Topics in Environmental Physiology and Medicine

Edited by Karl E. Schaefer

Short-Term Tests
for Chemical Carcinogens

Edited by

H.F. Stich
Head, Environmental Carcinogenesis Unit
British Columbia Cancer Research Centre

R.H.C. San
Chief, Carcinogen Testing Laboratory
British Columbia Cancer Research Centre

With 160 Illustrations

Springer-Verlag
New York Heidelberg Berlin

H.F. Stich
Head, Environmental Carcinogenesis Unit
British Columbia Cancer Research Centre
Vancouver, British Columbia
Canada

R.H.C. San
Carcinogen Testing Laboratory
British Columbia Cancer Research Centre
Vancouver, British Columbia
Canada

Library of Congress Cataloging in Publication Data
Main entry under title:

Short-term tests for chemical carcinogens.

  (Topics in environmental physiology and medicine)
  Includes index.
  1. Carcinogenicity testing—Congresses. 2. Car-
cinogens—Congresses. 3. Cocarcinogens—Congresses.
I. Stich, Hans F., 1927–   II. San, R.H.C.
[DNLM: 1. Carcinogens, Environmental—Analysis—
Congresses. 2. Biological assay—Congresses.
3. Mutagens—Analysis—Congresses. QZ202 S5593 1979]
RC268.65.S57   616.99′4071′0287   80-16545

© 1981 by Springer-Verlag New York Inc.
Softcover reprint of the hardcover 1st edition 1981

9 8 7 6 5 4 3 2 1

ISBN-13: 978-1-4612-5849-0   e-ISBN-13: 978-1-4612-5847-6
DOI: 10.1007/978-1-4612-5847-6

Contents

Viral Systems

1. A Biochemical Phage Induction Assay for Carcinogens 1

   R.K. Elespuru

2. Detection of DNA-Modifying Agents by Analyzing the Lesions Introduced into Purified PM2 DNA 12

   U. Kuhnlein, S.S. Tsang, and J. Edwards

3. Reactivation of Viruses 20

   A.J. Rainbow

DNA

4. Metabolic Activation of Nitroheterocyclic Compounds in Bacteria and Mammalian Cells 36

   D.R. McCalla

5. Utilization of the Alkaline Elution Assay as a Short-Term Test for Chemical Carcinogens 48

   J.A. Swenberg

6. DNA Synthesis Inhibition in Mammalian Cells as a Test for Mutagenic Carcinogens 59

   R.B. Painter

7. DNA Repair Synthesis (UDS) as an *in vitro* and *in vivo* Bioassay to Detect Precarcinogens, Ultimate Carcinogens, and Organotropic Carcinogens 65

   H.F. Stich, R.H.C. San, and H.J. Freeman

8.   Multi-well Assay for Unscheduled DNA Synthesis Using Human Diploid
     Fibroblasts   83

     G.R. Douglas, C.E. Grant, J.M. Wytsma, and A. Chan

9.   Sucrose Gradients: An Assay for DNA Damage   90

     B. Palcic and L.D. Skarsgard

10. The Testicular DNA-Synthesis Inhibition Test (DSI Test)   94

     J.P. Seiler

Microbial Systems

11. The *Salmonella* Mutagenicity Test: An Overview   108

     L. Haroun and B.N. Ames

12. Applications of the *Salmonella*/Microsome Assay   120

     V.F. Simmon

13. Determination of Genotoxic Activity Using DNA Polymerase-Deficient
     and -Proficient *E. coli*   127

     Z. Leifer, J. Hyman, and H.S. Rosenkranz

14. The Nucleotide-Permeable *Escherichia coli*: A Model System that
     Responds to DNA-Binding Carcinogens, Mutagens, and Antitumor Agents
     with DNA Excision Repair or with Inhibition of Replicative DNA
     Synthesis   140

     H.W. Thielmann

15. The Yeast *Saccharomyces cerevisiae*: An Assay Organism for
     Environmental Mutagens   161

     R.C. von Borstel

     Appendix I: Protocol for a Haploid Yeast Reversion Test for Assaying
     Mutagens   171

     R.C. von Borstel, M.M. Shahin, and R.D. Mehta

16. Induction and Genetic Characterization of Specific Locus Mutagens in the
     ad-3 Region in Two-Component Heterokaryons of *Neurospora
     crassa*   175

     F.J. de Serres

17. Neurospora and Environmentally Induced Aneuploidy   187

     A.J.F. Griffiths

Higher Plants

18. Plant Genetic Test Systems for the Detection of Chemical Mutagens   200

     W.F. Grant, A.E. Zinov'eva-Stahevitch, and K.D. Zura

Chromosomes

19. A Short-Term Cytogenetic Test for Genetic Instability in Humans   217

     T.C. Hsu, W.W. Au, L.C. Strong, and D.A. Johnston

20. The Sister Chromatid Exchange Test   236

     S. Wolff

21. The Micronucleus Assay. I. *In Vivo*   243

     J.A. Heddle and M.F. Salamone

22. The Micronucleus Assay. II. *In Vitro*   250

     J.A. Heddle, A.S. Raj, and A.B. Krepinsky

23. Automation in Cytogenetics   255

     J. Melnyk, K.R. Castleman, and G.W. Persinger

Mammalian Systems

24. Mutagenesis Studies in Diploid Human Cells with Different DNA-Repair
     Capacities   264

     J.J. McCormick and V.M. Maher

25. Liver Culture Indicators for the Detection of Chemical Carcinogens   277

     G.M. Williams

26. An *Escherichia coli* Differential Killing Test for Carcinogens Based on a
     uvrA recA lexA Triple Mutant   290

     M.H.L. Green and D.J. Tweats

27. Detection of Carcinogens Using the Fluctuation Test with S9 or with
     Hepatocyte Activation   296

     S.A. Hubbard, M.H.L. Green, and J.W. Bridges

Transformation

28. *In Vitro* Mammalian Cell Transformation for Identification of Carcinogens,
     Cocarcinogens, and Anticarcinogens   306

     C.H. Evans and J.A. DiPaolo

29. The Use of Cryopreserved Syrian Hamster Embryo Cells in a
    Transformation Test for Detecting Chemical Carcinogens   323

    R.J. Pienta, W.B. Lebherz III, and R.F. Schuman

30. Assay of Chemically Induced Transformation of Human Cells   338

    T. Kakunaga

31. Chemical-Viral Interactions: Enhancement of Viral Transformation by
    Chemical Carcinogens   350

    B.C. Casto

32. The Calcium Independence of Neoplastic Cell Proliferation: A Promising
    Tool for Carcinogen Detection   362

    A.L. Boynton, S.H.H. Swierenga, and J.F. Whitfield

Entire Animals

33. Induction of a Resistant Preneoplastic Liver Cell as a New Principle for a
    Short-Term Assay *in vivo* for Carcinogens   372

    E. Farber and H. Tsuda

34. Recent Achievements with *Drosophila* as an Assay System for Carcinogens   379

    E. Vogel

35. Strategy for Breeding Test Animals of High Susceptibility to Carcinogens   399

    F. Anders, M. Schwab, and E. Scholl

36. Methods for Human and Murine Sperm Assays   408

    A.J. Wyrobek

Cocarcinogens, Anticarcinogens, and Promoters

37. *In vitro* Assay for Tumor Promoters   420

    J.E. Trosko, L.P. Yotti, B. Dawson, and C.-C. Chang

38. Inhibition of Chemical Mutagenesis: An Application of Chromosome
    Aberration and DNA Synthesis Assays Using Cultured Mammalian Cells   428

    L. Wei, R.F. Whiting, and H.F. Stich

39. Detection of Cocarcinogens and Anticarcinogens with Microbial
    Mutagenicity Assays   438

    D.R. Stoltz

40. The Use of A Bacterial Assay to Identify Which Agents Modify
     Carcinogen-Induced Mutagenesis   449

     M.P. Rosin

Concepts

41. Quantitative Measures of Induced Mutagenesis   457
     F. Eckardt and R.H. Haynes

42. Tests for Potential Carcinogens: Unresolved Problems   474
     J. Ashby

43. Mutagenicity Testing: Problems in Application   483
     M.S. Legator and S.J. Rinkus

44. Short-Term Genetic Tests Extended to the Human   505
     M.L. Mendelsohn

     Index   511

Preface

The recent surge of interest in designing, validating, and implementing short-term tests for carcinogens has been spurred by the fairly convincing correlation between the carcinogenicity and mutagenicity of chemicals and physical agents and by the assumption that DNA alteration, mutations, and chromosome aberrations are somehow involved in neoplastic transformation. Moreover, it has been tacitly assumed that the mutagenic capacity alone of compounds would induce regulatory agencies to pass rules for their removal from the environment and would lead the public to avoid them. The actual response, however, is quite different.

Governmental departments shy away from making any decisions on the basis of *in vitro* test systems. The public at large is becoming irritated by daily announcements that many of their cherished habits could adversely affect their health. Industry appears to feel threatened and may reduce its search for new beneficial chemicals. The reluctance to accept wholeheartedly the mutagenicity tests for the detection of carcinogens is partly due to uncertainty about the involvement of mutations in neoplastic transformation, partly due to the present difficulty of extrapolating results from various endpoints obtained on numerous organisms to man, and partly due to a multitude of complex events that lead *in vivo* to the evolvement of benign or malignant tumors.

Following the initial rapid advances in the detection of environmental chemicals with carcinogenic and mutagenic properties, we seem to have arrived at a crossroads: We must now set new priorities for future research and must make an unbiased assessment of the *actual* hazard of a compound to man and the human population.

Forty-three experts were invited to assess the pros and cons of using short-term tests to detect the genotoxic and by implication carcinogenic potency of environmental chemicals and complex mixtures of compounds. It has become evident that no single bioassay can uncover all genotoxic agents. Thus this book covers a spectrum of tests that use a great variety of organisms and endpoints.

The possibility of using viral test systems is discussed in three papers. In the past, viruses, with their well-defined genomes and ease of handling, have not received the attention they seem to warrant. Seven papers focus on methods based on the interaction of genotoxic agents and carcinogens with the DNA of the target cells. Recent advances in microbial tests for mutagenicity were reviewed in seven papers. With the development of new tester strains that provide metabolic activation and improved handling procedures, microbes will undoubtedly find an even broader use in mutagenicity testing. A single paper defends the use of higher plants. The recent successful introduction of Tradescantia staminal hairs as a sensitive bioassay to detect airborne mutagens and carcinogens may lead to a wider recognition and application of various plant tests.

Chromosome aberrations, sister-chromatid exchanges, the micronucleus test, and the automation of cytogenic alterations are discussed in five papers. Anomalies of chromosome complements were found at high frequencies among congenital anomalies, stillbirth, and spontaneous or induced tumors. These chromosome anomalies may represent an endpoint which appears to be an integral part of several genetic disorders affecting human populations.

The most important aspect of mammalian tests including the use of cultured human cells are covered in four papers. The greatest contributions of these tests are in the area of metabolic activation of precarcinogens and the inactivation of ultimate carcinogens. Since human cells of various cancer-prone individuals can be used, it is possible to estimate the variations in response towards carcinogens and mutagens within human population groups. Neoplastic transformation *in vitro* is reviewed in five papers. There should be no question about the high relevance of these bioassays.

Scientists as well as regulators would like to see the introduction of an endpoint that is a definite part of tumor formation in mammals, including human populations. With the issue of relevance in mind, attempts are being made to design short-term tests in entire animals. Several of the newly developed *in vivo* assays incorporate the advantages of *in vitro* short-term tests with the completeness of bioassays using whole animals. These issues are summarized in four papers.

The emphasis on mutagens and carcinogens should not detract from the importance of modulating agents including anticarcinogens, desmutagens, cocarcinogens, promoters, antipromoters, sensitizers, electron scavengers, and DNA-repair inhibitors. Four papers deal with this important field. The final four papers are dedicated to the discussion of quantitative measurements of mutagenesis, the problem of application of short-term tests, and a host of unresolved issues.

This comprehensive review of short-term tests for genotoxicity should appeal to all interested in environmental carcinogenicity and mutagenicity. It will be helpful to all who actively work in this field as well as to regulators and administrators who must choose test systems that will provide reliable and relevant results for regulatory decisions.

H.F. Stich

Contributors

B.N. Ames *Chapter 11*
Department of Biochemistry
University of California
Berkeley, California 94720, U.S.A.

F. Anders *Chapter 35*
Genetisches Institut der Justus Liebig-
  Universität
Heinrich-Buff-Ring 58-62
6300 Giessen, Federal Republic of Ger-
  many

J. Ashby *Chapter 42*
Genetic Toxicology Section
Central Toxicology Laboratory
Imperial Chemical Industries Ltd
Alderley Park
Nr Macclesfield
Cheshire SK10 4TJ, England

W.W. Au *Chapter 19*
Section of Cell Biology
The University of Texas System Cancer
  Center
M.D. Anderson Hospital and Tumor Insti-
  tute
Texas Medical Center
Houston, Texas 77030, U.S.A.

A.L. Boynton *Chapter 32*
Animal and Cell Physiology Section
  Division of Biological Sciences

National Research Council of Canada
Ottawa, Ontario, Canada K1A 0R6

J.W. Bridges *Chapter 27*
Institute of Industrial and Environmental
  Health and Safety
University of Surrey
Guilford
Surrey GU2 5XU, England

K.R. Castleman *Chapter 23*
Jet Propulsion Laboratory
California Institute of Technology
Pasadena, California 94301, U.S.A.

B.C. Casto *Chapter 31*
Environmental Sciences Group
Northrop Services Inc.
P.O. Box 12313
Research Triangle Park, North Carolina
  27709, U.S.A.

A. Chan *Chapter 8*
Mutagenesis Section
Environmental and Occupational Toxicol-
  ogy Division
Department of National Health and Wel-
  fare
Tunney's Pasture
Ottawa, Ontario, Canada K1A 0L2

C.-C. Chang   *Chapter 37*
Department of Pediatrics and Human Development
College of Human Medicine
Michigan State University
East Lansing, Michigan 48824, U.S.A.

B. Dawson   *Chapter 37*
Department of Pediatrics and Human Development
College of Human Medicine
Michigan State University
East Lansing, Michigan 48824, U.S.A.

F.J. de Serres   *Chapter 16*
Associate Director for Genetics
Office of the Director
National Institute of Environmental Health Sciences
P.O. Box 12233
Research Triangle Park, North Carolina 27709, U.S.A.

J.A. DiPaolo   *Chapter 28*
Laboratory of Biology, Division of Cancer Cause and Prevention
National Cancer Institute
National Institutes of Health
Bethesda, Maryland 20205, U.S.A.

G.R. Douglas   *Chapter 8*
Mutagenesis Section
Environmental and Occupational Toxicology Division
Department of National Health and Welfare
Tunney's Pasture
Ottawa, Ontario, Canada K1A 0L2

F. Eckardt   *Chapter 41*
Gesellschaft für Strahlen-und Umweltforschung
Abteilung Strahlenbiologie
8042 Neuherberg, Federal Republic of Germany

J. Edwards   *Chapter 2*
Environmental Carcinogenesis Unit
British Columbia Cancer Research Centre
601 West 10th Avenue
Vancouver, B.C., Canada V5Z 1L3

R.K. Elespuru   *Chapter 1*
Biological Carcinogenesis Program
Frederick Cancer Research Center
P.O. Box B
Frederick, Maryland 21701, U.S.A.

C.H. Evans   *Chapter 28*
Tumor Biology Section
National Cancer Institute
National Institutes of Health
Bethesda, Maryland 20205, U.S.A.

E. Farber   *Chapter 33*
Department of Pathology
University of Toronto
Banting Institute
100 College Street
Toronto, Ontario, Canada M5G 1L5

H.J. Freeman   *Chapter 7*
Environmental Carcinogenesis Unit
British Columbia Cancer Research Centre
601 West 10th Avenue
Vancouver, B.C., Canada V5Z 1L3

C.E. Grant   *Chapter 8*
Mutagenesis Section
Environmental and Occupational Toxicology Division
Department of National Health and Welfare
Tunney's Pasture
Ottawa, Ontario, Canada K1A 0L2

W.F. Grant   *Chapter 18*
Genetics Laboratory, Department of Plant Sciences, Faculty of Agriculture
Macdonald Campus of McGill University
Ste Anne de Bellevue, P.Q., Canada H9X 1C0

M.H.L. Green   *Chapters 26, 27*
MRC Cell Mutation Unit
University of Sussex
Falmer
Brighton BN1 9QG, England

A.J.F. Griffiths   *Chapter 17*
Department of Botany
Biological Sciences 2125a

University of British Columbia
Vancouver, B.C., Canada V6T 1W5

L. Haroun   *Chapter 11*
International Agency for Research on Cancer
150 Cours Albert Thomas
69372 Lyon Cedex 2, France

R.H. Haynes   *Chapter 41*
Department of Biology
Faculty of Science
York University
4700 Keele Street
Downsview, Ontario, Canada M3J 1P3

J.A. Heddle   *Chapters 21, 22*
Department of Biology
Faculty of Science
York University
4700 Keele Street
Downsview, Ontario, Canada M3J 1P3

T.C. Hsu   *Chapter 19*
Section of Cell Biology
The University of Texas System Cancer
   Center
M.D. Anderson Hospital and Tumor Insti-
   tute
Texas Medical Center
Houston, Texas 77030, U.S.A.

S.A. Hubbard   *Chapter 27*
MRC Cell Mutation Unit
University of Sussex
Falmer
Brighton BN1 9QG, England

J. Hyman   *Chapter 13*
Department of Microbiology
New York Medical College, Basic Scien-
   ces Building
Valhalla, New York 10595, U.S.A.

D.A. Johnston   *Chapter 19*
Department of Biomathematics
The University of Texas System Cancer
   Center
M.D. Anderson Hospital and Tumor Insti-
   tute
Texas Medical Center
Houston, Texas 77030, U.S.A.

T. Kakunaga   *Chapter 30*
Cell Genetics Section
Laboratory of Molecular Carcinogenesis
Chemistry Branch
Building 37, Room 3E08
National Cancer Institute
Bethesda, Maryland 20205, U.S.A.

A.B. Krepinsky   *Chapter 22*
Department of Biology
Faculty of Science
York University
4700 Keele Street
Downsview, Ontario, Canada M3J 1P3

U. Kuhnlein   *Chapter 2*
Environmental Carcinogenesis Unit
British Columbia Cancer Research Centre
601 West 10th Avenue
Vancouver, B.C., Canada V5Z 1L3

W.B. Lebherz, III   *Chapter 29*
Chemical Carcinogenesis Program
Frederick Cancer Research Center
P.O. Box B
Frederick, Maryland 21701, U.S.A.

M.S. Legator   *Chapter 43*
Division of Environmental Toxicology
Department of Preventive Medicine and
   Community Health
The University of Texas Medical Branch
Galveston, Texas 77550, U.S.A.

Z. Leifer   *Chapter 13*
Department of Microbiology
New York Medical College, Basic Scien-
   ces Building
Valhalla, New York 10595, U.S.A.

V.M. Maher   *Chapter 24*
Carcinogenesis Laboratory—Fee Hall
Department of Microbiology and Depart-
   ment of Biochemistry
Michigan State University
East Lansing, Michigan 48824, U.S.A.

D.R. McCalla   *Chapter 4*
Department of Biochemistry
McMaster University
Health Sciences Center

1200 Main Street West
Hamilton, Ontario, Canada L8S 4J9

J.J. McCormick   *Chapter 24*
Carcinogenesis Laboratory—Fee Hall
Department of Microbiology and Department of Biochemistry
Michigan State University
East Lansing, Michigan 48824, U.S.A.

R.D. Mehta   *Appendix 1*
Department of Genetics
The University of Alberta
Edmonton, Alberta
Canada T6G 2E9

J. Melnyk   *Chapter 23*
Department of Developmental Cytogenetics
City of Hope National Medical Center
1500 East Duarte Road
Duarte, California 91010, U.S.A.

M.L. Mendelsohn   *Chapter 44*
Biomedical Sciences Division
Lawrence Livermore Laboratory
University of California
P.O. Box 5507
Livermore, California 94550, U.S.A.

R.B. Painter   *Chapter 6*
Laboratory of Radiobiology, School of Medicine
University of California
San Francisco, California 94143, U.S.A.

B. Palcic   *Chapter 9*
Medical Biophysics Unit
British Columbia Cancer Research Centre
601 West 10th Avenue
Vancouver, B.C., Canada V5Z 1L3

G.W. Persinger   *Chapter 23*
Department of Developmental Cytogenetics
City of Hope National Medical Center
1500 East Duarte Road
Duarte, California 91010, U.S.A.

R.J. Pienta   *Chapter 29*
Chemical Carcinogenesis Program
Frederick Cancer Research Center
P.O. Box B
Frederick, Maryland 21701, U.S.A.

A.J. Rainbow   *Chapter 3*
Department of Biology
McMaster University
1280 Main Street West
Hamilton, Ontario, Canada L8S 4K1

A.S. Raj   *Chapter 22*
Department of Biology
Faculty of Science
York University
4700 Keele Street
Downsview, Ontario, Canada M3J 1P3

S.J. Rinkus   *Chapter 43*
Division of Environmental Toxicology
Department of Preventive Medicine and Community Health
The University of Texas Medical Branch
Galveston, Texas 77550, U.S.A.

H.S. Rosenkranz   *Chapter 13*
Department of Microbiology
New York Medical College, Basic Sciences Building
Valhalla, New York 10595, U.S.A.

M.P. Rosin   *Chapter 40*
Environmental Carcinogenesis Unit
British Columbia Cancer Research Centre
601 West 10th Avenue
Vancouver, B.C., Canada V5Z 1L3

M.F. Salamone   *Chapter 21*
Department of Biology
Faculty of Science
York University
4700 Keele Street
Downsview, Ontario, Canada M3J 1P3

R.H.C. San   *Chapter 7*
Carcinogen Testing Laboratory
British Columbia Cancer Research Centre
601 West 10th Avenue
Vancouver, B.C., Canada V5Z 1L3

E. Scholl   *Chapter 35*
Genetisches Institut der Justus Liebig-
   Universität
Heinrich-Buff-Ring 58-62
6300 Giessen, Federal Republic of Ger-
   many

R.F. Schuman   *Chapter 29*
Chemical Carcinogenesis Program
Frederick Cancer Research Center
P.O. Box B
Frederick, Maryland 21701, U.S.A.

M. Schwab   *Chapter 35*
Genetisches Institut der Justus Liebig-
   Universität
Heinrich-Buff-Ring 58-62
6300 Giessen, Federal Republic of Ger-
   many

J.P. Seiler   *Chapter 10*
Swiss Federal Research Station for Fruit-
   Growing, Viticulture, and Horticulture
8820 Waedenswil, Switzerland

M.M. Shahin   *Appendix I*
Department of Genetics
The University of Alberta
Edmonton, Alberta, Canada T6G 2E9

V.F. Simmon   *Chapter 12*
Genex Corporation
6110 Executive Boulevard
Rockville, Maryland 20852, U.S.A.

L.D. Skarsgard   *Chapter 9*
Medical Biophysics Unit
British Columbia Cancer Research Centre
601 West 10th Avenue
Vancouver, B.C., Canada V5Z 1L3

H.F. Stich   *Chapters 7, 38*
Environmental Carcinogenesis Unit
British Columbia Cancer Research Centre
601 West 10th Avenue
Vancouver, B.C., Canada V5Z 1L3

D.R. Stoltz   *Chapter 39*
Toxicology Research Division

Food Directorate
Health Protection Branch
Health and Welfare Canada
Tunney's Pasture
Ottawa, Ontario, Canada K1A 0L2

L.C. Strong   *Chapter 19*
Section of Cell Biology
The University of Texas System Cancer
   Center
M.D. Anderson Hospital and Tumor Insti-
   tute
Texas Medical Center
Houston, Texas 77030, U.S.A.

J.A. Swenberg   *Chapter 5*
Department of Pathology
Chemical Industry Institute of Toxicology
P.O. Box 12137
Research Triangle Park, North Carolina
   27709, U.S.A.

S.H.H. Swierenga   *Chapter 32*
Division of Drug Toxicology
Health and Welfare Canada
Tunney's Pasture
Ottawa, Ontario, Canada K1A 0L2

H.W. Thielmann   *Chapter 14*
Institut für Biochemie
Deutsches Krebsforschungszentrum
Im Neuenheimer Feld 280
6900 Heidelberg 1, Federal Republic of
   Germany

J.E. Trosko   *Chapter 37*
Department of Pediatrics and Human
   Development
College of Human Medicine
Michigan State University
East Lansing, Michigan 48824, U.S.A.

S.S. Tsang   *Chapter 2*
Environmental Carcinogenesis Unit
British Columbia Cancer Research Centre
601 West 10th Avenue
Vancouver, B.C., Canada V5Z 1L3

H. Tsuda *Chapter 33*
Department of Pathology
University of Toronto
Banting Institute
100 College Street
Toronto, Ontario, Canada M5G 1L5

D.J. Tweats *Chapter 26*
Glaxo Group Research
Harefield
Uxbridge
Middlesex UB9 61S, England

E. Vogel *Chapter 34*
Department of Radiation Genetics and
  Chemical Mutagenesis
State University of Leiden
Wassenaarseweg 72
2333 AL Leiden,
The Netherlands

R.C. von Borstel *Chapter 15, Appendix I*
Department of Genetics
The University of Alberta
Edmonton, Alberta, Canada T6G 2E9

L. Wei *Chapter 38*
Environmental Carcinogenesis Unit
British Columbia Cancer Research Centre
601 West 10th Avenue
Vancouver, B.C., Canada V5Z 1L3

J.F. Whitfield *Chapter 32*
Animal and Cell Physiology Section
Division of Biological Sciences
National Research Council of Canada
Ottawa, Ontario, Canada K1A 0R6

R.F. Whiting *Chapter 38*
Canadian Centre for Occupational Health
  and Safety
McMaster University
1200 Main Street West
Hamilton, Ontario, Canada L8N 3Z5

G.M.Williams *Chapter 25*
Division of Experimental Pathology

Naylor Dana Institute for Disease Prevention
American Health Foundation
1 Dana Road
Valhalla, New York 10595, U.S.A.

S. Wolff *Chapter 20*
Laboratory of Radiobiology, School of
  Medicine
University of California
San Francisco, California 94143, U.S.A.

A.J. Wyrobek *Chapter 36*
Biomedical Sciences Division
Lawrence Livermore Laboratory
University of California
P.O. Box 5507
Livermore, California 94550, U.S.A.

J.M. Wytsma *Chapter 8*
Mutagenesis Section
Environmental and Occupational Toxicol-
  ogy Division
Department of National Health and Wel-
  fare
Tunney's Pasture
Ottawa, Ontario, Canada K1A 0L2

L.P. Yotti *Chapter 37*
Department of Pediatrics and Human
  Development
College of Human Medicine
Michigan State University
East Lansing, Michigan 48824, U.S.A.

A.E. Zinov'eva-Stahevitch *Chapter 18*
Genetics Laboratory, Department of Plant
  Sciences, Faculty of Agriculture
Macdonald Campus of McGill University
Ste Anne de Bellevue, P.Q., Canada H9X
  1C0

K.D. Zura *Chapter 18*
Genetics Laboratory, Department of Plant
  Sciences, Faculty of Agriculture
Macdonald Campus of McGill University
Ste Anne de Bellevue, P.Q., Canada H9X
  1C0

1

A Biochemical Phage Induction Assay for Carcinogens

ROSALIE K. ELESPURU

Introduction

Bacteriophage induction assays have been used periodically for many years as detection systems for agents with potential carcinogenic or cancer inhibiting properties (Lwoff, 1953; Endo *et al.*, 1963a,b; Price *et al.*, 1964; Geissler, 1967; Fleck, 1968, 1974). In the early seventies, a summary of the available experimental data (Heinemann, 1971) indicated that phage induction correlated reasonably well with antitumor properties but somewhat less well with the carcinogenic properties of the chemicals tested. Bacterial assays became much more practical after the introduction of mammalian enzyme activating systems and bacterial strains with increased sensitivity to chemicals (Ames *et al.*, 1975). Many carcinogens previously missed in bacterial assays could then be detected as mutagens in *Salmonella typhimurium* (McCann *et al.*, 1975). A bacteriophage lambda induction assay for carcinogens utilizing a sensitive strain of *E. coli* has recently been described (Moreau *et al.*, 1976; Moreau and Devoret, 1977). A number of carcinogens, including aflatoxins, polycyclic hydrocarbons, and aromatic amines, were shown to induce bacteriophage lambda in this system. These results encouraged the further development of phage induction assays for screening of chemicals.

Because phage induction is a response measurable in all or most of a population of bacteria exposed to inducing agents, biochemical assays of induction have been devised. Genes with easily measurable products have been placed under the control of bacteriophage lambda promoters and utilized in the study of induction (Smith and Oishi, 1976, 1978; Levine *et al.*, 1978). These assays, however, are not easily amenable to large-scale screening because of the use of expensive substrates and procedures that are impractical for routine use with hazardous chemicals. We have recently developed a more practical biochemical phage induction assay suitable for large-scale screening of chemicals (Elespuru and Yarmolinsky, 1979).

Table 1.1. *Bacterial Strains*

Strain	Immunity	Permeability	DNA repair	Phage replication	Lysis functions
BR 469	λt_{11}	+	$uvrB\Delta$	$- (P_R^-)$	$- (P_R^-)$
BR 513	λt_{11}	$envA$	$uvrB\Delta$	$- (P_R^-)$	$- (P_R^-)$
BR 293	434	+	$uvrB\Delta$	$+ (P_R^+)$	$- (S_7)$

Legend

t_{11} is a mutation in P_R (rightward promotor) of lambda preventing transcription and the expression of rightward functions (phage replication, cell lysis, turnoff, phage coat proteins) (Eisen *et al.*, 1966; Nijkamp *et al.*, 1970)

$envA$ is a mutation in the bacterial cell wall resulting in enhanced permeability (Normark *et al.*, 1969)

$uvrB\Delta$ is a deletion of a gene coding for a subunit of an enzyme in the UV-excision repair pathway (Seeburg, 1978)

S_7 is an amber mutation in a gene necessary for bacterial cell lysis (Harris *et al.*, 1967)

The strain of *E. coli* (BR 513, Table 1.1) constructed for the biochemical induction assay (BIA) is lysogenic for a bacteriophage lambda carrying a portion of the lactose operon, the *lacZ* gene, adjacent to the lambda leftward promoter. Upon induction (e.g., by chemicals or radiation), the lambda immunity repressor is cleaved, *lacZ* is transcribed, and its gene product, β-galactosidase, is produced. Normal phage replication, phage proteins, cell lysis, and leftward transcription control are inhibited by the presence of a mutation preventing transcription from the rightward promoter of the phage. Transcription of *lacZ* and synthesis of β-galactosidase are therefore prolonged after induction. Details of the genetic construction of this strain and its behavior upon induction have been previously described (Elespuru and Yarmolinsky, 1979). The assay of β-galactosidase using inexpensive substrates is a simple and straightforward procedure (Miller, 1972). The BIA may be performed as a spot test on agar or as a quantitative liquid incubation assay in a single test tube (Figs. 1.1 and 1.4).

Among the advantages of a biochemical screening assay for carcinogens are the following:

1. Rapidity—the assay is complete 3 to 5 hours after the addition of carcinogen.
2. Testing of complex samples: (a) samples containing microorganisms (e.g., fermentation broths or fecal samples) do not interfere with the assay. (b) samples in toxic solvents can often be tested because colony formation or full phage production is not necessary.
3. Simplicity—only one bacterial strain is necessary to detect a wide variety of DNA-chemical interactions; the genomic target size for phage induction is large, most likely the entire bacterial chromosome rather than a few base pairs or genes.
4. Safety—the carcinogen is confined to a single plate or tube to minimize exposure of operating personnel during performance of the assay.

Materials

Reagents

Bacteriological media are from Difco; inorganic and organic salts are from Fischer Scientific; sodium ampicillin (Polycillin N) is from Bristol Laboratories; ONPG, BNG, Fast Blue RR salt, NADP monosodium salt, G-6-P, chloramphenicol, β-mercaptoethanol, tris (as Trizma Base) are from Sigma.

Media and Buffers

LBE (per liter)—10 g Bactotryptone, 5 g yeast extract, 10 g sodium chloride, 5 ml 1M tris (hydroxymethyl) aminomethane.

After autoclaving, medium is supplemented with 4 ml of 50X medium E (Vogel and Bonner, 1956) and 10 ml of 20% glucose. For LBE amp agar, 15 g agar is added before autoclaving and 1 ml of a freshly prepared solution of sodium ampicillin (10 mg/ml) is added just before pouring plates. Plates should be poured on a level surface.

ZCM Buffer (per liter) (after Miller, 1972)—16.1 g $Na_2HPO_4 \cdot 7H_2O$ (or 8.5 g Na_2HPO_4 anhydrous); 5.5 g $NaH_2PO_4 \cdot H_2O$, 0.75 g KC1, 0.246 g $MgSO_4 \cdot 7H_2O$ (or 0.12 g $MgSO_4$ anhydrous), 2.7 ml β-mercaptoethanol, 25 mg chloramphenicol. Adjust pH to 7.0. Do not autoclave. Store cold.

Medium A (per liter) (Miller, 1972)—10.5 g K_2HPO_4, 4.5 g KH_2PO_4, 1 g $(NH_4)_2SO_4$, 0.5 g sodium citrate $\cdot 2H_2O$. Autoclave.

Carcinogens

Aromatic amines are obtained from the NCI Chemical Repository (IIT Research Institute, Chicago), aflatoxins and nitrosamines from W. Lijinsky, and other carcinogens from Sigma.

Bacterial Strains

Strains BR 513 and BR 469 were constructed for this assay (Elespuru and Yarmolinsky, 1979). Strain BR 293 was kindly supplied by M.B. Yarmolinsky.

Laboratory Supplies

Polystyrene plastic tubes (16 × 125 mm) are #2025 from Falcon; large (243 × 243 × 18 mm) bioassay plates are from A/S Nunc, Kamstrupvej 90, Kamstrup, DK-4000 Roskilde, Denmark, distributed in USA by Vangard International, Inc., 1111-A Green Grove Rd., Neptune, NJ 07753; "mini plates" (100 × 15 mm square) are #1012 from Falcon.

Methods

One-Tube Quantitative BIA

1. A culture of BR 513 is grown from a loopful of glycerol stock in 1–3 ml LBE overnight at 37° with mild shaking.
2. Next morning, the bacteria are diluted 100-fold into fresh LBE and grown at 37° for 2–3 hours to A_{600} 0.4. Bacteria are observed in the microscope for characteristic chain formation (chains of 2–4 cells) to avoid the use of contaminated cultures. Chains are generally not observed in stationary phase (overnight) cultures.
3. While the bacteria are growing, chemical solutions or test samples are distributed to tubes in aliquots of 10–50 μl (10–20 μl if chemicals are in toxic solvents such as DMSO or ethanol).
4. When the bacteria have reached A_{600} 0.4, they are diluted tenfold into fresh 37° LBE + ampicillin (10 $\mu g/ml$) (fresh ampicillin is added to LBE just before use, at a dilution of 1:100). If enzymatic activation is required, bacteria are diluted tenfold into 37° LBE-amp activation mix prepared just before use.

LBE-amp activation mix is (per ml):

LBE + amp (25 $\mu g/ml$)	0.40 ml
0.1M NADP	0.03 ml
0.1M G-6-P	0.10 ml
0.1M $MgCl_2$	0.10 ml
phosphate buffer, 0.1M pH 7.4	0.13 ml
rat liver S9 fraction	0.14 ml
	0.90 ml
(bacteria	0.10 ml)

Diluted bacteria are then distributed in 0.5 ml aliquots to tubes containing chemicals.

5. Bacterial suspensions are incubated with shaking at 37° for 3 hours.

6. 4.5 ml cold ZCM buffer is then added to each tube. Tubes may be stored in the cold overnight if desired at this point.
7. To initiate the enzyme assay, tubes and ONPG substrate are warmed to 28° in a water bath.
8. An aliquot of 1.0 ml ONPG (4 mg/ml in Medium A) is added to each tube. A stopwatch is started just before adding substrate. Each tube is vortexed quickly (one spin is sufficient; slow speed).
9. Tubes are incubated at 28° until they develop yellow color (10 min–several hours to an absorbance of ~0.5–1.0 at 420 nm).
10. To stop color development, 2.5 ml 1M Na_2CO_3 (sodium carbonate) is added. Tubes are quickly mixed by shaking or gentle vortexing. The exact time of the addition of Na_2CO_3 is recorded.
11. Yellow color is measured by absorbance (e.g., in B + L Spectronic 20 or Coleman Jr.) at 420 nm. The spectrophotometer may be adjusted to zero with a media control. Each tube is turned to obtain the lowest possible reading (to minimize optical artifacts such as scratches on the tubes).

12. Enzyme Units = $100 \, A_{420}/t_{hr}$

    where:
    A_{420} = absorbance at 420 nm
    t_{hr} = time required to develop yellow color in hours
    1 enzyme unit = the approximate value of an uninduced control tube (no liver S9) at zero time (see Elespuru and Yarmolinsky, 1979)

Spot Test

(See Table 1.2 for proportions for different-sized Petri dishes.)

1. A culture of BR 513 is grown from a loopful of glycerol stock (Elespuru and Yarmolinsky, 1979) in 1–3 ml LBE overnight at 37° with mild shaking.
2. Next morning bacteria are diluted 100-fold (or to A_{600} 0.05) into fresh LBE, and grown at 37° for 2–3 hours to A_{600} 0.4. Bacteria are examined in the microscope for characteristic chain formation.
3. The culture is pelleted by centrifugation (4000–5000 RPM, ~3 min). The supernatant is then decanted and fresh LBE

Table 1.2. *Spot Test: Proportions for Different-sized Petri Dishes*

	Petri dish diameter		
	100 mm	150 mm	243 mm
Number of spots/dish	10–20 (round) 36 (square)	30–50	100–200
Log phase bacteria (pelleted)	3 ml	7 ml	25 ml
LBE for resuspending pellet	.1 ml	.25 ml	1 ml
Agar (bottom layer)	~35 ml	~85 ml	~320 ml
Soft agar (2 top layers)	3 ml	7 ml	25 ml
S9 activation mix			
Standard Ames mix	.5 ml	1.2 ml	4.0 ml
or			
$MgCl_2$ 0.1M	20 μl	50 μl	.2 ml
G-6-P 0.1M	20 μl	50 μl	.2 ml
NADP 0.1M	10 μl	25 μl	.1 ml
S9 liver preparations	.01–1.0 ml	.025–2.5 ml	.1–10 ml
BNG	1 mg ⎤ .1 ml	2.5 mg ⎤ .25 ml	10 mg ⎤ 1.0 ml
Fast Blue RR salt	6 mg ⎦ DMSO	15 mg ⎦ DMSO	60 mg ⎦ DMSO

is added for resuspension of the pellet (do not vortex).

4. (optional) S9 activation mix is added to each tube (S9 mix is freshly made and stored cold; the mix is warmed to 37° just prior to use).

5. Melted soft agar (0.7% = 7 g/l) at 45° is added to each tube one at a time. The contents are poured on prewarmed (38°) Petri dishes containing LBE agar + 10 μg/ml ampicillin. Plates are made 1–4 days prior to use. The top layer is allowed to solidify at room temperature on a level surface.

6. Chemical solutions or mixtures are spotted on the plates using Pasteur or capillary pipets. Chemicals on discs may also be used. Plates should be kept warm on a warming plate if possible, and returned to the incubator within 15–20 minutes.

7. Plates are incubated at 38° (with lid off to allow drying of spots, if necessary) for 3 hours.

8. Color indicator (Fast Blue RR salt + BNG) is dissolved in DMSO just prior to use. Solution is aided by use of a Pasteur pipet. Soft agar (45°) is immediately added to the indicator solution and poured on the plate. (Discs, if present, should be removed prior to the addition of indicator overlay.) The agar is allowed to solidify. Color development takes about 10 minutes.

Bioautography

Purified chemical standards, mixtures, or extracts are chromatographed on thin layer plates (glass or plastic) containing silica gel or other adsorbant. A piece of wet filter paper is placed on the surface of a warmed (38°) 243 mm bioassay plate (containing LBE-amp agar) and smoothed to remove wrinkles and air bubbles. The thin layer plates are then placed face down on the filter paper slowly from one edge to prevent the accumulation of air bubbles. The plates are then incubated for 1 to 3 hours at 38° or overnight in the cold if longer diffusion times are required. After the incubation the thin layer plates and filter paper are removed from the agar surface. Bacteria grown for the biochemical phage induction spot test are poured in agar onto the warmed bioassay plate. The plate is further incubated at 38° for 3 hours and developed as described for the spot test.

Discussion

One-Tube Liquid Incubation Assay

The liquid incubation assay was designed so that phage induction and the enzyme assay are carried out in a single disposable tube (Fig. 1.1). Buffers and substrates are added consecutively with no necessity for the removal of any solution from the tubes. Measurement of the amount of enzyme present is made by insertion of the tubes in a suitable spectrophotometer (e.g., Coleman Jr. or Bausch & Lomb Spectronic 20). This method is intended to make the assay as simple and safe as possible.

Induction may be quantified, as shown in Figure 1.2, using the one-tube assay. Upon exposure to an optimal inducing dose of ENNG (a direct-acting carcinogen) there is an initial lag period of approximately one generation time, the latent period of the "SOS" response (Witkin, 1976; Defais et al., 1976). Thereafter, the amount of β-galactosidase increases linearly for several hours, well beyond the time when a lambda lysogen with rightward promoter functions would have lysed. Enzyme synthesis levels off after 5 or 6 hours, perhaps owing to the reestablishment of repression (in a cro^- background) or the depletion of essential nutrients. Enzyme induction may be measured conveniently after 2 or 3 hours for powerful inducing agents (Elespuru and Yarmolinsky, 1979). The level of sensitivity of the assay may be increased for weak inducers or suboptimal doses of strong inducers, by allowing enzyme synthesis to continue for longer times.

Enzyme produced from BR 513 (Figure

Tube Assay

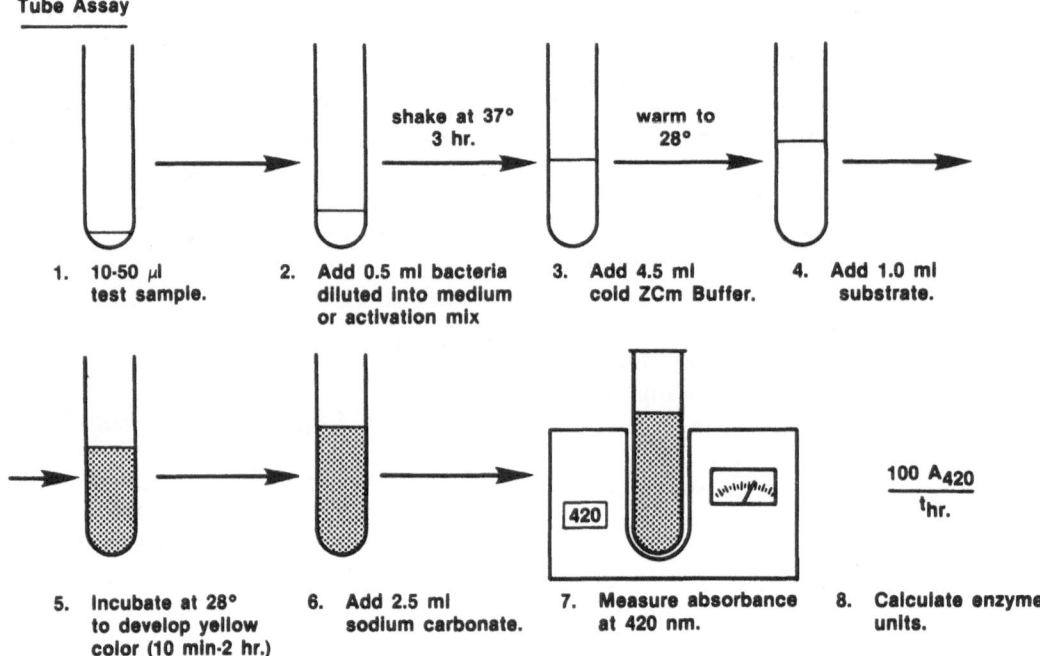

1. 10-50 µl test sample.

2. Add 0.5 ml bacteria diluted into medium or activation mix

   shake at 37° 3 hr.

3. Add 4.5 ml cold ZCm Buffer.

   warm to 28°

4. Add 1.0 ml substrate.

5. Incubate at 28° to develop yellow color (10 min-2 hr.)

6. Add 2.5 ml sodium carbonate.

7. Measure absorbance at 420 nm.

$$\frac{100\ A_{420}}{t_{hr.}}$$

8. Calculate enzyme units.

Figure 1.1. Schematic diagram of quantitative one-tube liquid incubation assay.

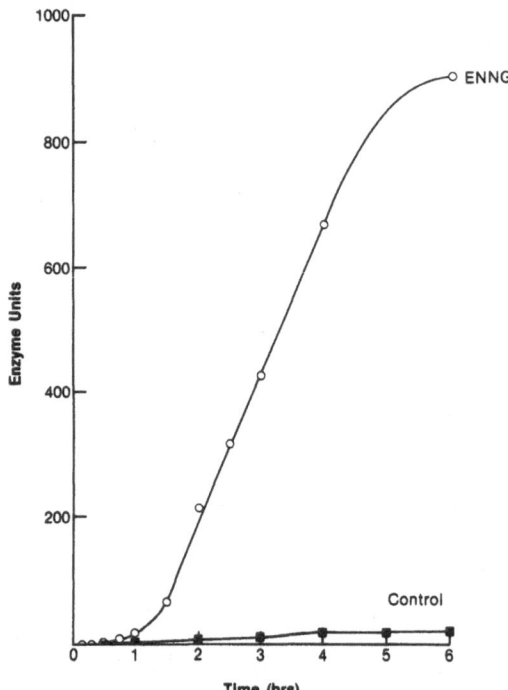

Figure 1.2. Kinetics of appearance of β-galactosidase after induction of BR 513 by an optimal inducing dose of ENNG (1.0 mM). Procedure was as described in Figure 1.1, except that at step 2 samples were taken at the times shown rather than at 3 hours.

1.2) arises from the transcription of a single copy of the *lacZ* gene per chromosome, owing to the absence of a phage replication function (Table 1.1). Enzyme production from a strain in which replication of the phage results in amplication of the *lacZ* gene is shown in Figure 1.3. The UV induction of the replication-proficient strain results in the appearance of approximately 10 times more enzyme than in the replication-deficient strain after 2 hours. The increased production of enzyme should result in an effective increase in sensitivity of the assay utilizable for the detection of lower levels of induction in a shorter length of time. Further development of the replication-proficient strain, including the addition of the mutation for enhanced permeability, is planned.

Activating enzymes can be added to the liquid incubation assay, with the loss of some sensitivity due to turbidity if crude enzyme preparations are used. The background absorbance increases severalfold if a standard rat liver S9 is used at a concentration of 10−20% in the assay.

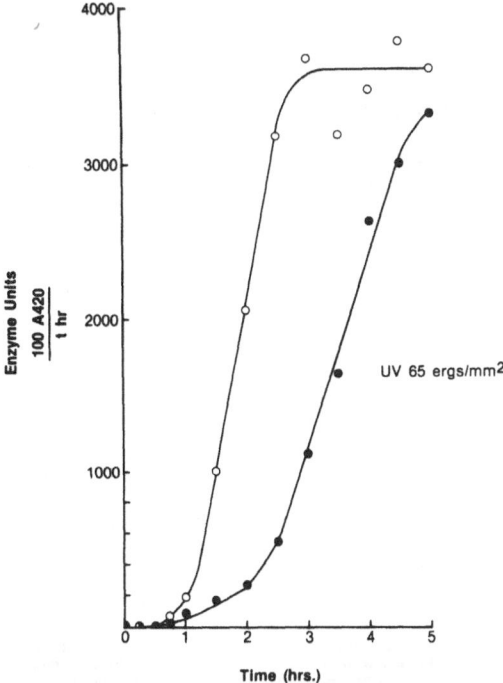

Figure 1.3. Comparison of induction by UV in lysogenic strains with (BR 293, open circles) or without (BR 469, closed circles) the capacity to replicate phage DNA. Procedure was as described in Figure 1.2. UV induction was with a GE germicidal lamp 8W at 50 cm (incident dose 13 ergs/mm²/sec). Bacteria were irradiated in 10 mM MgSO₄ followed by immediate dilution into LBE for gene expression.

Spot Test

A spot test for phage induction (Fig. 1.4) is performed by pouring bacteria in agar with activating enzymes onto Petri dishes, spotting test solutions or mixtures, and incubating the plates at 37° for 3 hours. Substrate is then added in an agar overlay, resulting in the appearance of an insoluble red dye over the areas of induction. Color development is complete within 10 minutes. The spot test is notable for its economy of time and materials, and for several useful applications. It is, however, only semiquantitative. Strong, weak, and intermediate levels of induction may be differentiated, but amounts of enzyme cannot be determined. Spot tests are as sensitive as the liquid incubation assay for the detection of small quantities

of inducing agents (i.e., inducing twice background amounts of enzyme in the liquid incubation assay or visible spots in the spot test after the same expression time, usually 3 hours). Gridded 15 × 100 mm square Petri dishes may be used for the test of as many as 36 chemical solutions on a single plate (Fig. 1.5). The same amount of S9 activation mix and cofactors is used for these dishes as is used for one Petri dish in the *Salmonella typhimurium* mutagenesis assay. Such economy of materials may be important when rare or expensive sources of activating enzymes are being tested (e.g., from a human tissue or a very small animal organ). If only a small quantity of test chemical is available, it will be present in higher concentration in a spot test than in a test in which it is distributed over the entire area of the dish (or throughout its volume).

Conventional mutagenesis and phage induction assays (including the liquid incubation assay) can be totally inhibited by toxic doses of chemicals. However, when chemicals are spotted on agar, a gradient of concentration is established within which a ring of bioactivity can usually be seen (Fig. 1.5). It is therefore possible to detect large as well as small amounts of an inducing agent in a spot test, e.g., 100 μg of aflatoxin B1 is detected as readily as 10 ng. Chemicals in most solvents, such as DMSO, methanol, or chloroform, can be spotted directly without adverse effects on the assay. In conventional plaque assays or mutagenesis assays, 100% DMSO and other solvents inhibit the growth of indicator bacteria or mutants, respectively, when spotted directly onto the plates.

Bioautography

The biochemical phage induction spot test can be adapted for bioautography (Fig. 1.4). Thin layer plates on which chemicals have been chromatographed are placed face down on an agar layer in large Petri dishes. Diffusion of the chemicals from the adsorbant into the agar results in a print of the chromatogram in the agar. Bioactive

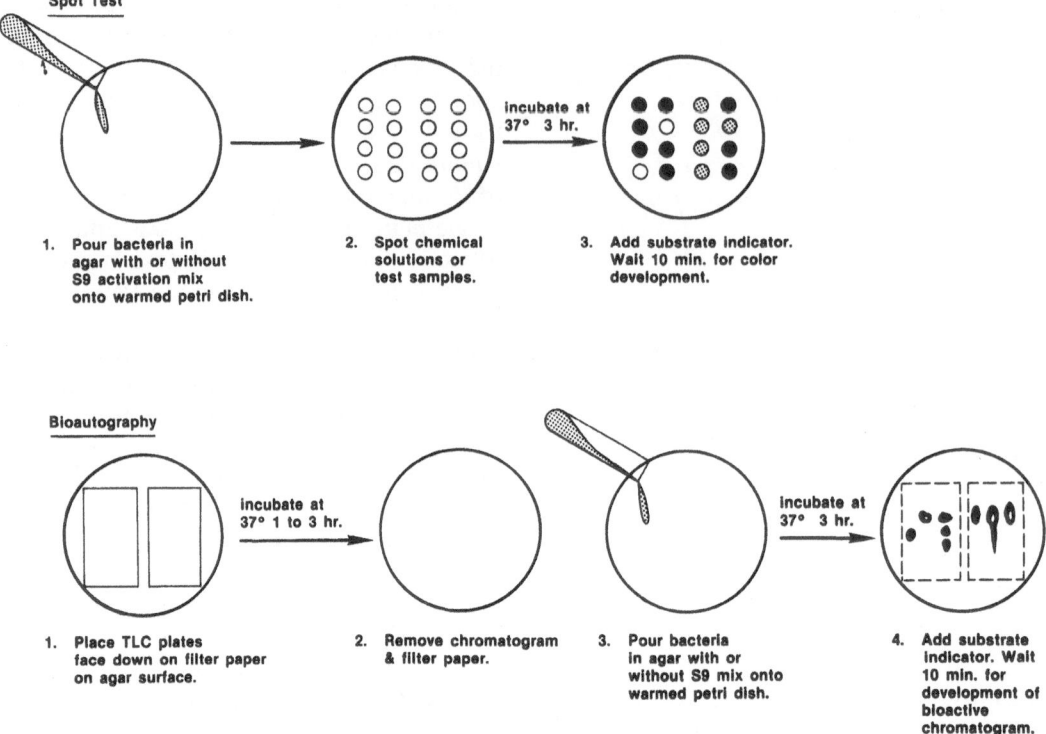

Spot Test

1. Pour bacteria in agar with or without S9 activation mix onto warmed petri dish.

2. Spot chemical solutions or test samples.

incubate at 37° 3 hr.

3. Add substrate indicator. Wait 10 min. for color development.

Bioautography

incubate at 37° 1 to 3 hr.

incubate at 37° 3 hr.

1. Place TLC plates face down on filter paper on agar surface.

2. Remove chromatogram & filter paper.

3. Pour bacteria in agar with or without S9 mix onto warmed petri dish.

4. Add substrate indicator. Wait 10 min. for development of bioactive chromatogram.

Figure 1.4. Schematic diagram of colorimetric spot test for phage induction. Adaptation of spot test for bioautography.

areas of the chromatogram appear as zones of red dye following the addition of inducible bacteria and enzyme substrate.

The bioautography of two complex mixtures and the resolution of bioactive components is shown in Figure 1.6. Figure 1.6a shows the chromatography of 6 different extracts from a fermentation broth containing the antitumor agent daunorubicin and related metabolites. A variety of phage-inducing areas can be seen. Figure 1.6b shows the bioautography of a natural mixture of aflatoxins and the purified components of the mixture. The location of the four aflatoxins on the silica gel and their transfer to the agar was monitored by fluorescence. Aflatoxins B1 and G1, constituting approximately 45% of the mixture each, are both carcinogenic and phage-inducing. One of the other components, aflatoxin B2, constituting 6% of the mixture, was equivocal as a carcinogen in a test on a

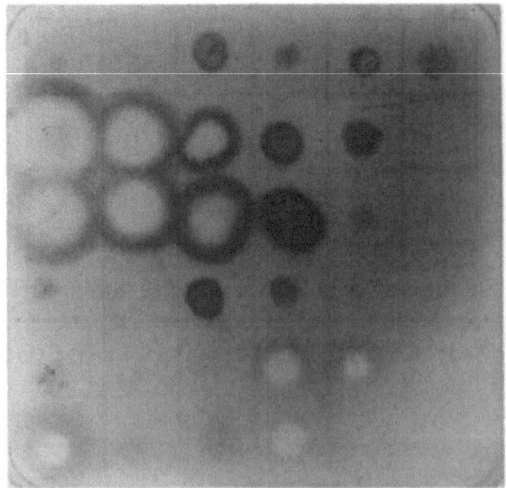

Figure 1.5. Spot test of carcinogens on gridded "mini plates" (100 mm), in the presence of arochlor-induced male hamster liver S9 activation mix (10 mg protein/plate). Procedure was as described in Figure 1.4 and Methods. Carcinogens and amounts spotted are as follows (see following table):

	1	2	3	4	5	6
A	AAF 100 μg	AAF 10 μg	N-OH-AAF 10 μg	N-OH-AAF 1 μg	N-Ac-AAF 10 μg	N-Ac-AAF 1 μg
B	AB1 10 μg	AB1 1 μg	AB1 0.1 μg	AB1 10 ng	AB1 1 ng	AB1 0.1 ng
C	AG1 10 μg	AG1 1 μg	AG1 0.1 μg	AG1 10 ng	AG1 1 ng	AG1 0.1 ng
D	DMBA 100 μg	DMBA 10 μg	MCA 100 μg	MCA 10 μg	B(a)P 100 μg	B(a)P 10 μg
E	2AA 10 μg	2AA 1 μg	DMSO Control	NPY 10 μg	NPY 5 μg	NPY 1 μg
F	DEN 10 μg	DEN 5 μg	DEN 1 μg	DMN 10 μg	DMN 5 μg	DMN 1 μg

AAF, B(a)P, and DMN were negative in this test; all other carcinogens were positive.

small number of animals (Lijinsky and Butler, 1966; Butler *et al.*, 1969). It appears to be a weak phage inducer. Aflatoxin G2, present as 3% of the mixture, is nega- tive for phage induction and has not been tested for carcinogenicity, possibly because of the lack of sufficient material for testing. Bioautography could be useful as a tool for

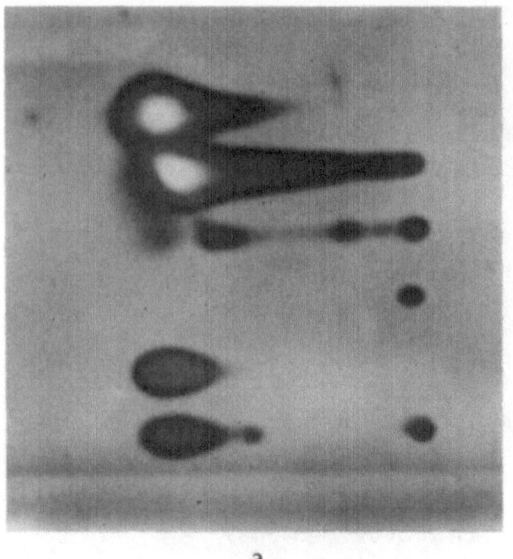

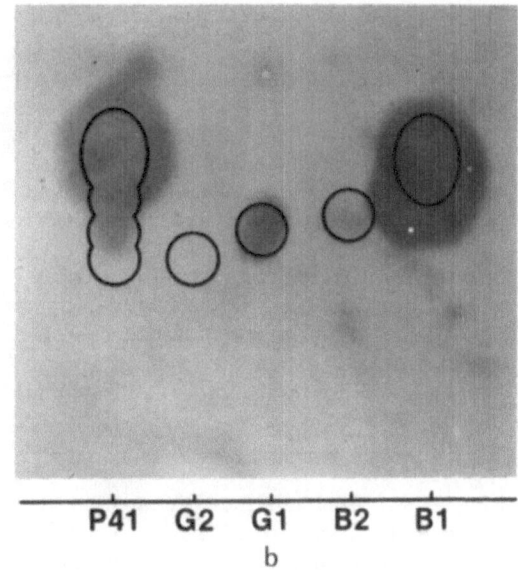

a   b

Figure 1.6. (a) Bioautography of extracts from daunorubicin fermentation broths. Concentrated ex- tracts were spotted onto plastic thin layer plates (Baker Flex 1B-F) containing silica gel and chroma- tographed in heptane:chloroform:methanol (5:5:1). The chromatograms were placed face down on agar for 2 hr prior to the addition of bacteria. The procedure for bioautography was then as described in Figure 1.4 and Methods. (b) Bioautography of aflatoxins B1, B2, G1, and G2 and the natural mixture (P41). Approximately 150 μg of sample was spotted as a solution in methanol on glass plates coated with 0.25 mm silica gel (EM Laboratories Silica Gel 60F-254) and chromatographed in chloro- form:methanol (97:3). The chromatograms were placed face down on agar for 1 hr prior to the addition of bacteria. Procedure for bioautography was then as described in Figure 1.4 and Methods. Superim- posed circles represent the location of each spot on the agar, as determined by fluorescence. Afla- toxins B1 and G1 were phage induction positive, while B2 was equivocal and G2 was negative.

identifying active impurities in samples of carcinogens or active components in complex industrial mixtures.

Abbreviations

ONPG, o-nitrophenyl-β-D-galactopyranoside; BNG, 6-bromo-2-naphthyl-β-D-galactopyranoside; G-6-P, glucose-6-phosphate; NADP, nicotinamide adenine dinucleotide phosphate; ENNG, N-ethyl-N'-nitro-N-nitrosoguanidine; AAF, 2-fluorenylacetamide; N-OH-AAF, N-hydroxy-2-fluorenylacetamide; N-ac-AAF, N-acetoxy-2-fluorenylacetamide; AB1, aflatoxin B1; AG1, aflatoxin G1; DMBA, 7,12-dimethylbenz(a)anthracene; MCA, 3-methylcholanthrene; B(a)P, benzo(a)pyrene; 2AA, 2-aminoanthracene; NPy, N-nitrosopyrrolidine; DEN, diethylnitrosamine; DMN, dimethylnitrosamine; DMSO, dimethylsulfoxide.

References

Ames, B.N., McCann, J., Yamasaki, E. (1975): Methods for detecting carcinogens and mutagens with the *Salmonella*/mammalian-microsome mutagenicity test. Mutat. Res. *31*:347–364.

Butler, W.H., Greenblatt, M., Lijinsky, W. (1969): Carcinogenesis in rats by aflatoxins B1, G1 and B2. Cancer Res. *29*:2206–2211.

Defais, M., Caillet-Fauquet, P., Fox, M.S., Radman, M. (1976): Induction kinetics of mutagenic DNA repair activity in *E. coli* following ultraviolet irradiation. Mol. Gen. Genet. *148*:125–130.

Eisen, H.A., Fuerst, C.R., Siminovitch, L., Thomas, R., Lanebert, L., Pereira Da Silva, L., Jacob, F. (1966): Genetics and physiology of defective lysogeny in K12 (λ): studies of early mutants. Virology *30*:224–241.

Elespuru, R.K., Yarmolinsky, M.B. (1979): A colorimetric assay of lysogenic induction designed for screening potential carcinogenic and carcinostatic agents. Environ. Mutagen. *1*:65–78.

Endo, H., Ishizawa, M., Kumiya, T. (1963a): Induction of bacteriophage formation in lysogenic bacteria by a potent carcinogen, 4-Nitroquinoline-1-oxide, and its derivatives. Nature (Lond.) *198*:195–196.

Endo, H., Ishizawa, M., Kamiya, T., Sonoda, S. (1963b): Relation between tumoricidal and prophage-inducing action. Nature (Lond.) *198*:258–260.

Fleck, W. (1968): Eine neue mikrobiologische Screening-Methode für die Suche nach potentiellen Carcinostatica und Virostatica mit Wirkung im Nucleinsäurestoffwechsel. Z.Allg.Mikrobiol. *8*:139–144.

Fleck, W.F. (1974): Development of microbiological screening methods for detection of new antibiotics. Postepy. Hig. Med. Dosw. *28*:479–498.

Geissler, E. (1967): Lysogenie und virale Kanzerogenese. Arch. Geschwulstforsch. *29*:355–372.

Harris, A.W., Mount, D.W.A., Fuerst, C.R., Siminovitch, L. (1967): Mutations in bacteriophage lambda affecting host cell lysis. Virology *32*:553–569.

Heinemann, B. (1971): Prophage induction in lysogenic bacteria as a method of detecting potential mutagenic, carcinogenic, carcinostatic and teratogenic agents. In Hollaender, A. (ed.): Chemical Mutagens: Principles and Methods for Their Detection. New York: Plenum, vol. 1, pp. 235–266.

Levine, A., Moreau, P.L., Sedgwick, S.G., Devoret, R., Adhya, S., Gottesman, M., Das, A. (1978): Expression of a bacterial gene turned on by a potent carcinogen. Mutat. Res. *50*:29–35.

Lijinsky, W., Butler, W.H. (1966): Purification and toxicity of aflatoxin G1. Proc. Soc. Exper. Biol. Med. *123*:151–154.

Lwoff, A. (1953): Lysogeny. Bacteriol. Rev. *17*:269–337.

McCann, J., Choi, E., Yamasaki, E., Ames, B.N. (1975): Detection of carcinogens as mutagens in the *Salmonella*/microsome test: Assay of 300 chemicals. Proc. Natl. Acad. Sci. USA *72*:5135–5139.

Miller, J.H. (1972): Experiments in Molecular Genetics. Cold Spring Harbor, New York: Cold Spring Harbor Laboratory, pp. 352–355.

Moreau, P., Bailone, A., Devoret, R. (1976): Prophage λ induction in *Escherichia coli* K12 envA uvrB: A highly sensitive test for potential carcinogens. Proc. Natl. Acad. Sci. USA *73*:3700–3704.

Moreau, P., Devoret, R. (1977): Potential carcinogens tested by induction and mutagenesis of prophage λ in *Escherichia coli* K12. In Hiatt, H., Watson, J.D., Winsten, J.A. (eds.): Origins of Human Cancer. Book C, Cold Spring Harbor, New York: Cold Spring Harbor Laboratory, pp. 1451–1472.

Nijkamp, H.J.J., Bvre, K., Szybalski, W. (1970): Controls of rightward transcription in coliphage λ. J. Mol. Biol. *54*:599–604.

Normark, S., Boman, H.G., Matsson, E. (1969): Mutant of *Escherichia coli* with anomalous cell division and ability to decrease episomally and chromosomally mediated resistance to ampicillin and several other antibiotics. J. Bacteriol. *97*:1334–1342.

Price, K.E., Buck, R.E., Lein, J. (1964): System for detecting inducers of lysogenic *Escherichia coli* W1709 (λ) and its applicability as a screen for antineoplastic antibiotics. Appl. Microbiol. *12*:428–435.

Seeburg, E. (1978): Reconstitution of an *Escherichia coli* repair endonuclease activity from the separated $uvrA^+$ and $uvrB^+$ / $uvrC^+$ gene products. Proc. Natl. Acad. Sci. USA *75*:2569–2573.

Smith, C.L., Oishi, M. (1976): The molecular mechanism of virus induction. I. A procedure for the biochemical assay of prophage induction. Molec. Gen. Genet. *148*:131–138.

Smith, C.L., Oishi, M. (1978): Early events and mechanisms in the induction of bacterial SOS functions: Analysis of the phage repressor inactivation process *in vivo*. Proc. Natl. Acad. Sci. USA *75*:1657–1661.

Vogel, M.J., Bonner, D.M. (1956): Acetylornithinase of *Escherichia coli*: Partial purification and some properties. J. Biol. Chem. *218*:97–106.

Witkin, E.M. (1976): Ultraviolet mutagenesis and inducible DNA repair in *Escherichia coli*. Bacteriol. Rev. *40*:869–907.

2

Detection of DNA-Modifying Agents by Analyzing the Lesions Introduced into Purified PM2 DNA*

URS KUHNLEIN, SIU SING TSANG, AND JANE EDWARDS

Principle

In this chapter we describe a simple short-term test for ultimate carcinogens (carcinogens which do not need metabolic activation). It is based on measuring directly the lesions introduced into DNA rather than the biological consequences of DNA damage. The double-stranded circular DNA of phage PM2 is incubated with the agent to be tested, separated from the unreacted chemical, and then tested for DNA lesions by methods which involve simple incubation steps and filtration through nitrocellulose filters. Nitrocellulose filters selectively retain denatured DNA or DNA containing single-stranded regions. This property can be used to measure single- and double-strand breaks (Center and Richardson, 1970), crosslinks, damages which locally denature

the DNA helix, damages which render the phosphodiester backbone sensitive to hydrolysis, and damages which increase the depurination or depyrimidination rate of the DNA. A diagram of the proposed testing scheme is presented in Figure 2.1. The assays for the different DNA lesions are based on the following principles:

Assay for lesions which locally denature the DNA. The modified PM2 DNA is filtered through a nitrocellulose filter at neutral pH. The filters will selectively retain DNA containing locally denatured molecules. Assuming a Poisson distribution, the average number of filter binding sites is $b = -\ln P$, where P is the fraction of DNA which is not retained by the filter.

Assay for crosslinks. The PM2 DNA is exposed to an alkali pH of 13.2 or higher and then neutralized. This treatment irreversibly denatures PM2 DNA unless the two strands of the DNA helix are held in register by interstrand crosslinks (Becker *et al.*, 1964; Pouwels *et al.*, 1969). The nitrocellulose filter will then retain the DNA

*Supported by the National Cancer Institute of Canada. Dr. U. Kuhnlein is a Research Scholar of the N.C.I. of Canada and S.S. Tsang is a Research Student of the N.C.I. of Canada.

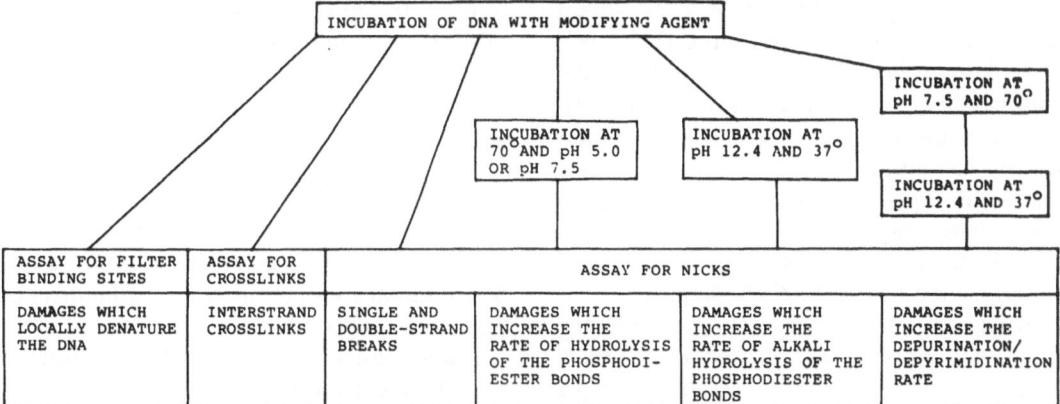

Figure 2.1. Diagram of proposed testing scheme for DNA modifications.

which has not been crosslinked, whereas crosslinked DNA will pass through the filter. Based on a Poisson distribution, the average number of crosslinks per DNA molecule (c) is ln (P/N). P and N are the fractions of crosslinked and untreated DNA, respectively, which bind to the filter. This equation takes into account that a small fraction of DNA without crosslinks (5%–10%) will also renature after the alkali treatment. The calculation assumes that one crosslink per DNA molecule is enough to prevent alkali denaturation.

Assay for single- and double-strand breaks (DNA nicking assay). Single- and double-strand breaks in DNA circles can also be determined with a simple filter assay. The DNA is exposed for 2 min to pH 12.4 and then neutralized. This treatment denatures nicked DNA circles but leaves covalently closed circles double-stranded (Center and Richardson, 1970; Pouwels *et al.*, 1968). The denatured DNA is then determined by filtration through nitrocellulose filters, which will retain the single-stranded DNA. The number of nicks per molecule (n) is calculated by assuming a Poisson distribution with the equation $n = -\ln P$, where P is the fraction of DNA which passes through the filter.

Assay for alkali-labile sites. DNA modifications which lead to an increased rate of alkali hydrolysis of the phosphodiester backbone can be measured by a simple modification of the nicking assay. Instead of neutralizing the DNA solution immediately after the alkali exposure, the DNA is left at pH 12.4 at room temperature or at 37° C. After this incubation step, the nicking assay is completed as usual.

Assay for heat-labile sites. PM2 DNA is rather stable when incubated at 70° C in the presence of high salt. Only about 10% of the DNA becomes nicked during incubation for 30 min at a pH between 6 and 10. A more acidic pH, however, increases the rate of hydrolysis significantly.

In order to test DNA for lesions which increase the heat sensitivity of the DNA phosphodiester bonds, we propose to incubate the modified DNA at pH 5.0 and pH 7.5. Alkali conditions would presumably reveal the same sites as incubation at pH 12.4 and 37° C. The resulting DNA breaks are determined with the standard DNA nicking assay.

Assay for DNA damages which increase the depurination/depyrimidination rate. In order to detect DNA damages which affect the stability of the glycosylic bond between base and deoxyribose, DNA is heated at 70° C at pH 7.5 to release the labile bases. The resulting apurinic or apyrimidinic sites are then hydrolyzed by incubating the DNA at pH 12.4 for 40 min at 37° C, and the nicks determined with the filter assay.

Interpretation of results. The order in which the tests are applied is important, since the result of each assay is necessary for the interpretation of the subsequent assays. First, the number of sites which lead to local denaturation of the DNA has to be known in order to evaluate the assay for crosslinks and to correct the values obtained in the nicking assay. If sites which locally denature the DNA are the major lesion, it is not possible to measure the other damages since the DNA will always bind to the nitrocellulose filter regardless of the presence of other DNA damages.

A high number of crosslinks, on the other hand, will obscure the presence of DNA breaks, but it will still be possible to measure the local denaturation sites.

An excess of DNA breaks will hide the presence of crosslinks, since denaturation of the DNA will create molecules which are partially single-stranded (tails and gaps). Such molecules will bind to the filter and not be recognized as being crosslinked.

Thus, the proposed scheme will generally be limited to detect the most dominant DNA lesions.

Methods

Growth and Maintenance of Phage PM2 and Pseudomonas Bal 31 (Espejo and Canelo, 1968)

Media and solutions. Bal Broth: 12 g $MgSO_4 \cdot 7H_2O$, 26 g NaCl, 8 g bacto-nutrient broth (Difco), 1 l distilled water. Sterilize, let cool, and add 10 ml 1 M $CaCl_2$, 3.5 ml 20% KCl, and 10 ml 1 M Tris-HCl (pH 7.5). Bal top agar: 1 l Bal-broth supplemented with 5 g bacto-agar (Difco). Bal plates and slants: 1 l Bal-broth supplemented with 23 g bacto-agar.

Preparation of PM2 phage stocks. Pseudomonas Bal 31 is grown at 28° C in an Erlenmeyer flask in Bal-broth stirred vigorously with a magnetic stirrer or by using a shaker bath. At a density of 2×10^7/ml, the bacteria are infected with PM2 at a multiplicity of 10^{-3} and incubated overnight. The bacterial debris and unlysed bacteria are removed by centrifuging for 10

min at 10,000 g. The supernatant is stored at 4° C. Typical phage yields are 10^{11}/ml.

Plaque assay for PM2. Pseudomonas Bal 31 is grown overnight at room temperature with aeration or in a shaker bath. An aliquot of 0.1 ml bacterial culture is mixed with 0.1 ml of PM2 (diluted in Bal broth) and 2.5 ml Bal top agar (heated to 50° C). The mixture is poured on Bal plates and left overnight at room temperature. Plaques appear after approximately 8 hours.

Maintenance of PM2 stocks and Bal 31 stocks. We routinely store PM2 at 4° C. However, the phage has a half-life of only about 1–2 weeks and new stocks have to be grown every half year. Further, since PM2 contains lipids, detergents and chloroform have to be avoided. Bal 31 can be kept on a slant at 4° C.

Preparation of ³H-labeled PM2 DNA (Espejo and Canelo, 1968; Espejo et al., 1969)

For the isolation of ³H-labeled PM2 DNA, bacteria are grown at 28° C in 1.5 l Bal broth supplemented with 10 mM Tris-HCl (pH 7.5) in a 3 l Erlenmeyer flask. The medium is stirred vigorously with a magnetic stirrer. When the cells reach a density of 5×10^8 per ml, 0.15 g of deoxyadenosine are added to the medium and 5 min later the cells are infected with PM2 at an MOI of 10. Five min after infection 1.5 mC of (³H)-methyl thymidine (specific activity 50 C/mmole) is added and the culture incubated overnight.

The bacterial debris is removed by centrifuging for 10 min at 10,000 g. The phage is then pelleted by centrifuging in a Beckman #21 rotor for 3 hr at 20,000 rpm and resuspended in 15 ml RB-buffer (1 M NaCl, 20 mM Tris-HCl; pH 8.0). It is easiest to let the phage pellets soak overnight before resuspension. The phage solution is then spun again at 10,000 g for 15 min to remove debris and at 50,000 rpm in a Beckman #50 rotor for 50 min to pellet the phage. The pellet is again resuspended in

15 ml RB-buffer with a Pasteur pipet. This cycle of low and high speed centrifugation is repeated once more. CsCl is then added to the phage solution to a density of 1.28 (0.3588 × weight of phage solution). The solution is distributed into two polyallomer tubes and centrifuged for 24 hr at 40,000 rpm and 20° C in a Beckman #50 rotor. Usually two bands are visible. The upper and major band contains the phage. It is collected with a syringe through the side of the tube.

The purified phage is dialyzed for 3 hr or longer against 1 l of BE-buffer (0.1 M NaCl, 1 mM EDTA, 20 mM Tris-HCl; pH 7.5). At room temperature sodium dodecyl sulfate (10%) is added by drop until the solution clears and the DNA extracted with an equal volume of phenol saturated with BE-buffer. The aqueous phase (aqueous 1) is re-extracted with 1/2 volume of BE saturated phenol (phenol 2) and the phenol phase with 1/2 volume of BE-buffer (aqueous 2). "Aqueous 2" is extracted with "phenol 2" and combined with "aqueous 1." With this extraction scheme every phase is extracted twice.

The extracted DNA is dialyzed for 12 hr periods twice against 1 l of 1M NaCl, 1mM EDTA, 10mM Tris-HCl (pH 7.5); twice against 1 l of 1 mM EDTA, 10 mM Tris-HCl (pH 7.5) and once against 1 l of 10 mM Tris (pH 7.5). The DNA is stored at 4° C.

Typical yields for a 1.5 l culture are 10–15 μmole DNA nucleotide with a radioactivity of 6000–8000 cpm/nmole. About 5% of the freshly prepared DNA is already nicked. Storage of DNA increases the percentage of nicked DNA by 5%–10% per month and seems to depend on the purity of the ^{3}H-thymidine.

Treatment with DNA-Modifying Agents and Re-isolation of the DNA

An aliquot of PM2 DNA (50 μM DNA nucleotide in 10 mM Tris-HCl, pH 7.5) is treated with the DNA-modifying agent in a reaction volume of 0.35 ml. After incubation (usually 30 min at 37° C) the DNA is re-isolated by one of the following methods: (1) The reaction is diluted by adding 0.65 ml of 10 mM Tris-HCl (pH 7.5) and extracted with 1 ml of phenol saturated with 10 mM Tris-HCl (pH 7.5) or by ether. The aqueous phase is collected and dialyzed overnight against buffer A (0.1 M NaCl, 0.01 M sodium citrate, 1 mM Tris-HCl; pH 7.5); (2) The reaction is chilled, 0.1 g of sucrose is added, and the solution layered between the eluant phase and resin phase of a 1 × 9 cm G75 Sephadex column which had been equilibrated with buffer A. The column is run with buffer A at a rate of 0.25 ml/min and fractions of 0.33 ml are collected. Ten μl of each fraction is assayed for radioactivity. The three fractions containing most activity are pooled. Freshly poured columns have to be run once with an aliquot of 5 mg Bovine serum albumin in order to saturate the sites which adsorb DNA.

Analysis of DNA Damages

The assays for DNA damages are carried out with 50 μl aliquots of the dialysate or the Sephadex pool which contains 0.1 M NaCl, 0.01 M sodium citrate, 1 mM Tris-HCl (pH 7.5), and abut 15 μM DNA.

Local denaturation sites. After adding successively 5 μl 100 mM Tris-HCl (pH 7.5), 0.15 ml of SE buffer (0.01% sodium dodecyl sulphate, 2.5 mM EDTA-NaOH; pH 7.0) and 5 ml of NT buffer (1 M NaCl, 50 mM Tris-HCl; pH 8.0), the solution is filtered through squares of nitrocellulose filter paper. The squares (2.4 × 2.4 cm) are best cut from 33 × 56 cm sheets (Schleicher and Schuell, type BA 85) and stored in NT buffer. For filtration a standard Millipore filter set-up can be used and the flow rate can be as high as 100 ml/min.

After filtering the sample, the tube is rinsed with 2 ml NT buffer. The filter is then washed with another 2 ml NT buffer and 5 ml of 0.3 M NaCl, 0.03 ml sodium citrate. The filters are dried and the radioactivity determined by liquid scintillation counting. Total DNA is determined by measuring the radioactivity of a DNA aliquot spotted on a blank filter.

Crosslinks. After adding 5 μl of 100 mM Tris-HCl (pH 7.5) and 0.15 ml SE buffer, the DNA is denatured with 0.4 ml of 0.15 M K_2HPO_4, 2.5 N KOH. After 2 min at room temperature, the reaction mixtures are neutralized with 0.2 ml of 0.5 M KH_2PO_4, 5 N HCl. Five ml NT buffer is then added and the filtration carried out as described for the determination of local denaturation sites.

Single- and double-strand breaks. Five μl of 100 mM Tris-HCl (pH 7.5) and 0.15 ml SE buffer are first added to the DNA aliquot. The DNA is then exposed to 0.2 ml of 0.3 M K_2HPO_4-KOH (pH 12.4) for 2 min at room temperature. The solution is neutralized with 0.1 ml of 1 M KH_2PO_4-HCl (pH 4.0); 0.2 ml of 5 M NaCl and 5 ml of NT buffer are added and the filtration through nitrocellulose filters carried out as described above.

Alkali-labile sites. The assay is the same as for DNA breaks except that the reactions are incubated for 40 min at 37° C after the addition of 0.2 ml of 0.3 M K_2HPO_4-KOH (pH 12.4).

Heat-labile sites at pH 5.0 and 7.5. For the determination of heat-labile sites at acid pH, the DNA aliquots are made pH 5.0 by adding 5 μl of 40 mM citric acid. After an incubation for 20 min at 70° C, the DNA is chilled; 0.15 ml of SE buffer is added and the single- and double-strand breaks determined as described above. For the determination of heat-labile sites at pH 7.5, the pH is adjusted by adding 5 μl of 100 mM Tris-HCl pH 7.5 and the incubation at 70° C is for 30 min.

Depurination/depyrimidination rate. The assay is the same as for DNA breaks except that the DNA solutions are incubated for 30 min at 70°C after adding 5 μl of 100 mM Tris-HCl (pH 7.5) and again for 40 min at 37° C after adding 0.2 ml of 0.3 M K_2HPO_4-KOH (pH 12.4).

Chemicals Tested

We have tested a series of DNA-modifying agents which introduce damages representative for each of the classes of DNA lesions. N-acetoxy acetylaminofluorene (N-acetoxy AAF) is known to render DNA susceptible to single-strand specific endonucleases, presumably by local denaturation of the double helix (Fuchs, 1975; Legerski et al., 1977). Such denaturation sites also bind DNA to nitrocellulose filters (Fig. 2.2 and Table 2.1). N-OH AAF, a carcinogen which needs to be activated in order to react with DNA, does not lead to filter binding. Cysteine, a hydroxyl radical-producing agent, also introduced filter binding sites into DNA, whereas nitrous acid, MNNG, and UV-light do not introduce these types of lesions (Table 2.1).

An agent known to crosslink DNA is nitrous acid (Becker et al., 1964). Figure 2.3 shows the amount of DNA which binds to the filter after alkali denaturation. After 60

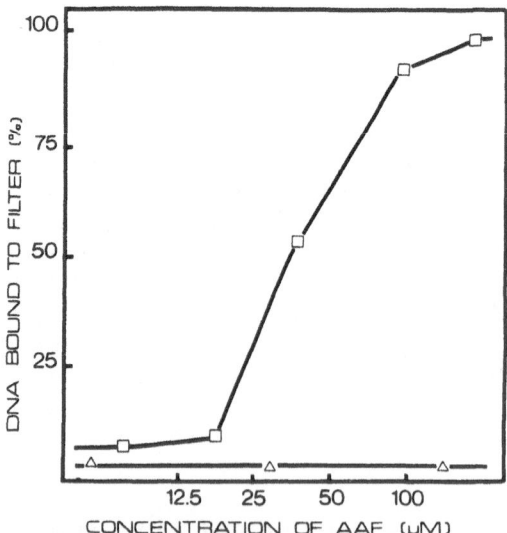

Figure 2.2. Filter binding of DNA after N-acetoxy AAF (□) and N-OH AAF (△) treatment. The reactions (50 μl) contained 10mM Tris-HCl (pH 7.5) and 5% DMSO. Filter binding was determined without re-extraction of the DNA.

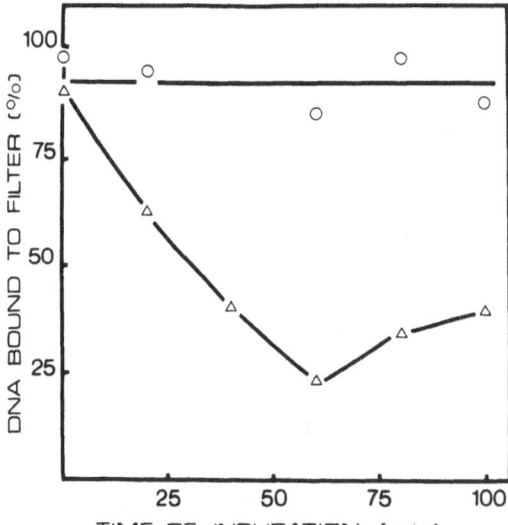

Figure 2.3. Crosslinking of DNA by nitrous acid. The reaction conditions were as described by Legerski *et al.* (1977) with a volume of 0.5 ml. 50 μl aliquots were diluted into 0.15 ml of SE buffer and directly assayed for crosslinks. △, DNA incubated with nitrous acid; ○, DNA incubated without nitrous acid but at the same ionic strength.

min of incubation with nitrous acid, only 25% of the DNA binds to the filter, indicating that 75% of the DNA contains crosslinks. MNNG and UV-light also introduce DNA crosslinks (Table 2.1).

Cysteine and ascorbate, both hydroxyl radical-producing agents, efficiently cleave PM2 DNA (Fig. 2.4). Hydroxyl radicals are produced during auto-oxidation of these compounds, a process which is inhibited by EDTA or catalase (Misra, 1974).

Lesions which render DNA heat-labile are introduced by nitrous acid (a deaminating agent) and at high doses of UV light (Table 2.1).

Among the methylating and ethylating agents, known to increase the depurination/depyrimidination rate, we have tested ethyl methane sulfonate (EMS) (Fig. 2.5) and N-methyl-N′-nitrosoguanidine (MNNG) (Table 2.1). UV light also increases the depurination/depyrimidination rate of DNA.

Disadvantages and Advantages of the Assay

It is clear that this assay cannot replace any of the standard short-term tests. First, it is limited to ultimate carcinogens because of interference of DNA endonucleases which are present in microsomal extracts. Further, the assay measures DNA damages without informing about the mutagenic or carcinogenic potential of a given lesion. Thus, MNNG is positive in this test not because of the O6-guanine alkylation but because of the more frequent N7-guanine alkylation, which is not thought to contribute significantly to the carcinogenicity of MNNG. The assay will also produce false positives, since many chemicals will react with purified DNA but are not able to reach the DNA of the tester organism.

However, the assay can complement biological tests in several ways. First, the toxicity of solvents or of the carcinogen itself does not interfere with the assay system. It is possible to react carcinogens with DNA in the presence of 50% DMSO or 50%

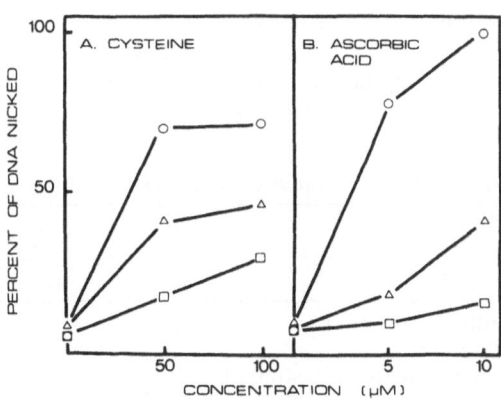

Figure 2.4. Nicking of DNA by cysteine (A) and ascorbic acid (B). The reaction mixtures (50 μl) contained 50 μM DNA, 10^5 M EDTA, 5×10^{-4} M CuSO$_4$, 10mM Tris-HCl (pH 7.5), and the concentrations of cysteine and ascorbic acid indicated. ○, no additions; □, 2 mM EDTA; △, 4 μ/ml catalase. Cysteine and ascorbic acid were neutralized with NaOH prior to the experiment.

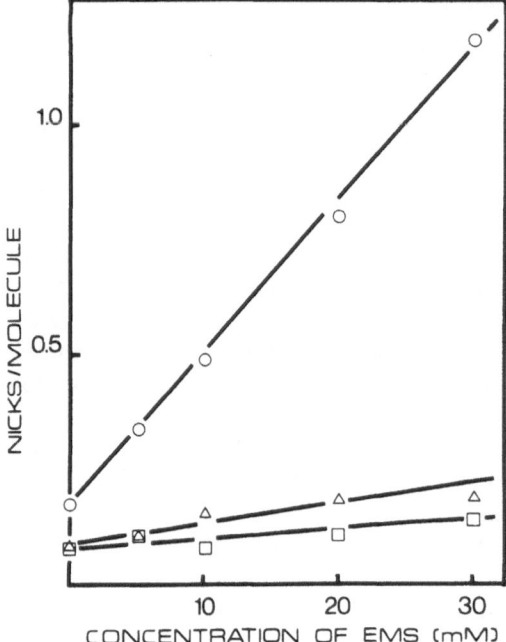

Figure 2.5. Depurination rate of EMS-treated DNA. The reaction mixtures contained 50 μM DNA, 10mM Tris-HCl (pH 7.5) and the concentrations of EMS indicated. After 30 min at 37° C the reaction mixtures were dialyzed overnight against 10mM Tris-HCl (pH 7.5) at 4° C. Then 1 part of 1 M NaCl, 0.1 sodium citrate was added to 9 parts of the DNA solutions and 50 μl aliquots of the mixtures were treated as follows: □,incubation for 15 min at 70° C; △, incubation overnight at room temperature after adding 0.15 ml SE buffer and 0.2 ml 0.3 M K₂HPO₄-KOH (pH 12.4); ○, incubation for 15 min at 70° C, followed by addition of 0.15 ml ES buffer and 0.2 ml 0.3 M K₂HPO₄-KOH (pH 12.4) and incubation overnight at room temperature. The nicking assays were then completed as indicated in Materials and Methods.

ethanol without affecting the subsequent analysis of the DNA damages, thus allowing the detection of carcinogens with a low solubility in aqueous media. The insensitivity of the assay to toxic substances is also important for the screening of complex mixtures of carcinogenic and toxic substances.

Second, the short-term test using purified DNA provides a tool to identify cellular protection mechanisms against carcinogens by comparing it with short-term tests using biological systems. For example, the nicking of purified DNA by cysteine can be detected at concentrations of less than 10^5 M, but cysteine does not mutate bacterial tester strains at concentrations as high as 5×10^{-2} M (Rosin and Stich, 1978). In

Table 2.1. *Classification of DNA Lesions*

	N-acetoxy AAF (53 μM)	Nitrous acid (60 min)	Cysteine (25 μM)	MNNG (100 μM)	UV light (980 J/m²)
Filter binding sites	123	4	53	2	0
Crosslinks	—[a]	139	—[b]	13	22
Single- and double-strand breaks	—	3[c]	247	1	2
Hydrolysis rate at pH 5.0 and 70° C	—	61	—	8	14
Hydrolysis rate at pH 7.5 and 70° C	—	1	—	2	32
Hydrolysis rate at pH 12.4 and 37° C	—	8	—	0	7
Depurination/depyrimidination rate	—	32	—	51	63

All reactions were carried out as described in the figure legends or the Methods section. After the reactions with the chemicals the DNA was re-extracted by Sephadex filtration and assayed for DNA modifications. All values are corrected for the values of untreated DNA.

[a]Values are not listed because of interference by local denaturation sites.

[b]Values are not listed because of interference by the high levels of DNA breaks.

[c]Values might be underestimated because of the presence of crosslinks.

human fibroblasts on the other hand, cysteine causes DNA damages (Stich *et al.*, 1978). The different response is probably due to the high level of catalase and superoxide dismutase in bacteria, which prevents the formation of hydroxyl radicals. The knowledge of such protection mechanisms is important for a thorough assessment of the carcinogenic potential of a chemical and for the identification of carcinogenic factors.

References

Becker Jr., E.F., Zimmerman, B.K., Geiduschek, E.P. (1964): Structure and function of crosslinked DNA, I. Reversible denaturation and *Bacillus subtilis* transformation. J. Mol. Biol. *8*: 377–391.

Center, M.S., Richardson, C.C. (1970): An endonuclease induced after infection of *Escherichia coli* with bacteriophage T7, I. Purification and properties of the enzyme. J. Biol. Chem. *245*: 6285–6291.

Espejo, R.T., Canelo, E.S. (1968): Properties of bacteriophage PM2, a lipid-containing bacterial virus. Virology *34*: 738–747.

Espejo, R.T., Canelo, E.S., Sinsheimer, R.L. (1969) DNA of bacteriophage PM2: a closed circular double-stranded molecule. Proc. Natl. Acad. Sci. USA *63*: 1164–1168.

Fuchs, R.P.P. (1975): In vitro recognition of carcinogen-induced local denaturation sites in native DNA by S_1 endonuclease from *Aspergillus oryzae*. Nature (Lond.) *257*: 151–152.

Legerski, R.J., Gray, H.B., Jr., Robberson, D.L. (1977): A sensitive endonuclease probe for lesions in deoxyribonucleic acid helix structure produced by carcinogenic or mutagenic agents. J. Biol. Chem. *252*: 8740–8746.

Misra, H.P. (1974): Generation of superoxide free radicals during the autooxidation of thiols. J. Mol. Biol. *249*: 2151–2155.

Pouwels, P.H., Knijnenburg, C.M., van Rotterdam, J., Cohen, J.A., Jansz, H.S. (1968): Structure of the replicative form of ϕX174, VI. Studies on alkali-denatured double-stranded ϕX174 DNA. J. Mol. Biol. *32*: 169–182.

Pouwels, P.H., van Rotterdam, J., Cohen, J.A. (1969): Structure of the replicative form of bacteriophage ϕX174, VII. Renaturation of denatured double-stranded ϕX174 DNA. J. Mol. Biol. *40*: 379–390.

Rosin, M.P., Stich, H.F. (1978): The inhibitory effect of cysteine on the mutagenic activity of several carcinogens. Mutat. Res. *54*: 73–81.

Stich, H.F., Wei, L., Lam, P. (1978): The need for a mammalian test system for mutagens: Action of some reducing agents. Cancer Letters *5*: 199–204.

3

Reactivation of Viruses

Andrew J. Rainbow

Introduction

The human population is constantly exposed to a large spectrum of chemicals and radiation which can cause alterations in the structure of DNA. Since there is a strong correlation between the ability of an agent to produce mutations in DNA and its ability to cause cancer, it has long been suspected that damage to DNA plays a causal role in human carcinogenesis. Normal human cells have been shown to possess a number of different mechanisms capable of repairing damage to DNA. Consequently, the level and fidelity of DNA repair may play a more central role in carcinogenesis than the type and extent of the DNA alterations produced following physical or chemical mutagens. Recent investigations which show that several rare human autosomal recessive genetic diseases that are accompanied by an increased cancer incidence are associated with defects in DNA repair add strong support to this idea (Marx, 1978; Setlow, 1978). Although

not all cancers arise from defects in repair, it seems reasonable to expect that unrepaired damage to DNA has a high carcinogenic potential. Although the cancer-prone syndromes, which include xeroderma pigmentosum (XP), Fanconi's anemia (FA), Bloom's (B) syndrome, and ataxia telengiectasia (AT), are rare, the heterozygous carriers are relatively frequent, amounting to between 0.5 and 1 percent of the population for most of the syndromes. Heterozygous carriers themselves also have a greater risk of malignancy than the general population, and a reduced DNA repair capacity has also be detected in some of them. The DNA repair capacity of the human cells is thus thought to play an important role in human carcinogenesis. Consequently, in order to evaluate the potential carcinogenicity of a physical or chemical agent on the human population, we must be acutely aware of those factors which will affect its response. The identification of highly sensitive groups within the population is, therefore, of considerable interest.

Irradiated phage particles have been used extensively to probe the DNA repair mechanisms of bacterial cells. Such investigations have included host-cell reactivation (HCR) (Rupert and Harm, 1966), Weigle or UV-induced reactivation (Weigle, 1953), carcinogen-induced reactivation (Sarasin et al., 1977) and SOS functions (Witkin, 1976). The existence of a number of repair-deficient bacterial mutants has added considerably to our understanding of DNA repair mechanisms in bacterial cells. In a similar way, irradiated or chemically treated viruses can be used to probe the DNA repair mechanisms of mammalian cells and, in particular, the human cell. The existence of a number of repair-deficient human syndromes may thus prove to be of considerable value in elucidating the role of DNA damage and its repair in human carcinogenesis.

The majority of investigations utilizing viruses as probes for cellular DNA repair have employed either herpesvirus, simian virus 40 (SV40), or human adenovirus (Ad). All are double-stranded DNA viruses which replicate in the nucleus of a susceptible cell. It is assumed that, for each virus type, the viral DNA is processed, in part at least, by the same enzyme systems which process the cellular DNA of the host. Several different types of investigation have been performed: (a) infection of cells with irradiated or chemically treated virus and an examination of virus survival which yields information concerning the capacity of the host-cell to repair the induced DNA lesions (host-cell reactivation [HCR]); (b) an examination of the survival of treated virus in cells which have been exposed to radiation or chemicals before infection. This reveals any enhancement in viral reactivation (enhanced virus reactivation [ER]); (c) an examination of viral expression for untreated virus in cells which have been exposed to radiation or chemicals before or during infection. This also yields information on cellular repair capacity; and (d) an examination of the mutation rate of viral genes following infection

of treated cells with treated virus, which will indicate the magnitude of inducible error-prone DNA repair modes in the cell.

Host-Cell Reactivation

HCR of irradiated virus has been investigated in several different cell types, using a variety of techniques including: plaque formation of herpesvirus (Takebe et al., 1974; Lytle, 1971); SV40 (Abrahams and van der Eb, 1976); and Ad (Rainbow and Mak, 1973; Day, 1974); intranuclear inclusion body formation of adenovirus (Rainbow and Mak, 1972; Stich et al., 1974): T antigen formation and transformation frequency of SV40 (Aaronson and Lytle, 1970); one cycle herpesvirus yield from mass cultures (Rabson et al., 1969; Coppey et al., 1978); V antigen formation of adenovirus (Rainbow, 1978); and repair of adenovirus DNA lesions (Rainbow, 1974, 1977a). Many of these techniques have been successfully applied in the detection of a defective repair mechanism for UV-induced DNA damage in fibroblasts from patients with XP. Most XP strains show some defect in their ability to perform excision repair of UV-induced cyclobutane pyrimidine dimers (Setlow et al., 1969; Cleaver and Trosko, 1970), whereas variant XP strains show normal excision but some defect in post-replication repair of UV-induced DNA damage (Lehmann et al., 1977). Cell hybridization studies have shown that the mutations leading to the XP phenotype fall into at least seven complementation groups—A, B, C, D, E, F, G—and the XP variant (Kraemer et al. 1975; Takebe et al., 1978; Bootsma et al., 1970, 1979; Robbins et al., 1974). In 1969, Rabson and coworkers reported a reduced virus yield in XP fibroblasts as compared to fibroblasts from normal donors following infection with irradiated herpesvirus (Rabson et al., 1969). Subsequently, a reduced HCR for XP fibroblasts was obtained using "T" antigen formation and transformation frequency of SV40 (Aaronson and Lytle, 1970). HCR of V antigen formation for UV-irradiated adenovirus has recently been examined in XP fibroblasts from several of the different complementation groups as well as those from an XP variant (Fig. 3.1, Table 3.1). Similar studies of XP fibroblasts from different complementation groups have previously been reported using plaque formation of UV-irradiated adenovirus (Day, 1974, 1975a), UV-irradiated herpesvirus (Takebe et al., 1978), and infectious center formation of UV-irradiated SV40 DNA (Abrahams and van der Eb, 1976).

Similar HCR values for the various XP complementation groups were obtained using V antigen expression or plaque formation of UV-irradiated Ad 2. Groups A and D have between 3 to 7% normal repair, group B 11%, group C 11 to 35%, group E 47%, group F 22%, and XP variants 57 to 64%. The relative amount of HCR found in these studies correlates well with the relative amount of UV-induced unscheduled DNA synthesis reported for the different XP complementation groups (Robbins et al., 1974; Bootsma et al., 1970; Takebe et al., 1978) except for group D, which shows far more UV-induced unscheduled DNA synthesis than its HCR ability. This discrepancy is, as yet, unresolved but may result from a larger UV-induced repair component for group D cells. HCR using either UV-irradiated adenovirus or UV-irradiated SV40 DNA was capable of detecting a repair deficiency in both XP and XP variant cells indicating that the repair of both these viruses depends, in part at least, on the excision repair and post-replication repair mechanisms of the host cell. Photoreactivation has also been reported to be involved in the HCR of some UV-irradiated virus systems (Lytle et al., 1976a,b) including human fibroblasts (Wagner et al., 1975).

HCR of UV-irradiated herpesvirus is also reduced in XP strains, although the 30% HCR value obtained for an XP group A strain is considerably higher than re-

ported using adenovirus or SV40 (Lytle, 1972). This may indicate a greater independence of host-cell functions for the larger

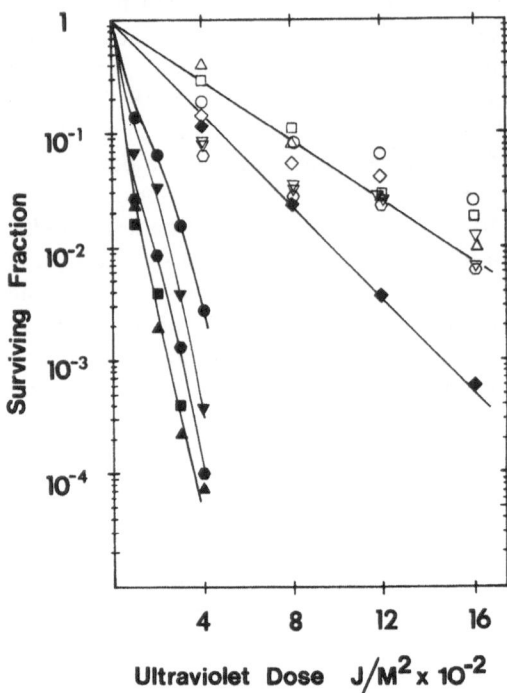

Figure 3.1. Survival of V antigen formation for UV-irradiated adenovirus type 2 in normal and xeroderma pigmentosum fibroblast strains. Adenovirus type 2 was prepared, diluted, irradiated, and used to infect fibroblast monolayers grown in eight-well chamber slides as described previously (Rainbow and Howes, 1977a). Virus was adsorbed for 2 hours, and at 48 hours after infection cell cultures were examined for the presence of antigen, using immunofluorescent staining. The number of V antigen positive cells was determined in duplicate at three serial dilutions for each treatment of the virus. The data points were then fitted to a straight line, using least squares analysis in order to obtain each survival point. Results were pooled for three separate experiments. Survival curves for all normal strains (open symbols) and each XP strain (closed symbols) were fitted by eye. Experiment 1: ○ CRL1119, ▽ A2, ◇ RE, ● XP12BE (A, 6%), ▼ XP2BE (C, 14%). Experiment 2: ○ GM964, ▽ A2 ■ XP5BE (D, 6%), ♦ XP4BE (variant 62%). Experiment 3: □ CRL1141, △ GM23, ▲ XP25RO (A, 6%) ● XP23OS (F, 22%).

The brackets following each XP strain show the complementation group and the percent HCR value obtained.

Table 3.1. *Host-Cell Reactivation of V Antigen Formation for Irradiated Adenovirus in Fibroblasts from Several Human Syndromes*

Fibroblast strain	Percent HCR[a]		References
	UV	Gamma	
Xeroderma pigmentosum[8b]	+++	++	Rainbow (1980) and unpublished data; Rainbow and Howes (1979)
Xeroderma pigmentosum variant[1]	++	+	Rainbow and Howes (1979)
Xeroderma pigmentosum heterozygotes[4]	++	+	Rainbow (1980)
Cockayne's syndrome[5]	+++	++	Rainbow and Howes (1980); Rainbow (unpublished data)
Fanconi's anemia[5]	++	++[c]	Rainbow and Howes (1977a, b)
Bloom's syndrome[4]	−	ND[d]	Krepinsky *et al.* (1979)
Bloom's syndrome (GM1492)[1]	++	ND	Krepinsky *et al.* (1979)
Progeria[3]	−	++	Rainbow and Howes (1979c); Rainbow (unpublished data)
Ataxia telangiectasia[1]	++	−	Rainbow (1978) and unpublished data
Huntington's chorea[4]	++	ND	Rainbow (unpublished data)

[a]D_{37} values of survival curve expressed as percentage of that obtained for normal strains (less than 50% normal, +++; 50-70%, ++; 70-85%, +; normal, −).

[b]Number of different cell strains tested.

[c]Determined for CRL1196 only.

[d]Not determined.

herpesvirus genome. The fact that some XP variant strains show normal HCR levels for UV-irradiated herpesvirus (Selsky and Greer, 1978; Lytle, 1978; Takebe *et al.*, 1978) further suggests that the repair of UV-damaged herpesvirus depends on cellular excision repair, but not post-replication repair. The fact that herpesvirus codes for its own DNA polymerase (Purifoy *et al.*, 1977) may allow it to be independent of the host-cell polymerases utilized for post-replication repair or other repair mechanisms for cellular DNA. HCR of UV-irradiated adenovirus in XP cells has also been studied by examining the fate of UV-induced alkaline labile lesions in the viral DNA after infection (Rainbow, 1977b) (Fig. 3.2). Repair of UV-induced alkaline labile damage in the viral DNA is detected 24 hours after infection of a normal fibroblast as well the XP4BE variant but is absent after infection of the XP10BE (group C) strain. Absent or substantially reduced repair of UV-induced alkaline labile DNA lesions in Ad 2 was also detected for the group D fibroblast strains XP5BE and XP7BE (data not shown). These results

suggest that in normal and XP variant cells, host-cell excision repair is involved in the repair of alkaline labile lesions in viral DNA. Since UV survival of adenovirus is known to reflect, in part, cellular post-replication repair, these results indicate that either the alkaline sucrose gradient technique is particularly insensitive or that post-replication repair does not remove alkaline lesions from UV-damaged DNA. The latter possibility is consistent with the current model of a bypass mechanism for post-replication repair (Lehmann, 1978).

A substantially reduced HCR for XP cells as compared to normal has also been detected for herpesvirus treated with N-acetoxy-2-acetylaminofluorene (Selsky and Greer, 1977) but not for herpesvirus treated with formaldehyde (Coppey and Nocentini, 1979), nitrogen mustard (Selsky and Greer, 1977) or X-rays (Lytle *et al.*, 1972). A reduced HCR for XP has been detected for adenovirus treated with gamma rays (Table 3.1) (Rainbow and Howes, 1979), nitrous acid (Day, 1975b), Benzo(a)pyrene diol-epoxide (Day *et al.*, 1978), Chlorpromazine plus near UV light

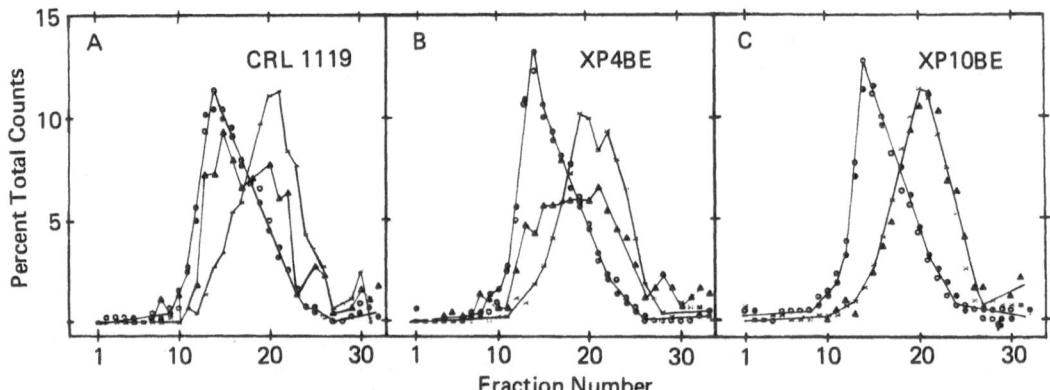

Figure 3.2. The fate of UV-induced alkaline labile DNA damage in adenovirus type 2 after infection of normal and xeroderma pigmentosum fibroblast strains. ^{3}H-labeled adenovirus type 2 was irradiated with 3.2×10^2 J/m^2 and used to infect fibroblasts at about 5 P.F.U. per cell. At 4 and 24 hours after infection samples of the infected cultures together with ^{14}C-labeled unirradiated marker virus were treated for 15 hours with alkaline lysing solution. The lysed samples were then analyzed by centrifugation on 17 ml alkaline sucrose density gradients in a similar manner to that described previously (Rainbow, 1977a). Gradient fractions were collected from the bottom of the tube. Radioactivity profiles for ^{14}C marker and ^{3}H-irradiated adenovirus are represented by the open circles and crosses, respectively, at 4 hours, and by close circles and triangles, respectively, at 24 hours after infection of fibroblast strains. A, CRL1119 (normal); B, XP4BE (variant); and C, XP10BE (group C).

(Day and DiMattina, 1977), and psoralen plus UV-light (Day *et al.*, 1975). The latter result is thought to reflect the fact that normal fibroblasts are incapable of repairing psoralen crosslinks in adenovirus DNA, but can repair psoralen-DNA mono-adducts, which are repaired poorly in XP cells.

HCR of gamma-irradiated adenovirus has also been studied by examining one cycle virus yields (Lee and Rainbow, 1977) and repair of single-strand viral DNA breaks (Rainbow, 1974). A reduced HCR of V antigen production for both UV and gamma-irradiated adenovirus type 2 has also been detected in fibroblasts from 4 XP heterozygotes representing 3 different complementation groups (Rainbow, 1980) (Fig. 3.3, Table 3.1). For a number of experiments, the HCR values obtained for each of the XP heterozygote strains ranged from 55 to 82% and 71 to 69% that obtained on normal strains for UV- and gamma-irradiated virus, respectively. Day has shown normal HRC of plaque formation for UV-irradiated Ad 2 on 4 XP heterozygous strains, 3 of which showed reduced HCR in the Ad 2 V antigen assay (Day, 1974). This difference in HCR using the different adenovirus assays presumably results from the fact that the HCR of V antigen formation measures the repair of viral DNA over a fixed time interval (48 hours) and this gives a measure of repair rate, whereas in the plaque assay, deficiencies in repair rate of the viral DNA may still result in eventual plaque formation when scored 16 to 18 days after infection. Although it remains to be seen whether the HCR for Vag production of UV-irradiated Ad 2 is capable of detecting repair deficiencies in all the other known heterozygous carriers, it appears to be a relatively simple technique with a good potential for identifying the cancer-prone heterozygote population. Normal levels of HCR in XP heterozygous strains have also been reported using UV-irradiated SV40 virus DNA (Abrahams and van der Eb, 1976), and herpesvirus treated with N-acetoxy-2-acetylamino-fluorene (N-AAAF) (Selsky and Greer, 1978). However, by examining the one-cycle viral yield following herpesvirus infection of fibroblasts, pretreated with UV,

Coppey *et al.* (1979a,b) have shown that the recovery of herpesvirus production capacity was significantly impaired in the XP heterozygous strain tested.

The effect of adding caffeine during the course of infection, on the survival of UV-irradiated virus has also been examined, using a number of different systems. HCR of plaque formation for UV-irradiated herpesvirus is reported to be inhibited by caffeine in normal human fibroblasts (Lytle, 1972) but not in XP variant cells (Selsky

and Greer, 1978) whereas no significant effect of caffeine was reported for HCR of herpesvirus treated with N-AAAF in normal XP or XP variant fibroblast strains. HCR of plaque formation for UV-irradiated adenovirus is significantly inhibited by caffeine in normal and XP variant cells (Day, 1974) but not in excision-deficient XP cells. As pointed out by Day, irradiated adenovirus is thus repaired by a caffeine-sensitive repair process in normal and XP variant cells which is either cellular excision repair or an excision-dependent repair process (Day, 1975c). Caffeine shows a greater effect on the survival of UV-irradiated adenovirus than on UV-irradiated herpesvirus in normal human cells, suggesting that a caffeine-sensitive recovery mechanism plays a greater role in the repair of adenovirus than in that of herpesvirus in normal human cells (Selsky and Greer, 1978). Addition of caffeine during infection also significantly inhibits the UV plaque survival of herpesvirus in CV-1 monkey kidney cells (Lytle, 1972), baby hamster kidney cells (Ross *et al.*, 1972), and primary rabbit kidney cells (Fogel *et al.*, 1979). Other types of treatment during viral infection also reduce HCR of UV-irradiated virus: acriflavin inhibits plaque survival of UV-irradiated herpesvirus in baby hamster kidney cells (Ross *et al.*, 1972) and hyperthermia inhibits inclusion body formation of UV-irradiated adenovirus in human fibroblasts (Lam and Stich, 1978).

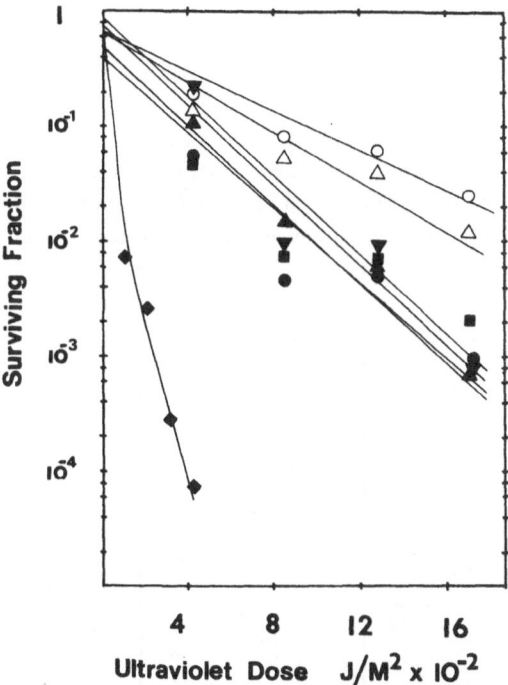

Figure 3.3. UV survival curves for V antigen formation of adenovirus type 2 in fibroblasts from xeroderma pigmentosum heterozygotes. Methods as for Figure 3.1. Survival curves for the normal and XP heterozygous strains were fitted by least squares regression analysis of the survival points. Results show a typical experiment using two normal strains, ○ CRL1119, △ RE, and four heterozygote strains, ● CRL1254 (mother of XP12BE), ▼ CRL1165 (mother of XP2BE), ▲ CRL1167 (father of XP2BE), ■ CRL1159 (mother of XP5BE). Percent HCR values obtained for this experiment: CRL1254 (59%); CRL1165 (54%);CRL1167 (55%), and CRL1159 (67%).

HCR for V antigen formation of irradiated adenovirus type 2 has been used extensively to examine the repair capacity of fibroblasts from a number of different human syndromes (Table 3.1) including FA (Rainbow and Howes, 1977a,b), AT (Rainbow, 1978), Cockayne's syndrome (CS) (Rainbow and Howes, 1980), progeria (P) (Rainbow and Howes, 1977c), Bloom's syndrome (B) (Krepinsky *et al.*, 1979), and Huntington's chorea (Rainbow, unpublished observations). HCR of UV-irradiated herpesvirus has also been examined for fibroblasts from 3 different Bloom's patients (Selsky *et al.*, 1978). Two strains showed normal HCR levels whereas one Bloom's strain GM1492 showed a reduced level of HCR, as was found using UV-irradiated adenovirus (Table 3.1). The DNA repair defect responsible for the reduced HCR in GM1492 is not clear at the present time. GM1492 cells themselves show colony survival after UV-irradiation similar to that obtained on normal strains (Krepinsky *et al.*, 1979).

Reduced HCR values for fibroblasts from Cockayne's syndrome have also been reported, using plaque formation of UV-irradiated adenovirus (Day and Ziolkowski, 1978), whereas UV-irradiated herpesvirus yields normal levels of HCR in CS fibroblasts (Ikenaga *et al.*, 1979). The FA line CRL1196 shows normal HCR levels for plaque formation of adenovirus treated with psoralen plus near UV light (Day *et al.*, 1975). Cells from FA patients have been reported to show reduced repair of crosslinks on their own DNA (Sasaki and Tonomura, 1973). Since psoralen plus near

UV light induces DNA crosslinks in adenovirus DNA, it has been suggested (Day *et al.*, 1975) that although FA and normal cells may differ in their ability to repair crosslinks in their own DNA, they may be alike in being unable to repair crosslinks in viral DNA. The HCR of 2 FA strains has also been investigated by examining V antigen formation of UV-irradiated Ad 12 at various times after infection (Rainbow and Howes, 1977b) (Fig. 3.4). HCR was substantially reduced in the 2 FA lines as compared to normal when V antigen scoring was done at 36 and 48 hours after infection, but not when assayed at 72 hours. This reduced rate of V antigen production for UV-irradiated Ad 12 in FA suggests a reduced rate of repair for some type of UV-induced viral DNA lesion for FA fibroblasts.

The detection of a reduced DNA repair capacity in fibroblasts from patients with CS, progeria, and Huntington's chorea, all non-cancer-prone conditions, indicates that

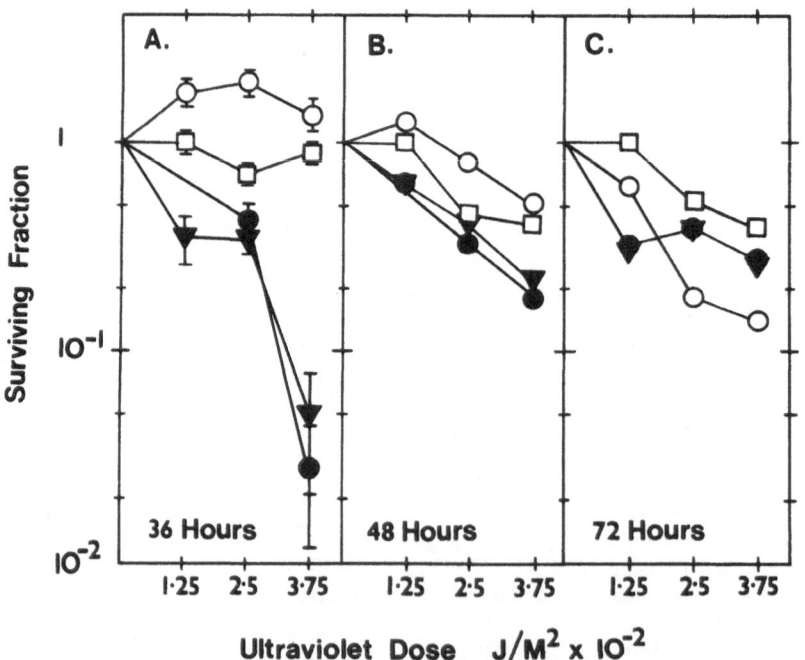

Figure 3.4. UV survival curves for V antigen formation of adenovirus type 12 in fibroblasts from two patients with Fanconi's anemia (closed symbols) and two normals (open symbols). Methods as for Figure 3.1. V antigen positive cells were scored at (A) 36 hours; (B) 48 hours; and (C) 72 hours after infection of □ GM726, ○ R.E., ▼ CRL1196, and ● GM646.

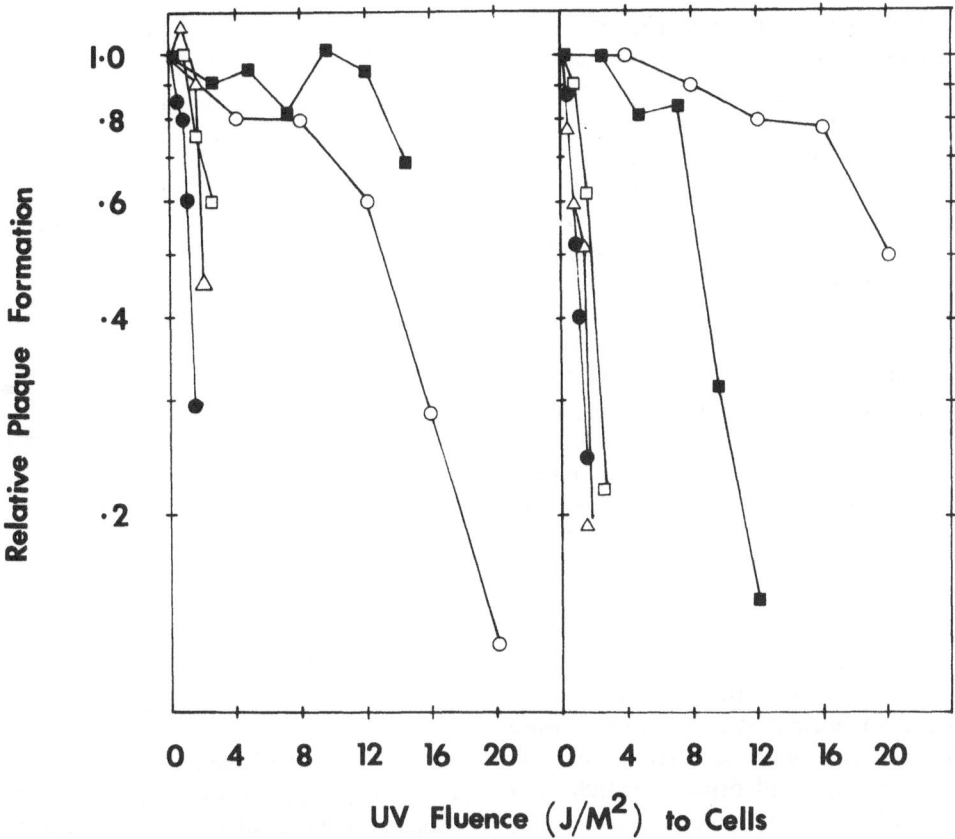

Figure 3.5. UV inactivation of cell capacity for herpes simplex virus plaque formation. Monolayers of normal human KD (○) and xeroderma pigmentosum (■) XP1BE; (△) XP7BE; (□) XP11BE; and (●) XP12BE fibroblasts were infected immediately following cell irradiation (left panel) or 4 days following cell irradiation (right panel). Absolute plaque formations for immediate infection were equal within 10% for all cells. Redrawn from Lytle *et al.* (1976).

a reduced DNA repair mechanism per se does not lead to elevated carcinogenesis.

Cellular Capacity for Virus Infection

Physical or chemical treatment of cell monolayers can decrease the cellular capacity to support virus infection. Such a decrease in capacity has been reported for herpesvirus following pretreatment of monkey kidney cells with either proflavin and light (Lytle and Hestler, 1976), angelicin or 8-methoxypsoralen and light (Coppey *et al.*, 1979a) or UV light (Lytle *et al.*, 1974; Coppey and Nocentini, *1976*; Bockstahler *et al.*, 1976). Capacity for virus infection is restored in monkey kidney cells when the

time between cell treatment and infection is increased. This is thought to result from the ability of repair mechanisms to remove damage from cellular DNA during this time period (Coppey *et al.*, 1979b).

The capacity of monkey kidney cells to support SV40 and herpesvirus infection has also been investigated following exposure of cells to several chemical carcinogens (Sarasin and Hanawalt, 1978; Lytle *et al.*, 1978). Lytle *et al.* (1976a,b) have found that pretreatment with UV light can also reduce the capacity of human fibroblasts to support herpesvirus infection (Fig. 3.4). In this investigation XP7BE, XP11BE, and XP12BE XP strains were found to be more sensitive to UV inactivation of capacity than XP1BE or normal strains (Fig. 3.5, left panel). Capacity curves for delayed infection (Fig. 3.5, right panel) showed a recovery of capacity for the normal strains, whereas all the XP strains tested showed a further decrease in capacity. Similar results have been reported for XP, XP variant, and XP heterozygote strains by Coppey et al. by examining the one-cycle herpesvirus yields in fibroblasts previously irradiated with UV (Coppey et al., 1979b; Coppey et al., 1978). A decreased cellular capacity for herpesvirus also exists in an XP strain as compared to normal strain following formaldehyde treatment (Coppey and Nocentini, 1979) or treatment with furocoumarins plus light (Coppey et al., 1979a). However, unlike the results obtained for UV treatment of cells, the recovery of cellular capacity from formaldehyde treatment was similar for both the normal and XP strains. XP strains also show a reduced capacity for adenovirus inclusion body formation as compared to normal strains when treated for 90 minutes with 4-nitroquinoline-1-oxide after virus adsorption (Stich and Laishes, 1975).

Enhanced Virus Reactivation

Several reports have shown that pretreatment of mammalian cells with physical or chemical agents can result in an enhanced survival for several UV-irradiated nuclear replicating DNA viruses such as herpesvirus (Bockstahler and Lytle, 1970), simian adenovirus (Bockstahler and Lytle, 1977), human adenovirus (Jeeves and Rainbow, 1979b), simian virus 40 (Sarasin and Hanawalt, 1978), or Kilham rat virus (Lytle, 1978). UV-enhanced reactivation (UVER) for UV-irradiated herpesvirus plaque formation has been reported following infection of several established cell lines, such as monkey kidney cells (Bockstahler et al., 1976), rat cells (Hellman et al., 1974), HeLa cells (Lytle et al., 1974), normal or XP human fibroblasts (Lytle et al., 1976a,b). UVER of UV-irradiated SV40 plaque formation has also been observed in monkey kidney cells (Sarasin and Hana-

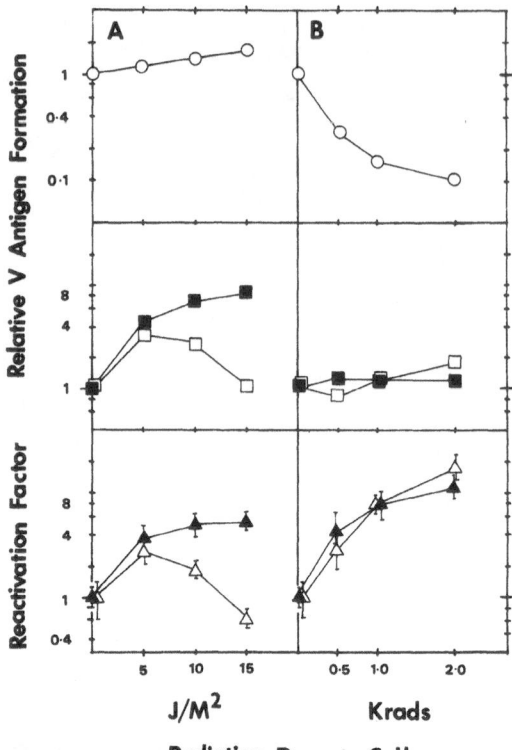

Figure 3.6. Enhanced reactivation of UV- and gamma-irradiated adenovirus type 2 in normal human fibroblasts for immediate infection of cells pretreated with (A) UV or (B) gamma rays. Top panels represent the survival of nonirradiated virus following immediate infection of fibroblast strain A 1 with either unirradiated ($\bigcirc$), UV-irradiated (10^3 J/m^2) ($\blacksquare$) or gamma-irradiated (2 Mrads) ($\square$) adenovirus. Bottom panels show the reactivation factors (ratio of values for irradiated virus to those for control virus) for UV-irradiated (10^3 J/m^2) ($\blacktriangle$) and gamma-irradiated (2 Mrads) virus ($\triangle$). Surviving fraction of virus on untreated cells was 4×10^{-3} and 9×10^{-2} for UV- and gamma-irradiated virus, respectively.

walt, 1978) and UVER of V antigen formation for UV-irradiated adenovirus has been found in human fibroblasts (Jeeves and Rainbow, unpublished data) (Fig. 3.6). X-ray ER of UV-irradiated herpesvirus plaque formation has been reported for both monkey kidney cells (Bockstahler and Lytle, 1977), and rat cells (Hellman et al., 1974), whereas γ-ray ER of V antigen formation for UV-irradiated adenovirus has been reported in human fibroblasts (Jeeves and Rainbow, 1979a).

Pretreatment of cells with several different chemical carcinogens has also been found to enhance the reactivation of UV-irradiated virus. ER of UV-irradiated herpesvirus plaquing was detected following pretreatment of CV-1 monkey kidney cells with aflatoxin B_1, N-acetoxy-2-AAF, N-hydroxy-2-AAF, and hydroxyurea but not following pretreatment with acetylaminofluorene or cycloheximide (Lytle and Goddard, 1979a,b; Lytle et al., 1978). ER of UV-irradiated herpesvirus has also been detected following pretreatment of primary rabbit kidney cells with caffeine, 5-bromodeoxyuridine, and hydroxyurea (Fogel et al., 1979). Pretreatment of CV-1 monkey kidney cells with metabolites of aflatoxin B_1, N-acetoxyacetylaminofluorene, methyl methanesulfonate, and ethyl methanesulfonate also enhanced the plaque survival of UV-irradiated SV40 (Sarasin and Hanawalt, 1978).

The phenomenon of enhanced virus reactivation in mammalian cells is similar to that observed by Weigle (1953) following pretreatment of bacterial cells. In the bacterial systems, the phenomenon involves the induction of an "error-prone" type of DNA that is highly mutagenic and is part of a number of so-called "SOS" functions (Radman, 1975). The precise mechanism leading to enhanced virus reactivation in mammalian cells is not known at the present time. It appears to be independent of excision repair (Lytle et al., 1976a,b) and multiplicity reactivation (Lytle, 1978). Several reports suggest that enhanced virus reactivation in CV-1 monkey kidney cells is induced and requires de novo protein synthesis (Lytle and Goddard, 1979a; DasGupta and Summers, 1978; Lytle et al., 1978; Sarasin and Hanawalt, 1978). The process may be initiated by DNA synthesis inhibition since agents which affect DNA synthesis all induce enhanced virus reactivation (Fogel et al., 1979; Lytle and Goddard, 1979a). This is of interest since DNA synthesis inhibition itself appears to be a good short-term test for chemical carcinogens (Painter, 1980).

Viral reactivation experiments have also been performed by following the course of infection as a function of time for one lytic cycle. This kind of investigation has shown that the kinetics of virus production after infection with UV-irradiated SV40 in UV-irradiated CV-1 monkey kidney cells is much faster than in untreated cells (Sarasin, 1978). This indicates that prior irradiation of the host cell leads to an increased kinetics in the expression of late viral genes from a UV-damaged template. A similar suggestion has been made for UV-irradiated adenovirus following pre-gamma irradiation of normal human cells (Jeeves and Rainbow, 1979a). Such an effect could result from the induction of a bypass repair mode allowing replication past UV-induced DNA lesions which would otherwise cause a block in the replication and transcription of the viral genome. The fact that UVER has been detected for the single-stranded DNA Kilham rat virus in rat nephroma cells suggests that bypass repair can be produced in some mammalian cells (Lytle, 1978).

Recent investigations demonstrate that radiation-enhanced reactivation also occurs for γ-irradiated adenovirus in normal human fibroblasts (Jeeves and Rainbow, 1979b) and X-irradiated herpesvirus in CV-1 monkey kidney cells (Lytle and Goddard, 1979b). Thus, UV or gamma irradiation of normal human fibroblasts prior to infection leads to an enhanced reactivation of V antigen formation for both UV- and γ-irradiated adenovirus (Fig. 3.6). The existence of an enhanced reactivation for both UV and gamma-irradiated virus suggests either that a single enhanced repair mode is capable of repairing broad spectrum of DNA damage or that at least two different repair modes are involved. Furthermore, it can be seen from Figure 3.6 that the capacity of unirradiated virus decreased following pre-γ-irradiation of the cells but increased slightly following pre-UV-irradiation. This suggests that UVER and γ-ray ER of irradiated adenovirus operate via different mechanisms in human cells. A similar suggestion has been made previously for monkey kidney cells since caffeine reduces UVER but not X-ray ER of UV-irradiated herpesvirus in these cells (Hellman *et al.*, 1976). Enhanced virus reactivation in mammalian cells may thus involve a number of different mechanisms, depending on the type of treatment given to the virus and host cell. This idea is consistent with results obtained for Weigle reactivation in repair-deficient *E. coli* strains, which suggests that there may be several pathways resulting in Weigle reactivation some or one of which may be mutagenic (Bresler *et al.*, 1978).

Error-Prone Reactivation

Recent reports indicate that enhanced virus reactivation is mutagenic in some mammalian systems, suggesting the induction of an "error-prone" DNA repair mode for mammalian cells. DasGupta and Summers (1978) showed that enhanced reactivation of herpesvirus was accompanied by an increased mutation frequency of the virus. They measured the frequency of forward mutations in the thymidine kinase gene of herpesvirus following one cycle of growth on nontreated and UV-irradiated Vero monkey kidney cells. Prior irradiation of the host cell resulted in an increased mutation frequency of the progeny virus for both UV-irradiated and unirradiated virus. Very similar results were obtained by Lytle and Goddard (Day, 1978) for herpesvirus in CV-1 monkey kidney cells. Sarasin (1979) carried out one-cycle growth experiments using a temperature-sensitive early mutant of SV40. He found increased reversion to wild type for UV-irradiated virus following pretreatment of CV-1 monkey kidney cells with UV light, adding further support for a UV-induced error-prone repair in monkey kidney cells. Day and Ziolkowski (1978), however, found little or no enhancement of the reversion frequency of adenovirus 5 ts2 following pretreatment of normal human fibroblasts 24 hours prior to infection with the UV-irradiated ts mutant. This latter finding

sheds some doubt on the universality of error-prone reactivation for all virus cell systems.

Conclusions

Viral reactivation in mammalian cells is a relatively new technique with considerable potential for aiding our understanding of human carcinogenesis. There are several unique aspects of viral reactivation which can be employed as short-term tests for physical and chemical carcinogenesis. Firstly, HCR of V antigen for irradiated or chemically treated adenovirus is a very sensitive assay for the DNA repair capacity of human cells, capable of detecting individuals in the population with an elevated sensitivity to carcinogens (Table 3.1). Adenovirus appears to have a greater sensitivity in detecting repair deficiencies than herpesvirus or SV40. Adenovirus also has the added advantage over SV40 in being able to plaque directly on human fibroblasts. The V antigen assay also has several advantages over the plaque assay for adenovirus since it is accomplished in a relatively short period of time (about 4 days) with relatively few cells and appears to be capable of detecting deficiencies in repair rate (Rainbow, 1978). Unlike plaque survival, which requires the integrity of the complete genome, V antigen survival reflects the DNA repair capacity required for DNA synthesis as well as the transcription and translation necessary for the synthesis of structural proteins. HCR levels for treated virus in XP strains as compared to normal strains can also be used to screen chemicals for potential carcinogens (Table 3.2). By using a battery of different cell types from patients with other cancer-prone syndromes the HCR technique can be extended to other susceptible groups in the population. The identification of chemicals which inhibit DNA repair can also be made by examining their ability to reduce HCR of irradiated virus.

Cellular capacity for virus infection and

Table 3.2. *Some Chemical and Physical Treatments Affecting the Reactivation of Double-Stranded DNA Viruses*

Treatment	Reactivation system examined[a]	Result[b]	References
UV	A (H, SV40, Ad)	+	Lytle *et al.* (1972); Takebe *et al.* (1978); Aaronson and Lytle (1970); Day (1974, 1975a); Rainbow (1980); Rainbow and Howes (1979); Rainbow (this work).
	B (H)	+	Lytle *et al.* (1976b); Coppey *et al.* (1979b)
	C (H, m, rt, rb, h)	+	Bockstahler *et al.* (1976); Hellman *et al.* (1974); Lytle *et al.* (1974); Lytle *et al.* (1976b)
	C (SV40, m)	+	Sarasin and Hanawalt (1978)
	C (Ad, h)	+	Jeeves and Rainbow (unpublished data)
Gamma rays	A (Ad)	+	Rainbow and Howes (1979)
	C (Ad, h)	+	Jeeves and Rainbow (1979a)
X-rays	A (H)	−	Lytle *et al.* (1972)
	C (H, m, rt)	+	Bockstahler and Lytle (1977); Hellman *et al.* (1974)
8-Methoxypsoralen plus light	B (H)	+	Coppey *et al.* (1979a)
Trimethylpsoralen plus light	A (Ad)	+	Day *et al.* (1975)
Chlorpromazine plus light	A (Ad)	+	Day and DiMattina (1977)
Angelicin plus light	B (H)	+	Coppey *et al.* (1979a)
Formaldehyde	B (H)	+	Coppey and Nocentini (1979)
	A (H)	−	Coppey and Nocentini (1979)
4-Nitroquinoline-1-oxide	B (Ad)	+	Stich and Laishes (1975)
Nitrous acid	A (Ad)	+	Day (1975b)
Benzo (a) pyrene diol-epoxide	A (Ad)	+	Day *et al.* (1978)
Caffeine	D (H, m, hm, rb)	+	Lytle (1972), Ross *et al.* (1972); Fogel *et al.* (1979)
	D (Ad, h)		Day (1975c); Rainbow and Mak (1973)
Acriflavin	C (H, rb)	+	Fogel *et al.* (1979)
	D (H, hm)	+	Ross *et al.* (1972)
Hyperthermia	D (Ad, h)	+	Lam and Stich (1978)
Aflatoxin B$_1$	C (H, m)	+	Lytle *et al.* (1978)
	C (SV40, m)	+	Sarasin and Hanawalt (1978)
Acetylaminofluorene	C (H, m)	−	Lytle *et al.* (1978)
N-acetoxy-2-acetylaminofluorene	C (SV40, m)	+	Sarasin and Hanawalt (1978)
	A (H)	+	Selsky and Greer (1977)
	C (H, m)	+	Lytle *et al.* (1978)
N-hydroxy-2-acetylaminofluorene	C (H, m)	+	Lytle *et al.* (1978)
Hydroxyurea	C (H, m)	+	Lytle and Goddard, (1979a)
	C (H, rb)	+	Fogel *et al.* (1979)
Cycloheximide	C (H, m)	−	Lytle and Goddard (1979a)
5-Bromodeoxyuridine	C (H, rb)	+	Fogel *et al.* (1979)
Methyl methanesulfonate	C (SV40, m)	+	Sarasin and Hanawalt (1978)
Ethyl methanesulfonate	C (SV40, m)	+	Sarasin and Hanawalt (1978)

[a]A, Reduced HCR of treated virus in XP cells as compared to normal cells; B, Reduced capacity for virus expression in XP cells as compared to normal cells following pretreatment of cells; C, Enhanced reactivation of UV-irradiated virus following pretreatment of cells; D, Reduced HCR of UV-irradiated virus following treatment of infected cells. Virus used: H, herpesvirus; Ad, adenovirus; SV40, simian virus 40. Cells used: h, 'normal' human (normal fibroblasts or established lines such as KB and HeLa); m, monkey; rb, rabbit; rt, rat; hm, hamster.

[b]+, effect observed; −, effect not observed.

its recovery is also a sensitive, although less direct, technique for detecting DNA repair deficiencies (Table 3.2). It may have certain advantages over host-cell reactivation in evaluating the response to chemical carcinogens which react poorly with viral DNA following treatment of the virion or when cellular activation of the carcinogen is required for its action. The main disadvantage is the indirect nature of the assay and our lack of detailed knowledge on the mechanism of reduced cellular capacity.

The enhanced virus reactivation assay itself shows considerable potential as a screening system for testing the carcinogenic potential of chemical compounds in mammalian cells (Lytle *et al.*, 1978) (Table 3.2). The sensitivity and the short assay time (3 days) are a distinct advantage over other mammalian systems. One major weakness is that the significance of enhanced virus reactivation is not clearly understood at the present time.

The detection of a DNA repair deficiency per se is only a first step in the identification of high-risk groups since not all individuals with reduced DNA repair capacity have an elevated risk of carcinogenesis. Such studies can be combined with viral assays for error-prone reactivation and, therefore, give further information as to the degree of error-proneness of the various repair modes which are brought into play following treatment of cells with carcinogens. Since error-prone DNA repair mechanisms are thought to play a role in physically and chemically induced carcinogenesis, an enhanced viral mutagenesis assay in human cells may prove to be an accurate indicator of chemical carcinogens.

Acknowledgments

The author wishes to thank Margaret Howes and Patrick Jeeves for participation in some of the experiments reported. Work from the author's laboratory was supported by the National Cancer Institute of Canada. Thanks also to Dr. C. David Lytle, Dr. Jacques Coppey, Dr. Alain Sarasin and Dr. Rufus Day, III for helpful discussions and for sharing their unpublished results.

References

Aaronson, S.A., Lytle, C.D. (1970): Decreased host-cell reactivation of irradiated SV40 virus in xeroderma pigmentosum. Nature (Lond.) *228*:359–361.

Abrahams, P.J., van der Eb, A.J. (1976): Host cell reactivation of ultraviolet-irradiated SV40 DNA in five complementation groups of xeroderma pigmentosum. Mutat. Res. *35*: 13–22.

Bockstahler, L.G., Lytle, C.D. (1970): Ultraviolet light-enhanced reactivation of a mammalian virus. Biochem. Biophys. Res. Commun. *41*:184–189.

Bockstahler, L.E., Lytle, C.D. (1977): Radiation-enhanced reactivation of nuclear replicating mammalian viruses. Photochem. Photobiol. *24*:477–482.

Bockstahler, L.E., Lytle, C.D., Stafford, J.E. (1976): Ultraviolet enhanced reactivation of a human virus: Effects of delayed infection. Mutat. Res. *34*: 189–198.

Bootsma, D., Mulder, M.P., Pot, F., Cohen, J.A. (1970): Different inherited levels of DNA repair replication in xeroderma pigmentosum strains after exposure to ultraviolet irradiation. Mutat. Res. *9*:507–516.

Bootsma, D. (1979): DNA repair deficiencies in man. In: Proc. 6th Int. Congr. Radiat. Res., S. Okada, M. Imanura, T. Terashima, and H. Yamaguchi (eds.). Tokyo: Toppan Printing Co., pp. 472–475.

Bresler, S.E., Kalinin, V.L., Shelegedin, V.N. (1978): W-reactivation and W-mutagenesis of gamma-irradiated phage lambda. Mutat. Res. *49*: 341–355.

Cleaver, J.E., Trosko, S.E. (1970): Absence of excision of ultraviolet induced cyclobutane dimers in xeroderma pigmentosum. Photochem. Photobiol. *11*:547–550.

Coppey, J., Nocentini, S. (1976): Herpesvirus and viral DNA synthesis in ultraviolet light-irradiated cells. J. Gen. Virol. *32*:1–15.

Coppey, J., Nocentini, S. (1979): Survival and herpes virus production of normal and xeroderma pigmentosum fibroblasts after treatment with formaldehyde. Mutat. Res. *62*: 355–361.

Coppey, J., Moreno, G., Nocentini, S. (1978): Herpes virus production by ultraviolet irradiated human skin cells: A marker of repair. Bull. Cancer *65*: 335–340.

Coppey, J., Averbeck, D., Moreno, G. (1979a): Herpes virus production in monkey kidney and human skin cells treated with angelicin or 8-methoxypsoralen plus 365 nm light. Photochem. Photobiol. *29*:797–801.

Coppey, J., Nocentini, S., Menezes, S., Moreno, G. (1979b): Herpes virus production as a marker of repair in ultra-violet irradiated human skin cells of different origin. Int. J. Radiat. Biol. *36*:1–10.

DasGupta, U.B., Summers, W.C. (1978): Ultraviolet reactivation of herpes simplex virus is mutagenic and inducible in mammalian cells. Proc. Natl. Acad. Sci. USA *75*:2378–2381.

Day, R.S. III (1974): Studies on repair of adenovirus 2 by human fibroblasts using normal, xeroderma pigmentosum, and xeroderma pigmentosum heterozygous strains. Cancer Res. *34*:1965–1970.

Day, R.S. III (1975a): Xeroderma pigmentosum variants have decreased repair of ultraviolet-damaged DNA. Nature (Lond.) *253*: 748–749.

Day, R.S. III (1975b): Human cells repair DNA damaged by nitrous acid. Mutat. Res. *27*:407–409.

Day, R.S. III (1975c): Caffeine inhibition of the repair of ultraviolet-irradiated adenovirus in human cells. Mutat. Res. *33*:321–326.

Day, R.S. III (1978): Viral probes for mammalian cell DNA repair: Results and prospects. In: DNA Repair Mechanisms, Hanawalt, P.C., Friedberg, E.C., Fox, C.F. (eds.). New York: Academic Press, pp. 531–534.

Day, R.S. III, Ziolkowski, C. (1978): Studies on UV-induced viral reversion, Cockayne's syndrome, and MNNG damage using adenovirus 5. In: DNA Repair Mechanisms, Hanawalt, P.C., Friedberg, E.C., Fox, C.F. (eds.). New York: Academic Press, pp. 535–539.

Day, R.S. III, Giuffrida, A.S., Dingman, C.W. (1975): Repair by human cells of adenovirus-2 damaged by psoralen plus near ultraviolet light treatment. Mutat. Res. *33*:311–320.

Day, R.S. III, DiMattina, M. (1977): Photodynamic action of chlorpromazine on adenovirus 5: Repairable damage and single-strand breaks. Chem. Biol. Interact. *17*: 89–97.

Day, R.S. III, Scudiero, D., DiMattina, M. (1978): Excision repair by human fibroblasts of DNA damaged by r-7, t-8-dihydroxy-t-9.10-oxy-7,8,9,10-tetrahydrobenzo(a)pyrene. Mutat. Res. *50*:383–394.

Fogel, M., Yamaniski, R., Rapp, F. (1979): Enhancement of host cell reactivation of ultraviolet-irradiated herpes simplex virus by caffeine, hydroxyurea and 5-bromodeoxyuridine. Int. J. Cancer *23*:657–662.

Hellman, K.B., Haynes, K.F., Bockstahler, L.E. (1974): Radiation-enhanced survival of a human virus in normal and malignant rat cells. Proc. Soc. Exp. Biol. Med. *145*:255–262.

Hellman, K.B., Lytle, C.D., Bockstahler, L.E. (1976): Radiation-enhanced reactivation of herpes simplex virus: Effect of caffeine. Mutat. Res. *36*:294–296.

Ikenaga, M., Sugita, T., Kozuka, T., Suehara, N., Inoue, M. (1979): Actions of radiation and radiomimetic chemicals on cultured cells from a patient with Cockayne syndrome. Proc. 6th Int. Cong. Rad. Res., Tokyo, Abstract E-9-9, p. 265.

Jeeves, W.P., Rainbow, A.J. (1979a): γ-ray enhanced reactivation of UV-irradiated adenovirus in normal human fibroblasts. Mutat. Res. *60*:33–41.

Jeeves, W.P., Rainbow, A.J. (1979b): Gamma-ray enhanced reactivation of gamma-irradiated adenovirus in human cells. Biochem. Biophys. Res. Commun. *90*:567–574.

Kraemer, K.H., deWeerd-Kastelein, E.A., Robbins, J.H., Keijer, W., Barret, S.F., Petinga, R.A., Bootsma, D. (1975): Five complementation groups in xeroderma pigmentosum. Mutat. Res. *33*:327–340.

Krepinsky, A.B., Rainbow, A.J., Heddle, J.A. (1979): Studies on the ultraviolet light sensitivity of Bloom's syndrome fibroblasts. Mutat. Res. *69*:357–368.

Lam, P., Stitch, H.F. (1978): Hyperthermia and host-cell reactivation of adenovirus 12. Can. J. Genet. Cytol. *20*:35–40.

Lee, P., Rainbow, A.J. (1977): Viral DNA synthesis and virus production in human KB cells infected with gamma irradiated adenovirus. Mutat. Res. *46*:135 (A42).

Lehmann, A.R. (1978): Replicative bypass mechanisms in mammalian cells. In: DNA Repair Mechanisms, Hanawalt, P.C., Friedberg, E.C., Fox, C.F. (eds.). New York: Academic Press, pp. 485–488.

Lehmann, A. R., Kirk-Bell, S., Arlett, C. F., Harcourt, S.A., deWeerd-Kastelein, E.A.,

Keijer, W., Hall-Smith, P. (1977): Repair of ultraviolet light damage in a variety of human fibroblast cell strains. Cancer Res. *37*: 904–910.

Lytle, C.D. (1971): Host cell reactivation in mammalian cells. I. Survival of ultraviolet irradiated herpes virus in different cell lines. Int. J. Radiat. Biol. *19*:329–337.

Lytle, C.D. (1972): Host-cell reactivation in mammalian cells. II. Effect of caffeine on herpes virus survival. Int. J. Radiat. Biol. *22*: 167–174.

Lytle, C.D. (1978): Radiation-enhanced virus reactivation in mammalian cells. Natl. Cancer Inst. Monogr. *50*:145–149.

Lytle, C.D., Goddard, J. G. (1979a): Enhanced virus reactivation in mammalian cells: Effects of metabolic inhibitors. Photochem. Photobiol. *29*:959–962.

Lytle, C.D., Goddard, J.G. (1979b): Low levels of radiation-enhanced reactivation of X-irradiated herpes simplex virus in mammalian cells. Proc. 6th Int. Congr. Radiat. Res., Tokyo, Japan, Abstract B-23-4, p. 156.

Lytle, C.D., Hestler, L.D. (1976): Photodynamic treatment of herpes simplex virus infection *in vitro*. Photochem. Photobiol. *24*: 443–448.

Lytle, C.D., Aaronson, S.A., Harvey, E. (1972): Host cell reactivation in mammalian cells. II. Survival of herpes simplex virus and vaccinia virus in normal human and xeroderma pigmentosum cells. Int. J. Radiat. Biol. *22*: 159–165.

Lytle, C.D., Benane, S.G., Bochstahler, L.E. (1974): Ultraviolet enhanced reactivation of herpes virus in human tumor cells. Photochem. Photobiol. *20*:91–94.

Lytle, C.D., Benane, S.G., Stafford, J.E. (1976a): Host cell reactivation in mammalian cells. V. Photoreactivation studies with herpes virus in marsupial and human cells. Photochem. Photobiol. *23*:331–336.

Lytle, C.D., Day, R.S. III, Hellman, K.B., Bockstahler, L.E. (1976b): Infection of UV-irradiated xeroderma pigmentosum fibroblasts by herpes simplex virus: Study of capacity and Weigle reactivation. Mutat. Res. *36*: 257–264.

Lytle, C.D., Coppey, J., Taylor, W.D. (1978): Enhanced survival of ultraviolet-irradiated herpes simplex virus in carcinogen-pretreated cells. Nature (Lond.) *272*:60–62.

Marx, J.L. (1978): DNA repair: New clues to carcinogenesis. Science *200*:518–521.

Painter, R.B. (1981): DNA synthesis inhibition in mammalian cells as a test for mutagenic carcinogens. In: Short-Term Tests for Chemical Carcinogens, Stich, H.F., San, R.H.C. (eds.). New York: Springer-Verlag, pp. 59–64.

Purifoy, D.J.M., Lewis, R.B., Powell, K.L. (1977): Identification of the herpes simplex virus polymerase gene. Nature (Lond.) *269*:621–623.

Rabson, A.S., Tyrrell, S.A., Legallais, F.Y. (1969): Growth of ultraviolet-damaged herpesvirus in xeroderma pigmentosum cells. Proc. Soc. Exp. Biol. Med. *132*:802–806.

Radman, M. (1975): SOS repair hypothesis: Phenomenology of an inducible DNA repair which is accompanied by mutagenesis. In: Molecular Mechanisms for Repair of DNA, Hanawalt, P.C., Setlow, R.B. (eds.). New York: Plenum Press, pp. 355–367.

Rainbow, A.J. (1974): Repair of radiation induced DNA breaks in human adenovirus. Radiat. Res. *60*:155–164.

Rainbow, A.J. (1977a): UV-induced alkaline labile DNA damage in human adenovirus and its repair after infection of human cells. Photochem. Photobiol. *25*:457–463.

Rainbow, A.J. (1977b): Repair of UV-induced alkaline labile damage in adenovirus after infection of several human fibroblasts. Mutat. Res. *46*:149 (A63).

Rainbow, A.J. (1978): Production of viral structural antigens by irradiated adenovirus as an assay for DNA repair in human fibroblasts. In: DNA Repair Mechanisms, Hanawalt, P.C. Friedberg, E.C., Fox, C.F. (eds.). New York: Academic Press, pp. 541–545.

Rainbow, A.J. (1980): A reduced capacity to repair irradiated adenovirus in fibroblasts from xeroderma pigmentosum heterozygotes. Cancer Res. (in press).

Rainbow, A.J., Howes, M. (1977a): Defective repair of ultraviolet and gamma-ray damaged DNA in Fanconi's anaemia. Int. J. Radiat. Biol. *31*:191–195.

Rainbow, A.J., Howes, M. (1977b): Reduced host-cell reactivation of UV-irradiated adenovirus in Fanconi's anaemia fibroblasts. Radiat. Res. *70*:686.

Rainbow, A.J., Howes, M. (1977c): Decreased repair of gamma-ray damaged DNA in progeria. Biochem. Biophys. Res. Commun. *74*:714–719.

Rainbow, A.J., Howes, M. (1979): Decreased repair of gamma-irradiated adenovirus in xeroderma pigmentosum fibroblasts. Int. J. Radiat. Biol.*36*:621–629.

Rainbow, A.J., Howes, M. (1980): A deficiency in the repair of UV and gamma-ray damaged DNA in fibroblasts from Cockayne's syndrome. Mutat. Res.

Rainbow, A.J., Mak, S. (1972): DNA strand breakage and biological function of human adenovirus after gamma irradiation. Radiat. Res. *50*:319–333.

Rainbow, A.J., Mak, S. (1973): DNA damage and biological function of human adenovirus after UV irradiation. Int. J. Radiat. Biol. *24*:59–72.

Robbins, J.H., Kraemer, K.H., Lutzner, M.A., Festoff, B.W., Coon, H.G.(1974): Xeroderma pigmentosum—an inherited disease with sun sensitivity, multiple cutaneous neoplasma, and abnormal DNA repair. Ann. Intern. Med. *80*:221–248.

Ross, L.J., Cameron, K.R., Wildy, P.(1972): Ultraviolet irradiation of herpes simplex virus: Reactivation processes and delay in virus multiplication. J. Gen. Virol. *14*:299–311.

Rupert, C., Harm, W. (1966): Reactivation after photobiological damage. Adv. Radiat. Biol. *2*:1–81.

Sarasin, A. (1978): Induced DNA repair processes in eucaryotic cells. Biochimie *60*: 1141–1144.

Sarasin, A. (1979): The use of DNA viruses as probe for studying DNA repair pathways in eucaryotic cells. In: Proc. 6th Int. Congr. Radiat. Res., S. Okada, M. Imamura, T. Terashima and H. Yamaguchi, (eds.). Tokyo: Toppan Printing Co., pp. 462–470.

Sarasin, A.R., Hanawalt, P.C. (1978): Carcinogens enhance survival of UV-irradiated simian virus 40 in treated monkey kidney cells: Induction of a recovery pathway. Proc. Natl. Acad. Sci. USA *75*:346–350.

Sarasin, A., Goze, A., Devoret, R., Monté, Y. (1977): Induced reactivity of UV-damaged phage gamma *E. coli* K12 host cells treated with aflatoxin B1 metabolites. Mutat. Res. *42*:205–214.

Sasaki, M.S., Tonomura, A. (1973): A high susceptibility of Fanconi's anaemia to chromosome breakage by DNA crosslinking agents. Cancer Res. *33*:1829–1836.

Selsky, C.A., Greer, S. (1978): Host-cell reactivation of UV-irradiated and chemically treated herpes simplex virus-1 by xeroderma pigmentosum, XP heterozygotes and normal skin fibroblasts. Mutat. Res. *50*:395–405.

Selksy, C., Weichselbaum, R., Little, J.B. (1978): Defective host-cell reactivation of UV-irradiated herpes simplex virus by Bloom's syndrome skin fibroblasts. In: DNA Repair Mechanisms, Hanawalt, P.C., Friedberg, E.C., Fox, C.F. (eds.). New York: Academic Press, pp. 555–558.

Setlow, R.B. (1978): Repair deficient human disorders and cancer. Nature (Lond.) *271*:713–717.

Setlow, R.B., Regan, J.D., German, J., Carrier, W.L. (1969): Evidence that xeroderma pigmentosum cells do not perform the first step in the repair of ultraviolet damage to their DNA. Proc. Natl. Acad. Sci. USA *64*:1035–1041.

Stich, H.F., Laishes, B. (1975): Carcinogens and DNA repair. In: Radiation Research: Biochemical, Chemical and Physical Perspectives Nygaard, O.F., Adler, H.I., Sinclair, W.K. (eds.). New York: Academic Press, pp. 727–734.

Stich, H.F., Stich, W., Lam, P. (1974): Susceptibility of xeroderma pigmentosum cells to chromosome breakage by adenovirus type 12. Nature (Lond.) *250*:599–601.

Takebe, H., Nii, S., Ishii, M.I., Ursumi, H. (1974): Comparative studies of host-cell reactivation of xeroderma pigmentosum, normal human and some other mammalian cells. Mutat. Res. *25*:383–390.

Takebe, H., Fujiwara, Y., Sasaki, M.S., Sato, Y., Kozuka, T., Nikaido, O., Ishizaki, K., Arase, S., Ikenaga, M. (1978): DNA repair and clinical characteristics of 96 xeroderma pigmentosum patients in Japan. In: DNA Repair Mechanisms, Hanawalt, P.C., Friedberg, E.C., Fox, C.F. (eds.). New York: Academic Press, pp. 617–620.

Wagner, E.K., Rice, M., Sutherland, B.M. (1975): Photoreactivation of herpes simplex virus in human fibroblasts. Nature (Lond.) *254*:627–628.

Weigle, J.J. (1953): Induction of mutations in a bacterial virus. Proc. Natl. Acad. Sci. USA *39*:628–636.

Witkin, E.M. (1976): Ultraviolet mutagenesis and inducible DNA repair in *E. coli*. Bacteriol. Rev. *40*:869–907.

4

Metabolic Activation of Nitroheterocyclic Compounds in Bacteria and Mammalian Cells

D.R. McCalla

Introduction

Several thousand nitroheterocyclic compounds, particularly derivatives of furan, thiazole, and imidazole, have been synthesized and subjected to some degree of biological evaluation. (See reviews by Paul and Paul 1964, 1966; Miura and Reckendorf, 1967; Grunberg and Titsworth, 1973; and McCalla, 1979.) Several of these compounds are or have been widely used as antibacterial and antiprotozoal agents, and development of new nitroheterocyclic compounds continues at a rapid pace. However, in the past decade it has become clear that some of these compounds are carcinogenic (Bryan, 1978), and an extensive literature now shows that essentially all nitroheterocycles are mutagenic to *Escherichia coli* and to the strains of *Salmonella typhimurium* that contain the pKM 101 plasmid (Tazima *et al.* 1975; McCann *et al.* 1975a,b; Klemenic and Wang, 1978). Indeed, in the list of 300 chemicals reported by McCann *et al.* (1975a) two ni-

trofurans appear to be the most potent mutagens of all the compounds tested.

This paper reviews some of the qualitative and quantitative differences between the activation of nitroheterocycles in test bacteria and mammalian cells and discusses some of the problems in using S9 preparations to activate these compounds when mammalian cells or mutant *Salmonella* or *E. coli* deficient in one type of activation system are used as test organisms.

Activation of Nitroheterocyclic Compounds by Bacteria and Mammalian Cells

Nitroheterocyclic compounds appear to be "direct-acting" mutagens since they induce mutations in the absence of added activation systems (Kada, 1973; McCalla and Voutsinos, 1974); however, there is good evidence that metabolic activation by endogenous enzymes is a prerequisite to induction of DNA damage and mutation by these compounds (see below). In contrast to the oxidative activation of aromatic amines and hydrocarbons, nitrosamines and many other compounds, activation of nitroheterocycles is a reductive process catalyzed by flavoprotein enzymes which transfer electrons from NADPH or NADH to the nitro group. Enzymes having nitro-reductase activity are widely distributed and occur in *E. coli* (Asnis, 1957; McCalla *et al.*, 1970, 1971, 1975), *S. typhimurium* (Rosenkranz and Speck, 1976; McCalla, unpublished) *Euglena gracilis* (McCalla and Voutsinos, 1975), and in many mammalian tissues and cultured cells. (See Olive and McCalla, 1977 for references.) Thus far two types of nitro-reductase activities, which work in fundamentally different ways and which produce different end products, have been identified. These activities were originally distinguished on the basis that type I was insensitive to oxygen while type II was inhibited by oxygen and catalyzed the net reduction of nitro compounds only under hypoxic conditions (Asnis, 1957). Recent work by Mason and his colleagues has shown that, even in the presence of oxygen, the type II reductases catalyze the addition of one electron to the nitro compound, thus producing the nitro radical anion (Fig. 4.1) (Mason and Holtzman, 1975a,b; Peterson *et al.*, 1979). If oxygen is present the radi-

cals are very rapidly reoxidized to the original nitro compound, forming superoxide in the process. The concentration of nitro radical anions remains too low to detect by electron spin-resonance spectroscopy (ESR). Since the rate of superoxide production by preparations containing type II reductases is greatly increased by addition of nitro compounds, the "futile" cycle of reduction followed by oxidation of the nitro compound can be looked upon as a catalytic process which produces superoxide— a compound which is known to be toxic. The cycle also consumes the oxygen dissolved in the medium and will thus eventually produce hypoxic conditions. The possible involvement of superoxide in the killing and mutagenesis by nitroheterocyclic compounds has not been explored.

In hypoxia, the radical anions are more stable and their concentrations can build up to the μM range, which allows a nonenzymatic disproportionation reaction to occur in which two molecules of the radical anion are converted to one molecule of the nitroso compound and one molecule of the original nitro compound. The overall process is thus equivalent to the two-electron reduction of one molecule of the nitro compound. The nitroso derivatives thus formed are further reduced to the level of the amine, but the process involved has not been studied.

In contrast, the oxygen-insensitive or

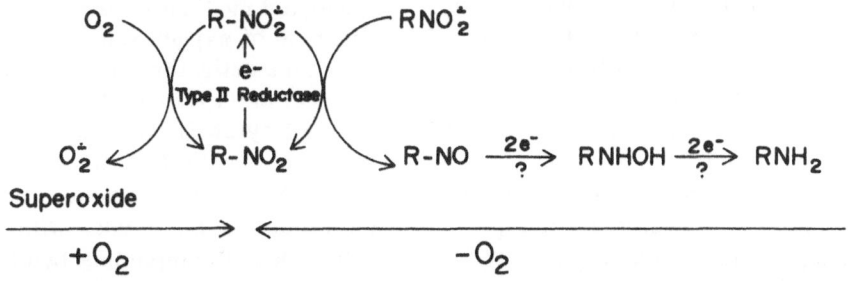

Figure 4.1. Pathway proposed by Peterson *et al.* (1979) for the reduction of nitrofurans by type II nitro-reductases.

R = the 2 substituted furan ring:

(See text for details.)

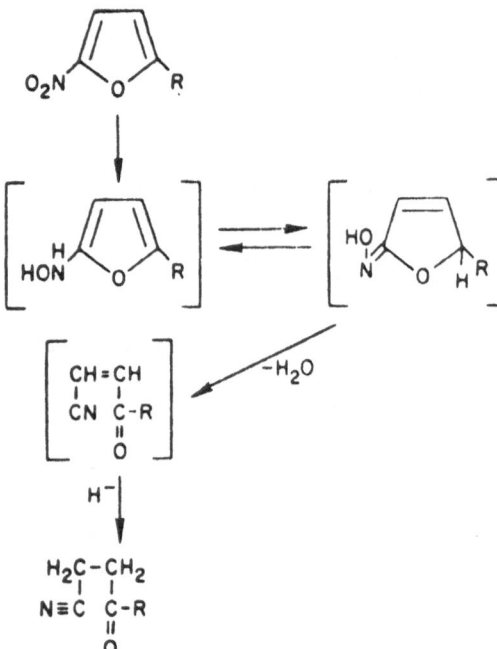

Figure 4.2. Pathway proposed by Gavin *et al.* (1966) for the reduction of nitrofurans by the type I reductases of bacteria. (See text for details.)

type I reductases must catalyze the addition of two or more electrons to the enzyme-bound nitro compound since the nitro radical anion cannot be detected by ESR even under hypoxic conditions (Peterson *et al.*, 1979). With nitrofurans, the end products formed by type I reductase appear to be open chain isomers of the aminofurans. Gavin *et al.* (1966) suggested that the reactions shown in Fig. 4.2 account for this process. At least with the type I reductases, a wide variety of nitrofurans having markedly different biological activities are reduced at nearly the same rate (McCalla *et al.*, 1970; Lu *et al.*, 1979). In *E. coli* and *S. typhimurium* most of the nitro-reductase activity is of type I but lower levels of type II activity are also present (Asnis, 1957; McCalla *et al.*, 1970).

The most direct evidence that type I nitro-reductases are really involved in the activation of nitro compounds comes from studies with nitrofuran-resistant bacteria which lack one or more reductase I compo-

nents (see below for details). In oxygenated media these mutants are resistant to the DNA-damaging (McCalla *et al.*, 1975) and mutagenic (McCalla and Voutsinos, 1975) effects of nitroheterocycles. However, these mutants retain their type II reductase activity and become more sensitive to nitro compounds under hypoxic conditions, presumably as a consequence of net reduction of the nitro compounds by type II reductase under these conditions. Neither the endogenous substrate nor the normal physiological role of any of the bacterial nitroreductases is known. It is of interest that nitro heterocycles with thiazole and imidazole rings are activated in a manner similar to nitrofurans and that mutants selected on the basis of their resistance to nitrofurazone are cross-resistant to other types of nitroheterocycles (Chessin *et al.*, 1978).

In contrast to bacteria, mammalian tissues contain only type II reductase activity. Enzymes which contribute to the nitroreductase activity of mammalian cells include NADPH-cytochrome-P450 reductase (Feller *et al.*, 1971; McCalla *et al.*, 1971; Boyd *et al.*, 1979) xanthine oxidase (Taylor *et al.*, 1951; Tatsumi *et al.*, 1976), and aldehyde oxidase (Wolpert *et al.*, 1973). The fact that mammalian cells are much more sensitive to the lethal (Mohindra and Rauth, 1976), DNA-damaging (Olive and McCalla, 1977), and mutagenic (McCalla *et al.*, 1978a) effects of nitroheterocyclic compounds in hypoxia than in oxygenated media is consistent with the involvement of oxygen-sensitive nitro-reductases in the activation of these compounds. Adams *et al.* (1976) have reported that the degree of cytotoxicity of a series of nitroheterocyclic compounds to aerobic Chinese hamster V79 cells depended on one-electron reduction potentials, with the most easily reduced compounds being the most toxic.

As noted above, the types I and II reductases produce different end products. These end products of reduction, i.e., amino heterocycles and open chain nitriles, have little or no toxicity (Ebitino *et al.*, 1962), a fact which implies that the "ultimate" agents are compounds having an oxidation state intermediate between that of the nitro compound and the amine (or open chain nitrile). The available evidence also suggests that the reactive intermediates formed in these two enzymatic pathways may also be different since when ^{14}C-nitrofurazone was reduced by reductase I, 20 to 30 times as much label was bound to added serum albumin as when reductase II catalyzed the reaction (McCalla et al., 1975). Nitrofurans activated by both types of enzymes lead to DNA damage (Lu et al., 1979) and mutations (McCalla et al., 1975), but the quantitative aspects have not be explored.

Thus, in terms of the use of bacterial mutagenicity tests to predict the hazard that nitroheterocyclic compounds pose to humans, we have a situation in which the activation of these agents in the usual test bacteria is appreciably different than it is in mammals, and it may well be that the microbial tests are giving the "right" qualitative answer for the wrong biochemical reasons. This situation is, of course, not unique to nitroheterocycles, since the S9 activating systems are known to produce different ultimate mutagens than are formed in intact cells from, for example, benzo(a)pyrene (King et al., 1975).

Type I Nitro-Reductases of *E. coli*

Table 4.1 summarizes the sensitivities of some wild type *E. coli* strains and their nitrofuran resistant (NFR) derivatives to nitrofurazone (McCalla et al., 1978b). Acquisition of resistance by strain AB1157 —and B/r (McCalla et al., 1970 and unpublished data), not shown—takes place in two steps. In the first of these, one distinct reductase component (Ia), representing about 70% of the total type I activity, is lost. In the second, another component (Ib) is lost, reducing the level of type I reductase to a few percent of that in the wild type strains. With each of these steps resistance to nitrofurazone is increased several-fold. The genes controlling these two enzymes (designated nfs A+ and nfs B+ respectively; (nfs = nitrofuran sensitivity) map close to gal (McCalla et al., 1978b). All of the first-step mutants derived from AB1157 are nfs A−, nfs B+. When the other "single" mutant, i.e., nfs A+·nfs B, was constructed, it turned out to have essentially the same sensitivity to nitrofurazone as AB1157, a property which accounts for lack of mutants of this genotype in the initial selection.

The major type I reductase components can be separated by gel electrophoresis

Table 4.1. *Pedigree and Properties of Some Nitrofuran-resistant (NFR) Mutants of* E. coli

Strain	AB 1157 $\xrightarrow{\text{NF}}$ NFR 402	$\xrightarrow{\text{NF}}$ NFR 502	$\xrightarrow{\text{AF2}}$ NFR 5022
Type I Reductase Complement	a,b,minor	−,b,minor	−,−minor	−,−(?)
nfs Genotype	nfs A+,B+	nfs A,B+	nfs A,B	nfs A B
Sensitivity to: NF μg/ml	5	15	35	35
AF2	0.75	1	1	10

Strain	WP2 uvr A $\xrightarrow{\text{NF}}$ NFR 343	$\xrightarrow{\text{AF2}}$ NFR 901
Type I Reductase Complement	a,−,minor	−,−minor	−,−(?)
nfs (Genotype)[a]	nfs A+,B	nfs A,B	
Sensitivity to: NF	5	45	50
AF2	0.3	1	5

The arrows represent mutational steps with either nitrofurazone (NF) or AF2 as selective agents as designated.

[a]Genotype inferred from the reductase complement.

(Fig. 4.3) or by chromatography on DEAE cellulose (not shown), using a buffered KCl gradient as the eluant. AB1157 contains two major components, one of which is absent from the $nfs\ A^+$, $nfs\ B^+$ strain and another which is missing in the $nfs\ A^+$, $nfs\ B$ strain. Double mutants ($nfs\ A$, $nfs\ B$) (not shown) lack both these components. Interestingly, the widely used test strains, *E. coli* WP2 and WP2 *uvr A*, which, like the $nfs\ A^+$, $nfs\ B$ strains, mutate to a high level of nitrofurazone resistance in a single step, contain only the Ia component of nitroreductase activity and thus appear to have lost the $nfs\ A^+$ gene in the course of their evolution from B/r. Further selection with the $nfr\ A$, $nfr\ B$ bacteria using nitrofurazone does not lead to the selection of strains having higher levels of resistance.

As noted above, the mutant $nfs\ A$, $nfs\ B$ strains are markedly resistant to nitrofurans, and with nitrofurazone few mutations are induced in aerated cultures. However, in the presence of AF2 (a nitrofuran derivative which is many times as potent as nitrofurazone) considerable mutation did take place, even in well-aerated cultures. This appears to be the result of activation of the AF2 by the small residual amount of type I reductase activity left in the resistant mutants. More resistant mutants (e.g., NFR 901) which proved to be essentially refractory to AF2 under aerated conditions were obtained when $nfs\ A$, $nfs\ B$ bacteria (NFR 502) were grown in medium containing AF2 (Fig. 4.1). Preliminary results suggest that some of the residual type I reductase was lost in this mutation step. When NFR 901 and similar strains were tested against nitrofurazone, they showed little if any further resistance to this agent. The fact that the third step of selection increased resistance only to a strongly mutagenic nitrofuran but not to the weaker agents indicates that either the amount of residual type I reductase remaining in NFR 343 is insufficient to produce toxic levels of nitrofurazone metabolites or that the residual enzyme is somewhat specific for the more powerful derivatives such as AF2. As

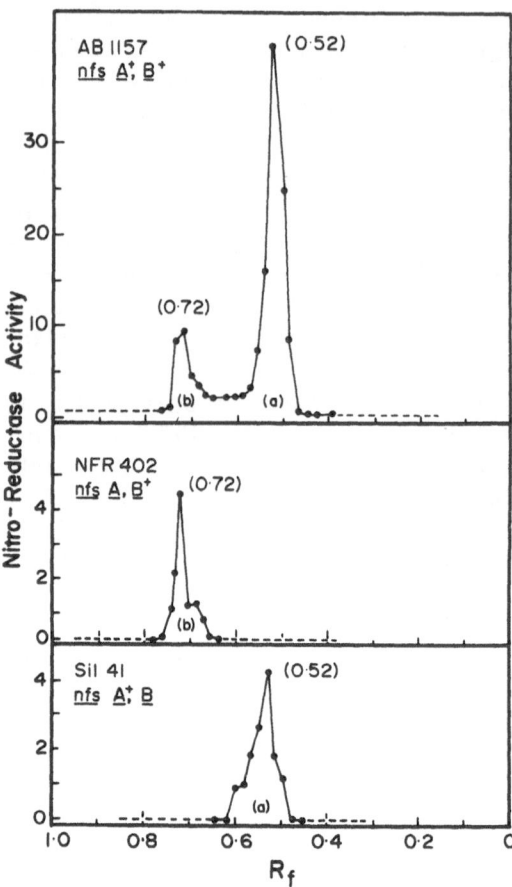

Figure 4.3. Gel electrophoresis profiles of type I nitrofurazone reductase activity in crude extracts of *E. coli* AB1157 and its NFR402 and SIL41 mutants. The "nitrofuran sensitivity" genotype of each strain is indicated on the figure. Electrophoresis was carried out in 8% total acrylamide (5% of which was *bis*-acrylamide), using the discontinuous buffer system No. 2330 of Jovin *et al.* (1970) at 0.5 mA per gel until the tracking dye approached the end of the tube. The gels were then removed from the glass tubes and stained with Coomassie Brilliant Blue dye or sliced into 1 mm sections, which were then assayed for reductase activity using procedures scaled down from those described by McCalla *et al.* (1970). Since different amounts of crude extract were applied to the various gels, the amount of apparent activity found in the different experiments is irrelevant.

noted above, all these mutant NFR strains contain the oxygen-sensitive type II reductases of their wild type parents, and the activity of these reductases can be "turned on" by hypoxia.

Activation of Nitroheterocycles by S9 Preparations?

The ability of S9 preparations to form mutagenic metabolites from nitroheterocyclic compounds has been studied in *E. coli* (Kada, 1974) and *S. typhimurium* (Rosenkranz and Speck, 1975, 1976), using mutants which lack reductase I activity, and positive results have been claimed. However, given the facts (1) that hypoxic conditions are required for net reduction of nitroheterocycles by the enzymes of the S9 fraction, (2) that S9 preparations "scavenge" oxygen from solution, especially in the presence of substrates like nitro compounds, and (3) that hypoxic conditions turn on endogeneous type II reductases, interpretation of these results is by no means clear, especially since in Kada's experiments no measures were taken to render the liquid media hypoxic and Rosenkranz and Speck employed standard plate assays with anaerobic bags to obtain hypoxic conditions.

We therefore undertook to determine whether or not S9 increased the mutagenicity of nitrofurazone toward strain NFR 343 under well-defined hypoxic con-

ditions. Enough S9 preparation was used to reduce half the nitrofurazone in 7 to 10 min while the bacteria alone would have taken over an hour to carry out this extent of reduction. It is evident from Table 4.2 that hypoxia alone was enough to induce a many-fold increase in the reversion of TRP$^-$ to TRP$^+$ in NFR 343 and that the inclusion of S9 mix, far from increasing the amount of mutation, considerably decreased it. Thus it appears that the nitrofurazone which was reduced outside the bacteria was relatively ineffective in inducing mutations compared to that reduced inside the bacteria. This is really not surprising since the volume occupied by the bacteria is very small compared to that of the medium, so that metabolites formed within a cell would have a much greater chance to react with cellular constituents than those generated in the extracellular medium. Thus, reduction of the nitrofuran by the S9 may well have the effect of decreasing substantially the amount of nitrofuran available to the endogenous bacterial reductases, which in turn decreases the induction of mutations, particularly after the first 15 minutes.

These results suggest an alternative explanation for the earlier apparently positive

Table 4.2.   *Mutagenicity of Nitrofurazone to* E. coli *NFR 343 under Various Conditions*

Experiment	Gas phase	Time (min)	S9	Induced mutation frequency
1	Air	40	−	2.1
	Argon	40	−	42.1
2	Argon	15	−	10.7
	Argon	15	+	2.6
	Argon	50	−	27.5
	Argon	50	+	2.3
	Air	50	+	2.2

Bacteria were grown overnight in penassay broth, harvested, and resuspended. The mutation assays were carried out by the treat and plate procedure described by Green and Muriel (1976). Bacteria (3×10^8/ml) were exposed to 84 μM nitrofurazone in 6 ml of 0.067 M phosphate butter containing 1 mg glucose. Where S9 mix (Ames *et al.*, 1975) was included, 1.6 ml was substituted for buffer. Hypoxia was achieved by flushing the media with argon and then tightly stoppering the tubes. Aerated samples were placed in 125 ml Erlenmeyer flasks which were shaken vigorously. All samples were treated at 37°C. After the times indicated, samples of the treated cultures were plated on solid medium containing limited tryptophan (Green and Muriel, 1976). Other samples of the same culture were diluted and plated on the same medium for determination of survival (over 80% in all treatments shown). The induced mutation frequency per 10^7 surviving cells was then calculated as previously described. The spontaneous mutation rate in untreated controls was 2.3 and 2.6 per 10^7 survivors in two successive experiments.

results, namely that the additional oxygen-scavenging power of the S9 fraction increased the extent or duration of hypoxia in the tubes (Kada, 1974) or on the plates (Rosenkranz and Speck, 1975, 1976) so that the endogenous bacterial type II reductases were more active (or active for a longer period) when S9 was added. Obviously, since mammalian cells contain only type II reductase activity, the same stricture applies to experiments such as those of Nakamura et al. (1977) in which mouse L5178 Y cells were exposed to AF2 with and without added S9 preparation without any reference to the development of hypoxic conditions. In order to interpret the results of experiments which employ cells containing mainly reductase II activity, it is essential to ensure that hypoxic conditions are rapidly achieved. This can best be done by bubbling liquid suspensions with oxygen-free argon or nitrogen. Removal of oxygen from solid media is likely to be slow where conventional anaerobic bags or jars are employed.

It should, however, be emphasized that what is at issue here is the relative effectiveness of the S9 and the endogenous enzymes under the conditions of the assay. The fact that added S9 preparations decreased the number of revertants is irrelevant to the role that enzymes of the S9 fraction, e.g., NADPH-cytochrome P450 reductase or xanthine oxidase, may play in activation of nitro compounds in mammalian cells.

Some Quantitative Aspects

Obviously the wide and successful employment of nitroheterocyclic compounds as antimicrobial agents in clinical and veterinary medicine reflects the selective toxicity of these agents to microbes. The exact extent of the difference depends on the cell types and species employed and is hard to determine from existing data. However, the

following comparison may serve as an illustration. Survival of E. coli WP2 is reduced to 10% by exposure for 20 minutes to 3 μM AF2 (Lu et al., 1979), whereas with cultured human fibroblasts this level of killing was reached only after exposure to 36 μM AF2 for 2 hours in hypoxia and aerobic human cells were several times more resistant than this (McCalla et al., 1978a). Similarly, bacteria are much more sensitive to the mutagenic effects of nitroheterocycles than are mammalian cells.

Why are the bacteria so much more sensitive? One factor is certainly the high levels of reductase activity in bacteria as compared to those in mammalian cells and tissues. From published data it appears that the specific activity, i.e., activity per mg total protein of nitro-reductase in crude extracts of rat liver assayed under anaerobic conditions, is at least 100 times lower than the specific activity of the E. coli reductases (McCalla et al., 1970; Olive and McCalla, 1977). The specific activity of the reductases in S. typhimurium are even higher than those in E. coli (unpublished results), which may explain the extreme sensitivity of Salmonella to nitroheterocyclic compounds. Further, while reduction of nitro compounds does take place under physiological conditions in intact mammals (Olivard et al., 1962; Wang et al., 1975), it is likely that the in vivo activity of the nitro-reductases is reduced by the oxygen present in the tissue.

A second factor that may be important is suggested by the observation that the mechanism of action and the intermediates produced by the type I reductases of bacteria and the type II reductases of mammalian cells appear to be different. This raises the possibility that we are dealing in the two cases with different types of DNA adducts, which may be formed in different yields or may differ in their biological effectiveness. Since we are now able to detect labeled DNA adducts when bacteria or mammalian cells are incubated with ^{14}C-ANFT (unpublished results), it should soon be possible to determine whether or not the adducts formed in the two situations are chemically identical.

A third factor is significant when considering data obtained with strains of *S. typhimurium* which contain the pKM 101 plasmid. It is worth recalling that in spot or plate tests, plasmid-free strains like TA 1535 give no evidence that they are mutated by nitrofurans (Yahagi *et al.*, 1974). More sensitive fluctuation (Green *et al.*, 1977) or "treat and plate" assays (Rosenkranz, 1977) do show that this strain is mutated, and it is thought that the negative results in the plate tests were due to excessive killing. In contrast to TA 1535, TA 100 is highly mutable by nitroheterocyclic compounds in plate and spot test (McCann *et al.*, 1975b; Yahagi *et al.*, 1976). From fluctuation assay data, it would appear that the presence of the plasmid increases mutability by nitrofurans by a factor of 100 to 1000. The mechanism of this effect is not understood, but, clearly, in the presence of the plasmid, damage that would have been repaired or tolerated in plasmid-free strains is expressed as mutations.

Meselson and Russel (1977) have shown that for several structurally different types of chemical carcinogens there is a consistent relation between carcinogenic potency to animals and mutagenic potency to *Salmonella*. If nitrofurans were to follow this relationship, compounds like FANFT and AF2 should be as carcinogenic as aflatoxin B. However, from the preceding discussion one would expect the nitro compounds to be less carcinogenic than might be predicted from their mutagenicity. This is, in fact, borne out by data from feeding experiments. Aflatoxin B_1, at levels as low as 15 ppb in the diet, induced liver tumors in rats (Butler and Barnes, 1966). In another study 5 ppm aflatoxin B_1 fed for 9 weeks induced liver tumors in 100% of the rats (Wogan and Newberne, 1967). In contrast, when FANFT was fed to rats for 30 weeks followed by 22 weeks of control diet, a high frequency of bladder tumors was induced only by levels of 1000 and 500 ppm; 10 and 5 ppm caused no abnormalities (Jacobs

et al., 1977). Similarly, 800 and 4000 ppm AF2 administered in the diet to female Wister rats for 18 months followed by 6 months of a control diet produced mammary tumors in 17 out of 48 and 37 out of 47 animals respectively (Takayama and Kuwabara, 1977). The general point that bacterial assays may exaggerate the potency of nitroheterocycles has been recognized by others, e.g., McCann and Ames, 1976, and Flamm, 1979, but it has not always been recognized that the difference may be 100- to 1000-fold.

Accepting that bacterial mutagenicity gives an exaggerated picture of the carcinogenic potency of nitroheterocycles relative to other compounds, one can ask if, within a series of nitrofuran compounds, there is a relation between mutagenicity and carcinogenic potency. The limited data available are summarized in Table 4.3. Certainly the three most potent carcinogens are powerful mutagens while the two compounds that have given negative results in carcinogenicity tests are much weaker mutagens, but more quantitative data are needed to permit a definite conclusion regarding the correlation. It is of interest that the relative mutagenicities of the four compounds which have been tested quantitatively in both *E. coli* WP2 uvrA and *S. typhimurium* TA 100 are similar over a potency range of 10,000-fold.

One other point that these data illustrate is the inadequacy of the carcinogenicity testing of nitrofurantoin—a compound which is widely used to treat human urinary tract infections. To the author's knowledge this compound has been tested in four separate experiments involving a total of only 70 treated animals. No significant excess tumors over the controls were observed in rats fed 3000 ppm nitrofurantoin in the diet for 44.5 weeks or 1,970 ppm for 16 weeks followed by 1000 ppm for 59 weeks. From the mutagenicity data these doses of nitrofurantoin would be expected to have an effect equivalent to that of somewhat less than 50 ppm of FANFT—a dose that produced mild hyperplasia but no tumors!

Table 4.3. *Comparison of the Carcinogenic and Mutagenic Potencies of Nitrofurans*

Chemical name	Common name or abbreviation	Carcinogenicity[a]			Relative mutagenic potency toward	
		Species[b]	Sex[c]	Relative potency	E. coli[d] WP2 uvrA	S. typhimurium[e] TA 100
2-(2-furyl)-3-(5-nitro-2-furyl)acrylamide	AF2	R	F	++	100	100
			M	+		
		M	F	+++		
		H	M	+++		
N-[4-(5-nitro-2-furyl)-2-thiazolyl]-formamide	FANFT	R	M	+++	40	80
			F	+++		
		M	F	+++		
			M	+++		
		H	M	+++		
		GP	M	−		
		D	F	+++		
2-amino-4-(5-nitro-2-furyl)thiazole	ANFT	R	F	+++	30	N.D.[f]
		M	F	+++		
N-(5-nitro-2-furfurylidine)-3-amino-2-oxazolidin-2-one	Furazolidone	R	F	+	10	N.D.
		M	F	+		
1-[(5-nitro-2-furfurylidene)-amino]hydantoin	Nitrofurantoin Furadantin	R	F	−	0.4	1
5-nitro-2-furaldehyde semicarbazone	Nitrofurazone	R	F	+	0.1	N.D.
			M	−		
5-nitrofuroic acid	−	R	F	−	0.003	0.001

[a]Taken from Cohen (1978).

[b]R, rat; M, mouse; H, hamster; GP, guinea pig; D, dog.

[c]F = female; M = male.

[d]Data normalized (AF2 = 100) from Lu *et al.* (1979).

[e]Data normalized (AF2 = 100) from McCann *et al.* (1975).

[f]Not determined.

Thus, the conclusion that nitrofurantoin is noncarcinogenic is premature.

Summary and Conclusions

1. Reductive activation of nitroheterocyclic compounds takes place at a much more rapid pace in *Escherichia coli* and *Salmonella typhimurium* than in mammalian cells and tissues.

2. The major nitro-reductase activity (type I) of the bacteria is insensitive to oxygen and yields open chain nitriles (isomeric with the corresponding amines) as the final products. The identity of the reactive intermediate which binds covalently to DNA and protein is unknown.

3. In contrast, the type II nitro-reductase activity of mammalian cells is sensitive to oxygen because the nitro radical anion (the first intermediate formed) is readily reoxidized by molecular oxygen.

Under hypoxic conditions the radical anions disproportionate in a nonenzymatic reaction to regenerate the original nitro compound and form the nitroso derivative which is further reduced to the amine. Again, the identity of the mutagenic intermediate remains unknown.

4. S9 preparations consume oxygen and can therefore lead to hypoxia, which in turn may account for the observations that addition of S9 to the media can increase the amount of mutation induced by nitroheterocycles in mammalian cells or in mutant bacteria containing only type II (oxygen-sensitive) nitro-reductase activity.

5. Mutagenicity assays, especially those employing plasmid-containing strains, give a misleading impression of the potential carcinogenicity of nitrofurans relative to other classes of compounds. With *S. typhimurium* TA 100 the extent of the difference may be as large as 1000-fold.

Tests with *E. coli* WP2 uvrA and *S. typhimurium* TA 100 give similar relative mutagenicities for several nitrofurans which differ widely in activity.

6. The conclusion that nitrofurantoin is noncarcinogenic is based on inadequate data.

Acknowledgments

The contributions of many co-workers over a period of several years are gratefully acknowledged. In particular I am indebted to Pierre Laneuville for the gel electrophoresis data. Financial support has been provided by the National Cancer Institute of Canada and the Natural Sciences and Engineering Research Council Canada.

Note added in proof: Since presenting this paper, W. Kutcher in our laboratory has observed the presence of type I (i.e., oxygen-insensitive) nitrofuran reductase activity in the soluble fraction of rat liver. This suggests that the action of nitrofurans

on aerated mammalian cells could be mediated by type I reductases, while the more potent action of these agents under hypoxic conditions is due to type II reductases.

References

Adams, G.E., Clarke, E.D., Jacobs, R.S., Stratford, I.J., Wallace, R.G., Wardman, P., Watts, M.E. (1976): Mammalian cell toxicity of nitro compounds; dependence upon reduction potential. Biochem. Biophys. Res. Comm. *72*: 824–829.

Ames, B.N., McCann, J., Yamasaki E. (1975): Methods for detecting carcinogens and mutagens with the *Salmonella*/mammalian microsome mutagenicity test. Mutat. Res. *31*: 347–364.

Asnis, R.E. (1957): The reduction of furacin by cell-free extracts of furacin-resistant and parent-susceptible strains of *Escherichia coli*. Arch. Biochem. Biophys. 66:208–216.

Boyd, M.R., Stiko, A.W., Sasame, H.A. (1979): Metabolic activation of nitrofurantoin—possible implications for carcinogenesis. Biochem. Pharmacol. *28*: 601–606.

Bryan, G.T. (ed.) (1978): Nitrofurans: Carcinogenesis—A Comprehensive Survey, Vol. 4. New York: Raven Press.

Butler, W.H., Barnes, J.M. (1966): Carcinoma of the glandular stomach in rats given aflatoxin. Nature (Lond.) *209*: 90.

Chessin, H., McLaughlin T., Mroczkowski, Z., Rupp, W.D., Low K.B. (1978): Radiosensitization, mutagenicity and toxicity of *Escherichia coli* by several nitrofurans and nitroimidazoles. Radiat. Res. *75*: 424–431.

Ebetino, F.F., Carroll, J.J., Gever, G. (1962): Reduction of nitrofurans I: aminofurans. J. Med. Pharm. Chem. *5*: 513–524.

Feller, D.R., Morita, M., Gillette, J.R. (1971): Reduction of heterocyclic nitro compounds in the rat liver. Proc. Soc. Exp. Biol. Med. *137*: 433–437.

Flamm, G. (1979): Discussion. In: V.K. McElheny, E. Abrahamson (eds.), Banbury Rept. 1: Assessing Chemical Mutagens: The Risk to Humans. Cold Spring Harbor Laboratory, Cold Spring Harbor, New York, p. 95.

Gavin, J.J., Ebetino, F.F., Freedman, R., Waterbury, W.E. (1966): The aerobic degradation of

1-(5-nitrofurfuryl ideneamino)-2-imidazalone (NF-246) by *Escherichia coli*. Arch. Biochem. Biophys. *113*:399–404.

Green, M.H.L., Muriel, W.J. (1976): Mutagen testing using the TRP⁺ reversion in *E. coli*. Mutat. Res. *38*:3–32.

Green, M.H.L., Rogers, A.M., Ward, A.C., McCalla, D.R. (1977): Use of a simplified fluctuation test to detect and characterize mutagenesis by nitrofurans. Mutat. Res. *44*: 139–143.

Grunberg, E., Titsworth, E.H. (1973): Chemotherapeutic properties of heterocyclic compounds: Monocyclic compounds with five-membered rings. Ann. Rev. Microbiol. *27*: 317–346.

Jacobs, J.B., Arai, M., Cohen, S.M., Friedell, G.H. (1977): A long-term study of reversible and progressive urinary bladder cancer lesions in rats fed N-[4-(5-nitro-2-furyl)-2-thiazolyl]formamide. Cancer Res. *37*: 2817–2821.

Jovin, T.M., Dante, M.L., Cramback, A. (1970): Multiphasic Buffer Systems Output, Document No. PB259311. Springfield, Virginia, National Technical Information Service.

Kada, T. (1973): *Escherichia coli* mutagenicity of furylfuramide. Jap. J. Genetics *48*: 301–305.

Kada, T. (1974): Metabolic activation and *Escherichia coli* mutagenesis of furylfuramide, a nitrofuran food additive. Ann. Rep. Natl. Inst. Genet. Jap. *24*:39–41.

King, H.W.S., Thompson, M.H., Brookes, P. (1975): The benzo(a)pyrene deoxyribonucleoside products isolated from DNA after metabolism of benzo(a)pyrene by rat liver microsomes in the presence of DNA. Cancer Res. *35*:1263–1269.

Klemenic, J.M., Wang, C.Y. (1978): Mutagenicity of nitrofurans. In: G.T. Bryan (ed.), Nitrofurans: Carcinogenesis—A Comprehensive Survey, Vol. 4. New York: Raven Press, pp. 99–130.

Lu, C., McCalla, D.R., Bryant, D.W. (1979): Action of nitrofurans on *E. coli*: Mutation and induction and repair of daughter-strand gaps in DNA. Mutat. Res. *67*:133–144.

Mason, R.P., Holtzman, J.L. (1975a): The role of catalytic superoxide formation in the O_2 inhibition of nitroreductase. Biochem. Biophys. Res. Commun. *67*:1267–1274.

Mason, R.P., Holtzman, J.L. (1975b): The

mechanism of microsomal and mitochondrial nitroreductase. Electron spin resonance evidence for nitroaromatic free radical intermediates. Biochemistry *14*: 1626–1632.

McCalla, D.R. (1979): Nitrofurans. In: F.E. Hahn (ed.), Antibiotics, Vol. 1, Mechanism of Action of Antibacterial Agents. Berlin: Springer-Verlag.

McCalla D.R., Voutsinos D. (1974): On the mutagenicity of nitrofurans. Mutat. Res. *36*:3–15.

McCalla, D.R., Voutsinos, D. (1975): Nitrofuran-reducing enzymes of Euglena. J. Protozool. *22*:130–134.

McCalla, D.R., Reuvers, A., Kaiser, C. (1970): Mode of action of nitrofurazone. J. Bacteriol. *104*:1126–1134.

McCalla, D.R., Reuvers, A., Kaiser, C. (1971): Activation of nitrofurazone in animal tissues. Biochem. Pharmacol. *20*:3532–3537.

McCalla, D.R., Olive, P., Tu, Y., Fan, M.L. (1975): Nitrofurazone-reducing enzymes in *E. coli* and their role in drug-activation *in vivo*. Can. J. Microbiol. *21*:1484–1491.

McCalla, D.R., Arlett, C.F., Broughton, B. (1978a): The action of AF2 on cultured hamster and human cells under aerobic and hypoxic conditions. Chem.-Biol. Interact. *21*: 89–102.

McCalla, D.R., Kaiser C., Green, M.H.L. (1978b): The genetics of nitrofurazone resistance in *Escherichia coli*. J. Bacteriol. *133*: 10–16.

McCann, J., Ames, B.N. (1976): Detection of carcinogens and mutagens in the *Salmonella*/microsome test: Assay of 300 chemicals: Discussion, Proc. Natl. Acad. Sci. USA *73*:950–954.

McCann, J., Choi, E., Yamasaki, E., Ames, B.N. (1975a): Detection of carcinogens as mutagens in the *Salmonella*/microsome test: Assay of 300 chemicals. Proc. Natl. Acad. Sci. USA *72*:5135–5139.

McCann, J. Springharn, N.E. Kobori, J., Ames, B.N. (1975b): Detection of carcinogens as mutagens: Bacterial tester strains with R factor plasmids. Proc. Natl. Acad. Sci. USA *72*:979–983.

Meselson, M., Russell, K. (1977): Comparisons of carcinogenic and mutagenic potency. In: H.H. Hiatt, J.D. Watson, J.A. Winsten (eds.), Origins of Human Cancer, Book C, Cold Spring Harbor Laboratory, New York, pp. 1473–1481.

Miura, K., Reckendorf, H.K. (1967): The nitrofurans. Prog. Med. Chem. 5:320–381.

Mohindra, J.K., Rauth, A.M. (1976): Increased cell killing by metronidazole and nitrofurazone of hypoxic compared to aerobic mammalian cells. Cancer Res. 36:930–936.

Nakamura, N., Suzuki, N., Okada, S. (1977): Mutagenicity of furylfuramide, a food preservative tested by using alanine-requiring mouse L5178 Y cells in vitro and in vivo. Mutat. Res. 46:355–364.

Olivard, J., Valenti, S., Buzard, J.A. (1962): The metabolism of 5-nitro-2-furaldehyde acetylhydrazone. J. Med. Pharm. Chem. 5: 524–531.

Olive, P.L., McCalla, D.R. (1975): Damage to mammalian cell DNA by nitrofurans. Cancer Res. 35:781–784.

Olive, P.L., McCalla, D.R. (1977): Cytoxicity and DNA damage to mammalian cells by nitrofurans. Chem. Biol. Interact. 16:223–233.

Paul, H.E., Paul, M.F. (1964): The nitrofurans—chemotherapeutic properties. Experimental Chemotherapy II:307–370.

Paul, H.E., Paul, M.F. (1966): The nitrofurans–chemotherapeutic properties. Experimental Chemotherapy IV:521–536.

Peterson, F.J., Mason, R.P., Hovsepian, J., Holtzman, J.L. (1979): Oxygen-sensitive and insensitive nitroreduction by Escherichia coli and rat hepatic microsomes. J. Biol. Chem. 254:4009–4014.

Rosenkranz, H.S. (1977): Studies on the mutagenicity of nitrofurans in Salmonella typhimurium. Biochem. Pharmacol. 26: 896–898.

Rosenkranz, H.S., Speck, W.T. (1975): Mutagenicity of metronidazole: activation by mammalian liver microsomes. Biochem. Biophys. Res. Commun. 66:520–525.

Rosenkranz, H.S., Speck W.T. (1976): Activation of nitrofurantoin to a mutagen by rat liver nitroreductase. Biochem. Pharmacol. 25: 1555–1556.

Takayama, S., Kuwabara, N. (1977): The production of skeletal muscle atrophy and mammary tumors in rats by feeding 2-(2-furyl)-3-(5-nitro-2-furyl)acrylamide. Toxicol. Lett. 1: 11–16.

Tatsumi, K., Kitamura, S., Yoshimura, H. (1976): Reduction of nitrofuran derivatives by xanthine oxidase and microsomes. Arch. Biochem. Biophys. 175:131–137.

Taylor, J.D., Paul, H.E., Paul, F.M. (1951): Metabolism of nitrofurans III. Studies with xanthine oxidase in vitro. J. Biol. Chem. 191: 223–231.

Tazima, Y., Kada, T., Murakami, A. (1975): Mutagenicity of nitrofuran derivatives, including furylfuramide, a food preservative. Mutat. Res. 32:55–80.

Wang, C.Y., Chiu, C.W., Kaiman, B., Bryan, G.T. (1975): Identification of 2-methyl-4-(5-amino-2-furyl)thiazole as the reduced metabolite of 2-methyl-4-(5-nitro-2-furyl) thiazole. Biochem. Pharmacol. 24:291–293.

Wogan, G.N., Newberne, P.M. (1967): Dose response characteristics of aflatoxin B_1 carcinogenesis in the rat. Cancer Res. 27: 2370–2376.

Wolpert, M.K., Althaus, J.R., Johns, D.G. (1973): Nitroreductase activity of mammalian liver aldehyde oxidase. J. Pharmacol. Exp. Ther. 185:202–213.

Yahagi, T., Nagao, M., Hara, K., Matsushima, T. Sugimura, T., Bryan, G.T. (1974): Relationships between the carcinogenic and mutagenic or DNA-modifying effects of nitrofuran derivatives, including 2-(2-furyl)-3-(5-nitro-2-furyl)acrylamide, a food additive. Cancer Res. 34: 2266–2273.

Yahagi, T., Matsushima, T., Nagao, M., Seino, Y., Sugimura, T., Bryan, G.T. (1976): Mutagenicities of nitrofuran derivatives on a bacterial tester strain with an R. factor plasmid. Mutat. Res. 40:9–14.

5

Utilization of the Alkaline Elution Assay as a Short-Term Test for Chemical Carcinogens

JAMES A. SWENBERG

Introduction

DNA damage and repair assays constitute one group of short-term tests that have recently been developed to help assess the carcinogenic and mutagenic potential of chemicals. The alkaline elution assay (AE) can be utilized to measure DNA damage and repair either *in vitro* or *in vivo*. This procedure was first described by Kohn *et al.* (Kohn and Grimek-Ewig, 1973; Kohn *et al.*, 1974, 1976) as a sensitive method for detecting single-strand breaks and alkali-labile sites in DNA of L1210 cells exposed to cancer chemotherapeutic agents. It has been subsequently used to evaluate DNA damage and repair induced *in vitro* (Brambilla *et al.*, 1978a, 1978b; Cavanna *et al.*, 1977; Fornace and Little, 1979; Swenberg *et al.*, 1976) and *in vivo* (Brambilla *et al.*, 1978a, 1978b; Eastman and Bresnick, 1979; Parodi *et al.*, 1978; Petzold and Swenberg, 1978) by a wide variety of chemical and environmental agents.

Principles of the Alkaline Elution Method

A detailed review of the theory and practice behind the alkaline elution assay has been published (Kohn, 1978). Briefly, mammalian cells or tissue homogenates are applied to a membrane filter, lysed, and washed. A mat of DNA, free of most membranes, protein, and RNA, remains on the filter. Alkaline-eluting solution (pH 12) is then slowly passed through the filter, causing double-stranded DNA to unwind and elute from the filter at a rate proportional to the length of the single strands (Kohn, 1978; Kohn *et al.*, 1974, 1976). Characteristic elution profiles can be obtained for several types of DNA damage, including single-strand breaks, alkali-labile sites, interstrand crosslinks and DNA-protein crosslinks. The shape of these profiles depends on the length of single-strand DNA, the formation of breaks due to alkali-labile sites during elution, and adsorption of DNA-bound protein to the filters. Large numbers of single-strand breaks cause rapid elution of DNA from the filters. Elution profiles obtained following treatment with agents that induce many alkali-labile sites in DNA will vary with the type of alkali-labile site induced. For instance, chemical hydrolysis of 7-alkylguanine proceeds very quickly under the alkaline conditions employed, whereas phosphotriesters require approximately 6 hours of elution. Interstrand or protein-DNA crosslinks impede the elution of DNA from the filter. This effect can be further magnified by irradiating both control and treated cells. Repair of damaged DNA can also be evaluated using the alkaline elution assay by examining elution profiles at different times after exposure to the test substance and observing a return to control-like profiles.

Test Systems

Initial studies using the *in vitro* alkaline elution assay utilized cell lines and relatively rapid flow rates (0.5 ml/min) for 30 minutes. Greater sensitivity was achieved by decreasing the flow rate by a factor of 10 and increasing the elution time to 6–20 hours. This minimized problems due to cytotoxicity since increased DNA elution was detectable at lower doses.

Several cell lines have been utilized in *in vitro* alkaline elution studies, including V-79, CHO, L1210, Wl-38 and 18BcR. S9 activation systems have been employed with V-79 and CHO cells successfully. The assay should work equally well with other cell lines and would probably be amenable to co-cultivation techniques employing primary hepatocytes for metabolic activation (Langenbach et al., 1978). For *in vitro* studies, DNA is usually prelabeled by incubating cultures with ^{14}C-thymidine.

Achieving proper metabolic activation remains one of the major problems encountered in any *in vitro* system for predicting

carcinogenic and mutagenic potential of chemicals. Microsomal and S9 activating systems favor oxidative metabolism, but lack some reductive metabolism and many of the natural detoxification pathways. These problems can be eliminated by evaluating the ability of a chemical to cause DNA damage *in vivo*. Several investigators have modified the alkaline elution method to detect chemically induced DNA damage in various tissues of rats and mice. Many tissues can be evaluated by labeling DNA of animals with ^{3}H-thymidine during the first three weeks of birth and exposing them to the test substance during the following three weeks (Eastman and Bresnick, 1979; Petzold and Swenberg, 1978). Adult replicating tissues and liver (after partial hepatectomy) can also be labeled with ^{3}H-thymidine (Thomas et al., 1978). Likewise, adult tissues of rodents (and presumably nonrodents) can be evaluated using microfluorometric methods (Brambilla et al., 1978a, 1978b; Parodi et al., 1978).

Equipment requirements will vary depending on the particular alkaline elution method employed. *In vitro* systems require standard tissue culture facilities for chemical carcinogens. The alkaline elution assay itself utilizes peristaltic pumps such as the eight-channel Gilson Minipuls peristaltic pumps with the slow speed control module. Manifold tubing of 1.30 mm internal diameter provides a flow range of 0.05 to 0.6 ml/min. The tubing should be changed routinely and the pumps recalibrated in order to achieve consistent and uniform results. Either Gelman syringe type filter holders or 25 mm filter holders consisting of an upper funnel section and lower collecting section are connected to a three-way stopcock by Teflon tubing. The stopcock allows removal of liquid in the tubing without disturbing the filter holder or disconnecting the peristaltic pump. Eluate can be collected in bulk or in fractions. The latter is considerably more informative since elution profiles are obtained. The Gilson Aliquogel fractionater with a special attachment to allow simultaneous collection of up to ten samples works well for this task. Techniques involving radiolabeled DNA require a scintillation counter, while a fluorometer is necessary for DNA quantitation using the diamino-benzoic acid microfluorometric assay.

Assay performance can be evaluated in several ways. Culture-to culture, animal-to-animal, and day-to-day variation should be monitored. Duplicate aliquots of each sample should be run to assure reproducibility. Filter-to-filter variation can be controlled by applying an internal standard of ^{3}H-thymidine-labeled cells for ^{14}C-thymidine studies and vice versa (Kohn, 1978). The data is then expressed graphically as the fraction of ^{14}C-DNA retained on the filter versus the fraction of ^{3}H-labeled reference DNA retained on the filter. The assay should always be run at a constant temperature and in a location that minimizes vibration (Eastman and Bresnick, 1979).

Protocols for the Alkaline Elution Assay

In vitro *Assay*

Stock cultures of the cell line employed are maintained in appropriate growth medium. Eagle's MEM supplemented with 10% fetal calf serum, 2mM L-glutamine, and 20 mM HEPES, pH 7.4, can be used for V-79 cells. DNA of cells in log phase growth is labeled by adding 2.0 ml of complete medium containing 0.1 μC ^{14}C-thymidine (^{14}C-dThd). Radioactive medium is removed 20–24 hours later and the cells are incubated for 4–20 hours in nonradioactive medium. Cells are then exposed to test chemicals, either in growth medium, or in the presence of an activation system such as S9 fraction (25 mg wet wt/ml), supplemented with 1 mM NADPH, 1.5 mM glucose-6-phosphate, 0.25 units/ml glucose-6-phosphate dehydrogenase, and 25 mM MgCl$_2$. Test chemicals are made up in saline or DMSO to achieve a 1% vehicle concentration. Following exposure to

chemicals, the cells are washed with 5mM phosphate-saline, pH 7.4 (PBS), and trypsinized. Trypsinized cell suspensions are diluted with cold medium without calf serum (4.0 ml) and placed on ice. Cells are then isolated by centrifugation an resuspended in growth medium. Aliquots are taken for cell counting, viability assessment, and alkaline elution. While cells can be frozen and assayed at a later time, it is preferable to assay them immediately after harvesting. This must be done when slow elution is used.

Alkaline Elution Procedure

Single 25 mm polyvinyl filters (2 μm) are carefully wetted with PBS and placed on a Gelman syringe filter holder (No. 4320-1). The filter holder is then assembled and filled with PBS, and a 20–50 ml syringe is attached as a reservoir. One ml of saline is placed in the reservoir and 0.2 ml aliquots containing approximately 0.5 × 10^6 cells (treated or control) are added. Aliquots (0.1 ml) containing 5 × 10^5 ^{3}H-dThd labeled cells that have been irradiated with 150 rads can be placed in the solution to serve as internal filter controls. The reservoirs are gently agitated to assure mixing and the cells are drawn onto the filter at a pumping rate of 0.6 ml per minute. The cells are lysed on the filters by adding 5 ml of a lysing solution containing 0.2% Sarkosyl, 2 M NaCl and 0.04 M EDTA, pH 10. The lysing solution is also pumped at 0.6 ml per minute. Following lysis, the filters are washed by pumping 5.0 ml of 1 mM EDTA solution (0.6 ml per minute) until it reaches the top of the filter holder. Single-stranded DNA is then eluted from the filter by pulling 12–40 ml of eluting solution containing 0.02 M EDTA (acid form) and enough tetrapropylammonium hydroxide to yield a pH of 12.0 to 12.2. The eluate is collected in ten fractions at a rate of 0.05 ml per minute and counted by liquid scintillation counting. The filters are removed and incubated in 0.5 ml of 1.0 N HCl for 1 hour at 70–80°C. After cooling to room temperature, 2.5 ml of 0.4 N NaOH is added and the preparations are allowed to stand for 30 minutes. The filters are then counted by liquid scintillation counting.

In vivo *Assay*

Beginning on day four of age, neonatal rats are labeled with ^{3}H-dThd for a period of 3 weeks, receiving approximately 0.75 μCi/g/dose intraperitoneally. The animals are dosed with the test substance during the subsequent three weeks. Animals are given additional ^{3}H-dThd for 2–3 days if the organs being assayed had rapid cell regeneration times. After exposure to chemicals for various periods of time, the animals are sacrificed by decapitation and the tissues repidly removed and placed in cold 0.02 M EDTA–0.9% NaCl solution. Samples of tissues are taken for histologic monitoring of toxicity if the toxic effects of the chemicals are unknown. Tissues are minced and pushed through stainless steel wire mesh screen or hand-homogenized in 20 ml of EDTA-NaCl buffer, pH 7.3–7.4 Specific homogenization techniques for each tissue so far evaluated have been published (Brambilla *et al.*, 1978a, 1978b; Eastman and Bresnick, 1979; Parodi *et al.*, 1978 Petzold and Swenberg, 1978; Thomas *et al.*, 1978). Tissue suspensions and homogenates are centrifuged at 1500 rpms for 2 to 4 minutes and the pellet is resuspended in 5–10 ml of EDTA-NaCl. Aliquots of this suspension are then applied to the alkaline elution assay as previously described.

Control Procedures

Most control procedures have already been alluded to; however, several of the most important aspects will be repeated to assure adequate coverage. The amount of DNA retained on the filter is affected by pH, time, and flow rate of eluting solution. Assay conditions should be adjusted so that 85–90% of the control DNA is retained. If duplicate filters vary by more than 3%, internal filter controls should be incorporated. Whenever possible, more than one control animal should be used for *in vivo* assays.

Statistical Applications

Procedures which collect a standard volume of eluate in bulk produce a single number for each assay. The percent of DNA retained on the filter from control and treated animals or cells can be compared by standard statistical methods. Data obtained from fractionated elutions is normally plotted as the log of the DNA remaining versus time. Such data can be compared and analyzed at a standard elution time. Furthermore, if data is plotted as the log of retained DNA versus log of reference DNA retained, data can be expressed as the amount of DNA retained when 50% of the reference DNA is eluted. These values can be compared using ordinary statistical methods. The best comparisons are for data obtained from a single experiment. Slight daily variations in procedure make cumulative data less amenable to statistical approaches. One paper has been published, however, on the attributes of this method (Parodi *et al.*, 1979).

Modifications of the Alkaline Elution Assay

Several modifications of the alkaline elution assay have been reported since its inception. The standard assay now utilizes a decreased flow rate and increased elution time in order to increase the sensitivity for detecting single-strand breaks and alkali-labile lesions in the DNA. By irradiating control and treated cells (Ewig and Kohn, 1978; Fornace and Little, 1979; Kohn, 1978; Ross *et al.*, 1978) or tissue homogenates (Thomas *et al.*, 1978) prior to filtration and examining the effect of proteinase K digestion prior to elution with alkali, the assay can distinguish between protein-DNA crosslinks and DNA-DNA crosslinks.

Considerable emphasis recently has been placed on developing methods for assaying unlabeled cells in tissues (Brambilla *et al.*, 1978a, 1978b; Cavanna *et al.*, 1977; Parodi *et al.*, 1978). These techniques permit analysis of DNA from adult animals, such as those on bioassay protocols and from tissues that are difficult to prelabel due to low levels of cell proliferation.

Evaluation of Alkaline Elution Assay Results

Several types of data can be obtained from the alkaline elution assay. In order to fully interpret the assay, dose response data for toxicity and elution, and time course data for repair are necessary. When increased elution is detected only at toxic doses, the test is generally considered to be negative. Elution profiles from such exposures usually consist of an initial rapid elution followed by an elution curve that nearly parallels the control. The alkaline elution test is considered positive if dose-related increases in elution occur at subtoxic doses. The shape of the elution profile provides some indication of the type of DNA damage, i.e., single-strand breaks and highly alkali-labile sites on the DNA give a first order elution. Elution profiles with accelerating elution suggest other alkali-labile sites that require additional exposure to alkali before becoming single-strand breaks. DNA repair can be readily visualized following exposure to positive compounds. Time courses for repair range from less than an hour to over a week. The alkaline elution assay is considered negative when no increase in elution is detected over an appropriate dose range. One must, however, always consider whether or not proper metabolic activation occurred. When treated cells or tissues have a dose-related

decrease in elution compared to controls, the agent should be suspected of inducing DNA crosslinks. The assay should be repeated with the appropriate modifications.

While data obtained from alkaline elution assays are clearly quantitative, the assay itself is considered qualitative with respect to predicting carcinogenic or mutagenic potential. One cannot say that because exposure to compound X produces a 50 percent increase in elution it is a more potent carcinogen or mutagen than compound Y, which results in only a 20 percent increase in elution. The alkaline elution assay detects many types of DNA damage. Some of these are pro-mutagenic, i.e., cause mutation if cell replication occurs prior to repair, while others are associated with toxicity but are not associated with mutation.

Correlation of the Alkaline Elution Assay with Carcinogenicity and Mutagenicity

Approximately 150 chemicals have been evaluated in one or more modifications of the alkaline elution assay. Although good bioassay data is not available on all these chemicals, results of the alkaline elution assay appear to correctly identify carcinogens from noncarcinogens 85 to 90 percent of the time (Swenberg and Petzold, 1979). Such statistics can be grossly misleading, however, since groups of chemicals with well-characterized metabolism and mutagenicity can be selected to improve the statistics. Conversely, compounds having poorly understood metabolism and uncharacterized mutagenic potential will most assuredly result in more false positive and false negative results. With this in mind, we must continue to evaluate chemicals that have completed proper bioassays. Such validation studies will eventually provide realistic correlation data. Future research must concentrate on understanding the mechanism of both correct and incorrect assay results.

Table 5.1. Alkaline Elution of DNA following in vitro Exposure of V79 Cells to Chemicals

Compounds	Exposure time-hr	Activation system[a]	Results and dose (mM)[b]									
			.001	.003	.01	.03	.1	.3	1.0	3.0	10.0	30.0
Alkylating Agents												
N-Methyl-N'-nitro-N-nitrosoguanidine (MNNG)	0.5	—	—	+	+	+	+					
N-Methyl-N'-nitro-N-nitrosoguanidine (MNNG)	1	R		—	—	—	+	+				
Methylnitrosourea (MNU)	1	—				—	+	+	+	+		
Ethylnitrosourea (ENU)	1	—				—	—	+	+	+		
Methyl Methanesulfonate (MMS)	1	—				—	+	+	+	+		
Ethyl Methanesulfonate (EMS)	1	—				—	—	—	+	+		
β-Propiolactone	1	—				—	+	—				
Propyleneimine	1	—					+	+	+	+	+	
Dimethylnitrosamine (DMN)	2	M					+	+	+	+	+	
Dimethylnitrosamine (DMN)	2	R					+	+	—	+		
Diethylnitrosamine (DEN)	2	M				—	—	—	—	+	+	
Diethylnitrosamine (DEN)	2	R				—	—	—	—	+	+	
Polycyclic Aromatic Hydrocarbons												
Benzo(a)pyrene [B(a)P]	2	R			—	+	+	+	+	+		
Benzo(e)pyrene [B(e)P]	4	R		—	—	+	+	+	+	+		
7, 12-Dimethylbenz(a)anthracene (DMBA)	2	R			—	+	+	+	—			
3-Methylcholanthrene (3MCA)	2	R			—	—	—	—				
Phenanthrene	1,2,4	R			—	—	—	—				
Anthracene	1,2,4	R			—	—	—	—				
Amines												
Benzidine	2	R			+	+	+	+				
N',N¹[phenyl-3-methyl phenyl]benzidine	2,4	R,M				—	—	—	—	—		
N',N¹-Diphenyl benzidine	2,4	—,R,M		—		—	—	—	—	—		
4-Aminoazobenzene	4	R				+	+	+	+	+		
4-Aminobiphenyl	2	R				—	—	+	+	+		
2-Acetylaminofluorene (2-AAF)	2	R				—	+	+	+			
N-Hydroxy-acetylaminofluorene (N-OH-AAF)	2	—			+	+	+	+	+			
N-Acetoxy-acetylaminofluorene (N-OAc-AAF)	2	—		—	+	+	+	+	+			
Aniline	1,2,4	R				—	+	+	+			
para-Rosaniline	2	R					+	+	+			
4,4'-Methylenedianiline (MDA)	1,2,4	R					—	+	+			
3,3',4,4'-Tetraaminobiphenyl (TAB)	2,4	R					—	+[c]	+			
2,4-Toluenediamine (TDA)	2,4	R					±	+[c]	±			
β-Naphthylamine	2	R					+	+	+			
α-Naphthylamine	2	R					+	+	+			

Table 5.1. *Alkaline Elution of DNA following in vitro Exposure of V79 Cells to Chemicals (Continued)*

Compounds	Exposure time-hr	Activation system[a]	Results and dose (mM)[b]									
			.001	.003	.01	.03	.1	.3	1.0	3.0	10.0	30.0
Drugs												
Estradiol, N.F.	2,4	–,R				–			–	–		
Hydrocortisone Acetate (U.S.P. 98.7%)	2,4	–,R				–		–	–	–		
Progesterone, U.S.P.	2,4	R						+^{c}	+^{c}	+^{c}		
Progesterone, U.S.P.	4	–						+^{c}	+^{c}			
Testosterone, N.F.	2,4	–,R				–		–	+^{c}	+^{c}		
Oxytetracycline, U-7881	4	R						–	–	–		
Solu-Medrol, U-9088	1	–					–		–			
Diethylstilbestrol (DES), U-1298	2	R				–		+^{c}	+^{c}	+^{c}		
Diazapam (Valium)	2	R				–		–	–	–		
Streptozocin	1	–				–		+	+	+	+	
1,3-Bis(2-chlorethyl-)-1-nitrosourea (BCNU)	1	–				–	–	+	+	+	+	
Cyclophosphamide	1,2,4	R						–	–	–	–	
Phenobarbital	2	–,R						–	–	–	–	
Aspirin	2	–,R				–		–	–	–	–	
Other Carcinogens/Mutagens												
Acrylonitrile	2	–,R					–		–	+	+	+
Aflatoxin B$_1$	2,4	R		+		+		+	+	+	+	
4-Nitroquinoline-1-oxide (4NQO)	1	R					+	+	+	+		
4-Nitroquinoline-1-oxide (4NQO)	1	–			–		–	–	–	–		
Safrole	1,2,4	R						–	–	–	–	
1-Hydroxysafrole	2	–,R					–	–	–	–		
1-Acetoxysafrole	2	–				+	+	+	+	+	+	
Ethionine	2	R						–	–	±		
Epichlorhydrin	2	–					+	–	–			
e-Caprolactone	4	R					+	+	+	+		
5-Nitro-2-furfrol Semicarbazon	1	–						+	+	+		
5-Nitro-2-furoic Acid	1	–						–	–			
Benzyl Chloride	2	–						+	+	+	+	
Dimethylcarbamyl Chloride	2	–						+	+	+	+	
1,2,3,4-Diepoxybutane	2	–						+	+	+	+	
1,2-Dimethylhydrazine	1	R						±	±	±		
Titanocene Dichloride	2	–						+	+	+	+	+
Miscellaneous Chemicals												
Dinitrophenol	2	–,R						–	–	–	–	
Methionine	2	–,R						–	–	–	–	

Table 5.1. Alkaline Elution of DNA following in vitro Exposure of V79 Cells to Chemicals (Continued)

Compounds	Exposure time-hr	Activation system[a]	Results and dose (mM)[b]									
			.001	.003	.01	.03	.1	.3	1.0	3.0	10.0	30.0
Caffeine	2	—								—	—	—
EDTA	1,2,4	—,R								—	—	—
Captan	2	—		—	+	+	+	+				
Captan	2	R					—		—			
DDT	2,4	R				—	—	—	—			
Dieldrin	2,4	R				—	—	—	—			
Other[d]	1,2,4	—,R								—		

[a]R = Rat; M = Mouse; (—) = None.

[b](+) = Damage observed; (—) = No damage observed.

[c]Toxic levels—cell viability low as determined by cellular ATP.

[d]Found negative at higher concentrations: ethanol, acetone, dimethylsulfoxide, propylene glycol, and benzene.

Table 5.2. Summary of DNA Damage Induced in vivo by Chemicals

Chemical	Dose range (mg/kg)	Routes	DNA elution[a]								
			Liver	Lung	Kidney	Brain	Thymus	Duodenum	Stomach	Bone marrow	Mammary
Dimethylnitrosamine[b]	10–40	IP	+++[f]	+	±[f]	−	±[f]	+[f]			
Diethylnitrosamine[b]	40–80	IP	+++[f]	−	±[f]	−	−	+			
N-Nitrosohexamethyleneimine[b]	25–250	IP,Oral	−[f]	−	−		−	−	−		
2-Acetylaminofluorene[c]	10–25	IP	+[f]	−	−		−				
N-Hydroxy-acetylaminofluorene[c]	20–40	IP	+[f]	−	−	−					
Benzidine-HCl[c]	250	IP,SC	+[f]	−	++					−	
Aflatoxin B1[d]	0.5–1	IP	++[f]	−	±		+				
1,2-Dimethylhydrazine[b]	20–100	SC	++[f]		±		−	++[f]			
Azoxymethane[b]	30–100	SC	+++[f]		±		−	++[f]			
4-Nitroquinoline-1-oxide[b]	25–100	SC	−	++[f]	±[f]	+		−			
4-Nitroquinoline-1-oxide[b]	100	IP	±[f]	+[f]	±[f]			−			
7,12-Dimethylbenz(a)anthracene[c]	75–200	Oral	−	±[f]	−		−	−	++		±[f]
Methylnitrosourea[e]	10–160	IP,IV	++	−	++	++[f]	±[f]				++[f]
Ethylnitrosourea[e]	50–100	IP,IV	+	−	±	+[f]	−				
Methyl Methanesulfonate[b]	100–200	IP	++	++	++[f]	+++[f]	++	+++	+++		
Ethyl Methanesulfonate[b]	100	IP	+		±[f]		+				
N-Methyl-N'-nitro-N-nitroso-guanidine[d]	250–500	Oral			−			++[f]	++[f]		
β-Propiolactone[c]	100–500	IP,Oral	++		++			+	±[f]		
Azaserine[b]	50–250	IP	++		++[f]	−				−	
Streptozocin[e]	500–100	IV				−				+[f]	
Cyclophosphamide[b]	5–80	IP	−		±	+[f]					
Ethyl Carbamate[b]	25–500	IP	−		−[f]						
Chloroform[d]	200,400	Oral	−								

a +++ = Increased elution > 50; ++ = Increased elution > 20; + = increased elution > 7.5; ± = increased elution > 4.0; − = increased elution < 4.0.
b Administered in saline, controls received a similar volume of saline.
c Administered in DMSO, control received DMSO.
d Administered in Corn Oil, control received Corn Oil.
e Administered in 0.06M Sodium Citrate-0.08M K_2HPO_4, pH 4.2; control received vehicle.
f Organ susceptible to tumor formation.

Compounds That Have Been Evaluated with the Alkaline Elution Assay

Several laboratories have published results of *in vitro* and *in vivo* alkaline elution assays. Table 5.1 summarizes data from *in vitro* testing in my laboratory during the past five years. Several other laboratories have published similar results with small numbers of chemicals (Cavanna *et al.*, 1977; Fornace and Little, 1979; Swenberg *et al.*, 1976). Kohn and co-workers have investigated the effect of exposures to many chemotherapeutics, alkylating agents, and X-irradiation (Erickson *et al.*, 1978; Ewig and Kohn, 1978; Kohn and Grimek-Ewig, 1973; Kohn *et al.*, 1974, 1976; Kohn, 1978; Ross *et al.*, 1978).

Table 5.2 summarizes *in vivo* alkaline elution results we have obtained (Petzold and Swenberg, 1978). Comparable results were reported by Brambilla *et al.* (1978a, 1978b), Parodi *et al.* (1978), and Eastman and Bresnick (1979).

Many additional proprietary drugs and chemicals have been evaluated with *in vitro* or *in vivo* alkaline elution assays. Bioassay data will become available on a number of these during upcoming years. Correlations of bioassay results and short-term tests on proprietary compounds frequently remain unpublished. This is particularly true when both sets of data are negative. Without such data, however, a great deal of information on the predictive value of short-term test is lost. It is hoped that a centralized data repository will be developed to provide a mechanism for such correlations.

References

Brambilla, G., Cavanna, M., Parodi, S., Sciaba, L., Pino, A., Robbiano, L. (1978a): DNA damage in liver, colon, stomach, lung and kidney of BALB/c mice treated with 1,2-dimethylhydrazine. Int. J. Cancer *22:* 174–180.

Brambilla, G., Cavanna, M., Parodi, S. (1978b): Evaluation of DNA damage and repair in mammalian cells exposed to chemical carcinogens. Pharmacol. Res. Comm. *10:* 693–717.

Cavanna, M., Parodi, S., Sciaba, L., Maura, A., Carlo, P., Cajelli, E., Brambilla, G. (1977): DNA damage and repair by alkaline elution in N-diazoacetylglycine amide-treated cells. Toxicol. Letters *1*:115–120.

Eastman, A., Bresnick, E. (1979): A technique for the measurement of breakage and repair of DNA alkylated *in vivo*. Chem.-Biol. Interact. *23*:369–377.

Erickson, L.C., Bradley, M.O., Kohn, K.W. (1978): Measurements of DNA damage in Chinese hamster cells treated with equitoxic and equimutagenic doses of nitrosoureas. Cancer Res. *38*:3379–3384.

Ewig, R.A., Kohn, K.W. (1979): DNA-protein cross-linking and DNA interstrand cross-linking by haloethylnitrosoureas in L1210 cells. Cancer Res. *38*:3197–3203.

Fornace, A.J., Little, J.B. (1979): DNA-protein crosslinking by chemical carcinogens in mammalian cells. Cancer Res. *39*:704–710.

Kohn, K.W. (1978): DNA as a target in cancer chemotherapy: Measurement of macromolecular DNA damage produced in mammalian cells by anticancer agents and carcinogens. Methods Cancer Res. *16*:291–345.

Kohn, K.W., Grimek-Ewig, R.A. (1973): Alkaline elution analysis, a new approach to the study of DNA single-strand interruptions in cells. Cancer Res. *33*:1849–1853.

Kohn, K.W., Friedman, C.A., Ewig, R.A.G., Iqbal, Z.M. (1974): DNA chain growth during replication of asynchronous L-1210 cells. Alkaline elution of large DNA segments from cells lysed on filters. Biochemistry *13*: 4134–4139.

Kohn, K.W., Erickson, L.C., Ewig, R.A.G., Friedman, C.A. (1976): Fractionation of DNA from mammalian cells by alkaline elution. Biochemistry *15*:4628–4637.

Langenbach, R., Freed, H.J., Huberman, E. (1978): Liver cell-mediated mutagenesis of mammalian cells with liver carcinogens. Proc. Natl. Acad. Sci. USA *75*:2864–2867.

Parodi, S., Taningher, M., Santi, L., Cavanna, M., Sciaba, L., Maura, A., Brambilla, G. (1978): A practical procedure for testing DNA damage *in vivo* proposed for a pre-screening of chemical carcinogens. Mutat. Res. *54*:39–46.

Parodi, S., Taningher, M., Cavanna, M., Brambilla, G., Boero, P. (1979): Evaluation of DNA damage: Problems related to signal/noise ratio in the perspective of a blind assay. In: L. Santi and S. Parodi (eds.), Short Term Tests for Prescreening of Potential Carcinogens, Istituto Scientifico Per Lo Studio e La Cura Dei Tumori, Genoa.

Petzold, G.L., Swenberg, J.A. (1978): Detection of DNA damage induced *in vivo* following exposure of rats to carcinogens. Cancer Res. *38*:1589–1594.

Ross, W.E., Ewig, R.A.G., Kohn, K.W. (1978): Differences between melphalan and nitrogen mustard in the formation and removal of DNA crosslinks. Cancer Res. *38:* 1502–1506.

Swenberg, J.A., Petzold, G.L. (1979): The usefulness of DNA damage and repair assays for predicting carcinogenic potential of chemi-' cals. In: B.E. Butterworth (ed.), Strategies for Short-Term Testing for Mutagens/Carcinogens. West Palm Beach, Florida: CRC Press, pp. 77–86.

Swenberg, J.A., Petzold, G.L., Harbach, P.R. (1976): *In vitro* DNA damage/alkaline eution assay for predicting carcinogenic potential. Biochem. Biophys. Res. Comm. *72:* 732–738.

Thomas, C.B., Osieka, R., Kohn, K.W. (1978): DNA cross-linking by *in vivo* treatment with 1-(2-chlorethyl)-3-(4-methylcyclohexyl)-1-nitrosourea of sensitive and resistant human colon carcinoma xenografts in nude mice. Cancer Res. *38*:2448–2454.

6

DNA Synthesis Inhibition in Mammalian Cells as a Test for Mutagenic Carcinogens

ROBERT B. PAINTER

Principle of Method

Current models for the mechanism of action of mutagens begin with alterations in the structure of DNA, i.e., DNA damage. The general concept is that if DNA damage of any kind occurs, there is a possibility that mutation may ensue, and therefore any agent that damages DNA is a potential mutagen. Because mutagenesis and carcinogenesis are closely correlated and because most, if not all, mutagens are also carcinogens, it is safe to assume that any agent that causes DNA damage is also capable of inducing the process that eventually results in human cancer.

With these guiding principles in mind, I proposed a rapid test based on measurements of DNA synthesis rate in human cells in culture (Painter, 1977a). Agents that damage DNA cause lesions that can inhibit the overall rate of DNA replication by any of three mechanisms: (1) slowing the rate of DNA fork displacement, i.e., inhibiting the movement of the many grow-ing points active at any one time in a human cell; (2) blocking the progression of DNA growing points already active in DNA replication (precocious termination); and (3) blocking the initiation of those replicating units (replicons) normally scheduled to operate. Each of these mechanisms has been shown to occur after various kinds of DNA damage, although, to my knowledge, intercalating agents are the only DNA-damaging agents that slow fork displacement rates (Painter, 1978). Precocious termination appears to be the major mechanism by which agents such as ultraviolet light and N-acetoxyacetoaminofluorene act (Regan and Setlow, 1973), and the inhibition of replicon initiation is the main mechanism for the inhibitions of DNA synthesis caused by X-rays (Makino and Okada, 1975; Painter and Young, 1975) and methyl methanesulfonate (Painter, 1977a; Dahle *et al.*, 1978).

Irrespective of the mechanism by which these agents inhibit DNA synthesis, they share the common feature that the DNA damage remains even after the damaging agent itself is gone. The rate of DNA synthesis continues to decrease as a function of time after the initial damage as more and more of the total genome becomes involved with the lesions, at least until repair processes intervene. It is this feature of continuing depression in rate of DNA synthesis that distinguishes the action of DNA-damaging agents from the action of substances that inhibit the rate of DNA synthesis but do not cause DNA damage. Nondamaging agents (e.g., the protein synthesis inhibitors cycloheximide and puromycin) cause their effects only when in contact with the cells; when they are removed, the rate of DNA synthesis increases (Painter, 1977b).

The difference in the rate of DNA synthesis after treatment is the basis of the mammalian cell DNA synthesis inhibition test. Cells (usually HeLa) are exposed to the agent for an arbitrarily chosen exposure time of 30 min for agents that do not require metabolic activation and 60 min for those that do. After the agent is removed and washed from the cells, the rate of DNA replication is measured immediately in treated and control cells. This is repeated at 0.5 and 1.5 hr after removal of the agent. Data are plotted as percentage of control rate of DNA synthesis as a function of time. A curve showing a decreasing rate of DNA synthesis between 0 and 0.5 hr or between 0.5 and 1.5 hr is considered to be diagnostic of a DNA-damaging agent. A curve showing an increasing rate of DNA synthesis between 0 and 0.5 hr is considered diagnostic of an agent that inhibits DNA synthesis but does not cause DNA damage. The damage to DNA caused by some agents seems to be repaired so rapidly that a curve showing a drop in rate of DNA synthesis during the first 0.5 hr followed by recovery during the period between 0.5 and 1.5 hr is sometimes observed (Fig. 6.1). If this observation is repeatable, the agent is still considered a DNA-damaging agent, e.g., ethyleneimine (10^{-4} M) and benzidine (6×10^{-4} M) (Painter, 1978). Thus far, no agent has been found that reproducibly

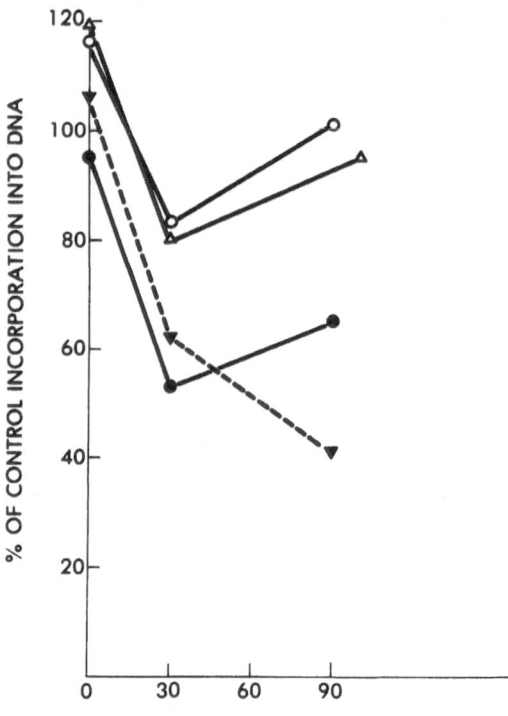

Figure 6.1. Inhibition of HeLa DNA synthesis by 9-aminoacridine (9A) and benzo(a)pyrene (BP). Both chemicals were dissolved in dimethyl sulfoxide and incubated with HeLa cells in the presence of S9 Mix (Ames *et al.*, 1975) for 1 hr. ○————○, 9A (2.6×10^{-6} M); △————△, 9A (7.8×10^{-6} M); ●————●, 9A (1.3×10^{-5} M); ▼-----▼, BP (4×10^{-5} M).

produces an increasing rate of DNA synthesis between 0 and 0.5 hr and a decreasing rate after 0.5 hr.

Test Organism and Testing Protocol

About 10^5 HeLa (or other rapidly growing mammalian) cells are inoculated into 35 mm Petri dishes containing 5 ml of Eagle's minimum essential medium and 0.01 μCi/ml of [^{14}C]thymidine (50 mCi/mmole). The cells are incubated in a humidified 5% CO_2 atmosphere for 24 hr or longer, the radioactive medium is removed from the cells, and a new medium is added that contains either the compound under test or an equivalent amount of carrier material; e.g., because some compounds must be dissolved in dimethyl sulfoxide, an equivalent amount of dimethyl sulfoxide is used in the medium for the control cultures. The compounds under test are incubated with the cells for 30 min; if, however, the compound must be activated by the liver microsomal extract S9 (Ames et al., 1975), the incubation is for 1 hr. The medium containing the compound is then removed from all cells, and the cells are washed twice with minimum essential medium. Fresh medium is added to all cultures except four; to these cultures (two controls and two treated) is added a medium containing 20 μCi/ml of [^{3}H]thymidine (50 Ci/mmole). After a 10-min incubation the radioactive medium is removed and the cells are washed rapidly with ice-cold SSC (0.15 M sodium chloride–0.015 M sodium citrate). The cells are scraped into cold SSC and filtered through Whatman GF/C filters wetted with ice-cold 4% perchloric acid; the filters are then washed with 4% perchloric acid, 70% alcohol, 95% alcohol, and 100% alcohol, dried, and counted with a liquid scintillation spectrometer.

One-half hour after removal of the drug, the medium on a second set of four cultures is removed and the cells are incubated with [^{3}H]thymidine and prepared for counting. At 1.5 hr after removal of the drug, the entire operation is repeated. Because the cells had previously incorporated [^{14}C] thymidine into parental DNA, the final counts are recorded as the ratio, ^{3}H/^{14}C, which is a measure of specific activity of the ^{3}H and therefore of the rate of DNA synthesis. The ^{3}H/^{14}C ratios for drug-treated cells are expressed as percentages of control incorporation into DNA as a function of time of incubation after end of treatment.

If nothing is known about the agent to be tested, the highest possible concentration is used first. If the agent inhibits DNA synthesis very strongly (greater than about 40%) or causes cell dislodgement, the concentration must be systematically reduced until a satisfactory zero-time inhibition is found. Previous experience or published reports on the agent or on similar agents sometimes aid in choosing a useful concentration of the agent for testing.

If the agent fails to inhibit DNA synthesis at the highest concentration, metabolic activation must be attempted; the agent is added to S9 Mix (Ames et al., 1975) and the mixture is incubated with the cells for 1 hr. S9 Mix is derived from a microsomal preparation from the liver of rats previously injected with Aroclor, a polychlorinated biphenyl (Ames et al., 1975), and is commercially available. The mixture is made up as follows:

Solution 1, prepared fresh each time, contains 8 ml of 0.125 M phosphate buffer, pH 7.4; 2 ml of a solution that is 0.04 M MgCl$_2$ and 0.165 M KCl; 0.05 ml of 1 M glucose-6-phosphate; and 0.4 ml of 4.2×10^{-2} M NADP.

Solution 2 contains 2 ml of freshly thawed liver extract plus 9 ml of Solution 1.

The final solution to be added to cells contains the agent diluted in McCoy's 5 A medium plus Solution 2. For unknown agents, the ratio is 7.5 ml of McCoy's to 2.5 ml of Solution 2; however, for most agents a 9:1 ratio is sufficient. Controls are made up in the same way but contain only the carrier (dimethyl sulfoxide, water, etc.). A positive control should be included with each test of unknowns. As an example of an agent that does not require metabolic activation, 4-nitroquinoline 1-oxide gives reproducible results. For an agent that requires metabolic activation, benzo(a)pyrene has proved valuable.

Evaluation of Results

To estimate the relative potency of the agent, a particular endpoint must be chosen. The concentration that produces a 40% inhibition of DNA synthesis (i.e., to 60% of control value) within 2.5 hr was originally used, but after screening of 50 to 60 compounds, it seems that the concentration necessary to inhibit DNA synthesis to 60% of control within only 1.5 hr is more advantageous.

The results with the first 25 agents tested (Table 6.1) were previously reported (Painter, 1977a,b; 1978); these agents had good correlation with the Ames test (Painter and Howard, 1978). About 30 more agents have been tested since those reports and the DNA synthesis inhibition test continues to detect those agents known to damage DNA (Table 6.2). As yet no false positives have been observed. Ascorbate is positive in the test (Galloway and Painter, 1979), as it is in the Ames test (Guttenplan, 1977) and in the unscheduled DNA synthesis test (Stich *et al.*, 1976). It seems, therefore, that ascorbate is capable of causing DNA damage, although normal levels of catalase, which protects intracellular DNA from ascorbate-induced damage,

may prevent this action *in vivo*. Very weak mutagens may not be detectable by the test—e.g., saccharin is negative although it has been reported to be positive in the sister chromatid exchange test (Wolff and Rodin, 1978).

The greatest advantage of the mammalian cell DNA synthesis inhibition test is its speed. Strongly positive compounds that do not require metabolic activation can be identified within about 3 hr after their addition to the cells. If the protocol of testing high concentrations without S9 followed by testing with S9 is used, even those compounds requiring metabolic activation can, theoretically, be identified within one day. Of course, as with any testing procedure, the test must be reproduced before it can definitely be considered positive. Negatives generally require more testing, and even after three or four runs at the highest possible concentration the compound can only be reported negative in this particular test.

Table 6.1. *Concentrations of Chemicals Required to Inhibit HeLa DNA Synthesis by 40% within 2.5 hr*

Chemical	S9 required?	Effective dose (M)
N-acetoxy-2-acetylaminofluorene (AAAF)	No	1.0×10^{-7}
Adriamycin	No	1.0×10^{-6}
Aflatoxin B_1	Yes	5.0×10^{-8}
Benzidine	No	6.0×10^{-4}
Benzo(a)pyrene	Yes	7.0×10^{-6}
Busulfan	No	$> 2.5 \times 10^{-2}$
Cycloheximide	—	Negative
Cyclophosphamide	Yes	6.0×10^{-4}
Dimethylnitrosamine	Yes	1.0×10^{-1}
Dimethyl sulfoxide (DMSO)	—	Negative
Ethyleneimine	No	1.0×10^{-4}
Ethyl methanesulfonate (EMS)	No	1.0×10^{-2}
Hydroxyurea	—	Negative
ICR-170	No	5.0×10^{-8}
Methyl methanesulfonate (MMS)	No	5.0×10^{-4}
N-methyl-N′-nitro-nitrosoguanidine (MNNG)	No	2.0×10^{-6}
Mitomycin C	No	1.0×10^{-6}
Natulan	—	Negative
Nickel chloride	—	Negative
4-Nitroquinoline 1-oxide	No	2.0×10^{-7}
Phorbol ester	—	Negative
Potassium cyanide	—	Negative
3′3′5′5′-Tetramethylbenzidine	—	Negative
Trenimon	No	2.0×10^{-7}
Triethylenemelamine (TEM)	No	1.0×10^{-4}

Table 6.2. *Chemicals Tested in the Mammalian Cell DNA Synthesis Inhibition Test Since Last Report (Painter, 1978)*

Chemical	S9 required?	Effective dose[a] (M)
Actinomycin D	No	1.0×10^{-6}
Amethopterin	—	Negative
9-Aminoacridine	Yes	1.3×10^{-5}
Aminopterin	—	Negative
Ascorbic acid	No	2.5×10^{-3}
Bis-chloroethylnitrosourea	No	1.0×10^{-4}
Captan	No	8.3×10^{-6}
Cholesterol	—	Negative
Cholesterol α epoxide	Yes	2.5×10^{-4}
Cis-platinum(II)diamine dichloride	No	1.0×10^{-6}
Dibenzo(a)acridine	—	Negative
1,2-Dibromo-3-chloropropane (DBCP)	No	1.0×10^{-2b}
1,3-Dibromopropanol	Yes	1.6×10^{-3}
1,3-Dichloropropanol	Yes	2.5×10^{-3}
Dieldrin	Yes	4.0×10^{-4}
Diethylsulfate	No	6.0×10^{-4}
Dimethylaminoazobenzene (butter yellow)	Yes	1.0×10^{-4}
7,12-Dimethylbenzanthracene (DMBA)	Yes	4.4×10^{-5}
7,12-Dimethyl-5-fluoro-benzanthracene (5FDMBA)	Yes	7.0×10^{-5}
7,12-Dimethyl-2-fluoro-benzanthracene (2FDMBA)	—	Negative
Dimethylsulfate	No	2.0×10^{-3}
Epibromohydrin	No	1.5×10^{-3b}
Epichlorohydrin	No	2.7×10^{-3b}
Ethylnitrosourea	No	1.0×10^{-4}
Mechloroethamine HCl	No	1.0×10^{-6}
Methylnitrosourea	No	5.0×10^{-5}
β-Naphthylamine	Yes	2.1×10^{-4}
Saccharin	—	Negative
Thiotepa	No	5.0×10^{-3}
Vincristine	—	Negative

[a] Dose required to inhibit DNA synthesis to 60% of control within 90 min after end of treatment.
[b] Performed with Chinese hamster ovary cells instead of HeLa cells.

Another advantage of the test is its cost. Other than the scintillation counter, which is an instrument found in nearly all biology and biochemistry laboratories, the materials are all easily obtained and inexpensive. And the growing of HeLa cell cultures and preparation of samples for the assay are procedures that do not require advanced training or specialized techniques.

The DNA synthesis inhibition test is a more reliable indicator of DNA damage than the unscheduled DNA synthesis test, which is completely insensitive to intercalating agents (Painter, 1978) and relatively poor at detecting agents that damage DNA in a manner similar to X-rays (Rasmussen and Painter, 1966; Painter and Young, 1972; Regan and Setlow, 1974). The DNA synthesis inhibition test is very sensitive to intercalating agents such as adriamycin (Painter, 1978) and actinomycin D (Table 6.2). Moreover, replicon initiation is uniquely sensitive to X-rays (Makino and Okada, 1975; Painter and Young, 1975); therefore, inhibition of overall DNA replication by X-rays is easily detected after doses of about 250 rads (Painter, 1978), whereas doses of 5000 rads or more are required before unscheduled DNA synthesis can routinely be observed (Rasmussen and Painter, 1966; Painter and Young, 1972; Fox and Fox, 1973).

The test is based on DNA damage and therefore is not meant to detect agents (such as asbestos) that induce cancers by other mechanisms. The test is also subject to the same uncertainties as other tests using *in vitro* metabolic activation; another microsomal preparation (e.g., from a tissue other than liver or from liver stimulated by another inducer) may yield different results. Combining this *in vitro* HeLa test with an *in vivo* (whole mouse) DNA synthesis inhibition test (Friedman and Staub, 1976; Amlacher and Rudolph, 1978), in which metabolic activation inherently occurs but which is more laborious to perform, may constitute a valuable screening method for DNA-damaging agents.

Acknowledgments

I thank Ricci Howard for performing these assays, Lynn Haroun, Drs. Bruce Ames, and Sidney Stolzenberg for furnishing chemicals, and Drs. Ron Hart and Bernie Daniels for HPLC-purified DMBA, 5FDMBA, and 2FDMBA. The work was supported by the U.S. Department of Energy.

References

Ames, BN., McCann, J., Yamasaki, E. (1975): Methods for testing carcinogens and mutagens with the Salmonella/mammalian-microsome mutagenicity test. Mutat. Res. *31*:347–364.

Amlacher, E., Rudolph, C. (1978): [Nuclear DNA synthesis rate and labelling index: effects of carcinogenic and non-carcinogenic chemicals on its behaviour in the organism of growing CBA-mice. The thymidine-incorporation-screening-system (TSS)] (In German). Exp Pathol. *16*:69–82.

Dahle, D.B., Griffiths, T.D., Carpenter, J.G. (1978): Inhibition of deoxyribonucleic acid synthesis and replicon elongation in mammalian cells exposed to methyl methanesulfonate. Mol. Pharmacol. *14*:278–289.

Fox, M., Fox, B.W. (1973): Repair replication in X-irradiated lymphoma cells *in vitro*. Int. J. Radiat. Biol. *23*:333–358.

Friedman, M.A., Staub, J. (1976): Inhibition of mouse testicular DNA synthesis by mutagens and carcinogens as a potential simple mammalian assay for mutagenesis. Mutat. Res. *37*:67–76.

Galloway, S.M., Painter, R.B. (1979): Vitamin C is positive in the DNA synthesis inhibition and sister-chromatid exchange tests. Mutat. Res. *60*:321–327.

Guttenplan, J.B. (1977): Inhibition by L-ascorbate of bacterial mutagenesis induced by two *N*-nitroso compounds. Nature (Lond.) *268*:368–370.

Makino, F., Okada, S. (1975): Effects of ionizing radiations on DNA replication in cultured mammalian cells. Radiat. Res. *62*:37–51.

Painter, R.B. (1977a): Inhibition of initiation of HeLa cell replicons by methyl methanesulfonate. Mutat. Res. *42*:299–304.

Painter, R.B. (1977b): Rapid test to detect agents that damage human DNA. Nature (Lond.) *265*:650–651.

Painter, R.B. (1978): DNA synthesis inhibition in HeLa cells as a simple test for agents that damage human DNA. J. Environ. Pathol. Toxicol. *2*:65–78.

Painter, R.B., Howard, R. (1978): A comparison of the HeLa DNA-synthesis inhibition test and the Ames test for screening of mutagenic carcinogens. Mutat. Res. *54*:113–115.

Painter R.B., Young, B.R. (1972): Repair replication in mammalian cells after X-irradiation. Mutat. Res. *14*:225–235.

Painter, R.B., Young, B.R. (1975): X-ray-induced inhibition of DNA synthesis in Chinese hamster ovary, human HeLa, and mouse L cells. Radiat. Res. *64*:648–656.

Rasmussen, R.E., Painter, R.B. (1966): Radiation-stimulated DNA synthesis in cultured mammalian cells. J. Cell Biol. *29*:11–19.

Regan, J.D., Setlow, R.B. (1973): Repair of chemical damage to human DNA. In: Chemical Mutagens; Principles and Methods for Their Detection, vol. 3, A. Hollaender (ed.). New York, Plenum Press, pp. 151–170.

Regan, J.D., Setlow, R.B. (1974): Two forms of repair in the DNA of human cells damaged by chemical carcinogens and mutagens. Cancer Res. *34*:3318–3325.

Stich, H.F., Karim, J., Koropatnick, J., Lo, L. (1976): Mutagenic action of ascorbic acid. Nature (Lond.) *260*:722–724.

Wolff, S., Rodin, B. (1978): Saccharin-induced sister chromatid exchanges in Chinese hamster and human cells. Science *200*:543–545.

7

DNA Repair Synthesis (UDS) as an in vitro *and* in vivo *Bioassay to Detect Precarcinogens, Ultimate Carcinogens, and Organotropic Carcinogens**

HANS F. STICH, RICHARD H.C. SAN, AND HUGH J. FREEMAN

Since Rasmussen and Painter (1966) first observed an unscheduled incorporation of tritiated thymidine (^{3}HTdR) by UV-irradiated cells, this method for assessing DNA repair synthesis has been used for several purposes. These include detection of mutagenic and carcinogenic activities of chemicals or complex mixtures, detection of DNA repair inhibitors, estimation of organotropic actions of carcinogens, and identification of individuals with DNA repair deficiencies. In this paper we will examine the usefulness of DNA repair synthesis as an economical and rapidly performed bioassay method for detecting DNA-damaging agents and, by implication, carcinogens and mutagens. We will deal with the potential usefulness of this technique as a tool to uncover the action of

organotropic carcinogens in various *in vivo* and *in vitro* systems. The mechanism of DNA repair has been comprehensively reviewed by others (Hanawalt *et al.*, 1978, 1979; Trosko and Chu, 1975; Trosko *et al.*, 1977), and an extensive examination of this aspect is beyond the scope of this paper.

The basic technique for detecting DNA repair synthesis (unscheduled DNA synthesis, UDS) involves initial exposure of nondividing cells to various concentrations of a carcinogen for 1 to 5 hours. Following carcinogen removal, the cells are exposed to ^{3}HTdR at concentrations of $1-10$ μCi/ml, fixed, and nonincorporated labeled thymidine extracted. Cells are then stained with a dye that will not mask the black silver grains in the autoradiographic preparation. Following processing for autoradiography, the unscheduled incorporation of tritiated thymidine is estimated from the average number of silver grains over each nucleus. To avoid misinterpretation, cells exhibiting DNA repair synthesis must be distinguished from cells undergoing regular

*Supported by research grants from the National Cancer Institute of Canada, Natural Sciences and Engineering Research Council of Canada, and the British Columbia Health Care Research Foundation. The authors wish to thank Miss Rosemary Johnson for typing this manuscript.

DNA replication. Usually, such S-phase nuclei can be readily identified by their high count of silver grains. However, cells entering or leaving S-phase may be lightly labeled and mistaken for "repairing" cells. In addition, higher concentrations of many carcinogens suppress semiconservative DNA replication, resulting in nuclei with only a few overlying silver grains. These nuclei could lead to an additional source of errors, and difficulties can be avoided by blocking cell proliferation with arginine-deficient medium (Stich and San, 1970), low serum content (about 1.5 to 2.5% serum), or hydroxyurea (5×10^{-2} to 10^{-3} M) (Cleaver, 1969) in culture medium.

Since the development of this method, a variety of modifications have been proposed. Whether one or more of these modifications is superior or inferior is controversial. Nevertheless, the research objectives, type of target cell and economic considerations ultimately determine their choice. The following application patterns for carcinogens and ^{3}HTdR were successfully used in screening over 200 compounds with ^{3}HTdR incorporation estimated by autoradiography or scintillation counting.

1. Cells are exposed to carcinogens for durations of 30 seconds to 5 hours and treated with ^{3}HTdR for 1–5 hours. Exposure times can be extended if the target cells require a longer period for activation of precarcinogens. For example, the binding of ^{3}H-benzo(a)pyrene to DNA increased about tenfold with prolongation of exposure from 12 hours to 24 hours (Brown et al., 1979). Alternatively, fast-reacting carcinogens (e.g., MNNG or 4 NQO) trigger UDS within minutes. To detect the peak of UDS, ^{3}HTdR must be added within a few minutes after exposure to these compounds.
2. Carcinogens and ^{3}HTdR are applied jointly for 0.5 to 5 hours. Attention must be given to a possible interaction between carcinogens and ^{3}HTdR. Moreover, a possible effect of the carcinogen on the penetration of ^{3}HTdR into the target cell must be taken into account.
3. Precarcinogen and S9 liver microsomal preparation precede ^{3}HTdR exposure.
4. Precarcinogen, S9 preparation, and ^{3}HTdR are applied jointly.
5. Precarcinogens are added to short-term cultures of hepatocytes (Williams, 1976, 1977, 1978). This approach avoids the use of S9 for metabolic activation of precarcinogens. ^{3}HTdR can be added with the precarcinogens.
6. Freshly prepared suspensions of mucosal cells of stomach, ileum, colon, liver, kidneys, etc. of rodents or humans may be exposed to carcinogen and ^{3}HTdR. This modification is particularly useful in detecting tissue-specific activation of precarcinogens.
7. Precarcinogens or ultimate carcinogens are administered (injections, force-feeding, etc.) to mice or rats followed by sacrifice at different time intervals after carcinogen treatment. The tissues are then removed, exposed to ^{3}HTdR, and prepared for sectioning and autoradiography (Stich and Kieser, 1974).

Foremost on the mind of those responsible for screening and monitoring programs is the *question of validation*. Unfortunately, to test a bioassay on a large series of compounds is a time-consuming, pedestrian task. Furthermore, validation remains controversial. The apparently simple task of random selection of chemicals for testing presents considerable difficulty and, to date, the validation of short-term tests is essentially based upon highly selected groups of compounds. Thus, claims regarding the reliability of specific methods for detection of a certain percentage of mutagens or carcinogens must be considered cautiously. More specifically, the use of UDS as a predictive test for chemical carcinogens is supported by the following observations:

1. Chemicals with isomers and derivatives of high, low, or noncarcinogenic activity were selected for validation. Their ca-

pacity to elicit UDS was compared to their carcinogenic properties. A good correlation was observed with 4-nitroquinoline-1-oxide (4NQO) and its derivatives (Stich, 1973; Stich *et al.*, 1971).

2. UDS-inducing capacities of triplets consisting of precarcinogen, proximate carcinogen, and ultimate electrophilic species revealed a close link between carcinogenicity and capacity to trigger a DNA repair synthesis (Fig. 7.1) (San and Stich, 1975).

3. As a rule, it is difficult to obtain valid animal data on the degree of carcinogenicity of a compound. The results can be particularly misleading when tumor frequencies of different organs, different species, different latency periods, and following different routes of agent administration are compared. Thus, claims of strict correlations between degree of carcinogenicity and level of mutagenicity may be dubious. 4NQO and its derivatives, however, represent a good example of an excellent correlation between carcinogenicity, DNA repair synthesis, and microbial mutagenicity (Nagao and Sugimura, 1976; Stich *et al.*, 1971).

4. Finally, the reliability of UDS as a bioassay for carcinogens was supported by testing a large number of known carcinogens and noncarcinogens (San and Stich, 1975; Martin *et al.*, 1978; Williams, 1976, 1979; Stich *et al.*, 1977). Of particular interest was the wide array of carcinogens that elicit a positive response (Table 7.1). Metals and H_2O_2-producing hydrazines, which are difficult to detect with microbial cells rich in catalase, triggered UDS in cultured mammalian cells.

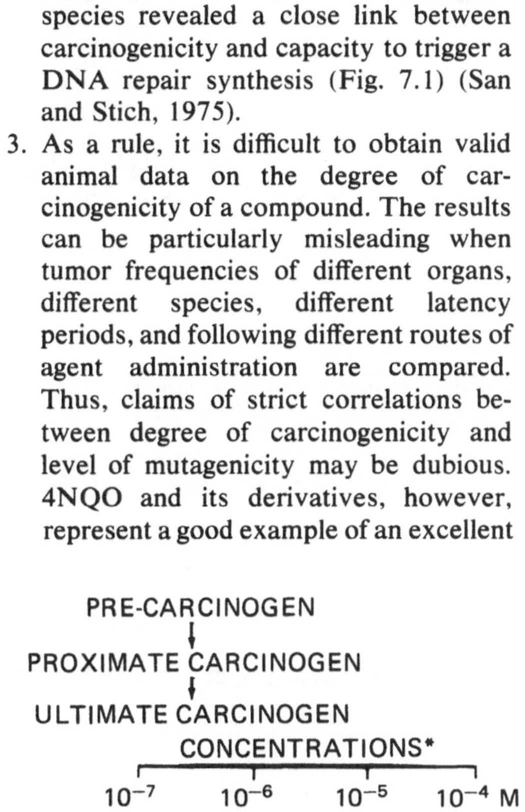

In spite of the good correlations between carcinogenicity of compounds and their capacity to induce UDS, *some questions remain unsolved.* For example, only a few of the 27 compounds known to be carcinogens for man have been examined in the UDS test (Rall's table in Maugh, 1978). The human carcinogen, diethylstilbestrol, produced a negative result in our laboratory. It is possible that hormones act through a non-DNA-damaging nonmutagenic mechanism. The lack of UDS with this agent correlates with the previous report of a negative *Salmonella* mutagenicity test (Glatt *et al.*, 1979), and suggests that the UDS result is not a "false negative."

A more difficult problem involves the handling of *complex mixtures* of compounds. These may contain carcinogens, anticarcinogens, promoters, and agents inhibiting DNA repair. The response of indicator cells may reflect an interplay among all these factors. For example, in the presence of a strong inhibitor of DNA repair, a mixture of compounds may not trigger a detectable UDS even though potent carcinogens or mutagens may be present. This phenomenon has been repeatedly encountered (Figs. 7.2 and 7.3). Of particular interest is the induction of chromosome aberrations by complex mixtures. As seen from Figures 7.2 and 7.3, results based on chro-

Figure 7.1. UDS-inducing capacity of pre-carcinogens, proximate carcinogens, and ultimate carcinogens in cultured human fibroblasts. (*Bars indicate concentration range within which unscheduled DNA synthesis was demonstrable in cultured human fibroblasts.)

Table 7.1. Chemical Carcinogens and Mutagens That Were Positive in the DNA Repair Test

Chemical	Cell type[a]	Metabolic activation[b]	Method of detection[c]	Reference
6-Acetoxy-benzo(a)pyrene	HF	−	A	Stich et al. (1974)
N-Acetoxy-4-acetylaminobiphenol	F	−	A	San & Stich (1975)
N-Acetoxy-2-acetylaminofluorene	F	−	A	Stich et al. (1972)
	HeLa	−	LSC	Martin et al. (1978)
N-Acetoxy-3-acetylaminofluorene	F	−	A	Stich et al. (1974)
N-Acetoxy-2-acetylaminophenanthrene	F	−	A	San & Stich (1975)
N-Acetoxy-4-acetylaminostilbene	F	−	A	San & Stich (1975)
N-Acetoxy-N-myristoyl-2-aminofluorene	F	−	A	San & Stich (1975)
3'-Acetoxy-safrole	F	−	A	San & Stich (1975)
2-Acetylaminofluorene	F	−	A	San & Stich (1975)
	HPC	(+)	A	Williams (1977)
	HeLa	+	LSC	Martin et al. (1978)
Aflatoxicol	F	+	A	Stich & Laishes (1975)
Aflatoxin B1	F	−	A	San & Stich (1975)
	F	+	A	Stich & Laishes (1975)
	HeLa	+	LSC	Martin et al. (1978)
	HPC	(+)	A	Williams (1976)
Aflatoxin G1	F	+	A	Stich & Laishes (1975)
	HeLa	+	LSC	Martin et al. (1978)
	HPC	(+)	A	Williams (1977)
Alderlin	HeLa	−	LSC	Martin et al. (1978)
Aldrin	F	−	A	Ahmed et al. (1977)
4-Aminobiphenyl	HPC	(+)	A	Williams (1978)
2-Amino-5-2-[(5-nitro-2-furyl)-1-(2-furyl)]-vinyl-1,3,4-oxadiazole (furamizole)	F	−	A	Tonomura & Sasaki (1973)
2-Amino-5-nitrophenol	F	−	A	[d]
Ascorbate + Cu	F	+	A	Stich et al. (1976)
Ascorbate + Mn	F	+	A	[e]
Benz(a)anthracene	HeLa	+	LSC	Martin et al. (1978)
Benz(a)anthracene-5,6-epoxide	F	−	A	Stich & San (1973)
Benzidine (p-diaminobiphenyl)	HPC	(+)	A	Williams (1978)
Benzo(a)pyrene-4,5-epoxide	HeLa	−	LSC	Martin et al. (1978)
Benzo(a)pyrene	HeLa	+	LSC	Martin et al. (1978)
	HF	+	A	Casto et al. (1976)
Benzo(e)pyrene	HeLa	+	LSC	Martin et al. (1978)
7-Bromomethylbenz(a)anthracene	HeLa	−	LSC	Martin et al. (1978)

Compound	Cell	Result	Method	Reference
Bromodeoxyuridine + fluorescent light	F	−	A	Kato (1977)
t-Butyl-hydroperoxide	F	−	A	Stich et al. (1974)
6-n-Butyl-4-nitroquinoline-1-oxide	HF	−	A	Stich et al. (1971)
6-t-Butyl-4-nitroquinoline-1-oxide	HF	−	A	Stich et al. (1971)
Captan	F	−	A	Ahmed et al. (1977)
Carbaryl	F	−	A	Ahmed et al. (1977)
6-Carboxy-4-nitroquinoline-1-oxide	HF	−	A	Stich et al. (1971)
Chlordane	F	−	A	Ahmed et al. (1977)
6-Chloro-4-nitroquinoline-1-oxide	HF	−	A	Stich et al. (1971)
2-Chloroethanol	F	−	A	d
Chloroethylene oxide	F	−	A	d
Chromate (potassium) K_2CrO_4	F	−	A	Whiting et al. (1979a)
Cyclophosphamide	HeLa	−	LSC	Martin et al. (1978)
Cysteine + Cu	F	−	A	Stich et al. (1978)
Cysteamine + Cu	F	−	A	Stich et al. (1978)
Decarbamoyl mitomycin C	F	−	A	Sasaki et al. (1977)
2,5-Diaminoanisole	F	−	A	d
Dianisidine	HeLa	+	LSC	Martin et al. (1978)
N-Diazoacetylglycine amide	F	−	A	Parodi et al. (1977)
Dibenz(a,c)anthracene	HeLa	+	LSC	Martin et al. (1978)
Dibenz(a,h)anthracene	HeLa	+	LSC	Martin et al. (1978)
Dibenz(c,h)anthracene 5,6-oxide	HeLa	+	LSC	Martin et al. (1978)
Di-n-butylnitrosamine	HeLa	+	LSC	Martin et al. (1978)
2,2'-Dichlorobenzidine	HeLa	+	LSC	Martin et al. (1978)
3,3'-Dichlorobenzidine	HeLa	+	LSC	Martin et al. (1978)
Dichromate ($K_2Cr_2O_7$)	F	−	A	Whiting et al. (1979a)
Dieldrin	F	−	A	Ahmed et al. (1977)
2,3-Diethyl-4-nitropyridine 1-oxide	F	−	A	Stich (1973)
Diethylnitrosamine	HeLa	+	LSC	Martin et al. (1978)
Diethylstilbestrol	HeLa	+	LSC	Martin et al. (1978)
2,4-D fluid	F	−	A	Ahmed et al. (1977)
2',3-Dimethyl-4-aminobiphenyl	HPC	(+)	A	Williams (1978)
N-Dimethyl-p-aminoazobenzene	HeLa	+	LSC	Martin et al. (1978)
1,2-Dimethylhydrazine + Mn	F	−	A	e
1,2-Dimethylhydrazine + Fe	F	−	A	e
1,2-Dimethylhydrazine + Cu	F	−	A	e
Dimethylnitrosamine	HeLa	+	LSC	Laishes & Stich (1973)
		+	LSC	Martin et al. (1978)
	HPC	(+)		Williams (1977)
2,3-Dimethyl-4-hydroxyaminopyridine-1-oxide	F	−	A	Stich (1973)

Table 7.1. *Chemical Carcinogens and Mutagens That Were Positive in the DNA Repair Test* (Continued)

Chemical	Cell type[a]	Metabolic activation[b]	Method of detection[c]	Reference
2,3-Dimethyl-4-nitropyridine-1-oxide	F	–	A	Stich (1973)
7,12-Dimethylbenz(a)anthracene	HPC	(+)	A	Williams (1977)
	HeLa	+	LSC	Martin et al. (1978)
1,1-Diphenyl-2-propynyl-N-cyclo-hexylcarbinol	F	–	A	San & Stich (1975)
Di-n-propylnitrosamine	HeLa	+	LSC	Martin et al. (1978)
Diquat	F	–	A	Ahmed et al. (1977)
Ethidium azide + light	L	–	LSC	Cantrell & Yielding (1977)
Ethyl methanesulfonate	F	–	A	San & Stich (1975)
	HeLA	–	LSC	Martin et al. (1978)
N-Ethyl-N-nitrosourea	HeLa	–	LSC	Martin et al. (1978)
3-Ethyl-4-nitropyridine-1-oxide	F	–	A	Stich et al. (1974)
3-Fluoro-4-nitroquinoline-1-oxide	HF	–	A	Stich et al. (1971)
Formaldehyde	HeLa	–	LSC	Martin et al. (1978)
2-(2-Furyl)-3-(5-nitro-2-furyl)acrylic acid amide (furylfuramide)	F	–	A	Tonomura & Sasaki (1973)
Glutathione + Cu	F	–	A	Stich et al. (1978)
Heptachlor	F	+	A	Ahmed et al. (1977)
Heptachlor-epoxy	F	+	A	Ahmed et al. (1977)
6-n-Hexyl-4-nitroquinoline-1-oxide	HF	–	A	Stich et al. (1971)
Hydrazine + Mn	F	–	A	Whiting et al. (1979b)
Hydrogen peroxide	F	–	A	Stich et al. (1978)
Hydrogen peroxide + Cu	F	–	A	[e]
Hydrogen peroxide + Fe	F	–	A	[e]
Hydrogen peroxide + Mn	F	–	A	[e]
N-Hydroxy-4-acetylaminobiphenyl	F	–	A	San & Stich (1975)
N-Hydroxy-2-acetylaminofluorene	F	–	A	Stich et al. (1972)
	HeLa	+	LSC	Martin et al. (1978)
N-Hydroxy-4-acetylaminostilbene	F	–	A	San & Stich (1975)
N-Hydroxy-2-acetylaminophenanthrene	F	–	A	San & Stich (1975)
3'-Hydroxy-safrole	F	–	A	San & Stich (1975)
4-Hydroxyaminoquinoline-1-oxide	F	–	A	Stich & San (1971)
6-Hydroxymethyl-2-2-(5-intro-2-furyl)-vinyl-pyridine (furpyrinol)	F	–	A	Tonomura & Sasaki (1973)
ICR-191	F	–	A	San & Stich (1975)
ICR-170	L	–	LSC	Meneghini (1974)

Compound	Cell	Result	Assay	Reference
Isoniazid + Mn	F	—	A	Whiting et al. (1979b)
Iproniazid + Mn	F	—	A	Whiting et al. (1979b)
Iproniazid + Cu	F	—	A;LSC	Whiting et al. (1979b)
8-Methoxy-psoralen + UVL	HeLa	—	LSC	Baden et al. (1972)
3-Methylcholanthrene	HF	+	A	Martin et al. (1978)
3-Methylcholanthrene	F	—	A	Casto et al. (1976)
3-Methylcholanthrene-11,12-epoxide	F	—	A	San & Stich (1975)
2-Methyl-4-dimethylaminoazobenzene	HPC	(+)	A	Williams (1977)
3'-Methyl-4-dimethylaminoazobenzene	HPC	(+)	A	Williams (1977)
4'-Methyl-4-dimethylaminoazobenzene	HPC	(+)	A	Williams (1977)
N-Myristoyloxy-N-acetyl-2-aminofluorene	F	—	A	San & Stich (1975)
N-Myristoyloxy-N-acetyl-2-aminofluorene				Bartsch et al. (1977)
N-Myristoyloxy-N-myristoyl-2-aminofluorene	F	—	A	San & Stich (1975)
N-Myristoyloxy-N-myristoyl-2-aminofluorene				Bartsch et al. (1977)
N-Methyl-N-nitrosourea	F	—	A	San & Stich (1975)
N-Methyl-N-nitrosourea	HeLa	—	LSC	Martin et al. (1978)
Methylguanidine (nitrosation products)	F	—	A	Lo & Stich (1975)
3-Methyl-4-hydroxyaminopyridine-1-oxide	F	—	LSC	Stich et al. (1974)
Methyl methanesulfonate	HeLa	—	A	Martin et al. (1978)
Methyl methanesulfonate	F	—	A	San & Stich (1975)
N-Methyl-N'-nitro-N-nitrosoguanidine(MNNG)	HPC	—	A	Williams (1976)
N-Methyl-N'-nitro-N-nitrosoguanidine(MNNG)	F	—	A	Stich et al. (1973a)
2-Methyl-4-nitroquinoline-1-oxide	HPC	—	A	Williams (1977)
2-Methyl-4-nitroquinoline-1-oxide	HeLa	—	LSC	Martin et al. (1978)
3,5,6,7, or 8-Methyl-4-nitroquinoline-1-oxide	HF	—	A	Stich et al. (1971)
3,5,6,7, or 8-Methyl-4-nitroquinoline-1-oxide	F	—	A	Stich et al. (1973a)
2-Methyl-4-nitropyridine-1-oxide	HF	—	A	Stich et al. (1971)
3-Methyl-4-nitropyridine-1-oxide	F	—	LSC	Stich et al. (1974)
3-Methyl-4-nitropyridine-1-oxide				Stich et al. (1973a)
Mitomycin C	Mouse P388	—	LSC	Oerstavik (1973)
Neocarzinostatin	L	—	A	Tatsumi et al. (1975)
Nialamide + Mn	F	—	A	Whiting et al. (1979b)
5-Nitro-2-furyl acrylic acid amide (nitrofurylacrylamide)	F	—	A	Tonomura & Sasaki (1973)
Nitrogen mustard	F	—	A	San & Stich (1975)
4-Nitro-o-phenylenediamine	F	—	A	[d]
2-Nitro-p-phenylenediamine	HeLa	—	LSC	Martin et al. (1978)
2-Nitro-p-phenylenediamine	HeLa	—	LSC	Martin et al. (1978)
4-Nitroquinoline	HF	—	A	Stich et al. (1971)

Table 7.1. Chemical Carcinogens and Mutagens That Were Positive in the DNA Repair Test (Continued)

Chemical	Cell type[a]	Metabolic activation[b]	Method of detection[c]	Reference
4-Nitroquinoline-1-oxide	HF	−	A	Stich & San (1970)
	F	−	A	Stich & San (1971)
	HeLa	−	LSC	Martin et al. (1978)
Patulin	F	−	A	[d]
1-Phenyl-1-(3,4-xylyl)-2-propynyl-cyclohexylcarbamate	F	−	A	San & Stich (1975)
β-Propiolactone	L	−	LSC	Lieberman et al. (1971a)
	F	−	A	Casto et al. (1976)
Selenate	F	−	A	Lo et al. (1978)
Selenite	F	+	A	Lo et al. (1978)
Selenite + glutathione	F	−	A	[e]
Sterigmatocystin	F	−	A	San & Stich (1975)
	F	+	A	Stich & Laishes (1975)
	HeLa	+	LSC	Martin et al. (1978)
Streptonigrin	F	−	A	San & Stich (1975)
	F	−	A	Stich et al. (1975b)
α-Terthienyl (+ light)	HeLa	+	LSC	Martin et al. (1978)
3,3',5,5'-Tetrachlorobenzidine	HeLa	+	LSC	Martin et al. (1978)
3,3',5,5'-Tetrafluorobenzidine	HeLa	+	LSC	Martin et al. (1978)
o-Tolidine	HeLa	+	LSC	Martin et al. (1978)
Triethylenemelamine (TEM)	L	−	A	Michel & Legator (1974)
Urethan	HeLa	+	LSC	Martin et al. (1978)

[a]HF = human fibroblasts; F = fibroblasts; HPC = hepatocyte primary culture; L = lymphocytes.

[b]+or−=with or without exogenous activation system added; (+) = endogenous activation.

[c]A = autoradiography; LSC = liquid scintillation counting.

[d]San and Stich (unpublished data).

[e]Whiting, Wei and Stich (unpublished data).

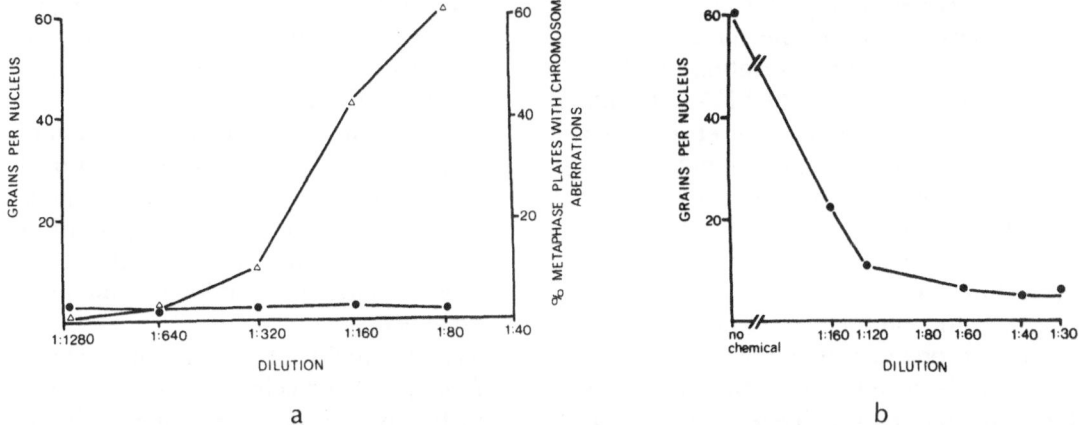

Figure 7.2. DNA- and chromosome-damaging activity in an industrial effluent: (a) unscheduled DNA synthesis (●) in cultured human fibroblasts and chromosome aberrations (△) in Chinese hamster ovary cells; (b) inhibitory effect on UV-induced DNA repair in cultured human fibroblasts.

mosome aberrations differ from those of UDS. A possible explanation is that DNA repair inhibition enhances the sensitivity of cells to the chromosome-damaging action of carcinogens in the mixture. A parallel exists in studies of xeroderma pigmentosum. Investigations using cells from these patients illustrated that sensitivity to chromosome-damaging agents is related to degree of repair deficiency (Stich *et al.*, 1973b; San *et al.*, 1977). In these cells car-

cinogens induce a high frequency of chromosome aberrations while the level of UDS is low. This pattern also results in a high sensitivity toward the lethal effect of carcinogens (Takebe *et al.*, 1972; Stich *et al.*, 1972, 1973a; Maher *et al.*, 1975).

Particular attention should be focused on the possibility of adapting UDS to estimate the action of carcinogens *in vivo*. In this regard, DNA repair offers a unique opportunity not fully explored to date. There are

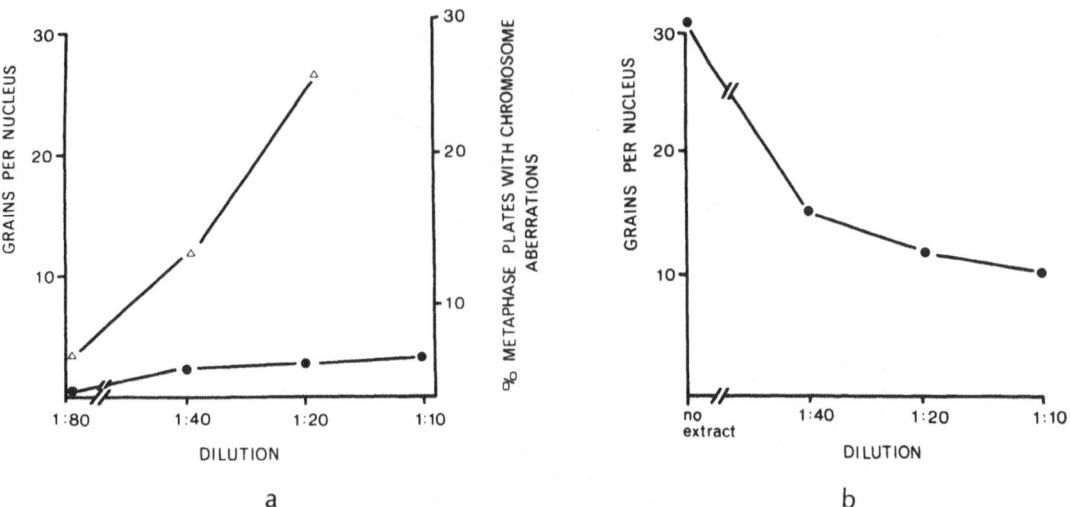

Figure 7.3. DNA and chromosome-damaging activity in a human fecal extract: (a) unscheduled DNA synthesis (●) in cultured human fibroblasts and chromosome aberrations (△) in Chinese hamster ovary cells; (b) inhibitory effect on UV-induced DNA repair in cultured human fibroblasts.

very few short-term tests for mutagens and carcinogens suitable for this purpose. Analyses of chromosome aberrations and the micronucleus test (Heddle, 1973; Heddle and Salamone, 1981; Heddle *et al.*, 1981; Schmid, 1975) have been used both with *in vitro* and *in vivo* systems. However, these methods are limited to rapidly dividing tissues. Various host-mediated assays can be used to detect carcinogens within an organism (Legator and Malling, 1971). However, these microbiological assays are limited since the indicator organisms are outside the cell and tissues. Thus, only precarcinogens which are activated within mammalian cells and subsequently released to the external surroundings are detectable. It appears that UDS and measurements of DNA fragmentation (Damjanov *et al.*, 1973; Laishes *et al.*, 1975; Stich *et al.*, 1975a,b; Koropatnick and Stich, 1976) are the most promising procedures applicable to all organs of a mammalian organism. A few examples will exemplify the various approaches that might be used in the design of an *in vivo* short-term test.

1. Short-term cultures of rat hepatocytes which retain their capacities to activate numerous precarcinogens (Williams, 1978, 1979) appear to be the first successful step toward using differentiated cells for UDS tests. This approach avoids the use of standard S9 preparations recognized to yield a pattern of carcinogen metabolites that differ from those occurring in the liver of an intact organism.

2. The possibility of monitoring UDS in freshly isolated cells from various mammalian tissues provides a means to detect organ-specific actions of ultimate carcinogens or an organ-specific activation of precarcinogens. Preliminary studies on freshly isolated mucosal cells of mouse stomach, ileum, and colon indicate UDS following exposure to ultraviolet irradiation (see Fig. 7.4) or N-methyl-N'-nitro-N-nitrosoguanidine(see Fig. 7.5). In addition, activation of the precarcinogen aflatoxin B_1 has been demonstrated(see Fig. 7.6). These results point to the feasibility of using freshly

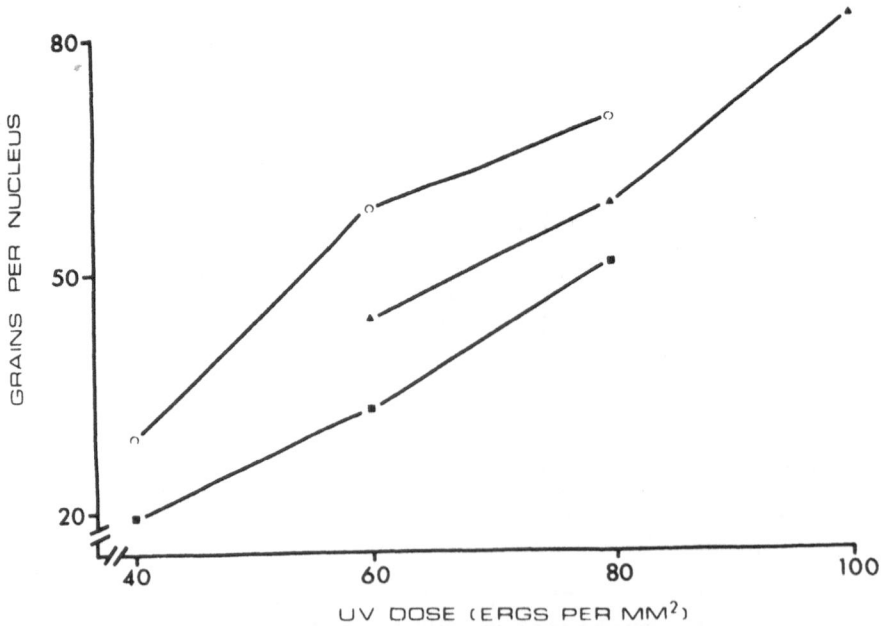

Figure 7.4. UV-induced unscheduled DNA synthesis in freshly isolated mucosal cells from mouse stomach (○), small intestine (■), and colon (▲) (San *et al.*, 1979).

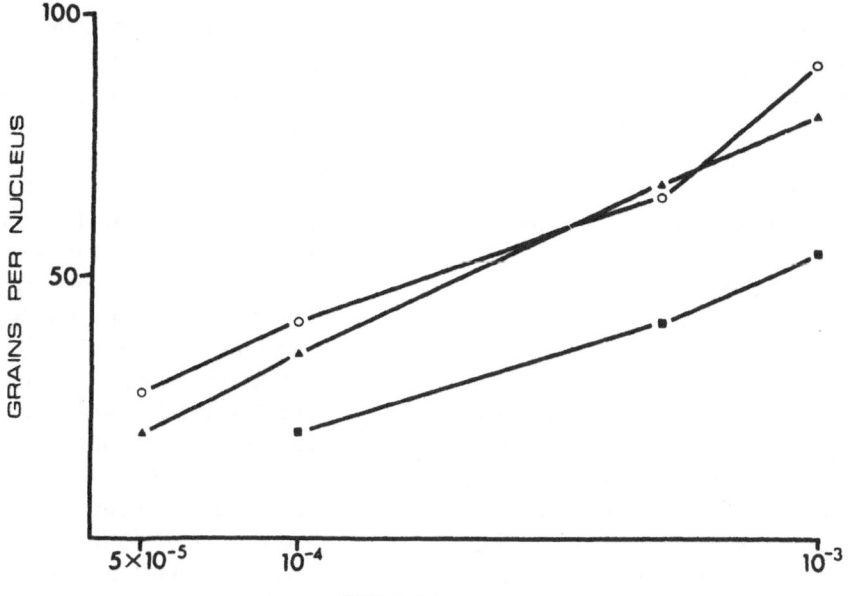

Figure 7.5. Unscheduled DNA synthesis in freshly isolated mucosal cells from mouse stomach (○), small intestine (■), and colon (▲) following a 2-hr exposure to MNNG and simultaneous labeling with 10 μCi/ml ³HTdR (San *et al.*, 1979).

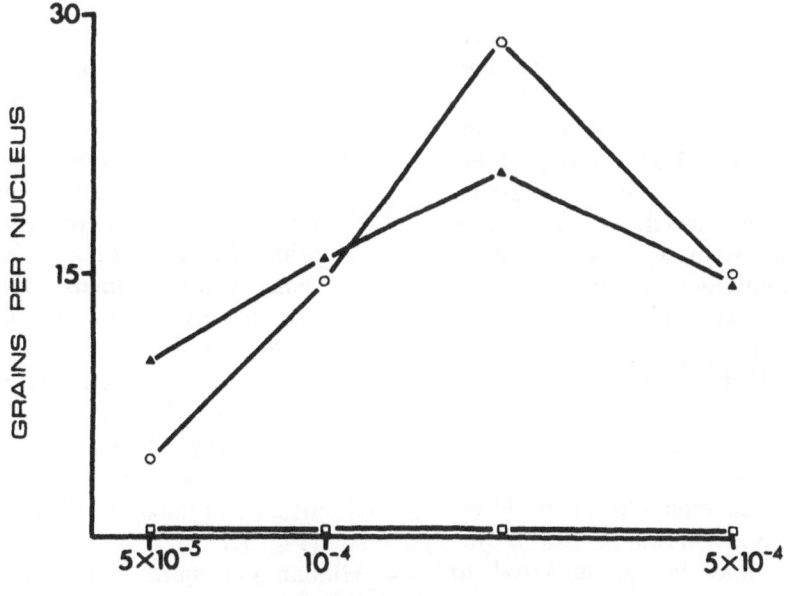

Figure 7.6. Activation of the precarcinogen aflatoxin B1 by freshly isolated mucosal cells from mouse stomach (○) and colon (▲), as measured by the ensuing unscheduled DNA synthesis. Negative control: cultured human fibroblasts (□) (San *et al.*, 1979).

isolated viable cells to uncover the organ-specific actions of carcinogens in differentiated tissues of the gastrointestinal tract.

3. Test systems using short-term cultures of hepatocytes or freshly prepared cell suspensions can be criticized on the grounds that the isolation procedures employed could change cell metabolism or permeability toward precarcinogens or ultimate carcinogens. Moreover, these techniques reveal little about the distribution of carcinogens within an organism or the accumulation of carcinogens in particular organs or tissues. To cope with these arguments, an *in vivo/in vitro* combination system was designed. The precarcinogen or carcinogen is administered to the animal (through injection, force-feeding, etc.) and specific tissue samples are taken at different intervals post-treatment. Basically, three different techniques can be used to follow the course of DNA repair.

(a) The tissue samples are divided into tiny (less than 1 mm^3) pieces, incubated with ^{3}HTdR (e.g., 10 μCi/ml) for 3–5 hours at 37° C and prepared for autoradiography (Lieberman *et al.*, 1973; Stich and Kieser, 1974; Stich *et al.*, 1975a). UDS is readily detectable in nondividing tissues. For rapidly proliferating tissue, particular care must be taken to distinguish DNA replication synthesis in dividing cells from UDS in nondividing ones.

(b) The carcinogen-induced DNA damage can be estimated by measurement of DNA fragmentation using alkaline sucrose gradient analysis (Damjanov *et al.*, 1973; Koropatnick and Stich, 1976; Abanobi *et al.*, 1977) or alkaline elution of filter-bound DNA (Swenberg, 1981). Carcinogens may be administered to test animals, tissues removed at various times after treatment, and DNA fragmentation estimated. Incorporation of radioactive precur-

sors has been widely used to detect DNA with these techniques. However, this method is restricted to rapidly proliferating tissues, where such precursors may be administered days or hours prior to experimentation. Prelabeling can be avoided if DNA and DNA fragments can be detected by reaction with ethidium bromide, a dye that intercalates duplex DNA with a 25-fold enhancement of its fluorescence. Neutralization of the alkaline gradient fractions allows formation of such duplex DNA by intramolecular hydrogen bonding (Morgan and Pulleybank, 1974). DNA from virtually all cells, both dividing and nondividing, may be detected in this manner. As a result, expensive and rather cumbersome prelabeling may be avoided.

The disappearance of fragmentation in DNA from tissues previously assaulted with carcinogens may also be measured. Thus, a form of DNA repair can be observed by following the gradual increase in size of DNA from small fragments (shortly after carcinogen administration) to nearly control-sized pieces.

The above-mentioned procedures (a and b) incorporate the advantages of *in vitro* short-term tests with the completeness of assays using whole animals. These tests are ideally suited to study the organotropic action of many carcinogens (Stich and Koropatnick, 1977), but the most promising feature of these tests is the possibility to apply them to human biopsies or tissues removed during surgery (Fig. 7.7).

4. Human peripheral blood lymphocytes have been used to estimate UDS following exposure to various carcinogens and mutagens (Connor and Norman, 1971; Evans and Norman, 1968; Frey-Wett-

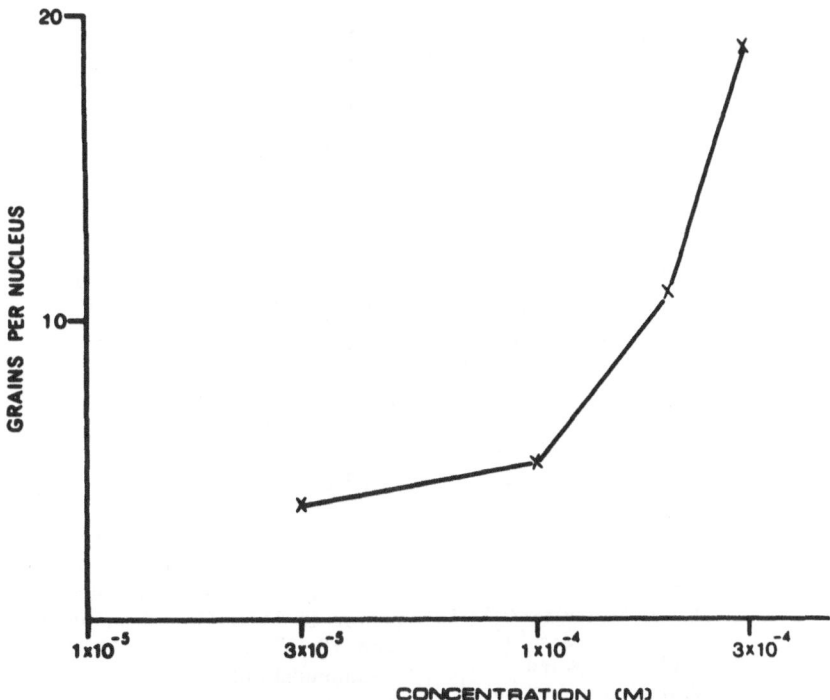

Figure 7.7. Activation of the precarcinogen aflatoxin B1 by freshly isolated mucosal cells from human small bowel, as measured by the autoradiographic detection of the ensuing unscheduled DNA synthesis (Freeman *et al.*, 1980).

stein *et al.*, 1969; Lieberman *et al.*, 1971a,b,c; Spiegler and Norman, 1970). It is surprising that these cells have not found a major use in uncovering repair deficiencies in human populations (Burk *et al.*, 1971).

5. A completely new usage of UDS was recently proposed by Miller and Miller (1979). Cats exposed to methylmercury doses which cause neurological dysfunctions showed UDS in their Purkinje cells and granular layer of the cerebellum. Autoradiographic examinations combined with DNase treatments revealed grains over the cytoplasm, suggesting UDS of mitochondrial DNA. These studies may open the door to a hitherto untouched field.

We would appear to be remiss if we neglected to mention some of the factors which could alter the level of UDS. Among them the genetically based repair deficiency of xeroderma pigmentosum patients

is the best known. The reduced repair capacity is not restricted to UV-induced DNA alterations as can be seen in Table 7.2. Thus, the use of UDS for screening of chemical agents for their mutagenic or carcinogenic properties requires an understanding of the DNA repair capacity of the indicator cells. The use of UDS is by no means restricted to human or rodent cells. For example, UDS can be readily measured in established fish cell lines exposed to carcinogens (Fig. 7.8). The only prerequisite is to have a standard which is not affected by different metabolic pathways in the indicator cells. Dose-response curves of UDS following UV-irradiation will suffice to show the DNA repair capacity of a cell population.

We would like to return to the important question of DNA repair inhibition. Compounds that inhibit DNA repair in a tissue or organ appear to yield an XP-like state. This phenomenon may frequently, if not always, occur when complex mixtures are

Table 7.2. *Response of Xeroderma Pigmentosum Cells to Carcinogens*[a]

Normal DNA repair	Reduced DNA repair
MNNG	4NQO + carcinogenic derivatives
MMS	3-Methyl-4NPO
EMS	MCA-epoxide (K-region)
NMU	BA-epoxide (K-region)
HN$_2$	N-Acetoxy-2-AAF
BHP	N-Hydroxy-2-AAF
ICR-191	N-Acetoxy-4-AAS
Nitrosation products of methyl-guanidine	N-Hydroxy-4-AAS
Isoniazid + Mn(II)	N-Acetoxy-4-AABP
Dimethylhydrazine + Mn(II)	N-Acetoxy-2-AAP
Dimethylhydrazine + Cu(II)	N-Hydroxy-2-AAP
	N-Myristoyloxy-2-AAF
	N-Acetoxy-3-AAF
	DMN, activated
	Aflatoxin B$_1$, activated
X-ray	UV

[a]Abbreviations: 4-AABP, 4-acetylaminobiphenyl; 2-AAF, 2-acetyl-aminofluorene; 2-AAP, 2-acetylaminophenanthrene; 4-AAS, 4-acetyl-aminostilbene; BA, benz(a)anthracene; BHP, butylhydroperoxide; DMN, dimethylnitrosamine; EMS, ethyl methanesulfonate; HN$_2$, nitrogen mustard; MCA, 3-methylcholanthrene; MMS, methyl methane-sulfonate; NMU, nitrosomethylurea; 4NPO, 4-nitropyridine 1-oxide; 4NQO, 4-nitroquinoline 1-oxide.

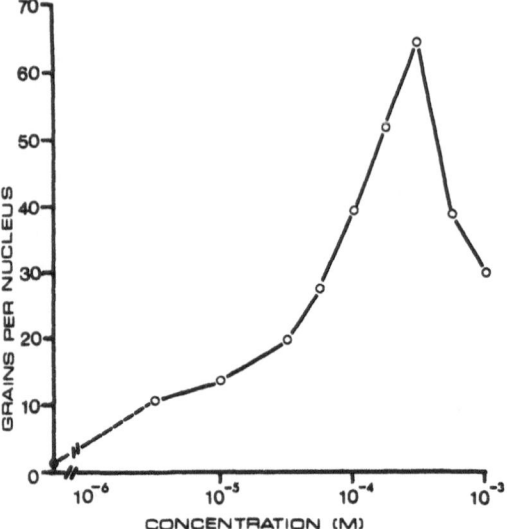

Figure 7.8. Unscheduled DNA synthesis in fish cells (RTG-2 line) following a 6-hr exposure to MNNG and simultaneous labeling with 10 μCi/ml ^{3}HTdR (D. Walton, personal communication).

examined because the chance that they include UDS inhibitors is very high. The significance of this observation can be found in the fact that in the presence of DNA repair inhibitors even small "tolerable" doses can produce serious chromosome-damaging or lethal effects. The simplest procedure to detect the presence of DNA repair inhibitors includes irradiation of cultured cells with a precise dose of UV radiation, thereafter exposing them to the tested chemical (or mixture of chemicals), sampling the tissue at various time intervals (e.g., 2, 4, and 8 hours post-UV-irradiation), and estimating the level of UDS (e.g., Figs. 7.2 and 7.3). To avoid misinterpretation of UDS results, a DNA repair inhibi-

tion test should be performed when complex mixtures are examined.

The DNA repair assays have many unique features lacking in microbial mutagenicity tests. The possibility to apply this assay in various *in vivo/in vitro* combination systems, on human cells and on human biopsy material may provide results of particular relevance to man. Because of its versatility, the UDS assay may help to bridge the gap between mutagenicity data obtained on supersensitive microbes and carcinogenicity data obtained on mammals, including man.

References

Abanobi, S.E., Popp, J.A., Chang, S.K., Harrington, G.W., Lotlikar, P.D., Hadjiolov, D., Levitt, M., Rajalaksmi, S., Sarma, D.S.R. (1977): Inhibition of dimethylnitrosamine-induced strand breaks in liver DNA and liver cell necrosis by diethyldithiocarbamate. J. Natl. Cancer Inst. *58*:263–270.

Ahmed, F.E., Hart, R.W., Lewis, N.J. (1977): Pesticide induced DNA damage and its repair in cultured human cells. Mutat. Res. *42*: 161-174.

Baden, H.P., Parrington, J.M., Delhanty, J.D.A., Pathak, M.A. (1972): DNA synthesis in normal and *Xeroderma pigmentosum* fibroblasts following treatment with 8-methoxypsoralen and long wave ultraviolet light. Biochim. Biophys. Acta *262*:247-255.

Bartsch, H., Malaveille, C., Stich, H.F., Miller, E.C., Miller, J.A. (1977): Comparative electrophilicity, mutagenicity, DNA repair induction, activity and carcinogenicity of some N- and O-acyl derivatives of N-hydroxy-2-aminofluorene. Cancer Res. *37*:1461-1467.

Brown, H.S., Jeffrey, A.M., Weinstein, B. (1979): Formation of DNA adducts in 10T¹/₂ mouse embryo fibroblasts incubated with benzo (a) pyrene or dihydrodiol oxide derivatives. Cancer Res. *39*:1673-1677.

Burk, P.G., Lutzner, M.A., Clarke, D.D., Robbins, J.H. (1971): Ultraviolet stimulated thymidine incorporation in *Xeroderma pigmentosum* lymphocytes. J. Lab. Clin. Med. *77*: 759-763.

Cantrell, C.E., Yielding, K.L. (1977): Repair synthesis in human lymphocytes provoked by photolysis of ethidium azide. Photochem. Photobiol. *25*:189-191.

Casto, B.C., Pieczynski, W. J., Janosko, N., DiPaolo, J.A. (1976): Significance of treatment interval and DNA repair in the enhancement of viral transformation by chemical carcinogens and mutagens. Chem.-Biol. Interact. *13*:105-125.

Cleaver, J.E. (1969): Repair replication of mammalian cell DNA: effects of compounds that inhibit DNA synthesis or dark repair. Radiat. Res. *37*:334-348.

Connor, W.G., Norman, A. (1971): Unscheduled DNA synthesis in human leucocytes. Mutat. Res. *13*:393-402.

Damjanov, I., Cox, R., Sarma, D.S.R., Farber, E. (1973): Patterns of damage and repair of liver DNA induced by carcinogenic methylating agents "*in vivo*". Cancer Res. *33*: 2122-2128.

Evans, R.G., Norman, A. (1968): Radiation stimulated incorporation of thymidine into the DNA of human lymphocytes. Nature (Lond.) *217*:455-456.

Freeman, H.J., San, R.H.C. (1980): Use of unscheduled DNA synthesis in freshly isolated human intestinal mucosal cells for carcinogen detection. Cancer Res. *40*: 3155-3157

Frey-Wettstein, M., Longmire, R., Craddock, C.G. (1969): Deoxyribonucleic acid (DNA) repair replication of ultraviolet (UV) irradiated normal and leukemia leukocytes. J. Lab. Clin. Med. *74*:109-118.

Glatt, H.R., Metzler, M., Oesch, F. (1979): Diethylstilbestrol and 11 derivatives. A mutagenicity study with *Salmonella typhimurium*. Mutat. Res. *67*:113-121.

Hahn, G.M., Jang, S-J., Parker, V. (1968): Repair of sublethal damage and unscheduled DNA synthesis in mammalian cells treated with monofunctional alkylating agents. Nature (Lond.) *220*:1142-1144.

Hanawalt, P.C., Friedberg, E.C., Fox, C.F. (eds.) (1978): DNA Repair Mechanisms. New York: Academic Press.

Hanawalt, P.C., Cooper, P.K., Ganesan, A.K., Smith, C.A. (1979): DNA repair in bacteria and mammalian cells. Ann. Rev. Biochem. *48*: 783-836.

Heddle, J.A. (1973): A rapid *in vivo* test for chromosomal damage. Mutat. Res. *18*: 187-190.

Heddle, J.A., Salamone, M.F. (1981): The micronucleus assay. I. *In vivo*. In: H.F. Stich, R.H.C. San (eds.), Short-Term Tests for Chemical Carcinogens. New York: Springer Verlag.

Heddle, J.A., Raj, A.S., Krepinsky, A.B. (1981): The micronucleus assay. II. *In vitro*. In: H.F. Stich and R.H.C. San (eds.), Short-Term Tests for Chemical Carcinogens. New York: Springer Verlag.

Kato, H. (1977): Mechanisms for sister chromatid exchanges and their relation to the production of chromosomal aberrations. Chromosoma *59*:179-191.

Koropatnick, D.J., Stich, H.F. (1976): DNA fragmentation in mouse gastric epithelial cells by precarcinogens, ultimate carcinogens and nitrosation products: an indicator for the determination of organotrophy and metabolic activation. Int. J. Cancer *17*:765-772.

Laishes, B.A., Stich, H.F. (1973): Repair synthesis and sedimentation analysis of DNA of human cells exposed to dimethylnitrosamine and activated dimethylnitrosamine. Biochem. Biophys. Res. Commun. *52*:827-833.

Laishes, B.A., Koropatnick, D.J., Stich, H.F. (1975): Organ-specific DNA damage induced in mice by the organotropic carcinogens 4-nitro-quinoline-1-oxide and dimethylnitrosamine. Proc. Soc. Exp. Biol. Med. *149*: 978-982.

Legator, M.S., Malling, H.V. (1971): The host-mediated assay, a practical procedure for evaluating potential mutagenic agents in mammals. In: A. Hollaender (ed.), Chemical Mutagens: Principles and Methods for Their Detection, Vol. 2. New York: Plenum Press, pp. 569–591.

Lieberman, M.W., Baney, R.N., Lee, R.E., Sell, S., Farber, E. (1971a): Studies on DNA repair in human lymphocytes treated with proximate carcinogens and alkylating agents. Cancer Res. 31:1297–1306.

Lieberman, M.W., Rutman, J.Z., Farber, E. (1971b): Thymidine labeling of pyrimidine isostichs from human lymphocyte DNA during repair after damage with N-acetoxy-acetylaminofluorene or nitrogen mustard. Biochem. Biophys. Acta 247:497–501.

Lieberman, M.W., Sell, S., Farber E. (1971c): Deoxydribonucleoside incorporation and the role of hydroxyurea in a model lymphocyte system for studying DNA repair in carcinogenesis. Cancer Res. 31:1307–1312.

Lieberman, M. W., Forbes, W., Donald, P. (1973): Demonstration of DNA repair in normal and neoplastic tissues after treatment with proximate chemical carcinogens and ultraviolet radiation. Nature (Lond.) 241:199–201.

Lo, L.W., Stich, H.F. (1975): DNA damage, DNA repair and chromosome aberrations of Xeroderma pigmentosum cells and controls following exposure to nitrosation products of methylguanidine. Mutat. Res. 30:397–406.

Lo, L.W., Koropatnick, J., Stich, H.F. (1978): The mutagenicity and cytotoxicity of selenite, "activated" selenite and selenate for normal and DNA repair-deficient human fibroblasts. Mutat. Res. 49:305–312.

Maher, V.M., Birch, N., Otto, Y.R., McCormick, J.J. (1975): Cytotoxicity of carcinogenic aromatic amides in normal and Xeroderma pigmentosum fibroblasts with different DNA repair capabilities. J. Natl. Cancer Inst. 54:1287–1294.

Martin, C.N., McDermid, A.C., Garner, R.C. (1978): Testing of known carcinogens and noncarcinogens for their ability to induce unscheduled DNA synthesis in HeLa cells. Cancer Res. 38:2621–2627.

Maugh, T.H. (1978): Chemical carcinogens: the scientific basis for regulation. Science 201:1200–1205.

Meneghini, R. (1974): Repair replication of opossum lymphocyte DNA: effect of compounds that bind to DNA. Chem.-Biol. Interact. 8:113–126.

Michel, T.M., Legator, M.S. (1974): DNA repair synthesis and chromosomal aberrations induced in vivo by triethylenemelamine. Mutat. Res. 24:41–45.

Miller, C.T., Miller, D.R. (1979): Biochemical toxicology of methylmercury. In: Effects of Mercury in the Canadian Environment, NRCC No. 16739, NRCC Associate Committee on Scientific Criteria for Environmental Quality, Ottawa, Canada.

Morgan, R.A., Pulleybank, D.E. (1974): Native and denatured DNA, cross-linked and palindromic DNA and circular covalently-closed DNA analyzed by a sensitive fluorometric procedure. Biochem. Biophys. Res. Commun. 61:396–403.

Nagao, M., Sugimura, T. (1976): Molecular biology of the carcinogen, 4-nitroquinoline-1-oxide. Adv. Cancer Res. 23:131–169.

Oerstavik, J. (1973): DNA repair synthesis in mouse P388 cells treated with mitomycin C. Acta Pathol. Microbiol. Scand. Sect. B, 81:711–718.

Parodi, S., Bolognesi, C., Pollack, R.L., Santi, L., Brambilla, G. (1977): Damage and repair of DNA in cultured mammalian cells with N-diazo-acetylglycine amide. Cancer Res. 37:4460–4466.

Rasmussen, R.E., Painter, R.B. (1966): Radiation-stimulated DNA synthesis in cultured mammalian cells. J. Cell Biol. 29:11–19.

San, R.H.C., Stich, H.F. (1975): DNA repair synthesis of cultured human cells as a rapid bioassay for chemical carcinogens. Int. J. Cancer 16:284–291.

San, R.H.C., Stich, W., Stich, H.F. (1977): Differential sensitivity of Xeroderma pigmentosum cells of different repair capacities toward the chromosome breaking action of carcinogens and mutagens. Int. J. Cancer 20:181–187.

San, R.H.C., Freeman, H.J., Stich, H.F. (1979): An in vitro system to quantitate the activation of organ-specific carcinogens in the gastrointestinal tract. Proc. Can Fed. Biol. Sci. 22:113.

Sasaki, M.S., Toda, K., Uzawa, A. (1977): Role of DNA repair in the susceptibility to chromosome breakage and cell killing in cultured human fibroblasts. In: Biochemistry of Cutaneous Epidermal Differentiation. Biochem. Cutaneous Epidermal Differ. Proc. Jpn U.S. Semin. (1976), pp. 167–180.

Schmid, W. (1975): The micronucleus test. Mutat. Res. 31:9–15.

Spiegler, P., Norman, A. (1970): Temperature

dependence of unscheduled DNA synthesis in human lymphocytes. Radiat. Res. *43*:187–195.

Stich, H.F. (1973): The link between oncogenicity of 4NQO and 4NPO derivatives, induction of DNA lesion and enhancement of viral transformation. Review. In: W. Nakahara, S. Takayama, T. Sugimura and S. Odashima (eds.), Topics in Chemical Carcinogenesis. Baltimore: University Park Press, pp. 17–28.

Stich, H.F., Kieser, D. (1974): Use of DNA repair synthesis in detecting organotropic actions of chemical carcinogens. Proc. Soc. Exp. Biol. Med. *145*:1339–1342.

Stich, H.F., Koropatnick, J. (1977): The adaptation of short term assays for carcinogens to the gastrointestinal system. In: E. Farber *et al.*(eds.), Pathophysiology of Carcinogenesis in Digestive Organs. Tokyo: University of Tokyo Press; Baltimore: University Park Press, pp. 121–134.

Stich, H.F., Laishes, B.A. (1975): The response of Xeroderma pigmentosum cells and controls to the activated mycotoxins aflatoxin and sterigmatocystin. Int. J. Cancer, *16*:266–274.

Stich, H.F., San, R.H.C. (1970): DNA repair and chromatid anomalies in mammalian cells exposed to 4-nitroquinoline-1-oxide. Mutat. Res. *10*:389–404.

Stich, H.F., San, R.H.C. (1971): Reduced DNA repair synthesis in Xeroderma pigmentosum cells exposed to the oncogenic 4-nitroquinoline-1-oxide. Mutat. Res. *13*:279–282.

Stich, H.F., San, R.H.C. (1973): DNA repair synthesis and cell survival of repair deficient human cells exposed to the K-region epoxide of benz(a)anthracene. Proc. Soc Exp. Biol. Med. *142*:155–158.

Stich, H.F., San, R.H.C., Kawazoe, Y. (1971): DNA repair synthesis in mammalian cells exposed to a series of oncogenic and non-oncogenic derivatives of 4-nitroquinoline-1-oxide. Nature (Lond.) *229*:416–419.

Stich, H.F., San, R.H.C., Miller, J.A., Miller, E.C. (1972): Various levels of DNA repair synthesis in Xeroderma pigmentosum cells exposed to the carcinogenic N-hydroxy- and N-acetoxy-2-acetylaminofluorene. Nature (Lond.) *238*:9–10.

Stich, H.F., San, R.H.C., Kawazoe, Y. (1973a): Increased sensitivity of Xeroderma pigmentosum cells to some chemical carcinogens and mutagens. Mutat. Res. *17*:127–137.

Stich, H.F., Stich, W., San, R.H.C. (1973b):

Chromosome aberrations in Xeroderma pigmentosum cells exposed to the carcinogens 4-nitro-quinoline-1-oxide and N-methyl-N'-nitro-nitrosoguanidine. Proc. Soc. Exp. Biol. Med. *142*:1141–1144.

Stich, H.F., Kieser, D., Laishes, B.A., San, R.H.C. (1974): The use of DNA repair in the identification of carcinogens, precarcinogens and target tissue. Proc. Canad. Cancer Conf. *10*:83–110.

Stich, H.F., Kieser, D., Laishes, B.A., San, R.H.C., Warren, P. (1975a): DNA repair of human cells as a relevant, rapid and economic assay for environmental carcinogens. In: Recent Topics in Chemical Carcinogenesis: Gann Monograph, *17*:3–15.

Stich, H.F., Lam, P., Lo, L.W., Koropatnick, D.J., San, R.H.C. (1975b): The search for relevent short term bioassays for chemical carcinogens: the tribulation of a modern Sisyphus. Can. J. Genet. Cytol. *17*:471–492.

Stich, H.F., Karim, J., Koropatnick, J., Lo, L. (1976): Mutagenic action of ascorbic acid. Nature (Lond.) *260*: 722–724.

Stich, H.F., San, R.H.C., Lam, P., Koropatnick, D.J., Lo, L. (1977): Unscheduled DNA synthesis of human cells as a short-term assay for chemical carcinogens. In: H.H. Hiatt, J.D. Watson, J.A. Winsten (eds.), Origins of Human Cancer, Book C, Cold Spring Harbor Conferences on Cell Proliferation, Vol. 4, Cold Spring Harbor, New York, Cold Spring Harbor Laboratory, pp. 1499–1512.

Stich, H.F., Wei, L., Lam, P. (1978): The need for a mammalian test system for mutagens: action of some reducing agents. Cancer Lett. *5*:199–204.

Swenberg, J.A. (1981): Utilization of the alkaline elution assay as a short-term test for chemical carcinogens. In: H.F. Stich and R.H.C. San (eds.), Short-Term Tests for Chemical Carcinogens. New York: Springer Verlag.

Takebe, H., Furuyama, J., Miki, Y., Kondo, S. (1972): High sensitivity of Xeroderma pigmentosum cells to the carcinogen 4-nitroquinoline-1-oxide. Mutat. Res. *15*:98–100.

Tatsumi, K., Sakane, T., Sawada, H., Shirakawa, G. (1975): Unscheduled DNA synthesis in human lymphocytes treated with neocarzinostatin. Gann *66*:441–444.

Tonomura, A., Sasaki, M.S. (1973): Chromosome aberrations and DNA repair synthesis in cultured human cells exposed to nitrofurans. Jap. J. Genet. *48*:291–294.

Trosko, J.E., Chu, E.H.Y. (1975): The role of DNA repair and somatic mutation in carcinogenesis. In: G. Klein, S. Weinhouse, A. Haddow (eds.), Advances in Cancer. New York: Academic Press, pp. 391–425.

Trosko, J.E., Chang, C., Glover, T. (1977): Analysis of experimental evidence of the relation between mutagenesis and carcinogenesis. Role of DNA repair in carcinogenesis. In: P. Dandel (ed.), Mecanismes d'Alteration et de Reparation du DNA: Relations avec la Mutagenese et la Cancerogenese Chimique, Paris, pp. 353–381.

Whiting, R.F., Stich, H.F., Koropatnick, D.J. (1979a): DNA damage and DNA repair in cultured human cells exposed to chromate. Chem.-Biol. Interact. 26:267–280.

Whiting, R.F., Wei, L., Stich, H.F. (1979b): Enhancement by transition metals of unscheduled DNA synthesis induced by isoniazid and related hydrazines in cultured normal and Xeroderma pigmentosum human cells. Mutat. Res. 62:505–515.

Williams, G.M. (1976): Carcinogen-induced DNA repair in primary rat liver cell cultures: a possible screen for chemical carcinogens. Cancer Lett. 1:231–236.

Williams, G.M. (1977): Detection of chemical carcinogens by unscheduled DNA synthesis in rat liver primary cell cultures. Cancer Res. 37:1845–1851.

Williams, G. (1978): Further improvements in the hepatocyte DNA repair test for carcinogens: detection of carcinogenic biphenyl derivatives. Cancer Lett. 4:69–75.

Williams, G.M. (1979): Review of in vitro test systems using DNA damage and repair for screening of chemical carcinogens. J. Assoc. Off. Anal. Chem. 62:857–863.

8

Multi-well Assay for Unscheduled DNA Synthesis Using Human Diploid Fibroblasts

George R. Douglas, Caroline E. Grant, Judith M. Wytsma, and Anthony Chan

Introduction

There is concern that man is being exposed to substances that have the potential to cause genetic disease. The development of a number of short-term tests for mutagenic activity (Kilbey *et al.*, 1977; Montesano *et al.*, 1976) has greatly facilitated the task of identifying mutagenic substances, and since it has been demonstrated that many mutagens are also carcinogens (McCann *et al.*, 1975), these substances may also be carcinogenic. Short-term tests for mutagenic activity can be divided into two groups: (1) tests that result in specific types of mutation events (i.e., point mutation or chromosome aberration) that are directly detected as phenotypic changes; and (2) tests that detect effects that correlate with genetic events. These latter effects may be precursor events which could ultimately, but not necessarily, lead to the formation of mutations. This category includes tests for induction of DNA damage and stimulation of unscheduled DNA synthesis (Stich *et al.*, 1976). Since they do not require the detection of a specific genotypic change which may require a rigid set of conditions, these *indicator* tests may be considered more universal in their ability to detect genetic activity. Thus they serve as an important adjunct to tests which measure mutations directly.

At present, microbial tests for mutagenic activity can, in general, be carried out more rapidly than comparable tests using mammalian cells. Accordingly, the process of screening individual compounds for mutagenic activity using combinations of microbial and mammalian tests could be delayed by less rapidly performed mammalian tests. Therefore, there is a need to develop more rapid mammalian assays to test for mutagenic activity. This requirement is intensified by the recent development of even more rapid microbial tests (McMahon *et al.*, 1979).

In response to this perceived requirement, a method has been developed for the detection of unscheduled DNA synthesis

(UDS) in human diploid fibroblasts, permitting the rapid processing of large numbers of samples. UDS, as determined using scintillation counting, was chosen as a suitable candidate for such a rapid multi-sample method because there is a minimum number of steps that require direct human observation, making this type of assay amenable to mass sampling procedures.

Materials and Methods

Cell Culture

Diploid human fibroblasts (IMR-90, obtained from the Institute for Medical Research, Camden, N.J.) are stored in liquid nitrogen at passage 13 in Eagles minimal essential medium (MEM) containing 10% fetal calf serum (FCS) (both supplied by GIBCO) and 10% glycerine. One ml/vial of cell suspension, at a density of approximately 5×10^5 cells/ml, is frozen in 2 ml plastic vials (Nunc). Before freezing, cultures are checked for *Mycoplasma* using direct culture and the uridine phosphorylase assay (Ho and Quinn, 1977).

In order to obtain fresh cultures for the UDS assay, one vial of cells is thawed rapidly in a 37° C water bath and added to a 75 cm² disposable culture flask (Falcon) with 20 ml of medium (CMEM), which contains MEM, 10% FCS, penicillin and streptomycin (100 units/ml and 100 μg/ml, respectively) (GIBCO), and gentamicin (5.0 μg/ml, Schering Corporation). The following day, the medium is changed to eliminate glycerine and any unattached cells. Cultures are maintained at 37° C, 5% CO_2, and high relative humidity. The medium is changed every two days until cells are confluent. All manipulations were carried out under gold fluorescent lights (General Electric, F40G0 6PK) in a biohazard laminar flow cabinet (Canadian Cabinets) which is vented to the outside. All pipetting is done with a Pipet-Aid (Drummond Scientific).

Confluent cells are subcultured at a 1:3 split ratio. The medium is removed and the cells washed with 5 ml of 0.25% trypsin solution (GIBCO). After removal of this wash, 10 ml of fresh trypsin solution is added and the cells observed with an inverted microscope. When approximately 50% of the cells have rounded up (1–3 minutes depending on room temperature, age of trypsin, and tightness of cells), the trypsin solution is removed and 10 ml of CMEM added. The cells are detached by shaking, vigorously pipetted to break up clumps, and then counted in a Coulter Counter (Coulter Electronics) if cell counts are required.

Prelabeling Cells with ^{14}C-Thymidine

Prelabeling the cells with methyl-^{14}C-thymidine (TdR) provides a means of later standardizing ^3H-TdR incorporation in the UDS assay to minimize variation due to DNA content between samples caused by uneven harvesting. Cells are seeded in a 150 cm² disposable culture flask (Corning) at 1.5×10^6 cells/flask in approximately 20 ml of CMEM. After 24 hours' incubation, the medium is removed and 20 ml fresh CMEM containing 0.35 μCi ^{14}C-TdR (53 μCi/mmole, Amersham Searle) are added. The cells are labeled for 24 hours. The ^{14}C-label is washed from the cells using three rinses of 10 ml phosphate-buffered saline (PBS, GIBCO). The cells are removed from the flask in 10 ml CMEM. After counting, the cells are diluted in CMEM to 6×10^4 cells/ml in a 25 cm² culture flask (Corning). 0.15 ml of the cell suspension is added to each well of a multi-well plate (96 × 0.35 ml capacity, flat-bottom, Linbro) so that each well contains 9×10^3 cells. Cells are added using a 2 ml pipet and are kept evenly suspended by inverting the flask (× 10) periodically during the addition. The plates are incubated at 37° C, 5% CO_2, and high relative humidity for 5–7 days before use in the UDS assay. Cells from passage 15 to 18 are used for the assay.

Treatment with Test Chemicals

Direct-acting carcinogens/mutagens. Medium in the wells is removed and replaced with 0.15 ml MEM without arginine with 2.5% FCS, fungazone (2.5 μg/ml), and kanamycin (100 μg/ml) (ADM) 20 hours before treatment with the test chemical (this medium was a gift from Dr. R.C. San). Two hours prior to treatment, this medium is replaced with 0.15 ml ADM containing 10 mM hydroxyurea (HU).

Separate concentrated stock solutions in the appropriate solvent are prepared for each treatment of the test chemical. The same volume of concentrated test chemical in solvent is added to each well of a multi-well plate. Accordingly, the final concentration of solvent in each well will be essentially the same. Depending on the solubility of the chemical, a 100 or 200× concentrated stock solution for each dose is prepared in disposable glass tubes with cork stoppers. The appropriate volume of the concentrated test chemical is added to another tube containing ADM without FCS with 10 mM HU, 5 μCi/ml methyl-^{3}H-TdR (Amersham Searle, 46 Ci/mmole) to give the desired final concentration of chemical, and thoroughly mixed with a pipet. After removing existing ADM/HU medium from the wells, the test chemical mixture, prepared as described, is added to the wells using a 2 ml disposable pipet (0.15 ml/well). Routinely, we use 12 wells (1 row) per treatment and each plate contains a solvent control. A positive control may be added as required. The cells are treated for 4 hours at 37° C, high relative humidity.

Carcinogens/mutagens requiring *in vitro* metabolic activation. Aroclor 1254-induced Sprague-Dawley rat liver homogenate (S9) is prepared according to the method outlined by Ames *et al.* (1975).

The cells are prepared and treated identically to those used for testing direct-acting chemicals until 2 hours before treatment is to begin. At this point, the ADM is replaced with 0.1 ml ADM without FCS containing 10 mM HU. Concentrated stock solutions (25×) of the test chemical in the appropriate solvent are made up in cork-stoppered disposable glass tubes. Each stock concentration is then diluted 1:9 parts ADM without FCS with 10 mM HU. The activation mixture contains: 1.6 mM HEPES buffer, pH 7.4 (Sigma); 3.8 mM MgCl$_2$ (Sigma); 24.8 mM KCl (Sigma); 3.8 mM glucose-6-phosphate (Sigma); 3.0 mM nicotinamide adenine dinucleotide phosphate (Sigma); 10 mM HU; 27.8% ADM without FCS; 22.5% S9; 22.5% ^{3}H-TdR (18.7 μCi/ml) in ADM without FCS with 10 mM HU.

The ADM/HU medium is not removed. Instead, 40 μl of the activation mix is added to the medium in each well with a Pipetman (Gilson). Then 10 μl of the diluted test chemical stock solution is added to each well. The cells are treated for 2 hours at 37° C and high relative humidity.

Cell Harvesting and Scintillation Counting

After removal of the mixture containing the test chemical, 0.1 ml of 0.1% protease (Sigma) in Earle's Balanced Salt Solution (GIBCO), prewarmed to 37° is added to the wells. The plates are incubated at 37° for 30 minutes and 0.1 ml of 10% ice-cold trichloroacetic acid (TCA, Fisher Scientific) is added to each well. The precipitated DNA is collected on glass fiber filter strips (Whatman) using a Mash II (Microbiological Associates) cell harvester which collects the precipitate 24 wells at a time. During harvesting the wells and the filter are washed 8–10 times with 5% TCA containing 3% sodium pyrophosphate (Baker). Before removal from the harvester, the filter is washed with absolute ethanol and then dried at 120° C for 20 minutes in an oven.

All procedures with the multi-well plates involving removal or addition of liquid, except the additions of cell suspension and test chemical, are done using a four-channel pipettor (Titertek). We use the bottom half of a Petri dish as a reservoir from which to fill the pipettor. Another Petri dish bottom can be used to collect waste liquid.

The filter discs containing the DNA precipitate are punched out of the filter strips into 7 ml glass mini-scintillation vials (Kimble) and 3 ml of aquasol (New England Nuclear) diluted 2 parts: 1 part xylene (Fisher Scientific) are added to each vial. Radioactivity is determined using a Mark III scintillation counter (Tracor) equipped with a paper tape output.

Data Analysis

The raw count data on paper tape is analyzed by a computer program which determines the ^{3}Hdpm/^{14}Cdpm ratios for each well, derives the mean ratio for each treatment and performs a Kolmogrov-Simirnov (K-S) test of significance comparing the difference between each treatment mean ratio and the control mean ratio. Based on our experience, the data from any treatment are not necessarily normally distributed. Therefore, the nonparametric K-S test is appropriate, whereas parametric statistics are not.

Chemicals

Chemicals, abbreviations, and suppliers other than those described previously are listed below:

Dimethylsulfoxide (DMSO), Aldrich, spectrophotometric grade, ICN K&K
4-Nitroquinoline oxide (4NQO), ICN K&K
Methyl methanesulfonate (MMS), Eastman
Ethyl methanesulfonate (EMS), Eastman
N-Methyl-N'-nitro-N-nitrosoguanidine (MNNG), ICN K&K
Benzo(a)pyrene (BP), Aldrich

Results and Discussion

With the use of multi-well plates it is not convenient to relate the level of ^{3}H-TdR incorporation directly to DNA content.

Therefore, UDS is estimated using the ratio of dpm ^{3}H-TdR/dpm ^{14}C-TdR. Since the ^{14}C-TdR label is incorporated more or less uniformly into all cells originally seeded into each well, this ratio serves as an excellent means of minimizing variation due to nonuniform recovery of DNA among wells. It is important to minimize variation in the number of cells per well seeded because normalization of DNA content may be rendered less effective. Furthermore, if the specific activity of ^{14}C-TdR labeled DNA varies from one experiment to the next, care should be exercised in comparing the ^{3}H/^{14}C ratios from different plates without reference to absolute ^{14}C levels. Accordingly, the inclusion of a control treatment in each plate permits more reliable comparisons among plates and increases the precision obtainable with this method.

The sensitivity of this method was evaluated using four direct-acting mutagens (4NQO, MNNG, MMS, EMS). The data shown in Figure 8.1 demonstrate that this method is capable of detecting response levels consistent with the range previously reported for these chemicals in a conventional UDS assay which uses scintillation counting (Martin *et al.*, 1978). The use of ADM, as shown earlier (Stich and San, 1970), tends to lower the background ^{3}H-TdR incorporation, which presumably is the result of residual semiconservative DNA replication (compare Figure 8.1a, b,c, with 8.1d). However, a higher background does not preclude the detection of a weak response for EMS predicted from published data (Martin *et al.*, 1978).

The incorporation of an *in vitro* metabolic activation system into the assay permits the detection of UDS caused by BP (Fig. 8.2). However, the weak and variable response found in earlier experiments with BP (data not shown) necessitated modifications to the protocol. In order to minimize an increase in background ^{3}H-TdR incorporation due to medium changes, medium is not removed from the wells immediately prior to treatment with a test chemical. Rather, the chemical and S9 mix are added to the existing medium to give the desired fi-

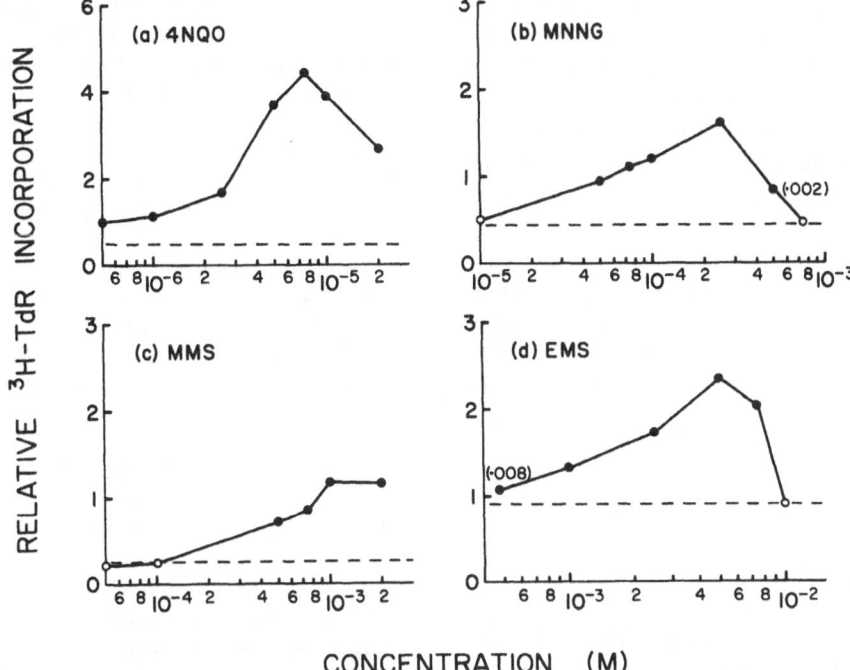

Figure 8.1. Mean relative ³H-TdR incorporation expressed as dpm ³H-TdR/dpm ¹⁴C-TdR, vs. concentration of different direct-acting mutagens. Dotted lines represent solvent control levels. Closed circles are mean ratios that are significantly different from control values at p ≤ 0.0002, except where bracketed numbers indicate other probabilities. Means depicted by open circles are not significantly different from control. Solvents were as follows: (a) 4NQO(DMSO), (b) MNNG (0.05 mM citrate buffer, pH 5.5), (c) MMS(MEM), (d) EMS (DMSO). IMR-90 cells were treated in multi-wells as described in Materials and Methods except (d) where medium contained arginine.

nal concentration. The chemical and S9 mix are added separately to minimize possible inactivation of short-lived metabolites by factors in the S9 before the treatment mixture comes into contact with the cells. The treatment time is reduced to 2 hours to minimize possible toxic effects of the S9 mix. While the sensitivity and potential of this method have been demonstrated, its utility as a screening tool must be validated by demonstrating appropriate responses with standard mutagens and carcinogens.

The resolving power of this multi-well method is clearly illustrated by the detection of statistical significance of seemingly small differences in ³H-TdR incorporation (Figs. 8.1d and 8.2). The method lends itself to the use of multiple replicates per treatment (in this case 12), which are required to detect such small differences.

Since human observation is reduced to a minimum and the method uses semi-au-

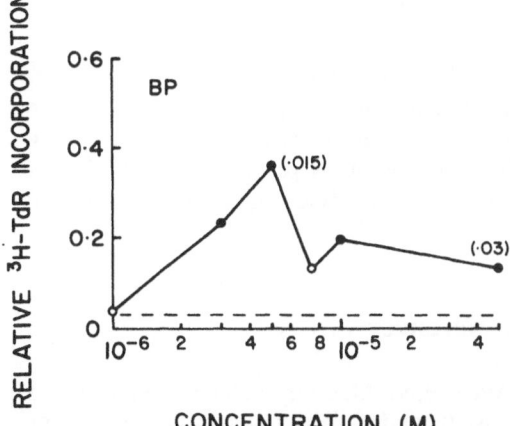

Figure 8.2. Mean relative ³H-TdR incorporation expressed as dpm ³H-TdR/dpm ¹⁴C-TdR, vs. concentration BP. Cells were treated with BP in presence of an *in vitro* metabolic activation mixture as described in Methods and Materials section. Symbols are as in Figure 8.1.

tomated processing, one person can conveniently perform an experiment with 24 different treatments (three plates, including controls) in 1 day, and have the data in final form the following morning. The limiting factor is the number of samples that fit into the scintillation counter and can be counted overnight.

As a test for genetic activity, UDS can be classified as an "indicator" test because it does not detect a specific type of mutation event. However, in spite of the ability of UDS to detect most mutagens and carcinogens (Martin et al., 1978; Stich et al., 1975; Williams, 1977), there are exceptions that may relate to the specific test organism used or to technical considerations such as the use of HU, which may cause artifacts (Andrae and Greim, 1979; Brandt et al., 1972; Clarkson, 1978). Nevertheless, the UDS method serves as an important adjunct or back-up test in a battery of tests for mutagenic activity.

While we have used human diploid fibroblast cells (IMR-90), the multi-well method for UDS could be adapted for use with metabolizing mammalian cells. It could also be easily adapted to detect inhibition of DNA synthesis (Painter, 1977).

Acknowledgments

We thank Dr. Richard San for his helpful comments and the gift of the ADM; Drs. K.C. Bora and E.R. Nestmann for reviewing the manuscript, and Dr. R.S. McCullough for advice on statistical analysis.

References

Ames, B.N., McCann, J., Yamasaki, E. (1975): Methods for detecting carcinogens and mutagens with Salmonella/mammalian-microsome mutagenicity test. Mutat. Res. 31: 347–364.

Andrae, U., Greim, H. (1979): Induction of DNA repair replication by hydroxyurea in human lymphoblastoid cells mediated by liver microsomes and NADPH. Biochem. Biophys. Res. Comm. 87:50–58.

Brandt, W.N., Flamm, W.G., Bernheim, N.J. (1972): The value of hydroxyurea in assessing repair synthesis of DNA in HeLa cells. Chem. Biol. Interact. 5:327–339.

Clarkson, J.M. (1978): Enhancement of repair replication in mammalian cells by hydroxyurea. Mutat. Res. 52:273–284.

Ho, T.Y., Quinn, P.A. (1977): Rapid detection of Mycoplasma in tissue culture by SEM. Scanning Electron Microscopy/1977, 2: 291–299.

Kilbey, B.J., Legator, M., Nichols, W., Ramel, C. (eds.) (1977): Handbook of Mutagenicity Test Procedures. Amsterdam, Elsevier/North Holland Biomedical Press.

Martin, C.N., McDermid, A.C., Garner, R.C. (1978): Testing of known carcinogens and noncarcinogens for their ability to induce unscheduled DNA synthesis in HeLa cells. Cancer Res. 38:2621–2627.

McCann, J., Choi, E., Yamasaki, E., Ames, B.N. (1975): Detection of carcinogens as mutagens in the Salmonella/microsomal test: Assay of 300 chemicals. Proc. Natl Acad. Sci. USA 72:5135–5139.

McMahon, R.E., Cline, J.C., Thompson, C.Z. (1979): Assay of 855 test chemicals in ten tester strains using a new modification of the Ames test for bacterial mutagens. Cancer Res. 39:682–693.

Montesano, R., Bartsch, H., Tomatis, L. (eds.) (1976): Screening Tests in Chemical Carcinogenesis, IARC Scientific Publications No. 12, Lyon, International Agency for Research on Cancer.

Painter, R.B. (1977): Rapid test to detect agents that damage human DNA. Nature (Lond.), 265:650–651.

Stich, H.F., San, R.H.C. (1970): DNA repair and chromatid anomalies in mammalian cells exposed to 4-nitroquinoline-1-oxide. Mutat. Res. 10:389–404.

Stich, H.F., Lam, P., Lo, L.W., Koropatnick, D.J., San, R.H.C. (1975): The search for relevant short-term bioassays for chemical carcinogens: the tribulation of a modern Sisyphus. Can. J. Genet. Cytol. 17:471–492.

Stich, H.F., San, R.H.C., Lam, P., Koropatnick, D.J., Lo, L.W., Laishes, B.A. (1976): DNA fragmentation and DNA repair as an *in vitro* and *in vivo* assay for chemical precarcinogens, carcinogens and carcinogenic nitrosation products. In: R. Montesano, H. Bartsch, and L. Tomatis (eds.), Screening Tests in Chemical Carcinogenesis, IARC Scientific Publications No. 12, Lyon, International Agency for Research on Cancer, pp. 617–636.

Williams, G. (1977): Detection of chemical carcinogens by unscheduled DNA synthesis in rat liver primary cell cultures. Cancer Res. *37*: 1845–1851.

9

Sucrose Gradients: An Assay for DNA Damage*

B. Palcic and L.D. Skarsgard

Principle of the Method

Exposure of cells to various physical and chemical agents often leads to damage in the cellular genetic material, DNA. Ionizing radiation, UV light, fluorescent light, microwaves, ultrasound, and heat have all been shown to induce DNA breaks. Many chemicals, including pharmacological agents, have also been found to cause a variety of injuries in DNA. Unrepaired or improperly repaired DNA damage may lead to a mutation, and, since mutagenesis and carcinogenesis are believed to be closely related, it can be assumed that agents which damage DNA are also capable of inducing human cancer.

A rapid and reliable method for the assay of DNA damage is the method of sucrose gradients. In this paper we shall limit the discussion to alkaline sucrose gradients where whole cells can be lysed on the top of

*This work is supported by the British Columbia Cancer Foundation and the National Cancer Institute of Canada.

a gradient, so that no extraction of DNA from cells is necessary.

The method was first applied to bacterial cells (McGrath and Williams, 1966) and was later modified for mammalian cells (Lett et al., 1967). Cellular DNA is first uniformly labeled by growing cells in medium containing either ^{3}H-TdR or ^{14}C-TdR. The cells are then exposed to the agent under investigation, washed, and lysed on the top of an alkaline sucrose gradient. The lysing solution on the top of the gradient is of pH >12 and contains detergents and salts. Cells placed into the lysing solution are immediately lysed; membranes, proteins, and RNA are denatured and dissolved. DNA is also denatured and becomes single-stranded in high pH solution. DNA single strands are freed of associated protein, RNA, and lipoprotein. This is followed by centrifugation of the DNA in a preparative ultracentrifuge and from the sedimentation patterns the molecular weight of the DNA molecules can be calculated. A comparison of the molecular weights of treated and untreated cells yields the number of damaged DNA sites.

Testing Protocol

Cells

Cultured mammalian cells are seeded into plastic Petri dishes or flasks at 1.25 x 10⁴ cells/cm² in growth medium. After 24 hours, the medium is replaced with growth medium containing either ^{14}C-TdR or ^{3}H-TdR (0.05 μCi/ml, specific activity 50 mCi/mmole, and 0.25 μCi/ml, specific activity 40 Ci/mmole, respectively). Approximately 0.75 ml/cm² of labeled medium is used. The label is removed after 24 hours and the cells are further incubated for a few hours in growth medium. The cells are then harvested by trypsinization and washed. At this stage they are exposed under prescribed conditions to the agent under investigation. The described labeling procedure yields approximately 0.05 decays/second/cell.

Gradients

A gradient former is used to prepare 5–20% linear sucrose gradients in centrifuge tubes. Gradient solutions contain 0.3 M NaOH, 0.01% sodium dodecyl sulfate (SDS), 0.001 M ethylenediaminetetraacetate (EDTA) and appropriate concentrations of sucrose. A layer of lysing solution containing 0.5 M NaOH, 0.2% SDS and 0.01 M EDTA is placed on the top of each gradient. For example, if 17 ml Beckman centrifuge tubes are employed (SW 27.1 rotor), 0.5 ml of lysing solution is used.

Lysing Cells

A microsyringe is used to dispense cells into the lysing layer. A volume of 0.02 ml containing 1×10^4 cells is carefully delivered into the lysing solution. The cells are then lysed for 6 hours at 20°C.

Centrifugation

Centrifugation is performed at 20°C in a preparative ultracentrifuge using a swinging bucket rotor. The time of centrifugation t and the angular speed of centrifugation ω are chosen such that the peak of the DNA distribution sediments to a position one-third to two-thirds of the total sedimentation distance d.

Collecting and Counting

After centrifugation, fractions are collected from each gradient in scintillation vials. For example, 17 ml Beckman tubes are typically collected as 25 fractions. Scintillation cocktail is then added and radioactivity is determined by measuring the decay rate in each vial, using a liquid scintillation counter.

Calculation of the Number of DNA Breaks

Weight average molecular weight M_ω and number average molecular weight M_n can be calculated from the following equations:

$$M_\omega = \left(\frac{\beta\, d_f}{\omega^2\, t\, a}\right)^k \frac{\Sigma c_i\, (i - ^1/_2)^k}{\Sigma c_i} \qquad (1)$$

$$M_n = 0.5\, M_w \qquad (2)$$

where a and k are constants 0.0528 and 2.5, respectively. β and d_f take different values depending on the conditions of sedimentation (temperature, solutions, choice of tubes). For example, for 17 ml Beckman tubes, 20°C, sedimentation in the above solutions, $d_f = 0.385$ cm and $\beta = 6.51 \times 10^{10}$ (rpm)² × hours × cm⁻¹. The relative number of counts c_i corrected for background is determined for each fraction.

The number of DNA breaks, n, can be obtained:

$$n = \frac{M_n,c}{M_n,t} - 1 \qquad (3)$$

where M_n,c represents the number average molecular weight of the DNA of the untreated control cells, and M_n,t the number average molecular weight of the DNA of the treated cells.

Further details of the above procedures can be found elsewhere (Palcic and Skarsgard, 1972, 1974).

Agent Tested

As an example of the application of this technique, we present data from our studies with misonidazole. Misonidazole (RO-07 0582) is a heterocyclic nitro compound which selectively sensitizes hypoxic cells to ionizing radiation (Asquith et al., 1974). It has been suggested for use as an adjunct to radiotherapy and limited clinical trials are underway. The drug is mildly toxic to mammalian cells; it is considerably more toxic to hypoxic cells than to cells treated with the drug in the presence of molecular oxygen (Moore et al., 1976; Hall and Roizin-Towle, 1975). Misonidazole mutagenicity is readily detected by the Ames test; the yield of histidine revertants is about ten times higher than background at a dose of 100 μg per plate (Josephy et al., 1980).

Figure 9.1 shows the DNA profiles of treated and untreated cells. Chinese hamster cells of the $CH2B_2$ cell line were exposed to 15 mM misonidazole for 1 hour at 37°C in the absence of oxygen (N_2 atmosphere). The DNA of the exposed cells sedimented a much shorter distance than that of control cells, indicating a smaller size distribution of the DNA molecules. The number average molecular weight of control DNA was 2.0×10^8 daltons, and of the treated cells, was 1.25×10^8 daltons. There

were 0.6 breaks induced per molecule of DNA or approximately 15,000 DNA breaks per cell. Longer incubation times resulted in more DNA breaks. For example, a 3-hour incubation induced five breaks per DNA molecule. If the cells were exposed to the drug in the presence of O_2 (aerobic conditions), no DNA breaks were observed.

It should be noted, however, that the cells must be washed free of misonidazole prior to lysis, since it was observed that the presence of misonidazole in the lysing solution led to increased DNA breaks irrespective of the treatment which the cells received.

Evaluation of Results

The technique of alkaline sucrose gradients allows one to quickly determine whether a particular agent causes DNA breaks in mammalian cells, including human cells grown in vitro. The technique requires a preparative ultracentrifuge, gradient former, gradient collector, and scintillation counter, as well as the necessary equipment for culturing mammalian cells. If this instrumentation is available, the assay is rather inexpensive. The results can be obtained in 2–3 days.

The technique measures the sum of various types of damages: single-strand breaks, double-strand breaks, alkali labile points (i.e., base and sugar damage which under alkali conditions leads to DNA backbone rupture). The technique cannot distinguish between these different lesions. A serious disadvantage of the technique is that not all DNA damage is detected. For example, UV-induced thymine dimers cannot be directly measured in this manner. This may also be the case for many chemicals that are known mutagens and/or carcinogens. Although techniques exist which overcome this problem to some extent (e.g., pretreating DNA with extracts of Micrococcus luteus prior to lysis of the

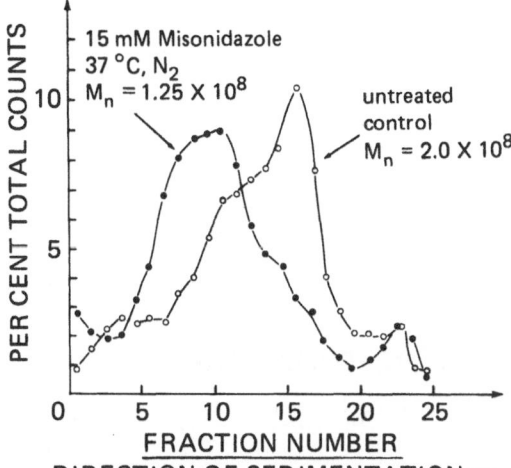

Figure 9.1. Exposure of cells to misonidazole under hypoxic conditions. ^{14}C-TdR-labeled CH2B$_2$ cells were exposed to 15 mM misonidazole in growth medium (—●——●—) or to medium alone (—○——○—) under hypoxic conditions (N$_2$ atmosphere, less than 5 ppm O$_2$ present) for 1 hour at 37°C. The cells were then thoroughly washed and lysed on the top of an alkaline sucrose gradient (17 ml tubes, Beckman SW 27.1 rotor). After 6 hours' lysis at 20°C, the gradients were spun for 11 hours at 14,000 rpm.

DNA on the gradient; examining of size of newly synthesized DNA rather than parental DNA), their use considerably complicates this method.

Another possible disadvantage of the method is the fact that a considerable amount of DNA damage must be inflicted in the cells before it is detectable. In the example shown in Figure 9.1, it is estimated that at least 2000 DNA breaks per cell must occur before one can distinguish between treated and untreated DNA profiles.

Essentially similar techniques which have the potential for greater sensitivity have been described; these are the hydroxylapatite chromatography method (Rydberg, 1975) and the alkaline elution method (Kohn and Grimek-Ewig, 1973). These techniques also detect DNA single-strand breaks exposed by alkali treatment but are potentially ten times more sensitive than the alkaline sucrose gradient method de-

scribed here. All these methods have a common shortcoming in that they detect, under normal use, only one class of DNA lesions, those which are labile under alkali conditions. False negative results might therefore be expected if these methods are used to assay DNA damaged produced by carcinogens or mutagens which lead primarily to other types of lesions.

References

Asquith, J.C., Watts, M.E., Patel, K., Smithen, C.E., Adams, G.E. (1974): Electron affinic sensitization of hypoxic bacterial and mammalian cells *in vitro* by some nitroimidazoles. Radiat. Res. *60*:108–118.

Hall, E.J., Roizin-Towle, L. (1975): Hypoxic sensitizers: radiobiological studies at the cellular level. Radiology *117*:453–457.

Josephy, P.D., Palcic, B., Skarsgard, L.D. (1980): Synthesis and properties of reduced derivatives of misonidazole. Cancer Clin. Trials (in press).

Kohn, K.W., Grimek-Ewig, R. (1973): Alkaline elution analysis, a new approach to the study of DNA single-strand interruptions in cells. Cancer Res. *33*:1849–1853.

Lett, J.T., Caldwell, I., Dean, C.J., Alexander, P. (1967): Rejoining of X-ray induced breaks in the DNA of leukaemia cells. Nature *214*: 790–792.

McGrath, R.A., Williams, R.W. (1966): Reconstruction *in vivo* of irradiated *E. coli* DNA; the rejoining of broken pieces. Nature *212*: 534–535.

Moore, B.A., Palcic, B., Skarsgard, L.D. (1976): Radiosensitizing and toxic effects of the 2-nitroimidazole Ro 07-0582 in hypoxic mammalian cells. Radiat. Res. *67*:459–473.

Palcic, B., Skarsgard, L.D. (1972): The effect of oxygen on DNA single-strand breaks produced by ionizing radiation in mammalian cells. Int. J. Radiat. Biol. *21*:417–433.

Palcic, B. Skarsgard, L.D. (1974): Absence of ultra fast processes of repair of single-strand breaks in mammalian DNA. Int. J. Radiat. Biol. *27*: 121–133.

Rydberg, B. (1975): The rate of strand separation in alkali of DNA of irradiated mammalian cells. Radiat. Res. *61*:274–287.

10

The Testicular DNA-Synthesis Inhibition Test (DSI Test)

J.P. SEILER

Introduction

The success of the bacterial short-term tests for the detection of mutagenic and potentially carcinogenic chemicals accentuated the need for a simple and quick mammalian *in vivo* screening system, which could be used in parallel to the bacterial assay in order to assess the probability of such an effect in mammals. Most mammalian *in vivo* systems make use of a singular genetic endpoint, namely chromosome breakage in one or another form. However, as experiments with Drosophila have shown (Vogel and Leigh, 1975), chromosome breakage is an insensitive indicator of genetic damage, and an endpoint common to chromosome breakage and other genetic effects would thus be desirable. An example of such a common endpoint could be the covalent binding of a chemical to DNA.

Another approach to this problem could be based on the fact that the binding of chemical carcinogens to DNA inhibits its *in vitro* replication (Berthold *et al.*, 1978; Zieve, 1973; Moore and Strauss, 1979;

Farber, 1968). This property of chemical mutagens and carcinogens is now easily amenable to measurement. Friedman and Staub (1976) reported first upon the applicability of such a test to carcinogens of different chemical classes, and subsequently we enlarged their study by validating it with over 300 chemical compounds (Seiler, 1978). On one hand these first data were rather encouraging, as most carcinogens could be detected by this method, whereas noncarcinogens generally proved to be inactive. On the other hand certain technical problems became rapidly apparent and had to be solved in order to enhance the reproducibility of the observed effects. Although still amenable to perfection, we think the technical side of this test is developed far enough to warrant its application in a tier system of carcinogen screening.

Method

We use 25–30 g male mice of an ordinary outbred colony for this test, but hamsters, rats, or guinea pigs can also be used. Routinely the test is performed as follows.

On the eve of the test day the test animals receive an injection of 18.5 kBq (0.5 μCi) (methyl-14C)-thymidine which is given subcutaneously in the back of the animals. For better reproducibility we use a Hamilton syringe of 500 μl capacity equipped with a repeating dispenser and inject 10 μl of undiluted 14C-thymidine solution (The Radiochemical Centre, Amersham, 1.85 MBq/ml (50 μCi/ml)). In filling the syringe care has to be taken that all air bubbles trapped in it or in the injection needle are expelled before injecting the animals; if these precautionary measures are observed the reproducibility of the injection will be better than $\pm$ 2.5%.

The next day these prelabeled animals are randomly assigned to the various test and control groups, and they receive then the test substances by some suitable route (i.p., p.o., etc.). Three hours after the administration of the test compounds the animals are injected again subcutaneously with labeled thymidine, but this time with 10 μl $= 0.37$ MBq (10 μCi) (methyl-3H)-thymidine. Exactly 45 minutes later the incorporation of tritiated thymidine is stopped by an i.p. injection of 3.0 mg unlabeled thymidine in 0.2 ml physiological saline, which provides for a 5×10^4-fold dilution of the radioactive thymidine. About half an hour later the mice are sacrificed and their testes excised. The testes are then transferred into 5% trichloroacetic acid (TCA) and homogenized in a Potter-type homogenizer with three to four strokes of the pestle at 1000–1200 rpm; alternatively, they can be stored frozen in distilled water at $-20°$ C until treated further. After centrifugation an aliquot of the supernatant is retained for the determination of total tritium present in the testes, and the pellet is resuspended in fresh 5% TCA by violent agitation on a vortex mixer and adding—dropwise at the beginning—the TCA solution. This washing is repeated twice; then the homogenate is washed twice with methanol and again twice with 5% TCA. After this last washing

the pellet is again suspended in 5% TCA. and is now heated for 30 minutes in a water bath at 95° C for DNA hydrolysis. After cooling and centrifugation, aliquots of the supernatant are taken for the determination of radioactivity and for the measurement of the DNA content, which is done photometrically by the diphenylamine method (Ashwell, 1957). Radioactivity is measured in a commercial scintillator (Aquasol-2, New England Nuclear) by liquid scintillation counting. The result is obtained by computing the channel ratio 3H/14C and by correcting this ratio for the spillover of 14C-counts into the 3H-channel, which can be done most easily by subtracting the channel ratio of a pure 14C standard.

This modified method deviates from the original one (Friedman and Staub, 1976; Seiler, 1977), which measured the thymidine incorporation as 3H counts/min per μg DNA and compared the results obtained from the treated animals with those of the concurrent controls. This formerly used method leads to larger and more irregular standard deviations as well as to large daily fluctuations in the control values (see Fig. 10.1a), since not only errors in the injection of tritiated thymidine but also errors in the DNA determination, pipetting errors, and differences in body weight, testicular mass, and thus in the amount of thymidine available per testicular cell, may be introduced into the result. The modified method as described here, on the other hand, is dependent only on the exact application of the two differently labeled thymidines, and a direct comparison of the incorporation capacity in the same animal before and after treatment is possible. For control reasons, however, we are still determining DNA content routinely in our tests. By a strict adherence to the time limits between the pulse of tritiated thymidine and the chase of "cold" thymidine, the ratio of the two labels in the testicular DNA becomes constant (at least as long as the same batches of labeled thymidine with constant specific activities are used), and the former necessity of using a relatively large number of concurrent controls is greatly reduced.

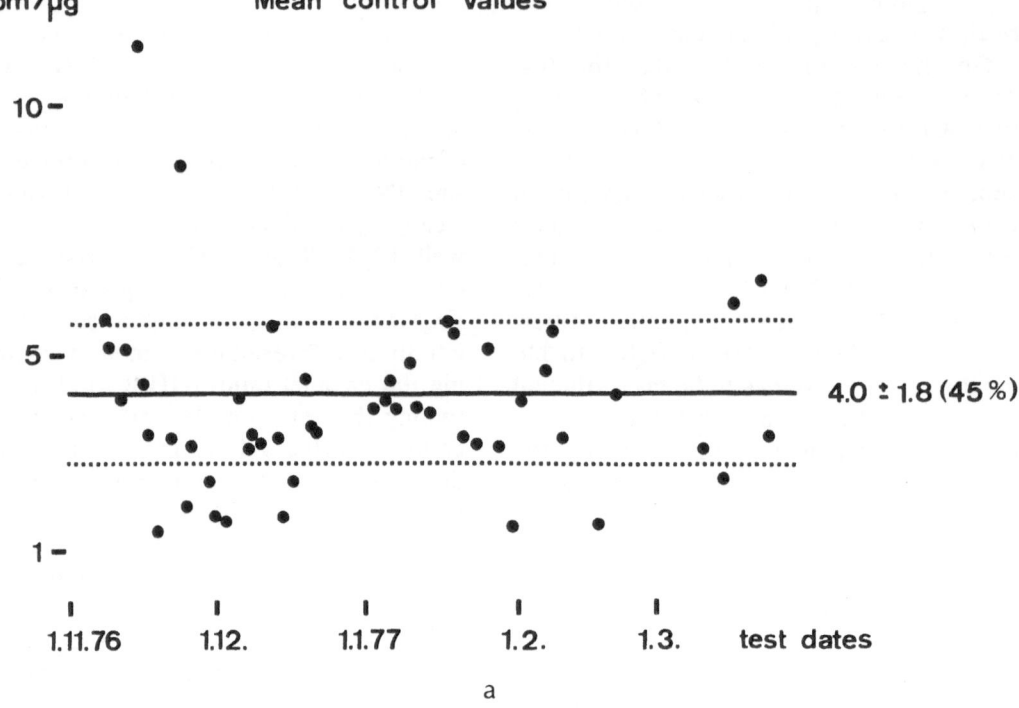

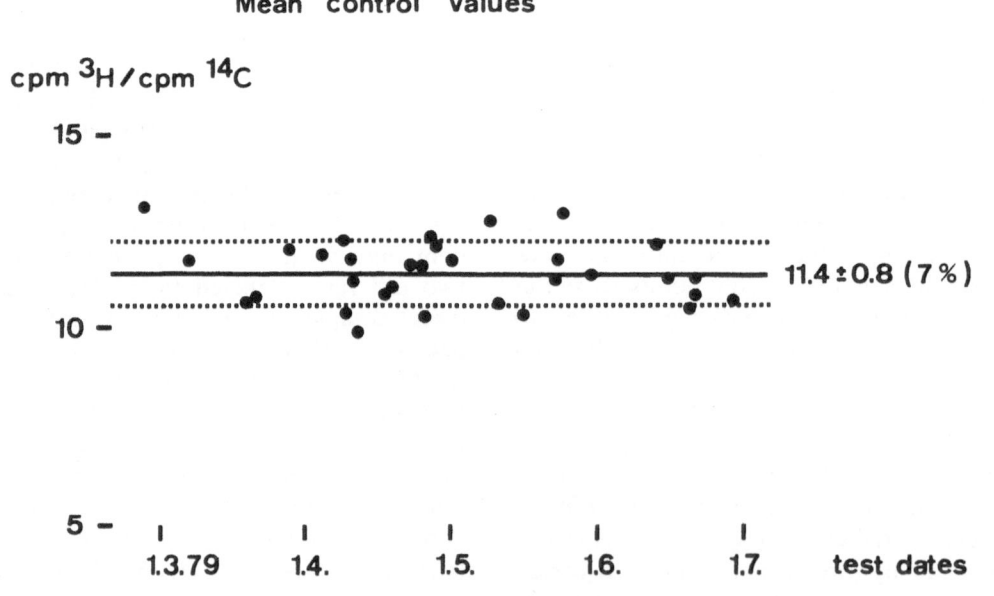

Figure 10.1. Daily fluctuations of control values in the DSI test. (a) Values expressed as cpm 3H per μg DNA; (b) values expressed as cpm 3H per cpm 14C.

In this way the results on control values shown in Figure 10.1b have been obtained, and it is obvious that the observed deviations from a mean value of 11.4 can well be tolerated in such an *in vivo* experiment. For clarity the data are finally transformed to express "percent thymidine incorporation" relative to the control or to "percent DNA synthesis inhibition" (% DSI).

Statistical significance of the results is best assessed by a nonparametric test (Wilcoxon-rank-test or equivalent); a t-test, the test of choice for many problems, cannot be used here, as the variances in treated and control groups will be unequal, which is also to be expected theoretically.

Results and Their Interpretation

Firstly, the suitability of a short-term test is judged by its record in detecting mutagens and carcinogens and in discerning them from nonmutagens and noncarcinogens. In this respect the DSI test more than fulfilled the expectations we had set on its performance (Table 10.1); it proved to have a predictive value analogous to that of the Ames test. However, a closer look at the data indicated that, although sensitive in picking up mutagens and carcinogens, it could have a rather large share of false positives. Theoretically, this is to be expected of a test of this type, as we may assume that every mutagen or carcinogen has to possess some activity toward DNA synthesis, whereas not every biologically active substance needs to be a mutagen or a carcinogen. However, as we shall see later, the system is able to recognize and thus to eliminate many of the "false" positives.

A second point to be considered in evaluating short-term tests is their response toward the different chemical groups of carcinogenic substances. To be useful, a test has to detect not only a particular type of carcinogen, but it should be capable of detecting all kinds of carcinogenic chemicals. Thus a test system cannot be evaluated only according to its potency in detecting direct-acting alkylating agents like methylmethanesulfonate (MMS) etc. but also according to its ability to detect as many as possible of the wide array of carcinogenic substances.

The DSI test detects carcinogens among all classes of known carcinogenic chemicals. Table 10.2 gives some idea of the range of compounds which can be detected. The interpretation of this table is very straightforward; those compounds inhibiting thymidine incorporation are carcinogens, and the ones leaving DNA synthesis unimpaired are not.

Considering the above-mentioned points, namely a good statistical correlation between long-term biological activity and activity in the short-term test, as well as a good response to all classes of chemical carcinogens, the DSI test can confidently be said to at least reach the standard set for a useful short-term test system.

In any short-term test system the inves-

Table 10.1. Compilation of DSI Test Results and Predictive Value of the Test

Substances tested[a]	134			
Proven Carcinogens	74;	DSI-positive	67	(= 91%)
"Noncarcinogens"	60;	DSI-positive	17	(= 28%)

("Noncarcinogens" include 13 antimetabolites, polycyclic aromatic hydrocarbons with initiating activity and weak mutagens, all of which are DSI-positive; false positives, excluding these compounds: 4/47 = 8.5%.)

[a]Compounds referred to in this table are those commercially available from the lists of carcinogens and noncarcinogens given by McCann *et al.* (1975), Heddle and Bruce (1977), Ishidate and Odashima (1977), Purchase *et al.* (1978), Seino *et al.* (1978), and Poirier and Weisburger (1979).

Table 10.2. *Selected Examples of DSI Test Results on Carcinogens and Noncarcinogens from Different Chemical Classes*

Compound	Dose (mg/kg)	% DSI[a]	Carcinogenicity (reference)	LD$_{50}$, remarks
A. Aromatic Amines				
2-Acetylamino- fluorene	50 i.p. 100 150 200	7.2 22.5 + 21.4 + 39.1 +	+ (McCann *et al.* 1975)	1020 mg/kg p.o. mouse
Benzidine	20 p.o. 50 100 200	3.8 7.1 19.0 + 28.9 +	+ (McCann *et al.* 1975)	
3,3',5,5'- Tetramethyl- benzidine	20 p.o. 50 100 200	1.7 9.6 5.2 0	− (Holland *et al.* 1974)	
B. Polycyclic Aromatic Hydrocarbons				
Benzo(a)pyrene	100 p.o. 200 p.o. 400 p.o. 500 p.o.	13.3 + 16.1 + 25.6 + 38.2 +	+ (IARC 1973b)	500 mg/kg i.p. mouse
Pyrene	100 p.o. 200 p.o. 500 p.o.	10.3 10.7 26.2 +	− (McCann *et al.* 1975)	toxic dose
9,10-Dimethyl- anthracene	75 p.o. 150 p.o. 300 p.o. 400 p.o.	8.4 27.0 + 30.7 + 52.9 +	+ (Purchase *et al.* 1978)	
C. Alkylating Agents				
Methylmethane- sulfonate	20 i.p. 50 80 120	6.7 22.5 + 43.7 + 68.1 +	+ (IARC 1974c)	150 mg/kg i.p. mouse
Dimethyl- sulfoxide	1100 p.o. 2200 4400	1.4 11.0 0	− (McCann *et al.* 1975)	21 g/kg p.o. mouse
Nitrogen mustard	2 p.o. 10 50 250	10.0 6.9 42.2 + 67.8 +	+ (IARC 1975)	
1,2-Dibromo- ethane	100 p.o. 200 300 500	0 27.0 + 49.2 + 79.4 +	+ (IARC 1977)	250 mg/kg p.o. mouse
Ethyleneimine	0.1 i.p. 0.3 1.0	21.7 + 28.9 + 74.0 +	+ (IARC 1975)	
Ethylamine	5 i.p. 15 50	0 0 5.6	−	400 mg/kg p.o. rat

Table 10.2. *Selected Examples of DSI Test Results on Carcinogens and Noncarcinogens from Different Chemical Classes* (Continued)

Compound	Dose (mg/kg)	% DSI[a]	Carcinogenicity (reference)	LD$_{50}$, remarks
D. Nitrosamines and Related Compounds				
N-Nitroso-dimethyl-amine	10 i.p.	9.4	+ (IARC 1978)	36 mg/kg i.p. rat
	20	19.8 +		
	30	31.0 +		
	40	41.5 +		
N-Nitroso-diphenyl-amine	200 p.o.	5.2	− (Purchase *et al.* 1978)	
	500 p.o.	0		
	1000 p.o.	8.1		
N-Nitroso-methylurea	20 i.p.	33.5 +	+ (IARC 1978)	
	40	62.0 +		
	80	81.6 +		
	160	90.5 +		
Urethane	500 p.o.	3.3	+ (IARC 1974c)	
	1000	18.0 +		
	1500	27.8 +		
Methylcarbamate	500 p.o.	0	− (IARC 1976b)	
	1000 p.o.	0		
	1500 p.o.	0		
E. Miscellaneous Compounds				
D,L-Ethionine	200 p.o.	0	+ (McCann *et al.* 1975)	1000 i.p. mouse
	750 p.o.	1.2		
	2000 p.o.	19.3 +		
L-Methionine	200 p.o.	1.3		
	750 p.o.	0		
	2000 p.o.	0.5		
Actinomycin D	0.3 i.p.	6.3	+ (IARC 1976a)	5 mg/kg i.p. mouse
	1	12.4 +		
	3	18.9 +		
	10	37.7 +		
Diethylstil-bestrol	100 p.o.	19.0 +	+ (IARC 1974b)	
	200	21.7 +		
	300	31.0 +		
	500	20.9		
Cadmium acetate	10 p.o.	18.2 +	+ (IARC 1973a)	88 mg/kg p.o. rat
	20	40.5 +		
	40	47.0 +		
	60	45.6 +		
Sodium acetate	200 p.o.	4.0	− (McCann *et al.* 1975)	
	500	10.5		
	1000	3.7		
Glucose	500 p.o.	8.0	− (Heddle and Bruce 1977)	
	1000	0		
	2000	13.5		
	4000	0		
Sodium glutamate	500 p.o.	5.5	− (Heddle and Bruce 1977)	
	1000 p.o.	0		
	2000 p.o.	5.4		

Table 10.2. *Selected Examples of DSI Test Results on Carcinogens and Noncarcinogens from Different Chemical Classes* (Continued)

Compound	Dose (mg/kg)	% DSI[a]	Carcinogenicity (reference)	LD$_{50}$, remarks
Ascorbic acid	500 p.o.	7.5	− (Heddle and Bruce 1977)	520 i.v. mouse
	1000 p.o.	0		
	2000 p.o.	3.5		
L-Adrenaline	0.1 i.p.	0	− (Heddle and Bruce (1977)	1.5 mg/kg s.c. mouse
	0.4	10.1		
	1.0	0.2		
	2.0	27.8 +		lethal dose

[a]Statistical significance assessed with the V test (Carnal and Riedwyl, 1972), + denotes significant difference to controls at $p \leqslant 0.05$.

tigation of test results which are in discordance with long-term assays and theoretical expectations can yield a deeper insight into the mechanisms of the test and the mode of action of the compound in question. This is especially true of the DSI test, and in the following we will show how such "wrong" results can have a biochemical explanation.

Several possibilities may be envisaged as to how false negative or false positive answers can be generated. A false negative result, e.g., will arise when (1) the chemical in question causes tumors through nongenetic mechanisms; or when (2) a breakdown product is the cancer-causing agent, but which is not produced in large enough amounts in the organism. A false positive answer on the other hand can be caused (3) by the cytotoxicity of the test compound or (4) by its interference with nucleic acid metabolism without directly producing DNA damage, e.g., by enzyme inhibition. In an analogous way, as the insufficient production of a reactive metabolite can lead to a false negative, (5) the appearance of anomalous metabolites or the saturation of detoxification mechanisms at very high dosages can be the cause of false positives. In Table 10.3 these possibilities are exemplified and we shall try to point out how such false answers may be recognized.

Aminotriazole can be regarded as an example for the first point. This compound is a goitrogenic chemical producing thyroid tumors at high doses. The appearance of tumors is, however, dependent on preneoplastic conditions, e.g., hyperplasia of the thyroid, and it is therefore assumed that the tumors arise by a nongenetic mechanism (McCann and Ames, 1976).

The second point can be exemplified by the drug isoniazid. Although this compound produces tumors in mice, it is inactive in the DSI test. Its tumorigenic activity in mice is most probably mediated through the major metabolite hydrazine, which by itself is quite capable of inhibiting the thymidine incorporation into mouse testicular DNA. The dose of hydrazine needed to produce an effect in the DSI test, however, is much greater than the amount of hydrazine which can be liberated from the maximally tolerated dose of isoniazid, thus leading to a false negative answer in testing isoniazid.

The third point is probably the most widely applicable one. Cytotoxic processes can, of course, also affect DNA synthesis, thus yielding false positive results. Although a compound such as 2,4-dinitrophenol is to be regarded as a noncarcinogen, it interrupts cellular metabolic processes and in the end also DNA synthesis, which, however, is inhibited only at excessive doses leading to acute poisoning and death (see Table 10.2).

A simple sign for the existence of a false positive through toxicity is the body temperature of the mice. Theoretically, a low-

Table 10.3. *Carcinogens and Noncarcinogens Evoking False Negative and False Positive DSI Responses*

Compound	Dose (mg/kg)	% DSI[a]	Carcinogenicity (reference)	LD$_{50}$, remarks
3-Amino-1,2,4-triazole	300 p.o. 600 p.o. 1000 p.o. 1200 p.o.	0 10.9 0 17.2 +	+ (IARC 1974c)	1100 mg/kg p.o. rat toxic dose
Isoniazid	20 p.o. 40 p.o. 50 p.o. 60 p.o. 75 p.o.	2.8 0 0 6.6 8.6	+ (IARC 1974a)	130 mg/kg i.p. mouse $^1/_4$ died
Hydrazine	100 200 400	0 16.2 + 47.1 +	+ (IARC 1974a)	
2,4-Dinitro-phenol	10 p.o. 20 30	8.1 8.9 54.8 +	− (Purchase *et al.* 1978)	30 mg/kg p.o. rat $^3/_8$ dead, others dying after 3 hr
Hydroxyurea	100 i.p. 150 i.p. 200 i.p.	7.8 48.7 + 93.6 +	− (Heddle & Bruce 1977)	500 mg/kg i.p. mouse
Phenanthrene	75 p.o. 150 p.o. 300 p.o. 500 p.o. 600 p.o.	0 4.5 26.8 + 36.8 + 33.4 +	− (McCann *et al.* 1975) (Poirier & Weisburger 1979)	700 mg/kg p.o. mouse

[a]For explanation see Table 10.2.

ering of the body temperature by 10°C should lead to a depression of thymidine incorporation by a factor of 2, which, indeed, could be demonstrated experimentally (B.E. Matter and P. Donatsch, personal communication).

Less easy to recognize are such compounds which inhibit DNA synthesis by inhibiting enzyme activities, by shifting cellular equilibria, or by altering membrane properties. When such effects are to be detected an additional assumption has to be made. It can be argued that in the case of an enzyme inhibitor, the rate of DNA synthesis in an affected organ is proportional to the concentration of the inhibitor within that organ. By looking at the detoxification and excretion kinetics of this agent in the target organ, a time interval can be deter-mined within which the DSI result should return to normal control values. Only agents reacting covalently with DNA and blocking in this manner the DNA synthesis are expected to inhibit thymidine incorporation after they have been inactivated or excreted (Seiler, 1979). Figure 10.2 and Table 10.4 demonstrate these opposite kinds of behavior: Methylmethanesulfonate (MMS), N-nitrosoethylenethiourea, and ethidium bromide, respectively, are inactivated or excreted rather rapidly, but the inhibition of thymidine uptake into testicular DNA is measurable even some days later. On the other hand, inhibitors and toxicants like hydroxyurea, caffeine and acetyl salicylic acid are losing inhibitory power at the rate at which they are inactivated or excreted. This approach is

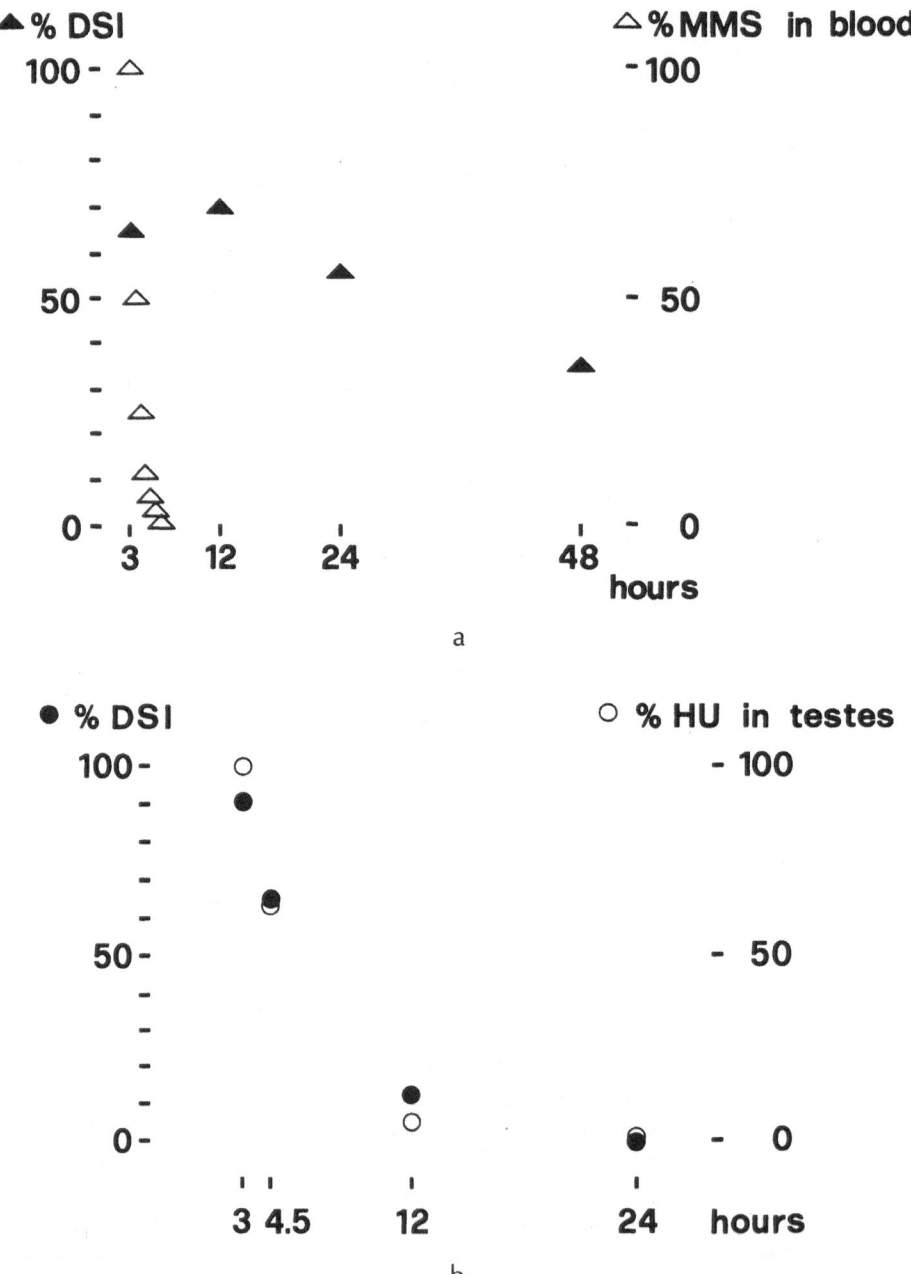

Figure 10.2. Time dependence of DSI value and pharmacokinetics of test substance. (a) Methyl-methanesulfonate (100 mg/kg i.p.), kinetics determined as alkylating potency in blood; (b) hydroxy-urea (200 mg/kg i.p.), kinetics determined as concentration of HU in testicular homogenate.

analogous to that used by Painter (1977) in his *in vitro* test system.

Note added in proof: In cases, where the detoxification and excretion kinetics of a compound are unknown and not easily determined, another approach seems feasible. Since testicular cells can be kept alive for some time in tissue culture media, noxious chemicals may be washed out and enzyme inhibition may be reduced by cul-

Table 10.4. *Kinetics of DSI Response and Pharmacokinetics of DNA-damaging Agents and General Toxicants*

Compound	Dose	Time (hr)	% DSI[a]	% of agent remaining active within organism
N-nitroso-ethylene-thiourea	100 i.p.	3	72.9 +	0 (T $^1/_2$ ca. 20 min)
		8	60.8 +	
		24	44.6 +	
		50	14.9	
Ethidium bromide	60 i.p.	2	46.0 +	80 (free compound)
		6	34.0 +	30
		22	46.1 +	0
		50	41.7 +	
		120	65.8 +	
Acetyl-salicylic acid	500 p.o.	3	24.0 +	100
		30	0	0 (12% after 24 h)
		50	2.8	0
Caffeine	100 p.o.	3	48.0 +	100
		8	17.4 +	65
		24	0.3	18
		50	0	0

[a]For explanation see Table 10.2.

tivating the cells for some time before the addition of 3H-thymidine. Also, the influence of reduced body temperature on the incorporation of 3H-thymidine will be eliminated by this combination method. The general procedure, which is not yet fully developed, consists of the following steps:

1. *In vivo* treatment of mice with 14C-thymidine
2. *In vivo* treatment with test compound
3. Sacrifice of animals and preparation of a testicular cell suspension
4. Incubation of cells *in vitro* for a short time (0.5–2 hr)
5. After removal of (wash) medium, addition of fresh medium containing 3H-thymidine
6. Incubation for about 1 hr for the incorporation of 3H-thymidine
7. After washing the cells in several steps, preparation of DNA and determination of its radioactivity

In this manner the DSI-value for, e.g., 2,4-dinitrophenol at 30 mg/kg p.o. of 54.8% (see Table 10.3) could be reduced to a nonsignificant deviation of 7.1% from the concurrent control in such a combined *in vivo/in vitro* experiment [Seiler, J.P. (1980): Testicular DNA synthesis inhibition—an *in vivo* test system for the detection of mutagenic and carcinogenic chemicals. 10th Annual Symposium on the Analytical Chemistry of Pollutants, Dortmund (FRG), May 28–30].

As is shown with the example of hydroxyurea, agents which inhibit some enzymatic step in the DNA synthesis can be powerful inhibitors in the DSI test. At first glance the case of 5-fluorouracil (5-FU) presented in Table 10.5 seemed to be an exception to this rule, as no inhibition of thymidine incorporation could be demonstrated when testing in the routine manner. 5-FU, however, is a peculiar agent in that it specifically inhibits thymidine synthesis; this in turn leads to a larger incorporation of the exogenously supplied radioactive thymidine, thus mimicking an even enhanced DNA synthesis. By measuring DNA synthesis as incorporation of desoxycytidine the actual inhibition of DNA synthesis can be demonstrated.

Into the same category of compounds

Table 10.5.   *DSI Response of 5-Fluorouracil Measured as Thymidine or Desoxycytidine Incorporation*

Compound	Dose (mg/kg)	% thymidine incorporation[a]	% desoxycytidine incorporation[a]
5-Fluorouracil	100	201.5 +	17.5 +
	300	215.1 +	14.1 +
	500	200.2 +	13.3 +
MMS	100	49.1 +	66.4 +
None (control)	—	100.0	100.0

[a]For explanation see Table 10.2.

can be placed those substances which do not inhibit thymidine incorporation but which alter the distribution pattern of the radioactive precursor by either changing precursor pool sizes, membrane properties, or vascular parameters. The measurement of total tritium in the testicular homogenate can provide clues to this possible explanation. Table 10.6 shows such a case in which a seemingly positive DSI test can be explained by the drastically reduced amount of 3H-thymidine available to the testicular cells.

The fifth possibility in yielding false positives is probably the most difficult to investigate and to differentiate from true positives. In many cases such compounds will be structurally related to chemical carcinogens, as is shown by the example of phenanthrene in Table 10.3. This compound is a nonmutagenic and noncarcinogenic polycyclic aromatic hydrocarbon (PAH) related structurally to other PAH's which are carcinogens, e.g., benzo(a)-pyrene. Bücker *et al.* (1979) have shown, however, that, by inhibiting the enzyme epoxide hydratase in rat liver homogenate, phenanthrene turned out to be mu-

tagenically active; Scribner (1973) even described a weak initiating activity of phenanthrene and anthracene on mouse skin. It can therefore be expected that most, if not all, PAH's will eventually act as mutagens if the protective mechanism of epoxide hydration is blocked. A positive result in the DSI test at the high doses needed to produce an effect in this test system even with the carcinogenic PAH's (see, e.g., Table 10.2), can then be explained by the saturation of this detoxification pathway with a concomitant "spillover" of reactive metabolites which otherwise would have been inactivated before reaching a target molecule such as DNA.

Summary and Outlook

The DSI test has —as should be evident by now—major advantages besides a few disadvantages. To begin with the latter, the DSI test exhibits a rather large share of false positives, which may cause unnecessary alarm, if such results are dispersed uncritically. On the other hand, the inher-

Table 10.6.   *Availability of 3H-Thymidine and DSI Response*

Compound	Dose (mg/kg)	% DSI[a]	% 3H in testicular homogenate	% DSI corrected
Serine	100 p.o.	0.8	102.1	2.8
	200 p.o.	11.9	97.0	9.2
	400 p.o.	24.0+	80.2	5.3

[a]For explanation see Table 10.2.

ent versatility of this test system offers the possibility of investigating positive and negative results further. Thus, at least a certain category of false positives, i.e., the cytotoxic and enzyme-inhibiting compounds, can be discriminated from the real positives. Another weakness of the test is the need for sometimes relatively high dosages, or, in other words, its relative insensitivity. The DSI test may thus not be able to detect a carcinogenic impurity in a technical chemical, as, e.g., the Ames test is capable of. Furthermore the high-dose treatment may lead to false positives by the saturation of some detoxification mechanism. Although these limitations of the DSI test have to be kept in mind, the advantages of this test system, which shall be shortly mentioned once more, in my opinion far outweigh them. The DSI test in its short-term form is a rapid *in vivo* test with the endpoint in a genetically meaningful organ and on an important biochemical process. It is a reliable test insofar as it detects carcinogens of all chemical classes with a low percentage of error. It is a flexible test system allowing the investigation of unexpected results in the same system by changing one or another parameter, e.g., the time course of the study, or the incorpo-

rated nucleoside. It is a relevant test as it incorporates the pharmacokinetics and the metabolic potential of a complete, intact mammalian organism. One has to bear in mind, however, that it is an indirect method to determine the carcinogenic or mutagenic potential of a chemical substance. Figure 10.3 describes this statement: The final and direct proof of carcinogenicity is the induction of neoplasms in an animal (or in man); while the induction of mutations or the inhibition of DNA synthesis need not be directly related to the process of carcinogenicity (as they may represent only one of several underlying causes) they nevertheless have some indirect causal relationship to carcinogenesis. As far as the DSI test is developed up till now, it would seem most suitable for use in a tier system, in conjunction with other short-term tests exhibiting different endpoints. In this situation the DSI test can give very valuable indications on the behavior of chemicals in the intact mammalian organism.

Furthermore, the DSI test has the potential for some presently less well explored applications. Thymidine incorporation is obviously not restricted to testicular tissue but can be measured in other organs as

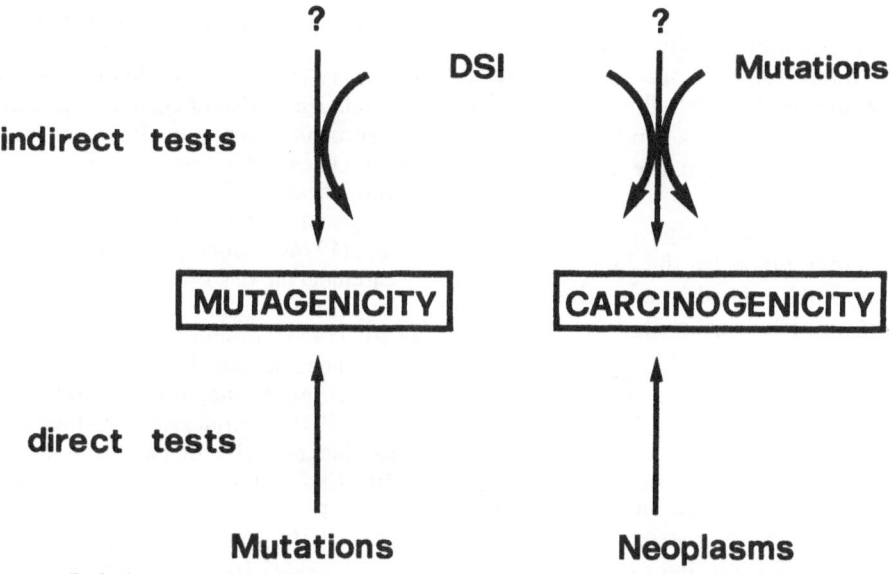

Figure 10.3.   Relationship of the DSI test to carcinogenicity and mutagenicity.

well. Mirvish *et al.* (1978), e.g., used this test in the rat to investigate the response of esophageal epithelium to different nitrosamines and found a significant relationship between the dose producing a 50% inhibition of thymidine incorporation and the tumor dose. From their data the DSI test seems able to predict the dose necessary for the induction of tumors in an organ; this indication of potency by the DSI test has already been pointed out by Friedman and Staub (1976), but more data on other substances and classes of compounds are needed in order to prove this point conclusively. Other possibilities could include the investigation of interactions between chemicals (Mirvish *et al.*, 1978) or the subacute administration and testing of substances. It will certainly be exciting to watch the progress of the DSI test as such approaches are tried further, and also to observe its performance, when more compounds of defined physiological activity, drugs and pesticides, e.g., are assayed in this system.

Acknowledgments

First of all I have to thank my co-workers Dr. F. Linder, Mr. E. Barben, Mr. Th. Thurnher, Miss M. Herensperger, and Miss K. Schneider for the data they generated. I thank furthermore my colleagues Dr. L. Stalder and Dr. H.-P. Bosshardt for their support of this work, and Dr. B. Matter (Sandoz Ltd., Basle) for valuable discussions.

I gratefully acknowledge also the financial assistance given by the Swiss Federal Office of Public Health and the World Wildlife Fund (Switzerland).

References

Ashwell, G. (1957): Colorimetric analysis of sugars. In: S.P. Colowick and N.O. Kaplan (eds.), Methods in Enzymology, Vol. 3. New York: Academic Press, pp. 73–105.

Berthold, V., Thielmann, H.W., Geider, K. (1978): Carcinogens inhibit DNA synthesis with isolated DNA polymerases from *Escherichia coli.* FEBS Letters 86:81–84.

Bücker, M., Glatt, H.R., Platt, K.L., Avnir, D., Ittah, Y., Blum, J., Oesch, F. (1979): Mutagenicity of phenanthrene and phenanthrene K-region derivatives. Mutat. Res. *66:* 337–348.

Carnal, H., Riedwyl, H. (1972): On a one-sample distribution-free test statistic. V. Biometrika 59:465–467.

Farber, E. (1968): Biochemistry of carcinogenesis. Cancer Res. 28:1859–1869.

Friedman, M.A., Staub, J. (1976): Inhibition of mouse testicular DNA synthesis by mutagens and carcinogens as a potential simple mammalian assay for mutagenesis. Mutat. Res. 37: 67–76.

Heddle, J.A., Bruce, W.R. (1977): Comparison of tests for mutagenicity or carcinogenicity using assays for sperm abnormalities, formation of micronuclei, and mutations in *Salmonella.* In: H.H. Hiatt, J.D. Watson and J.A. Winsten (eds.), Origins of Human Cancer, Book C, Cold Spring Harbor Laboratory, New York, pp. 1549–1557.

Holland, V.R., Saunders, B.C., Rose, F.L., Walpole, A.L. (1974): A safer substitute for benzidine in the detection of blood. Tetrahedron *30:*3299–3302.

IARC (1973a): Monographs on the evaluation of carcinogenic risk of chemicals to man, Vol. 2. Lyon: Int. Agency for Research on Cancer.

IARC (1973b): Monographs on the evaluation of carcinogenic risk of chemicals to man, Vol. 3. Lyon: Int. Agency for Research on Cancer.

IARC (1974a): Monographs on the evaluation of carcinogenic risk of chemicals to man, Vol. 4. Lyon: Int. Agency for Research on Cancer.

IARC (1974b): Monographs on the evaluation of carcinogenic risk of chemicals to man, Vol. 6. Lyon: Int. Agency for Research on Cancer.

IARC (1974c): Monographs on the evaluation of carcinogenic risk of chemicals to man, Vol. 7. Lyon: Int. Agency for Research on Cancer.

IARC (1975): Monographs on the evaluation of carcinogenic risk of chemicals to man, Vol. 9. Lyon: Int. Agency for Research on Cancer.

IARC (1976a): Monographs on the evaluation of carcinogenic risk of chemicals to man, Vol. 10. Lyon: Int. Agency for Research on Cancer.

IARC (1976b): Monographs on the evaluation of carcinogenic risk of chemicals to man, Vol.

12. Lyon: Int. Agency for Research on Cancer.

IARC (1977): Monographs on the evaluation of carcinogenic risk of chemicals to man, Vol. 15. Lyon: Int. Agency for Research on Cancer.

IARC (1978): Monographs on the evaluation of carcinogenic risk of chemicals to man, Vol. 17. Lyon: Int. Agency for Research on Cancer.

Ishidate, M., Jr., Odashima, S. (1977): Chromosome tests with 134 compounds on Chinese hamster cells in vitro —a screening for chemical carcinogens. Mutat. Res. 48:337–354.

McCann, J., Ames, B. (1976): Detection of carcinogens as mutagens in the Salmonella/microsome test: Assay of 300 chemicals: Discussion. Proc. Natl. Acad. Sci. USA 73:950–954.

McCann, J., Choi, E., Yamasaki, E., Ames, B. (1975): Detection of carcinogens as mutagens in the Salmonella/microsome test: Assay of 300 chemicals. Proc. Natl. Acad. Sci. USA 72:5135–5139.

Mirvish, S.S., Chu, C., Clayson, D.B. (1978): Inhibition of (3H)thymidine incorporation into DNA of rat esophageal epithelium and related tissues by carcinogenic N-nitroso compounds. Cancer Res. 38:458–466.

Moore, P., Strauss, B.S. (1979): Sites of inhibition of in vitro DNA synthesis in carcinogen- and UV-treated ΦX174 DNA. Nature (Lond.) 278:664–666.

Painter, R.B. (1977): Rapid test to detect agents that damage human DNA. Nature (Lond.) 265:650–651.

Poirier, L.A., Weisburger, E.K. (1979): Selection of carcinogens and related compounds tested for mutagenic activity. J. Natl. Cancer Inst. 62:833–840.

Purchase, I.F.H., Longstaff, E., Ashby, J., Styles, J.A., Anderson, D., Leferre, P.A., Westwood, F.R. (1978): An evaluation of 6 short-term tests for detecting organic chemical carcinogens. Br. J. Cancer 37:873–903.

Scribner, J.D. (1973): Tumor initiation by apparently noncarcinogenic polycyclic aromatic hydrocarbons. J. Natl. Cancer Inst. 50:1717–1719.

Seiler, J.P. (1977): Inhibition of testicular DNA synthesis by chemical mutagens and carcinogens. Preliminary results in the validation of a novel short term test. Mutat. Res. 46:305–310.

Seiler, J.P. (1978): Inhibition of testicular DNA synthesis by chemical mutagens and carcinogens as a valid mammalian short-term testing procedure. Mutat. Res. 53:260.

Seiler, J.P. (1979): Phenoxyacids as inhibitors of testicular DNA synthesis in male mice. Bull. Environ. Contam. Toxicol. 21:89–92.

Seino, Y., Nagao, M., Yahagi, T., Hoshi, A., Kawachi, T., Sugimura, T. (1978): Mutagenicity of several classes of antitumor agents to Salmonella typhimurium TA98, TA100, and TA92. Cancer Res. 38:2148–2156.

Vogel, E., Leigh, B. (1975): Concentration-effect studies with MMS, TEB, 2,4,5-triCl-PDMT and DEN on the induction of dominant and recessive lethals, chromosome loss and translocations in Drosophila sperm. Mutat. Res. 29:383–396.

Zieve, F.J. (1973): Effects of the carcinogen N-acetoxy-2-fluorenylacetamide on the template properties of deoxyribonucleic acid. Mol. Pharmacol. 9:658–669.

11

The Salmonella *Mutagenicity Test: An Overview**

LYNNE HAROUN AND BRUCE N. AMES

Introduction

The *Salmonella* test is currently the most widely used short-term test for screening environmental mutagens and carcinogens. It is in use in more than 2000 government, industrial and academic laboratories throughout the world. Test results on over 2600 chemicals have been reported in the Environmental Mutagen Information Center (EMIC) bibliography. In addition to its use in chemical screening programs, the assay has been used in a variety of other ways, providing toxicological information about a chemical that cannot easily be obtained from conventional animal tests. For example, the assay has been used to study the metabolic activation of chemical car-

* Sections of this review have been adapted from McCann and Ames (1977) and Ames (1979). A modified version of this paper will appear in S.O. Krezoski *et al.* (eds.), Microsomes, Drug Oxidations, and Chemical Carcinogenesis (4th International Symposium on Microsomes and Drug Oxidations), Academic Press: New York (in press).

cinogens. Differences among S9 fractions prepared from various tissues and species can be determined, and the effect of additions to the "S9 mix" such as antioxidants, EDTA, or enzyme inhibitors, can easily be investigated. In addition, the test has been used to help isolate and identify mutagens present in complex mixtures, to study the effect of comutagens, and as an adjunct in the development of useful, non-mutagenic chemicals. (Applications of the *Salmonella* assay have recently been reviewed by Hollstein *et al.*, 1979.)

The methods for testing in *Salmonella* have been described in detail (Ames *et al.*, 1973a, 1973b, 1975b; McCann *et al.*, 1975b), and the test system recently reviewed (Hollstein *et al.*, 1979; McCann and Ames, 1977; Sugimura *et al.*, 1977a). In this overview, we give a current assessment of the *Salmonella* assay for use in chemical screening programs and discuss several recent modifications of the assay that have increased its usefulness as a screening tool and limitations of the test.

Validation Against Known Animal Carcinogens and Noncarcinogens

Ideally, one would like to validate the efficiency of a test for use as a predictive tool for the detection of carcinogens by testing a large number of known human carcinogens. There is, however, little data available on chemicals that cause cancer in man. Almost all the organic chemicals that are known or suspected human carcinogens are mutagenic in *Salmonella* (Table 11.1); however, this is not sufficient to validate the test, both because the number of chemicals is small and because it must be shown that the test gives a negative response with noncarcinogens. We therefore examined about 300 chemicals that had been reported as carcinogens or noncarcinogens in animal experiments. About 90% (156/174) of the chemical carcinogens were mutagenic in *Salmonella*, while 87% (94/108) of the noncarcinogens were nonmutagens (McCann and Ames, 1976; McCann *et al.*, 1975a). In most cases, the test discrimi-

nated very well between carcinogens and noncarcinogens. This is especially striking with regard to results with closely related chemicals, where one is a potent carcinogen and the other is extremely weak or a noncarcinogen (Fig. 11.1). The relative mutagenicity of these chemicals illustrates the ability of the test to discriminate between minor differences in chemical structure that also alter carcinogenicity.

The test has been independently validated in three other studies (Table 11.2). They report a 72–91% correlation between mutagenicity in *Salmonella* and carcinogenicity in animal tests. However, the actual percentage of carcinogens that will be detected in a screening program cannot be correctly predicted from the above studies, unless the chemicals tested accurately reflect the distribution of carcinogens in the real world, or those to be tested in a given screening program. While compounds from a wide variety of chemical classes were selected for these studies, they probably were not a statistically representative sampling of the various classes of carcinogens as they are distributed in the environment. It is recognized that the ability of the *Salmonella* assay to detect carcinogens from any given chemical class may be lower (or higher) than the combined correlation values for all chemicals reported in these studies. The success rate in *Salmonella* is markedly lower for some classes of carcinogens (e.g., the chlorinated hydrocarbons) and there are some classes of carcinogens that will not be detected. For example, promoters and carcinogens such as griseofulvin, that do not directly interact with DNA, will not be detected as mutagens in *Salmonella*.

In setting up a screening program, it is thus important to recognize the strengths and limitations of any given short-term test. No one assay detects all the recognized carcinogens, and consequently, the use of a battery of tests is recommended. The correlation between mutagenicity and carcinogenicity should be determined separately for each class of chemical carcinogens.

Table 11.1. Organic Chemicals Known or Suspected as Human Carcinogens: Results of Mutagenicity Tests in Salmonella[a]

Mutagens	Nonmutagens
Aflatoxins	Benzene
4-Aminobiphenyl	Diethylstilbestrol
Auramine dye mixture	
Benzidine	
Chlornaphazine	
Bis (chloromethyl) ether	
Chloroprene	
Cigarette smoke condensates	
Coal tar	
Cyclophosphamide	
Melphalan	
Mustard gas	
2-Naphthylamine	
4-Nitrobiphenyl	
Soot	
Vinyl chloride	

[a]Reprinted from McCann and Ames (1977). Copyright 1977 by Cold Spring Harbor Laboratory.

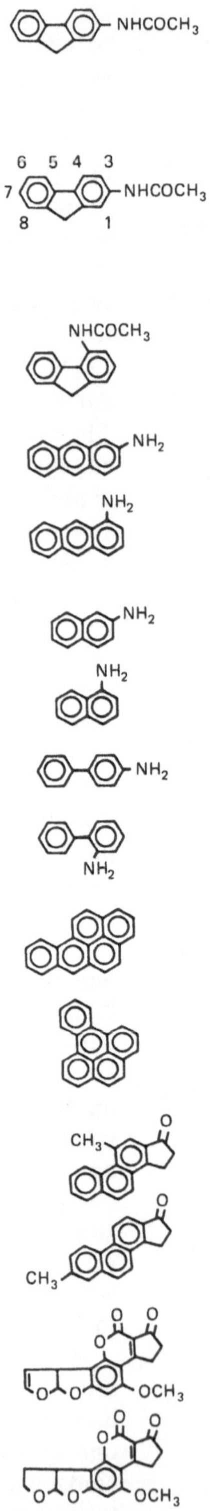

	CARC	REV/nMOLE	RATIO
2-ACETYLAMINOFLUORENE (AAF)	+	108	
1-HYDROXY-2-AAF	0	< 0.02	> 5400
3-HYDROXY-2-AAF	0	< 0.02	> 5400
5-HYDROXY-2-AAF	0	< 0.04	2700
7-HYDROXY-2-AAF	c0	< 0.3	> 3600
4-ACETYLAMINOFLUORENE	0	0.3	360
2-AMINOANTHRACENE	+	510	
1-AMINOANTHRACENE	w+	22	23
β-NAPHTHYLAMINE	+	8.5	
α-NAPHTHYLAMINE	c0	0.42	20
4-AMINOBIPHENYL	+	31	
2-AMINOBIPHENYL	+	0.51	61
BENZO (a) PYRENE	+	121	
BENZO (e) PYRENE	w+	0.6	202
15,16-DIHYDRO-11-METHYL-CYCLOPENTA (a) PHENANTHRENE-17-ONE	+	84	
15,16-DIHYDRO-3-METHYL-CYCLOPENTA (a) PHENANTHRENE-17-ONE	0	< 0.03	> 2800
AFLATOXIN B$_1$	+	7057	
AFLATOXIN B$_2$	w+	2.1	3360

Figure 11.1. Relative mutagenicity of carcinogen/noncarcinogen (weak carcinogen) pairs. Explanation of symbols: (+) carcinogen; (0) noncarcinogen; (w+) weak carcinogen; (c0) noncarcinogen in most studies, but with some reports of weak or marginal activity; (?) inadequate data available for classification but generally regarded as noncarcinogen based on structural considerations. For discussion, see Donahue *et al.* (1978) and McCann and Ames (1976). Reprinted from McCann and Ames (1977), copyright 1977 by Cold Spring Harbor Laboratory.

Table 11.2. *Summary of Four Validation Studies in* Salmonella *with Organic Chemical Carcinogens and Noncarcinogens*

Carcinogens			Noncarcinogens			
Number tested	Mutagenicity in *Salmonella* (+)	(−)	Number tested	Mutagenicity in *Salmonella* (+)	(−)	Reference
174	90%	10%	108	13%	87%	McCann *et al.* (1975a); McCann and Ames (1976)
167	85%	15%	86	26%	74%	Sugimura *et al.* (1977b)
58	91%	9%	62	6%	94%	Purchase *et al.* (1978)
38	72%	28%	16	19%	81%	Heddle *et al.* (1977)

[a]Reported values recalculated to exclude inorganic and physical carcinogens.

Ideally, it would be useful to identify carcinogen/noncarcinogen pairs from each chemical class that could then be used to validate all the short-term tests. This would not only determine the ability of a test to detect a wide variety of carcinogens, but also its ability to discriminate between carcinogens and noncarcinogens. In a recent study, Rinkus and Legator (1979) characterized 465 known or suspected carcinogens, dividing them into 39 chemical categories. Of these, 271 compounds had been tested in *Salmonella*. The correlation with mutagenicity in *Salmonella*. The correlation with mutagenicity in *Salmonella* was determined for each category; values ranged from 22 to 94%. The combined correlation factor for all chemicals tested was 77% (210/271). We have reanalyzed their study and find that the percentage of the organic carcinogens detected as mutagens should be 82% (Ames and McCann, 1980).

Mutagens Subsequently Found to be Carcinogens

While it is important to recognize the limitations of the validation studies, the usefulness of short-term tests will ultimately be determined by their ability to accurately predict the carcinogenic potential of untested chemicals. There is an increasing number of cases where chemical carcinogens were detected in short-term tests before being tested in animals. These include furylfuramide (AF-2), used as a food additive in Japan from 1964 to 1965 (Sugimura *et al.*, 1977b); ethylene dibromide, an industrial chemical and grain fumigant; and ethylene dichloride, the precursor of vinyl chloride (Ames, 1979). Two examples, Tris-BP and hair dyes, are discussed below.

Tris-BP (*tris*(2,3-dibromopropyl) phosphate) was the main flame retardant used in children's polyester pajamas from 1972 to 1977. Tris-BP, its metabolite dibromopropanol, and an impurity, dibromochloropropane (a carcinogen and human sterilant), are mutagens in *Salmonella* (Blum and Ames, 1977; Prival *et al.*, 1977). Fifty million children wore sleepwear that contained this material, added on at about 5% of the weight of the fabric. We argued that Tris-BP would pose a serious hazard to these children, since nonpolar chemicals such as tris are readily absorbed through the skin (Blum and Ames, 1977). Since its detection as a mutagen in *Salmonella*, it has been shown to be active in a number of short-term tests: it damages human DNA *in vitro* (Gutter and Rosenkranz, 1977), induces heritable mutations in *Drosophila* (Valencia *et al.*, unpublished, cited in Blum and Ames, 1977), and causes unscheduled DNA synthesis in human cells in tissue culture (Stich, unpublished, cited in Blum and Ames, 1977). Recent tests at the National Cancer Institute have shown that Tris-BP is a carcinogen in both rats and mice (National Cancer Institute, 1978a). In skin-painting studies, it induced cancer in mice (Van Duuren *et al.*, 1978) and testicular atrophy and sterility in rabbits (Osterberg *et al.*, 1977). It was recently shown that a mutagenic metabolite of Tris-BP, dibromopropanol, is present in the urine of children who wore tris-treated pajamas (Blum *et al.*, 1978).

Hair dyes also contain mutagens. In a study at our laboratory, about 90% (150/169) of the commercial oxidative-type (hydrogen peroxide) hair dye formulations tested were mutagenic; of the 18 components of these hair dyes (mostly aromatic amines), 9 were mutagenic (Ames *et al.*, 1975a). Many semipermanent hair dyes were also mutagenic. Hair dye components are known to be absorbed through the skin, and a variety of these ingredients have now been shown to be mutagenic in other short-term tests. One of the mutagenic hair dye components, 4-methoxy-*m*-phenylenediamine (MMPD), was recently shown to be a carcinogen in feeding studies at NCI (National Cancer Institute, 1978b). Several of the other chemicals are also being tested at NCI and appear to be carcinogens. The hair dye manufacturers in the United States have removed this compound from their products. However, at least one manufacturer has replaced MMPD with a closely related chemical, 4-ethoxy-*m*-phenylenediamine. This compound has not been tested for carcinogenicity, but a recent report indicates that it, too, is a mutagen in *Salmonella* (Prival *et al.*, 1980).

False Positives and Negatives

In general, most classes of chemical mutagens/carcinogens are detected in *Salmonella*. However, it is important to identify those classes that are not detected in the assay. The false positives and negatives reported in the McCann validation study have been previously discussed (McCann and Ames, 1976). We believe that most false negatives are due to inadequacies in the *in vitro* activation system and that most false positives are due to inadequate carcinogenicity tests, i.e., they are not true false positives. In calculating the correlation between carcinogenicity and mutagenicity, we thus feel it is important to examine the false positives and negatives separately. The problems in classifying a compound as a noncarcinogen are discussed below.

False Positives

Over 150 chemicals negative in animal cancer tests (noncarcinogens) have been tested in *Salmonella* (Hollstein *et al.*, 1979); the majority of these were negative. Most noncarcinogens positive in the test were weak mutagens and may in fact be weak carcinogens so far undetected in the animal cancer tests. A few of the false positives (5/14) (discussed in McCann and Ames, 1976) were strongly mutagenic and are mutagenic in other short-term tests as well (Hollstein *et al.*, 1979). These chemicals may possibly be false positives, but it would be premature to make that classification based on the animal cancer test data currently available. We think that most false positives are due to the limitations of the animal cancer tests and not to the fact that there are mutagens that are not carcinogens. The NCI has published criteria describing an adequate cancer test (Sontag *et al.*, 1975). The application of such criteria in evaluating noncarcinogenicity would mean that few chemicals can be considered noncarcinogens with a high degree of certainty. Recently, our laboratory has been involved in an extensive analysis of the world's cancer literature. One of the purposes of this study is to develop a measure of the thoroughness of a negative cancer test. This is expressed as a "no greater than" potency value that evaluates a test in terms of the weakest carcinogen that it could have detected. Given the enormous variation in the design and size of animal cancer tests, it is essential to be able to give a quantitative estimate of the sensitivity of any test in which a compound is found to be a noncarcinogen.

The limitations of animal cancer tests are clearly illustrated with the testing of 7-hydroxy-2-acetylaminofluorene (7-OH-2-AAF). This compound was classified as a noncarcinogen based on the animal tests (Morris *et al.*, 1960; Weisburger, personal communication, cited in Donahue *et al.*, 1978). When the sample of 7-OH-2-AAF that was used in the animal tests was tested in *Salmonella*, it was found to be a weak mutagen (McCann *et al.*, 1975a); this activity was later shown to be due to an impurity, the carcinogen 2-acetylaminofluorene (2-AAF), present in the sample (Donahue *et al.*, 1978). (The amount of 2-AAF in the sample of 7-OH-2-AAF was about 0.25%.) Thus the animal cancer test was not sensitive enough to detect a small amount of a potent carcinogen.

If this difference in sensitivity between *Salmonella* and animal tests is generally applicable to the detection of weak carcinogens, a similar explanation may apply to some of the other false positives in *Salmonella*. This may be the case for some of the noncarcinogens that are weak mutagens in the assay. These could be weak carcinogens that the animal cancer tests were not able to detect owing to the limitations in sensitivity of the tests.

In addition to the limitations of animal cancer tests, there are other potential sources of false positives in *Salmonella*. For example, there could be enzymes in *Salmonella* capable of metabolizing chemicals to mutagenic intermediates that are not present in mammals. However, there are many bacteria present in the human gut, including *Escherichia coli*, a close relative of *Salmonella*, that may have the same or similar enzymes. The nitrocarcinogens represent the largest class of chemicals activated to mutagens by bacterial enzymes (McCann and Ames, 1976). As these enzymes are present in the liver and bacterial flora of the gut, the most likely effect of *in situ* activation by *Salmonella* is to greatly increase the relative mutagenic potency of these compounds. At present there are no known compounds specifically activated by enzymes unique to *Salmonella*.

Impurities can also have a significant effect in mutagenesis testing. There is a 10^6-fold range in the mutagenic potency of compounds detected in the test (McCann and Ames, 1977) and, as a result, a small amount of a potent mutagen present as an impurity could cause a significant mutagenic response. As discussed above, this was clearly shown in the testing of 7-OH-2-AAF, where the presence of less than a 1% impurity (2-AAF) was detected in the *Salmonella* assay.

False Negatives

The general problems that can lead to false negatives in the assay have been discussed elsewhere (McCann and Ames, 1977) and are only briefly reviewed here.

1. *Chemical toxicity.* The limitations of the *Salmonella* assay in terms of maximum tolerated dose are defined by the toxicity of the chemical. In general, we have found that the expression of mutagenic activity occurs at concentrations below the toxic level, while toxicity is associated with excessive DNA damage. However, the test could fail to detect chemicals that are toxic to the bacteria for reasons unrelated to their mutagenic properties (e.g., antibiotics). Modifications of the standard test protocol, including liquid suspension assays, pulse exposures (Brem *et al.*, 1974; Rosenkranz *et al.*, 1976), and the fluctuation test (Green *et al.*, 1976) have been used to increase the sensitivity of the assay for the detection of toxic chemicals.

2. *The* in vitro *metabolic activation system.* We believe that most false negatives are due to technical inadequacies of the *in vitro* metabolic activation system. For example, the rate of conversion of the premutagen to its active form may be quite low, or specific enzymes or cofactors necessary for activation may not be present in the S9 mix.

Recent modifications of the *Salmonella* assay that enable the detection of mutagens previously reported as negative in *Salmonella* are discussed below.

3. *Tester strain specificity.* Some compounds may not be detected in the assay because they interact only with specific base sequences in the DNA not found in the present set of tester strains. We are planning to sequence the DNA of the tester strains in the region of the histidine mutations to identify the base sequences at the site of the mutations. Additional strains are being screened so that a complete set of tester strains will be available.

Some compounds, such as the carcinogen mitomycin C (Kondo *et al.*, 1970; McCann *et al.*, 1975b) and malondialdehyde (Mukai and Goldstein, 1976), require an intact *uvrB* repair system to be detected as mutagens and are thus detected only with strains not usually used for routine testing (TA92 and TA94). A number of methylating agents (e.g., dimethylnitrosamine, cycasin, and N-methyl-N'-nitro-N-nitrosoguanidine) are much more active on strain *hisG46* (Matsushima *et al.*, 1978). For specific classes of chemicals, additional strains may be useful for screening.

4. *The pharmacokinetics of the active metabolite.* Chemicals that are metabolized to highly reactive intermediates (e.g., free radicals) may react with the S9 or other cellular components before reaching the DNA. Other compounds, such as actinomycin D, may be negative because they do not accumulate intracellularly in the bacteria (Benedict *et al.*, 1977).

5. *Carcinogens that are not mutagens.* Not all carcinogens interact with DNA. Promoters, hormones, and compounds such as griseofulvin that do not interact directly with the DNA are not detected in the *Salmonella* assay. Griseofulvin is known to interfere with microtubule formation during mitosis, which could result in chromosome abnormalities. It may be an example of a class of carcinogens that will be detected only in eucaryotic organisms.

Recent Modifications of the *Salmonella* Assay

The methods for mutagen testing in *Salmonella* were described in detail in 1975 (Ames *et al.*, 1975b). Since that time, several modifications have been introduced, and as a result several compounds that were previously reported as false negatives are now positive in the assay. As our understanding of both the mechanism of carcinogenesis and the metabolism of chemical carcinogens increases, many of the problems resulting in false negatives will be eliminated. Here we discuss a few of the more important modifications that have significantly increased the usefulness of the assay as a screening tool.

1. *Liquid preincubation.* The liquid preincubation method (Yahagi *et al.*, 1977) is a modification of the pour plate assay in which the bacteria, compound, and S9 (or phosphate buffer) are preincubated for 20–30 minutes before being poured on the plate. Several compounds that are weak or not detected in the pour plate assay are detected using this method. These include dimethylnitrosamine (DMN), diethylnitrosamine, dimethylaminoazobenzene, and several carcinogenic pyrrolizidine alkaloids (lasiocarpine, senkirkine, and heliotrine). The mutagenic activity of several other compounds, including aflatoxin B_1, benzidine, benzo(a)pyrene, and methyl methanesulfonate, has been determined using both methods; in all cases the preincubation assay is of equal or greater sensitivity than the pour plate assay (reviewed in Matsushima *et al.*, 1980).

Based on studies with DMN, it had been postulated that the liquid preincubation assay was more sensitive because the metabolites of DMN reacted with the agar, present at all times in the plate incorporation assay (Bartsch *et al.*, 1976). However, recent studies suggest that the increased sensitivity of the preincubation assay is due to the fact that the bacteria, S9, and compound are incubated at higher concentrations than are present in the plate incorpo-ration assay (Prival *et al.*, 1979).

2. *The liver homogenate (S9)*. Liver homogenates (the S9) prepared from Aro-clor-induced rats are recommended for rou-tine screening. Other tissues and species have been used as a source of S9 (reviewed in Hollstein *et al.*, 1979), but until recently no tissue- or species-specific compounds had been identified. Recently, several mu-tagen/carcinogens that were negative (or very weakly positive) when tested with rat liver preparations were positive when tested with an S9 from hamster liver. These include phenacetin, lasiocarpine, pe-tasitenine (T. Matsushima, personal com-munication), dimethylnitrosamine (T. Mat-sushima, personal communication; Prival *et al.*, 1979), *para*-rosaniline, and Fyrol FR2 (V. Simmon, personal com-munication).

It is clear that the liver homogenate is able to metabolize most premutagens to their active, mutagenic forms. However, it cannot faithfully duplicate whole animal ab-sorption, distribution, metabolism, and ex-cretion of a chemical. Some of the false positives and negatives may be due to these limitations. Further work in this area will presumably show what additions or modifi-cations of the assay are needed to make it correspond more closely to the whole animal (e.g., the addition of riboflavin or glycosidase, discussed below). In spite of the theoretical limitations of the S9, there are few false positives or negatives in *Sal-monella*. In addition, many experiments in-vestigating the metabolism of carcinogens would be very costly or difficult in whole animal studies, but are easily done with *Sal-monella* or the other short-term tests. There have been several studies comparing chemical activation by different species and tissues (reviewed in Hollstein *et al.*, 1979), including the use of an S9 prepared from human tissue. Quercetin (discussed below), hair dye components, and Tris-BP (un-published results, this laboratory), mu-tagens to which there has been extensive human exposure, have all been shown to be activated by an S9 prepared from human liver.

3. *Riboflavin*. Several azo and diazo dyes are mutagenic only when tested in the pres-ence of riboflavin. These include Congo Red, Eriochrom blue black B, and the car-cinogens Ponceau R and Trypan blue. The mutagenic activity of some dyes, while not dependent on riboflavin, is increased (e.g., Ponceau 3R), although the activity of sev-eral other azo dyes is decreased (e.g., di-methylaminoazobenzene [DAB]) (Sugi-mura *et al.*, 1977b; Matsushima *et al.*, 1980).

The riboflavin may directly reduce the azo bond (G. Jaffe, personal com-munication) or act as a cofactor of the azo-reductase present in the liver homogenate.

4. *Glycosidases*. Quercetin, kaempferol, and a number of other naturally occurring flavonols are mutagens in *Salmonella* (Bjel-danes and Chang, 1977; Sugimura *et al.*, 1977c). These compounds are present in many edible plants and are used in drugs and food supplements (Brown and Die-trich, 1979). However, most of the fla-vonols present in plants are in the form of glycosides (over 70 different glycosides of quercetin have been identified), and as such are not mutagenic. Orally ingested glyco-sides are usually hydrolyzed by the micro-flora of the lower bowel. As neither *Sal-monella* nor liver have active glycosidases, the glycosides of natural mutagens are inac-tive in *Salmonella* unless the appropriate glycosidase is added. We have now devel-oped a model for the human gut flora that can be added directly to the *Salmonella* test. An enzyme preparation, fecalase, was made from a sonicate of human feces, and has been found to have a broad spectrum of glycosidase activity (Tamura *et al.*, 1980). Two commercially available glycosides of quercetin, rutin (quercetin-3-rutinoside, Fig. 11.2) and quercitrin (quercetin-3-L-rhamnoside) were mutagenic in the pres-ence of fecalase. Cycasin (the glucoside of methylazoxymethanol), a known car-cinogen, is also positive in *Salmonella* when tested with fecalase.

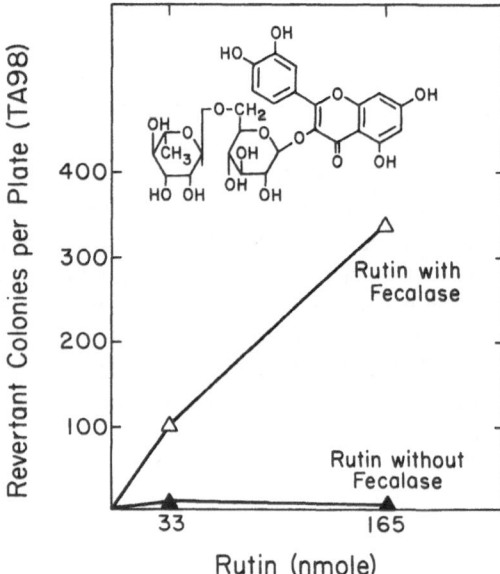

Figure 11.2. Mutagenicity of rutin in the standard plate incorporation test. Spontaneous revertant colonies have been subtracted. Liver homogenate from Aroclor-induced rats and fecalase were added as described (Tamura *et al.*, 1980). Fecalase is an enzyme preparation (2 mg/ml protein in pH 7.4 phosphate buffer) made by sonicating human feces, centrifuging, filtering the supernatant through a Sephadex gel, and then filter sterilizing.

A rat cecal extract (Brown and Dietrich, 1979) has also been used in the assay as a source of glycosidase activity, and several commercial enzyme preparations are available. These include β-glucosidase (from almonds), hesperidinase (a mixture of glycosidases isolated from *Aspergillus niger*) (Matsushima *et al.*, 1980), and preparations from the snail *Helix pomatia* (Brown and Dietrich, 1979).

5. *Chemical interactions: comutagens.* The *Salmonella* test and other short-term assays should be useful for investigating a variety of interactions between chemicals, including comutagenic and antimutagenic interactions. For example, recent work in Japan (reviewed in Matsushima *et al.*, 1980) has shown that two β-carboline derivatives, harman and norharman, are comutagens. While they are not mutagenic in *Salmonella*, they significantly alter the mutagenic response of several carcinogens.

The carcinogens, aniline, and *o*-toluidine (both nonmutagens), and DAB (a weak mutagen) are all strong mutagens when tested in the presence of norharman. The mutagenicity of several other compounds, including benzo(a)pyrene, acetylaminofluorene, and Trp-P-1, is either decreased or increased in the presence of norharman, depending on the amount of S9 used in the assay. The mechanism by which norharman is acting is presently unknown. It has been shown both to intercalate between the base pairs of double-stranded DNA (possibly changing the susceptibility of DNA to attack by mutagens) and to modify the enzymatic metabolism of compounds (possibly altering the balance between activation and deactivation pathways).

Conclusions

An advantage of the *Salmonella* test is that the standard assay can easily be added to or modified to include new methods. (The standard assay is that described in Ames *et al.*, 1975b. All compounds are tested in the pour plate assay on strains TA1535, TA1537, TA98, and TA100, both in the presence and absence of an S9 mix. The S9 homogenate is prepared from the liver of Aroclor-induced rats; 50 μl of the homogenate is added per plate.) While most carcinogens are detected in the standard assay (Table 11.2), a few additional ones will be detected by some of the various modifications described. What then should be recommended for a routine assay? As indicated above, several compounds have now been identified that are positive only when specific modifications of the standard assay are used: for example, DMN is positive only with the preincubation modification; *para*-rosaniline only with hamster liver; and mitomycin C only on strains TA92 and TA94. Riboflavin and glycosidase are necessary for the detection of azo and diazo dyes, respectively. We would like to discuss the use of each of these modifications for routine screening.

1. *Liquid preincubation.* It is not yet possible to identify the specific classes of compounds that will be positive only in the preincubation assay. However, as the work of the Japanese shows that the preincubation method is more sensitive than the pour plate assay for some chemicals (discussed above), we think that it is useful for routine screening. This procedure is only a slight modification of the plate test and does not significantly increase the time or cost in running the assay.

2. *Hamster liver S9.* Although several compounds have been identified that are positive only when tested with hamster liver S9, there is not yet enough information to recommend substituting hamster S9 in place of rat S9 for routine testing. In comprehensive screening programs, it would be desirable to test compounds with both hamster and rat liver.

3. *Additional tester strains.* TA92 and TA94 are useful for detecting crosslinking agents, although only two compounds, mitomycin C and malondialdehyde, have been identified that are positive on these strains but not on the standard set of strains. These strains are available from this laboratory, although at present there is not enough information to determine if it is worth the extra cost to use them for routine screening. It is not possible to predict the number of compounds that would be missed if these strains were not routinely used. The addition of two strains would increase the size of the assay by 50%.

4. *Riboflavin.* Based on the evidence presented above, all azo and diazo dyes should be tested both in the presence and absence of riboflavin.

5. *Glycosidase.* Glycosides and all extracts from plants should be tested both in the presence and absence of a source of glycosidases.

Testing a compound using the conditions outlined above would generally be considered adequate. However, there are other possible modifications and, depending on the compound, further testing might include assays with additional *Salmonella* tester strains, varying the concentration of the S9 homogenate, using S9 prepared from a different tissue or species, or testing in the presence of norharman. Additional testing will increase the probability of detecting a potential carcinogen, although it is difficult to determine if it will be worth the extra time and cost. Whether further testing is initiated, the significance of a negative result in *Salmonella* or any other short-term test must be considered in the light of the chemical class of the compound and the ability of the test to detect compounds in that class. The use of a battery of short-term tests designed to detect as wide a range of chemicals as possible will help to minimize the problem of false negatives.

Acknowledgments

We would like to thank M. Hollstein for helpful criticism of the manuscript. This work was supported by Department of Energy contract EY-76-S-03-0034 PA156.

References

Ames, B.N. (1979): Identifying environmental chemicals causing mutations and cancer. Science *204*:587–593.

Ames, B.N. Durston, W.E., Yamasaki, E., Lee, F.D. (1973a): Carcinogens are mutagens: A simple test system combining liver homogenates for activation and bacteria for detection. Proc. Natl Acad. Sci. USA *70*:2281–2283.

Ames, B.N., Lee, F.D., Durston, W.E. (1973b): An improved bacterial test system for the detection and classification of mutagens and carcinogens. Proc. Natl Acad. Sci. USA *70*:782–786.

Ames, B.N., Kammen, H.O., Yamasaki, E. (1975a): Hair dyes are mutagenic: Identification of a variety of mutagenic ingredients. Proc. Natl Acad. Sci USA *72*:2423–2427.

Ames, B.N., McCann, J. (1980): Validation of the *Salmonella* test: A reply to Rinkus and Legator. Cancer Res.

Ames, B.N., McCann, J., Yamasaki, E.(1975b):

Methods for detecting carcinogens and mutagens with the *Salmonella*/mammalian microsome mutagenicity test. Mutat. Res. *31*:347–364.

Bartsch, H., Camus, A., Malaveille, C. (1976): Comparative mutagenicity of N-nitrosamines in a semi-solid and in a liquid incubation system in the presence of rat or human tissue fractions. Mutat. Res. *37*:149–162.

Benedict, W.F., Baker, M.S., Haroun, L., Choi, E., Ames, B.N. (1977): Mutagenicity of cancer chemotherapeutic agents in the *Salmonella*/microsome test. Cancer Res. *37*:2209–2213.

Bjeldanes, L.F., Chang, G.W. (1977): Mutagenic activity of quercetin and related compounds. Science *197*:577–578.

Blum, A., Ames, B.N.(1977): Flame-retardant additives as possible cancer hazards. Science *195*:17–23.

Blum A., Gold, M.D., Ames, B.N., Kenyon, C., Jones, F.R., Hett, E.A., Dougherty, R.C., Horning, E.C., Dzidic, I., Carroll, D.I., Stillwell, R.N., Thenot, J.-P. (1978): Children absorb tris-BP flame retardant from sleepwear; urine contains the mutagenic metabolite, 2,3-dibromopropanol. Science *201*:1020–1023.

Brem, H., Stein, A.B., Rosenkranz, H.S. (1974): The mutagenicity and DNA-modifying effect of haloalkanes. Cancer Res. *34*:2576–2579.

Brown, J.P., Dietrich, P.S. (1979): Mutagenicity of plant flavonols in the *Salmonella*/mammalian microsome test: activation of flavonol glycosides by mixed glycosidases from rat cecal bacteria and other sources. Mutat. Res. *66*:223–240.

Donahue, E.V., McCann, J., Ames, B.N. (1978): Detection of mutagenic impurities in carcinogens and noncarcinogens by high-pressure liquid chromatography and the *Salmonella*/microsome test. Cancer Res. *38*:431–438.

Green, M.H.L., Muriel, W.J., Bridges, B.A. (1976): Use of a simplified fluctuation test to detect low levels of mutagens. Mutat. Res. *38*:33–42.

Gutter, B., Rosenkranz, H.S. (1977): The flame retardant tris(2-3-dibromopropyl)-phosphate: alteration of human cellular DNA. Mutat. Res. *56*:89–90.

Heddle, J.A., Bruce, W.R. (1977): Comparison of tests for mutagenicity or carcinogenicity using assays for sperm abnormalities, formation of micronuclei, and mutations in *Salmonella*. In: H.H. Hiatt, J.D.Watson, and J.A.Winsten (eds.), Origins of Human Cancer, Book C, Cold Spring Harbor Conferences on Cell Proliferation, Vol. 4, pp. 1549–1557, Cold Spring Harbor Laboratory, Cold Spring Harbor, New York.

Hollstein, M., McCann, J., Angelosanto, F.A., Nichols, W.W (1979): Short-term tests for carcinogens and mutagens. Mutat. Res. *65*:133–226.

Kondo, S., Ichikawa, H., Iwo, K., Kato, T. (1970): Base-change mutagenesis and prophage induction in strains of *Escherichia coli* with different DNA repair capacities. Genetics *66*:187–217.

Matsushima, T., Shirai, A., Sawamura, M., Umezawa, K., Sugimura, T. (1978): Difference in sensitivity of *Salmonella typhimurium* strains to alkylating agents. Mutat. Res. *54*:243.

Matsushima, T., Sugimura, T., Nagao, M., Yahagi, T., Shirai, A., Sawamura, M. (1980): In: K. Norpoth and R.C. Garner (eds.), Short-Term Mutagenicity Systems for Detecting Carcinogens. Springer-Verlag, Berlin.

McCann, J., Ames, B.N. (1976): Detection of carcinogens as mutagens in the *Salmonella*/microsome test: Assay of 300 chemicals. Discussion. Proc. Natl Acad. Sci. USA *73*:950–954.

McCann, J., Ames, B.N. (1977): The *Salmonella*/microsome mutagenicity test: predictive value for animal carcinogenicity. In: H.H. Hiatt, J.D. Watson, and J.A. Winsten (eds.), Origins of Human Cancer, Book C. Cold Spring Harbor Conferences on Cell Proliferation, Vol. 4, pp. 1431–1450, Cold Spring Harbor Laboratory, Cold Spring Harbor.

McCann, J., Choi, E., Yamasaki, E., Ames, B.N. (1975a): Detection of carcinogens as mutagens in the *Salmonella*/microsome test: Assay of 300 chemicals. Proc. Natl Acad. Sci. USA *72*:5135–5139.

McCann, J., Spingarn, N.E., Kobori, J., Ames, B.N. (1975b): Detection of carcinogens as mutagens: bacterial tester strains with R factor plasmids. Proc. Natl Acad. Sci. USA *72*:979–983.

Morris, H.P., Velat, C.A., Wagner, B.P., Dahlgard, M., Ray, F.E. (1960): Studies of carcinogenicity in the rate of derivatives of aromatic amines related to N-fluorenylacetamide. J. Natl Cancer Inst. *24*:149–180.

Mukai, F.H., Goldstein, B.D. (1976): Mutagenicity of malonaldehyde, a decomposition product of peroxidized polyunsaturated fatty acids. Science *191*:868–869.

National Cancer Institute (1978a): Bioassay of Tris (2,3-Dibromopropyl)phosphate for Possible Carcinogenicity, Carcinogenesis Technical Report Series No. 76, US Department of Health, Education and Welfare Publication No. (NIH) 78-1326.

National Cancer Institute (1978b): Bioassay of 2,4-Diaminoanisole Sulfate for Possible Carcinogenicity, Carcinogenesis Technical Report Series No. 84, US Department of Health, Education and Welfare Publication No. (NIH) 78-1334.

Osterberg, R.E., Bierbower, G.W., Hehir, R.M. (1977): Renal and testicular damage following dermal application of the flame retardant tris(2,3-dibromopropyl)phosphate. J. Toxicol. Environ. Health *3*:979–987.

Prival, M.J., McCoy, E.C., Gutter, B., Rosenkranz, H.S. (1977): Tris(2,3-dibromopropyl)phosphate: mutagenicity of a widely used flame retardant. Science *195*:76–78.

Prival, M.J., Mitchell, V.D., Gomez, Y.P. (1980): Mutagenicity of a new hair dye ingredient: 4-ethoxy-*m*-phenylenediamine. Science *207*:907–908.

Prival, M.J., Sheldon, A.T., Jr, King, V.D. (1979): The effect of species used as the source for S-9 on the mutagenicity of dialkylnitrosamines in the *Salmonella* plate assay. 10th Annual Meeting of the Environmental Mutagen Society, Abstract No. Cb-4, pp. 67–68.

Purchase, I.F.H., Longstaff, E., Ashby, J., Styles, J.A., Anderson, D., Lefevre, P.A., Westwood, F.R. (1978): An evaluation of 6 short-term tests for detecting organic chemical carcinogens. Br. J. Cancer *37*:873–959.

Rinkus, S.J., Legator, M.S.(1979): Chemical characterization of 465 known or suspected carcinogens and their correlation with mutagenic activity in the *Salmonella typhimurium* system. Cancer Res. *39*:3289–3318.

Rosenkranz, H.S., Gutter, B., Speck, W.T. (1976): Mutagenicity and DNA-modifying activity: a comparison of two microbial assays. Mutat. Res. *41*:61–70.

Sontag, J., Page, N.P., Saffiotti, U. (1975): Guidelines for Carcinogen Bioassay in Small Rodents, Division of Cancer Cause and Prevention, National Cancer Institute, Bethesda, Maryland.

Sugimura, T., Kawachi, T., Matsushima, T., Nagao, M., Sato, S., Yahagi, T. (1977a): A critical review of submammalian systems for mutagen detection. In: D. Scott, B.A. Bridges, and F.H. Sobels (eds.), Progress in Genetic Toxicology, pp. 125–140, Elsevier/North-Holland Biomedical Press, Amsterdam.

Sugimura, T., Nagao, M., Kawachi, T., Honda, M., Yahagi, T., Seino, Y., Sato, S., Matsukura, N., Matsushima, T., Shirai, A., Sawamura, M., Matsumoto, H. (1977b): Mutagen-carcinogens in food, with special reference to highly mutagenic pyrolytic products in broiled foods. In: H.H.Hiatt, J.D. Watson, and J.A. Winsten (eds.), Origins of Human Cancer, Book C, Cold Spring Harbor Conferences on Cell Proliferation, Vol. 4, pp. 1561–1577, Cold Spring Harbor Laboratory, Cold Spring Harbor.

Sugimura, T., Nagao, M., Matsushima, T., Yahagi, T., Seino, Y., Shirai, A., Sawamura, M., Natori, S., Yoshihira, K., Fukuoka, M., Kuroyanagi, M.(1977c): Mutagenicity of flavone derivatives. Proc. Japan. Acad. *53*:194–197.

Tamura, G., Gold, C., Ferro-Luzzi, A., Ames, B.N. (1980): Fecalase: a model for the activation of dietary glycosides to mutagens by intestinal flora. Proc. Natl Acad. Sci. USA *77*:4961–4965.

Van Duuren, B.L., Loewengart, G., Seidman, I., Smith, A.C., Melchionne, S. (1978): Mouse skin carcinogenicity tests of the flame retardants tris(2,3-dibromopropyl)phosphate, tetrakis(hydroxymethyl)phosphonium chloride, and polyvinyl bromide. Cancer Res. *38*:3236–3240.

Yahagi, T., Nagao, M., Seino, Y., Matsushima, T., Sugimura, T., Okada, M. (1977): Mutagenicities of N-nitrosamines on *Salmonella*. Mutat. Res. *48*:121–129.

12

Applications of the Salmonella/*Microsome Assay*

Vincent F. Simmon

Introduction

The *Salmonella*/microsome assay has been described in detail (Ames *et al.*, 1973, 1975b; McCann *et al.*, 1975b). I will present some rationale for using this procedure as well as the results obtained in testing some chemicals.

On the surface, there would appear to be some discrepancies with respect to the accuracy of the *Salmonella*/microsome assay in detecting chemical carcinogens as mutagens. This accuracy has been variously reported as 64–90% (McCann *et al.*, 1975a; Simmon, 1979) for retrospective testing of chemicals of known carcinogenic activity. The higher accuracies were obtained when variations of the standard procedure were included (e.g., assays in suspension, preincubation, assays in desiccators), and with the use of metabolic activation preparations other than those obtained from rats previously induced with Aroclor 1254 (e.g., using livers of mice or

hamsters and using phenobarbital or 3-methylcholanthrene as inducers). The significance of the high accuracy of this mutagenicity assay is that the majority of known organic chemical carcinogens are mutagens. From another viewpoint, most organic chemical carcinogens can either covalently bind to DNA (in their activated form) or reduce the fidelity of DNA replication. The preponderance of evidence would suggest that the association of carcinogenic and mutagenic activities is not fortuitous. Nevertheless, the significance of this observation appears to escape the awareness of the detractors of this test system.

There is considerable evidence and awareness regarding the application of the *Salmonella*/microsome assay to achieving an understanding of the molecular mechanisms of carcinogenesis. For example, much of our current understanding of the metabolism of polycyclic aromatic compounds has evolved from the isolation and testing of individual metabolites using *Sal-*

monella (Lehr *et al.*, 1977; Wood *et al.*, 1978).

Perhaps the most telling fact is that in the three-volume series, *Origins of Human Cancer*, Ames is the most cited author (Hiatt *et al.*, 1977).

From a pragmatic point of view, mutagenic assays with *Salmonella* have forecast accurately the carcinogenic activity (or transforming activity) of chemicals such as vinylidene chloride (Bartsch *et al.*, 1975), 1,2-dibromoethane (Ames, 1971), AF-2 (McCann *et al.*, 1975b), tris(2,3-dibromopropyl)phosphate (Blum and Ames, 1977; Prival *et al.*, 1977), and complex mixtures such as hair dyes (Ames *et al.*, 1975a), drinking water concentrates (Loper *et al.*, 1978), substances in urine (Legator *et al.*, 1975) and particulate diesel exhaust (Huisingh *et al.*, 1979).

Using the standard *Salmonella*/microsome procedure, a minimum accuracy of 65% would seem reasonable. In most instances, false positive responses arise from the relative insensitivity of *in vivo* assays as well as a reliance upon carcinogenic data which is woefully inadequate by comparison with current bioassay methodology (McCann and Ames, 1976). In a limited (25 chemicals) prospective, double-blind study, 10 of 13 carcinogens, 3 of 4 suspect carcinogens, and 3 of 8 noncarcinogens were mutagenic (Dunkel and Simmon, 1980), using the standard assay procedure. Thus, the accuracy of predicting carcinogenic activity for carcinogens and suspect carcinogens (as defined by the Chemical Advisory Group) was greater than 76%.

From the economic viewpoint, detecting environmental carcinogens by their mutagenic activity in the *Salmonella*/microsome assay is extremely advantageous. Assuming 70% accuracy and a cost of $500 per chemical or substance tested, 70 of 100 carcinogens could be detected at a cost of $50,000. Using an *in vitro* transformation assay costing approximately $5,000 per sample, only 10 carcinogens could be detected, assuming

100% accuracy for this procedure. The same $50,000 would pay for the preliminary subchronic experiments of a single bioassay; the total bioassay would cost in excess of $250,000. Thus, in the real world with real limitations, the *Salmonella*/microsome assay is of great value.

One class of chemicals which is not detected, or is only weakly detected as mutagenic, using the standard agar overlay procedure, is the alkyl halides. During the last few years, a number of alkyl halides have been found to be carcinogenic in animals and humans. Previously, many of these chemicals, for example, vinyl chloride, were thought to be innocuous. We have found that a simple modification of the standard *Salmonella*/microsome test can be effectively used to detect carcinogenic alkyl halides as mutagens by containing the volatile vapors (Simmon *et al.*, 1977) (see Fig. 12.1).

Results and Discussion

Table 12.1 presents the results of assays with the alkyl halide epichlorohydrin ($CH_2-CH-CH_2Cl$) and indicates the $\diagdown O \diagup$ sensitivity of the desiccator assay procedure.

These results were obtained by incubation of the plates seeded with *S. typhimurium* TA100 in the top agar for 7 hours

Table 12.1. *Sensitivity of the Desiccator Assay*

Epichlorohydrin (nl/desiccator)	ppb	Average *Salmonella typhimurium* TA100 revertants per plate
0	0	112
25	2.8	146
50	5.6	172
100	11.1	207
250	27.8	263
1000	111	538

DESICCATOR ASSAY

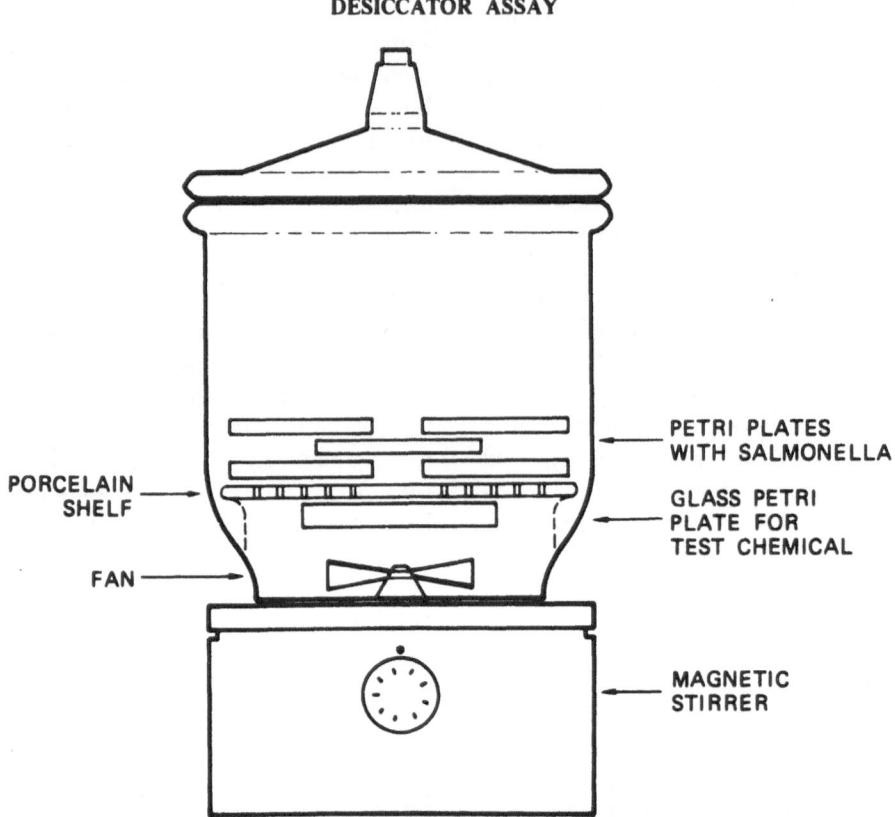

PETRI PLATES
WITH SALMONELLA

GLASS PETRI
PLATE FOR
TEST CHEMICAL

PORCELAIN
SHELF

FAN

MAGNETIC
STIRRER

Figure 12.1.   A diagram of the desiccator assay. Petri plates seeded with *Salmonella* and with their lids removed, were placed right side up in a 9-liter desiccator. The test chemical (liquid) was added to the glass Petri plate beneath the porcelain shelf. For gaseous chemicals, a known volume of gas was drawn into a partially evacuated desiccator, after which the desiccator was equilibrated to atmospheric pressure with sterile air. The fan (and magnetic stirrer) provide dispersion of vapors.

in the presence of various concentrations of epichlorohydrin (Simmon, 1977). Bridges (1978) has recently reported that this sensitivity may be increased if the bacteria are spread on the surface of the agar. The results presented are the average of five plates per dose. The beginning of the dose-response was approximately 3 ppb, which is exquisitely sensitive. Epichlorohydrin is the most mutagenic compound we have tested in the desiccator assay. It should be noted that the boiling point of epichlorohydrin is 117° C. We have found the desiccator procedure to be more sensitive than the standard agar incorporation for most alkyl halides, even when the boiling point is in excess of 175° C (e.g., benzyl chloride, b.p. 179° C). Thus, the desiccator technique is not to be used just for low molecular

weight, volatile compounds. We believe that the decreased sensitivity of the standard agar incorporation is due to the hydrophobic nature of these compounds rather than their volatility. These water-insoluble, or slightly soluble, compounds are probably readily lost from the thin agar overlay when incubated at 37° C.

We have previously reported that methylene chloride is mutagenic when assayed in desiccators, but is not mutagenic in the standard assay (Simmon *et al.*, 1977). Table 12.2 presents results of desiccator assays with methylene chloride.

Methylene chloride is considerably less mutagenic than epichlorohydrin, but is much more mutagenic than vinyl chloride. The minimum detectable concentration of vinyl chloride is about 3–5% (30–50 ppt)

Table 12.2.   *Reversion of* Salmonella typhimurium *TA100 by Methylene Chloride*[a]

Methylene chloride (μl/ desiccator)	ppm	Average *Salmonella typhimurium* TA100 revertants per plate
0	0	146
50	5.6	218
100	11.1	307
200	22.2	423
400	44.4	681
600	66.6	994
800	88.9	1352

[a]Assays were conducted in desiccators at 37°C for 8 hours.

(Malaveille *et al.*, 1975; Rannug *et al.*, 1974). Methylene chloride is a widely used industrial solvent and is present in the household in paint strippers and aerosol cans. Methylene chloride has replaced vinyl chloride and fluorocarbons in aerosol cans and has replaced trichloroethylene as a decaffeinating agent. Interestingly, methylene chloride increases reverse mutation in TA100 and TA98, but is only very weakly mutagenic to TA1535 (Simmon, unpublished data).

We have observed another differential mutagenic effect with n-propyl iodide and i-propyl iodide, both of which are mutagenic in *S. typhimurium* strains TA1535 and TA100 as well as in *E. coli* WP2 (uvrA) (Fig. 12.2). i-Propyl iodide was equally mutagenic to both TA1535 and TA100 (left panel) and n-propyl iodide was equally mutagenic to both these strains (right panel). However, i-propyl iodide was much more mutagenic than n-propyl iodide. Chemically, the secondary halide would be expected to be a stronger alkylating agent

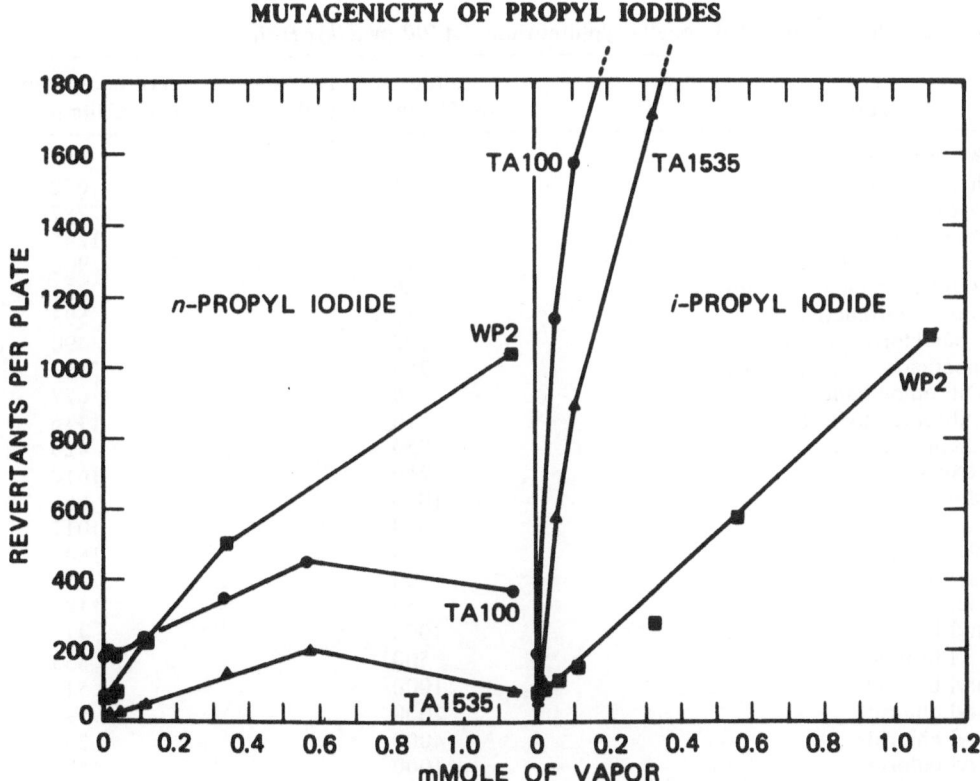

Figure 12.2.   The mutagenic activity of n-propyl iodide and i-propyl iodide with *S. typhimurium* strains TA1535 and TA100 and with *E. coli* WP2 (uvrA). Each point represents the average of three plates. Plates were exposed to the test chemical for 8 hours at 37°C, then removed from the desiccator and incubated for 2 days before counting the revertant colonies.

than the primary halide, and therefore a more potent mutagen. While this expected effect occurred in both *Salmonella* strains, the two compounds were, unaccountably, equally mutagenic to *E. coli* WP2.

The mutagenic activities of a number of alkyl halides are presented in Table 12.3. In general, there is a good correlation between mutagenic activity and chemical reactivity. The bromide and iodide analogues tend to be more mutagenic than the chlorides. For example, methylene bromide is considerably more mutagenic than methylene chloride (Table 12.2), methyl bromide is more mutagenic than methyl chloride, and 2-iodopropane is more mutagenic than 2-chloropropane. Of the four trihalomethanes tested, only chloroform (data not presented) was not mutagenic. The trihalomethanes are a by-product of water chlorination.

The desiccator procedure has also been quite useful in testing volatile anesthetics, many of which are halogenated. In collaboration with investigators from Stanford University School of Medicine, we have tested a number of anesthetics. These include halothane, methoxyfluorene, isofluorene, enfluorane, fluoroxene, trichloroethylene, nitrous oxide, divinyl ether, and cyclopropane (Baden *et al.*, 1976, 1977, 1979). Three of these, trichloroethylene, divinyl ether, and fluoroxene, were found to be mutagenic. None of these three is currently used as an anesthetic in the United States.

The desiccator technique is applicable to chemicals other than alkyl halides—lactones, sultones, and epoxides. The results with ethylene oxide and propylene oxide are presented in Tables 12.4 and 12.5, respectively. Ethylene oxide is a chemosterilant which has found wide acceptance, particularly in hospitals. At less than toxic

Table 12.3. *Reversion of* Salmonella typhimurium *TA100 by Alkyl Halides*[a]

Test chemical	(nl/desiccator) (or % atmosphere)	Average number of revertants per plate
Methyl bromide	(.05)	518
Methyl chloride	(5)	642
Methyl iodide	50	287
Dibromomethane (methylene bromide)	3	1245
Bromochloromethane	10	906
Tribromomethane (bromoform)	100	431
Dibromochloromethane	10	593
Bromodichloromethane	100	490
Bromoethane	750	7260
1,2-Dibromoethane	20	677
1,1-Dibromoethylene	10	1336
1,2-Dichloroethane	750	426
Iodoethane	250	1610
2-Chloropropane	1000	656
1-Chloropropene	500	1017
3-Chloropropene	500	1640
1-Iodopropane	500	450
2-Iodopropane	50	1140
i-Butyl bromide	1000	412
s-Butyl bromide	500	835
t-Butyl bromide	1000	513
n-Butyl chloride	2000	594
s-Butyl chloride	4000	232
t-Butyl chloride	1000	576

[a]Assays were conducted in sealed desiccators at 37°C for 8 hours. Values are the average of three plates and are taken from the linear portion of a dose response curve.

Table 12.4. *Reversion of* Salmonella typhimurium *TA1535 and TA100 by Ethylene Oxide*[a]

Ethylene oxide (cc injected/desiccator)	Average revertants per plate	
	TA1535	TA100
0	17	163
0.1	16	171
1	27	234
5	119	405
10	312	778

[a]Assays were conducted in desiccators (sealed) at 37°C for $7^{1}/_{2}$ hours. Values are the average of three plates.

concentrations, ethylene oxide is a potent mutagen. The concentrations given are maxima; it is possible that the sample of ethylene oxide we tested contained an inert carrier gas. This sample was judged to be free of alkyl halide(s) by gas chromatography/mass spectrometry.

In summary, the *Salmonella*/microsome assay has been extremely useful for identifying mutagens and potential carcinogens. The desiccator technique increases the sensitivity of the assay and extends the range of carcinogens that are detected as mutagens in *Salmonella*.

Table 12.5. *Reversion of* Salmonella typhimurium *TA1535 and TA100 by Propylene Oxide*[a]

Propylene oxide (ml added/desiccator)	Average revertants per plate	
	TA1535	TA100
0	28	136
0.5	53	187
1	68	213
2.5	147	302
5	289	471
7.5	422	595
10	508	744
25	1115	1206

[a]Assays were conducted in desiccators (sealed) containing liquid volumes of propylene oxide. Incubation was at 37° C for 8 hours. Values are the average of three plates.

References

Ames, B.N. (1971): The detection of chemical mutagens with enteric bacteria. In: A. Hollaender (ed.), Chemical Mutagens: Principles and Methods for Their Detection, Vol. 1. New York: Plenum Press, pp. 267–282.

Ames, B.N., Durston, W.E., Yamasaki, E., Lee, F.D. (1973): Carcinogens are mutagens: A simple test system combining liver homogenates for activation and bacteria for detection. Proc. Natl. Acad. Sci. USA *72* 2423–2427.

Ames, B.N., Kammen, H.O., Yamasaki, E. (1975a): Hair dyes are mutagenic: Identification of a variety of mutagenic ingredients. Proc. Natl. Acad. Sci. USA *72*:2423–2427.

Ames, B.N., McCann, J., Yamasaki, E. (1975b): Methods for detecting carcinogens and mutagens with the *Salmonella*/mammalian microsome mutagenicity test. Mutat. Res. *31*:347–364.

Baden, J.M., Brinkenhoff, M., Wharton, R.S., Hitt, B., Simmon, V.F., Mazze, R.I. (1976): Mutagenicity of volatile anesthetics: Halothane. Anesthesiology *45*:311–318.

Baden, J.M., Kelley, M.J., Wharton, R.S., Hitt, B.A., Simmon, V.F., Mazze, R.I. (1977): Mutagenicity of halogenated ether anesthetics. Anesthesiology *46*:346–350.

Baden, J., Kelley, M.J., Mazze, R.I., Simmon, V.F. (1979): Mutagenicity of inhalation anesthetics; nitrous oxide, cyclopropane, trichloroethylene and divinyl ether. Br. J. Anaesth. *51*:417–421.

Bartsch, H., Malaveille, C., Montesano, R., Tomatis, L. (1975): Tissue-mediated mutagenicity of vinylidene chloride and 2-chlorobutadiene in *Salmonella typhimurium*. Nature (Lond.) *255*: 641–643.

Blum, A., Ames, B.N. (1977): Flame-retardent additives as possible cancer hazards. The main flame retardant in children's pajamas is a mutagen and should not be used. Science *195*:17–23.

Bridges, B.A. (1978): Detection of volatile liquid mutagens with bacteria: Experiments with dichlorvos and epichlorohydrin. Mutat. Res. *54*:367–371.

Dunkel, V.C., Simmon, V.F. (1980): Mutagenic activity of chemicals previously tested for carcinogenicity in the National Cancer Institute Bioassay Program. In: R. Montesano, H. Bartsch, L. Tomatis (eds.), Molecular and Cellular Aspects of Carcinogen Screening

126   Vincent F. Simmon

Tests. Lyon: IARC Scientific Publications No. 27.

Hiatt, H.H., Watson, J.D., Winsten, J.A. (eds.) (1977): Origins of Human Cancer, Cold Spring Harbor Laboratory, Cold Spring Harbor, New York.

Huisingh, J., Bradow, R., Fungers, R., Claxton, L., Zweidinger, R., Tejada, S., Bumgarner, J., Duffield, F., Waters, M., Simmon, V.F., Hard, C., Rodriguez, C., Snow, L. (1979): Application of bioassay to the characterization of diesel particle emissions. In: M.D. Waters, et al. (eds.), Application of Short-Term Bioassays in the Fractionation and Analysis of Complex Environmental Mixtures, Environmental Science Research Series, Vol. 15, pp. 381–418.

Legator, M.S., Connor, T.H., Stoeckel, M. (1975): Detection of mutagenic activity of metronidazole and niridazole in body fluids of humans and mice. Science 188:118–119.

Lehr, R.E., Schaefer-Ridder, M., Jerina, D.M. (1977): Synthesis and properties of the vicinal transdihydrodials of anthracene, phenanthrene and benzo(a)anthracene. J. Org. Chem. 42:744–746.

Loper, J.C., Lang, D.R., Smith, C.C. (1978): Mutagenicity of complex mixtures from drinking water. In: R.L. Jolley, H. Gorchev, and D.H. Hamilton, Jr. (eds.), Water Chlorination: Environmental Impact and Health Effects, Vol. 2. Ann Arbor, Michigan, Ann Arbor: Science Publishers, pp. 433–450.

Malaveille, C., Bartsch, H., Barbin, A., Camus, A.M., Montesano, R., Croisy, A., and Jacquignon, P. (1975): Mutagenicity of vinyl chloride, chloroethylene oxide, chloroacetaldehyde and chloroethanol. Biochem. Biophys. Res. Commun. 63:363–370.

McCann, J., Ames, B.N. (1976): Detection of carcinogens as mutagens in the Salmonella/microsome test: Assay of 300 chemicals. Discussion. Proc. Natl. Acad. Sci. USA 73:950–954.

McCann, J., Choi, E., Yamasaki, E., Ames, BN. (1975a): Detection of carcinogens as mutagens in the Salmonella/microsome test: Assay of 300 chemicals. Proc. Natl. Acad. Sci. USA 72:5135–5139.

McCann, J., Spingarn, N.E., Kobori, J., Ames, D.N. (1975b): Detection of carcinogens as mutagens: Bacterial tester strains with R factor plasmids. Proc. Natl. Acad. Sci. USA 72: 979–983.

Prival, M.J., McCoy, E., Gutter, B., Rosenkranz, H.S. (1977): Tris (2,3-dibromopropyl) phosphate: mutagenicity of a widely used flame retardant. Science 195: 76–78.

Rannug, U., Johansson, A., Ramel, C., and Wachtmeister, C.A. (1974): Mutagenicity of vinyl chloride after metabolic activation. Ambio 3:194–197.

Simmon, V.F. (1977): Structural correlations of carcinogenic and mutagenic alkyl halides. In: I.M. Asher and C. Zervos (eds.), Structural Correlations of Carcinogenesis and Mutagenesis: A Guide to Testing Priorities. Proceedings of Second FDA Office of Science Summer Symposium, pp.163–171, Office of Science, Food and Drug Administration.

Simmon, V.F. (1979): In vitro mutagenicity assays of chemical carcinogens and related compounds with Salmonella typhimurium. J. Natl. Cancer Inst. 62:893–899.

Simmon, V.F., Kauhanen, K., Tardiff, R.G. (1977): Mutagenic activities of chemicals identified in drinking water. In: D. Scott, B.A. Bridges, and F.H. Sobels (eds.), Progress in Genetic Toxicology, Amsterdam: Elsevier/North-Holland Biomedical Press, pp. 249–258.

Wood, A.W., Levin, W., Thomas, P.E., Ryan, D., Karle, J.M., Yagi, H., Jerina, M.D., Conney, A.R. (1978): Metabolic activation of dibenzo (a,h) anthracene and its dihydrodiols to bacterial mutagens. Cancer Res. 39:1967–1973.

13

Determination of Genotoxic Activity Using DNA Polymerase-Deficient and -Proficient E. coli

Zev Leifer, Julie Hyman, and Herbert S. Rosenkranz

Microbial assays are gaining acceptance as valuable procedures for the recognition of environmental chemicals possessing the potential for inducing cancer. Recent studies have shown that assays based upon the preferential inhibition of DNA repair-deficient bacteria, both as adjuncts to other assays (such as the *Salmonella* mutagenicity procedure) as well as in their own right, have a high probability of detecting carcinogens (Poirier and deSerres, 1979; Rosenkranz and Poirier, 1979) as DNA-modifying agents. One of the assays which has undergone extensive evaluation and validation is the one based upon the preferential inhibition of DNA polymerase I-deficient *E. coli* (Slater *et al.*, 1971; Rosenkranz and Leifer, 1980). The present report is concerned with a description of the various modes in which this assay can be run, with the advantages and limitations of each, and includes a comparison of responses in this assay and the potential for inducing cancers in animals.

The Spot-Test (Disc-Diffusion Assay)

As originally devised, the DNA repair test made use of a pair of isogenic *E. coli* strains which differed in that one member of the pair was deficient in DNA polymerase I. Exposure of such cells to direct-acting agents which possessed the ability to modify the cellular DNA resulted in the preferential killing or the inhibition of the DNA polymerase-deficient strain (*E. coli* pol $A_1{}^-$) (Fig. 13.1). This was presumably due to the fact that DNA polymerase I plays an essential role in the DNA repair process. On the other hand, agents which interfered with structures other than the DNA affected the two strains to the same extent. Based upon these observations, a simple disc-diffusion assay was devised which enabled the facile and rapid screening of large numbers of chemicals (Slater *et al.*, 1971). The assay could be run in either of two modifications: (a) the bacteria were spread directly onto the surface of the agar

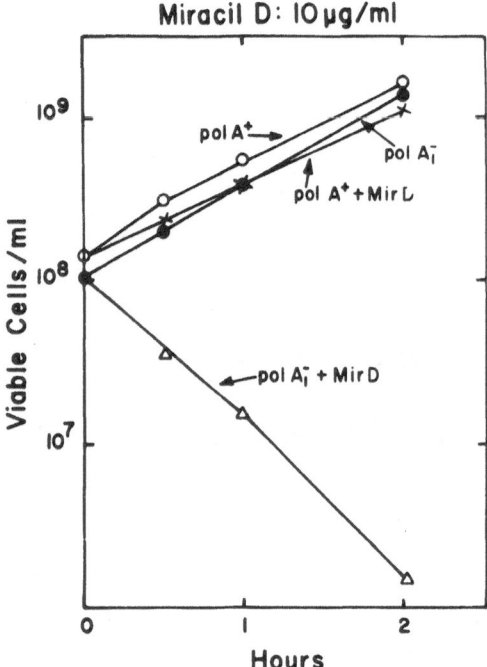

Figure 13.1. Preferential killing of *E. coli* pol A_1^- by Miracil D. Bacteria were brought to the exponential growth phase at which time portions of the cultures were supplemented with Miracil D (final concentration: 10 μg/ml). The cultures were incubated with aeration at 37° C and at intervals portions of the cultures were withdrawn and processed for the determination of the number of viable cells.

(1.5%) which then also received discs impregnated with the test chemicals (Fig. 13.2), or (b) the bacteria were added to 2 ml of molten top agar (0.75%) which was then poured onto agar plates and, upon solidification of the top agar, discs containing the test chemicals were placed on the surface of the plates (Fig. 13.3). From an esthetic point of view, it is evident that the procedure in which the bacteria are in the agar overlay gives a "nicer" picture. The edges of the zone's growth inhibition are smoother and therefore are more readily measurable (Fig. 13.3). Actually there is another reason that the latter procedure is preferable. Because the procedure depends upon diffusion in agar, which in turn depends upon the agar concentration, it is obvious that the lower agar content of the top

agar in the second procedure will result in better diffusion.

It was shown that this procedure is compatible with a microsomal activation mixture of the type used in the *Salmonella* mutagenicity assay (Slater *et al.*, 1971). Originally the enzyme and cofactors were placed in a central well; however, more recently experiments have indicated that better results were obtained when the microsomal activation mixture was incorporated together with the bacteria into the agar overlay (unpublished results).

Using this procedure, a number of chemicals, both mutagens and carcinogens, were shown to possess DNA-modifying activity (Table 13.1). The details of the procedure are described elsewhere (Slater *et al.*, 1971; Rosenkranz and Leifer, 1980).

Although many chemicals have been tested by the disc-diffusion procedure, its chief disadvantage is related to the fact that it depends upon diffusion. Chemicals which, because of poor solubility or large size, do not diffuse may inhibit neither strain and therefore may give a *"no test"* result. It should be noted that such a finding cannot be interpreted as a negative result; in the disc-diffusion method, a negative result is defined as one in which a chemical inhibits the two tester strains to the same extent. Because many environmentally important chemicals yield such "no test" results, procedures to circumvent this limitation were sought; some of these are discussed below.

The Plate Incorporation Assay

A plate incorporation procedure has been shown to possess great versatility and discriminatory power in the *Salmonella* mutagenicity assay developed by Ames and his associates (1975). It was thought, therefore, that incorporation of bacteria *and* of test chemicals into an agar overlay followed by incubation and the determination of the number of viable cells could form the basis of a plate incorporation assay using DNA

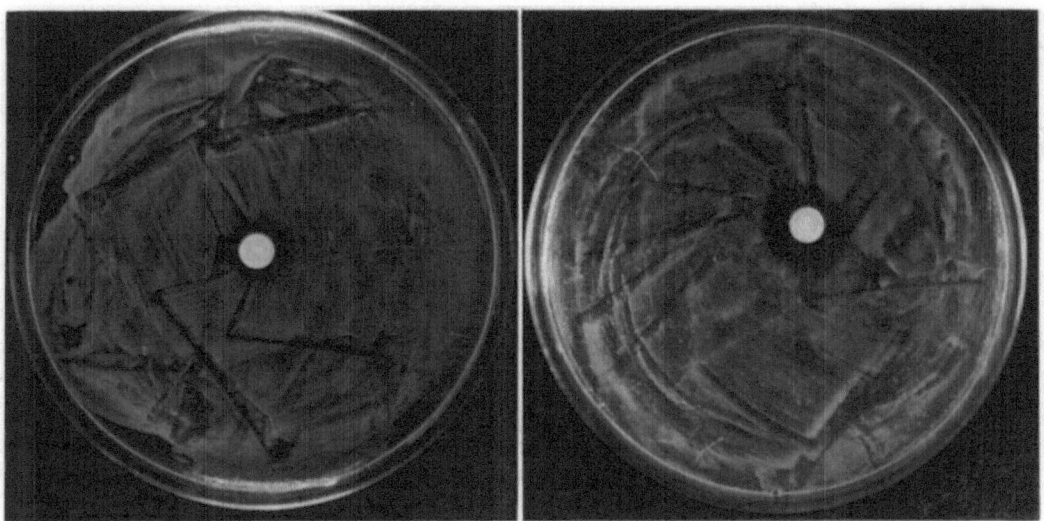

Figure 13.2.   Effect of N-hydroxyurethan on the growth of *E. coli* pol A$^+$ (left) and pol A$_1^-$ (right). The bacteria were spread on the surface of the agar plates.

repair-proficient and -deficient microorganisms. Such a method would have the added advantage of permitting the expression of results as fractional survival, a biologically and statistically meaningful value (as opposed to diameters of zones of growth inhibition). A procedure to carry out such an assay is described below.

Cultures of *E. coli* pol A$^+$ and its DNA polymerase-deficient derivative pol A$_1^-$ are grown to the middle of the exponential growth phase in medium HA + T where upon they are diluted in fresh medium HA + T to approximately 2,000–10,000 cells per ml. Replicate 0.1 ml amounts of the cultures together with dilutions of the test chemicals are added to tubes containing 2 ml of molten agar (0.75% HA + T

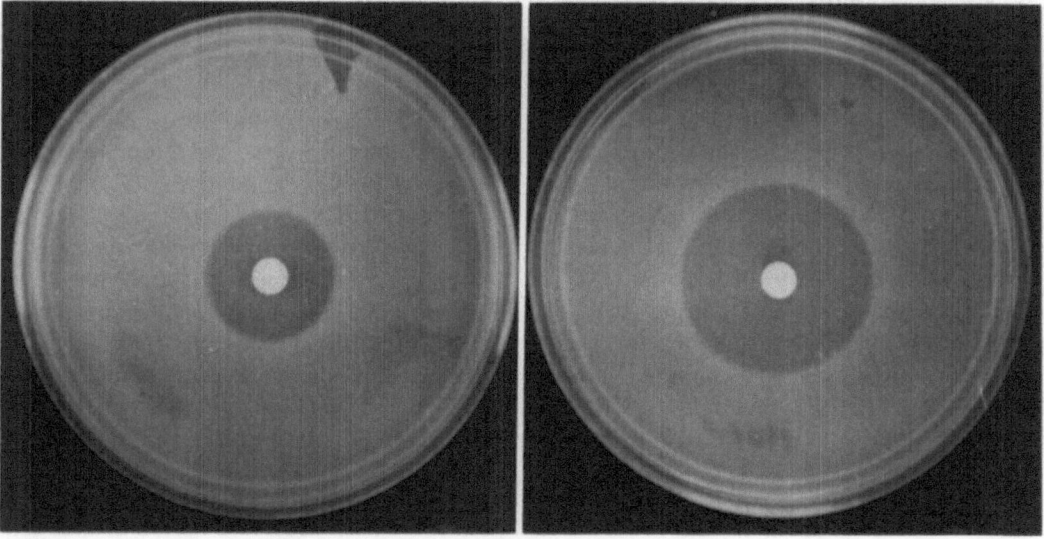

Figure 13.3.   Effect of N-hydroxyurethan on the growth of *E. coli* pol A$^+$ (left) and pol A$_1^-$ (right). The bacteria were incorporated into the agar overlay.

Table 13.3. *Effects of Various Agents on the Growth of a DNA Polymerase-Deficient Strain (Standard Assay)*

			Size of zone of inhibition (mm)	
Group	Agent	Amount	Pol A$^+$	Pol A$_1^-$
I	Penicillin	3 units	9	8
	Erythromycin	15 μg	9	9
	Cycloserine	50 μg	62	62
	Chloramphenicol	30 μg	30	30
	Streptomycin	10 μg	26	26
	Kanamycin	30 μg	18	18
II	Methyl methanesulfonate	10 μl	44	60
	Ethyl methanesulfonate	10 μl	0	20
	N-Methyl-N-Nitrosourea	0.5 μmoles	45	85
	N-Ethyl-N-Nitrosourea	0.5 μmoles	0	13
	N-Methyl-N-Nitrosourethan	0.1 μmoles	2	46
	N-Ethyl-N-Nitrosourethan	0.5 μmoles	0	16
	Formaldehyde	10 μl	59	62
	N-Hydroxyurethan	20 μmoles	12	21
	Nitrosofluorene	0.5 μmoles	0	15
	N-Hydroxylaminofluorene	0.5 μmoles	0	12
	1,2-Dimethylhydrazine	250 μg	0	12
	NFTF[a]	60 μg	25	38
	1-phenyl-3,3-dimethyl-triazine	250 μg	24	47
	Natulan	250 μg	16	22
	Chloroquine	0.2 μmoles	9	15
	Acridine Orange	250 μg	7	9
	Miracil D	1 μmole	7	19
	Auramine O	250 μg	9	14
	p-Rosaniline	250 μg	6	10

[a]NFTF, N-[4-(5-nitro-2-furyl)thiazolyl] formamide.

agar, 43° C). Following mixing, the contents of the tubes are poured onto the surface of agar plates (1.5% HA+T agar). After solidification of the top agar, the plates are incubated at 37° in the dark for two days and colony-forming units are enumerated. Results are expressed as per cent survival relative to that of the solvent controls. An expression of the preferential killing of the pol A$_1^-$ strain is obtained by dividing the survival of the pol A$_1^-$ by the survival of the pol A$^+$ strain, i.e., the survival index. A survival index of 1.00 (0.96–1.00) is taken as evidence of a negative result; values in the range of 0.86–0.95 are taken as borderline or weakly positive, while a value of less than 0.85 indicates a preferential killing of the pol A$_1^-$ strain (i.e., a positive result).

The assay was shown to be quite effective in demonstrating the DNA-modifying ability of methyl methanesulfonate and ethyl methanesulfonate (Table 13.2). Moreover, the concentration dependence of the survival index was also demonstrated (Table 13.2). Similarly, 3-bromo-1-propanol, a chemical related to the flame retardant "tris" tris(2,3-dibromopropyl) phosphate (Carr and Rosenkranz, 1978; Prival *et al.*, 1976) was also shown to be a DNA-modifying agent (Table 13.2). In view of the dose dependence of the survival indices, it seemed that a measure of relative genotoxic activity could be obtained by determining the concentration of chemical giving a survival index of 0.5. The genotoxic activities of a series of halogenated propanols are recorded in Table 13.3.

Table 13.2. *Genotoxic Activity Determined by the Plate Incorporation Procedure*

Chemical	Amount per plate	Pol A$^+$		Pol A$_1^-$		Survival index
		Survivors per plate	Percent of control	Survivors per plate	Percent of control	
None	0	458	100	523	100	1.00
Methyl methanesulfonate	1 μl	466	102	362	69	0.68
	5 μl	463	101	<0.5	<0.1	<0.001
Ethyl methanesulfonate	30 μl	479	105	338	65	0.62
	50 μl	471	103	215	41	0.40
3-Bromo-1-propanol	0	381	100	217	100	1.00
	64 μmoles	358	93	161	74	0.79
	128 μmoles	353	93	153	71	0.76
	640 μmoles	42	11	8	4	0.34

The assay, while it performs well with direct-acting agents capable of modifying cellular DNA chemically, does not seem to be adequate for other DNA-modifying agents such as proflavin, 9-aminoacridine, and chloroquine, which are known intercalating agents (Table 13.4). These chemicals, however, do give a positive test in the disc-diffusion and liquid assay modifications (Rosenkranz and Leifer, 1980, and unpublished results) of the procedure. It is conceivable that in the plate incorporation test, owing to the continuous contact and the possibility of DNA repair through pathways other than DNA polymerase, the DNA lesions induced by intercalating agents are repaired.

Thus far, the plate incorporation assay has not performed well when coupled to metabolic activation mixtures (Table 13.5) similar to the one which has been shown to perform well under similar conditions in the

Table 13.3. *Relative Genotoxic and Mutagenic Activities of a Series of Halogenated Propanols[a]*

Chemicals	Genotoxicity, agar plate		Genotoxicity, liquid suspension		Mutagenicity, *Salmonella* assay	
	Specific activity (μmoles per plate)	Relative activity	Specific activity (M)	Relative activity	Specific activity (revertants per nmole)	Relative activity
1-Bromo-2-propanol	314	2.0	1.0×10^{-1}	3.4	0.096	20.5
3-Bromo-1-propanol	419	1.5	1.1×10^{-2}	32	0.112	23.9
1-Chloro-2-propanol	638	1.0	3.5×10^{-1}	1.0	0.005	1.0
3-Chloro-1-propanol	357	1.8	1.4×10^{-1}	2.6	0.000	0.0
2,3-Dibromo-1-propanol	148	4.3	5.0×10^{-2}	7.1	0.046	9.8
2,3-Dichloro-1-propanol	7	90	6.2×10^{-5}	5710	0.347	73.8
Tris (2,3-dibromopropyl) phosphate			2.1×10^{-3}	169		

[a]Specific genotoxic activities are defined as the concentration required to give a survival index of 0.5. Relative genotoxic activities were calculated by setting the value of 1-chloro-2-propanol as 1.0. The mutagenic activity of the halogenated propanols for *Salmonella typhimurium* have been reported previously (Carr and Rosenkranz, 1978). Relative mutagenic activities are expressed in relation to the activity of 1-chloro-2-propanol.

Table 13.4.   *Assay of Intercalating Agents by the Plate Incorporation Method*

Compound	Amount per plate	Pol A^+		Pol A_1^-		Survival index
		Survivors per plate	Percent of control	Survivors per plate	Percent of control	
None	0	472	100	288	100	1.00
9-Aminoacridine	100 μg	434	92	239	83	0.90
Proflavin	100 μg	464	98	298	103	1.05
Chloroquine	100 μg	453	96	286	99	1.03

Salmonella mutagenicity assay (Ames *et al.*, 1975).

It would thus seem that the plate incorporation test has a number of shortcomings which at this time do not make it suitable for routine use. On the other hand, for many direct-acting agents, it may well be useful for determining specific genotoxic activities which may be of value in comparative studies and in structure-activity investigations. Certainly, in the context of the halogenated propanols, the plate incorporation procedure has demonstrated the DNA-modifying activity of 3-chloro-1-propanol which had been demonstrated to be nonmutagenic in the *Salmonella* assay and to be nonresponsive in the disc-diffusion procedure as well. This finding in itself is quite important as 3-chloro-1-propanol had been considered as an alternate monomer for new flame retardants devoid of genetic and genotoxic activities (Carr and Rosenkranz, 1978).

Liquid Assay

Because of some of the limitations of the agar incorporation procedure and in light of the fact that the original discovery of the preferential killing of DNA polymerase-deficient bacteria by DNA-modifying agents was done by determination of the preferential killing rate of *E. coli* pol A_1^- (Fig. 13.1) in suspension, it seemed that a simplification of the suspension procedure, whereby the effect of a test agent on bacterial viability after a single exposure interval could be measured, would be useful. It was

demonstrated, moreover, that such a procedure was compatible with metabolic activation by microsomal enzymes (Table 13.6) (Rosenkranz *et al.*, 1976; Rosenkranz and Poirier, 1979). However, the original procedure was uneconomical because it required large quantities of microsomal preparations and cofactors. Attempts were therefore made to scale the procedure down as well as to optimize it. The current method is outlined below.

Overnight cultures of *E. coli* pol A^+ or pol A_1^- in medium HA + T (Slater *et al.*, 1971) are diluted in fresh medium HA + T to a bacterial density of approximately 1500 cells per ml and replicate 100-μl amounts of these cultures are distributed into tubes (or vials), which also receive 10-μl dilutions of the test chemicals (either in water or dimethylsulfoxide). The tubes are incubated in a 37° C water bath for 2 hours whereupon they are cooled to 23° C and each receives 2 ml of molten agar (0.75% HA + T agar at 43° C). They are mixed, their contents poured onto agar plates (25 ml, 1.5% HA + T agar), and, upon solidification of the top agar, these are incubated at 37° in the dark for two days and surviving bacteria enumerated. All determinations are made in replicate and each experiment includes solvent, positive (ethyl methanesulfonate), and negative (chloramphenicol) controls (Table 13.7). Results are expressed as percent survival compared to the solvent control as described for the plate incorporation assay. The survival index is as defined above: a value of 1.00 (0.96–1.0) indicates a lack of preferential killing of the pol A_1^- strain (i.e., a negative result); values in the range of

Table 13.5. *Assay of Compounds Requiring Metabolic Activation by the Plate Incorporation Method*[a]

Experiment	Compound	Amount per plate	Pol A+				Pol A$_1$$^-$			
			Survivors per plate (−activation)	Percent of control	Survivors per plate (+activation)	Percent of control	Survivors per plate (−activation)	Percent of control	Survivors per plate (+activation)	Percent of control
I	None	0	971	100	1008	100	650	100	805	100
	2-Anthramine	200 µg	962	99	1091	108	695	107	766	95
	Survival Index:							1.08		0.88
II	None	0	1063	100	1124	100	532	100	543	100
	2-Fluorenamine	250 µg	832	78	1037	92	420	79	614	113
	Survival Index:							1.01		1.23
	Nitrosomorpholine	500 µg	949	89	1029	92	567	107	505	93
	Survival Index:							1.20		1.01

[a]Microsomes were derived from the livers of rats induced with Aroclor 1254.

Table 13.6. *Comparison of Mutagenic, Carcinogenic, and DNA-Modifying Activity of Test Chemicals*

Chemicals	Carcinogenicity	Mutagenicity (*Salmonella*)	Pol A$_1^-$ assay[a]	
Aromatic Amines				
N-Acetoxy-N-2-Fluorenyl-acetamide	UC	+	+	
4-Aminoazobenzene	PC	+	−	
2-Aminobiphenyl	NC	−	−	
4-Aminobiphenyl	PC	+	+	(S9)
Aniline	NC	−	−	
1-Anthramine	NC	+	±	(S9)
2-Anthramine	PC	+	+	(S9)
Auramine O	UC	−	+	
Bis-(p-dimethylamino)di-phenylmethane	PC	−	+	
o-Chloraniline	NT	−	+	
p-Chloraniline	PC	−	+	
N,N-Dimethyl-4-amino-azobenzene	PC	−	−	
2,3-Dimethyl-4-amino-biphenyl	PC	+	+	(S9)
2-Fluorenamine	PC	+	+	(S9)
N-2-Fluorenylacetamide	PC	+	+	(S9)
N-4-Fluorenylacetamide	NC	+	−	
N-Hydroxy-N-2-fluorenylacet-amide	PC	+	+	
1-Hydroxy-2-fluorenylacet-amide	NC	−	−	
3-Hydroxy-2-fluorenylacet-amide	NT	−	−	
5-Hydroxy-2-fluorenylacet-amide	NT	+	+	(S9)
7-Hydroxy-2-fluorenylacet-amide	NC	+	+	(S9)
3-Methoxy-4-aminoazobenzene	PC	+	+	(S9)
2-Methyl-4-dimethylamino-azobenzene	PC	−	−	
3′-Methyl-4-dimethylamino-azobenzene	PC	−	−	
1-Napthylamine	NC	+	+	(S9)
2-Naphthylamine	PC	+	+	(S9)
1-Naphthylhydroxylamine	UC	+	+	
2-Naphthylhydroxylamine	UC	+	+	
α-Naphthylisothiocyanate	NC	−	−	
4-Nitrobiphenyl	PC	+	+	
2-Nitrofluorene	PC	+	+	
2-Nitronaphthalene	PC	+	+	
p-Rosaniline	PC	−	+	
o-Toluidine	PC	−	+	
p-Toluidine	PC	−	−	(S9)
4-o-Tolylazo-o-toluidine	PC	+	+	
Aromatic Polycyclic Hydrocarbons				
Anthracene	NC	−	−	
Benz(a)anthracene	PC	+	−	
Benzo(a)pyrene	PC	+	+	(S9)
Benzo(e)pyrene	PC	−	−	

Table 13.6. *Comparison of Mutagenic, Carcinogenic, and DNA-Modifying Activity of Test Chemicals* (Continued)

Chemicals	Carcinogenicity	Mutagenicity (*Salmonella*)	Pol A$_1^-$ assay[a]
7-Bromoethyl-12-methyl-benz(a)anthracene	UC	+	+
Chrysene	PC	−	−
4,5-Dihydrobenzo(a)pyrene-4,5-epoxide	UC	+	+
7,12-Dimethylbenz(a)anthracene	PC	+	± (S9)
3-Hydroxybenzo(a)pyrene	PC	+	+ (S9)
7-Hydroxymethyl-12-methyl-benz(a)anthracene	PC	−	+ (S9)
Phenanthrene	NC	−	−
Heterocyclic Compounds			
1′-Acetoxysafrole	UC	+	+
Acridine orange	PC	−	+
Aflatoxin B$_2$	PC	−	−
3-Amino-1,2,4-triazole	PC	−	−
7,9-Dimethylbenz(c)-acridine	PC	+	+ (S9)
4-Hydroxylaminoquinoline-N-oxide	UC	+	+
1′ Hydroxysafrole	PC	−	−
N-[4-(5-Nitro-2-furyl)-thiazolyl]formamide	PC	+	+
4-Nitroquinoline-N-oxide	PC	+	+
Safrole	PC	−	+
Alkylating Agents			
Benzyl chloride	UC	+	+
Bromobenzene	NT	−	±
Butane sultone	UC	+	+
ε-Caprolactone	NT	−	−
1,2,3,4-Diepoxybutane	UC	+	+
1,2-Epoxybutane	NC	+	+
Ethyl-p-toluenesulfonate	UC	+	+
Glycidaldehyde	UC	+	+
Glycidol	NC	+	+
Methylazoxymethanol acetate	PC	+	+
Methyl iodide	UC	+	+
Propane sultone	UC	+	+
β-Propiolactone	UC	+	+
Propyleneimine	UC	+	+
Uracil mustard	UC	+	+
Nitrosamines, Hydrazines, and Related Substances			
1,2-Dimethylhydrazine	PC	−	+
Diphenylnitrosamine	NC	−	−
Hydrazine sulfate	PC	+	+
N-Methyl-N′-Nitro-N-nitrosoguanidine	UC	+	+
Natulan	PC	−	+
N-Nitrosodiethyl-amine	PC	−	+ (S9)
N-Nitrosoethylurea	UC	+	+
1-Phenyl-3-dimethyl-triazene	PC	+	+

Table 13.6. *Comparison of Mutagenic, Carcinogenic, and DNA-Modifying Activity of Test Chemicals* (Continued)

Chemicals	Carcinogenicity	Mutagenicity (*Salmonella*)	Pol A₁⁻ assay[a]
Amides, Ureas, and Acylating Agents			
Acetamide	PC	−	−
Dimethylcarbamyl chloride	UC	−	+
Methyl carbamate	NC	−	−
Succinic anhydride	NT	−	−
Thioacetamide	PC	−	−
Thiourea	PC	−	−
Urethan	PC	−	−
N-Hydroxyurethan	PC	−	+
Antimetabolites			
5-Bromodeoxyuridine	NC	−	No Test
Ethionine	PC	−	No Test
5-Fluorodeoxyuridine	NC	−	−
5-Iododeoxyuridine	NC	−	No Test
Methotrexate	NC	−	No Test
Inorganics			
Beryllium sulfate		−	No Test
Hydroxylamine hydrochloride		−	+
Lead acetate		−	No Test
Titanocene dichloride		−	No Test
Promoters			
Phorbol		−	No Test
12-O-Tetradecanoyl-phorbol-13-acetate		−	No Test
1,8,9-Trihydroxy-anthracene		+	+

Abbreviations: NC, noncarcinogen; PC, procarcinogen; UC, ultimate carcinogen; NT, not tested; S9, indicates that DNA-modifying activity was detected in the presence of microsomal enzymes from rat liver.

[a] For the pol A₁⁻ assay, chemicals were first tested in the standard disc-diffusion assay and results recorded as positive (preferential inhibition of pol A₁⁻ strain) or negative (equal inhibition of both tester strains). Chemicals inhibiting neither strain ("no test" result) were then tested in the modified liquid suspension assay. Chemicals not showing preferential inhibition of the pol A₁⁻ strain were retested in the presence of S9 and cofactors. In the above table a "no test" result indicates that the test agent was tested only by the plate diffusion assay, and that in that assay neither strain was inhibited.

[b] Although, in our hands, dimethylcarbamyl chloride was not mutagenic in the standard *Salmonella* assay, a mutagenic response was obtained when it was tested under conditions which prevented evaporation (Rosenkranz *et al.*, 1980).

0.86–0.95 are taken as borderline or weakly positive results, and values below 0.85 are taken as evidence of a preferential killing of the pol A₁⁻ strain (i.e., a positive result).

The assay is compatible with metabolic activation by microsomal enzymes (Table 13.8). We have found that addition of 1 µl of a mixture containing, per ml: 12 mg S9 (microsomes) from Aroclor-induced rat livers, 2.33 mg NADP, 1.07 mg D-glucose-6-phosphate, 0.06 ml 0.1 M MgCl₂, 0.25 ml

Table 13.7. *Preferential Inhibition of DNA Polymerase-Deficient* E. coli *by DNA-Modifying Agents*

Additions	Amount per tube (mg)	Pol A⁺		Pol A₁⁻		Survival index
		Survivors per plate	Percent of control	Survivors per plate	Percent of control	
1-Bromo-2-	0	271	100.0	226	100.0	1.00
propanol	0.08	277	102.2	229	101.3	1.00
	0.16	268	98.9	221	97.8	0.98
	0.31	279	103.0	215	95.1	0.96
	0.62	262	96.7	180	79.7	0.82
	0.93	255	94.1	156	69.0	0.73
	1.24	252	93.0	123	54.4	0.58
	1.55	243	89.7	89	39.4	0.43
	1.86	230	84.9	56	24.8	0.29
Ethyl methane-	0	160	100.0	151	100.0	1.00
sulfonate	0.28	158	98.8	124	82.1	0.83
	0.56	155	96.9	101	66.9	0.69
	0.84	147	91.9	65	43.1	0.47
	1.12	144	90.8	38	25.2	0.27
	1.40	141	88.1	15	9.9	0.11
Chlorampheni-	0	160	100.0	151	100.0	1.00
col	0.005	163	101.9	154	102.0	1.00
	0.010	153	95.6	157	104.0	1.09
	0.020	143	89.4	152	100.7	1.13

0.1 M KCl and 0.07 ml phosphate buffer (1.0 M, pH 7.4) to the tube containing the bacteria (100 μl) and the test chemical (10 μl) was optimal. When the assay included metabolic activation, the incubation period was shortened to 1 hour at 37° (water bath) to minimize the growth-promoting effects of the S9 preparation.

For the comparison of genotoxic potency, we have found it convenient to titrate the indicator strains with a graded series of test chemicals and to express the survival index as a function of chemical concentration (Table 13.7). The specific genotoxic activity then is the concentration of test chemical giving a survival index of 0.50 which is determined from the linear portion of the dose-response curve.

The specific genotoxic activities of a series of halopropanols, related to the flame

Table 13.8. *Microsome-mediated Activation of Chemicals to DNA-Modifying Agents*[a]

Additions	Amount per tube (μg)	Microsomes	Percent survivors		Survival index
			Pol A⁺	Pol A₁⁻	
None	0	−	100.0	100.0	1.00
None	0	+	107.7	112.2	1.04
Ethyl methanesulfonate	1400	−	36.9	<0.21	<0.006
Chloramphenicol	20	−	100.0	103.8	1.04
2-Aminofluorene	100	−	87.3	91.7	1.05
2-Aminofluorene	100	+	87.0	39.5	0.45
3-Methylcholanthrene	50	−	99.0	98.1	0.99
3-Methylcholanthrene	50	+	67.4	29.5	0.44
7,12-Dimethylbenz(a)anthracene	50	−	100.0	94.7	0.95
7,12-Dimethylbenz(a)anthracene	50	+	51.7	23.0	0.45
Tris (2,3-dibromopropyl)phosphate	10	−	100.0	100.0	1.00
Tris (2,3-dibromopropyl)phosphate	10	+	79.3	23.2	0.29

[a]Microsomes were derived from the livers of rats induced with Aroclor 1254.

retardant tris (2,3-dibromopropyl) phosphate, are given in Table 13.3. Such calculations will ultimately permit a comparison of DNA-modifying potency with mutagenic and carcinogenic potencies (see, for example, Table 13.3).

In routine testing, chemicals showing no toxicity for either strain should be tested to the limit of their solubility. In addition to the features enumerated above, this assay includes the possibility of removing the test chemical either by filtration or by centrifugation following the initial period of exposure and thereby eliminating residual toxicity.

Genotoxic and Mutagenic Potencies

Interest has been expressed in assessing carcinogenic risk based on the mutagenic potency of test chemicals. Although this is an area of some controversy at this time, some quantitative correlations have been made (Ames, 1979; Meselson and Russell, 1977). In building up a data base to evaluate the potential significance of genotoxic potency, we took advantage of the data obtained with the halogenated propanols to compare the results of the two DNA modification assays described here to each other and to the mutagenicity assays of these same chemicals (Table 13.3). It was thus found that, while the plate incorporation assay required large quantities of chemicals, its discriminatory power was not as great as that of the liquid assay procedure, although the order of relative activities was essentially similar. Mutagenic activities followed essentially the order found with the DNA-modifying assays, with the exception that, while 3-chloro-1-propanol was consistently nonmutagenic for *Salmonella* (Carr and Rosenkranz, 1978), it displayed considerable genotoxic activity.

These findings, which indicate the usefulness of performing the DNA-modifying assays in tandem with mutagenicity determinations, also suggest that future studies based upon the differential killing of DNA repair-deficient microorganisms be done at varying doses so as to build up a data base that could eventually be used to express genotoxic activity, which could then be compared to mutagenic and carcinogenic potencies.

The Ability of the pol A_1^- Assay to Detect Carcinogens and Mutagens

As part of an effort to assess the effectiveness of the pol A_1^- assay to predict carcinogenicity, a group of 82 chemicals composed of known carcinogens and noncarcinogens (Table 13.9) was assayed in the *Salmonella* mutagenicity assay and the *E. coli* pol A_1^- test (Rosenkranz and Leifer, 1980; Rosenkranz and Poirier, 1979; Simmon, 1979). The results of these studies are summarized in Tables 13.6 and 13.10. For the carcinogens, they indicate that 82% of the chemicals (for which there is carcinogenicity and liquid suspension assay data) were positive in both the *Salmonella* and pol A_1^- assays, and 79% and 65% were positive only in the pol A_1^- and *Salmonella* assays, respectively. Eighteen percent of the carcinogens were inert in both assays. Obviously some carcinogens were detectable only by the pol A_1^- assay, e.g., acridine orange, auramine O, p-rosaniline, dimethyl-

Table 13.9.   *Classes of Compounds Tested for Mutagenicity, Carcinogenicity, and DNA-Modifying Activity*

Class	Number
Aromatic Amines	33
Aromatic Polycyclic	
Hydrocarbons	11
Heterocyclic Compounds	10
Alkylating Agents	13
Nitrosamines, Hydrazines and	
Related Compounds	8
Amides, Ureas, and	
Acylating Agents	7
Noncarcinogens	16
Ultimate Carcinogens	21
Procarcinogens	45

Table 13.10.   *Analysis of the Response of Carcinogens and Noncarcinogens in Microbial Assays*

	Noncarcinogens		Ultimate carcinogens		Procarcinogens		All carcinogens	
	N	Percent	N	Percent	N	Percent	N	Percent
1) *Salmonella* alone	7	44	19	90	24	53	43	65
2) Pol A alone	6	38	21	100	31	69	52	79
3) Both *Salmonella and* pol A	5	31	19	90	22	49	41	62
4) Either *Salmonella or* pol A	8	50	21	100	33	73	54	82
	16		21		45		66	

hydrazine, safrole, 1'-acetoxysafrole, etc., and some only by the *Salmonella* assay, e.g., 4-aminoazobenzene and benz(a)anthracene. It would thus appear that the pol A_1^- assay may be useful for detecting potential carcinogens. Moreover, because there are some chemicals which are uniquely detected by only one or the other of the assay systems, it would be advisable to run this assay as part of a battery of test systems.

Acknowledgments

These studies were supported by the National Cancer Institute (NO1-CP-33395 and NO1-CP-65855).

References

Ames, B.N. (1979): Identifying environmental chemicals causing mutations and cancer. Science *204*:587–593.

Ames, B.N., McCann, J., Yamasaki, E. (1975): Methods for detecting carcinogens and mutagens with the *Salmonella*/mammalian-microsome mutagenicity test. Mutat. Res. *31*: 347–364.

Carr, H.S., Rosenkranz, H.S. (1978): Mutagenicity of derivatives of the flame retardent tris (2,3-dibromopropyl) phosphate: halogenated propanols. Mutat. Res. *57*:381–384.

Meselson, M., Russell, K. (1977): Comparisons of carcinogenic and mutagenic potency. In: H.H. Hiatt, J.D. Watson and J.A. Winsten (eds.), Origins of Human Cancer, Book C,

Cold Spring Harbor Conferences on Cell Proliferation, Vol. 4. Cold Spring Harbor Laboratory, New York, pp. 1473–1481.

Poirier, L.A., de Serres, F.J. (1979): Initial National Cancer Institute studies on mutagenesis as a prescreen for chemical carcinogens: an appraisal. J. Natl. Cancer Inst. *62*:919–926.

Prival, M.J., McCoy, E.C., Gutter, B., Rosenkranz, H.S. (1976): Tris (2,3-dibromopropyl)phosphate: mutagenicity of a widely used flame retardent. Science *195*: 76–78.

Rosenkranz, H.S., Leifer, Z. (1980): Determining the DNA-modifying activity of chemicals using DNA-polymerase-deficient *Escherichia coli*. In: F.J. de Serres (ed.), Chemical Mutagens: Principles and Methods for Their Detection, Vol. 6. New York: Plenum Press, pp. 109–147.

Rosenkranz, H.S., Poirier, L.A. (1979): An evaluation of the mutagenicity and DNA modifying activity in microbial systems of carcinogens and noncarcinogens. J. Natl. Cancer Inst. *62*:873–892.

Rosenkranz, H.S., Gutter, B., Speck, W.T. (1976): Mutagenicity and DNA-modifying activity: a comparison of two microbial assays. Mutat. Res. *41*:61–70.

Rosenkranz, H.S., McCoy, E.C., Biuso, L., Speck, W.T. (1980): Short-term microbial assays in the assessment of carcinogenic risk. In: A.N. Rowan and C.J. Stratmann (eds.), The Use of Alternatives in Drug Research. London: Macmillan Press, pp. 103–145.

Simmon, V.F. (1979): *In vitro* mutagenicity assays of chemicals, carcinogens and related compounds with *Salmonella typhimurium*. J. Natl. Cancer Inst. *62*:893–899.

Slater, E.E., Anderson, M.D., Rosenkranz, H.S. (1971): Rapid detection of mutagens and carcinogens. Cancer Res. *31*:970–973.

14

The Nucleotide-Permeable Escherichia coli: *A Model System that Responds to DNA-Binding Carcinogens, Mutagens, and Antitumor Agents with DNA Excision Repair or with Inhibition of Replicative DNA Synthesis*

HEINZ WALTER THIELMANN

Principle of Method

Living *E. coli* wild type cells have been reported as capable of performing several types of DNA repair: some of them are relatively error proof (such as DNA excision repair), and others are error prone (for references, see Witkin, 1976; Smith, 1978; Kimball, 1978). The system being presented here, the ether-permeabilized *E. coli* cell, predominantly or exclusively performs DNA excision repair.

The system is characterized by the following features: after treatment of exponentially growing *E. coli* cells with diethylether, the cells are not viable anymore; they exhibit, however, high rates of DNA synthesis, which depend solely upon an external supply of ATP and the four deoxynucleoside triphosphates (Vosberg and Hoffmann-Berling, 1971). After the ether treatment, the *E. coli* membrane becomes permeable not only to the DNA precursors but also to carcinogens or DNA-binding agents of high molecular weight such as bleomycin.

Ether-permeabilized (nucleotide-permeable) *E. coli* (wild type or mutants) is exposed to carcinogens, mutagens, antineoplastic agents or compounds of potential carcinogenic hazard, and DNA repair polymerization is measured as a function of carcinogen (or drug) concentration used for pretreatment.

The test procedure may serve two purposes. First, ether-permeabilized wild type cells can routinely be used for assessing repair-inducing ability of a compound to be tested. Second, by using mutants defective in DNA pathways it is possible to analyze which enzymic activity is involved in the excision of damaged nucleotides.

Test Organism: The Ether-Permeabilized *E. coli* Cell

Growth and ether treatment of cells. Cells were grown in 400 ml casamino-enriched minimal medium at 37° C under aeration to a density of 3×10^8 cells/ml.

The cells were harvested by centrifugation (15 min, 4000 × *g*) at 4°C, washed with 20 ml starvation buffer, pelleted, suspended in 2 ml buffer A, and gently agitated at 4°C for 45 sec with 2 ml diethylether in a partition tube. The cell suspension was separated, pelleted through a 5-ml cushion of buffer A containing 32% (w/v) sucrose, resuspended in buffer A to give 5×10^{10} cells/ml, and frozen in solid CO_2.

The following bacterial strains were used:

Designation of strain	Derived from *E. coli*	Relevant genetic markers
Hfr C6[a]	K-12	Hfr Cavalli *endA*
Hfr C7[a]	K-12	Hfr Cavalli *endAuvr*
H 540[b]	K-12/C hybrid	Hfr Cavalli *endA*
H 512[a]	C	F^- *uvrA*$_{103}$ *endA*
H514[a]	C	F^- *thyA*$_{46}$ *uvrA*$_{103}$ *endA*
H 551[a]	K-12	Hfr Cavalli *uvrB*
H 560[c]	K-12/C hybrid	F^+ *polAendA*
BT 1026[d]	K-12/C hybrid	F^+ H 560 *thy endA polA*$_1$ *polC*
H 10265[e]	K-12/C hybrid	F^+ H 560/BT 1026 *thyA polA*$_1$ *polB*$_1$
KMBL 1788[f]	K-12	*polC*$_{1026}$ *endA*
KMBL 1778[f]	K-12	F^- *endA*$_{101}$
		F^- *polA*$_{107}$ *endA*$_{101}$

[a]Dürwald and Hoffmann-Berling, 1968; [b] gift of Hoffmann-Berling; [c] Vosberg and Hoffmann-Berling, 1971; [d] Wechsler *et al.*, 1973; [e] Hirota *et al.*, 1972; [f] Glickman *et al.*, 1973.

Equipment and reagents. Liquid scintillation spectrometer, centrifuge performing 4000 × *g* at 4°C. Sterile Erlenmeyer flask (content 1 l) cotton-stoppered, with glass tube ending in a porous glass disc for growing bacteria under aeration. Test tubes (10 ml); one stoppered partition tube (15 ml). Filtering apparatus: The filtering surface consists of a porous glass disc to which vacuum is applied. Nitrocellulose millipore filters (Schleicher & Schuell Selectron BA 85/1, 25 mm diameter) were used. The filters were prewetted with 10% trichloroacetic acid containing 1 mg/ml RNA (the mixture having been preheated for 15 min at 90°C). The four deoxyribonucleoside triphosphates (dNTPs) and [³H] dTTP or [α-³²P] dTTP.

Starvation buffer: 67 mM KCl, 17 mM NaCl, 10 mM Tris · HCl (pH 8.1), 0.4 mM $MgSO_4$, 1 mM $CaCl_2$. Buffer A: 50 mM Tris · HCl (pH 7.4), 80 mM KCl, 8 mM magnesium acetate, 2.5 mM EGTA, 17% (w/v) sucrose. Buffer B: 0.1 M KCl, 54 mM Tris · HCl (pH 7.6), 7.5 mM magnesium acetate, 2.5 mM EGTA, 1.25 mM ATP, 6.3% (w/v) sucrose. For solubilization of water-insol. carcinogens (e.g., 7-bromomethyl-benz[*a*]anthracene), a soln. of equal vol. of dimethyl sulfoxide and Cremophor EL[R] (a polyethoxylated ricinus oil from Badische Anilin & Soda-Fabrik AG, Ludwigshafen) was used.

Checking procedure of the organism. Permeabilized wild type and *uvr* mutant cells were first checked for repair polymerization after application of increasing doses of uv light. Briefly, aliquots of 5×10^8 cells were suspended in 8 ml of starvation buffer, irradiated with uv light (0, 100, 200, 400, 1000 ergs/mm²[100 J/m²]) in a Petri dish in the dark, centrifuged (15 min at 4000 × *g*) at 4°C, and resuspended in 0.4 ml buffer B. DNA synthesis, performed during 20 min at 37°C, was then measured. With increasing uv-irradiation, wild type cells exhibited an increase of DNA synthesis, *uvr* mutant cells a decrease. Temperature-sensitive *pol* mutants were tested for DNA synthesis at both the permissive and the nonpermissive temperatures.

Storage. After permeabilization, cells were stored at −80° C.

Testing Protocol

Treatment of permeabilized cells with carcinogens and incorporation of deoxyribonucleotides. To follow carcinogen-stimulated [³H]dTMP or [³²P]dTMP incorporation, 2.5×10^9 cells were incubated with various concentrations of carcinogens or drugs in 0.4 ml buffer B for 20 min at 30° C. The chemicals to be tested were freshly dissolved in buffer (10 mM Tris·HCl, pH 7.6), except 7-bromomethylbenz[a]anthracene, BCNU, CCNU, and the aminofluorene derivatives, which were solubilized in Cremophor EL^R/Me₂SO (1/1). Me₂SO should not have exceeded 5% in the assays. DNA synthesis was then initiated by altering the temperature to 37° C and adding the dNTPs (dissolved in 0.1 ml 10 mM Tris·HCl, pH 7.4) to give final concentrations of $20 \mu M$ dATP, dCTP and [³H]dTTP (spec. act. 0.5 Ci/mmol). At the required times (0, 10, 20, 30 min), 0.1 ml aliquots were removed, precipitated with 10% ice-cold trichloroacetic acid containing 1 mg/ml RNA (the mixture having been preheated for 15 min at 90° C) and collected on prewetted millipore filters. The filters were washed three times with trichloroacetic acid/RNA, dried (45 min, 75° C and counted (scintillation fluid of 0.4% butyl-PBD in toluene) in a liquid scintillation spectrometer.

Since [³H]dTMP incorporation was found to be approximately linear during the 30-min incubation, the above procedure was simplified for rapid screening of chemicals by means of various *E. coli* mutants. To quantify repair polymerization, aliquots of 5×10^8 permeabilized cells were preincubated with various concentrations of carcinogen or chemical for 20 min at 30° C, and DNA synthesis was allowed to proceed for 30 min at 37° C. Since the concentration of the test compound causing maximum DNA repair synthesis is unknown, a series of concentrations (usually increasing by a factor of 2) had to be applied. Incorporation of the radioactive precursor into the DNA was allowed for 30 min, and the extent of DNA synthesis was measured. During the preincubation period practically none of the compounds to be tested had reacted to completion; therefore, fractions of them were still present when DNA synthesis started. However, this only exceptionally caused interference with DNA synthesis.

Controls. Control assays were always performed in duplicate; they did not receive carcinogens or chemicals but solvent or solubilizer instead.

Modifications. There are additional possibilities for enhancing the sensitivity of the test system. Thus, in routine testing, excess carcinogen or drug was not washed out of the cells at the end of the preincubation period. This might have caused some depression of repair polymerization owing to possible pH changes. Therefore, when repair was at the limit of detectability, pelleting (4000 × g) and resuspension of the cells prior to the addition of the dNTPs improved the sensitivity of the assay for repair response.

Evaluation of the Results

Mutants defective in DNA-pathways. The repair-inducing potency of a compound is assessed by determining [³H]dTMP incorporation as a function of increasing carcinogen concentrations, as outlined above. These two parameters were plotted on the same graph. Essentially two types of curves could be obtained: (1) those which showed only inhibition of replicative DNA synthesis (e.g., as seen after application of *cis*-Pt(II)diammine chloride); and (2) those which showed an initial depression of [³H]dTMP incorporation (trough) followed by an increase of incorporation or an incorporation maximum. Thus in the curves of type (2), DNA polymerization due to repair was superimposed on a steadi-

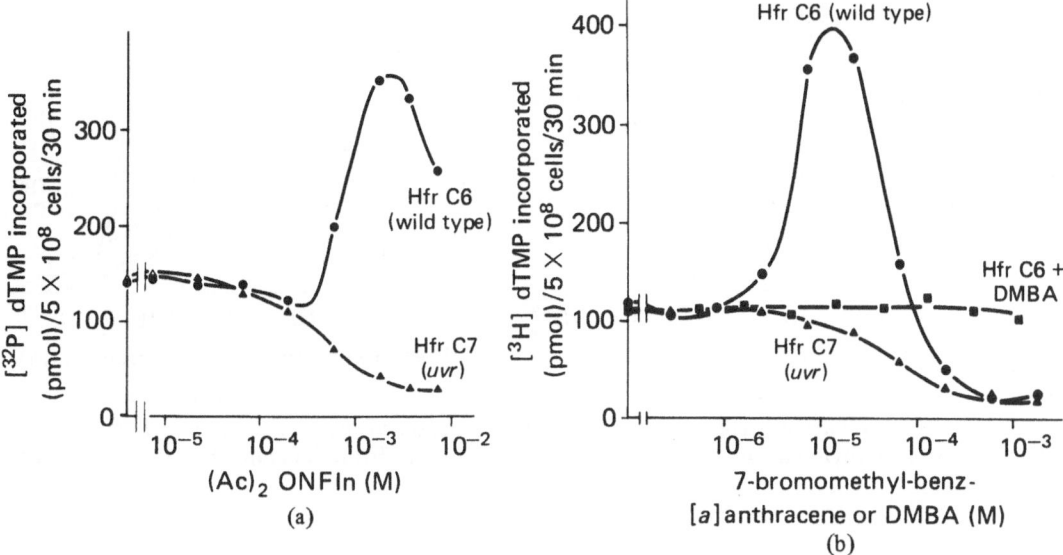

Figure 14.1. Stimulated [^{32}P]dTMP incorporation in nucleotide-permeable Hfr (wild type) and HfrC7 (*uvr*) cells as a function of (Ac)$_2$ONFln (a), 7-bromomethyl-benz[*a*]anthracene and DMBA concentrations (b). From Thielmann, Eur. J. Biochem. *61*, 501 (1976).

ly decreasing curve which represented increasing inhibition of replicative DNA synthesis. Repair and replication could be distinguished from each other by subtracting the [^{3}H]dTMP value of the trough from the maximum value of reparative DNA synthesis. There were exceptional cases where high levels of replicative DNA synthesis masked small repair maxima. In these cases repair polymerization could be assessed more accurately by using permeabilized cells with inhibited DNA replication (as achieved by mitomycin C pretreatment; Lindqvist and Sinsheimer, 1967).

Figure 14.1 shows a typical example of stimulation of DNA synthesis in *E. coli* wild type cells as a consequence of increasing concentrations of (Ac)$_2$ ONFln. It can be seen that treatment with ~2 mM of this ultimate carcinogen caused an increase of more than 200% of DNA synthesis above the background level measured in the carcinogen-free control. This effect was not found in *uvrA*, *uvrB* and *uvrC* mutant cells, whereas *uvrD* cells exhibited similar amounts of carcinogen-stimulated [^{32}P] dTMP incorporation as wild type cells (Fig. 14.2).

Wild type and *uvr* cells responded to the ultimate carcinogen 7-bromomethyl-benz[*a*]anthracene in a similar fashion as to (Ac)$_2$ONFln (Fig. 14.1). Evidence suggests that the genes *uvrA*, *uvrB*, and *uvrC* control early (endonucleolytic) steps in DNA repair (Waldstein *et al.*, 1974; Seeberg and Strike, 1976; Seeberg, 1978). The *uvrD* gene appears to be involved in later steps of excision repair (Ogawa *et al.*, 1968).

The carcinogen 7,12-dimethyl-benz[*a*] anthracene did not induce DNA repair in *uvr*-proficient cells (Fig. 14.1). This compound binds to DNA only after metabolic activation by microsomal enzymes (Sims and Grover, 1974). *E. coli* does not seem to possess enzymic equivalents thereof (Geissler, 1962; Slater *et al.*, 1971).

Figure 14.3 shows that the 5'-3' exonucleolytic activity of DNA polymerase I (exonuclease VI) is involved in the repair of (Ac)$_2$ONFln-damaged DNA. The ability of (Ac)$_2$ONFln to stimulate DNA synthesis was measured using a mutant (KMBL 1789, *polA*$_{107}$; Glickman *et al.*, 1973) lacking the 5'-3' exonucleolytic, but not the polymerase activity of DNA poly-

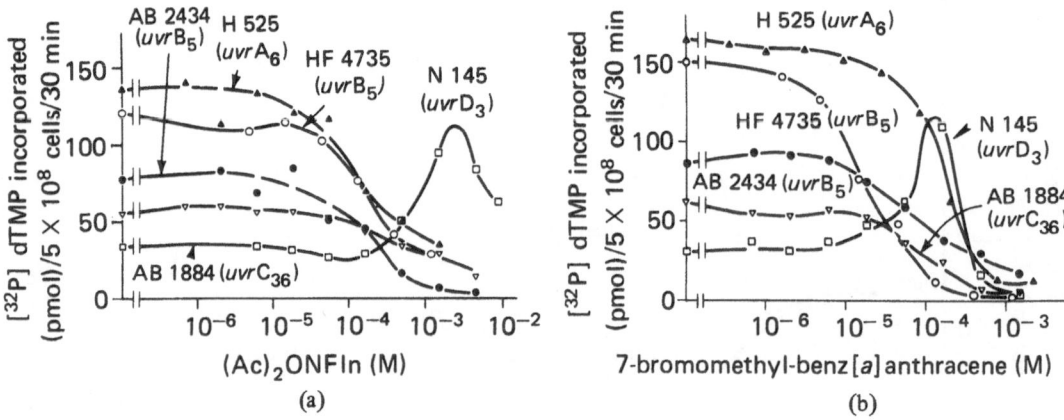

Figure 14.2. [^{32}P] dTMP incorporation in nucleotide-permeable *uvr* cells as a function of (Ac)$_2$ONFln (a) and 7-bromomethyl-benz[a]anthracene (b) concentrations. From Thielmann, Eur. J. Biochem. *61*, 501 (1976).

merase I. At the same time the wild type isogen (KMBL 1788) was tested as a control. In the *polA*$_{107}$ mutant, repair synthesis was cut back by more than 66%. Kornberg (1974) has shown that the polymerizing ac-

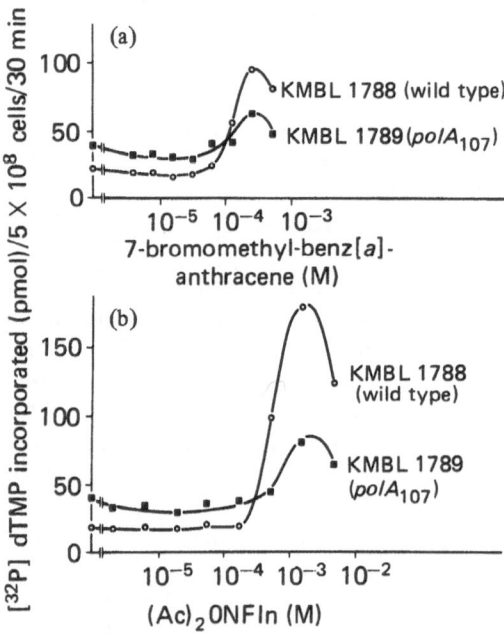

Figure 14.3. Involvement of the 5′-3′ exonucleolytic activity of DNA polymerase I in the DNA repair induced by (Ac)$_2$ONFln (a) and 7-bromomethyl-benz[a]anthracene (b). From Thielmann, Eur. J. Biochem. *61*, 501 (1976).

tion of DNA polymerase I is coupled with its 5′-3′ exonuclease activity. The enzyme synthesizes in the 5′-3′ direction as it hydrolyzes in the same direction, thus translating the nick and eliminating possible DNA damages at the 5′ end of the nick. Since the *polA*$_{107}$ mutant lacks the exonuclease VI activity, nick translation was not possible and, as a consequence, DNA polymerization did not take place. The residual stimulated [^{3}H]dTMP incorporation observed in *polA*$_{107}$ cells could be facilitated by nuclease activity other than the 5′-3′ of polymerase I. Thus, primer termini might have been created which were extended by DNA polymerases, a process which may not have contributed to the removal of damaged nucleotides from DNA.

The contributions of the respective DNA polymerases to gap filling are exemplified by the directly acting carcinogen 7-methyl-benz [a] anthracene-5, 6-oxide. Figure 14.4 shows that repair polymerization in the *polA*$_1$ mutant H560 was only ~1% of that of control cells at 37°C, suggesting that under the assay conditions DNA polymerase I was the main polymerizing enzyme in the excision repair process. The residual K-region epoxide-stimulated [^{3}H]dTMP incorporation observed in H 560 cells was probably facilitated by DNA

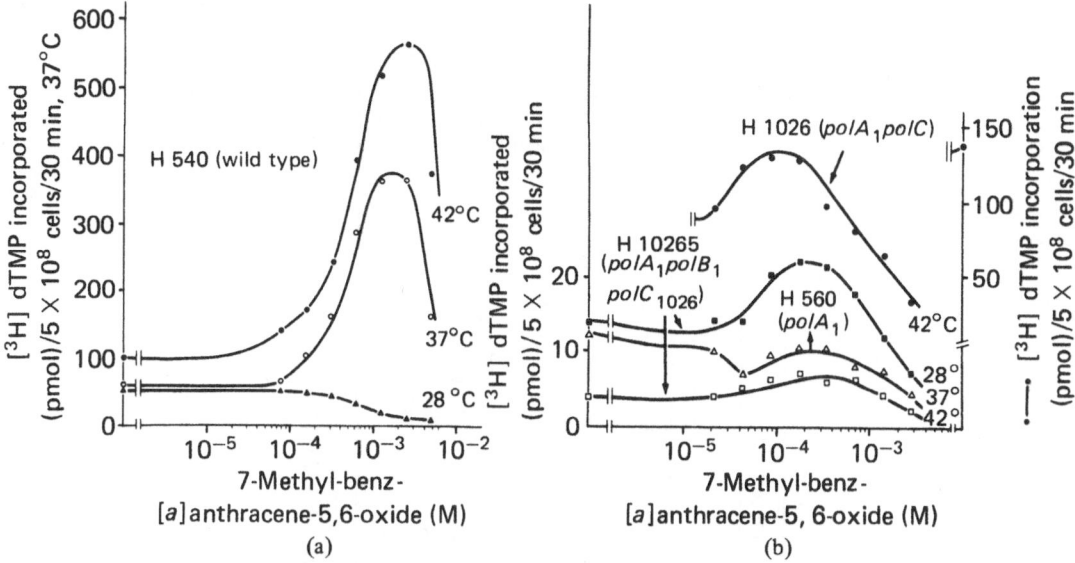

Figure 14.4. DNA polymerases involved in repair polymerization induced by 7-methylbenz[*a*]-anthracene-5,6-oxide. DNA synthesis was measured in wild type cells (H540)(a), at the respective temperatures, to serve as a control for the polymerase-deficient mutants(b). From Thielmann and Gersbach, *Z. Krebsforsch.* *92*, 157 (1978a).

polymerases II and III. To distinguish between the possible contributions in *polA*$_1$ cells, DNA repair synthesis can be determined in the *polA*$_1$, *polC* mutant BT 1026 at the restrictive temp. of 42° C, and in the *polA*$_1$ *polB*$_1$ *polC*$_{1026}$ mutant H 10265 at the permissive temperature of 28° activity (45 pmol) was found in the *polA*$_1$ *polC* mutant (Fig. 14.4) as compared with the wild type control strain H 540 tested at 42° C (450 pmol; Fig. 14.4), indicating that the contribution of DNA polymerase II is small. In H 10265 cells, DNA polymerase III catalysed some repair synthesis (8 pmol) at the permissive temperature of 28° C, and, as expected, practically none at the restrictive temperature of 42° C. In H 560 control cells, no K-region epoxide-stimulated [³H]dTMP incorporation could be measured at 28° C (Fig. 14.4), possibly because inhibition of replicative DNA synthesis masked reparative synthesis.

Tables 14.1 to 14.5 summarize results obtained with carcinogenic epoxides, alkylating agents, 2-acetylaminofluorene- and bromomethyl-aryl-derivatives, crosslinking carcinogens and intercalating agents, OsO$_4$ and H$_2$O$_2$ and antitumor agents.

K-region epoxides and 1,2,3,4-diepoxybutane. All epoxides tested, except phenanthrene-9,10-oxide, causes DNA excision repair in ether-permeabilized *E. coli* wild type cells (Table 14.1). The repair-inducing potencies of the respective oxides assessed by their ability to stimulate DNA polymerization decreased in the order: 7-methylbenz [*a*] anthracene-5, 6-oxide ≃ 7,12-dimethyl-benz [*a*] anthracene-5, 6-oxide 7 chrysene-5, 6-oxide ≃ benzo [*a*] pyrene-4,5-oxide. This order correlates well with the relative mutagenic and carcinogenic activities of the compounds (Sims and Grover, 1974; Levine *et al.*, 1974; Wood *et al.*, 1976; Siebert et al., 1980). Removal of damaged nucleotides requires exonuclease VI, since only little repair synthesis was observed in the *polA*$_{107}$ mutant lacking this enzyme.

It is of interest to consider the types of DNA modification which might provide substrates for the repair endonucleases that initiate the excision of epoxide-induced DNA damage. Exact data on the products of binding of K-region epoxides to double-stranded DNA are lacking. Still, *in vitro* model reactions suggest that K-region

Table 14.1. DNA Repair Polymerization Induced by Epoxides in Nucleotide-Permeable E. coli Cells[a]

Epoxide	[³H]dTMP incorporation (pmol)				
	Wild type (H 540)	uvrA (H 512)	uvrB (H 551)	Wild type (KMBL1788)	polA₁₀₇ (KMBL1789)
7,12-dimethyl-benz[a]anthracene-5,6-oxide	>290(>1.2 mM)	0[b](40 µM–5 mM)	0[b](40 µM–5 mM)	>400(>5 mM)	~16(~5 mM)
7-methyl-benz[a]anthracene-5,6-oxide	>420(>1 mM)	0 (40 µM–5 mM)	0 (40 µM–5 mM)	385(2 mM)	10(2 mM)
Benzo[a]pyrene-5,6-oxide	30(65 µM)	—	0 (5–600 µM)	320(0.1 mM)	<5(~0.1 mM)
Chrysene-5,6-oxide	36(0.15 mM)	0[b](4–500 µM)	0[b](4–500 µM)	41(32 µM)	0[b](4–500 µM)
Phenanthrene-9,10-oxide	8(~0.1 mM)	0(4–500 µM)	0[b](4–500 µM)	10(0.2 mM)	0(4–500 µM)
1,2,3,4-diepoxybutane	>520(>80 mM)	>85(>300 mM)	>125(>150 mM)	320(110 mM)	~10(>150 mM)
TPA-6α,7α-oxide	50(80 µM)	40(120 µM)	65(0.3 mM)	50(0.2 mM)	<12(0.2 mM)
TPA-6β,7β-oxide	40(0.1 mM)	10(60 µM)	0(6–700 µM)	0(6–700 µM)	0(6–700 µM)

[a] Epoxide concentrations at which maximal stimulation of DNA synthesis was reached are given in brackets. In cases where no stimulation was observed, the epoxide concentration range used is given. From Thielmann and Gersbach. Z. Krebsforsch. 92, 157 (1978a).

[b] No repair but inhibition of replicative DNA synthesis.

epoxides bind covalently to purine ring nitrogens (N-7 of Gua (Friesel and Hecker, 1977; Thielmann, 1977) or N-3 of Ade and/or to the extranuclear amino group of Gua residues (Blobstein *et al.*, 1975; Jeffrey *et al.*, 1976)). If the first binding type had predominated in ether-permeabilized cells, one would have found that, according to previous results (Thielmann *et al.*, 1975), DNA excision repair was independent of *uvrA* and *uvrB* gene products. However the current experiments showed that with the K-region epoxides used, no excision repair at all was inducible in *uvrA* and *uvrB* mutant cells. Therefore, it was concluded that the arene oxides predominantly bound to the N² of guanine residues or other sites, thus causing steric distortions and localized denaturation of the double helix as documented for uv-irradiated DNA (Salganik *et al.*, 1967; Denhardt and Kato, 1973; Gupta and Mitra, 1974; Woodworth-Gutai *et al.*, 1977; Heflich *et al.*, 1977). It appears reasonable to assume that areas of steric distortion or local denaturation represent the signals for the uv-endonuclease to act.

In contrast to the arene oxides, 1,2,3,4-diepoxybutane did stimulate [³H]dTMP incorporation in *uvrA* and *uvrB* mutant cells. The rationalization of this result in terms of DNA modification must to some extent remain ambiguous, since not all the DNA adducts are known.

In general, the aliphatic epoxides resemble the alkyl sulfates, alkyl sulfonates and nitroso compounds in alkylating Gua residues in preference to Ade residues, and Ade preferentially to Cyt residues (Singer, 1975). Further hints as to the possible binding product come from a study (Lawley and Jarman, 1972) in which the reaction of propylene oxide with DNA has been investigated *in vitro*. Lawley and Jarman identified 7-(2-hydroxypropyl)Gua and 3-(2-hydroxypropyl)Ade as the principal products. It would not be surprising if 1,2,3,4-diepoxybutane alkylated Gua (N-7) and Ade (N-3). Since Lawley and Jarman also observed spontaneous loss of 2-hydrox-

ypropylated Gua bases from DNA at neutral pH, one can assume that a (nonenzymic) generation of AP-sites occurred in the system. Single-strand breaks as a consequence of depurination have been observed (Verly and Brakier, 1969). Conceivably, DNA glycosylase-catalyzed hydrolysis of Ade residues also took place. Both processes would explain repair polymerization in the *uvr* strains tested, since alkylated and AP-sites in DNA are substrates for *uvr*-independent endonucleases (Friedberg *et al.*, 1969; Thielmann *et al.*, 1975; Radman, 1976; Gates and Linn, 1977). Considering that the repair level measured in wild type cells was about fourfold higher than that measured in *uvr* mutants, it appears that the *uvrA*- and *uvrB*-controlled uv-endonuclease is involved in the excision of diepoxide-modified bases. Goldschmidt *et al.* (1968) treated 2-dGuo with glycidaldehyde and showed that in fact the N² and the N-1 of Gua acted as nucleophiles to open the epoxide ring.

6,7-Epoxides of the tumor promoter TPA. TPA exerts its tumor-promoting activity without metabolic activation (Hecker, 1978). However, when applied repeatedly to mouse skin at high doses, TPA produces tumors in low yield after a long latency period (Hecker, 1968). This low tumorigenic activity may be attributed to 6,7-epoxides of TPA, one of which is formed by autoxidation (Schmidt and Hecker, 1975). Since it is conceivable that the 6,7-epoxides of TPA can bind to DNA, they were tested for repair-stimulating activity (Table 14.1).

The 6α,7α-epoxide of TPA stimulated repair in wild type and *uvr* mutant cells to similar extents, which suggests involvement of *uvr*-independent incision steps.

The DNA sites to which the epoxide binds are not known. Still, on the basis of these data, attachment to the N-7 of Gua or the N-3 of Ade is conceivable. Covalent fixation of alkyl substituents on the N-7 of Gua and the N-3 of Ade are known to cause depurination (Lawley and Brookes, 1963). AP-sites can be repaired in the absence of uv-endonuclease (Thielmann *et al.*, 1975). The $6\beta,7\beta$-epoxide was less active in wild type cells and practically inactive in *uvr* mutants. The weaker effects of this compound are explainable by its lower reactivity. Thus, on silica gel, TPA-$6\alpha,7\alpha$-oxide-20-acetate underwent an acid-catalyzed opening of the epoxide ring, whereas the corresponding $6\beta,7\beta$-derivative did not (R. Schmidt and E. Hecker, personal communication). The probable explanation is that TPA, for steric reasons, is more accessible to reactants from the β than from the α direction.

The repair-stimulating properties of the TPA-6,7-epoxides lend support to the assumption that the low tumorigenic activity observed in long-term carcinogenesis experiments could be caused by the conversion of TPA to its 6,7-epoxides.

Alkylating, carboxyethylating and acylating carcinogens. MeNOUr and Me(NO)(NO$_2$)Gdn predominantly methylate N-7 of Gua residues (60%), phosphate oxygens (~25%), N-3 of Ade (8%) and the extranuclear atoms of purine and pyrimidine bases (Loveless and Hampton, 1969; Goth and Rajewsky, 1974; Sun and Singer, 1975). Treatment of DNA with Et-NOUr leads to the formation of phosphotriesters (~60%), et^{7}Gua and et^{6}Gua (10% each), et^{7}Ade and other ethylation products (Goth and Rajewsky, 1974; Singer, 1975). 1-(*p*-Chlorophenyl)-3-methyltriazene methylates DNA *per se* (Preussmann *et al.* 1974), whereas the 3,3-dimethyl-derivative has to be hydroxylated by microsomal enzymes to become reactive toward DNA. *E. coli* is not able to perform this metabolizing step under the experimental conditions. β-Propiolactone yields mainly 7-(2-carboxyethyl)Gua residues (Roberts and Warwick, 1963). For chemical reasons maleic hydrazide should acylate the same nucleophilic sites in DNA as the alkylating agents do, although the relative proportions of the DNA-adducts might be different.

It is assumed that the base modifications mentioned above account for the repair effects observed in ether-permeabilized cells (Table 14.2). In particular, alkylation of the N-7 position of Gua and N-3 position of Ade residues readily leads to the formation of AP-sites at neutral pH (Lawley and Brookes, 1963). In *E. coli* these sites are recognized by AP-endonucleases (Verly and Rassart, 1975; Radman, 1976; Ljungquist, 1977). Furthermore, hydrolysis of m^{3}Ade residues from DNA has been shown to be mediated by 3-methyl-adenine-DNA glycosylase (Lindahl, 1976), a process which contributes to the generation of AP-sites. There is evidence that in ether-permeabilized *E. coli* cells, excision of alkylation-damaged nucleotides is initiated by both the AP-endonucleases and a *p*-chloromercurio-benzoate-sensitive DNA glycosylase activity (Thielmann, unpublished).

Aromatic carcinogens: 2-acetylaminofluorene- and bromomethyl-aryl derivatives. DNA repair caused by (Ac)$_2$ONFln and 7-bromomethyl-benz[*a*]anthracene was strongly dependent on *uvrA*, *uvrB* and *uvrC* gene products (see Fig. 14.2 and Table 14.3). By contrast, benzyl bromide, a simple analogue of 7-bromomethyl-benz[*a*]anthracene, formed DNA-adducts which were excised by a *uvr*-independent mechanism. It appears plausible that benzyl bromide arylalkylates purine ring nitrogens, labilizing the *N*-glycosidic bond and leading to AP-sites. DNA glycosylases may also be involved in generating AP-sites. Arylalkylated bases and AP-sites are known to be substrates for the endonucleases II and III and for AP-endonucleases (Friedberg *et al.*, 1969; Kirtikar and Goldthwait, 1974; Verly and Rassart, 1975; Radman, 1976; Ljungquist, 1977). Table 14.3 furthermore shows that the proximate carcinogens (Ac)NOHFln and (Ac)NHFln do not elicit excision repair. It is thus not surprising that the proximate carcinogens 4-nitroquinoline-1-oxide and the aflatoxins B$_1$, B$_2$, G$_1$ and G$_2$ proved to be inactive.

OsO$_4$ and H$_2$O$_2$. In ether-permeabilized *E. coli* cells, the repair-inducing activity of OsO$_4$ is clearly demonstrable (Fig. 14.5) even at very low concentrations (0.1–1

Table 14.2. *DNA Repair Polymerization Induced by Alkylating Carcinogens in Nucleotide-Permeable E. coli Cells*[a]

Carcinogen	[³H]dTMP incorporation (pmol)				
	Wild type (H 540)	uvrA (H 512)	uvrB (H 551)	Wild type (KMBL1788)	polA107 (KMBL1789)
N-methyl-N-nitrosourea	150 (10 mM)	113 (12 mM)	100 (10 mM)	450 (13 mM)	42 (13 mM)
N-ethyl-N-nitrosourea	25 (8 mM)	36 (10 mM)	—	—	—
1-(p-chlorophenyl)-3-methyl-triazene	480 (6 mM)	36[b] (5 mM)	160 (~6 mM)	280 (4 mM)	40 (~8 mM)
1-(p-chlorophenyl)-3,3-dimethyl-triazene (0.12–62.5 mM)	<5	0	0	0	0
N-methyl-N'-nitro-N-nitrosoguanidine	430 (2 mM)	215 (4 mM)	75 (3 mM)	390 (4 mM)	20 (4 mM)
N-methyl-N'-nitro-N-nitrosoguanidine + N-acetyl-cysteine (7 mM)	670 (4 mM)	>360 (>8.5 mM)	>120 (>8.5 mM)	>460 (>8.5 mM)	>56 (>8.5 mM)
β-propiolactone[b]	140[c] (15 mM)	83[c] (20 mM)	30 (19 mM)	255 (15 mM)	25 (20 mM)
Maleic hydrazide	36 (300 μM)	70 (300 μM)	28 (300 μM)	26 (30 μM)	0 (0.2–450 μM)

[a] Carcinogen concentrations at which maximal stimulation of DNA synthesis was reached are given in brackets. In cases where no stimulation was observed, the concentration range used for testing is given. From Thielmann and Gersbach, Z. Krebsforsch. 92, 177 (1978b).
[b] Preincubation: 10 min at 27°C.
[c] Cells were pretreated with mitomycin C in vivo to inhibit DNA replication.

Table 14.3. *DNA Repair Polymerization Induced by Carcinogenic 2-Acetylaminofluorene and Bromomethyl-aryl Derivatives, and Selected Indirectly Acting Carcinogens*[a]

Carcinogen	[³H]dTMP incorporation (pmol)				
	Wild type (H 540)	uvrA (H 512)	uvrB (H 551)	Wild type (KMBL1788)	$polA_{107}$ (KMBL1789)
N-acetoxy-2-acetylaminofluorene	166(2 mM)	0[b](35 μM–4.4 mM)	0[b](26 μM–13.3 mM)	110(0.6 mM)	12(2 mM)
N-hydroxy-2-acetylaminofluorene	0(16 μM–2.2 mM)	—	—	—	—
2-acetylaminofluorene	0(18 μM–2.2 mM)	—	—	—	—
7-bromomethyl-benz[a]-anthracene	245(13 μM)	—	0(2–780 μM)	95(220 μM)	36(220 μM)
Benzyl bromide	285(2.5 mM)	105(2.6 mM)	55(2.3 mM)	380(1.3 mM)	29(1.3 mM)
4-nitroquinoline-1-oxide	0[c](0.5 μM–1.1 mM)	0(0.5 μM–1.1 mM)	0(0.5 μM–1.1 mM)	0(4 μM–0.5 mM)	0(4 μM–0.5 mM)
Aflatoxins B₁, G₁	0(0.2 μM–1 mM)	—	—	—	—
Aflatoxins B₂, G₂	0(0.2–76 μM)	—	—	—	—

[a]Carcinogen concentrations at which maximal stimulation of DNA synthesis was reached are given in brackets. In cases where no stimulation was observed, the concentration range used for testing is given. From Thielmann and Gersbach, Z. Krebsforsch. 92, 177 (1978b).

[b]No repair but inhibition of replicative DNA synthesis.

[c]No repair but 20% inhibition of replication at highest concentration tested.

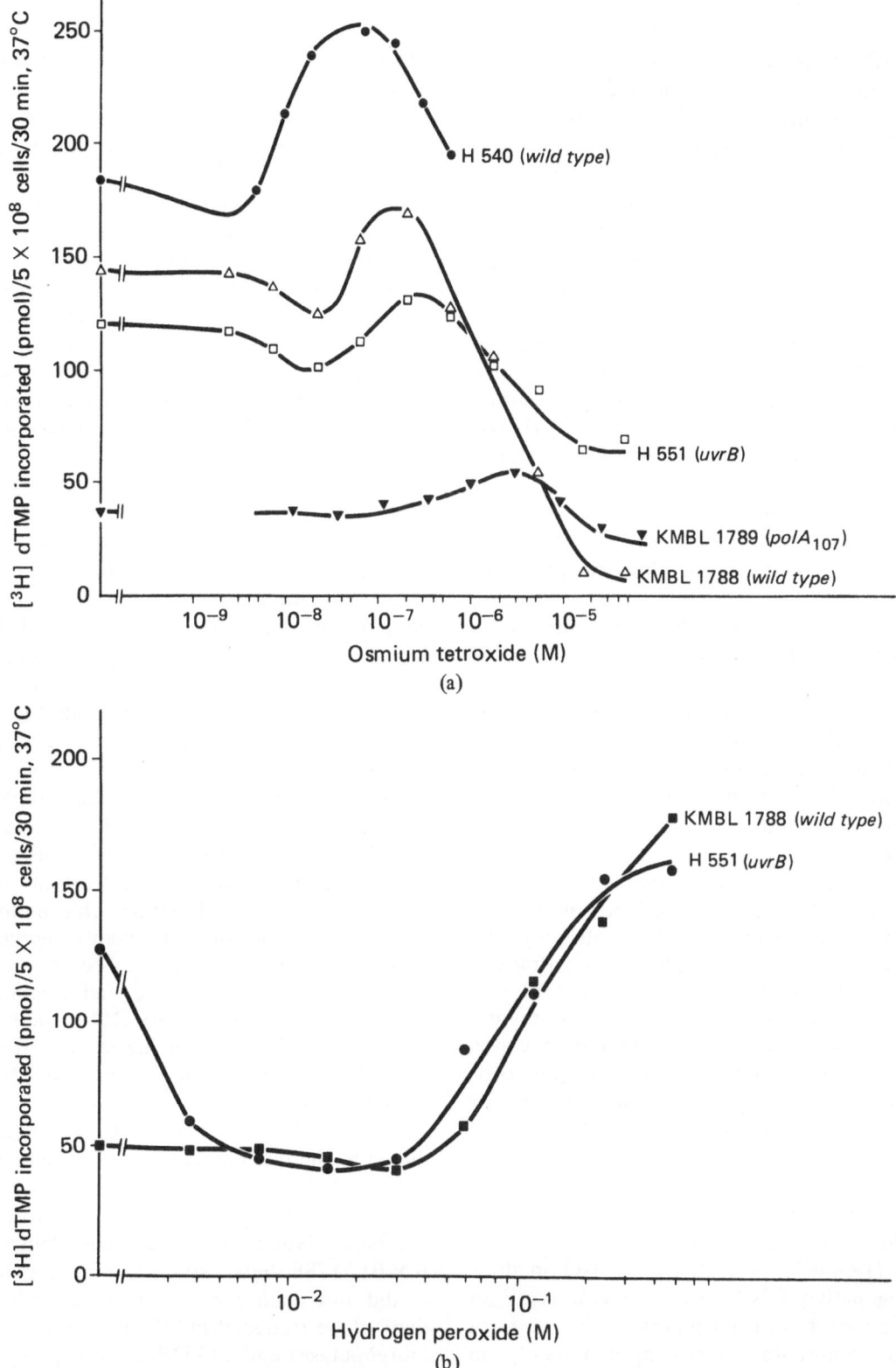

Figure 14.5. Induction of repair polymerization in nucleotide-permeable wild type, *uvr* and *polA*₁₀₇ cells as a function of osmium tetroxide (a), or hydrogen peroxide (b) concentrations. Osmium tetroxide (0.1 mg) was freshly dissolved in 50 μl methanol before 950 μl of 20mM Tris · HCl pH 7.6 was added. This stock solution was further diluted. From Thielmann and Gersbach, Z. Krebsforsch. **92**, 177 (1978b).

μM), but proved to be inhibitory to DNA synthesis at concentrations exceeding approx. 1 μM. Since DNA synthesis was stimulated to the same extent in *uvrB* and wild type cells, it can be concluded that the excision system removing OsO_4-damaged nucleotides does not involve uv-repair machinery (see also Hariharan and Cerutti, 1974).

OsO_4 reacts with DNA in aqueous solution to form mainly 5,6-dihydroxy-dihydrothymine and apyrimidinic sites, but causes no Ade damage or strand breaks (Hariharan and Cerutti, 1974). The 5,6-dihydroxy-dihydrothymine residues which tend to form apyrimidinic sites probably result from hydrolysis of oxo-osmium (VI) heterocyclic esters. The question remains as to the chemical nature of the DNA-OsO_4 products which give rise to excision repair at these low amounts of OsO_4. It can be envisaged that the stable heterocyclic osmate esters are either recognized *in situ* by *uvr*-independent DNA glycosylases/endonucleases or are hydrolyzed off the DNA, thus forming apyrimidinic sites which in turn could be substrates for endonucleases (Gates and Linn, 1977).

Similar results were obtained with H_2O_2. Figure 14.5 shows that DNA synthesis attained comparable levels in both *uvrB* and wild type cells. The concentrations required, however, were as high as 50–400 mM. Apparently under the experimental conditions (absence of transition metals), generation of HO· radicals, thought to be the ultimate reactants, was low. The repair polymerization measured in wild type and *uvrB* cells was approximately equal. Evidently the excision of the DNA-H_2O_2 products proceeds well in the absence of the *uvrB* gene products.

H_2O_2 is less selective than OsO_4 in altering native DNA. Thus several types of reaction have been reported: saturation of the 5,6-double bonds predominantly in pyrimidine bases, slow liberation of all four (unaltered) bases, and breakage of the DNA backbone (Rhaese and Freese, 1968).

DNA-crosslinking carcinogens. The bifunctional *N-mustard* is known to cause interstrand crosslinks in duplex DNA. Monofunctional alkylation, however, is approximately 25 times as frequent as crosslinking (Kohn *et al.*, 1966).

When exposed to *N-mustard*, wild type cells show highly stimulated DNA repair synthesis (Table 14.4). In *uvrA* and *uvrB* mutant cells, approximately one-third to one-half of this amount of repair polymerization is measured (Table 14.4), suggesting that *uvrA* and *uvrB* gene functions are involved in the recognition of some of the DNA damage caused by *N-mustard*. This result is in agreement with the observation that the uv-sensitive strain Bs$_1$ lacks a mechanism for unlinking DNA crosslinked with *N-mustard* (Kohn *et al.*, 1965). It is therefore assumed that the residual repair seen in *uvr* mutant cells is due to a type of alkylation (possibly monovalent) that generates AP-sites. This latter type of alkylation seems to predominate in the case of quinacrine half mustard, since high repair levels were seen in both *uvrA* and wild type cells. The DNA repair synthesis caused by quinacrine half mustard can be attributed to the alkylating chloroethyl group alone; control experiments showed that the basic heterocyclic compounds quinacrine, 9-amino-acridine, and acridine, which are able to intercalate DNA, did not act as repair inducers in the system. The same holds true for the intercalating dye ethidium bromide.

The antibiotic *mitomycin C*, when reduced to the quinone form, is also capable of crosslinking duplex DNA (Szybalski and Iyer, 1967) in a similar fashion to *N-mustard*. However, ether-permeabilized KMBL 1788 cells exposed to mitomycin C did not exhibit excision repair unless diaphorase (reduced-NAD: lipoamide oxidoreductase) and NADPH were present (Table 14.4). In contrast, *uvrB* cells responded to the same treatment with inhi-

Table 14.4. *DNA Repair Polymerization Induced by DNA-Crosslinking Carcinogens and Intercalating Agents in Nucleotide-Permeable* E. coli *Cells*[a]

Carcinogen or intercalating agent	[³H]dTMP incorporation (pmol)						
	Wild type (H 540)	uvrA (H 512)	uvrB (H 551)	Wild type (HfrC 6)	uvr (HfrC 7)	Wild type (KMBL1788)	polA$_{107}$ (KMBL1789)
N-mustard	350(0.7 mM)	115(4.3 mM)	140(0.6 mM)	400(1.4 mM)	210(1.4 mM)	270(0.3 mM)	65(0.8 mM)
Quinacrine half mustard	420(1 µM)	235(2 µM)	—	—	—	590(4.6 µM)	33(6 µM)
Quinacrine (600 nM–10 mM)	0[b]	0[b]	0[b]	—	—	0[b]	0[b]
Acridine (0.7 µM–0.35 mM)	0[c]	0[c]	—	—	—	0[c]	0[c]
9-amino-acridine	0[b](2 nM–3.5 mM)	0[b](0.4 nM–13 µM)	—	—	—	—	—
Ethidium bromide	0[d](20 nM–2.5 µM)	0[d](5 nM–2.5 µM)	—	—	—	—	—
Mitomycin C[e]	—	—	0(1 mM)	—	—	150(1 mM)	—

[a]Carcinogen concentrations at which maximal stimulation of DNA synthesis was reached are given in brackets. In cases where no stimulation was observed, the concentration range used for testing is given. From Thielmann and Gersbach, Z. Krebsforsch. 92. 177 (1978b).
[b]Inhibition of replicative DNA synthesis at concns ≥ 10 µM.
[c]Slight inhibition of replication at concns ≥0.3 mM.
[d]Drastic inhibition of replication at concns ≥ 1 µM.
[e]After reduction (15 min, 37° C) by 0.8 U/ml diaphorase in the presence of 0.1 mM NADPH.

bited DNA replication, indicating that excision of the damaged nucleotides requires the intact *uvrB* function.

Mitomycin does not bind to DNA *per se* but requires reduction of the *p*-quinone structure (NADPH plus quinone reductase) and loss of the tertiary methoxy group to form an aromatic indole system which, in turn, activates the aziridine ring and the methyl urethane side chain (Szybalski and Iyer, 1967). The *N*-7 position of Gua is not involved in mitomycin C binding (Tomasz, 1970), as is the case for *N*-mustard. Given the two reactive centers in mitomycin, the molecule seems to fit best between complementary strands, when crosslinking two O^6 groups of the nearest but opposite guanines (Szybalski and Iyer, 1967). Still, the majority of ligands is bound monovalently (Szybalski and Iyer, 1967).

Antitumor agents. The influence upon DNA synthesis of some antitumor agents, whose major mode of action is likely to be covalent or noncovalent (i.e., electrostatic, coordinative, intercalative) DNA-binding, is described in the following experiments (Table 14.5). Substances known to bind noncovalently (such as *cis*-Pt(II)diammine derivatives, adriamycin, daunomycin, H 33258), inhibited DNA replication without detectably stimulating DNA repair. Apparently these compounds simply create an obstacle for the replication apparatus. Substances which react covalently with DNA (bleomycin, BCNU, CCNU and others) stimulated DNA repair and, to various extents, inhibited DNA replication. Two representatives of the respective groups, the *cis*-Pt(II)diammine complexes and the halogenated nitrosoureas, will be discussed in more detail.

The products of *in vivo* DNA-platinum binding which are responsible for the attenuation of DNA synthesis are not known with certainty. Still, potentiometric chloride determinations have shown that *cis*-Pt(NH_3)_2Cl_2 forms a stable complex with Gua residues in DNA using the N-7 and O^6 sites of Gua as chelating ligands (Macquet

and Theophanides, 1975). As a consequence, the hydrogen bond between O^6 of Gua and N^4-H of Cyt is abolished, which leads to a partial denaturation of the double helix. It is conceivable that this chelate complex formation accounts for the biological effects observed. DNA interstrand crosslinks, although possibly very infrequent, may also play a role (Pascoe and Roberts, 1974).

In the concentration of 3 μM to 1 mM there was no repair detectable in wild type cells, but rather an inhibition of replication. The effectiveness of *cis*-Pt(NH_3)_2Cl_2 in inhibiting DNA synthesis was tenfold that of *cis*-Pt(NH_3)_2(NO_2)_2. This difference correlates well with the rates of substitution of the anionic ligands by the chelating sites, since kinetic studies have shown that chloride is a much better leaving group than nitrite (Leh and Wolf, 1976).

At neutral pH the halogenated nitrosoureas BCNU and CCNU decompose to form mainly 2-chloroethyl and vinyl carbonium ions, and the corresponding isocyanates, $O = C = NR$ (Wheeler, 1976). CCNU in which the N' is substituted by a 2-chloroethyl group generates, on hydrolysis, 2-chloroethylisocyanate; this ester in turn hydrolyzes via carbamic acid to form carbon dioxide and 2-chloroethylamine, the latter possibly also being an alkylating agent (Wheeler, 1976).

Application of BCNU and CCNU (≤ 1mM) to ether-permeabilized wild type cells and *uvrB* mutants led to inhibition of replicative DNA synthesis (Thielmann and Gersbach, 1978a,b). BCNU and CCNU, however, at concentrations greater than 3 mM stimulated [^{3}H]dTMP incorporation (Table 14.5).

Conclusions

The data presented show that ether-permeabilized *E. coli* cells are a sensitive system responding to ultimate carcinogens, mutagens, and alkylating antitumor agents

Table 14.5. DNA Repair Polymerization Induced by Antineoplastic Agents in Nucleotide-Permeable E. coli Cells[a]

Antineoplastic agent	[³H]dTMP incorporation (pmol)				
	Wild type (H 540)	uvrA (H 512)	uvrB (H 551)	Wild type (KMBL1788)	polA107 (KMBL1789)
cis-Pt(NH₃)₂Cl₂(3 μM–1.8 mM)	0[b]	—	0[b]	—	—
cis-Pt(NH₃)₂(NO₂)₂ (1 μM–0.6 mM)	0[b]	—	0[b]	—	—
CCNU (Lomustine)	25(2 mM)	—	22(2.4 mM)	—	—
BCNU (Nitrumon)	15(10 mM)	—	23(7 mM)	—	—
Adriamycin (adriablastin) (7.1 nM–0.9 mM)	—	—	—	0[b]	—
Daunomycin (daunoblastin) (1 μM–2.4 mM)	0[b]	0[b]	—	0[b]	—
H 33258 (0.6 μM–1.3 mM)	0[b]	0[b]	—	—	—
Triethylenethiophospho-ramide (Thiotepa)	110(1.3 mM)	65(0.6 mM)	46(1 mM)	36(1.8 mM)	24(1.8 mM)
Triaziquone (Trenimon)	—	45(0.3 mM)	0[b](10 μM–0.1 mM)	50(0.1 mM)	20(0.12 mM)
Cyclophosphamide (Endoxan)	40(5 mM)	70(5 mM)	0[b](0.14 μM–18 mM)	38(6 mM)	28(6 mM)
Ifosfamide (Holoxan)	45(10 mM)	60(9 mM)	70(12 mM)	—	—
4-hydroperoxy-cyclo-phosphamide	120(0.2 mM)				
Bleomycin (30 μM–0.6 mM)	>468(>0.6 mM)	>268(>0.6 mM)	>208(>0.6 mM)	488(0.3 mM)	>96(>0.6 mM)
Bleomycin+5 mM dithiothreitol	792(0.2 mM)	368(75 μM)	232(20 μM)	640(40 μM)	204(40 μM)

[a] Drug concentrations at which maximal stimulation was reached are given in brackets. In cases where only replication was inhibited, the concentration ranges used are given. From Thielmann and Gersbach, Z. *Krebsforsch.* 92, 177 (1978b).

[b] No repair but inhibition of replicative DNA synthesis.

with DNA excision repair. This repair can be easily quantified by determining the ability of the cells to incorporate [³H]dTMP into DNA.

Carcinogens, mutagens, or antitumor agents known to bind covalently to DNA elicited excision repair. Neither the pre- or proximate carcinogens used, nor the intercalating or chelate complex forming agents, however, caused repair polymerization under the conditions applied. The data indicate that only substances which possess the chemical potential of transferring electrophilic groups to DNA are capable of inducing excision repair. The applicability of the ether-permeabilized *E. coli* cell as an accurate indicator for repair in mammalian cells would, therefore, require that the pre- or proximate carcinogens or mutagens be metabolized to the same reactive products which arise from their metabolism in mammals. In preliminary experiments, however, only marginal activation of the proximate carcinogen Me₂NNO was achieved by pretreating it with Udenfriend's hydroxylating mixture (Udenfriend *et al.*, 1954). Preincubation of the compounds with microsomes (Ames *et al.*, 1973) might be a better alternative.

All carcinogens, mutagens, and antitumor agents that gave rise to excision repair in wild type cells either induced considerably less (bleomycin) or no repair at all in a mutant lacking exonuclease VI. This enzyme, therefore, appears to play a central role in degrading stretches of damaged strands in double-stranded DNA.

Steric distortions in duplex DNA arising from attachment of arylating or arylalkylating carcinogens seem to be the crucial signals recognized by the uv-endonuclease. For example, (Ac)₂ONFln, 7-bromomethyl-benz[a]anthracene, and mitomycin C required intact *uvrA* and *uvrB* functions.

Carcinogens, mutagens, or antitumor agents which predominantly alkylate nucleophilic centers in duplex DNA (such as the N-3 of Ade or the N-7 of Gua) were capable of inducing DNA excision repair in wild type strains and in both *uvrA* and *uvrB* mutant cells. Typical examples were MeNOUr, Me(NO)(NO₂)Gdn, and maleic hydrazide. These agents probably have in common that they give rise to spontaneous (Lawley and Brookes, 1963) or enzymically catalyzed (Lindahl, 1979) depurination. AP-sites are substrates of the AP-endonucleases (Verly and Rassart, 1975; Radman, 1976; Ljungquist, 1977), which are *uvr*- and ATP-independent enzymes.

There is an intermediate group of carcinogens which consistently induced more DNA repair synthesis in wild type cells than in *uvr* mutant cells. N-mustard, benzyl bromide, and 1-(*p*-chlorophenyl)-3-methyltriazene are cases in point. Under the assay conditions, these carcinogens appear to be very reactive and, as a consequence, not very selective with respect to their action on DNA. Thus, a broad spectrum of DNA modifications was to be expected. Some of these modifications (e.g., O⁶-alkylation of Gua) are likely to alter duplex DNA sterically. Other chemical modifications (e.g., N-7 alkylation of Gua, N-3 alkylation of Ade) probably have no steric consequences, since the alkyl groups can be easily accommodated in one of the DNA grooves. DNA glycosylases and AP-endonucleases seem to deal with these types of DNA alteration.

Intercalating agents (quinacrine, adriamycin, ethidium bromide) and chelate complex-forming agents (*cis*-Pt(II)diammine complexes) proved to be effective in inhibiting DNA replication without detectably stimulating repair synthesis. By contrast, some ultimate carcinogens (e.g., β-propiolactone, quinacrine mustard) inhibited DNA replication at low concentrations and stimulated repair synthesis at higher concentrations. Present knowledge regarding the interaction of the tested compounds with DNA suggests that *repair polymerization* might be indicative of *covalent* binding of the repair inducer, whereas *inhibition of replication* points to *covalent and/or noncovalent* binding of the causative agent.

The question remains as to how sensitive the ether-permeabilized *E. coli* cell is for monitoring covalent binding of carcinogenic residues via repair polymerization. Experiments are currently in progress to correlate the number of methyl, benz [*a*] anthryl-7-methyl, and *N*-2-fluorenylacetamido residues excised from the genome with the number of nucleotides inserted. These suggest that, under the conditions described, permeabilized cells insert 30–120 nucleotides per carcinogenic residue removed, depending on the type of carcinogen. Thus, if one allows repair polymerization in the presence of deoxynucleoside triphosphates of high specific activity, even rare DNA-modifying events (provided they are excised) induce sufficient repair synthesis to be detected.

Abbreviations

Abbreviations for nucleotides follow CBN Recommendations; see Eur. J. Biochem. *15* (1970) 203–208. dNTPs are the four deoxyribonucleoside triphosphates dATP, dCTP, dGTP and dTTP; Gua, guanine; dGuo, 2'-deoxyguanosine; Ade, adenine; et^{6}Gua, O^6-ethylguanine; et^{7}Gua, 7-ethylguanine; m^{3}Ade, 3-methyl-adenine; EGTA, ethyleneglycol-bis-(ffl'aminoethyl)-*N*,*N*'-tetraacetic acid; butyl-PBD, 2-(4-*t*-butyl-phenyl)-5-(4-biphenyl)-1,2,3-oxadiazole. MeNOUr, *N*-methyl-*N*-nitrosourea; EtNOUr, *N*-ethyl-*N*-nitrosourea; Me(NO) (NO$_2$) Gdn, *N*-methyl-*N*'-nitro-*N*-nitrosoguanidine; Me$_2$NNO, dimethylnitrosamine; (Ac)$_2$ONFln, *n*-acetoxy-2-acetylaminofluorene; (Ac)NOHFln, *N*-hydroxy-2-acetylaminofluorene; (Ac) NHFln, 2-acetylaminofluorene; TPA, 12-*O*-tetradecanoyl-phorbol-13-acetate; BCNU, *N*,*N*'-bis(2-chloroethyl)-*N*-nitrosourea; CCNU, *N*-(2-chloroethyl)-*N*'-cy-clohexyl-*N*-nitrosourea; H 33258 ("Hoechst 33258"), 2,2(4-hydroxy-phenyl)-6-benzimidazolyl-6-(1-methyl-4-piperazyl)-benzimidazol-trihydrochloride; AP-sites, apurinic sites.

Enzymes

DNA polymerase I (EC 2.7.7.7); endonuclease II (EC 3.1.4.-); exonuclease VI (EC 3.1.4.-); reduced-NAD: lipoamide oxidoreductase (EC 1.6.4.3).

References

Ames, B.N., Durston, W.E., Yamasaki, E., Lee, F.D. (1973): Carcinogens are mutagens: a simple test system combining liver homogenates for activation and bacteria for detection. Proc. Natl. Acad. Sci. USA *70*:2281–2285.

Blobstein, S.H., Weinstein, I.B., Grunberger, D., Weisgras, J., Harvey, R.G. (1975): Products obtained after *in vitro* reaction of 7,12-dimethylbenz[*a*]anthracene-5,6-oxide with nucleic acids. Biochemistry *14*:3451–3458.

Brookes, P., Lawley, P.D. (1960): The reaction of mustard gas with nucleic acids *in vitro* and *in vivo*. Biochem. J. *77*:478–484.

Denhardt, D.T., Kato, A.C. (1973): Comparison of the effect of ultraviolet radiation and ethidium bromide intercalation on the conformation of superhelical ϕX174 replicative form DNA. J. Mol. Biol. *77*:479–494.

Dürwald, H., Hoffmann-Berling, H. (1968): Endonuclease I-deficient and ribonuclease I-deficient *Escherichia coli* mutants. J. Mol. Biol. *34*:331–346.

Friedberg, E.C., Hadi, S.M., Goldthwait, D.A. (1969): Endonuclease II of *Escherichia coli*. J. Biol. Chem. *244*:5879–5889.

Friesel, H., Hecker, E. (1977): Reaction of arene oxides with nucleosides. Cancer Lett. *3*:169–175.

Gates III, F.T., Linn, S. (1977): Endonuclease from *Escherichia coli* that acts specifically upon duplex DNA damaged by ultraviolet light, osmium tetroxide, acid, or X-rays. J. Biol. Chem. *252*:2802–2807.

Geissler, E. (1962): Über die Wirkung von Nitrosaminen auf Mikroorganismen. Naturwissenschaften *49*:380–381.

Glickman, B.W., van Sluis, C.A., Heijneker, H.L., Rörsch, A. (1973): A mutant of *Escherichia coli* K12 deficient in the 5' - 3' exonucleolytic activity of DNA polymerase I. Mol. Gen. Genet. *124*:69–82.

Goldschmidt, B.M., Blazej, T.P., van Duuren, B.L. (1968): The reaction of guanosine and deoxyguanosine with glycidaldehyde. Tetrahedron Lett. *13*:1583–1586.

Goth, R., Rajewsky, M.F. (1974): Molecular and cellular mechanisms associated with pulse-carcinogenesis in the rat nervous system by ethylnitrosourea: ethylation of nucleic acids and elimination rates of ethylated bases from DNA of different tissues. Z. Krebsforsch. *82*:37–64.

Gupta, R.D., Mitra, S. (1974):Strand separation of DNA induced by ultraviolet irradiation *in vitro*. Biochim. Biophys. Acta *374*:145–158.

Hariharan, P.V., Cerutti, P.A. (1974): Excision of damaged thymine residues from gamma-irradiated poly(dA-dT) by crude extracts of *Escherichia coli*. Proc. Natl. Acad. Sci. USA *71*:3532–3536.

Hecker, E. (1968): Cocarcinogenic principles from the seed of *Croton tiglium* and from other Euphorbiaceae. Cancer Res. *28*:2338–2349.

Hecker, E. (1971): Isolation and characterization of the cocarcinogenic principles from croton oil. In: Methods in Cancer Research. Busch, H. (ed.), Vol. 6, New York: Academic Press, pp. 439–484.

Hecker, E. (1978): Structure-activity relationships in diterpene esters irritant and cocarcinogenic to mouse skin. In: Mechanisms of Tumor Promotion and Cocarcinogenesis. Slaga, T.J., Sivak, A., Boutwell, R.K. (eds.), Vol. II. New York: Raven Press. pp. 11–48.

Heflich, R.H., Dorney, D.J., Maher, V.M., McCormick, J.J. (1977): Reactive derivatives of benzo[a]pyrene and 7,12-dimethyl-benz[a]anthracene cause S$_1$ nuclease sensitive sites in DNA and "UV-like" repair. Biochem. Biophys. Res. Commun. *77*:634–641.

Hirota, Y., Gefter, M., Mindich, L. (1972): A mutant of *Escherichia coli* defective in DNA polymerase II activity. Proc. Natl. Acad. Sci. USA *69*:3238–3242.

Jeffrey, A.M., Blobstein, S.H., Weinstein, I.B., Beland, F.A., Harvey, R.G., Kasai, H., Nakanishi, K. (1976): Structure of 7, 12-dimethyl-benz[a]anthracene-guanosine adducts. Proc. Natl. Acad. Sci. USA *73*:2311–2315.

Kimball, R.F. (1978): The relation of repair phenomena to mutation induction in bacteria. Mutat. Res. *55*:85–120.

Kirtikar, D.M., Goldthwait, D.A. (1974): The enzymatic release of O^6-methylguanine and 3-methyladenine from DNA reacted with the carcinogen *N*-methyl-*N*-nitrosourea. Proc. Natl. Acad. Sci. USA *71*:2022–2026.

Kirtikar, D.M., Dipple, A., Goldthwait, D.A. (1975): Endonuclease II of *Escherichia coli*: DNA reacted with 7-bromomethyl-12-methylbenz[a]anthracene as a substrate. Biochemistry *14*:5548–5553.

Kohn, K.W., Steigbigel, N.H., Spears, C.L. (1965): Cross-linking and repair of DNA in sensitive and resistant strains of *E. coli* treated with nitrogen mustard. Proc. Natl. Acad. Sci. USA *53*:1154–1161.

Kohn, K.W., Spears, C.L., Doty, P. (1966): Interstrand crosslinking of DNA by nitrogen mustard. J. Mol. Biol. *19*:266–288.

Kornberg, A. (1974): DNA Synthesis. San Francisco: H.W. Freeman & Co., pp. 94–98.

Lawley, P.D., Brookes, P. (1963): Further studies on the alkylation of nucleic acids and their constituent nucleotides. Biochem. J. *89*: 127–138.

Lawley, P.D., Jarman, M. (1972): Alkylation of propylene oxide of deoxyribonucleic acid, adenine, guanosine and deoxyguanylic acid. Biochem. J. *126*:893–900.

Leh, F.K., Wolf, W. (1976): Platinum complexes: a new class of antineoplastic agents. J. Pharmacol. Sci. *65*:315–328.

Levine, A.F., Fink, L.M., Weinstein, I.B., Grunberger, D. (1974): Effect of *N*-2-acetylaminofluorene modification on the conformation of nucleic acids. Cancer Res. *34*:319–327.

Lindahl, T. (1976): New class of enzymes acting on damaged DNA. Nature (Lond.) *259*:64–66.

Lindahl, T. (1979): DNA glycosylases, endonucleases for apurinic/apyrimidinic sites, and base excision-repair. In: Progress in Nucleic Acid Research and Molecular Biology. Cohn, W.E., (ed.), Vol. 22, New York: Academic Press, pp. 135–193.

Lindqvist, B.H., Sinsheimer, R.L. (1967): The process of infection with bacteriophage ΦX174. XV. Bacteriophage DNA synthesis in abortive infections with a set of conditional lethal mutants. J. Mol. Biol. *30*:69–80.

Ljungquist, S. (1977): A new endonuclease from *Escherichia coli* acting at apurinic sites in

DNA. J. Biol. Chem. *252*:2808-2814.

Loveless, A., Hampton, C.L. (1969): Inactivation and mutation of coliphage T_2 by N-methyl- and N-ethyl-N-nitrosourea. Mutat. Res. *7*:1-12.

Macquet, J.-P., Theophanides, T. (1975): DNA-platinum interactions *in vitro* with *trans-* and *cis*-Pt(NH$_3$)$_2$Cl$_2$. Bioinorg. Chem. *5*:59-66.

Ogawa, H., Shimada, K., Tomizawa, J.-I. (1968): Studies on radiation-sensitive mutants of *E. coli*. Mol. Gen. Genet. *101*:227-244.

Pascoe, J.M., Roberts, J.J. (1974): Interactions between mammalian cell DNA and inorganic platinum compounds-I. Biochem. Pharmacol. *23*:1345-1357.

Preussmann, R., Ivankovic, S., Landschütz, C., Gimmy, J., Flohr, E., Griesbach, U. (1974): Carcinogene Wirkung von 13 Aryldialkyltriazenen an BD-Ratten. Z. Krebsforsch. *81*:285-310.

Radman, M. (1976): An endonuclease from *Escherichia coli* that introduces single polynucleotide chain scissions in ultraviolet-irradiated DNA. J. Biol. Chem. *251*:1438-1445.

Rhaese, H.-J., Freese, E. (1968): Chemical analysis of DNA alterations. Biochim. Biophys. Acta. *155*:476-490.

Roberts, J.J., Warwick, G.P. (1963): The reaction of β-propiolactone with guanosine, deoxyguanylic acid and RNA. Biochem. Pharmacol. *12*:1441-1442.

Salganik, R.J., Drevich, V.F., Vasyunina, E.A. (1967): Isolation of ultraviolet-denatured regions of DNA and their base composition. J. Mol. Biol. *30*:219-222.

Schmidt, R., Hecker, E. (1975): Autoxidation of phorbol esters under normal storage conditions. Cancer Res. *35*:1375-1377.

Seeberg, E., Strike, P. (1976): Excision repair of ultraviolet-irradiated deoxyribonucleic acid in plasmolyzed cells of *Escherichia coli*. J. Bacteriol. *125*:787-795.

Seeberg, E. (1978): Reconstitution of an *Escherichia coli* repair endonuclease activity from the separated *uvrA*$^+$ and *uvrB*$^+$/*uvrC*$^+$ gene products. Proc. Natl. Acad. Sci. USA *75*:2569-2573.

Siebert, D., Marquardt, H., Friesel, H., Bartsch, H., Hecker, E. (1980): PAH and possible metabolites: convertogenic activity in yeast and tumor initiating activity in mouse skin. J. Cancer Res. Clin. Oncol.

Sims, P., Grover, P.L. (1974): Epoxides in polycyclic aromatic hydrocarbon metabolism and carcinogenesis. Advan. Cancer Res. *20*:165-274.

Singer, B. (1975): The chemical effects of nucleic acid alkylation and their relation to mutagenesis and carcinogenesis. In: Progress in Nucleic Acid Research and Molecular Biology. Cohn, W.E. (ed.), Vol. 15, New York: Academic Press, pp. 219-284.

Slater, E.E., Anderson, M.D., Rosenkranz, H.S. (1971): Rapid detection of mutagens and carcinogens. Cancer Res. *31*:970-973.

Smith, K.C. (1978): Multiple pathways of DNA repair in bacteria and their roles in mutagenesis. Photochem. Photobiol. *28*:121-129.

Sun, L., Singer, B. (1975): The specificity of different classes of ethylating agents toward various sites of Hela cell DNA *in vitro* and *in vivo*. Biochemistry *14*:1795-1802.

Szybalski, W., Iyer, V.N. (1967): The mitomycins and porfiromycins. In: Antibiotics. Gottlieb, D., Shaw, P.D. (eds.), Vol. 1, Berlin: Springer-Verlag, pp. 211-245.

Thielmann, H.W. (1976): Carcinogen-induced DNA repair in nucleotide-permeable *Escherichia coli* cells. Eur. J. Biochem. *61*:501-513.

Thielmann, H.W. (1977): Detection of strand breaks in ΦX 174 RFA and PM2 DNA reacted with ultimate and proximate carcinogens. Z. Krebsforsch. *90*:37-69.

Thielmann, H.W., Gersbach, H. (1978a): Carcinogen-induced DNA repair in nucleotide-permeable *Escherichia coli* cells. Z. Krebsforsch. *92*:157-176.

Thielmann, H.W., Gersbach, H. (1978b): The nucleotide-permeable *Escherichia coli* cell, a sensitive DNA repair indicator for carcinogens, mutagens and antitumor agents binding covalently to DNA. Z. Krebsforsch. *92*:177-214.

Thielmann, H.W., Vosberg, H.-P., Reygers, U. (1975): Carcinogen-induced DNA repair in nucleotide-permeable *Escherichia coli* cells. Eur. J. Biochem. *56*:433-447.

Tomasz, M. (1970): Novel assay of 7-alkylation of guanine residues in DNA. Application to nitrogen mustard, triethylenemelamine and mitomycin C. Biochim. Biophys. Acta *213*:288-295.

Udenfriend, S., Clark, C.T., Axelrod, J., Brodie, B.B. (1954): Ascorbic acid in aromatic hydroxylation. J. Biol. Chem. *208*:731-739.

Verly, W.G., Brakier, L. (1969): The lethal action of monofunctional and bifunctional alkylating

160   Heinz Walter Thielmann

agents on T7 coliphage. Biochim. Biophys. Acta *174*:674–685.

Verly, W.G., Rassart E. (1975): Purification of *Escherichia coli* endonuclease specific for apurinic sites in DNA. J. Biol. Chem. *250*:8214–8219.

Vosberg, H.-P., Hoffmann-Berling, H. (1971): DNA synthesis in nucleotide-permeable *Escherichia coli* cells. J. Mol. Biol. *58*:739–753.

Waldstein, E.A., Sharon, R., Ben-Ishai, R. (1974): Role of ATP in excision repair of ultraviolet radiation damage in *Escherichia coli*. Proc. Natl. Acad. Sci. USA *71*:2651–2654.

Wechsler, J.A., Nüsslein, V., Otto, B., Klein, A., Bonhoeffer, F., Hermann, R., Gloger, L., Schaller, H. (1973): Isolation and characterization of thermosensitive *Escherichia coli* mu-

tants defective in deoxyribonucleic acid replication. J. Bacteriol. *113*:1381–1388.

Wheeler, G.P. (1976): A review of studies on the mechanism of action of nitrosoureas. In: Cancer Chemotherapy. Sartorelli, A.C. (ed.), Vol. 30, Washington, D.C.: Amer. Chem. Symp. Series (Gould, R.F., ed.), pp. 87–119.

Witkin, E.M. (1976): Ultraviolet mutagenesis and inducible DNA repair in *Escherichia coli*. Bacteriol. Rev. *40*:869–907.

Wood, A.W., Wislocki, P.G., Chang, R.L., Levin, W., Lu, A.Y., Yagi, H., Hernandez, O., Jerina, D.M., Conney, A.H. (1976): Mutagenicity and cytotoxicity of benzo[*a*]pyrene benzo-ring epoxides. Cancer Res. *36*:3358–3366.

Woodworth-Gutai, M., Lebowitz, J., Kato, A.C., Denhardt, D.T. (1977): Ultraviolet light irradiation of PM2 superhelical DNA. Nucleic Acids Res. *4*:1243–1256.

15

The Yeast Saccharomyces cerevisiae: *An Assay Organism for Environmental Mutagens*

R.C. VON BORSTEL

Genetic Properties of *Saccharomyces cerevisiae*

The yeast *Saccharomyces cerevisiae* has come to be an important vehicle for transferring the vast amount of knowledge learned from the molecular biology of prokaryotes to the study of the cells of higher organisms. Yeast is a true eukaryote, well understood genetically, with both haploid and diploid cell cycles. Meiosis takes place in the diploid cell without the prior differentiation of elaborate fruiting bodies or other sexual paraphernalia.

Saccharomyces cerevisiae has 17 chromosomes, yet the total DNA content is but five times that of *Escherichia coli*. Yeast can be handled with standard microbiological techniques, yet genetically and molecularly it is a typical eukaryote with a true nucleus, true chromosomes with telomeres and centromeres, and interstitial noncoding DNA within genes which has to be processed out of mRNA and certain tRNAs.

Like other eukaryote organisms, yeast has mitochondria.

Since *Saccharomyces* functions perfectly well either aerobically or anaerobically, the mitochondria are accessible to genetic study. Spontaneous, antibiotic resistant mutants of mitochondrial DNA were first found in *Saccharomyces* by Thomas and Wilkie (1968). By 1975 about 12 mitochondrial genes had been identified (Plischke *et al.*, 1976). A technique for mutating mitochondrial genes (Putrament *et al.*, 1973; Putrament *et al.*, 1978), and another for selecting mutants (Tzagaloff *et al.*, 1975), led to a complete saturation of the genetic map of the mitochondrial genome within three years (Dujon *et al.*, 1977; Borst and Grivell, 1978). With the advent of the general use of restriction enzymes, restriction maps of the mitochondrial DNA became available (Bernardi, 1978; Fonty *et al.*, 1978), and the rules governing genetic recombination of the mitochondrial DNA are being worked out (Strausberg *et al.*, 1978). This explosion of activity made it possible to explore relations of the mitochondria and the cell nucleus in a detail not possible with any other organism.

Genetic transformation of yeast cells with exogenous DNA has now been demonstrated (Hinnen et al., 1978). Restriction enzymes are used to cut up nuclear DNA of yeast. It is then attached to plasmids, and clones of different segments of yeast chromosomes are established in *Escherichia coli*. The DNA of a particular genetic region of yeast can thus be amplified through multiplication of the plasmids. Yeast spheroplasts are exposed to the plasmid DNA, and the DNA becomes incorporated in yeast chromosomes. Genetic analyses have been made of how and where the exogenous DNA recombines—mostly it returns to or beside its former genetic location, but occasionally its attachment is elsewhere in the genome. This transfer of DNA from the eukaryote to a prokaryote and back again to the eukaryote, where it can be analyzed genetically, is not yet possible with any other higher organism.

From the above, we can infer that the yeast *Saccharomyces cerevisiae* has properties that make it open to genetic studies of the nucleus (tetrads and haploid and diploid cell cycles) and the mitochondria (aerobic and anaerobic metabolism), and that molecular techniques are available (genetic transformation) which make it unparalleled among eukaryotes for most kinds of genetic analysis.

Now let us briefly examine some properties which make it unparalleled among eukaryotes for mutational analysis.

Radiobiology of Yeast

In dose-action experiments on yeast, it was found early that diploid cells are much more resistant than haploid cells to lethal effects of ionizing radiation. It was also observed that yeast cells undergoing DNA synthesis are more resistant to radiation killing than cells which are in interphase.

Mutagen sensitivity during the G_1 phase and resistance during the S-phase are not limited to lethality; they have also been observed for mitotic recombination (Fig. 15.1) (Pittman, 1956), and for mutation of haploid cells with certain chemicals (Figs. 15.2 and 15.3) (Zimmermann and Schwaier, 1963).

When yeast cells are irradiated and held for 48 hours before plating, the survival of the cells is substantially increased. This phenomenon is called liquid-holding recovery. Liquid-holding recovery is observed after treatment of haploid and diploid cells with ultraviolet radiation, and after treatment of diploid cells with ionizing radiation. Cells without mitochondrial DNA do not exhibit liquid-holding recovery unless glucose is present in the holding medium, so it is believed that energy metabolism is required for repair of the damage (Nunes de Langguth and Gelos, 1972).

During the last decade, the repair of induced lesions has been analyzed genetically in yeast (Game and Cox, 1973; Brendel and Haynes, 1973). At the present time nearly 100 repair-deficient mutant loci have been found (Lemontt, 1980). Four repair pathways are known for ultraviolet-radiation-induced damage: photorepair (Resnick, 1969), excision repair (Game and Cox, 1973), double-strand-break repair (Game and Mortimer, 1974; Ho and Mortimer, 1976; Resnick, 1976), and mutagenic repair (Lemontt, 1971; Lawrence and Christensen, 1976). The latter two pathways are the principal systems for repairing lesions induced by ionizing radiation (Game and Mortimer, 1974). The familiar radiation-resistance of a/α diploid cells exists because these heterozygotes have enhanced double-strand-break repair capacity (Ho, 1975). At least one other pathway is used for lesions induced by methyl methanesulfonate (Prakash and Prakash, 1977). Spontaneous lesions have been shown to be repaired by these and possibly one or two other pathways as well (Hastings et al., 1976; von Borstel and Hastings, 1977).

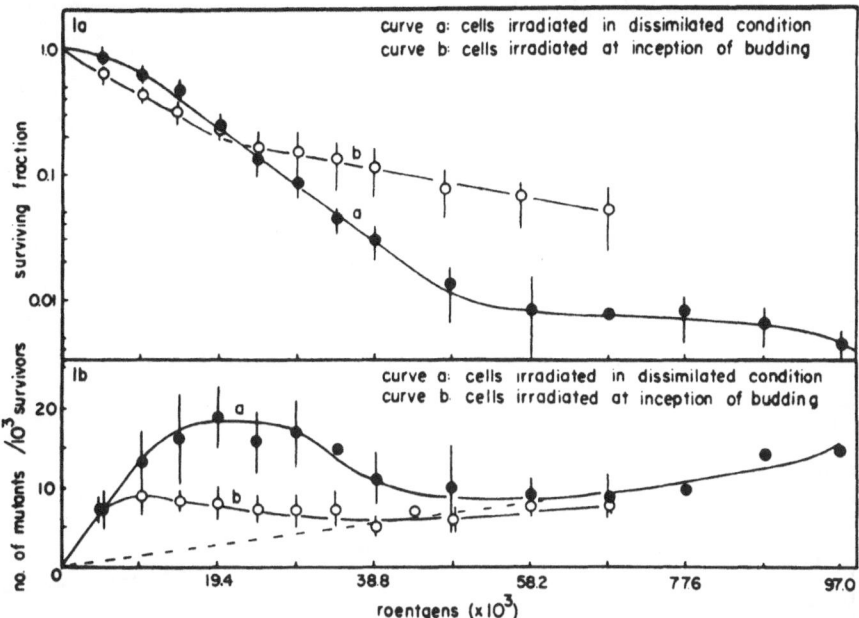

Figure 15.1. The curves for survival (1a) and induced recombinants (1b) of diploid mitotic cells of *Saccharomyces cerevisiae:* Stationary phase cells (0); rapidly growing cells (O). The cells in the inter-budding stages are more sensitive to γ-radiation, and the budding cells are more resistant. The recombinants from induced mitotic crossing-over of cells in the sensitive fraction reach a peak and then decline in frequency as these cells are killed; induced mitotic crossing-over begins to increase again in the resistant fraction (Pittman, 1956).

Enzymes used for one type of lesion in one pathway are used in other pathways for different types of lesions. For example, an ethyl methanesulfonate-induced lesion uses at least one enzyme in the double-strand-break repair pathway for the mutagenic repair pathway (Prakash and Prakash, 1977). Also, the double-strand-break repair

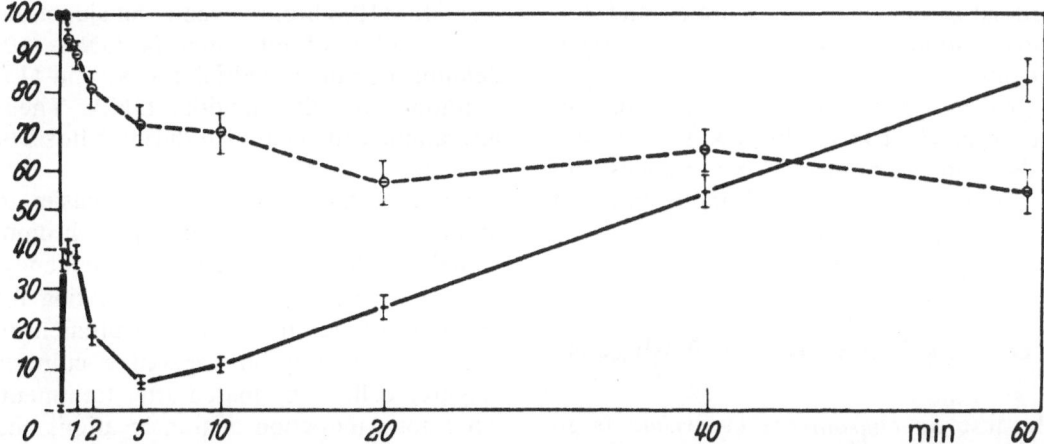

Figure 15.2. Reversion frequencies of *ade6-45* after different exposures to 0.003% concentration of *N*-nitroso-*N*-methylacetamid (pH 7.5). Ordinate: survival (----) and reversions/10^6 survivors (———); Abscissa: exposure time in minutes. Cells in this interbudding stage are more sensitive and cells during budding are more resistant, both to killing and to mutation induction (Zimmermann and Schwaier, 1963).

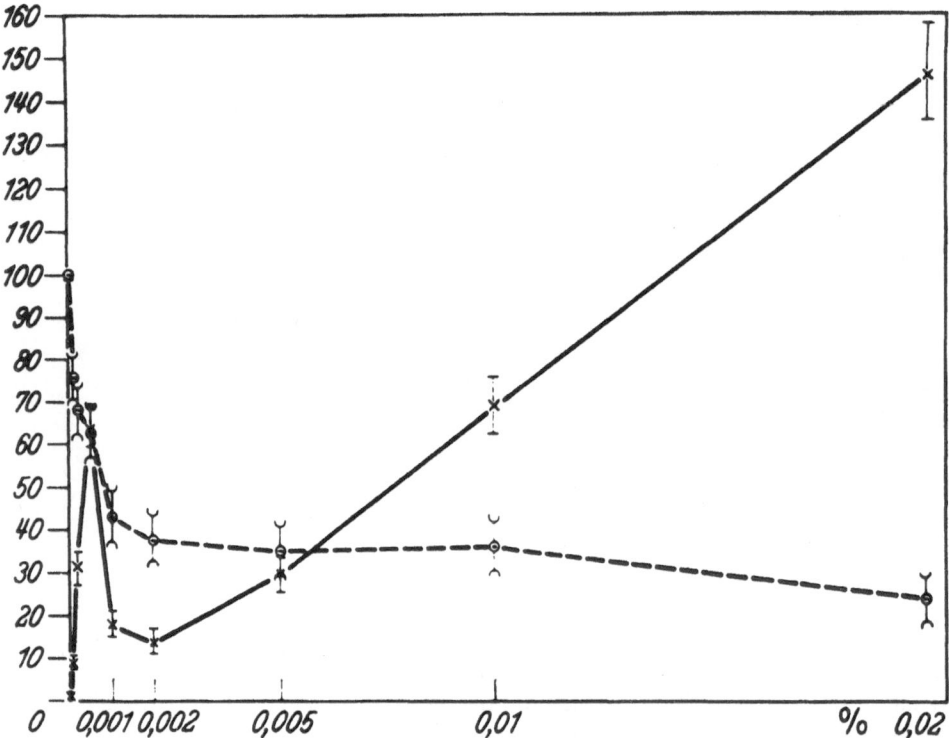

Figure 15.3.   Reversion frequencies of *ade6–45* after treatment with different concentrations of *N*-nitroso-*N*-methylacetamid with a standard exposure time of 30 minutes (pH 7.5). Ordinate: survival (---) and reversions/10⁶ survivors (———); Abscissa: concentration in % (Zimmermann and Schwaier, 1963).

pathway needs at least one step of the excision repair pathway (*RAD3*) for repairing crosslinks induced with psoralen and 360 nm radiation (Jachymczyk and von Borstel, unpublished data).

It is now apparent that most lesions induced in yeast DNA by radiation, chemicals, and spontaneous occurrences are reparable by one or more mutagenic and nonmutagenic pathways.

Yeast as a Test System for Mutagens

Because *Saccharomyces cerevisiae* is so well known genetically, it is possible to identify the following genetic alterations: (1) cell killing, (2) dominant lethality, (3) forward mutations, (4) frameshift mutations, (5) base substitution mutations identifiable as either a transversion or a transition, (6)

gene conversion, (7) intragenic recombination, (8) intergenic recombination, (9) translocation, (10) chromosome nondisjunction, (11) total loss of mitochondrial DNA, (12) deletion of mitochondrial DNA, and (13) mutations in mitochondrial DNA. These alterations can be studied during mitosis or meiosis.

Some of these alterations are easier to identify than others. For example, chromosomal translocations are rare in frequency, and each one requires genetic testing for certainty of identification; measurement of dominant lethality in vegetative cultures requires cells to be mated after treatment. Therefore, induction of translocations and dominant lethality would not be a suitable endpoint at this time for simple rapid assays. The same is true for distinguishing total and partial loss of mitochondrial DNA. All of the rest of the assayable endpoints in yeast can be carried out in one step, although usually not in the same strain.

Before we discuss the specific strains usable for testing of environmental mutagens in yeast, the differences between stationary phase and growing vegetative cells should be discussed because it is now evident that, for highest efficiency, tests of environmental mutagens should be carried out on both phases.

The Cell Cycle in Relation to Mutagen Sensitivity

As was pointed out before, cells fall into two populations with respect to sensitivity to ionizing radiation and to certain chemicals. The cells in S-phase are resistant both to killing and to mutation induced by X-radiation. The cells in G_1 phase are relatively sensitive. These differences to X-radiation disappear if the double-strand-break repair pathway is rendered nonfunctional by mutation of one of the steps.

It is of considerable interest that some compounds, like hycanthone (von Borstel and Igali, 1975) and pyrvinium pamoate (Galindo-Prince and von Borstel, unpublished data), are much more mutagenic to stationary phase cells than to growing cells, and other compounds, like cyclophosphamide and diethylnitrosamine, are much more mutagenic to growing cells than to stationary phase cells. The mystery disappeared when it was shown that when the 9000 g supernatant fraction containing microsomes of mammalian liver cells (S9 preparation) is present during mutagenization, then pyrvinium pamoate is not mutagenic in the stationary phase (Galindo-Prince and von Borstel, unpublished data), and cyclophosphamide and dimethylnitrosamine are mutagenic in the stationary phase. It is clear that the cytochrome P-450 system detoxifies pyrvinium pamoate and converts cyclophosphamide and dimethylnitrosamine into mutagens.

In summary, cell stages may be involved with intrinsic sensitivity differences owing to differences in availability or capacity of repair pathways, but the most marked responses are the result of the lack or presence of endogenous metabolic activation.

Endogenous Activation of Premutagens

Oxygen is used in two quite different aerobic metabolisms in yeast although, anaerobically, yeast can get along without either of them. The first is where oxygen is reduced to H_2O during the formation of ATP by catalysis by the respiratory pathway in mitochondria. The second is where oxygen is used in the synthesis of such compounds as sterols and the degradation of such compounds as toxic chemicals. Cytochrome P-450 associated with microsomes activates the oxygen used for this second set of reactions. It is important to note that enzyme systems are used in association with cytochrome P-450 to activate the oxygen, and this is very likely the source of species variation in reactions or in variation brought about by cell differentiation.

Gillette (1966) attributes the following reactions to microsomal cytochrome P-450 from the mammalian liver:

Aromatic hydroxylation	N-Oxidation
Aliphatic hydroxylation	Sulfoxidation
N-Dealkylation	Dehalogenation
S-Dealkylation	Azoreduction
O-Dealkylation	Nitroreduction
Deamination	Peroxidation
Desulfuration	Epoxidation

As is known from the work of Miller and Miller (1971), many of these reactions can convert nascent compounds into active mutagens and carcinogens. Callen and Philpot (1977) have demonstrated in yeast that different premutagens can be activated by either N-hydroxylation (β-naphthylamide and ethyl carbamate), dealkylation (dimethylnitrosamine), epoxidation (aflatoxin B_1), sulfoxidation (dimethyl sulfoxide), or hydrolysis (cyclophosphamide).

The cytochrome P-450 is induced naturally in growing cells using glucose as a carbon source (Wiseman et al., 1975; Callen and Philpot, 1977). Phenobarbital also has been used to induce cytochrome P-450 (Wiseman and Lim, 1975). When cells enter the stationary phase, the cytochrome P-450 rapidly disappears. Therefore, in order to obtain maximum endogenous activation of premutagens, it is essential to carry out the experiment with growing cells or cells just entering the stationary phase.

Callen and Philpot (1977) also found that strain variations exist in levels of cytochrome P-450. Strain D5 has approximately nine times as much as strain D4. Other strains, including strain XV185-14C, more closely resemble strain D4 (Callen, unpublished data).

Callen (1978) points out that differences do exist between yeast and mammals for the levels of some types of metabolic activation. Certainly it is well known that levels differ for various mammalian species as well as for tissues. Also, the cytochrome P-450 tends to be in the reduced form in yeast, and in the oxidized state in mammalian cells. Consequently, the cytochrome P-450 and its adjunct enzymes in yeast cannot be regarded as an exact analogue of the human liver. Nevertheless, it is a near enough analogue so that growing yeast can be used as a first screen for mutagens without adding an exogenous S9 microsomal fraction from mammalian liver for activation of premutagens.

Strains Used for Testing of Environmental Mutagens

An array of strains has been constructed for testing of presumptive or possible environmental mutagens. These are of two general types, diploids and haploids, and each has certain advantages.

The diploid strains are shown in Table 15.1. The D3 strain (Zimmermann et al., 1966) is simply heterozygous for the ade2 locus, and its endpoint is homozygosity for ade2 as most often induced by an intergenic recombination between the ade2 locus and the centromere during mitosis. Theoretically, red colonies could arise by gene conversion, nondisjunction, or a mutation.

Strain D3 is less sensitive to environmental mutagens than some of the other strains which have been developed, but more chemicals have been tested with this strain than with any other strain of yeast.

Strain D4 (Zimmermann and Schwaier, 1967) was developed to measure induced intragenic recombination.

Strain D5 (Zimmermann, 1973) uses both mitotic gene conversion and intergenic mitotic recombination as endpoints. This strain fortuitously has an endogenous cytochrome P-450 activity approximately nine times higher than strain D4 and some of the haploid strains. It is probably the strain of choice (as yeast is the organism of choice) for developing nations, or for field conditions where S9 preparations and synthetic media are too difficult or too expensive to use. The tests with strain D5 use yeast extract, peptone and glucose medium for both growth in liquid and colony formation in plates; synthetic media are not required.

Strain D6 (Parry and Zimmermann, 1976) is a strain developed to use nondisjunction as an endpoint, but also indicates mitotic intergenic recombination, gene conversion, and mutation. This strain seems to be one of the more sensitive yeast strains available for general use. It requires constant monitoring because it breaks down easily.

Strain D7 (Zimmermann et al., 1975) was developed to have the best properties of D4 and D5. It detects a slightly different array of mutagens than does strain D6, and it is quite stable and easy to use.

Strain 6117 (Sora et al., 1979) is a nondisjunction testing strain with the same ad-

Table 15.1. *Diploid Strains and Genotypes of Saccharomyces cerevisiae Commonly Used for Assaying Chemical Carcinogens*

Strain	Genotype	Genetic endpoint	Reference
D3	$\dfrac{CYH4\ ade2\ his8}{+\quad +\quad +}$	Mitotic recombination	Zimmermann *et al.*, 1966
D4	$\dfrac{\alpha\ gal2\ ade2\text{-}2\ trp5\text{-}12\ \ leul\ +}{a\ +\ \ ade2\text{-}1\ trp5\text{-}27}$	Intragenic recombination	Zimmermann & Schwaier, 1967
D5	$\dfrac{\alpha\ trpl\ ade2\text{-}40\ MAL1\ +}{a\ +\ ade2\text{-}119\ +\ MAL4}$	Mitotic recombination, gene conversion and mutation	Zimmerman, 1973
D6	$\dfrac{\alpha\ his4\ tifade3\ leul\ trp5\ cyh2\ met13\ ade2\text{-}40}{a\ +\quad +\quad +\quad +\quad +\quad +\quad ade2\text{-}40}$	Mitotic nondisjunction	Parry & Zimmermann, 1976
D7	$\dfrac{\alpha\ ade2\text{-}40\ trp5\text{-}12\ ilv1\text{-}92}{a\ ade2\text{-}119\ trp5\text{-}27\ ilv1\text{-}92}$	Intragenic recombination, induced mitotic crossing-over, gene conversion and reverse mutations	Zimmermann *et al.*, 1975
6117	$\dfrac{\alpha\ +\ +\ cyh2\ met13\ tyr3\ lys5\ ade5\text{-}7\ ade2\text{-}1}{a\ leul\ trp5\ +\ \ +\ \ +\ \ +\ \ +\quad ade2\text{-}1}$ $\dfrac{ura4\ can1}{+\quad +}$	Mitotic gene conversion	Sora *et al.*, 1979b
JD1	$\dfrac{\alpha\ his4ABC\ +\ \ +\ \ +\ \ trp5\text{-}U6}{a\ his4C\ \ ade2\text{-}1\ ser7\ his8\ trp5\text{-}U9}$	Intragenic recombination	Davies *et al.*, 1975
CM-1194	$\dfrac{\alpha\ ade2\text{-}119\ trp5\text{-}27\ 1\text{-}131\ ilv1\text{-}92}{a\ ade2\text{-}40\ trp5\text{-}12\,cyc1\text{-}45\ ilv1\text{-}92}$	Intragenic recombination, mitotic recombination, gene conversion and mutation	Moore, 1978
CM-1293	$\dfrac{\alpha\ ade2\text{-}119\ trp5\text{-}27\ cyc1\text{-}131\ ilv1\text{-}92\ +}{a\ ade2\text{-}40\ trp5\text{-}12\ cyc1\text{-}45\ ilv1\text{-}92\ can1}$	Intragenic recombination, mitotic recombination, gene conversion and mutation	Moore & Schmick, 1979

vantages as strain D6. It has not been as well validated as strain D6.

Strain JD1 (Davies *et al.*, 1975) tests for intragenic recombination at two different loci. Indeed, it seems to be an excellent strain, with not quite the same spectrum of specificity as strain D7.

Strains CM-1194 and CM-1293 resemble strain D7 with slight changes to extend the usefulness of the strain. It has not been well validated, but CM-1293 was shown to respond to saccharin (Moore and Schmick, 1979).

The haploid strains are shown in Table 15.2. In general, the haploid strains are used for testing for mutations and killing only, and they probably identify a slightly different array of mutagens than strains D6 and D7. The protocol for assaying mutations by XV185-14C, the strain developed by S.-K. Quah and me, is given in the Appendix. It was developed from a series of suppressible mutants from R.K. Mortimer's laboratory, the *hom3-10* mutant from G.E. Magni's laboratory, and the *his1-7* mutant from S. Fogel's laboratory. The strains have been designed to be quite useful for detecting different types of mutations, frameshifts, transversions of different types, transitions or just general base substitutions. The other strains from our laboratory and those from Larimer's laboratory derive from strain XV185-14C.

The haploid strains, taken all together, contain the following alleles: *ade2-1*, *arg4-17*, *lys1-1*, and *trp5-48* which are ochre mutants, *trp1-1* and *tyr7-1* which are amber mutants, and *ade1-10* and *hom3-10* which are believed to be frameshift mutants because they exhibit a higher spontaneous rate during meiosis than during mitosis (Magni and von Borstel, 1962; Magni, 1963). Also, *hom3-10* responds to frameshift suppressors (G. Lucchini Bonomini and G.E. Magni, unpublished data). The mutants *his1-7* and *arg4-27* are missense mutants revertible by back mutation and intragenic missense suppression.

Strain XΛ4-8C does not carry an arginine-requiring mutant so it can be used for assaying for canavanine resistance, every mutant of which is recessive and at one locus only. If a strain carries the red color markers *ade1* or *ade2*, it can also be used to assay forward mutations at five adenine loci encoding enzymes on the adenine pathway prior to where the red adenine markers act. These mutants show up as white, adenine-requiring colonies among a background of red colonies. The temperature-sensitive *ade2-912* (in strain XV1000-1A) facilitates study of the forward mutations in the loci which act early in the adenine pathway.

The lack of liquid-holding recovery in XV185-14C, the presence of *rad3* in XV423-2A, and the presence of *rad2* in XΛ4-8C possibly widen the sensitivity ranges. Strain XV185-14C is somewhat temperature-sensitive, exhibiting a syndrome similar to those of some of the cell division cycle mutants of Hartwell (1975); this may also aid in its general sensitivity to mutagens.

A haploid, triple-rad strain *rad1 rad50 rad18*, which uses lethality only as the endpoint, has been used by J.M. Parry (unpublished data). It is a more sensitive strain than D6 but, as expected, does give more false positives.

The haploid strains which carry either *ade1* or *ade2*, or the diploid strains which carry one of them homozygously, are useful for recognition of respiratory-deficient mutants which usually have defective mitochondrial DNA. These petite strains, as they are called, take on the characteristic red coloration rather slowly. Certain compounds, such as intercalating agents like the ethidium salts, can be recognized as possible mutagens more rapidly this way than by any other.

The use of yeast mitochondria as an assay for mutagens has not been explored systematically. This needs to be done, because mitochondrial DNA responds to some mutagens, like $MnSO_4$, which are difficult to detect with standard yeast assays.

With respect to efficiency for carcinogen detection, yeast seems to be somewhat bet-

Table 15.2. *Haploid Strains and Genotypes of Saccharomyces cerevisiae Commonly Used For Assaying Chemical Carcinogens*

Strain	Genotype	Genetic endpoint	Reference
XV185-14C	a *trp5-48 arg4-17 lys1-1 ade2-1 his1-7 hom3-10*	Reversion of ochre, missense and putative frameshift	Quah & von Borstel, cited in Shahin & von Borstel, 1977, 1978
XV1000-1A	a *ade2-912 arg4-17 his1-7 hom3-10 trp5-48*	Forward mutation, reversion of ochre, missense and frameshift	von Borstel, unpublished
6126/16c	a *his4-1 ade1-10 arg4-27 tyr7-1 trp1-1*	Reversion of ochre, amber, missense and putative frameshift	Sora *et al.,* 1979
XΛ4-8C	α *ade2-1 his1-7 hom3-10 lys1-1 trp5-48 rad2*	Reversion of ochre, missense and frameshift	Larimer *et al.,*
XV423-2A	α *ade2-1 his1-7 hom3-10 lys1-1 trp5-48 rad3-12*	Reversion of ochre, missense and frameshift	Grant *et al.,* 1979
Repair-sensitive	*rad2 rad18 rad50*	Survival	Sharp & Parry (in press)

ter than the Salmonella test devised by Bruce Ames, but a higher concentration of mutagen is required for an individual test. Yeast has been positive with every human carcinogen so far tested.

In conclusion, any three yeast strains taken together would provide almost total information about any mutagen if both stationary and growing cells are tested. The strains I would choose would be D6, XV185-14C, and D7 or CM-1293. Undoubtedly we can reach a point where one or two strains may be all that would be needed. If the strains are well aerated and used under growing and nongrowing conditions, it is likely that the exogenously added S9 microsomal fraction will be found to be unnecessary for general testing. We are not ready to proclaim this advanced state of the art yet, but strains are being planned and constructed which should improve yeast even further as a simple organism for testing environmental mutagens.

APPENDIX I

Protocol for a Haploid Yeast Reversion Test for Assaying Mutagens

R.C. von Borstel, M.M. Shahin, and R.D. Mehta

Preparation of Cell Suspensions, Treatment Conditions, and Plating

Stationary phase cells. Freshly grown cells on YEPD-agar plates (yeast extract 1%, peptone 2%, glucose 2%, agar 1.5%) are inoculated into liquid YEPD medium and incubated at 30° C in a shaking water bath until they reach stationary phase. The cells are then harvested by centrifugation and washed twice in phosphate buffer (pH 7.0). Cell concentrations are adjusted to $5 \times 10^7 - 1 \times 10^8$ cells/ml. At zero time, the compound to be tested is added to the cell suspension to give the desired final concentration of the test chemical. Control and treated cell suspensions are continually agitated at 30° C. At appropriate intervals, samples are withdrawn, centrifuged, and washed twice in phosphate buffer. For survival estimation, treated and control samples are diluted and plated onto YEPD-agar plates to give about 100–300 colonies per plate. The plates are incubated for 4–5 days at 30° C, and colony forming ability is used as the criterion for survival. For reversion studies involving the metabolic activation system, the following components are used: S9 microsomal fraction (Litton Bionetics), NADP, glucose-6-phosphate, $MgCl_2$, KCl, phosphate buffer (pH 7.4). The S9 mix was prepared as described by Ames et al. (1975) and was added to the control and treatment tubes to give a final concentration of the S9 microsomal fraction as 55 or 110 μl/ml of cell suspension.

Growing cells. An inoculum prepared from the 48-hour-old culture grown on YEPD-agar plates at 30° C is added to 4 ml of liquid YEPD medium dispensed into test tubes (150 mm $\times$ 25mm) to give a cell concentration of 1×10^7 cells/ml. Cultures are then incubated at 30° C on a water bath shaker for 4 hours. At this time, the desired concentrations of the test compound are added to the test tubes. Incubation of control and treated samples is continued for another 24 and/or 48 hours. Samples are withdrawn at appropriate time intervals after addition of the compounds, and are handled in the same manner as described for stationary phase cells.

Determination of Spontaneous and Induced Reversion Frequency

For the determination of the spontaneous or the induced reversion frequency, undiluted samples (1×10^6–5×10^7 cells/ml) are plated (0.2 ml/plate) on medium containing 0.67% Difco yeast nitrogen base without amino acids, 2% glucose, 2% agar, plus all required supplements except histidine, threonine, or arginine. Reversion at the *his1*, *hom3* or *arg4* loci produced cells capable of growth on medium deficient in histidine, threonine, or arginine, respectively. The *hom3* mutant is auxotrophic either for homoserine or for both methionine and threonine.

Preparation of the Solution of the Test Compounds

All compounds are dissolved in different volumes of DMSO (dimethylsulfoxide) depending upon the solubility of a given compound, except for some compounds which are solubilized in water.

Criteria Used

1. Compounds which show mutation frequency double or higher than that of the controls for at least one of the three markers tested are recorded as mutagenic and are marked as (+).
2. When the mutation frequency is found to be consistently higher than the controls—at least by a factor of 1.5 of the control for at least one marker—such compounds are recorded as weakly mutagenic and are marked as (±).
3. The compounds which show less than 1.5 times the control mutation frequencies for all the markers tested are re-

corded as nonmutagenic and are marked as (−).
4. Compounds, which in some cases show only a slight increase (generally ≤ 1.2 times the control frequency) in induced mutation frequency or in other cases if the mutation frequency is higher but data are found to be inadequate to support the conclusion, are recorded as doubtful ones and are marked (?).

References

Ames, B.N., McCann, J., Yamasaki, E. (1975): Methods for detecting carcinogens and mutagens with the *Salmonella*/mammalian microsome mutagenicity test. Mutat. Res. *31*: 347–364.

Bernardi, G. (1978): Intervening sequences in the mitochondrial genome. Nature (Lond.) *276*:558–559.

Borst, P., Grivell, L.A. (1978): The mitochondrial genome of yeast. Cell *15*: 705–723.

Brendel, M., Haynes, R.H. (1973): Interactions among genes controlling sensitivity to radiation and alkylation in yeast. Mol. Gen. Genet. *125*:197–216.

Callen, D.F. (1978): A review of the metabolism of xenobiotics by microorganisms with relation to short-term test systems for environmental carcinogens. Mutat. Res. *55*:153–163.

Callen, D.F., Philpot, R.M. (1977): Cytochrome P450 and the activation of promutagens in *Saccharomyces cerevisiae*. Mutat. Res. *45*: 309–324.

Davies, P.J., Evans, W.E., Parry, J.W. (1975): Mitotic recombination induced by chemical and physical agents in the yeast *Saccharomyces cerevisiae*. Mutat. Res. *29*: 301–314.

Dujon, B., Colson, A.M., Slonimski, P.P. (1977): The mitochondrial genetic map of *Saccharomyces cerevisiae*: computation of mutations, genes, genetic and physical maps. In: Mitochondria 1977, Genetics and Biogenesis of Mitochondria, W. Bandlow, R.J. Schweyen, K.Wolf, F. Kaudewitz (eds.). Walter de Gruyter, Berlin, New York, pp. 579–669.

Fonty, G., Goursot, R., Wilkie, D., Bernardi, G. (1978): The mitochondrial genome of wild-type yeast cells. VII. Recombination in crosses. J. Mol. Biol. *119*:213–235.

Game, J.C., Cox, B.S. (1973): Synergistic interaction between *RAD* mutations in yeast. Mutat. Res. *20*:35–44.

Game, J.C., Mortimer, R.K. (1974): A genetic study of X-ray sensitive mutants in yeast. Mutat. Res. *24*:281–292.

Gillette, J.R. (1966): Biochemistry of drug oxidation and reduction of enzymes in hepatic endoplasmic reticulum. In: Advances in Pharmacology, Vol. 4. S. Garattini and P.A. Shore (eds.). New York, Academic Press, pp. 219–261.

Grant, E.L., von Borstel, R.C., Ashwood-Smith, M.J. (1979): Mutagenicity of cross-links and monoadducts of hirocoumarins (psoralen and angelicin) induced by 360-nm radiation in excision-repair-defective and radiation-insensitive strains of *Saccharomyces cerevisiae*. Environ. Mutagenesis *1*:55–63.

Hartwell, L.H. (1974): *Saccharomyces cerevisiae* cell cycle. Bacteriol. Rev. *38*:164–198.

Hastings, P.J., Quah, Siew-Keen, von Borstel, R.C. (1976): Spontaneous mutation by mutagenic repair of spontaneous lesions in DNA. Nature (Lond.) *264*:719–722.

Hinnen, A., Hicks, J.B., Fink, G.R. (1978): Transformation of yeast. Proc. Natl. Acad. Sci. USA *75*:1929–1933.

Ho, K. (1975): Induction of DNA double-strand breaks by X-rays in a radio-sensitive strain of the yeast *Saccharomyces cerevisiae*. Mutat. Res. *30*:327–334.

Ho, K., Mortimer, R.K. (1976): X-ray-induced dominant lethality and chromosome breakage and repair in a radiosensitive strain of yeast. In: Molecular Mechanisms for Repair of DNA, Part B, P.C. Hanawalt and R.B. Setlow (eds.). New York, Plenum Press, pp. 545–547.

Larimer, F.W., Ramey, D.W., Lijinsky, W., Epler, J.L. (1978): Mutagenicity of methylated *N*-nitrosopiperidines in *Saccharomyces cerevisiae*. Mutat. Res. *47*:155–161.

Lawrence, C.W., Christensen, R. (1976): UV mutagenesis in radiation-sensitive strains of yeast. Genetics *82*:207–232.

Lemontt, J.F. (1971): Mutants of yeast defective in mutation induced by ultraviolet light. Genetics *68*:21–33.

Lemontt, J.F. (1980): Genetic and physiological factors affecting repair and mutagenesis in yeast. In: DNA Repair and Mutagenesis in Eukaryotes, F.J. de Serres, W.M. Generoso, and M.D. Shelby (eds.). New York, Plenum Press.

Magni, G.E. (1963): The origin of spontaneous mutations during meiosis. Proc. Natl. Acad. Sci. USA *50*:975–980.

Magni, G.E., von Borstel, R.C. (1962): Different rates of spontaneous mutation during mitosis and meiosis in yeast. Genetics *47*:1097–1108.

Miller, J.A., Miller, E.C. (1971): Chemical carcinogenesis mechanism and approaches to its control. J. Natl. Cancer Inst. *59*:1651–1658.

Moore, C.W. (1978): Bleomycin-induced mutation and recombination in *Saccharomyces cerevisiae*. Mutat. Res. *58*:41–49.

Moore, C.W., Schmick, A. (1979): Recombinogenicity and mutagenicity of saccharin in *Saccharomyces cerevisiae*. Mutat. Res. *67*:215–219.

Nunes de Langguth, E., Gelos, U. (1972): Dark recovery in a petite mutant of *Saccharomyces cerevisiae* irradiated with ultraviolet light. Experentia *28*:1413–1415.

Parry, J.M., Zimmermann, F.K. (1976): The detection of monosomic colonies produced by mitotic chromosome nondisjunction in the yeast *Saccharomyces cerevisiae*. Mutat. Res. *35*:49–66.

Pittman, D.D. (1956): The relation of population heterogeneity to the plateau phenomenon in radiation dose-response curves. J. Bacteriol. *71*:500–501.

Plischke, M.E., von Borstel, R.C., Mortimer, R.K., Cohn, W.E. (1976): Genetic markers and associated gene products in *Saccharomyces cerevisiae*. In: Handbook of Biochemistry and Molecular Biology, 3rd ed., G. Fasman (ed.), Vol. 2. Nucleic Acids, Cleveland, Ohio, CRC Press, pp. 767–832.

Prakash, L., Prakash, S. (1977): Isolation and characterization of MMS-sensitive mutants of *Saccharomyces cerevisiae*. Genetics *86*:33–55.

Putrament, A., Baranowska, H., Prazmo, W. (1973): Induction by manganese of mitochondrial antibiotic resistance mutations in yeast. Mol. Gen. Genet. *126*:357–366.

Putrament, A., Baranowska, H., Ejchart, A., Prazmo, W. (1978): Manganese mutagenesis in yeast. In: Methods in Cell Biology, Vol. 20, D.M. Prescott (ed.), New York, Academic Press, pp. 25–34.

Resnick, M.A. (1969): Genetic control of radiation sensitivity in *Saccharomyces cerevisiae*. Genetics *62*:519–531.

Resnick, M.A. (1976): The repair of double-strand breaks in chromosomal DNA of yeast.

174   R.C. von Borstel, M.M. Shahin, and R.D. Mehta

In: Molecular Mechanisms for Repair of DNA, Part B, P.C. Hanawalt, and R.B. Setlow (eds.). New York, Plenum Press, pp. 549–556.

Shahin, M.M., von Borstel, R.C. (1978): Comtagenic and lethal effects of α-benzene hexachloride, dibutyl phthalate and trichloroethylene in *Saccharomyces cerevisiae*. Mutat. Res. *48*:173–180.

Shahin, M.M., von Borstel, R.C. (1978): Comparisons of mutation induction in reversion systems of *Saccharomyces cerevisiae* and *Salmonella typhimurium*. Mutat. Res. *53*: 1–10.

Sharp, D., Parry, J.M. (1980): The use of repair-deficient strains of yeast to assay the activity of 40 coded compounds. In: International Program for the Evaluation of Short-term Tests for Carcinogenicity, F.J. de Serres, J. Ashby, P. Brookes, B. Bridges, I.F.H. Purchase, and T. Sugimura (eds.). Amsterdam, Elsevier.

Sora, S., Panzeri, L., Lucchini Bonomini, G., Carbone, M.L. (1979): Conversione genica mitotica e crossing-over mitotico. In: Mutagenesi Ambientale Metodiche di Analisi, Vol. I, Test *in vitro*, G.E. Magni (ed.). Consiglio Nazionale delle Ricerche AQ/1/18-34, Rome, Italy, pp. 141–168.

Strausberg, R.L., Vincent, R.D., Perlman, P.S., Butow, R.A. (1978): Asymmetric gene conversion at inserted segments on yeast mitochondrial DNA. Nature (Lond.) *276*: 577–583.

Thomas, D.Y., Wilkie, D. (1968): Recombination of mitochondrial drug-resistance factors in *Saccharomyces cerevisiae*. Biochem. Biophys. Res. Commun. *30*:368–372.

Tzagaloff, A., Aka, A., Needleman, R.B. (1975): Assembly of the mitochondrial membrane system: Isolation of nuclear and cytoplasmic mutants of *Saccharomyces cerevi-siae* with specific defects in mitochondrial functions. J. Bacteriol. *122*:826–831.

von Borstel, R.C., Hastings, P.J. (1977): Mutagenic repair pathways in yeast. In: Research in Photobiology, A. Castellani (ed.). London, Plenum Press, pp. 683–687.

von Borstel, R.C., Igali, S. (1975): Mutagenicity testing of antischistosomal thioxanthenones and indazoles on yeast. J. Toxicol. Environ. Health *51*:281–291.

Wiseman, A., Lim, T-K. (1975): Induction of cytochrome P-450 in yeast by phenobarbital. Biochem. Soc. Trans. *3*:974–977.

Wiseman, A., Lim, T-K., McCloud, C. (1975): Relationship of cytochrome P-450 to growth phase of brewer's yeast in 1% and 20% glucose medium. Biochem. Soc. Trans. *3*:276–278.

Zimmermann, F.K. (1973): A yeast strain for visual screening for the two reciprocal products of mitotic crossing over. Mutat. Res. *21*: 263–269.

Zimmermann, F.K., Schwaier, R. (1963): Eine ungewöhnlich Dosiswirkungs-Beziehung der N-Nitros-N-Methylacetamid induzierten Mutationsraten bei *Saccharomyces cerevisiae*. Z. Verebungl. *94*:261–268.

Zimmermann, F.K., Schwaier, R. (1967): Induction of mitotic gene conversion with nitrous acid, 1-methyl-3-nitrosoguanidine and other alkylating agents in *Saccharomyces cerevisiae*. Mol. Gen. Genet. *100*:63–76.

Zimmermann, F.K., Schwaier, R., von Laer, U. (1966): Mitotic recombination induced in *Saccharomyces cerevisiae* with nitrous acid, diethylsulfate and carcinogenic, alkylating nitrosamide. Z. Vererbungl. *98*:230–246.

Zimmermann, F.K., Kern, R., Rasenberger, H. (1975): A yeast strain for simultaneous detection of induced mitotic crossing over, mitotic gene conversion and reverse mutation. Mutat. Res. *28*:381–388.

16

Induction and Genetic Characterization of Specific Locus Mutations in the ad-3 Region in Two-Component Heterokaryons of Neurospora crassa

FREDERICK J. DE SERRES

Principle of Method

General Utility

Many assay systems have been developed which seem especially well suited for use in mass-screening programs where the objective is to screen large numbers of unknowns and to identify those chemicals with carcinogenic and mutagenic potential. The *ad-3* forward mutation test system is not suitable for such use.

In my laboratory we have been particularly interested in developing an assay system that could further characterize those chemicals which give a positive response for more detailed study and evaluation. It is not only important to determine to what extent such chemicals might provide a carcinogenic risk but also to what extent they might be a genetic hazard to the human population. What type of genetic damage is produced and what are the consequences of successful transmission through the germ line? In addition, we need to know why

some mutagenic chemicals act as carcinogens and others (perhaps) do not; is there any correlation with the production of a particular type of genetic alteration?

In this context, the *ad-3* test system in Neurospora is particularly useful in (1) not only providing confirmatory data that chemicals identified as active in mass-screening programs as being active in producing specific locus mutations in eukaryotic organisms, but also (2) providing data on relative potency and in providing an assay which can characterize the genetic damage produced, ranging from point mutations within the locus to multilocus deletions (involving not only the locus of interest but those in immediately adjacent regions as well), and (3) providing information as to how this spectrum of damage might be modified both quantitatively and qualitatively in excision-repair-deficient strains. This is particularly important so that we can develop a data base that is appropriate for risk estimation—not only for exposure of the normal individuals in the population but the repair-deficient individuals as well.

In man, gene mutations occur both by alteration of the DNA within the locus by point mutation as well as physical removal of the locus (as well as other loci in the immediately adjacent regions) by multilocus or chromosome deletion. In haploid eukaryotic organisms, multilocus deletions are lethal and genetic analysis is usually restricted to point mutations. Although Neurospora is a haploid organism, by using two-component heterokaryons between appropriately marked strains both point mutations and multilocus deletions can be recovered and characterized genetically. Characterization of specific locus mutants is not possible on a routine basis in most eukaryotic organisms; since they can arise by a variety of genetic mechanisms their characterization may reveal effects which are not obvious in the overall induction curve. In a forward-mutation system each overall curve is, in effect, a summation of a series of component curves. The methods developed for the genetic analysis of gene mutations in the *ad-3* region of Neurospora permit rapid genetic characterization of individual *ad-3* mutants and of the development of dose-response curves for many different classes of point mutations and multilocus deletions involving the *ad-3A* and *ad-3B* loci (de Serres and Malling, 1971; de Serres, 1980). These two loci control sequential steps in the purine biosynthetic pathway and in Neurospora they are closely linked. The *ad-3* test system has been used not only to evaluate the mutagenicity of chemical carcinogens but to evaluate their genetic effects with regard to the type of mutations produced. In Neurospora chemical carcinogens produce point mutations with a high frequency of base-pair substitutions (Malling and de Serres, 1969; Ong and de Serres, 1972, 1973). In addition, by comparing *ad-3* mutations induced in wild-type strains versus excision-repair-deficient strains we have shown that the *ad-3* mutations frequencies can be markedly altered quantitatively as well as qualitatively (for review, see de Serres, 1980).

Perhaps the most important use of the *ad-3* test system in Neurospora is (1) to provide confirmatory data for the mutagenic activity in a eukaryotic organism, and (2) to provide data on the relative frequencies of point mutations and multilocus deletions. Whereas recessive point mutations may not be expressed for many generations, there are data from the analysis of specific locus mutations in mice which show that multilocus deletions can show immediate expression in the F_1 generation (Russell, 1971; Russell and Raymer, 1979). Thus, exposure of the human population to chemicals may not only provide a risk with regard to the induction of cancer, but genetic damage in germ cells may provide a substantial genetic risk if high frequencies of multilocus deletions are produced. In addition, there are data which show that point mutations can be converted to multilocus deletions in excision-repair-deficient strains, which indicates that selected subgroups may be at an even higher genetic risk (de Serres, 1980) than the major portion of the human population.

Description of Test Method

A general and detailed description of the *ad-3* test system has been published previously (de Serres and Malling, 1971) and will be described only briefly here. In two-component heterokaryons heterozygous for mutations at the *ad-3A* and *ad-3B* loci (Fig. 16.1), both point mutations and multilocus deletions can be recovered. *Ad-3* mutants are recovered on the basis of the accumulation of a reddish purple pigment in the vacuoles of the mycelium under culture conditions where wild-type conidia form colonies which are colorless. *Ad-3* mutant colonies are isolated, subcultured, and subjected to a series of heterokaryon tests to characterize them genetically. For the purposes described above it is considered sufficient to perform only those tests required (1) to determine the genotype of the *ad-3* mutants and (2) to distinguish point mutations from multilocus deletions.

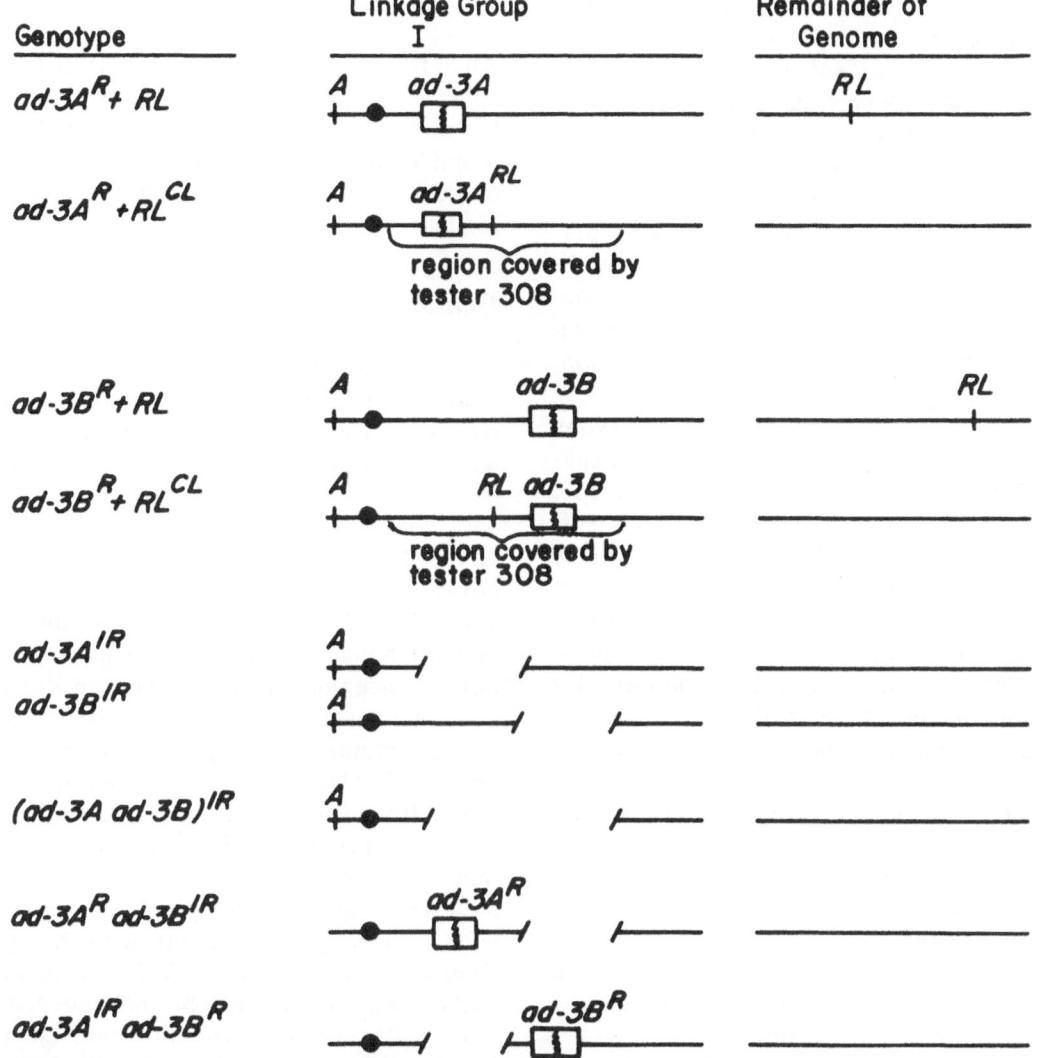

Figure 16.1. Some of the genetic alterations possible in *ad-3* mutants induced in a two-component heterokaryon which give a negative dikaryon test.

Assays for Induction of *Ad-3* Mutations

Test Organisms

Storage and culture techniques. Stock cultures of heterokaryons 12 and 59 are kept on silica gel at 4° C. Subcultures are made by making a conidial suspension from 8 to 10 crystals of such a stock culture (see p. 317, de Serres and Malling, 1971). These conidia are plated on minimal medium to make it possible to select single-colony isolates of each heterokaryon. This procedure ensures a low spontaneous background of *ad-3* mutants (e.g., about 0.4×10^6). This procedure, in addition to selection of mature cultures with bright orange conidia, ensures that the heterokaryotic fraction of conidia is at least 20–25%.

Equipment. The usual equipment for plating and incubating cultures of microorganisms is required. All plates are incubated at 30°C. The unusual aspect of this assay is that all treated cells are incubated in 10 liters of medium in a 12-liter Florence flask. The addition of 0.125% agar provides a "soft agar" which allows each viable heterokaryotic conidium to form a small colony about 1–2 mm in diameter. The number of such flasks required depends on the number of treatments as well as the number of *ad-3* mutants desired for genetic analysis. An additional incubator is required for those genetic tests (dikaryon and trikaryon) which are performed at 35°C. (See later section, Genetic Analysis of *Ad-3* Mutants.)

Strains. Two different heterokaryons have been used for evaluation of the genetic effects of known chemical carcinogens: (1) heterokaryon 12, which is repair sufficient, and (2) heterokaryon 59, which is excision-repair deficient (Table 16.1). The genotypes of the individual components of each heterokaryon must be checked prior to making silica gel stock cultures. By using single-colony isolates from such stock cultures in each experiment no further check is usually required.

Testing protocol. Two different methods of treatment can be used: (1) tests on growing cultures or (2) tests on nongrowing conidial suspensions. If the chemical requires DNA replication or metabolic activation for mutagenic activity (e.g., base analogs and mycotoxins) then assays must be done on growing cultures.

In general, tests are made on suspensions of conidia treated for 2 hours at various concentrations of the test chemical. This approach is used to determine those chemical concentrations which will give approximately 80, 40, 20, and 10% survival. Two flasks are inoculated with untreated conidia and two flasks are inoculated with treated conidia for each survival level to give a total of 10^6 surviving heterokaryotic colonies in each of the ten flasks. This procedure will provide data for a dose-response curve for mutation-induction, so that a second experiment can be performed (if necessary) to provide a sample of mutants (usually 100–150) adequate for genetic analysis.

Statistical Considerations

All of the procedures required for statistical analysis of data for dose-response curves for survival and mutation-induction can be found in Smith and de Serres (1969); copies of this publication will be supplied upon request to the second author.

Evaluation of Results

Since the spontaneous frequency of *ad-3* mutants is so low (0.4 per 10^6 colonies) the recovery of no mutants and five or more mutants in a pair of flasks in any of the four treated series usually means that the chemical is mutagenic. Occasionally a spontaneous mutation will occur early during the growth of a single-colony isolate from our stock cultures so that a significant number of *ad-3* mutants will be found in control flasks. Since this is an extremely rare event and complicates the genetic analysis of induced mutants, such experiments are best discarded. Another minor complication arises from variation in the shape of the dose-response curve for mutation-induction which may show a peak response at high levels of survival rather than at low levels. In early experiments with X-rays (Webber and de Serres, 1965) we found that point mutations and chromosome deletions may show different induction kinetics. Since this may also be true with chemicals, *ad-3* mutants should either be isolated from comparable forward-mutation frequencies (100×10^{-6} survivors) or at three to four different forward-mutation frequencies if dose-response curves are desired. The rigor to which samples of *ad-3* mutants are to be subjected to genetic analysis can be determined only by human exposure levels and the need for this type of data for risk-estimation.

Genetic Analysis of *Ad-3* Mutants

Principle of method. In general, genetic analysis of *ad-3* mutant is accomplished

Table 16.1. *Genetic Composition of Each Component of Two-Component Heterokaryons of* Neurospora crassa *Used To Recover Specific Locus Mutants in the ad-3 Region*

Component	Strain number		Linkage group									
			I				III		IV		V	VI
Heterokaryon 12												
I	74-OR60-29A	A	hist-2	+	ad-3B	nic-2	+	ad-2	+	+	inos	+
II	74-OR31-16A	A	+	+	+	+	al-2	+	cot-1	+	+	pan-2
Heterokaryon 59												
I	74-OR276-40A	A	hist-2	+	ad-3B	nic-2	+	ad-2	+	uvs-2	inos	+
II	74-OR244-3A	A	+	+	+	+	al-2	+	cot-1	uvs-2	+	pan-2

through a series of heterokaryon tests with special tester strains. The objective is to determine genotype: *ad-3A*, *ad-3B*, *ad-3A ad-3B*, *ad-3B nic-2* or *ad3A ad-3B nic-2*. The latter two classes usually arise by multilocus deletion. The *ad-3A* and *ad-3B* loci are about 0.1 map units with the *nic-2* locus some 5 to 8 map units distal. Genotype is determined by heterokaryon tests on Petri plates with a special set of four testers. Positive results are usually obtained in 24–48 hours. The next two tests, the dikaryon test and the trikaryon test, are designed to determine whether the mutation in the *ad-3* is a point mutation or a multilocus deletion. If a point mutation has been induced in component II of either heterokaryon 12 or 59, then this component should grow on minimal medium supplemented with adenine and calcium pantothenate. If the plates are incubated at 35° C, the *cot-1* marker produces a small dense colony rather than a large spreading type. Where such small dense colonies are isolated, the resulting culture will have albino conidia rather than orange conidia because of the *al-2* marker. In a negative dikaryon test, no small dense colonies are found in 500 to 1000 colonies scanned.

A negative dikaryon test can be due to the fact that (1) a multilocus deletion has been induced in the *ad-3* region or (2) that a point mutation has been induced at the *ad-3A* or *ad-3B* locus *in addition to* a separate site of recessive lethal damage elsewhere in the genome. To distinguish between these two alternatives the strain under test is combined with three different testers, each carrying a different multilocus deletion (e.g., *ad-3A*, *ad-3B* and *ad-3A*, *ad-3B nic-2*), in a series of trikaryon tests. The data from these latter tests make possible a precise distinction between the above alternatives so that all mutants can be assigned to one class or the other. If a sufficient number of mutants has been analyzed, computer programs for statistical analysis (Smith and de Serres, 1969) make it possible to generate dose-response curves for each of these different classes of genetic alterations with as-

sociated 95% confidence intervals. Regression analysis of the data for each curve makes it possible to determine their slopes.

Heterokaryon Tests to Determine Genotype

Storage and culture techniques. All tester strains are stored on silica gel (de Serres and Malling, 1971). For use, strains are subcultured on minimal medium with the appropriate supplements.

Tester strains. Heterokaryon tests to determine genotype are performed with the strains indicated in Table 16.2. All presumed *ad-3* mutants should give a negative test with one or more of the first three strains and a positive test with the *ad-2 inos* strain used as a positive control.

Heterokaryon tests. These tests are performed on the surface of solidified minimal medium in Petri plates supplemented with 0.5% sucrose and 0.5% sorbose to retard the growth of combinations of strains giving a positive test. Each plate is marked on the bottom with a code number for each strain as well as a spot for the unknown being tested. Tester strains are usually added first (~0.1 ml spot/plate) as a spot on the surface of the agar and allowed to dry prior to the addition of the inoculum of the strain to be tested. The four tester strains are added near the periphery of the plate at positions corresponding to 3, 6, 9, and 12 on a clock face. A spot containing the unknown alone is added to the center of the plate as well as being added (~0.1 ml spot/plate) *carefully* to the four spots containing the testers. Care needs to be taken *not* to cross-contaminate

Table 16.2.   Strain Number and Genotype of Tester Strains Used in Heterokaryon Tests To Determine Genotype of ad-3 *Mutants*

Strain number	Genotype
74A-Y68-M13	*ad-3A*
74A-Y112-M2	*ad-3B*
74-OR33-3A	*hist-2 nic-2 al-2*
74-OR60-44A	*ad-2 inos*

Table 16.3. *Genotypic Characterization of Presumptive* ad-3 *Mutants in Heterokaryon Tests on Minimal Medium at 30°C*

Initial genotype	Response with each tester strain				Inferred genotype
	ad-3A	ad-3B	hist-2 nic-2 al-2	ad-2 inos	
ad-3 (1)	$-^a$	+	+	+	ad-3A
ad-3 (2)	+	$-^a$	+	+	ad-3B
ad-3 (3)	+	−	−	+	ad-3B nic-2
ad-3 (4)	−	−	+	+	ad-3A ad-3B
ad-3 (5)	−	−	−	+	ad-3A ad-3B nic-2
ad-3 (6)	+	+	+	+	unknown

^{a}ad-3A and ad-3B mutants are sometimes "leaky" and show residual growth on minimal medium; these − responses should be compared with the growth of each unknown ad-3 mutant alone.

the testers with each other. All plates are incubated at 30° C and scored at 12-hour intervals for up to 72 hours, although most positive tests will have occurred by 48 hours.

Occasional "leaky ad-3 mutants will show residual growth with the ad-3A or ad-3B tester which should match that shown alone. In some samples the percentage of leaky mutants can be quite high (~35%). Occasional mutants will grow with all testers as well as alone; these are either ad-3 mutants too leaky to test or unknowns. The frequency of such mutants is very low in most experiments.

The genotypes of each presumptive ad-3 mutant which can be inferred on the basis of the type of overall response with the four strains used as testers is given in Table 16.3.

Dikaryon Test

Rationale. The dikaryon test is performed to determine whether the ad-3 mutation induced in component II of heterokaryon 12 or 59 is a point mutation or some other type of genetic alteration. If a point mutation has been induced in either the ad-3A or ad-3B locus, Component II should grow as a homokaryon on minimal medium supplemented with adenine and calcium pantothenate. If these plates are incubated at 35° C these conidia will form a

tiny dense colony because of the cot-1 marker. This provides a simple visual check to show that a point mutation has been induced in the ad-3 region. If the test is negative and no cot colonies can be found, this can be due to the fact that the induced mutation in the ad-3 is a point mutation but a separate site of massive lethal damage has been induced elsewhere in the genome. Alternatively, the induced mutation in the ad-3 region could be a multilocus deletion. These latter two alternatives can be distinguished in trikaryon tests; see later section, Trikaryon Test.

Test procedure. Each new ad-3 mutant is plated out on minimal medium supplemented with adenine and calcium pantothenate. Usually one tube of medium is inoculated directly from the culture tube and a 1 ml aliquot is removed to make a tenfold dilution in a second tube. These plates are incubated at 35° C for 48 hours so that they can be scored under a dissecting microscope for the presence of tiny dense cot colonies. At least three single colonies are isolated onto slants of the same medium in test tubes and incubated at 35° C for 48 hours. If the single-colony isolate remains restricted in growth (to verify the presence of the cot marker) the tubes are moved to room temperature under normal lighting conditions. At this temperature cot is nonrestrictive and a normal appearing culture should result which has albino conidia (due to the al-2 marker). An outline of the

Table 16.4. *Results and Conclusions from Dikaryon Tests on* ad-3 *Mutant Induced in Two-Component Heterokaryons*

Event	Result
Positive test	
1. Conidial plating of original culture at 35°C	Appearance of *cot* colonies among large spreading colonies
2. Single-colony isolate of *cot* colonies incubated at 35°C for 48 hr	Single-colony isolates remain small and dense on agar slants
3. Single-colony isolates transferred to 20–24°C under normal lighting	Single-colony isolates grew up into normal cultures with albino (not orange) conidia
Negative Test	
1. Conidial plating of original culture at 35°C	No *cot* colonies in either original or repeated platings

events expected in both positive and negative tests is given in Table 16.4.

Trikaryon Test

Rationale. The objective of the trikaryon test (actually tetrakaryons, but because two components from different sources are identical the resulting cultures are trikaryons) is to provide further characterization of *ad-3* mutations and to distinguish between point mutations which have arisen *in combination with* a separate site of recessive lethal damage elsewhere in the genome from multilocus deletions in the *ad-3* region. With careful selection of strains for use as testers in such assays a precise characterization of many different genotypes is possible. In Figure 16.2 the array of different types of damage which can be distinguished among *ad-3A* mutants is illustrated. With this assay it is possible to distinguish between, for example, point mutations at the *ad-3A* locus with a separate site of recessive lethal damage elsewhere in the genome ($ad\text{-}3A^R + RL$); point mutations at the *ad-3A* locus with a closely linked site of recessive lethal damage ($ad\text{-}3A^R + RL^{CL}$) and multilocus deletion of the *ad-3A* locus ($ad\text{-}3A^{IR}$).

Tester strains. The three strains used as testers are kept as silica gel stock cultures which are subcultured for different batches of trikaryon tests on newly induced mutants. The strain numbers, genotypes, and code numbers of these strains are given in Table 16.5. For further information on the extent of genetic damage in each strain, see Figure 3, de Serres, 1968.

Test performance. Subcultures of each tester and all *ad-3* mutants giving a negative dikaryon test are subcultured on minimal medium supplemented with adenine, niacin and calcium pantothenate. When these cultures have conidiated, trikaryons are made on the same medium with 1% sorbose, 0.05% glucose, and 0.05% fructose to inhibit growth in unslanted 13 × 100 mm test tubes.

To three sets of tubes, 0.1 ml drops of a 1×10^6 conidia/ml suspension of each test-

Table 16.5. *Two-Component Heterokaryons Used as Tester Strains in Trikaryon Tests*[a]

Strain number	Genetic alteration in component II in *ad-3* region	Code
12-7-215	$ad\text{-}3A^{IR}$	21
12-5-182	$ad\text{-}3B^{IR}$	38
12-1-18	$(ad\text{-}3A\ ad\text{-}3B\ nic\text{-}2)^{IR}$	308

[a]See de Serres, 1968.

(1) $ad\text{-}3A^R + RL$

Strain	Component	Genotype of each component
Tester 308	I	A hist-2 $ad\text{-}3A^R$ $ad\text{-}3B^R$ nic-2 + ; $ad\text{-}2$; + ; inos; +
	II	A + ($ad\text{-}3A$ $ad\text{-}3B$ nic-2)^{IR} al-2; + ; cot-1; + ; pan-2
New $ad\text{-}3A$ Mutant	I	A hist-2 $ad\text{-}3A^R$ $ad\text{-}3B^R$ nic-2 + ; $ad\text{-}2$; + ; inos; + +
	II	A + $ad\text{-}3A^R$ + + al-2; + ; cot-1; + ; pan-2 RL
	II	A + ($ad\text{-}3A$ $ad\text{-}3B$ nic-2)^{IR} al-2; + ; cot-1; + ; pan-2 +
	II'	A + $ad\text{-}3A^R$ + + al-2; + ; cot-1; + ; pan-2 RL

Isolate will grow as a *cot* colony at 35°C on minimal medium supplemented with adenine + pantothenate but not minimal medium + pantothenate.

(2) $ad\text{-}3A^R + RL^{Cl}$

Strain	Component	Genotype of each component
Tester 308	I	A hist-2 $ad\text{-}3A^R$ $ad\text{-}3B^R$ + nic-2 + ; $ad\text{-}2$; + ; inos; +
	II	A + ($ad\text{-}3A$ $ad\text{-}3B$ RL nic-2)^{IR} + ; + ; cot-1; + ; pan-2
New $ad\text{-}3A$ Mutant	I'	A hist-2 $ad\text{-}3A^R$ $ad\text{-}3B^R$ + nic-2 + ; $ad\text{-}2$; + ; inos; +
	II'	A + $ad\text{-}3A^R$ + RL + al-2; + ; cot-1; + ; pan-2
	II	A + ($ad\text{-}3A$ $ad\text{-}3B$ RL nic-2)^{IR} al-2; + ; cot-1; + ; pan-2
	II	A + $ad\text{-}3A$ + RL + al-2; + ; cot-1; + ; pan-2

Isolate will not grow as a *cot* colony on either minimal medium + pantothenate or minimal medium + pantothenate + adenine. However, isolate will grow as an adenine-requiring *cot* colony in combination with Tester 21.

(3) $ad\text{-}3A^{IR}$

Strain	Component	Genotype of each component
Tester 308	I	A hist-2 $ad\text{-}3A^R$ $ad\text{-}3B^R$ nic-2 + ; $ad\text{-}2$, + ; inos; +
	II	A + ($ad\text{-}3A$ $ad\text{-}3B$ nic-2)^{IR} al-2; + , cot-1; + ; pan-2
New $ad\text{-}3A$ Mutant	I'	A hist-2 $ad\text{-}3A^R$ $ad\text{-}3B^R$ nic-2 + ; $ad\text{-}2$; + , inos; +
	II'	A + ($ad\text{-}3A$)^{IR} + + al-2; + ; cot-1; + ; pan-2
	II	A + ($ad\text{-}3A$ $ad\text{-}3$ nic-2)^{IR} al-2; + ; cot-1; + ; pan-2
	II'	A + ($ad\text{-}3A$)^{IR} + + al-2; + ; cot-1; + ; pan-2

Isolate will not grow as a *cot* colony on minimal medium + pantothenate or minimal medium + pantothenate + adenine. Neither will it grow as a *cot* colony on any medium in combination with Tester 21.

Figure 16.2. Distinction between various types of genetic alterations giving rise to *ad-3A* mutants by means of trikaryon tests with selected testers: (1) point mutation at the *ad-3A* locus with a separate site of recessive lethal damage elsewhere in the genome (*ad-3A+ RL*), (2) point mutation at the *ad-3A* locus with a closely linked, but separate site of recessive lethal damage (*ad-3A + RL*^{CL}, and (3) a multilocus deletion covering the *ad-3A* locus (*ad-3A*^{IR}).

er are added. A comparable inoculum of mutants to be tested is then added to a set of tubes previously inoculated with each of the three testers. In this way, conidia will germinate, fuse, and form a compact mycelium on the surface of the unslanted agar. After 3–4 days these tubes are flooded with the same supplemented medium with 0.75% fructose + 0.75% glucose + 1.2% agar. This medium is kept at 45° C and dispensed 2 ml per tube from a buret. The tubes are slanted after adding this medium.

After trikaryons have conidiated they are plated on minimal medium supplemented with adenine and pantothenate + 1% sorbose + 0.05% glucose and 0.05% fructose + 1.5% agar dispensed 12 ml into 20 × 150 mm test tubes. Trikaryons are plated in the same manner as dikaryons and both plates from each culture are incubated at 35° C for 2 days. After this incubation period plates are scanned under a dissecting microscope to isolate small dense *cot* colonies. As indicated in Table 16.6 *cot* colonies can be either adenine-requiring (ad⁻) or adenine-independent (ad⁺). In some cases *cot* colonies are not expected

because the newly induced *ad-3* mutant will not complement the *ad-3* mutant in Component II of any of the three strains used as testers. *Cot* colonies (at least six) are isolated from each trikaryon onto divided Petri plates prepared with minimal medium + 1.0% sorbose + 0.05% glucose + 0.05% fructose + calcium pantothenate on one side and the same medium + adenine on the other side. These plates are incubated for 2 days at 35° C and scored for their adenine requirement (ad⁻ or ad⁺). Trikaryons giving a negative test are usually replated using conidia from a different part of the original culture.

Consistency Checks on the Genetic Analyses

After completing the heterokaryon tests to determine genotype and the dikaryon and trikaryon tests, a tabulation of the data provides a simple check for consistency. It is clear that the trikaryon test is also a test which can be used to determine genotype and if a mutant scores as *ad-3A* in the het-

Table 16.6. *Recovery and Adenine Requirements of* cot *Colonies from Trikaryons*

Genotype from heterokaryon test	Tester strain[a]						Inferred genotype
	21		38		308		
	M+P	M+AP	M+P	M+AP	M+P	M+AP	
ad-3A	—	—	ad⁺	ad⁺	—	—	*ad-3A^{IR}*
"	—	ad⁻	ad⁺	ad⁺	—	ad⁻	*ad-3A^R+RL*
"	—	ad⁻	ad⁺	ad⁺	—	—	*ad-3A^R+RL^{CL}*
ad-3B	ad⁺	ad⁺	—	—	—	—	*ad-3B$^{IR\ x}$*
"	ad⁺	ad⁺	—	ad⁻	—	ad⁻	*ad-3B^R+RL*
"	ad⁺	ad⁺	—	ad⁻	—	—	*ad-3B^R+RL^{CL}*
ad-3A ad-3B	—	—	—	—	—	—	*(ad-3A ad-3B)^{IRxx}*
"	—	ad⁻	—	—	—	—	*ad-3A^{R}ad-3B^{IR}*
"	—	—	—	ad⁻	—	—	*ad-3A^{IR}ad-3B^R*
"	—	ad⁻	—	ad⁻	—	ad⁻	*ad-3A$^{R\ ad-3B4}$+RL*
"	—	ad⁻	—	ad⁻	—	—	*ad-3A^{R}ad-3B^R+RL^{Cl}*

x = (*ad-3B nic-2*)^{IR} will give same type of data

xx = (*ad-3A ad-3B nic-2*)^{IR} will give same type of data

— = no *cot* isolate found

ad⁻ = adenine-requiring

ad⁺ = adenine-independent

[a]21, 38, and 308 were plated on minimal medium + pantothenate and minimal medium + pantothenate + adenine at 35°C.

Table 16.7. *Agents Found To Be Mutagenic in Heterokaryon 12 of* Neurospora crassa

---

(1) *Radiations*

X-rays	730 Mev. Protons
UV	^{60}Co γ rays
450 Mev. Protons	^{137}Cs γ rays
101 Mev. Carbon ions	^{32}P β rays
39 Mev. Helium ions	^{85}Sr γ rays
442 Mev. Protons	

(2) *Chemicals*

Actinomycin D	Leucanthone analog (IA-III)
AF-2 2(2-furyl)-3-(5-nitro-2-furyl) acrylamide	Methylmethanesulfonate
Aflatoxin B$_1$	Metronidazole (1-(2-hydroxyethyl)-2-methyl-5-nitroimidazole
2-amino-N^6-hydroxyadenine	Mitomycin C
2-aminopurine	Myleran
Captan	N-acetoxy acetylaminofluorene
1,2-dibromoethane	Natulan
1,2,5,6-diepoxyhexane	N-ethyl-N'-nitro-N-nitrosoguanidine
1.2.7,8-diepoxyoctane	N-ethyl-N-nitrosourea
1,2,4,5-diepoxypentane	N-hydroxy acetylaminofluorene
9, 10-dimethylbenzanthracene	6-N-Hydroxylamino purine
Diethylnitrosamine	Niridazole (1-(5-nitro-1-thiazolyl)-2-imidazolidinone)
Dimethylnitrosamine	
Epichlorohydrin	N-methyl-N'-nitro-N-nitrosoguanidine
Ethyleneimine	N-methyl-N-nitrosourea
Ethylmethanesulfonate	Nitrofurazone (5-nitro-2-furoldehydesemi-carbazone)
F30066 (N-idoptopyl-3-(5-nitro-2-furyl)-acrylamide	4-nitroquinoline-1-oxide
FANFT (2-formylamino-4-(5-nitro-2-furyl) thirazole)	Nitrous acid
	NO-benzthiozuron
P-fluorophenylalanine	O-methyl hydroxylamine
H-193 (5-nitro-2-(p-carbamoylstyryl)furan)	1-phenyl-3-monomethyltriazene
Hycanthone	Proflavin + light
Hycanthone analog (IA-II)	β-propiolactone
Hycanthone analog (IA-IV)	Sodium bisulfite
4-(Hydroxyamino) quinoline-1-oxide	SQ18506 (trans-5-amino-3-[2-(5-nitro-2-furyl)-vinyl]- 1,2,4-oxadiazole)
Hydroxylamine	
ICR-170	Sterigmatocystin
Leucanthone	2,3,5,6-tetra-ethyleneimino-1,4,-benzoquinone
Leucanthone analog (IA-I)	Trenimon

---

erokaryon tests, then it should also score as an *ad-3A* mutant in the trikaryon tests. The number of mutants that will require trikaryon tests varies markedly from sample to sample within an experiment as well as between experiments.

Dose-Response Curves for Individual Classes of ad-3 Mutants

The genetic analysis of *ad-3* mutants makes it possible to resolve the overall dose-response curve into its different components (Smith and de Serres, 1969).

The first major subdivision is point mutations (*ad-3^R*) versus multilocus deletions (ad-3^{IR}). Point mutations can be further subdivided into *ad-3A^R* and *ad-3B^R*; multilocus deletions into *ad-3A^{IR}*, *ad-3B^{IR}*, *ad-3B nic-2^{IR}*. The mutations which involve the *nic-2* locus usually result from large multilocus deletions and have been shown to occur after X-irradiation (Webber and de Serres, 1965) and chemical treatment (Ong and de Serres, 1972, 1980).

Mutagenic Agents Tested

A list of the different types of radiations and chemicals which share given positive results in the *ad-3* test system in heterokaryon 12 is given in Table 16.7.

Summary and Conclusions

This genetic characterization of *ad-3* mutants thus makes it possible to distinguish point mutations from multilocus deletions of varying sizes. The genetic analysis makes it possible to produce a data base more appropriate for risk estimation than the overall frequencies of *ad-3* mutations. The high percentages of *ad-3* mutants resulting from hycanthone treatment, for example (Ong and de Serres, 1980), suggest that the genetic effects of chemotherapy may warrant serious consideration if these same genetic effects are produced in human germ cells. The Neurospora data show that 82% of the *ad-3* mutants recovered in these experiments are large multilocus deletions which also cover the *nic-2* locus. A more general application of the *ad-3* test system will make it possible to obtain a more thorough genetic characterization of specific locus mutations and make possible a rapid assessment of both immediate and long-term effects of the genetic damage produced in eukaryotic organisms.

References

de Serres, F.J. (1968): Genetic analysis of the extent and type of functional inactivation in irreparable recessive lethal mutations in the *ad-3* region of *Neurospora crassa*. Genetics *58*: 69–77.

de Serres, F.J. (1980): Mutation-induction in repair-deficient strains of *Neurospora*. In: W.M. Generoso, M.D. Shelby, F.J. de Serres (eds.), DNA Repair and Mutagenesis in Eukaryotes. New York: Plenum Press.

de Serres, F.J., Malling, H.V. (1971): Measurement of recessive lethal damage over the entire genome and at two specific loci in the *ad-3* region of a two-component heterokaryon of *Neurospora crassa*. In: A. Hollaender (ed.), Chemical Mutagens: Principles and Methods for Their Detection, Vol. 2. New York: Plenum Press, pp. 311–342.

Malling, H.V., de Serres, F.J. (1969): Mutagenicity of alkylating carcinogens. Ann. N.Y. Acad. Sci. *163*: 788–800.

Ong, T., de Serres, F.J. (1972): Mutagenicity of chemical carcinogens in *Neurospora crassa*. Cancer Res. *32*: 1890–1893.

Ong, T., de Serres, F.J. (1973): Genetic characterization of *ad-3* mutants induced by chemical carcinogens, 1-phenyl-3-monomethyltriazene and 1-phenyl-3,3-dimethyltriazene, in *Neurospora crassa*. Mutat. Res. *20*: 17–223.

Ong, T., de Serres, F.J. (1980): Genetic analysis of adenine-3 mutants induced by hycanthone, lucanthone and their indazole analogs in *Neurospora crassa*. Environ. Mutagen.

Russell, L.B. (1971): Definition of functional units in a small chromosome segment of the mouse and its use in interpreting the nature of radiation-induced mutations. Mutat. Res. *11*: 107–123.

Russell, L.B., Raymer, G.D. (1979): Analysis of the albino-locus region of the mouse. III. Time of death of prenatal lethals. Genetics *92*: 205–213.

Smith, D.B., de Serres, F.J. (1969): Computer Programs for Statistical Analysis of Forward-Mutation Experiments at Specific Loci in *Neurospora crassa*, Quarterly Progress Report to the National Aeronautics and Space Administration, 1 January–31 March, 1968, Oak Ridge National Laboratory ORNL-TM-2544.

Webber, B.B., de Serres, F.J. (1965): Induction kinetics and genetic analysis of X-ray-induced mutations in the *ad-3* region of *Neurospora crassa*. Proc. Natl. Acad. Sci. USA *53*: 430–437.

17

Neurospora and Environmentally Induced Aneuploidy

A.J.F. GRIFFITHS

Introduction

Research in environmental mutagenesis has traditionally stressed mutagenic agents which cause gene (point) mutations, as opposed to chromosomal mutations. There are good reasons for this emphasis. First, in the area of germinal mutation, of all diseases which can be attributed to specific genetic causes, the major proportion in any one generation is due to genetic changes involving single genes. Carter (1977) has estimated that the frequency of disorders due to "mutant genes of large effect" is about 12 per thousand live births, composed of 9.0 autosomal dominant, 2.5 autosomal recessive, and 0.5 X-linked conditions. However, the role of environmental mutagens in the maintenance and/or change in this large component of society's genetic load is by no means established, and is still the subject of much debate (Sobels, 1977). Second, in the area of somatic mutation, the demonstration that most carcinogens will

cause gene mutation in microbes (McCann *et al.*, 1975) has strengthened the notion that their carcinogenesis involves gene mutation (McCann and Ames, 1976; deSerres, 1976). Once again, however, the precise nature of this involvement of the mutagens in the etiology of the disease (cancer) is not understood.

The present discussion is to draw attention to the importance of chromosomal mutation in environmental mutagenesis, both at the germinal and somatic levels. The type of chromosomal mutation which assumes the position of major relevance is aneuploidy. Aneuploidy is the term that describes a karyotype with an abnormal number of chromosomes, differing from the wild type by parts of chromosome sets. Aneuploids showing excess numbers of chromosomes are hyperploid, and those showing a reduced number are hypoploid. Usually the excess or reduction is in terms of one or a small number of chromosomes.

In the area of germinal change, the results of several surveys (Carter, 1977;

Sankaranarayanan, 1979) have revealed that chromosomal mutation ranks close to gene mutation as a cause of human ill health. For example, Carter (1977) estimates that the total frequency of unbalanced chromosomal anomalies, predominantly aneuploids, is about 4.0 per thousand live births, in other words, about one-third of those due to gene mutation. Furthermore, the frequencies of chromosomal mutations in spontaneous abortions is startling: Carter estimates that unbalanced autosomal chromosomal anomalies (mainly aneuploids) in first trimester abortions may be as high as 50%. Thus, it is evident that aneuploidy is not at all uncommon, and that those particular aneuploids which are seen in the surviving fractions, mainly Down syndrome (DS), represent a considerable emotional and financial load. As with the area of germinal gene mutation, the contribution of environmental factors to these frequencies is not known, and has scarcely begun to be investigated.

Particularly significant is the fact that in contrast to germinal gene mutation, germinal chromosomal mutation (aneuploidy) is almost entirely *de novo*, the entire mutant population arising anew each generation. The phenotypic frequencies of diseases caused by germinal gene mutation (especially recessives) are determined in a much more complex relation, by fundamental parameters such as mutation rate, allele frequency, and selection coefficients. First, two recessive mutant alleles from different parents have to unite to produce a mutant phenotype. Second, allele frequencies in any one generation are the result not of mutation in the recent past but of the intricate interplay of mutation and selection, probably acting over hundreds of previous generations. Germinal chromosomal mutations are quite different. They usually act as dominant genetic factors instantly revealing themselves in the phenotype with a high degree of penetrance, usually expressed as severe and incapacitating ill health.

There are several consequences of these basic attributes of aneuploids. In the first place, they are easily identified unambiguously, and subsequently quantified and studied. In the second place, there is usually no problem of transmission of the condition to offspring. The problems of one generation are theirs alone. Implicit here is the idea that if the germinal chromosomal mutation rates could be controlled, the impact upon the population would be immediate and complete. There is no question of a "genetic time bomb" as in the case for germinal gene mutation.

Chromosomal mutation has been shown to be intimately connected also with the area of somatic mutation. In the context of mammalian disease, this association has been made particularly strongly in the case of cancer. Numerous articles have reviewed the well-established occurrence of chromosomal mutations, particularly aneuploids, in the cells of cancerous tissue (Koeller, 1972; de Grouchy and Turleau, 1974, Nowell, 1974). Recent advances in chromosome banding technology have permitted the identification of specific chromosomal involvements in these aneuploid conditions. A few examples will illustrate these associations. Sonta and Sandberg (1978), in a study of primary intestinal tumors, found that of 15 tumors studied, 12 showed cells with abnormal chromosomes, usually trisomy for chromosomes 8, 13, 15, 17, or 21. Hypoploids were found, too, but were less frequent. Mitelman *et al.*, (1976a,b) reviewed the situation for human leukemia. For chronic myeloid leukemia a total of 67 patients were reviewed. All of these showed from 20–100% of their cells (bone marrow or spleen) with chromosome abnormality in addition to the usual Philadelphia chromosome. Thirty-three cases had an extra Philadelphia chromosome, and there were 28 cases of trisomy for the long arm of 17. For acute myeloid leukemia a total of 57 patients were reviewed. Most of these showed aneuploidy; 21 for chromosome 8, 16 for chromosome 9, and 10 for chromosome 21. Once again hyperploidy was more common that hypoploidy. These authors state, "There can be little doubt that the cytogenetic abnormalities are correlated at the molecular level to mechanisms important in the induction of the malignant state" (Mitelman *et al.*, 1976b).

Particularly striking are the results from cancers induced by chemical carcinogens. Chan *et al.* (1979) studied murine thymomas induced by benzo(a)pyrene and 7, 12-dimethylbenz(a)anthracene (DMBA), as well as X-rays and an endogenous leukemia virus. They found that in a total of 89 tumors studied, 76 showed trisomy 15. Weiner *et al.* (1978) found trisomy 15 "in almost all leukemia cells" in murine T-cell leukemias induced by DMBA. On the other hand, Uenaka *et al.* (1978) found that trisomy 2 was the most common in rat leukemias induced by N-nitroso-N-butylurea. Some analogous information is available from studies on induced tumors in humans. Mitelman *et al.* (1978) studied bone marrow cell chromosomes in individuals who had been exposed occupationally to solvents, insecticides, or petroleum products. In the exposed group 82.6% showed aneuploidy, especially monosomy for chromosomes 5 or 7, and trisomy for chromosomes 8 or 21. Thus, not only is aneuploidy found in association with cancer but environmentally induced cancers also show these associations.

The interrelationship between cancer and aneuploidy shows a further intriguing complexity in the association of some germinally derived aneuploid conditions, such as DS, with increased risk of cancer. The epidemiological relationship is reviewed by Miller (1970). For example, DS individuals shown an approximately 15-fold increased incidence of leukemia. The disease progression, once again, shows the addition of extra chromosomes other than that responsible for DS itself (chromosome 21), which also persists. Goh *et al.* (1978), in speculating on the way in which these additional trisomies arise, suggest that nondisjunction occurs more frequently in cells that are already chromosomally unbalanced. They state ". . . the mechanism of

the chromosomal evolution found in DS leukemia patients may be similar to that of the acute transformation of chronic myelocytic leukemia, i.e., a nondisjunction that resulted in the phenotypic expression of acute leukemia." In one study using cell lines from DS and from normal individuals it was found that DS-derived cells are more susceptible to chromosome breakage by the leukemogen DMBA (O'Brien *et al.*, 1971), but in another study, hematopoietic cell lines from DS and normal individuals showed no difference in their response to mitomycin-C and caffeine, known clastogens in tissue culture (Banerjee *et al.*, 1977).

Obviously the precise relation between aneuploidy and the progress to cancer is by no means established. The most obvious need is to discover which is cause and which is effect. Nevertheless, the induction of aneuploidy in cancer is obviously highly relevant from the environmental standpoint. Thus, it has been shown that there is a need for research on the specific environmental induction of aneuploidy in the areas of germinal and of somatic environmental mutagenesis.

The Neurospora Test System

In order to discover which chemicals and other environmental agents might represent a hazard to human populations in promoting aneuploidy, a test system is necessary. Several features are desirable in such a system. First, because of the large numbers of environmental agents which might be tested, the test system, like those for gene mutation, should be simple and inexpensive, and take up as little space as possible. Second, the test organism must be a eukaryote with relatively standard chromosomal organization, and a well-mapped and investigated genome. These two items virtually dictate the choice of a fungal system. Third, the phenomenon of aneuploidy should be well-documented either from meiotic ("germinal") or mitotic ("somatic") cells. Fourth, there should be a selective system for the identification of aneuploids, in order to quantify their occurrence. Fifth, the system should show aneuploid frequencies which are sensitive to exposure to a selection of externally added environmental agents.

Several fungi have well-documented aneuploid detection systems, and these have started to be used as test organisms in this connection. These fungi are *Aspergillus nidulans* (mitotic aneuploidy, Bignami *et al.*, 1974), *Neurospora crassa* (meiotic aneuploidy, Griffiths and Delange, 1977), *Saccharomyces cerevisiae* (meiotic and mitotic aneuploidy, Parry, 1977), and *Sordaria brevicollis* (meiotic aneuploidy, Bond and McMillan, 1979). All these systems have their own advantages and disadvantages, and these will be reviewed elsewhere (Griffiths, 1980). The present report concentrates on the Neurospora system, which I have developed. It should be noted that a variety of other test systems have been proposed in nonfungal organisms, for example, Drosophila, mammals, and higher plants. These will be found reviewed in de Serres (1979).

Neurospora is a haploid ($n = 7$) eukaryote. It has been used extensively in standard transmission and biochemical genetics. Its meiotic chromosomes look, and behave, like typical eukaryotic chromosomes. (In fact tetrad analysis in Neurospora was instrumental in confirming many of the canons of the chromosome theory of inheritance.)

Pittenger (1954) first showed that pseudowild types of Neurospora are in fact derived from disomic ($n + 1$) aneuploid meiotic products. These aneuploids are detectable in systems in which a pair of homologous chromosomes at meiosis have several different auxotrophic markers in trans arrangements. The meiotic products (ascospores) are plated on minimal selective medium, and the plates are later scanned for colonies. Most meiotic products will bear auxotrophic alleles and will produce no col-

onies. Disomics, however, through intergenic complementation, will grow and produce a visible pseudowild type colony. Pittenger showed that the aneuploid condition is unstable, and breaks down, probably soon after the first mitotic nuclear divisions begin, into two nuclear types, each corresponding to one member of the disomic chromosome pair. The situation is diagrammed in Figure 17.1. By having several parental chromosomes marked, Pittenger was able to show that the original aneuploids were disomic for one chromosome pair only, or perhaps for a small number of pairs. In later studies, Smith (1973) showed that, in some crosses at least, Neurospora pseudowild types could be multiply disomic, in fact, approaching diploidy in some cases.

Threlkeld and Stoltz (1970) performed tetrad analyses in order to discover the meiotic mechanisms responsible for producing disomic ascospores. They used parents with two different complementing *pan*-2 alleles. On limiting pantothenate, *pan*-2 mutations result in a pale ascospore. Disomic ascopores are black and can be spotted in the ascus. The entire contents of the disomic-containing ascus can then be grown up individually and tested. Briefly, these workers found that the disomics

DISOMIC, n + I

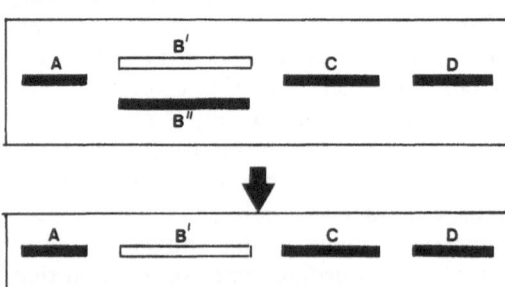

HETEROKARYON (PSEUDOWILD TYPE)
Figure 17.1. Possible mechanism for breakdown of disomic meiotic products to heteronuclear heterokaryons.

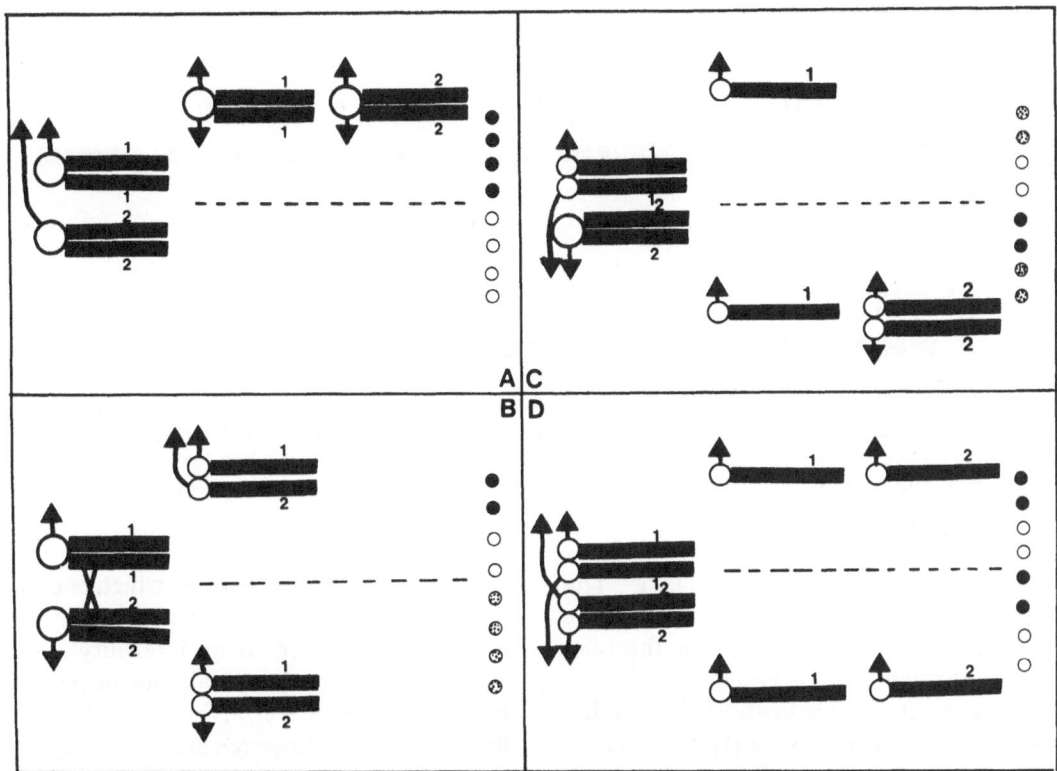

Figure 17.2. Main ways of generating aneuploid products of meiosis in Neurospora, based on the work of Threlkeld and Stoltz (1970). (A) Nondisjunction at the first meiotic division. (B) Nondisjunction at one of the second meiotic divisions, accompanied by a crossover. (C) Precocious division of one centromere at the first meiotic division. (D) Precocious division of both centromeres at first meiotic division. In (C) and (D) only representative second meiotic divisions are shown. 1 and 2 are complementing heteroalleles of the *pan*-2 gene. In the asci, black is disomic (n + 1), white is nullisomic (n − 1), and stippled is haploid (n). (See text for further explanation.)

could be a result of either first or second division meiotic nondisjunction, or precocious centromere division at the first division of meiosis. Some examples are illustrated in Figure 17.2.

Griffiths and Delange (1977) constructed two parental strains of opposite mating type, in a cross designed specifically to test for agents which can produce aneuploids at meiosis. The genotypes of the two strains are shown in Figure 17.3. The marker alleles serve the following functions.

1. *arg-1, ad-3A, ad-3B,* and *nic-2* are auxotrophic mutations serving as selective markers for disomy. Colonies growing on minimal medium are phenotypically arg⁺ ad⁺ nic⁺. These colonies have been tested extensively, and shown to be pseudowild types composed of two nuclear types corresponding to the marked parental chromosomes or recombinants thereof. No multiple recombinants appear because of the tight linkage relationships. The background frequency of pseudowild types is always low, showing that a normal meiosis is occurring.

2. *tol* is a recessive suppressor of mating type incompatibility, and permits the growth of disomics heterozygous for the mating type chromosome.

3. *C/c, D/d* and *E/e* are three separate he-

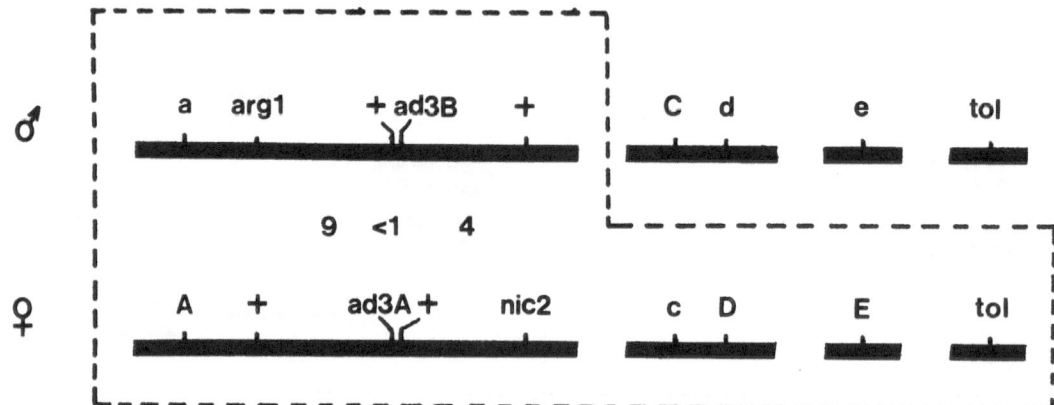

Figure 17.3. The genotypes of the two parents in the Neurospora cross from which prototrophic disomics are selected. A typical disomic is shown enclosed by a broken line. (See text for explanation of gene symbols.) The numbers between the chromosomes represent map distances, in map units.

terokaryon incompatibility loci. The presence of heterozygosity at any one of these loci, in a disomic for the relevant chromosome, is enough to prevent the growth of the disomic into a heterokaryotic pseudowild type. Thus, the presence of these markers in the standard cross tends to select for single, or possibly low number, multiple disomics, and is meant to make the system more closely model the human hyperploidy conditions. These same markers also serve to guard against the chance fusion of the germination tubes of adjacent ascospores of complementing auxotrophic genotypes.

Experimental protocol is quite simple and is diagrammed in Figure 17.4. The protoperithecial (female) parent is inoculated into tubes of crossing medium and the morphogenesis of the sexual apparatus occurs in about one week. At this time, suspensions of cells from the male parent are spread over the female cultures to effect fertilization. Six hours later, the crosses are exposed to the environmental agent. The surfaces of the cultures are flooded with the chemical agents either as a solution, emulsion, or suspension. Solvents have seldom been used because standard solvents have a radical and adverse effect on the fertility of the crosses. For an untried agent, a wide range of final concentrations or exposures

is attempted, and the highest which is compatible with adequate cross-fertility is eventually analyzed. If no infertility is encountered in the first trial, the doses are increased until a sterility threshold is obtained. When ascospores are shot, they are suspended and counted, either automatically in a cell counter, or by colony-forming ability on supplemented medium. An aliquot of known density is plated on selective medium and the pseudowild frequency is obtained. A minimum of three to five replicates at the threshold dose is aimed for. Untreated controls are run concurrently with each treatment batch.

Results

Fifty-seven agents have been tested to date in the Neurospora system: a total of 56 chemicals plus gamma radiation. The selection of these agents was on the basis of a combination of an "off the shelf" approach, and a consideration of mutagenicity or clastogenicity in other organisms. In the initial phase of the experiments, there was a need to demonstrate the sensitivity of the pseudowild type frequency to externally added agents. At this stage, some obvious choices, such as colchicine, were tried but found to have no effect on fertility even at massive doses, and these were put aside. The sensitivity of the system was eventu-

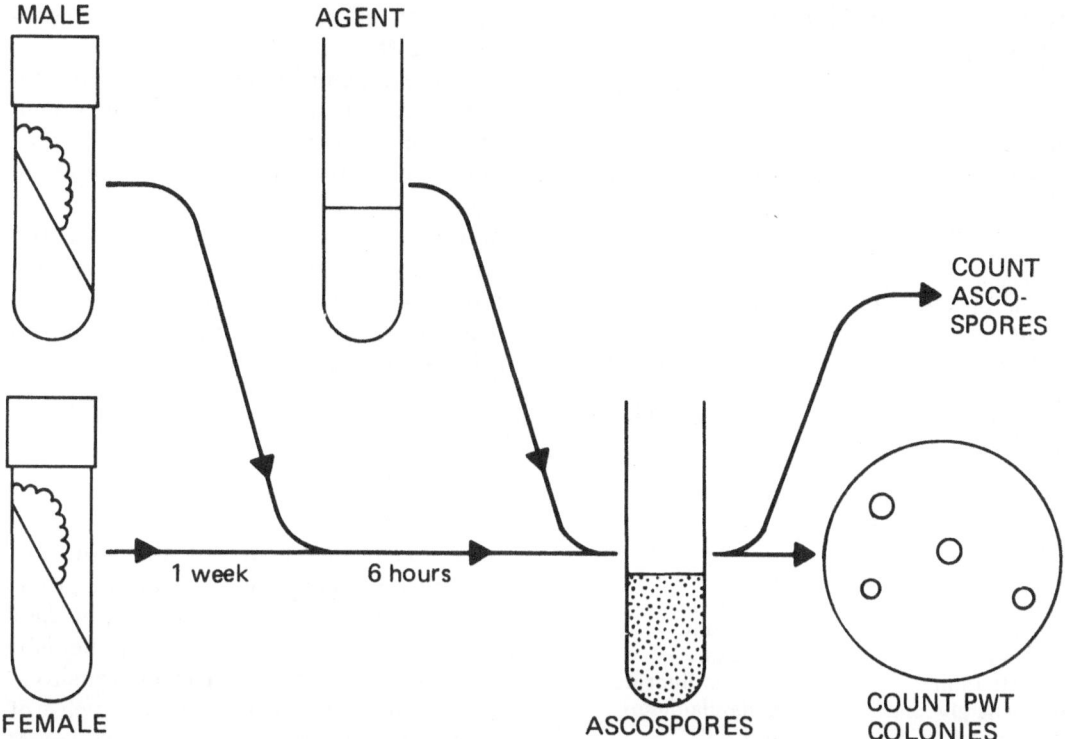

Figure 17.4. Protocol for testing agents for ability to produce aneuploids (detected as pseudowild-type, PWT, colonies).

ally demonstrated by the use of p-fluoro-phenylalanine (Griffiths and Delange, 1977), and this remains probably the most consistently potent chemical agent. At later stages of the experiments, the selection of agents was more on the basis of trying to discover the general incidence of aneu-ploid-producing potency. Carcinogenic and mutagenic agents were also included in the list.

The average control frequency of pseu-dowild types is about 6 per 10^5 ascospores, with a range rarely extending above 1.0 per 10^4. The frequencies obtained for the test agents were compared both with the overall range of control values, and with the simul-taneously run control values, using analy-ses of variance. Agents showing significant differences from controls (positives) are shown in Table 17.1. Agents showing no

Table 17.1. *Agents Testing as Positive in the Neurospora System*

Agent	Mean PWT frequency ($\times 10^{-5}$)	Concurrent control
Amethopterin	22.9	8.4
Caffeine	24.8	7.1
2,4-diaminoanisole sulfate	21.1	7.4
p-fluorophenylalanine	28.7	4.9
Gamma rays	14.3	4.1
Sulfacetamide	29.9	6.7
Trifluralin	19.6	5.8
Trimethoprin	9.5	5.1
Mean of all controls to date	—	5.5

Table 17.2.   *Agents Testing as Negative in the Neurospora System*

actidione	8-hydroxyquinoline
adriamycin	isopropanol
alloxan	maleic hydrazide
ametryne	mercuric chloride
aminopterin	methanol
m-aminophenol	mitomycin C
atrazine	metronidazole
bromacil	I-naphthol
4-chlororesorcinol	p-nitrobiphenyl
m-cresol	4-nitro-o-
croton oil	phenylenediamine
cyanazine	nitroquinoline
cyclic AMP	nitrosoguanidine
2,5-diaminoanisole	nystatin
sulfate	orthene
2,4-diaminotoluene	phenylmercuric
2,3-dibromopropanol	nitrate
diethylstilbestrol	potassium fluoride
dimethylsulfoxide	resorcinol
enterovioform	saccharin
eptan	sarcosine
ethanol	simazine
DL ethionine	sodium fluoride
ethylene glycol	sulfanilamide
fluorodeoxyuridine	terbacil
B-fluoropyruvic acid	vinblastine sulfate

significant difference from controls (negatives) are shown in Table 17.2.

In general, among the positive chemical agents, no response to increased final concentrations has been demonstrated. The data for p-fluorophenylalanine is shown in Figure 17.5. An optimum treatment concentration is suggested by these data, but this is rarely impressive for the chemicals. In the same figure, the data for gamma ray treatment is graphed, and this does show a response to increased dose.

Discussion

It has been demonstrated that, in this Neurospora test system, some agents prove to be positive, and some negative, in their ability to produce aneuploids at meiosis. Thus, the system is demonstrably sensitive, and to rather a wide range of types of agents. About 14% of the agents tested showed positive effects. Although the

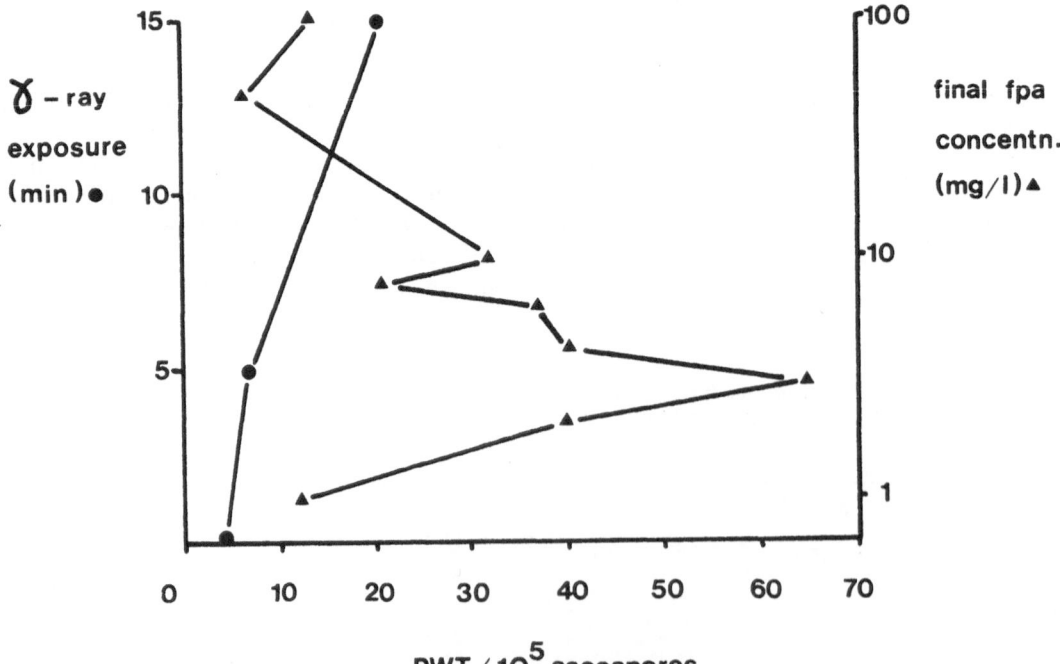

Figure 17.5. The relation between pseudowild (PWT) frequency dose, for γ-rays, and p-fluorophenylalanine (fpa). γ-Rays were from a ^{60}Co source delivering 40 r/sec.

agents were not selected on a random basis, it is clear that positive agents are not uncommon.

The significance of these results can only be assessed in relation to the entire battery of tests for detecting aneuploid-producing agents. It is beyond the scope of this presentation to review all this data, but some comments can be made as they relate to the discussion of the Neurospora results.

False positives and negatives. The general problems inherent under this heading have been reviewed recently by McCann and Ames (1978), and many of those problems are relevant in the present context. I suspect that one of the main specific errors in the present work is that the meiotic apparatus is not being consistently and/ or adequately challenged by the chemical agents. This is manifested in the rather large doses necessary to obtain decreases in fertility, and in the variability of pseudo-wild type frequencies between replicates (Griffiths, 1979). The probable cause is that the chemical agents are not entering the perithecia or the ascogenous hyphae. Failure of entry could reflect either an inherent meiotic insularity, or the way the chemicals are administered. Since most organic solvents interfere with fertility or viability, we have had to resort to the use of emulsions and suspensions in some cases. Nevertheless, since threshold doses were obtained, there is reason to believe that some level of biological potency is attained, although perhaps not at the crucial stages under study. The least offensive solvent is DMSO and at present an attempt is being made to incorporate DMSO into the protocol. The failure to consistently challenge the meiotic system is probably also the reason for the lack of differential response to dose in the chemical treatments. In contrast, the frequencies due to gamma ray treatment do show a correlation with dose. It should be mentioned that a dose response for chemicals causing meiotic aneuploidy has been observed in other fungal systems (Parry, 1977; Bond and McMillan, 1979). This could reflect inherent differences between the species involved.

The magnitude of the effects. Not only in the Neurospora system, but also in the yeast, Sordaria, Aspergillus, and mammalian systems for aneuploidy testing, there is only a relatively small effect of positive agents in the optimal dose range. This is true both of meiotic and mitotic systems. In Neurospora a six- to tenfold increase of means over control means is the best that has been achieved. This same order of magnitude of response is encountered in the other systems: for example, in yeast (Parry, 1977) p-fluorophenylalanine increased the frequency of meiotic disomics a maximum of eightfold, and in Sordaria (Bond and McMillan, 1979) about a tenfold increase for the same chemical was observed for meiotic nondisjunction. These results stand in contrast to the million-fold increases observed for some chemicals in the Salmonella test for point mutagens (McCann and Ames, 1978). Such "supermutagens" have not yet been encountered for aneuploidy. The reason for this is not known. It is possible that the meiotic target processes are fundamentally more sensitive, and abort completely at higher doses. This is certainly true for Neurospora; fertility is possible only within a narrow range of environmental conditions, whereas cell proliferation is far less intolerant of the same environmental abuse.

Carcinogens versus noncarcinogens. If the present list of agents is examined for carcinogens (by comparing, for example, with the list of McCann et al., 1975) about 13 will be found. It is probably too early in the development of this kind of research to attempt generalization, but it is apparent that agents which are positive in promoting aneuploidy are not necessarily and, furthermore, not usually, carcinogens or mutagens. (Notice also that ethionine, a carcinogen which is not mutagenic in the Ames test, is not active in the Neurospora aneuploidy system.)

Interorganismal comparisons. A satisfying degree of concurrence is seen if the results from different organisms are compared. Table 17.3 shows some comparisons compiled from notes made at a recent

Table 17.3. *Some Agents Which Have Been Tested in Several Systems*[a]

	Neurospora (mei)[b]	Aspergillus (mit)	Sordaria (mei)	Yeast (mei or mit)	CHO cells (mit)	Drosophila (mei)	Mouse (mei)	Vole (mei)	Human (mei)
Benomyl		+					−		
Caffeine	+			+					
Colcemid						+	−		
Colchicine		+			+				
Cytoxan					+				+
Dimethylsulfoxide	−	−			+				
Ethylmethanesulfonate				+		−	+		
p-fluorophenylalanine	+	+	+	+				+	
Fluorodeoxyuridine	−	−							
Methyl benzomyl carbamide		+						+	
Methyl mercury					+	+	−		
Methylmethanesulfonate		+		+			+	+	
Mitomycin C	−			+			−		
Saccharin	−			+			−		
Sulfacetamide	+			+					
Sulfanilamide	−	−		−					
Trifluralin	+		+			+			
Trimethoprin	+			+					
X-rays				+			+	+/−	+
γ-rays	+			+					

[a]Based on notes taken at a workshop. The formal version of this table will be published in Environmental Health Perspectives.
[b]mei stands for meiotic, and mit for mitotic systems.

workshop (de Serres, 1979). These correlations increase the likelihood that it is the same, or similar, biological target processes that are being affected in all these organisms. At the same time, credence is given to the notion that there is a class of agents which can generally increase the frequency of aneuploidy. p-Fluorophenylalanine appears, so far, to be a good example of such an agent.

Meiotic versus mitotic aneuploidy systems. At first glance, one might surmise that a meiotic aneuploidy detection system could be more relevant to germinal aneuploidy in man, and a mitotic test more revelant to somatic aneuploidy in human tissue. This may well prove to be the case eventually, and different sets of specificities may emerge, but it is presently interesting to note that several agents are active in

both meiotic and mitotic systems. For example, from Table 17.3, it can be seen that this is true of p-fluorophenylalanine, in several systems, and of sulfacetamide and trimethoprin in the Neurospora and yeast systems.

Germinal risks in populations. Theoretically, at the germinal level, an increased population exposure to agents which produce aneuploidy, should be detectable in the most important human aneuploid condition, Down syndrome. There should be one or both of two main effects. First, the overall frequency of DS in the population could increase. Second, because of the well-known maternal age effect associated with DS, the mean age of mothers of DS individuals might be expected to decrease, if the maternal age effect were assumed to be basically due to increased exposure in older mothers. Lowry *et al.* (1976) examined the data on DS in British Columbia for the period 1952–1972 and Griffiths *et al.* (1979) extended this analysis to 1975. Thus, these studies spanned a total of 23 years. No significant change in the overall frequency of DS was detected. A significant decline in the mean age of mothers of DS individuals was observed. However, examination of the frequencies within specific maternal age groups revealed no increases, and hence the overall decline of DS maternal age was best accounted for by the overall downshift of maternal age in the population as a whole, rather than any possible increased exposure to environmental agents. Thus, for BC in this time period there is no evidence of a deteriorating germinal situation. The same might not be true, however, for other geographical regions, and other historical comparisons. Furthermore, specific occupational or regional exposures may constitute a germinal aneuploidy hazard. There is evidence that this was the case in the occupational exposure of workmen to dibromochloropropane (Kapp *et al.*, 1979), based on studies of fluorescent Y chromosomes in sperm samples. The situation for X-ray exposure is still in doubt (Sankaranarayanan, 1979).

Somatic risks in populations. The interrelationship between cancer and aneuploidy is not at all well understood at the cellular level. Three observations may have to be reconciled: (1) Carcinogens are point mutagens and hence point mutagenesis is presumably involved in carcinogenesis; (2) most cancers show aneuploidy at some stage, and hence aneuploid chromosomal mutation is involved in carcinogenesis; and (3) most agents producing aneuploidy in various test systems to date are not carcinogen. Obviously more work is necessary before these relationships become clear. At the very least, further research on aneuploid-producing agents is needed to strengthen the correlations between aneuploid-, cancer-, and mutation-producing potencies.

The Neurospora test system, described above, is being remodeled to increase its usefulness. This work is proceeding along three major lines. First, a breeding program is being used to try to improve the sensitivity of the strains. This has been a difficul task because the original parental strains exhibit an extraordinarily high level of fertility for multiply marked strains in this species, and this high fertility has always been lost in derivatives. Second, exploratory crosses will be made to determine the effects of pretreating agents with a mammalian metabolic activating system. Third, a project to incorporate better solvents into the protocol will be continued. In addition, more studies of interspecies comparisons are planned, stemming from uniform samples: this has been instigated as part of a cooperative program involving several laboratories.

References

Banerjee, A., Jung, O., Huang, C.C. (1977): Response of hematopoietic cell lines derived from patients with Down's Syndrome and from normal individuals to mitomycin C and caffeine. J. Natl. Cancer Inst. *59*:37–40.

Bignami, M., Morpurgo, G., Pagliani, R., Carere, A., Conti, G., di Giuseppe, G. (1974): Non-disjunction and crossing-over induced by pharmaceutical drugs in *Aspergillus nidulans*. Mutat. Res. 26:159–170.

Bond, D.J., McMillan, L. (1979): The effect of p-fluorophenylalanine on the frequency of aneuploid meiotic products in *Sordaria brevicollis*. Mutat. Res. *60*:221–24.

Carter, C.O. (1977): The relative contribution of mutant genes and chromosome abnormalities to genetic ill-health in humans. In: D. Scott, B.A. Bridges, F.H. Sobels, (eds.), Progress in Genetic Toxicology, Amsterdam: Elsevier/North-Holland Biomedical Press, pp. 1–14.

Chan, F.P., Ball, J.K., Sergovich, F.R. (1979): Trisomy No. 15 in murine thymomas induced by chemical carcinogens, X-irradiation, and an endogenous murine leukemia virus. J. Natl. Cancer Inst. *62*:605–610.

de Grouchy, J., Turleau, C. (1974): Clonal evolution in the myeloid leukemias. In: J. German (ed.), Chromosomes and Cancer. New York: John Wiley and Sons, pp. 287–312.

de Serres, F.J. (1976): Mutagenicity of chemical carcinogens. Mutat. Res. 41:43–50.

de Serres, F.J. (1979): Chairman of a workshop: Systems to detect induction of aneuploidy by environmental mutagens. Environ. Health Perspect. 31: (whole volume).

Goh, K., Lee, H., Miller, G. (1978): Down's syndrome and leukemia: mechanism of additional chromosomal abnormalities. Am. J. Ment. Defic. 82:542–548.

Griffiths, A.J.F. (1979): A Neurospora prototroph selection system for studying aneuploid production. Environ. Health Perspect. 31: 75–80.

Griffiths, A.J.F. (1980): Microbial systems for detecting aneuploidy. In: A. Hollaender and F.J. de Serres (eds.), Chemical Mutagens: Principles and Methods for Their Detection, Vol. 7. New York: Plenum Press (in preparation).

Griffiths, A.J.F., Delange, A.M. (1977): p-Fluorophenylalanine increases meiotic nondisjunction in a Neurospora test system. Mutat. Res. 46: 345–354.

Griffiths, A.J.F., Lowry, R.B., Renwick, D.H.G. (1979): Down's syndrome and maternal age in British Columbia 1972–1975. Environ. Health Perspect. 31:9–12.

Kapp, R.W., Picciano, D.J., Jacobson, C.B. (1979): Y-Chromosomal nondisjunction in dibromochloropropane-exposed workmen. Mutat. Res. 64:47–51.

Koeller, P.C. (1972): The Role of Chromosomes in Cancer Biology. New York: Springer-Verlag.

Lowry, R.B., Jones D.C., Renwick, D.H.G., Trimble, B.K. (1976): Down syndrome in British Columbia, 1952–73; incidence and mean maternal age. Teratology 14:29–34.

McCann, J., Ames, B.N. (1976): Detection of carcinogens as mutagens in the Salmonella/microsome test: Discussion. Proc. Natl. Acad. Sci. USA 73:950–954.

McCann, J., Ames, B.N. (1978): The Salmonella/microsome mutagenicity test: predictive value of animal carcinogenicity. In: W.G. Flamm and M.A. Mehlman (eds.), Advances in Modern Toxicology, Vol. 5. New York: John Wiley and Sons, pp.87–108.

McCann, J., Choi, E., Yamasaki, E., Ames, B.N.(1975): Detection of carcinogens as mutagens in the Salmonella/microsome test: Assay of 300 chemicals. Proc. Natl. Acad. Sci. USA 72:5135–5139.

Miller, R.W. (1970): Neoplasia and Down syndrome. Ann. N.Y. Acad. Sci. 171: 637–644.

Mitelman, F., Levan, G., Nilsson, P.G., Brandt, L. (1976a): Nonrandom karyotypic evolution in chronic myeloid leukemia. Int. J. Cancer 18:24–30.

Mitelman, F., Nilsson, P.G., Levan, G., Brandt, L. (1976b): Nonrandom chromosome changes in acute myeloid leukemia. Chromosome banding examination of 30 cases at diagnosis. Int. J. Cancer 18:31–38.

Mitelman, F., Brandt, L., Nilsson, P.G. (1978): Relation among occupational exposure to potential mutagenic/carcinogenic agents, clinical findings, and bone marrow chromosomes. in acute nonlymphocytic leukemia. Blood 52: 1229–1237.

Nowell, P.C. (1974): Chromosome changes and the clonal evolution of cancer. In: J. German (ed.), Chromosomes and Cancer. New York: John Wiley and Sons, pp. 267–286.

O'Brien, R.L., Poon, P., Kline, E., Parker, J.W. (1971): Susceptibility of chromosomes from patients with Down's Syndrome to 7,12—dimethylbenz (a) anthracene—induced aberrations in vitro. Int. J. Cancer 8:202–210.

Parry, J.M. (1977): The detection of chromosome non-disjunction in the yeast Saccharomyces cerevisiae. In: D. Scott, B.A. Bridges, F.H. Sobels (eds.), Progress in Genetic Toxicology. Amsterdam: Elsevier/North-Holland Biomedical Press, pp. 223–232.

Pittenger, T.H. (1954): The general incidence of pseudo-wild types in Neurospora crassa. Genetics 39:326–342.

Sankaranarayanan, K. (1979): The role of nondisjunction in aneuploidy in man. Mutat. Res. 61:1–28.

Smith, D.A. (1973): Unstable diploids of Neurospora and a model for their somatic behaviour. Genetics, 76:1–17.

Sobels, F.H. (1977): Extrapolation from experimental test systems for evaluation of genetic risks in man. In: D. Scott, B.A. Bridges, F.H. Sobels (eds.), Progress in Genetic Toxicology. Amsterdam: Elsevier/North-Holland Biomedical Press, pp. 175–184.

Sonta, S., Sandberg, A.A. (1978): Chromosomes and causation of human cancer and leukemia. XXX. Banding studies of primary intestinal tumours. Cancer 41:164–173.

Threlkeid, S.F.H., Stolz, J.M. (1970): A genetic analysis of non-disjunction and mitotic recombination in *Neurospora crassa*. Genet. Res. Camb. *16*:29–35.

Uenaka, H., Ueda, N., Maeda, S., Sugiyama, T. (1978): Involvement of chromosome No. 2 in chromosome changes in primary leukemia induced in rats by N-nitroso-N-butylurea. J. Natl. Cancer Inst. *60*:1399–1404.

Weiner, F., Spira, J., Ohno, S., Haran-Ghera, N., Klein, G. (1978): Chromosome changes (trisomy 15) in murine T-cell leukemia induced by 7,12-dimethylbenz(a)anthracene (DMBA). Int. J. Cancer *22*:447–453.

18

Plant Genetic Test Systems for the Detection of Chemical Mutagens

WILLIAM F. GRANT, A.E. ZINOV'EVA-STAHEVITCH, AND
K.D. ZURA

Introduction

Higher plants provide valuable systems for screening and monitoring environmental chemicals. Although plant screening and monitoring assay systems have been in existence for many years, they are only beginning to receive the recognition which these sensitive and reliable systems warrant (de Serres, 1978). A general lack of familiarity with plant mutagenesis research, and the belief that plant and animal cells diverge so greatly in their physiology and phylogeny as to make comparisons meaningless, has led to a dearth of interest in, and funding for, plant mutagenesis. Recent symposia on the subject (Hart *et al.*, 1978; de Serres, 1978) may help improve the situation.

Numerous test systems already exist in higher plants. In a recent compilation, Shelby (1976) lists 233 plants which have been used in various aspects of mutagenesis research. Of these, *Vicia faba*, *Allium cepa*, *Hordeum vulgare*, *Trades-*

cantia, *Arabidopsis*, *Zea mays*, *Nicotiana*, *Pisum sativa*, *Glycine max*, *Lycopersicon*, and *Crepis capillaris* have been used most frequently (Nilan and Vig, 1976). In the area of mutagen monitoring, there are at present no organisms as useful as the *Tradescantia* stamen hair (Schairer *et al.*, 1978) and the *Osmunda regalis* (Klekowski, 1978) systems for the detection of air and water pollutants, respectively.

Recent studies have shown that, for a specific chemical agent, comparable results in terms of genetic abnormalities are obtained in plant and animal systems. For example, a recent survey of studies on the effects of pesticides (Grant, 1978) has shown that excellent correlations exist between the frequency of both chromosomal abnormalities and C-mitoses in plant and animal systems. A similar conclusion was drawn in studies on the effect of eight chemicals in several systems, including *in vitro* and *in vivo* mammalian systems, bacteria, *Drosophila*, and plant systems. The plant systems showed excellent correlations with the mammalian systems (Clive and Spector, 1978).

Plant assays have in their favor relative ease of handling, amenability to diverse growth and testing conditions, and relatively low cost, compared to animal systems (Nilan and Vig, 1976; Grant, 1978). It is evident that higher plant systems offer an attractive alternative to mammalian and nonmammalian animal assays in mutagen screening programs. Consequently, consideration should be given to the incorporation of higher plants into the tier testing system.

This paper outlines some of the most common procedures for screening for genetic aberrations in plants. Its purpose is twofold. Primarily, it provides an introduction to the techniques currently available in plant mutagenicity testing, particularly for those individuals who have worked exclusively with microbial and animal systems. Secondarily, it is hoped that it will serve as an inducement to expanding and refining existing plant systems, and providing incentive for the development of new testing procedures for mutagenicity.

Chromosome Aberrations in Plant Root Tips as a Test and Monitoring System

The standard method for the detection of chromosome aberrations in plants involves treatment and analysis of root-tip meristematic cells (Kihlman, 1971). Recent studies have shown that this classical assay for induced chromosome aberrations can be used effectively to monitor environmental chemicals (Grant, 1978; Klekowski, 1978; Tomkins and Grant, 1976). Root tips can be obtained in large numbers, are easy to handle, and display a degree of sensitivity which makes them well-suited to studies utilizing low chemical doses. Plants which have large chromosomes, including onion (*Allium* species), corn (*Zea mays*), broad bean (*Vicia faba*), barley (*Hordeum*

vulgare), pea (*Pisum sativum*), spiderwort (*Tradescantia* species), hawk's beard (*Crepis* species), and lily (*Lillium* species) are ideal for this type of study.

The Allium Test

The classical test for studying the effects of chemicals on plant chromosomes and cell division was developed by Levan (1938, 1949) using the common onion, *Allium cepa* ($2n = 16$). Other members of the genus, including *A. carinatus*, *A. flavum*, *A. fistulosum* (Welsh leek), *A. proliferum*, and *A. sativum* (garlic), have been used in mutagenesis studies, but none, except the tree or Egyptian onion, *A. proliferum* ($2n = 16$), have been widely used. While *A. cepa* is readily available, *A. proliferum* has its own attractive features. The plants reproduce from bulbils (rather than seed), which give rise to a large population genetically identical to the parent bulb. Furthermore, the chromosomes of *A. proliferum* being more asymmetric than those of *A. cepa* are easier to distinguish. The general procedure for handling these species is the same as for bulbs of *A. cepa*.

Schedule of Treatment for Allium Species

1. Select young onion bulbs of uniform size so that the base will fit a round-mouth container (glass jar) or the holes in a rack when several bulbs are to be treated simultaneously in a single tank.
2. Denude bulbs, removing the loose outer scales, and scrape the bulbs so that the apices of the root primordia are exposed. The bulbs are now ready for use.
3. Submerge bulbs in tap water to about one-quarter the depth of the bulb. Change water daily or aerate. A small aerator sold for aquaria will provide sufficient aeration for several small containers. The bottles or tanks should be kept in the dark for the period of the

experiment. After approximately 2 to 4 days at 20° C, the roots will grow to a length of 1–3 cm.

4. Since chemicals vary in their mode of action on the genetic material (for example, some chemical agents are most active when present at the time of DNA synthesis), the duration of the mitotic cycle and the period of DNA synthesis must be determined. The duration of the mitotic cycle in onion has been reported as follows: 20.5 hr at 19° C; 18.8 hr at 20° C and 23 hr at 21° C with G_1, S and G_2 being 3.3, 12, 3.7 hr, respectively and mitosis itself 4 hr (Kihlman, 1971).

5. For treatment, transfer bulbs to a similar vessel containing the solution to be tested. Treatment times from 2 to 24 hr are most common, with a series such as 2, 4, 8, 12, 24 hr being used to determine toxicity and threshold levels.

6. Recovery period: Since chromosome aberrations may be produced by different chemicals at different stages in the cell cycle, recovery periods prior to fixation should include a time period of one complete cell cycle. Time periods which have been used include 2, 4, 8, 12, 24, 28, and 72 hr after treatment.

7. Pretreatment: After the recovery period, harvest root tips and, depending on the experiment, either pretreat prior to fixation or fix immediately. For pretreatment, a 0.05 to 0.1% aqueous solution of colchicine is most commonly used to obtain the large number of metaphases which are required for the scoring of interchanges and deletions. At 20° C, a 1–2 hr pretreatment period should suffice. An alternate to colchicine is a 0.002 M 8-hydroxy-quinoline solution with a treatment period of 1–2 hr. If only anaphase aberrations are scored, pretreatment should be omitted since colchicine and hydroxyquinoline destroy the spindle, preventing the mitotic cycle from continuing beyond metaphase.

8. Fixation: Fix in absolute or 95% alcohol-glacial acetic acid (3:1) from 30 min to 24 hr. After fixation, the material can be maintained in 70% alcohol in a refrigerator, and the procedure resumed with the next step when convenient.

9. Staining: The usual procedures for aceto-carmine, aceto-orcein or Feulgen reagent may be followed. The latter procedure, specific for DNA, is generally preferred because of the excellent contrast obtained. The procedure is as follows: (a) Hydrolyze in 1N HCl at 60° C for about 5 min (optimum time, between 4 and 12 min, will vary with specific conditions); and (b) stain in Feulgen reagent for 2 hr.

10. Maceration: Treat with 5% pectinase for 1–3 hr. If roots are left too long in pectinase, they will become overly soft and difficult to handle.

11. Slide preparation: On a slide remove meristematic region and squash in a drop of 45% acetic acid.

12. Mount with a coverslip and make temporary seal with paraffin-gum arabic mixture, clear fingernail polish, or rubber cement.

13. Control: Set up a control experiment in which all the above steps are followed, substituting distilled water for the chemical compound. Variation in type and frequency of chromosome aberrations between the two sets will give a measure of the activity of the compound.

14. Permanent slides: A temporary slide will deteriorate after a few days even when kept in a Petri dish and stored in a refrigerator. To make permanent, use dry ice (Conger and Fairchild, 1953), liquid carbon dioxide, or a freezing apparatus to lower the temperature below −40° C (Sharma and Sharma, 1972). When the preparation is frozen, the coverslip is removed with a razor blade or scalpel. The root tip material adheres to the slide, which subsequently

is immersed in two changes of absolute alcohol and mounted in Euparal with a clean coverslip.

Onion Seed

Sax and Sax (1968) have used *A. cepa* seed to test for mutagenic chemicals. Their technique is as follows:

1. Pour 10 cc of test solution over 5 layers of paper toweling (or filter paper) in a Petri dish.
2. Sow about 100 onion seeds on the moist paper.
3. When the radicles (primary roots) are 0.5–1.0 cm long, usually in 3 days, fix in alcohol-acetic acid (3:1) for 24 hr.
4. After fixation prepare aceto-carmine smears or store in 70% alcohol until convenient.
5. Analyze 10 root tips, each with an average of more than 100 anaphases.

Detection of Chromosome Aberrations Using Vicia faba *Root Tips*

Vicia faba has long been used for cytological and radiobiological studies (Read, 1959). This species possesses six pairs of chromosomes, which are designated according to centromere position as either M (median) or S (subterminal). The single pair of M chromosomes is more than twice the length of the S chromosomes (mean 2.3:1) and possesses a large satellite on the short arm. Döbel *et al.* (1973) have developed new karyotypes with chromosomal structural changes in which all of the chromosomes may be distinguished from one another.

Cultural conditions. Since the seed of *V. faba* are large and do not lend themselves to germination in a Petri dish, various techniques have been developed for their germination and culture (Kihlman, 1971). The

technique practised in our Laboratory will be described (Zura and Grant, 1979):

1. Germination: About 30 seeds are germinated in well-aerated, thermally regulated (20° C) running tap water, using a 500-ml beaker and a small thermostatically controlled heater.
2. After 3–4 days, when the primary roots reach approximately 1 cm, the seedlings are planted in a bed of moist perlite. The bed consists of a rectangular wooden or plexiglass frame (16 × 14 cm) with plastic screening secured to the bottom. Perlite is spread over the screening to a depth of 2.5 cm. The seedlings (four per bed) are planted so that the primary root of each seedling protrudes down through the screen, allowing the seedling to obtain nutrients from the solution below. The bed is placed on top of a rectangular glass receptacle (butter dish) whose exterior surface is painted black.
3. Fill the dish to within 0.5 cm of the rim with Hoagland's nutrient solution adjusted to pH 6. Plants used for negative and positive controls are cultured under identical conditions at pH 6. Freshly prepared solutions should be used, as many chemicals are unstable in solution. The pH of the treatment solutions must also be carefully controlled as Zura and Grant (1980) have shown recently that low pH media are genetically active, particularly if weak acids are present in solution. Consequently, when chemical agents which possess weak acid properties are tested at low pH (approaching, or below, the pK of the chemical), the assay can be interpreted critically only if the prospective mutagen-induced effects can be differentiated from those induced by low pH *per se*. This criterion can be met by choosing, as controls, buffering agents with acid-base properties and concentrations similar to those of the chemical being tested.
4. The root tips of the seedlings are im-

mersed in the nutrient solution, which is aerated continuously with an air pump. Several containers can be maintained continually in controlled environmental conditions: 16 hr photoperiod, $24 \pm 1°$ C light, $19 \pm 1°$ C dark. In many cases, it may be necessary to use growth chambers for temperature control and in order to carry out the treatment and recovery in the dark if this is required.

5. Decapitate primary roots when they reach a length of approximately 4 cm.

6. Maintain plants in nutrient solution until the secondary roots reach 3–4 cm, at which time they are ready for treatment. Of the four seedlings in each bed, two or three will reach this stage at the same time.

Other practices to follow in order to standardize the procedure have been documented by Kihlman (1971, 1975). Chromosome aberrations encountered and procedures for scoring slides have been given by various authors, including Kihlman (1971) and Nicoloff and Gecheff (1976).

Tests for Somatic Chromosome Aberrations in Microspores and Pollen Tubes in Tradescantia

Mitosis in the male gametophyte of *Tradescantia* has been used to monitor the effect of chemicals and radiation on chromosomes (Ma and Kahn, 1976; Rushton, 1969; Smith and Lotfy, 1954). Somatic divisions of either the microspore or generative nuclei are scored. The study of the microspore nuclei has the advantage that, in the first postmeiotic division, the microspores undergo a relatively synchronous division, permitting examination of a large population of cells at the same stage.

To test for chromosome aberrations in generative nuclei in response to chemical agents, Smith and Lotfy (1954) carried out the following variation of the coated slide method:

1. Harvest fresh pollen in the morning and place in a desiccator at 6° C for 4 hr.

2. Treat dried pollen with chemical to be tested.

3. Sow immediately on culture medium (1% agar, 12% lactose and 0.01% colchicine), which has been spread lightly on a cover glass to form a thin film.

4. Invert cover glass over a glass ring attached to a slide to form a Van Tieghem cell, which may be humidified with a small piece of moist absorbent paper.

5. Place cultures in an incubator at 25° C for 24 hr. A number of metaphases of the dividing generative nuclei should be present; controls reach metaphase on an average of 2–4 hr sooner.

6. Remove cover slips and place in acetic-alcohol (1:3) for a few seconds.

7. Invert cover slip in a large drop of aceto-carmine on a slide for several minutes.

8. Remove cover slip with a pair of tweezers.

9. Pass through the following dehydration series:
   (1) 70% alcohol:glacial acetic acid (3:1)
   (2) absolute alcohol:glacial acetic acid (3:1)
   (3) absolute alcohol:glacial acetic acid (9:1)
   (4) absolute alcohol
   A prolonged immersion (more than 15 sec) in absolute alcohol may cause the development of crystals in the medium, which will make examination of the slide difficult.

10. Mount slide in Euparal or other mounting medium and examine microscopically.

Alternatively, the pollen can be cultured by using the hanging drop method (Sharma and Sharma, 1972) or the floating cellophane technique (La Cour and Fabergé,

1943). The culture medium can be varied by substituting sucrose (3–30% solution) for lactose, or gelatin (1–3%) for agar. Extracts from the style, placenta, or stigma have also been added to the medium (Sharma and Sharma, 1972).

Sister Chromatid Exchanges as a Test System

Sister chromatid exchanges (SCEs) consist of reciprocal exchanges between sister chromatids which have been differentiated so that they can be distinguished from one another. In an autoradiographic study on *Vicia faba*, Taylor *et al.* (1957) first demonstrated the occurrence of SCEs, which were observed as switches in tritiated thymidine label from one chromatid to its sister. The literature on SCEs has been reviewed by DuPraw (1970), Peacock (1963), Kihlman and Kronborg (1975), and Kato (1977).

The relatively poor resolution of the autoradiographic technique made an accurate estimation of the number of exchanges unreliable. However, with the development of new techniques the use of SCEs in the testing of mutagens has become a valid test system (Beck and Obe, 1975; Perry and Evans, 1975). A number of SCEs occur in preparations under normal culture or growth conditions, but it has been amply demonstrated that the frequency of SCEs can be increased by agents known to produce chromosomal aberrations, for example, X-rays (Gatti *et al.*, 1974) and UV-irradiation (Wolff *et al.*, 1974). A much greater increase in the frequency of SCEs has been observed following treatments with some chemical mutagens, such as caffeine and mitomycin C (Kato, 1974). While considerable information is available on the chemistry involved in the differentiation of sister chromatids, the biological significance or consequences of SCEs are unknown. The technique which has been

most successful in demonstrating SCEs in plants is that of fluorescence plus Giemsa (FPG). The incorporation of 5-bromouracil (an analogue of thymine) into the DNA component of the chromosomes affects their staining properties. A chromatid which has incorporated BrdU into both strands of its DNA stains more weakly with the FPG technique than a chromatid which has incorporated BrdU into only one strand. Such a unifilarly substituted chromatid does not stain as strongly as a completely unsubstituted chromatid (containing only thymine). After one cell cycle, in the presence of BrdU the two sister chromatids cannot be distinguished by FPG differential staining, and the SCEs which may have occurred cannot be detected. However, if the BrdU is omitted during the following cell cycle, one of the chromatids will contain thymine in both DNA strands whereas the second will be unifilarly substituted with BrdU. Alternatively, if BrdU is supplied during the second cycle, one of the sister chromatids will become bifilarly substituted whereas the second will remain unifilarly substituted. Under these circumstances, sister chromatids can be differentiated by using fluorochrome staining (e.g., "33258 Hoechst") and observed using UV light, or by the FPG technique. The FPG technique gives a much clearer, permanent resolution of the exchanges than fluorescence alone, which rapidly fades.

The first detailed procedure described for SCEs in plants has been that for *Vicia faba* by Kihlman and Kronborg (1975): their procedure is as follows:

1. Lateral roots (procedures for the growth of root tips have been outlined in a previous section) are exposed for 22 hr to an aqueous solution of 100 μM 5-bromodeoxyuridine (BrdU), 0.1 μM 5-fluorodeoxyuridine (FdUrd) and 5 μM uridine (Urd).

   In a recent modification of the technique, the time of exposure to BrdU was reduced to 16–18 hr in order to minimize the risk of obtaining bifilarly substituted chromatids (Kihlman *et al.*, 1978).

2. Replace BrdU solution with one containing 100 μM thymidine (dThd) and 5 μM Urd for 21 hr.
3. Pretreat with 0.05% colchicine for 3 hr.
4. Fix overnight in cold 1:3 acetic-methanol. In our Laboratory, we fix for 2 hr because longer fixation does not improve the preparation and renders the material more difficult to squash.

    Note: Do not expose roots to light during steps 1 to 4. Treatments should be carried out in the dark at 20° C. Kihlman and Kronborg (1975) used glass tubes (containing 40 ml of solution) covered on the outside with black tape; the main root passes through a small hole in a black rubber disk resting on the top of the tubes.

5. After fixation, rinse roots in 0.01 M citric acid ($C_6H_8O_7 \cdot H_2O$)-sodium citrate ($Na_3C_6H_5O_7 \cdot 2H_2O$) buffer, pH 4.7.
6. Incubate roots for 75 min at 27° C with 0.5% pectinase dissolved in the same buffer. In our Laboratory, we incubate for 2 hr at 22° C.
7. Squash root tips in 45% acetic acid on alcohol-cleaned slides, coated (subbed) with a 10:1 mixture of gelatin and chrome alum (chromium potassium sulfate).
8. Using dry ice (Conger and Fairchild, 1953), liquid carbon dioxide, or a freezing apparatus (Sharma and Sharma, 1972), lower the temperature to below −40° C, freeze preparation, and remove cover slip.
9. Process slide through ethanol series from absolute, 95, 85, 70, 50, and 30% to distilled water.
10. Place slide in a moist chamber and incubate for 50 min at 27° C in a RNase solution prepared as follows: 1 mg RNase (Sigma, RNase-A from bovine pancreas) dissolved in 10 ml 0.5 × SSC (0.075 M NaCl + 0.0075 M

$Na_3C_6H_5O_7 \cdot 2H_2O$); 200 μl are placed onto the tissue and the preparation covered with a fresh cover slip. The RNase step is not absolutely necessary but improves the contrast between sister chromatids.

11. Rinse in 0.5 × SSC.
12. Stain for 20 min with a solution of "33258 Hoechst" prepared as follows: 1 mg of the fluorochrome dissolved in 1 ml ethanol and 0.1 ml of this solution added to 200 ml 0.5 × SSC.
13. After staining, rinse preparations and mount on 0.5 × SSC; temporarily seal cover slips with gum mastic, clear fingernail polish, or rubber cement.
14. Observe preparation using a fluorescence microscope. With a Zeiss UV-microscope exciter, filter combinations BG38 and BG 12 and barrier filters 43/54 may be used. Differentiation of sister chromatids may be improved by exposing the preparations to long-wave UV light for 30 min (Kihlman *et al.*, 1978).
15. Store slides over distilled water in a moist chamber for 4 days at 4° C and then incubate for 60 min at 55° C in 0.5 × SSC. The latter step may be omitted if the preparation is stored for 4 days at 25° C.
16. Rinse slides thoroughly in 0.017 M phosphate buffer, pH 6.8.
17. Stain for 6 to 7 min in a solution containing 3% Giemsa (Gurr's R66) in the same buffer. The staining time is critical.
18. Rinse preparations, first in phosphate buffer, then in distilled water, and air dry.
19. Dip the dry preparations in xylene and mount in a mounting medium such as permount or Canada balsam.

In our Laboratory, a modified technique of Kihlman and Kronborg's (1975) schedule has been adopted by Zura and Grant for evaluating the mutagenicity of pesticides. The modifications were found desirable because the procedure is shortened considerably and the routine removal of the cover slip is rendered unnecessary. Essentially, all staining procedures are applied to intact, fixed root tips with squashing reserved for the very last step. Only those slides which are to be kept indefinitely need to be made permanent. Following step 6 in the above schedule our procedure is as follows:

7. Rinse in 0.5 × SSC and stain with 33258 Hoechst prepared according to Kihlman and Kronborg's schedule (step 12).
8. Maintain root tips on 0.5 × SSC overnight at 4° C.
9. Incubate 60 min at 55°C in 0.5 × SSC.
10. Rinse material in 0.017 M phosphate buffer (pH 6.8) and stain 7–20 min in a solution containing 3% Giemsa (Gurr's R66) in the same buffer.
11. Incubate in 0.1 N HCl for 2 min at 60° C. This maceration step does not have serious adverse effects on the differentiation of the chromatids but considerably improves the ease of squashing.
12. Squash the stained root tip meristems in a drop of 45% acetic acid and analyze microscopically using visible light; the addition of UV light improves the differentiation.

Sister chromatid exchanges have also been successfully induced in *Allium cepa* (Schvartzman and Cortes, 1977) and it might be expected that species with large chromosomes, such as *Tradescantia*, may also be suitable for the development of sister chromatid exchange assay systems.

Other plants that may be amenable for use in the chemical induction of sister chromatid exchanges include several in which Giemsa banding has already been demonstrated. These include: *Anacyclus* (Ehrendorfer *et al.*, 1977), *Anemone* (Marks, 1974; Marks and Schweizer, 1974), *Avena* (Yen and Filion, 1977), *Crepis* (Komatsu and Tanaka, 1978), *Fritillaria*

(Schweizer, 1973), *Gibasis* (Kenton, 1978), *Hepatica* (Marks and Schweizer, 1974), *Hordeum* (Linde-Laursen, 1975; Vosa, 1977), *Leopoldia* (Bentzer and Landström 1975), *Lotus* (Shankland and Grant, 1976), *Nigella* (Klasterska and Natarajan, 1975; Marks, 1975), *Ornithogalum* (Stack *et al.*, 1974; Schweizer, 1977), *Phaseolus* (Schweizer, 1976), *Scilla* (Grielhuber and Speta, 1976; Schweizer, 1977), *Secale* (Gill and Kimber, 1974a; de Vries and Sybenga, 1976; Kranz, 1976; Singh and Röbbelen, 1976; Nakata *et al.*, 1977), *Trillium* (Schweizer, 1973; Takehisa and Utsumi, 1973), *Triticum* (Gill and Kimber, 1974b; Hadlaczky and Belea, 1975), *Tulbaghia* (Vosa, 1977), *Tulipa* (Filion and Blakey, 1979), and *Zea* (Hadlaczky and Kalman, 1975; Sachan and Tanaka, 1977).

Cytogenetic Test Systems

Hordeum vulgare *as a Test System*

Cultivated barley is excellent for testing for mutagenesis as considerable knowledge has been accumulated on both the cytology and the genetics of the group (Nilan, 1964, 1974). Barley is diploid ($2n = 14$) with relatively large chromosomes which are easily distinguished. Since the plant is normally self-fertilized, recessive mutations are phenotypically expressed. Hybridization is easily carried out, which makes it possible to test for recessiveness. Chlorophyll-deficient mutants have been used widely in barley because many can be detected readily in the seedling stage. In addition to barley, several other species have been used for both cytological and genetic testing of the effects of putative mutagens. These include tomato (Grant and Harney, 1960), pea (Blixt, 1972), maize (Spasojevic, 1975), cotton (Endrizzi and Brown, 1968), *Solanum* (Lamm, 1945), and *Tradescantia* (Sparrow *et al.*, 1968).

An example from the author's Laboratory will illustrate an approach for testing for mutagenicity in barley (Wuu and Grant, 1966).

1. Treat 50–200 seeds per treatment with concentrations of 0, 500, 1000, and 1500 ppm of the test chemical for 6, 12, and 24 hr. The maximum concentration is determined by the solubility of the chemical in water and by its toxicity. Ideally, as with pesticides, the range of treatments should include the concentration recommended by the manufacturer. However, if this concentration is too toxic, then the concentration or the duration of treatment may be reduced.

   In order to determine the efficiency of the assay and for comparative purposes, positive controls should be used. For example, Ehrenberg *et al.* (1961) demonstrated that a dose of 5500 R X-ray was capable of inducing approximately 50% sterility in barley. The highly mutagenic alkylating agent, ethylmethanesulfonate, is also widely used as a positive control in comparing cytological and genetic effects.

2. The effects of the chemical on germination, growth rate, teratogenicity of seedlings, and root and shoot growth should be tested.

3. Cytological studies of the root tip cells should be carried out in order to determine the types and frequencies of the chromosomal aberrations which are induced. Each barley plant can develop up to six primary tillers so that a mutation in a cell on one primary tiller will not be expressed in any other primary tiller. Thus, chimeras may be expected at the C_1 plant level.

4. Score the morphological mutants in the C_1 generation.

5. Determine the fertility of the C_1 plants. An index of fertility can be obtained by examining either quartet cells for micronuclei or the pollen of intact anthers for unfilled and shriveled pollen grains. The latter is carried out by allowing pollen to stain in a drop of lactophenol green on a slide for a few hours. Cotton blue, Müntzing's mixture of glycerol and 1% aceto-carmine, or Owzarzak's methyl green phloxine may be substituted for the fast green in lactophenol. Unstained pollen is probably inviable. A more accurate measure of pollen fertility is germination rate of the pollen tubes, procedures for which have been outlined by Sharma and Sharma (1972).

6. Study C_1 pollen mother cells (PMCs) for meiotic chromosome aberrations.

7. Collect seed from individual C_1 plants and grow out for C_2 studies.

8. Cytogenetic studies on the C_2 generation include the following: Examination of root tip cells for chromosome aberrations, screening for morphological mutants, examination of PMCs for meiotic chromosome aberrations, and determination of fertility levels. The C_2 generation is used to assess the degree of genetic effects and can be compared with the toxic effects observed in the C_1 generation.

Testing for Chromosome Aberrations in Meiotic Material

Chromosome aberrations may be determined at metaphase and anaphase in meiotic cells of most plants whose root tips are amenable to cytological analysis (Wuu and Grant, 1967). Chromosome aberrations in meiosis may be determined following treatment of the seed and growing out of the plants to the stage of bud formation. Alternatively, cuttings may be treated in, or the plants may be sprayed with, the test solution. Further evidence of chromosomal abnormalities may be obtained by examining cells in the quartet stage for micronuclei resulting from chromosome breakage or lagging. The results are comparable to the micronucleus test in animals (Bruce and Heddle, 1979). Techniques for studying chromosomes in gametophytic material are reviewed by Sharma and Sharma (1972).

Genetic Test Systems

Arabidopsis thaliana *as a Test System*

A member of the Cruciferae, *Arabidopsis thaliana* possesses certain features which have made it an excellent plant, comparable to some microorganisms, for testing the effects of mutagens. It is a diploid ($2n = 10$), self-fertilized species. Its short life-cycle (28 days) and small size permit a large number of plants to be grown in standard test tubes under culture conditions. A single plant may produce more than 50,000 seeds. A number of marker loci are available for mutation studies. Rédei (1974) has reviewed the literature on *Arabidopsis*.

Rédei (1974) and Rédei and Acedo (1976) have described the procedures used to detect different classes of mutants in *Arabidopsis*. The main mutation detection system scores lethal or chlorophyll defective embryos in 11- to 13-day-old siliques of the plants which develop from mutagen-treated seeds, or chlorophyll-deficient mutants in M_2 seedlings.

For the detection of chlorophyll-deficient or thiamine mutants, Rédei and Acedo (1976) suggest:

1. Treat 10^5 seeds and suspend in cold viscose agar solution.
2. Disperse seed at the rate of 100 seeds per 5 ml of solution per 13 cm pot.
3. Grow M_1 generation plants under continuous illumination for 6–7 weeks.
4. Harvest M_1 seed in bulk.
5. Plant M_2 seed at the rate of 100 per 13 cm pot.
6. Chlorophyll-deficient mutants may be tabulated; identify and stake thiamine auxotrophs and feed with a dilute solution containing thiamine twice weekly until maturity.
7. Each suspected thiamine mutant can be progeny-tested for verification of the mutation.

Specific Locus Test Systems

A number of specific locus systems have been developed for mutagenicity testing. These include chlorophyll-deficient systems in barley, maize, and *Arabidopsis* (Constantin, 1978), in which either the seed is treated or plants are exposed throughout their entire life-cycle to the putative mutagen. When seed is treated, the M_1 generation plants can be used to assess morphophysiological features such as reduction in seedling height, survival to maturity, and fertility. The M_2 generation is used to assess genetic effects—most commonly the frequency of chlorophyll-deficient phenotypes.

Specialized Systems for the Detection of Mutations

Three specialized high resolution systems have been developed: (1) the *waxy* locus in *Zea mays* and *Hordeum vulgare*, (2) *yg*-2 locus in *Zea mays*, and (3) the self-incompatibility locus (*S*)—which is found in many plants. Phenotypes in all three systems are expressed in the gametophytic stage. Several hundred-thousand pollen grains with their own genotype can be scored, and hence, provide a genetic analysis with a resolving power comparable to that obtained using microorganisms.

The *waxy* locus. The maize *wx* locus assay can be used to measure forward mutation frequency ($Wx{\rightarrow}wx$), reverse mutation frequency ($wx{\rightarrow}Wx$), and the frequency of intragenic recombination. Alleles of the *waxy* locus in barley are being developed for screening and monitoring mutagens (Rosichan *et al.*, 1979). The *waxy* (*wx*) allele is recessive to *starchy* (*Wx*). When subjected to an iodine solution, kernels carrying the *Wx* allele stain blue-black; *wx/wx* kernels stain red. The starch type of a pollen grain is controlled by the genetic constitution of that pollen grain, not by that of the parental sporophyte. Hence, a genetic reversion of *wx* to *Wx* can be detected by scoring for pollen grains from plants that are homozygous for *wx*.

The procedure used by Plewa (1978) for detecting the mutagenic effects of pesticides in *Zea mays* is as follows:

1. Divide material into treatments and controls. For treatments, incorporate pesticide in the soil following manufacturer's instructions.
2. Plant kernels homozygous for *wx* in soil.
3. Grow plants to early anthesis; harvest tassels.
4. Agitate individual tassels in 70% ethanol to remove any foreign contaminant pollen from the surface of the tassel.
5. Remove approximately 15 unopened florets from the tassel and agitate in a Petri dish filled with 70% ethanol.
6. Dissect out anthers from unopened florets.
7. Place in a stainless steel cup of a VirTis microhomogenizer containing 0.6 ml of gelatin-iodine stain; mince anthers with scissors and homogenize for 30 sec.
8. Strain homogenate through cheesecloth onto the surface of a large microscope slide and place cover slip on pollen suspension.
9. To make a second slide add 0.4 ml of stain into steel cup, rehomogenize anthers for 20 sec and repeat step 8.
10. Examine solidified pollen suspension under a dissecting microscope and count blue-black staining pollen grains.
11. Estimate total number of pollen grains per slide by counting the number of pollen grains with 20 randomly chosen 1 mm^2 areas and multiply by the appropriate factor.
12. Examine several slides and for each plant calculate *wx* to *Wx* reversion frequency by dividing the total number of *Wx* pollen by the total estimated number of pollen grains analyzed.
13. Determine the mean and the standard error for each treatment group.

The *yg*-2 allele in *Zea mays*. The *yg*-2 allele is a chlorophyll deficiency that is detected and scored in seedling leaves (Conger and Carabia, 1977). The gene is used as a test assay system as follows:

1. Place kernels (heterozygous, *Yg*-2/*yg*-2) in water and aerate at 25° C for 72 hr.
2. Transfer kernels to a test tube containing chemical to be tested (adjusted to pH 7) and aerate at 25° C for 5 to 8 hr. For controls, transfer kernels to a test tube containing 0.01 M phosphate buffer (at pH 7) and aerate at 25° C for the same time as treated kernels.
3. Wash kernels and place in running water overnight.
4. Plant kernels in flats or pots.
5. In approximately 3 to 4 weeks, count *yg*-2 sectors per leaf on the fourth and fifth leaves, and compare the results in treated and untreated material.

Self-incompatibility test systems. Gametophytic self-incompatibility assay systems have been developed in *Oenothera*, *Nicotiana*, and *Petunia* (Nettancourt, 1977); recently Mulcahy and Johnson (1978) have described a method for the construction of a largely autonomous incompatibility system which would provide continuous monitoring of the environment. All these systems are based on the principle that pollen cannot fertilize a plant bearing the same *S* allele. For example, pollen carrying an S_1 allele cannot fertilize a plant homozygous for S_1, can effect 50% fertilization in plants with the genotype S_1S_2, and can completely fertilize plants not containing the S_1 allele. In practice, estimated numbers of pollen grains from treated or control plants are placed on the stigma of the donor plant to effect self-fertilization. Fruit- and seed-set are used as criteria of compatibility; pollen-style inhibition, as a criterion of incompatibility (Pandey, 1964). Since the incompatible style acts as a sieve excluding all but mutant pollen, scoring is limited to the small number of pollinations involving mutant pollen grains.

Tests Employing Somatic Mutations

The Tradescantia Assay Systems for the Detection of Mutations

One of the unique plants for testing for mutagenicity is *Tradescantia*, in which many genetic end points can be tested. These include chromosome aberrations in root tips, stamen hairs, and microspores, as well as somatic mutations in petals and stamen hairs. The *Tradescantia* genetic system, developed by the late Arnold Sparrow and his colleagues at Brookhaven for studying the effects of ionizing radiation, demonstrated that stamen hairs were sensitive to as little as 0.25 rad of X-rays. This uniquely sensitive stamen hair system is now being used in a mobile laboratory to monitor for air pollutants (Schairer *et al.*, 1978).

Clones of *Tradescantia*, heterozygous for flower color, are used for the analysis of somatic mutations in flower petals and stamen hairs. Clones of *Tradescantia* are easily maintained by vegetative propagation. These flower continuously throughout the year if maintained in growth chambers (Underbrink *et al.*, 1973). Clones 02, 0106, and 4430, the most commonly used, are diploid ($2n = 12$) hybrids between pink- and blue-flowering parents, with blue dominant to pink. The visible marker used in the test system is the phenotypic change in pigmentation from blue to pink in either petals or stamen hairs. Although mutant homozygous recessive pink sectors occur spontaneously in the petals, their frequency greatly increases following exposure to mutagens. In young developing floral tissue, the mutations are evident as isolated pink cells which appear within 5 to 18 days after exposure to a mutagen. In the mature flow-

er, they appear as groups of cells (Mericle and Mericle, 1967).

The stamen hair system is extremely sensitive. A linear relationship between mutation frequency and dosage is found over a wide range of radiation levels, with no evidence of a threshold effect even at levels as low as 250 mrad of X-rays (Schairer *et al.*, 1978). Mature stamen hair cells not only mutate to normal-size pink cells but also to giant or dwarf pink, as well as blue and colorless cells. For facility, only pink cells are scored routinely.

Two methods for testing for mutagenicity using *Tradescantia* are as follows:

Method A (Tomkins and Grant, 1972)

1. Select immature flower buds on a number of plants (or fresh cuttings of as uniform age as possible) and wrap each bud with a strip of absorbent cotton 2.5 cm by 7.5 cm.
2. Apply 5 ml of test solution to the absorbent cotton.
3. After 24 hr, remove absorbent cotton.
4. In 7–15 days, examine flower petals under low magnification, dissect out stamen hairs, and score both for pink sectors.

Method B (Underbrink *et al.*, 1973)

1. Prepare fresh cuttings with young inflorescences which contain developing flower buds.
2. Treat a selected number of cuttings with a given concentration of chemical to be tested.
3. Grow cuttings after treatment in aerated Hoagland's nutrient solution, which is changed weekly. Score pink sectors in petals (using a dissecting microscope). From day 7 through day 16 posttreatment, dissect out stamen hairs and score pink mutants.

Subsequent procedures and statistical

treatments are given in Underbrink *et al.* (1973).

The Soybean Assay System for the Detection of Mutations

Chlorophyllous-achlorophyllous spots or sectors of a number of plants, including soybean, tobacco, maize, pea, and tomato, lend themselves to tests for mutagenesis. Such phenotypic changes may arise from point mutations, somatic crossing over, gene conversion, chromosome losses, or minute chromosome deletions.

Vig (1975) has developed an assay system for somatic mutagenesis in soybean involving spots on the leaves, which are generally confined to the two simple leaves and the first compound leaf. The system exhibits incomplete dominance: $Y_{11}Y_{11}$ genotypes are dark green in color, heterozygous $Y_{11}y_{11}$ genotypes are light green, and homozygous $y_{11}y_{11}$ genotypes are golden yellow. In some heterozygous plants there arise twin spots indistinguishable from $Y_{11}Y_{11}$ and $y_{11}y_{11}$. It has been postulated that these arise from somatic crossing over. In addition, single spots have been observed, and it has been suggested that these are the result of point mutations.

The experimental procedure for mutagenesis testing as carried out by Vig (1973) is as follows:

1. Soak seeds in the specific concentrations of the chemical to be tested.
2. Wash seeds with water.
3. Plant in sand.
4. The first compound leaf will appear in 3–5 weeks; analyze the two simple leaves and the compound leaf for the types and frequencies of spots.

Vig (1978) has recently listed a number of species of higher plants which are promising for the study of mutagen-induced mosaicism, including *Gossypium, Nicotiana, Petunia, Salvia, Lycopersicon, Zea, Hordeum,* and *Tradescantia*.

Other Plant Test Systems

In brief, higher plants offer considerable potential for the development of new genetic test systems (Nilan, 1978). For example, test systems are being developed with haploid plants (Christianson and Chiscon, 1978) and an *Adh-1* allelic system in *Zea mays* (Freeling, 1978). Furthermore, considerable effort has been devoted to the selection of cultivated and wild plants for screening and monitoring pollutants (Feder, 1978; Klekowski, 1978; Koranda and Robison, 1978; Tomkins and Grant, 1976; Wolverton and McDonald, 1978).

Tumors and fasciations which occur widely in higher plants (White, 1948) offer other potentially important test systems. These cancerlike growths are analogous to cancers which occur in animals in that both plant and animal tumors express the same physical manifestation in response to different stimuli, such as bacteria or X-rays. Tumors and fasciations should be investigated from the point of view of chemical mutagenesis, as they may not only provide test systems but may also contribute further information to the relationship between mutagenesis and carcinogenesis in general. Additional plant systems for screening for mutagens have been described by Ehrenberg (1971) and Nilan and Vig (1976).

Concluding Remarks

It should be evident that higher plants provide excellent systems for mutagenicity testing. Numerous plant assays have been developed, the most important of which are outlined in this paper. The results in plant systems correlate well with comparable animal studies. Plant systems cover a wide range of resolutions. Furthermore, plants are inexpensive, easy to handle and amenable to a wide range of experimental conditions. For these reasons, it is recommended that plant systems be given their due recognition by committees concerned with regulating test systems.

Acknowledgments

The technical assistance of Ms. Rosemary Haraldsson of my Laboratory has been greatly appreciated. The authors thank Dr. M.J. Plewa of the Institute for Environmental Studies, University of Illinois, Urbana, for his constructive comments on the allelic systems in *Zea mays*. Financial support from the Natural Sciences and Engineering Research Council of Canada for studies in genetic toxicity of environmental chemicals is gratefully acknowledged.

References

Beck, B., Obe, G. (1975): The human leucocyte test system. VI. The use of sister chromatid exchanges as possible indicators for mutagenic activities. Humangenetik 29:127–134.

Bentzer, B., Landström, T. (1975): Polymorphism in chromosomes of *Leopoldia comosa* (Liliaceae) revealed by Giemsa staining. Hereditas 80:219–232.

Blixt, S. (1972): Mutation genetics in *Pisum*. Agri Hort. Genet. 30:1–293.

Bruce, W.R., Heddle, J.A. (1979): The mutagenic activity of 61 agents as determined by the micronucleus, *Salmonella* and sperm abnormality assays. Can. J. Genet. Cytol. 21:319–334.

Christianson, M.L., Chiscon, M.O. (1978): Use of haploid plants as bioassays for mutagens. Environ. Health Perspect. 27:77–83.

Clive, D., Spector, J.F.S. (1978): Comparative chemical mutagenesis: An overview. In: F.J. de Serres (ed.), Proceedings of Comparative Chemical Mutagenesis Workshop. NIEHS. Research Triangle Park, N.C.

Conger, A.D., Fairchild, L.M. (1953): A quick-freeze method for making smear slides permanent. Stain Technol. 28:281–283.

Conger, B.V., Carabia, J.V. (1977): Mutagenic effectiveness and efficiency of sodium azide versus ethyl methanesulfonate in maize: Induction of somatic mutations at the yg_2 locus by treatment of seeds differing in metabolic state and cell population. Mutat. Res. 46:285–296.

Constantin, M.J. (1978): Utility of specific locus systems in higher plants to monitor for mutagens. Environ. Health Perspect. 27:69–75.

de Serres, F.J. (1978): Introduction: Utilization of higher plant systems as monitors of environmental mutagens. Environ. Health Perspect. 27:3–6.

de Vries, J.M., Sybenga, J. (1976): Identification of rye chromosomes: the Giemsa banding pattern and the translocation tester set. Theoret. Appl. Genet. 48:35–43.

Döbel, P., Rieger, R., Michaelis, A. (1973): The Giemsa banding patterns of the standard and four reconstructed karyotypes of *Vicia faba*. Chromosoma 43:409–422.

DuPraw, E.J. (1970): DNA and Chromosomes. New York: Holt, Rinehart and Winston.

Ehrenberg, L. (1971): Higher plants. In: A. Hollaender (ed.), Chemical Mutagens: Principles and Methods for Their Detection, Vol. 2. New York: Plenum Press, pp. 365–386.

Ehrenberg, L., Gustafsson, A., Lundqvist, U. (1961): Viable mutants induced in barley by ionizing radiations and chemical mutagens. Hereditas 47:243–282.

Ehrendorfer, F., Schweizer, D., Greger, H., Humphries, C. (1977): Chromosome banding and synthetic systematics in *Anacyclus* (Asteraceae-Anthemideae). Taxon. 26:387–394.

Endrizzi, J.E., Brown, M.S. (1968): Cytological and genetical studies of the hybrid of *Gossypium herbaceum* L. and *G. triphyllum* Hoch. Genetica 39:272–288.

Feder, W.A. (1978): Plants as bioassay systems for monitoring atmospheric pollutants. Environ. Health Perspect. 27:139–147.

Filion, W.G., Blakey, D.H. (1979): Differential Giemsa staining in plants. VI. Centromeric banding. Can. J. Genet. Cytol. 21:373–378.

Freeling, M. (1978): Maize *Adhl* as a monitor of environmental mutagens. Environ. Health Perspect. 27:91–97.

Gatti, M., Pimpinelli, S., Olivieri, G. (1974): The frequency and distribution of isolabelling in Chinese hamster chromosomes after exposure to X-rays. Mutat. Res. 23:229–238.

Gill, B.S., Kimber, G. (1974a): The Giemsa C-banded karyotype of rye. Proc. Natl. Acad. Sci. USA 71:1247–1249.

Gill, B.S., Kimber, G. (1974b): Giemsa C-banding and the evolution of wheat. Proc. Natl. Acad. Sci. USA 71:4086–4090.

Grant, W.F. (1978): Chromosome aberrations in plants as a monitoring system. Environ. Health Perspect. 27:37–43.

Grant, W.F., Harney, P.M. (1960): Cytogenetic effects of maleic hydrazide treatment of tomato seed. Can. J. Genet. Cytol. 2:162–174.

Grielhuber, J., Speta, F. (1976): C banded karyotypes in the Scilla hohenackeri group. S. persica and Puschkinia (Liliaceae). Plant Syst. Evol. 126:149–188.

Hadlaczky, Gy., Belea, A. (1975): C-banding in wheat evolutionary cytogenetics. Plant Sci. Lett. 4:85–88.

Hadlaczky, Gy., Kalman, L. (1975): Discrimination of homologous chromosomes of maize with Giemsa staining. Heredity 35:371–374.

Hart, R.W., Kraybill, H.F., de Serres, F.J. (1978): A Rational Evaluation of Pesticidal vs. Mutagenic/Carcinogenic Action. U.S. Dept. Health, Education and Welfare, DHEW Publ. No. (NIH) 78–1306. 122 pp.

Kato, H. (1974): Induction of sister chromatid exchanges by chemical mutagens and its possible relevance to DNA repair. Exp. Cell Res. 85:239–247.

Kato, H. (1977): Spontaneous and induced sister chromatid exchanges as revealed by the BUdR-labeling method. Int. Rev. Cytol. 49:55–97.

Kenton, A. (1978): Giemsa C-banding in Gibasis (Commelinaceae). Chromosoma 65:309–324.

Kihlman, B.A. (1971): Root tips for studying the effects of chemicals on chromosomes. In: A. Hollaender (ed.), Chemical Mutagens: Principles and Methods for Their Detection, Vol. 2. New York: Plenum Press, pp. 489–514.

Kihlman, B.A. (1975): Root tips of Vicia faba for the study of the induction of chromosomal aberrations. Mutat. Res. 31:401–412.

Kihlman, B.A., Kronborg, D. (1975): Sister chromatid exchanges in Vicia faba. I. Demonstration by a modified fluorescent plus Giemsa (FPG) technique. Chromosoma 51:1–10.

Kihlman, B.A., Natarajan, A.T., Andersson, H.C. (1978): Use of the 5-bromodeoxyuridine-labelling technique for exploring mechanisms involved in the formation of chromosomal aberrations. I. G$_2$ experiments with root-tips of Vicia faba. Mutat. Res. 52:181–198.

Klasterska, I., Natarajan, A.T. (1975): Distribu-

tion of heterochromatin in the chromosomes of Nigella damascena and Vicia faba. Hereditas 79:154–156.

Klekowski, E.J., Jr. (1978): Screening aquatic ecosystems for mutagens with fern bioassays. Environ. Health Perspect. 27:99–102.

Komatsu, H., Tanaka, R. (1978): Morphological changes of C-bodies in the mitotic cycle of Crepis vesicaria ssp. taraxacifolia. Proc. Jpn. Acad. Ser. B. 54:228–233.

Koranda, J.J., Robison, W.L. (1978): Accumulation of radionuclides by plants as a monitor system. Environ. Health Perspect. 27:165–179.

Kranz, A.R. (1976): Karyotype analysis in meiosis: Giemsa banding in the genus Secale L. Theoret. Appl. Genet. 47:101–107.

La Cour, L., Fabergé, A.C. (1943): The use of cellophane in pollen tube technic. Stain Technol. 18:196.

Lamm, R. (1945): Cytogenetic studies in Solanum sect. tuberosum. Hereditas 31:1–128.

Levan, A. (1938): The effect of colchicine on root mitosis of Allium. Hereditas 24:471–486.

Levan, A. (1949): The influence on chromosomes and mitosis of chemicals, as studied by the Allium test. Proc. Eighth Int. Congr. Genet. Stockholm. Hereditas, Suppl. pp. 325–337.

Linde-Laursen, Ib. (1975): Giemsa C-banding of the chromosomes of 'Emir' barley. Hereditas 81:285–289.

Ma, T.-H., Khan, S.H. (1976): Pollen mitosis and pollen tube growth inhibition by SO$_2$ in cultured pollen tubes of Tradescantia. Environ. Res. 12:144–149.

Marks, G.E. (1974): Giemsa banding of meiotic chromosomes in Anemone blanda. Chromosoma 49:113–119.

Marks, G.E. (1975): The Giemsa-staining centromeres of Nigella damascena. J. Cell Sci. 18:19–25.

Marks, G.E., Schweizer, D. (1974): Giemsa banding differences in Anemone and Hepatica. Chromosoma 44:405–416.

Mericle, L.W., Mericle, R.P. (1967): Genetic nature of somatic mutations for flower color in Tradescantia, clone 02. Radiat. Bot. 7:449–464.

Mulcahy, D.L., Johnson, C.M. (1978): Self-incompatibility systems as bioassays for mutagens. Environ. Health Perspect. 27:85–90.

Nakata, N., Yasumuro, Y., Sasaki, M. (1977):

An acetocarmine-Giemsa staining of rye chromosomes. Jpn. J. Genet. 52:315-318.

Nettancourt, D. de. (1977): Incompatibility in Angiosperms. Berlin, Heidelberg, New York: Springer-Verlag.

Nicoloff, H., Gecheff, K. (1976): Methods of scoring induced chromosome structural changes in barley. Mutat. Res. 34:233-244.

Nilan, R.A. (1964): The Cytology and Genetics of Barley, 1951-1962. Monogr. Suppl. No. 31, Res. Studies, Washington State Univ. Press.

Nilan, R.A. (1974): Barley (Hordeum vulgare) In: R.C. King (ed.), Handbook of Genetics. Vol. 2. Plants, Plant Viruses and Protists. New York: Plenum Press, pp. 93-110.

Nilan, R.A. (1978): Potential of plant genetic systems for monitoring and screening mutagens. Environ. Health Perspect. 27:181-186.

Nilan, R.A., Vig, B.K. (1976): Plant test systems for detection of chemical mutagens. In: A. Hollaender (ed.), Chemical Mutagens: Principles and Methods for Their Detection, Vol. 4. New York: Plenum Press, pp. 143-170.

Pandey, K.K. (1964): Elements of the S-gene complex. I. The S_{FI} alleles in Nicotiana. Genet. Res. 5:397-409.

Peacock, W.J. (1963): Chromosome duplication and structure as determined by autoradiography. Proc. Natl. Acad. Sci. USA 49:793-801.

Perry, P., Evans H.J. (1975): Cytological mutagen-carcinogen exposure by sister chromatid exchange. Nature (London) 258:121-125.

Plewa, M.J. (1978): Activation of chemicals into mutagens by green plants: A preliminary discussion. Environ. Health Perspect. 27:45-50.

Read, J. (1959): Radiation Biology of Vicia faba in Relation to the General Problem. Oxford: Blackwell, 270 pp.

Rédei, G.P. (1974): Arabidopsis thaliana. In: R.C.King (ed.), Handbook of Genetics. Vol. 2. Plants, Plant Viruses and Protists, New York: Plenum Press, pp. 151-180.

Rédei, G.P., Acedo, G. (1976): Biochemical mutants in higher plants. In: D. Dudits, G.L. Farkas, P. Maliga (eds.), Cell Genetics in Higher Plants, Budapest: Akademiae Kiado, p. 39.

Rosichan, J., Arenaz, P., Nilan, R.A. (1979): A high resolution plant mutagen monitoring system. Genetics 92:s108.

Rushton, P.S. (1969): The effects of 5-fluorodeoxyuridine on radiation-induced chromatid aberrations in Tradescantia microspores. Radiat. Res. 38:404-413.

Sachan, J.K.S., Tanaka, R. (1977): Variation and pattern of C-banding in Zea chromosomes. Nucleus 20:61-64.

Sax, K., Sax, H.J. (1968): Possible mutagenic hazards of some food additives, beverages and insecticides. Jpn. J. Genet. 43:89-94.

Schairer, L.A., Van't Hof, J., Hayes, C.G., Burton, R.M., de Serres, F.J. (1978): Exploratory monitoring of air pollutants for mutagenicity activity with the Tradescantia stamen hair system. Environ. Health Perspect. 27:51-60.

Schvartzman, J.B., Cortes, F. (1977): Sister chromatid exchanges in Allium cepa. Chromosoma 62:119-131.

Schweizer, D. (1973): Differential staining of plant chromosomes with Giemsa. Chromosoma 40:307-320.

Schweizer, D. (1976): Giemsa and fluorochrome banding of polytene chromosomes in Phaseolus vulgaris and P. coccineus. In: K. Jones, P.E. Brandham (eds.), Current Chromosome Research. Amsterdam: Elsevier/North-Holland Biomedical Press, pp. 51-56.

Schweizer, D. (1977): R-banding produced by DNase I digestion of chromomycin-stained chromosomes. Chromosoma 64:117-124.

Shankland, N.E., Grant, W.F. (1976): Localization of Giemsa bands in Lotus pedunculatus chromosomes. Can. J. Genet. Cytol. 18:239-244.

Sharma, A.K., Sharma, A. (1972): Chromosome Techniques: Theory and Practice. Baltimore: University Park Press.

Shelby, M.D. (1976): Chemical Mutagenesis in Plants and Mutagenicity of Plant-related Compounds. Oak Ridge National Laboratory, ORNL/EMIC-7 327 pp.

Singh, R.J., Röbbelen, G. (1976): Giemsa banding technique reveals deletions within rye chromosomes in addition lines. Z. Pflanzenzüchtg. 76:11-18.

Smith, H.H., Lotfy, T.A. (1954): Comparative effects of certain chemicals on Tradescantia chromosomes as observed at pollen tube mitosis. Am. J. Bot. 41:589-593.

Sparrow, A.H., Schairer, L.A., Marimuth, K.M. (1968): Genetic and cytologic studies of Tradescantia irradiated during orbital flight. BioScience 18:585-590.

Spasojevic, V. (1975): Cytogenetical effect of

Tuberite on *Zea mays*. Arkiv. Poljopr. Nanke 28:119–125.

Stack, S.M., Clarke, C.R., Cary, W.E., Muffly, J.T. (1974): Different kinds of heterochromatin in higher plant chromosomes. Cell Sci. 14:499–504.

Takehisa, S., Utsumi, S. (1973): Heterochromatin and Giemsa banding of metaphase chromosomes in *Trillium kamtschaticum* Pallas. Nature (Lond.) 244:286–287.

Taylor, J.H., Woods, P.S., Hughes, W.L. (1957): The organization and duplication of chromosomes as revealed by autoradiographic studies using tritium-labeled thymidine. Proc. Natl. Acad. Sci. USA 43:515–529.

Tomkins, D.J., Grant, W.F. (1972): Comparative cytological effects of the pesticides menazon, metrobromuron and tetrachloroisophthalonitrile in *Hordeum* and *Tradescantia*. Can. J. Genet. Cytol. 14:245–256.

Tomkins, D.J., Grant, W.F. (1976): Monitoring natural vegetation for herbicide-induced chromosomal aberrations. Mutat. Res. 36:73–84.

Underbrink, A.G., Schairer, L.A., Sparrow, A.H. (1973): *Tradescantia* stamen hairs: A radiobiological test system applicable to chemical mutagenesis. In: A. Hollaender (ed.), Chemical Mutagens: Principles and Methods for their Detection, Vol. 3. New York: Plenum Press, pp. 171–207.

Vig, B.K. (1973): Somatic crossing over in *Glycine max* (L.) Merrill: Mutagenicity of sodium azide and lack of synergistic effect with caffeine and mitomycin C. Genetics 75:265–277.

Vig, B.K. (1975): Soybean (*Glycine max*): A new test system for study of genetic parameters as affected by environmental mutagens. Mutat. Res. 31:49–56.

Vig, B.K. (1978): Somatic mosaicism in plants with special reference to somatic crossing over. Environ. Health Perspect. 27:27–36.

Vosa, C.G. (1977): Heterochromatic patterns and species relationship. Nucleus 20:33–41.

White, O.E. (1948): Fasciation. Bot. Rev. 14:319–358.

Wolff, S., Bodycote, J., Painter, R.B. (1974): Sister chromatid exchanges induced in Chinese hamster cells by UV irradiation of different stages of the cell cycle: The necessity for cells pass through S. Mutat. Res. 25:73–81.

Wolverton, B.C., McDonald, R.C. (1978): Bioaccumulation and detection of trace levels of cadmium in aquatic systems by *Eichhornia crassipes*. Environ. Health Perspect. 27:161–164.

Wuu, K.D., Grant, W.F. (1966): Morphological and somatic chromosomal aberrations induced by pesticides in barley. Can. J. Genet. Cytol. 8:481–501.

Wuu, K.D., Grant, W.F. (1967): Chromosomal aberrations induced by pesticides in meiotic cells of barley. Cytologia 32:31–41.

Yen, S.-T., Filion, W.G. (1977): Differential Giemsa staining in plants. V. Two types of constitutive heterochromatin in species of *Avena*. Can. J. Genet. Cytol. 19:739–743.

Zura, K.D., Grant, W.F. (1980): Manuscript in preparation.

19

A Short-Term Cytogenetic Test for Genetic Instability in Humans

T.C. Hsu, William W. Au, Louise C. Strong, and
Dennis A. Johnston

Introduction

It has been postulated by various investigators that 70 to 90% of all human cancers can be attributed to environmental causes (Boyland, 1969; Higginson, 1976). These environmental causes may be due to the agents that persons are exposed to in the atmosphere or agents that are ingested. However, not all people exposed to the similar environment develop cancer. Genetic factors may be responsible for some of the variations in susceptibility from individual to individual. The genetic factors may be (1) the ability of an individual to metabolize (activate or inactivate) mutagenic agents as well as promoters, cocarcinogens etc.; (2) the ability of an individual to repair the induced damage; and (3) the immunocompetence of an individual to recognize the potential cancer cells.

One of the potential mechanisms concerns individuals with unstable genetic constitution, especially those who can be identified as having elevated rate of chromosome breakage. These are patients with ataxia telangiectasia, Bloom's syndrome, Fanconi's anemia, and xeroderma pigmentosum. German (1972) classified them as patients with "chromosome instability syndrome." Patients with the first three types of syndromes are characterized by having a higher rate of spontaneous chromosome breakage than normal individuals. The instability appears to be constitutional and can be observed in different cell types. In addition to having chromosome instability, these three groups of patients are predisposed to develop neoplasms (German, 1972).

Other patients expressing spontaneous chromosome aberrations have been reported. In surveys of families showing the "cancer family syndromes" (Lynch et al., 1977), spontaneous chromosome instability has been observed in probands as well as in some family members without evidence of cancer (Wurster-Hill et al., 1974; Cervenka et al., 1977; Law et al., 1977; Wurster-Hill et al., 1979; Cheng et al., 1979). Thus, chromosome abnormalities appear to be intimately associated with predisposition to malignancy. Whether the aberrations are due to genetic defects per se or due to interactions with specific external agents remains to be determined.

Xeroderma pigmentosum patients do not have increased spontaneous rate of chromosome breakage. However, they are susceptible to sunlight-induced skin abnormalities manifested eventually as skin carcinomas and melanomas (Robbins *et al.*, 1974). This sensitivity of the target tissue to sunlight can be demonstrated *in vitro* by cytogenetic analysis. Fibroblast but not lymphocyte cultures from these patients showed increased rate of UV-induced chromosome breakage compared with normal cells (Parrington *et al.*, 1971). The chromosome abnormalities from these four types of patients are so distinct that in some cases cytogenetic analysis can be used to confirm the clinical diagnosis. For example, the diagnosis of Bloom's syndrome is confirmed when quadriradials are observed in lymphocyte cultures (German, 1974).

Chromosome instability syndromes result from homozygocity for the respective autosomal gene. The frequency of occurrence was estimated to be 1 in 40,000 for ataxia telangiectasia (Swift, 1976), 1 in 360,000 for Fanconi's anemia (Swift, 1976), and 1 in 250,000 for xeroderma pigmentosum (Robbins *et al.*, 1974). Bloom's syndrome patients are not as frequently found as others. Although the homozygous patients are rarely found and they have recognizable phenotypic and cytogenetic characteristics, the heterozygotes have no distinct clinical features. However, the heterozygotes may be predisposed to develop malignancy as well. The evidence was obtained from observation of excess cancer in families of homozygous probands of ataxia telangiectasia (Swift *et al.*, 1976), Fanconi's anemia (Swift, 1971; Swift, 1976), and xeroderma pigmentosum (Swift and Chase, 1979). It has been estimated by Swift (1971) that 5% of all individuals dying of acute leukemia may be Fanconi's anemia

heterozygotes. Again, Swift and Chase (1979) suggested that xeroderma pigmentosum heterozygotes constitute at least 0.1% of the population and "account for a substantial proportion of genetic predisposition to sun-induced nonmelanoma skin cancer." It does not require further emphasis on the importance of identifying these heterozygotes as well as others who are also predisposed to develop malignancy.

Auerbach and Wolman (1978) discovered that cells heterozygous for Fanconi's anemia were particularly sensitive to the induction of chromosome breakage by diepoxybutane (DEB). Subsequently, it was suggested by Auerbach *et al.* (1979) that such a cytogenetic method could be used for prenatal detection of the Fanconi anemia gene.

Another group of patients with possible genetic instability that has been found in the human population is that with congenital chromosome abnormalities. It was shown by Seabright (1976) that cultured cells obtained from patients with chromosome abnormality, whether or not extra chromatin was present, developed an excess of X-ray–induced breakage. Similar X-ray sensitivity has also been identified in patients with Down syndrome (Lambert *et al.*, 1976; Goh *et al.*, 1978) and D-deletion retinoblastomas (Weichselbaum *et al.*, 1976; Little *et al.*, 1978). However, the evidence for disposition to develop cancer in chromosomally abnormal patients is not clear except in those with D-13 deletion (Wilson *et al.*, 1973; Francke, 1976) and those with Down syndrome (Goh *et al.*, 1978).

It is possible that subtle differences exist among phenotypically normal individuals in terms of the rate of spontaneous chromosome breakage, but to detect such differences may require the scoring of thousands of metaphases in multiple samples for statistical analysis. An alternative is to apply a clastogen such that the differences between "stable" and "unstable" genotypes may be exaggerated to facilitate an evaluation.

During the last 2 years, we have been analyzing the phenomenon of genetic instability or stability of individuals. We wish to determine whether chromosomal instability, as measured by spontaneous or clastogen-induced chromosome breakage, might reflect susceptibility to carcinogenesis. For our patient population, we selected patients with known heritable cancer-predisposing syndromes, those with unusual familial aggregation of cancer, and those with constitutional chromosome abnormalities. Also included were patients with retinoblastoma and nevoid basal cell carcinoma syndrome, disorders which include an exceptional risk of environmentally (radiation) induced tumors (Strong, 1977).

Since a survey like ours includes individuals with heterogeneous genetic background and with different repair deficiencies for affected individuals, we assumed that any clastogen was as good as another. We have been using gentian violet for the induction of chromosome breakage for the following reasons:

1. It is a potent clastogen in mammalian cells in culture (Au *et al.*, 1978).
2. It induces reparable DNA damage in bacterial assay (Au *et al.*, 1979).
3. Unlike most clastogens, gentian violet does not significantly inhibit mitosis in a 5-hr treatment experiment. In addition, gentian violet does not induce extensive chromosome tangling and stickiness; therefore, the induced breaks and exchanges are easy to score for quantitation.
4. A solution of gentian violet in water is stable for at least 2 weeks. This is more convenient than weighing out a small amount of compound every time we test a blood sample.

In this report, we present the findings of our initial survey. It includes 14 normal individuals, cancer patients, members of families with high risk to develop cancer, and members of families with constitutional chromosome abnormalities.

Materials and Methods

Selection of Individuals for Analysis

Fourteen healthy volunteers were used as our normal controls. The male-to-female ratio was 4:10 and the age range was 23 to 54. Four blood samples, once a week, were obtained from each of eight persons and three samples from each of six persons.

Patients and families were selected for study because of (a) known heritable cancer-predisposing condition (bilateral retinoblastoma, Fanconi's anemia, and nevoid basal cell carcinoma syndromes); (b) unusual familial aggregation of cancer (family VI, Tables 19.1 and 19.6); or (c) known constitutional chromosomal abnormality and cancer in a relative with the same chromosomal anomaly (Blattner *et al.*, 1980) (family VIII, Tables 19.1 and 19.3). Other cancer patients before chemotherapy and patients with constitutional abnormalities were also selected. The age, diagnosis, and treatment prior to study are listed in Table 19.1. Relatives of the probands indicated as not affected at the time of study had undergone clinical and, in some cases, radiographic evaluation to seek out retinoblastoma, nevoid basal cell carcinoma syndrome, hematologic disease, or other neoplasia. With the exception of the ring-15 patient, only one blood sample was obtained and analyzed per person.

Gentian Violet Solution

Gentian violet was dissolved in water to obtain a stock solution of 100 μg/ml. This solution was stored in the refrigerator and every 2 weeks new stock solution was made.

Table 19.1. *Clinical Information on Members Selected for Family Studies*

Family number	Patient	Diagnosis	Cytotoxic treatment/age (year)	Age (year) at study
I	Proband	FA	androgens, corticosteroids/7+	12
	Father	heterozygote, FA	—	41
	Mother	heterozygote, FA	—	40
II	Proband	ALL in remission	XRT, CT/5+	6
	Father	ALL, HD in nieces and nephews	—	42
	Mother	FH negative for cancer	—	27
III	Proband	NBCCS	—	49
	Spouse	—	—	45
	Son	—	—	23
	Daughter I	—	—	26
	Daughter II	—	—	21
	Daughter III	—	—	16
IV	Proband	B-RTB	XRT/1	10
	Mother	—	—	46
V	Proband	U-RTB	—	3
	Father	—	—	31
	Mother	—	—	25
VI	Proband	B-RTB	—	5 mos.
	Father	—	—	35
	Mother	B-RTB	—	30
	Maternal aunt	—	—	31
VII	Proband	ALL in remission	XRT, CT/8+	9
	Monozygotic twin	—	—	9
	Father	—	—	32
	Mother	—	—	31
VIII	Proband	46,XX, t(13;18) t18; Inv 9	—	4
	Sib	45,XX, t(13;18)	—	6
	Father	46,XY, Inv 9	—	unknown
	Mother	45,XX t(13;18)	—	34
	Control	—	—	unknown
IX	Proband	46,XX, r(15)	—	15 mos.
	Father	46,XY	—	30
	Mother	46,XX	—	29
Patients	—	Adrenocortical Adenocarcinoma, FH Gardner's syndrome	—	8
		Astrocytoma, FH Astrocytoma	—	1.5

ALL = acute lymphocytic leukemia; B-RTB = bilateral retinoblastoma; CT = chemotherapy; FA = Fanconi's anemia; FH = family history; HD = Hodgkin's disease; NBCCS = nevoid basal cell carcinoma syndrome; U-RTB = unilateral retinoblastoma; XRT = radiation therapy.

Blood Cultures and Scoring Procedure

Phytohemagglutinin-stimulated whole-blood cultures were set up in Gibco chromosome medium 1A (10 drops of blood into 5 ml of medium) for 3 and 4 days. Half of the cultures were harvested on day 3. Gentian violet (final concentration 5 μg/ml) was added to cultures during the last 5 hr of incubation, and colcemid (0.04 μg/ml) was added during the last hour. Untreated control cultures from normal controls and from patients were harvested after 1 hr colcemid treatment. During harvesting, the cells were centrifuged, treated with 8 ml of hypotonic solution (0.075M KC1) for 10 minutes, and fixed in Carnoy's fixative (3 methanol:1 glacial acetic acid) 3 times. Air-dried slides were made and coded.

Unstained slides were scanned to ascertain a good supply of mitosis with good chromosome spreading. If the mitotic rate was low from the 3-day samples, the remaining cultures were harvested on day 4 using the same protocol. Air-dried preparations were stained in Giemsa (4% in 0.01 M phosphate buffer) for approximately 5 minutes, rinsed, air-dried, and mounted in permount.

The coded slides were scored for spontaneous and induced chromosome breakage. Fifty metaphases were scored per experimental point. Our scoring and recording procedures were described previously (Au et al., 1978). All types of chromosome aberrations were recorded in our raw data. In the final quantitation of damage, all aberrations were converted into the number of lesions per metaphase. Chromosome gaps and breaks were calculated as 1 lesion each and exchange and isochromatid breaks were 2 lesions each. Metaphases with more than 10 breaks were calculated as 10 lesions each.

Statistical Analysis of Data

Fourteen normal subjects were analyzed in triplicate or quadruplicate for percent of cells with breaks, exchanges and average number of lesions per metaphase with or without the treatment with gentian violet (Table 19.2). These normal subjects were characterized by sex and for differences between individuals, using both parametric (t-tests and one-way analysis of variance) and nonparametric tests (Mann-Whitney and Kruskal-Wallis tests). No significant sex or other differences among the normal subjects were observed on any tests (p > 0.1 for all tests), indicating that all normal subjects could be pooled into a single population. The six measurements (% breaks, %

exchanges, and lesions/metaphase with and without gentian violet) were tested for a Gaussian distribution using a Kolmogorov-Smirnov goodness of fit test, which showed that % breaks and lesions/metaphase were distributed with a Gaussian distribution (p > 0.1 for all cases). Log normality was also tested with the same result (p > 0.26 for all cases), indicating a logarithm might be more appropriate. Exchange data had too many zero exchanges to be Gaussian (p > 0.003 for all cases). The means, standard deviations, and 95% tolerance ranges are given at the bottom of Table 19.2. The ranges are given with 2.5% probability of the patient value exceeding the tolerance level. Tolerance ranges were also calculated for the log-transformed data and given in the anti-log form as well to facilitate comparisons. Tolerance ranges for exchanges are given using percentiles since the data is not Gaussian. Since there are 48 readings, the 95th percentile is approximately at the second highest value (e.g., 12 for exchanges on gentian violet-treated cells).

Results

The finding of the spontaneous and induced chromosome damage in the 14 normal controls is shown in Table 19.2. Variations in chromosome damage were observed among the individuals as well as between samples of the same individual. However, statistical analysis of the data showed that there was no individual variation in similarly treated normal controls (i.e., whether it was spontaneous or induced). There was also no difference between the sexes of the donors. From this pool of data, it was determined that a spontaneous breakage rate of more than 0.26 lesion/metaphase and an induced rate of more than 2.58 lesions/metaphase would constitute significant differences from the rates among the controls.

Table 19.3 shows the breakage rate of a Fanconi's anemia family. The proband had higher spontaneous and induced rates of

Table 19.2. *Spontaneous and Gentian Violet-induced Chromosome Breakage in Normal Control Individuals*

Individual	Control			Gentian violet (5 µg/ml)		
	% metaphases with breaks	% metaphases with exchanges	Number of lesions per metaphase	% metaphases with breaks	% metaphases with exchanges	Number of lesions per metaphase
1 ♂	16	2	0.24	60	12	1.76
	10	0	0.10	56	12	1.86
	12	0	0.16	60	0	1.32
	8	0	0.12	40	12	1.24(1.54)[c]
2 ♀	20	0	0.24	48	8	0.92
	10	0	0.12	36	0	1.46
	12	0	0.14	48	12	2.12
	30	0	0.36	68	4	2.24(1.68)
3 ♀	10	0	0.10	40	0	1.48
	14	0	0.16	44	0	0.96
	16	0	0.20	24	4	0.36(0.94)
	10	0	0.10	72	0	2.14
4 ♂	8	0	0.08	40	4	1.04
	4	0	0.04	36	0	0.52(1.22)
	18	0	0.20	72	0	1.32
	12	0	0.16	54	6	1.92
5 ♂	8	0	0.08	44	0	0.96(1.40)
	12	0	0.12	48	4	1.36
	6	0	0.06	60	4	2.12
	6	0	0.08	40	0	0.84(1.44)
6 ♀	12	0	0.12	40	0	0.72
	4	0	0.04	76	0	2.32
	4	0	0.04	32	0	0.52
	6	0	0.06	52	0	1.32(1.24)
7 ♀	12	0	0.12	68	0	1.36
	8	0	0.10	40	4	0.56
	8	0	0.08	64	0	1.40(1.10)
	12	0	0.14	40	4	0.84
8 ♀	8	0	0.08	Mitotic Inhibition		
	10	0	0.12			
	12	0	0.12			
	16	0	0.16			
9 ♀	8	0	0.08	44	4	1.00
				72	0	1.60(1.14)
				48	4	1.20
				60	0	1.80
10 ♀	16	0	0.16	60	0	1.16(1.38)

11 ♂	12	0	0.20	56	0	1.96
	8	0	0.08	48	0	1.24
	4	0	0.04	72	12	2.32
	0	0	0.00	44	4	0.88(1.60)
12 ♀	16	0	0.16	52	4	1.04
	4	0	0.04	52	8	1.48
	6	0	0.06	64	8	0.96
	12	0	0.12	70	16	2.96(1.62)
13 ♀	24	0	0.24	80	8	2.52
	8	0	0.08	60	0	1.44
	8	0	0.08	48	4	1.08
	4	0	0.08	46	4	1.06(1.52)
14 ♀	4	0	0.04	56	8	1.76
	4	0	0.04	40	0	0.80
	20	0	0.24	76	12	2.16
	12	0	0.18	48	0	0.76(1.37)
Mean	10.67	0.04	0.12	53.21	3.79	1.40
s.d.	5.59	0.29	0.07	13.22	4.56	0.59
95% upper tolerance (two-sided)	21.85	0[a]	0.26	79.65	12[a]	2.58
Mean log[b]	1.02	0.01	0.05	1.72	0.46	0.37
s.d. log[b]	0.20	0.07	0.03	0.11	0.46	0.11
95% upper tolerance (two-sided)	1.42	N.D.	0.11	1.94	N.D.	0.59
95% upper tolerance (retransformed by antilogs)	25.30	N.D.	0.29	86.10	N.D.	2.89

N.D. = not determined.

[a] Calculated as the mean and standard deviation of the logarithms of the data after adding one $[y = \log_{10}(X + 1)]$.

[b] Approximate 95th percentile.

[c] Number in parenthesis is the mean.

Table 19.3. Spontaneous and Gentian Violet-induced Chromosome Breakage in a Fanconi's Anemia Family

Individual	Control			Gentian violet (5 μg/ml)		
	% metaphases with breaks	% metaphases with exchanges	Number of lesions per metaphase	% metaphases with breaks	% metaphases with exchanges	Number of lesions per metaphase
Family I						
Proband	30	0	0.56[a]	88	26	6.28[a]
Father	8	0	0.10	50	10	1.18
Mother	20	0	0.18	46	0	1.24

[a]Significantly higher than normal control.

chromosome breakage than the other individuals studied. However, the spontaneous and induced rates of the parents were no different from those of the controls.

The findings from a family with multiple lymphoproliferative neoplasm is shown in Table 19.4. The father of this family was considered a potential carrier since members of his blood relatives had high incidences of childhood leukemia and lymphoma (Fig. 19.1). The mother had no history of cancer in her family. In this family, only the mother had increased rate of induced breakage.

Table 19.5 shows the findings in a patient with nevoid basal cell carcinoma and his relatives. The proband had an elevated rate of induced breakage. One of the daughters had increases in both spontaneous and induced aberration rates while the rest of the members were similar to the controls.

Three different retinoblastoma patients and their relatives were obtained for this survey. The results are shown in Table

19.6. None of the probands or relatives showed increased spontaneous or induced chromosome instability.

Results from other cancer patients are shown in Table 19.7. Two melanoma patients had increased rate of induced breakage. One of them had spontaneous instability. Induced instability was observed in patients with osteosarcoma and astrocytoma but not in a patient with adrenal cortical adenocarcinoma. A family was selected for this survey because one of the monozygotic twin offspring developed leukemia. This proband had a significant increase in induced chromosome breakage over the normal control. However, his twin and other relatives had breakage rates within the normal limit.

The findings of patients with constitutional chromosome abnormalities are shown in Table 19.8. None of these patients have any history of neoplasia. The first family studied was composed of members with chromosome abnormalities involving Nos. 9, 13, and 18 (Fig. 19.2).

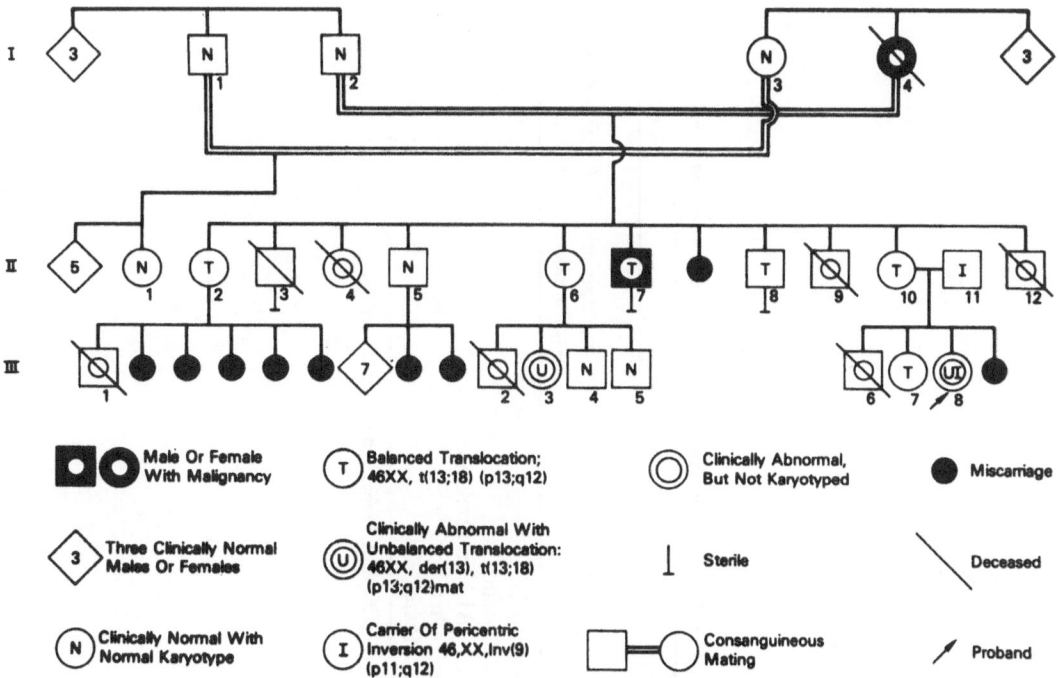

Figure 19.1. Pedigree of family II with multiple lymphoproliferative neoplasms—squares are males and circles are females.

Table 19.4. Spontaneous and Gentian Violet-induced Chromosome Breakage in Hematologic Cancer Patients and Families

Individual	Control			Gentian violet (5 μg/ml)		
	% metaphases with breaks	% metaphases with exchanges	Number of lesions per metaphase	% metaphases with breaks	% metaphases with exchanges	Number of lesions per metaphase
Family II						
Proband (ALL in remission)	8	0	0.18	54	4	2.08
Father	10	0	0.20	76	4	2.16
Mother	4	0	0.06	66	12	2.62[a]

[a]Significantly higher than normal control.

Table 19.5. *Spontaneous and Gentian Violet-induced Chromosome Breakage in a Nevoid Basal Cell Carcinoma Family*

Individual	Control			Gentian violet (5 μg/ml)		
	% metaphases with breaks	% metaphases with exchanges	Number of lesions per metaphase	% metaphases with breaks	% metaphases with exchanges	Number of lesions per metaphase
Family III						
Proband	20	0	0.20	70	6	2.74[a]
Wife	16	0	0.16	74	10	1.60
Son	8	0	0.12	42	4	1.08
Daughter I	24	0	0.28[a]	82	18	3.38[a]
Daughter II	16	0	0.16	84	8	1.68
Daughter III	8	0	0.10	64	12	2.42

[a]Significantly higher than normal control.

Table 19.6. Spontaneous and Gentian Violet-induced Chromosome Breakage in Retinoblastoma Families

Individual	Control			Gentian violet (5 μg/ml)		
	% metaphases with breaks	% metaphases with exchanges	Number of lesions per metaphase	% metaphases with breaks	% metaphases with exchanges	Number of lesions per metaphase
Family IV						
Proband (B-RTB)	20	0	0.20	58	2	1.58
Mother	12	0	0.12	72	8	1.72
Family V						
Proband (U-RTB)	10	0	0.12	48	0	0.98
Father	4	0	0.04	56	2	0.94
Mother	0	0	0.00	32	0	0.92
Family VI						
Proband (B-RTB)	8	0	0.08	50	0	1.44
Father	10	0	0.12	40	0	0.64
Mother (B-RTB)	6	0	0.08	68	12	2.54
Maternal aunt	10	0	0.12	60	8	1.72

Table 19.7. *Spontaneous and Gentian Violet-induced Chromosome Breakage in Other Patients*

Individual	Control			Gentian violet (5 μg/ml)		
	% metaphases with breaks	% metaphases with exchanges	Number of lesions per metaphase	% metaphases with breaks	% metaphases with exchanges	Number of lesions per metaphase
Family VII						
Proband (ALL in remission)	12	0	0.12	98	6	2.90[a]
Father	16	0	0.16	68	2	1.08
Mother	12	0	0.12	46	0	0.72
Monozygotic twin (normal)	20	0	0.20	84	6	1.90
Adrenal cortical adenocarcinoma	0	0	0.00	32	0	0.60
ALL	12	0	0.14	84	12	5.16[a]
Astrocytoma	12	0	0.18	52	14	2.60[a]
Lymphoma	20	0	2.24[a]	60	20	2.52
Melanoma I	24	0	0.40	64	12	2.80[a]
Melanoma II	8	0	0.08	76	12	2.72[a]
Osteosarcoma	18	0	0.22	76	4	2.92[a]

[a]Significantly higher than normal control.

Table 19.8. Spontaneous and Gentian Violet-induced Chromosome Breakage in Chromosomally Abnormal Patients

	Control			Gentian violet (5 μg/ml)		
Individual	% metaphases with breaks	% metaphases with exchanges	Number of lesions per metaphase	% metaphases with breaks	% metaphases with exchanges	Number of lesions per metaphase
Family VIII						
Proband						
46,XX,t(13;18)+18;Inv 9	12	0	0.12	76	12	3.68[a]
Sib 45,XX,t(13;18)	8	0	0.08	80	4	1.96
Father 46,XY, Inv 9	0	0	0.00	88	6	2.96[a]
Mother 45,XX,t(13;18)	4	0	0.04	76	24	4.12[a]
Control	20	0	0.20	64	0	1.16
Family IX						
Proband (sample I)	12	0	0.12	82	10	3.96[a]
46,XX,r(15)						
(sample II)	10	0	0.12	76	16	3.48[a]
Father 46,XY	8	0	0.08	48	6	1.38
Mother 46,XX	6	0	0.08	36	0	0.80
Down syndrome I						
47,XX,21	4	0	0.04	64	8	1.90
Down syndrome II						
47,XY,21	10	0	0.12	74	16	3.08[a]
Down syndrome III						
47,XX,21	8	0	0.12	48	0	1.04

[a]Significantly higher than normal control.

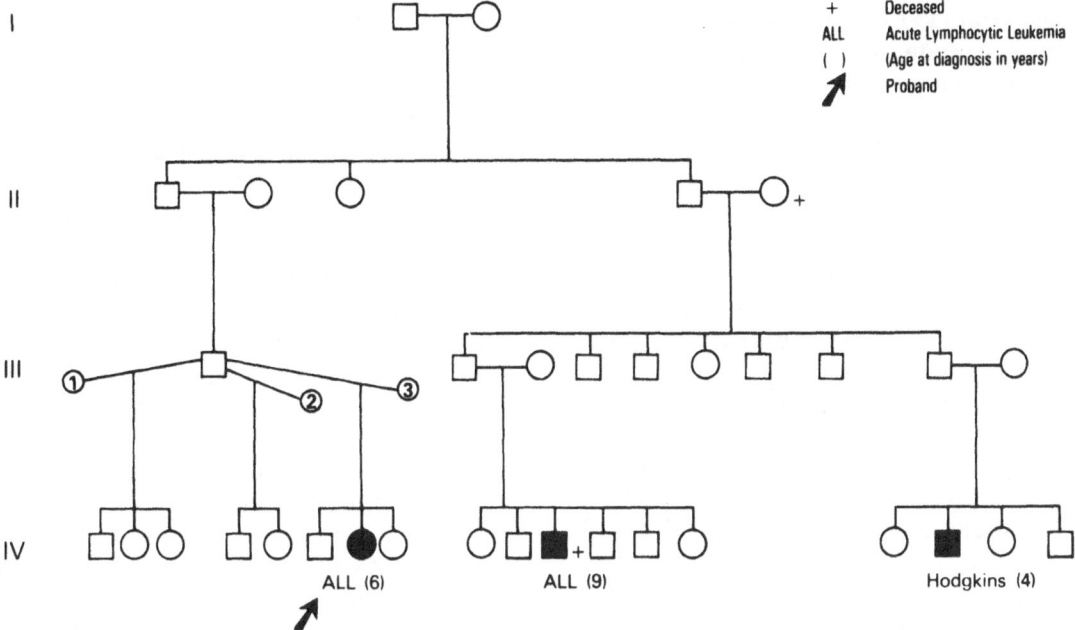

Figure 19.2. Pedigree of family VIII with constitutional chromosome abnormalities—squares are males and circles are females.

Three of these four members had increased rates of induced breakage. A relative by marriage to this family had a normal rate. The detailed cytogenetic analysis of this family will be published elsewhere. A second family studied involved a daughter with ring chromosome 15. This abnormal chromosome was formed spontaneously since it was not found in the parents. In addition, it was assumed to be a constitutional abnormality since the ring chromosome was observed in both blood and fibroblast cultures (David Ledbetter, personal communication). In two blood samples obtained from the proband, the induced breakage rate was significantly higher than that of normal persons. In three Down syndrome patients, only one showed an elevated induced breakage rate.

Discussion

It has been shown by investigators that both congenital and acquired chromosome abnormalities are intimately associated with neoplasia (Rowley, 1977; Mitelman and Levan, 1978). In several cases, chromosome abnormality had been observed before the manifestation of malignancy. Examples of these are abnormalities observed in preleukemic patients (Nowell et al., 1976; Nowell and Finan, 1978) and in patients with high risk to develop cancer (German, 1972). Certain patients, e.g., Down syndrome and Fanconi's anemia heterozygotes, were shown to be preferentially susceptible to induction of chromosome breakage by specific agents (Lambert et al., 1976; Auerbach and Wolman, 1978). This finding involving the Fanconi's anemia heterozygotes is particularly important because these heterozygotes as well as others (e.g., ataxia telangiectasia, xeroderma pigmentosum) are phenotypically indistinguishable from normal persons, but they may contribute substantially to the incidence of neoplasia (Swift, 1976; Swift et al., 1976; Swift and Chase, 1979).

Cytogeneticists have been interested in studying the genetic stability or instability of various individuals, especially their response to insult by clastogens. Littlefield and Goh (1973) analyzed the spontaneous breakage rate of 10 males and 21 females. They found variations between different persons and between cultures of the same individual. Similar variations were observed in different individuals by Pero *et al.* (1978) on the response to DNA-damaging agents. Our study of 14 normal controls with repeated samples from each did not show a significant difference between individuals whether the cultures were treated with a clastogen or not.

From this preliminary survey of ours, several indications have been obtained. In a family with Fanconi's anemia, the proband had spontaneous and gentian violet-induced chromosome instability. Unlike the study of Auerbach and Wolman (1978) using diepoxybutane, gentian violet did not induce significant increase chromosome breakage in the heterozygotes.

In a family with hematologic cancer, there does not appear to be a consistent response among the affected and supposedly normal family members. From a cytogenetic point of view, acute leukemia appears to be a very heterogeneous disease. Approximately 50% of the acute leukemia patients have chromosome abnormality and the remaining ones have normal karyotype (Berghe *et al.*, 1977; Mitelman and Levan, 1978). In our future studies, we plan to analyze the chromosome constitution of the acute leukemic patients in order to determine whether abnormal karyotype plays a role in the individual's response to gentian violet.

Nevoid basal cell carcinoma is a dominantly inherited disease. In the one family (Table 19.5), the proband responded with an increase in induced rate of breakage. Among the offspring, only one had spontaneous and induced instability. We plan to obtain repeated samples from this family to determine if the instability, whether spontaneous or induced, persists and whether this corresponds to the clinical course of this disease. No significant response to gentian violet-induced instability was observed in a series of retinoblastoma patients and relatives.

Eight cancer patients were studied in this survey before the application of therapy. Six of them responded positively to gentian violet-induced instability (Table 19.6). Whether this phenomenon of induced instability predisposed these individuals to develop primary neoplasms is difficult to determine. However, it is important to inquire if these patients showing induced genetic instability would have higher risks to develop second primary tumors. Studies in patients surviving various forms of cancers have provided some interesting findings. In survivors of heritable retinoblastomas, the occurrence of a second primary osteosarcoma appears to be related to the dose of radiation therapy applied to the first tumor (Sagerman *et al.*, 1969; Abramson *et al.*, 1976). In a study involving survivors of Ewing's sarcoma, Strong *et al.* (1979) found a significant increase in cancer risk in irradiated patients. All new tumors arose in heavily irradiated areas. In addition, intensive chemotherapy was found to enhance the risk of developing new tumors (Strong *et al.*, 1979). These findings, as well as others, led these authors to suggest that "persons who survive one cancer may be at risk for additional cancers due to common predisposing factors as well as factors related to treatment of the first cancer."

Among patients with chromosome abnormalities, these patients with structural abnormality responded more frequently to induced instability than those with numerical abnormality (Table 19.8). In the first family studied, the mother with the 13/18 translocation had the highest rate of induced breakage. Her brother, not studied in our survey, having the same translocation developed leukemia recently. Some relationships between cytogenetic abnormality and predisposition to develop cancer seem to exist. Indeed, the association of 13-deletion with retinoblastoma (Wilson *et al.*, 1973; Francke, 1976; Sparks *et al.*, 1979) and Down syndrome with leukemia (Goh *et al.*, 1978) have been reported. If the patients with chromosome abnormality are indeed sensitive to induced breakage, then they may have an increased risk to develop cancer induced by environmental factors.

Although our data are preliminary and insufficient, they provide enough encouragement for continuation of the survey, both with gentian violet and with other clastogens (with different mechanisms of actions), for substantiation of these possible conclusions: (1) genetic instability can be demonstrated by the cytogenetic response to clastogens, (2) genetic instability is associated with a higher risk to develop neoplasia, and (3) an assay system can be developed to detect individuals with a higher susceptibility to potentially hazardous professions and environment.

Acknowledgments

Supported in part by research grants ES-01304 from National Institute of Environmental Health Sciences, GM-19513 from National Institute of General Medical Sciences, CA-11430 from National Cancer Institute, and ENV 76-82241 from National Science Foundation.

We thank Dr. William A. Blattner, Environmental Epidemiology branch, National Cancer Institute, for providing us with blood samples and the family pedigree of Family VIII for our study; Miss Cheryl J. Collie for her technical assistance, and Miss Mary B. Cline for preparation of the manuscript.

References

Abramson, D.H., Ellsworth, R.M., Zimmerman, L.E. (1976): Nonocular cancer in retinoblastoma survivors. Trans. Am. Acad. Ophthalmol. Otolaryngol. *81*:454–457.

Au, W., Pathak, S., Collie, C.J., Hsu, T.C. (1978): Cytogenetic toxicity of gentian violet and crystal violet on mammalian cells in vitro. Mutat. Res. *58*:269–276.

Au, W., Butler, M.A., Bloom, S.E., Matney, T.S. (1979): Further study of the genetic toxicity of gentian violet. Mutat. Res. *66*: 103–112.

Auerbach, A.D., Wolman, S.R. (1978): Carcinogen-induced chromosome breakage in Fanconi's anemia heterozygous cells. Nature *271*:69–71.

Auerbach, A.D., Warburton, D., Bloom, A.D., Chaganti, R.S.K. (1979): Prenatal detection of the Fanconi anemia gene by cytogenetic methods. Am. J. Hum. Genet. *31*:77–81.

Berghe, H. van den, Borgström, G.H., Brandt, L., Catovsky, D., Chapelle, A. de la, Golomb, H.M., Hossfeld, D.K., Lawler, S., Lindsten, J., Louwagie, A., Mitelman, F., Reeves, B.R., Rowley, J.D., Sandberg, A.A., Teerenhovi, L., Vuopio, P. (1977): Chromosomes in leukemia. Cytogenet. Cell Genet. *19*:321–325.

Blattner, W.A., Kistenmacher, M.L., Tsai, S., Punnett, H.H., Giblett, E.R. (1980): Clinical manifestation of familial 13:18 translocation. J. Med. Genet.

Boyland, E. (1969): The correlation of experimental carcinogenesis and cancer in man. Prog. Exp. Tumor Res. *11*:222–234.

Cervenka, J., Anderson, R.S., Nesbit, M.E., Krivit, W. (1977): Familial leukemia and inherited chromosomal aberration. Int. J. Cancer *19*:783–788.

Cheng, W.S., Mulvihill, J.J., Greene, M.H., Pickle, L.W., Tsai, S., Whang-Peng, J. (1979): Sister chromatid exchanges and chromosomes in chronic myelogenous leukemia and cancer families. Int. J. Cancer *23*:8–13.

Francke, U. (1976): Retinoblastoma and chromosome 13. Cytogenetics *16*:131–135.

German, J. (1972): Genes which increase chromosomal instability in somatic cells and predispose to cancer. Prog. Med. Genet. *8*: 61–101.

German, J. (1974): Bloom's syndrome. II. The prototype of human genetic disorders predisposing to chromosomal instability and cancer. In: J. German (ed.), Chromosomes and Cancer. New York: John Wiley and Sons, pp. 601–617.

Goh, K., Lee, H., Miller, G. (1978): Down's syndrome and leukemia: Mechanism of additional chromosome abnormalities. Am. J. Ment. Defic. *82*:542–548.

Higginson, J. (1976): Present trends in cancer epidemiology. Can. Cancer Conf. 8:40–75.

Lambert, B., Hansson, K., Bui, T. H., Funes-Cravioto, F., Lindsten, J. (1976): DNA repair and frequency of X-ray and UV light induced chromosome aberrations in leukocytes from patients with Down's syndrome. Ann. Hum. Genet. 39:293–305.

Law, I.P., Hollinshead, A.C., Whang-Peng, J., Dean, J.H., Oldham, R.K., Herberman, R.B., Rhode, M.C. (1977): Familial occurrence of colon and uterine carcinoma and of lymphoproliferative malignancies. II. Chromosomal and immunologic abnormalities. Cancer 39:1229–1236.

Little, J.B., Weichselbaum, R.R., Nove, J., Albert, D.M. (1978): X-ray sensitivity of fibroblasts from patients with retinoblastoma and with abnormalities of chromosome 13. In: P.C. Hanawalt, E.C. Friedberg, C.F. Fox (eds.), DNA Repair Mechanisms. New York: Academic Press, pp. 685–690.

Littlefield, L.G., Goh, K.O. (1973): Cytogenetic studies in control men and women. I. Variations in aberration frequencies in 29,709 metaphases from 305 cultures obtained over a three-year period. Cytogenet. Cell Genet. 12: 17–34.

Lynch, H.T., Guirgis, H.A., Lynch, P.M., Lynch, J.F., Harris, R.E. (1977): Familial cancer syndromes: A survey. Cancer 39: 1867–1881.

Mitelman, F., Levan, G. (1978): Clustering of aberrations to specific chromosomes in human neoplasms. III. Incidence and geographic distribution of chromosome aberrations in 856 cases. Hereditas 29:207–232.

Nowell, P., Finan, J. (1978): Chromosome studies in preleukemic states. IV. Myeloproliferative versus cytogenetic disorders. Cancer 42: 2254–2261.

Nowell, P., Jensen, J., Gardner, F., Murphy, S., Chaganti, R.S.K., German, J. (1976): Chromosome studies in "preleukemia." III. Myelofibrosis. Cancer 38:1873–1881.

Parrington, J.M., Delhanty, J.D.A., Baden, H.P. (1971): Unscheduled DNA synthesis. UV-induced chromosome aberrations and SV40 transformation in cultured cells from Xeroderma pigmentosum. Ann. Hum. Genet. 35: 149–160.

Pero, R.W., Bryngelsson, C., Mitelman, F., Kornfält, R., Thulin, T., Norden, A. (1978): Interindividual variation in the responses of cultured human lymphocytes to exposure

from DNA damaging chemical agents. Mutat. Res. 53:327–341.

Robbins, J.H., Kraemer, K.H., Lutzner, M.A., Festoff, B.W., Coon, H.G. (1974): Xeroderma pigmentosum. An inherited disease with some sensitivity, multiple cutaneous neoplasms, and abnormal DNA repair. Ann. Intern. Med. 80:221–248.

Rowley, J.D. (1977): Mapping of human chromosomal regions related to neoplasia: Evidence from chromosomes 1 and 17. Proc. Natl. Acad. Sci. USA 74:5729–5733.

Sagerman, R.H., Cassady, J.R., Tretter, P. (1969): Radiation induced neoplasia following external beam therapy for children with retinoblastoma. Am. J. Radiat. 105:529–535.

Seabright, M. (1976): Patterns of induced aberrations in humans with abnormal autosome compliment. In: P.L. Pearson, K.R. Lewis (eds.), Chromosomes Today, Vol. 5. New York: John Wiley and Sons, pp. 293–298.

Sparks, R.S., Muller, H., Klisak, I., Abram, J.A. (1979): Retinoblastoma with 13q- chromosomal deletion associated with maternal paracentric inversion of 13q. Science 203: 1027–1029.

Strong, L.C. (1977): Genetic and environmental interactions. Cancer 40:1861–1866.

Strong, L.C., Herson, J., Osborne, B.M., Sutow, W.W. (1979): Risk of radiation-related subsequent malignant tumors in survivors of Ewing's sarcoma. J. Natl. Cancer Inst. 62: 1401–1406.

Swift, M. (1971): Fanconi's anemia in the genetics of neoplasia. Nature (Lond.) 230: 370–373.

Swift, M. (1976): Malignant disease in heterozygous carriers. In: D. Bergsma (ed.), Cancer and Genetics, The National Foundation–March of Dimes, Birth Defects: Original Article Series, Vol. 12. New York: Alan R. Liss, pp. 133–144.

Swift, M., Chase, C. (1979): Cancer in families with xeroderma pigmentosum. J. Natl. Cancer Inst. 62:1415–1421.

Swift, M., Sholman, L., Perry, M., Chase, C. (1976): Malignant neoplasms in the families of patients with ataxia-telangiectasia. Cancer Res. 36:209–215.

Weichselbaum, R.R., Nove, J., Little, J.B. (1976): Skin fibroblasts from a D-deletion type retinoblastoma patient are abnormally X-ray sensitive. Nature (Lond.) 226: 726–727.

Wilson, M.G., Towner, J.W., Fujimoto (1973):

Retinoblastoma and D-chromosome deletion. Am. J. Hum. Genet. 25:57–61.

Wurster-Hill, D.H., Cornwell, G.G., McIntyre, O.R. (1974): Chromosomal aberrations and neoplasm—a family study. Cancer 33:72–81.

Wurster-Hill, D.H., Cornwell, G.G., McIntyre, O.R. (1979): Chromosome aberrations of myloid and lymphoid cells in cancer patients and family members without evidence of cancer. Cancer Detect. Prevent. 2:125–126.

20

The Sister Chromatid Exchange Test*

Sheldon Wolff

Principle of Method

Mutagenic carcinogens such as alkylating agents, or physical mutagens such as ultraviolet light, induce lesions in DNA. When DNA containing such lesions replicates during S-phase, exchanges often occur between the two daughter molecules (Wolff et al., 1974). At the cytological level this becomes manifest as exchanges between the two sister chromatids of a metaphase chromosome. To visualize these exchanges, however, the cells must be treated so that the two sister chromatids will be different from one another. This can be accomplished by exploiting the fact that DNA is a double molecule that replicates semiconservatively, i.e., it has two complementary polynucleotide strands that separate from one another, and each acts as a template for the synthesis of new complementary strands. This means that

*This work was supported by the U.S. Department of Energy

each newly synthesized DNA molecule consists of one old polynucleotide strand that is conserved and a brand new strand.

The differentiation between sister chromatids of a given chromosome was first accomplished by Taylor, Woods, and Hughes (Taylor et al., 1957), who made cells with chromosomes containing sister chromatids physically different from one another. They allowed the cells to replicate their DNA for one cell cycle in the presence of tritiated thymidine to produce chromatids in which the DNA contained one normal polynucleotide strand and one complementary radioactive polynucleotide strand. When cells with such chromosomes were allowed to replicate again, this time in the absence of tritiated thymidine, the radioactive strand separated from the normal strand and both acted as templates for normal polynucleotide chains. Consequently, after these two rounds of replication, the chromosomes contained one chromatid that was radioactive and another that was not. These two sister chromatids could be differentiated from one another by autoradiographic procedures, and exchanges between the two physically different sister chromatids could be observed as exchanges of radioactive segments between them.

Autoradiography is an inherently cumbersome technique with a relatively low level of resolution that precludes the observation of small sister chromatid exchanges (SCEs). Thus, in recent years new techniques have supplanted autoradiography (Zakharov and Egolina, 1972; Latt, 1973; Ikushima and Wolff, 1974; Perry and Wolff, 1974; Wolff and Perry, 1974; Kato, 1974). With these techniques it is possible for chromosomes to have sister chromatids that are chemically, rather than physically, different from one another. These chemically different chromatids will stain differentially, leading to excellent cytological preparations that give far greater precision and resolution for SCEs than is possible with autoradiography. Because of the ease of preparation of the material and the ease of scoring the cytogenetic effects manifested as sister chromatid exchanges, these new techniques have made the study of SCEs a favored way to determine whether or not a material damages DNA (Latt, 1974; Kato and Shimada, 1975; Perry and Evans, 1975; Solomon and Bobrow, 1975; Stetka and Wolff, 1976a,b; Takehisa and Wolff, 1977). In fact, it has been found that SCEs are often produced by concentrations of a chemical that are one or two orders of magnitude lower than the concentration necessary to cause chromosome aberrations, which have heretofore been used as the measure of cytogenetic damage.

The production of chromosomes with chemically different sister chromatids depends upon the incorporation of chemical analogues of thymidine into DNA. Usually, 5-bromodeoxyuridine (BrdUrd) is the chemical analogue of choice, although other analogues such as 5-iododeoxyuridine can also be used. If chromosomes replicate for one round of DNA replication in the presence of BrdUrd and then another round in its absence, chromosomes are

produced in which one chromatid is unsubstituted and the other is singly substituted with BrdUrd, thus forming chromosomes with chemically different sister chromatids. Alternatively, if BrdUrd is present for both rounds of replication, the resultant chromosomes will have one chromatid that is singly substituted, whereas its sister will be bifilarly substituted with BrdUrd. In either case, the sister chromatids can now be made to stain differently from one another.

This differential staining can be accomplished if the preparations are stained with Giemsa alone (Zakharov and Egolina, 1972; Ikushima and Wolff, 1974; Korenberg and Freedlender, 1974), with fluorescent dyes (Latt, 1973; Perry and Wolff, 1974; Kato, 1974; Lin and Alfi, 1976), or with fluorescent dyes followed by exposure to Giemsa or other stains (Perry and Wolff, 1974; Kim, 1974; Goto et al., 1975). In such preparations SCEs can be readily scored.

Test Cells

The methods for making harlequin chromosomes containing sister chromatids that stain differently from one another work with a variety of cells. Thus, experiments have been carried out with mammalian, plant (Kihlman and Kronborg, 1975), insect (Wienberg, 1977; Gatti et al., 1979), avian (Bloom and Hsu, 1975), and fish (Kligerman and Bloom, 1976) cells. They also work for cells that have been exposed to BrdUrd in vivo (Allen and Latt, 1976; Schneider et al., 1976; Tice et al., 1976; Allen et al., 1977). It is possible to use any of these systems to determine whether or not a chemical can attack DNA and induce SCEs. To obtain optimally stained preparations slight modifications are used for each separate system and even for each separate cell line. Because the use of Chinese hamster ovary (CHO) cells offers many advantages, the methods and protocols for their use will be given as prototypes for the study of SCEs.

Among the reasons for choosing CHO cells are that they have a low level of DNA excision repair, which tends to increase the sensitivity of the system, and that they are transformed and thus grow readily in tissue culture. Also they have a short cell cycle time (about 12 hr) so that an experiment may be carried out in 24 hours, and the mitotic cells are loosely attached to the culture vessels so that slides

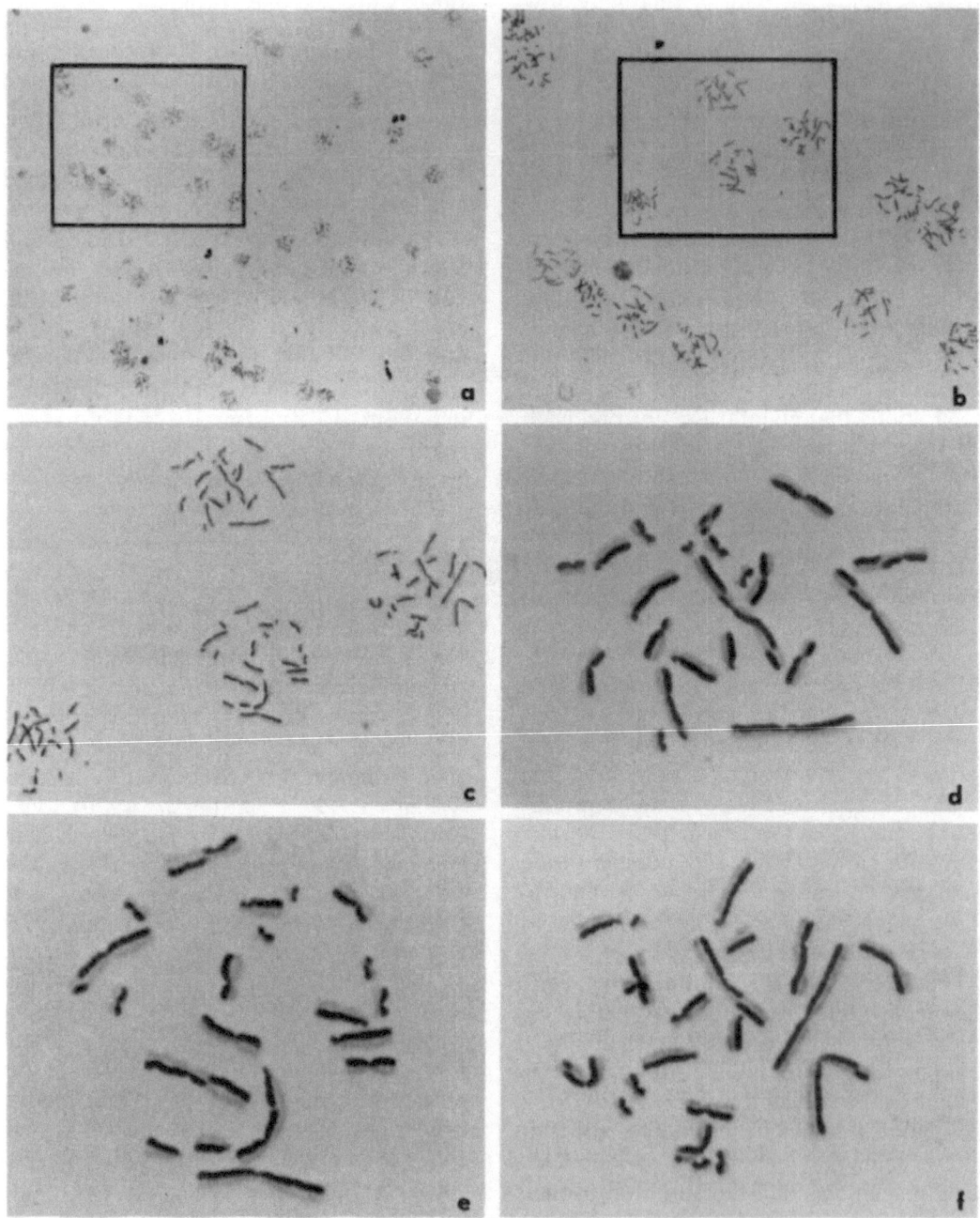

Figure 20.1. Section of slide showing metaphase CHO cells. (a) Low magnification; (b) higher magnification of inset from (a); (c) higher magnification of inset from (b); (d) upper cell from (c); (e) right hand cell from (c); (f) lower cell from (c).

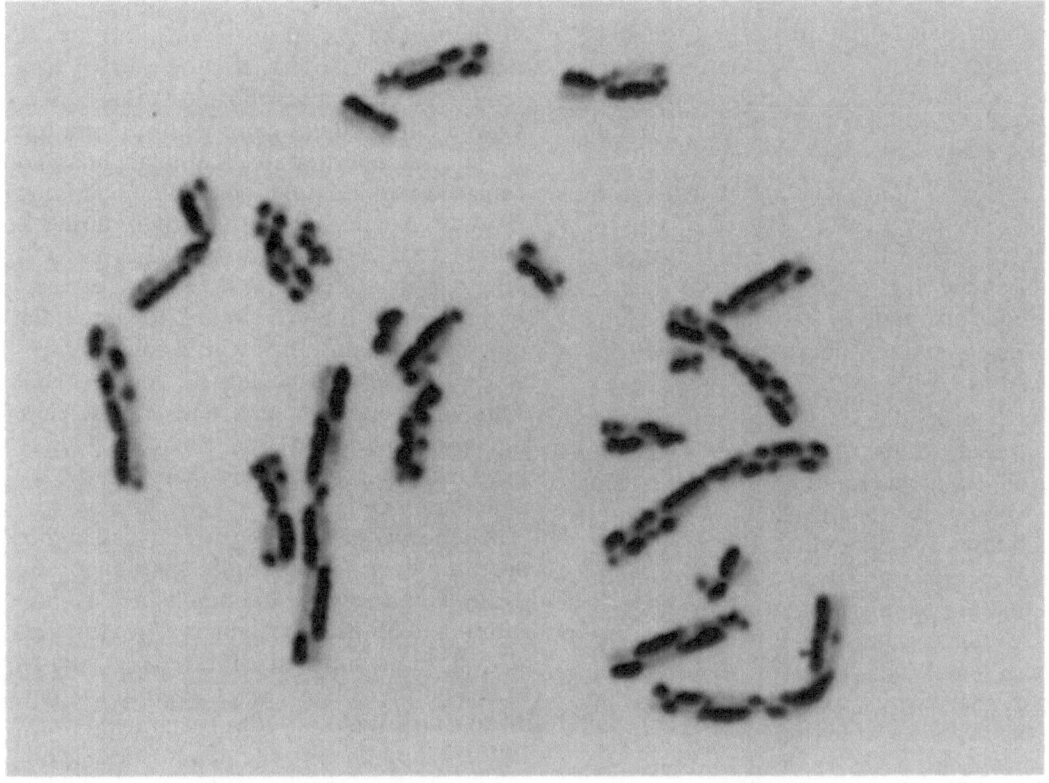

Figure 20.2. Harlequin stained CHO cell showing multiple SCEs after treatment with a potent mutagenic carcinogen.

containing virtually only metaphase cells can be prepared. This is illustrated in Figure 20.1, which shows at increasing magnifications a typical section of a slide made of CHO cells. If the cultures had been treated with a potent alkylating agent the cells obtained could all look very much like the cell illustrated in Figure 20.2.

A successful technique for CHO cells in tissue culture consists of seeding 75 cm² plastic tissue culture flasks with 4×10^5 cells in 10 ml of McCoy's 5A medium supplemented with 10% fetal calf serum, 10 units of penicillin, and 10 μg/ml of streptomycin. These cultures are gassed with 5% CO_2, sealed, and then incubated at 37°. If the experiments are carried out in a CO_2 incubator, the cultures need not be gassed. Some 24 hr later the cells will have divided twice and the culture will be in exponential growth phase. At this time the medium is supplemented with BrdUrd. A final con-

centration of 10 μM BrdUrd has been found sufficient to give preparations with well-differentiated chromatids and also to ensure that the cells are on the plateau of a curve relating SCEs to concentrations of BrdUrd in the medium (Perry and Wolff, 1974). Once BrdUrd has been added, the cells should either be kept in complete darkness or be handled only under yellow or red photographic safelights, although sometimes incandescent light can be used. In no case should the cells be exposed to fluorescent light, which contains wavelengths that can cause photolysis of BrdUrd-containing DNA. After 22 hr of culture, Colcemid is added to a final concentration of 2×10^{-7} M to accumulate mitotic cells. The loosely attached metaphase CHO cells are collected by the mitotic selection method of Terasima and Tolmach (1961). After gently rocking the cultures, the medium is poured into centrifuge tubes and centrifuged at $800 \times g$ for 4–10 min. After the medium is poured off, the pellet, which contains virtually only mitotic cells, is exposed to a hypotonic solution consisting of 0.075 M KCl to spread the chromosomes. The cells are then fixed in methanol:acetic acid (3:1). After two or three changes of fixative, the pellet of cells is suspended in a small amount of fixative and slides are made by placing drops of the suspension on clean microslides, which are allowed to air-dry.

If cells other than CHO are used, it may be necessary to trypsinize the flasks at the time of cell collection. In this case, and also when cells are grown in suspension (e.g., human lymphocytes), the slides will contain metaphase cells in a background of cells that are mainly in interphase.

The slides can be stained by a variety of methods (see Wolff, 1980 for review). One method that works extremely well is to stain the slides for at least 25 min in a solution of Hoechst 33258 in M/15 Sorensen's buffer, pH 6.8, for at least 20 min. The slides are then washed, dried, and mounted in the same buffer with a coverslip. The wet slides are then exposed to 360 nm wavelength light. This can be conveniently done by placing them for various periods of time in the sun, under white- or black-light fluorescent bulbs, or exposing them to light from a high-pressure mercury lamp. The slides can then be stained in 3% Giemsa made in the same buffer. If the chromosomes are too light, they must be exposed to less light. If they are too dark, they should be exposed to more light before staining with Giemsa. Preparations can then be washed and mounted in Gurr's DEPEX or Canada balsam.

Testing Protocol

Treatment of cells with a direct-acting mutagenic carcinogen is accomplished by simply adding the compound to the cultures. In this way, cells in all stages of the cell cycle are exposed. In most cases, the chemical can be left in the medium for the entire culture period. If, however, it is a long-lived substance whose removal is necessary, the medium can merely be poured off after an appropriate treatment time and replaced with complete medium containing BrdUrd. Since SCEs are formed during S (Wolff *et al.*, 1974), damaged S cells will give rise to SCEs, as will cells exposed in other parts of the cell cycle if they contain long-lived lesions that will produce SCEs when the cells move into S. Since CHO cells are relatively deficient in excision repair, the induction of SCEs by such long-lived lesions is more likely than it is in normal human cells.

When it is necessary to treat cells with procarcinogens that require metabolic activation, this can be accomplished by the addition of S9 Mix (Malling, 1971; Ames *et al.*, 1975), composed of rat liver microsomes and an NADPH generating system (Stetka and Wolff, 1976b; Natarajan *et al.*, 1976). When this is done, the medium from exponentially growing cultures should be replaced with medium in which the serum has either been omitted or has been reduced to 2% (Stetka and Wolff, 1976b). After treatment with the chemical in S9 Mix for at least 1 hr, this medium can be poured off and the cells washed before being recultured in normal medium containing BrdUrd. It should be noted that with some chemicals a longer period of exposure in the presence of S9 Mix might be necessary for metabolic activation (Takehisa and Wolff, 1978). Furthermore, some chemicals will inhibit the cell cycle. In these cases it may be necessary to culture the cells for longer than 24 hr to obtain two rounds of DNA replication.

In any experiment both positive and negative controls should be run. A positive control could conveniently consist of treatment with 10^{-3} M cyclophosphamide in the presence of S9 Mix, whereas the negative controls should consist of a culture with no chemical exposure and a culture exposed to the solvent in which the test chemical is dissolved.

When large numbers of SCEs are induced by a chemical, adequate statistical results can be obtained by scoring 20–50 cells. If, however, the baseline level of SCEs is increased by only 30–40%, far more reliable results are obtained when 100 cells are scored at each point.

Because SCEs are usually distributed among the chromosomes according to the Poisson formula (Wolff and Perry, 1974), Poisson errors can be used on the points, that is, the results can be recorded as the mean number of SCEs per cell $\pm$ standard error of the mean. This would be recorded as the number of SCEs divided by the number of cells $\pm$ the square root of the number of SCEs divided by the number of cells. The difference between two means can be calculated by a t-test. The value of t is computed by dividing the difference between the two means by the square root of the sum of the variances of the means. The variance of each mean is the square of the standard error of the mean.

Large numbers of chemicals have been tested and an ever-increasing number are appearing in the literature. The SCE test is able to detect not only strong carcinogenic mutagens but also such compounds as saccharin (Wolff and Rodin, 1978) that are both weak mutagens and weak carcinogens.

References

Allen, J.W., Latt, S.A. (1976): In vivo BrdU-33258 Hoechst analysis of DNA replication kinetics and sister chromatid exchange formation in mouse somatic and meiotic cells. Chromosoma 58:325–340.

Allen, J.W., Shuler, C.F., Mendes, R.W., Latt, S.A. (1977): A simplified technique for in vivo analysis of sister chromatid exchanges using 5-bromodeoxyuridine tablets. Cytogenet. Cell Genet. 18:231–237.

Ames, B.N., McCann, J., Yamasaki, E. (1975): Methods for testing carcinogens and mutagens with the Salmonella/mammalian-microsome mutagenicity test. Mutat. Res. 31: 347–364.

Bloom, S.E., Hsu, T.C. (1975): Differential fluorescence of sister chromatids in chicken embryos exposed to 5-bromodeoxyuridine. Chromosoma 51:261–267.

Gatti, M., Santini, G., Pimpinelli, S., Olivieri, G. (1979): Lack of spontaneous sister chromatid exchanges in somatic cells of Drosophila melanogaster. Genetics 91:255–274.

Goto, K., Akematsu, T., Shimazu, H., Sugiyama, T. (1975): Simple differential Giemsa staining of sister chromatids after treatment with photosensitive dyes and exposure to light and the mechanism of staining. Chromosoma 53:223–230.

Ikushima, T., Wolff, S. (1974): Sister chromatid exchanges induced by light flashes to 5-bromodeoxyuridine and 5-iododeoxyuridine substituted Chinese hamster chromosomes. Exp. Cell Res. 87:15–19.

Kato, H. (1974): Spontaneous sister chromatid exchanges detected by a BUdR-labelling method. Nature (Lond.) 251:70–72.

Kato, H., Shimada, H. (1975): Sister chromatid exchanges induced by mitomycin C: a new method of detecting DNA damage at chromosomal level. Mutat. Res. 28:459–464.

Kihlman, B.A., Kronborg, D. (1975): Sister chromatid exchanges in Vicia faba. I. Demonstration by a modified fluorescent plus Giemsa (FPG) technique. Chromosoma 51:1–10.

Kim, M.A. (1974): Chromatidaustausch und Heterochromatinveranderungen menschlicher Chromosomen nach BUdR-Markierung. Humangenetik 25:179–188.

Kligerman, A.D., Bloom, S.E. (1976): Sister chromatid differentiation and exchanges in adult mudminnows (Umbra limi) after in vivo exposure to 5-bromodeoxyuridine. Chromosoma 56:101–109.

Korenberg, J., Freedlender, E.F. (1974): Giemsa technique for the detection of sister chromatid exchanges. Chromosoma 48: 355–360.

Latt, S.A. (1973): Microfluorometric detection of deoxyribonucleic acid replication in human metaphase chromosomes. Proc. Natl. Acad. Sci. USA 70:3395–3399.

Latt, S.A. (1974): Sister chromatid exchanges, indices of human chromosome damage and repair: Detection by fluorescence and induction by mitomycin C. Proc. Natl. Acad. Sci. USA 71:3162–3166.

Lin, M.S., Alfi, O.S. (1976): Detection of sister chromatid exchanges by 4'-6-diamidino-2-phenylindole fluorescence. Chromosoma 57: 219–225.

Malling, H.V. (1971): Dimethylnitrosamine: formation of mutagenic compounds by interaction with mouse liver microsomes. Mutat. Res. *13*:425–429.

Natarajan, A.T., Tates, A.D., van Buul, P.P.W., Meijers, M., de Vogel, N. (1976): Cytogenetic effects of mutagens/carcinogens after activation in a microsomal system in vitro. I. Induction of chromosome aberrations and sister chromatid exchanges by diethylnitrosamine (DEN) and dimethylnitrosamine (DMN) in CHO cells in the presence of rat-liver microsomes. Mutat. Res. *37*:83–90.

Perry, P., Evans, H.J. (1975): Cytological detection of mutagen-carcinogen exposure by sister chromatid exchange. Nature (Lond.) *258*: 121–125.

Perry, P., Wolff, S. (1974): New Giemsa method for the differential staining of sister chromatids. Nature (Lond.) *251*:156–158.

Schneider, E.L., Chaillet, J.R., Tice, R.R. (1976): In vivo BUdR labeling of mammalian chromosomes. Exp. Cell Res. *100*:396–399.

Solomon, E., Bobrow, M. (1975): Sister chromatid exchanges—a sensitive assay of agents damaging human chromosomes. Mutat. Res. *30*:273–278.

Stetka, D.G., Wolff, S. (1976a): Sister chromatid exchange as an assay for genetic damage induced by mutagen-carcinogens. I. In vivo test for compounds requiring metabolic activation. Mutat. Res. *41*:333–342.

Stetka, D.G., Wolff, S. (1976b): Sister chromatid exchange as an assay for genetic damage induced by mutagen-carcinogens. II. In vitro test for compounds requiring metabolic activation. Mutat. Res. *41*:343–350.

Takehisa, S., Wolff, S. (1977): Induction of sister chromatid exchanges in Chinese hamster cells by carcinogenic mutagens requiring metabolic activation. Mutat. Res. *45*:263–270.

Takehisa, S., Wolff, S. (1978): The induction of sister-chromatid exchanges in Chinese hamster ovary cells by prolonged exposure to 2-acetylaminofluorene and S-9 Mix. Mutat. Res. *58*:103–106.

Taylor, J.H., Woods, P.S., Hughes, W.L. (1957): The organization and duplication of chromosomes as revealed by autoradiographic studies using tritium-labeled thymidine. Proc. Natl. Acad. Sci. USA *43*: 122–128.

Terasima, T., Tolmach, L.J. (1961): Changes in X-ray sensitivity of HeLa cells during the division cycle. Nature (Lond.) *190*: 1210–1211.

Tice, R., Chaillet, J., Schneider, E.L. (1976): Demonstration of spontaneous sister chromatid exchanges in vivo. Exp. Cell Res. *102*: 426–429.

Wienberg, J. (1977): BrdU-Giemsa-technique for the differentiation of sister chromatids in somatic cells of *Drosophila melanogaster*. Mutat. Res. *44*:283–286.

Wolff, S. (1980): Sister chromatid exchange in mammalian cells. In: Handbook of DNA Repair Techniques, Friedberg, E.C., and Hanawalt, P.C. (eds.), Vol. 5, New York: Marcel Dekker, Inc.

Wolff, S., Perry, P. (1974): Differential Giemsa staining of sister chromatids and the study of sister chromatid exchanges without autoradiography. Chromosoma *48*:341–353.

Wolff, S., Rodin, B. (1978): Saccharin-induced sister chromatid exchanges in Chinese hamster and human cells. Science *200*:543–545.

Wolff, S., Bodycote, J., Painter, R.B. (1974): Sister chromatid exchanges induced in Chinese hamster cells by UV irradiation of different stages of the cell cycle: The necessity for cells to pass through S. Mutat. Res. *25*: 73–81.

Zakharov, A.F., Egolina, N.A. (1972): Differential spiralization along mammalian mitotic chromosomes. I. BUdR-revealed differentiation in Chinese hamster chromosomes. Chromosoma *38*:341–365.

21

The Micronucleus Assay. I. In Vivo

JOHN A. HEDDLE AND MICHAEL F. SALAMONE

Introduction

The micronucleus assay was developed to detect *in vivo* chromosomal breakage more easily than traditional cytogenetic techniques (Heddle, 1973; Schmid, 1973, 1975, 1976; Matter and Schmid, 1971). It was not designed to be a short-term test for carcinogenicity and as such has shown only moderate success in this role, which has sometimes been thrust upon it. As a method for detecting chromosomal breakage, however, it has been quite successful and has largely replaced the traditional scoring of metaphases in bone marrow cells. We have tested more than 100 agents in our laboratory by one or more variations of this assay (Heddle and Bruce, 1977a, 1977b; unpublished data): approximately an equivalent number, including some of the same compounds, have been tested in other laboratories (see, for example: Matter and Grauwiler, 1975; Jenssen and Ramel, 1978; Trzos *et al.*, 1978; Maier and Schmid, 1976; Miller, 1973). Positive results

with the assay demonstrate an *in vivo* genetic activity, namely chromosome breakage in bone marrow. Clearly this does not conclusively show that the agent tested is a mutagen or carcinogen, but experience with the assay suggests that this will usually be the case. Negative results are much less meaningful and should not preclude further testing because they can arise for a variety of reasons: (1) there is no *in vivo* chromosomal damage (whether or not intragenic mutations or cancers are produced); (2) chromosomal damage is tissue-specific and the bone marrow is not affected; (3) the sample size is too small, or (4) the treatment and sampling protocol are inadequate.

The Underlying Biology

Micronuclei arise from chromosomal fragments that are not incorporated into daughter nuclei at the time of cellular division.

244   John A. Heddle and Michael F. Salamone

They should be detectable in any dividing cell population in which chromosomal fragments are lost. The polychromatic erythrocytes (PCE) found in bone marrow are a convenient cell population for the study of micronuclei (von Ledebur and Schmid, 1973). These cells are the immediate end-product of a series of cellular divisions, although they themselves do not divide. Their polychromatic staining characteristic is the result of ribosomal RNA which is present for about 24 hours after the cell is formed. As PCEs develop into mature erythrocytes, they lose their ribosomal RNA and the accompanying polychromatic staining property. In mammals, mature erythrocytes do not contain nuclei; the nuclei are expelled ~8–12 hr after the last mitosis preceding the formation of an erythrocyte. For reasons that are not understood, micronuclei are not expelled or, at least, many remain. From this information, one can deduce two aspects of the change in micronucleus frequency in the PCE population after the animal has been treated with a clastogen: (1) treatments within 8–12 hr of sampling will not be effective in producing micronuclei, and (2) even a very brief pulse of micronucleus production would produce an elevated frequency of micronuclei for about 24 hr. Normally the production itself will last for several hours at least, so that the micronuclei will be present longer than 24 hr, especially if more than one cell cycle is affected or differential mitotic delay occurs. Since most mutagens interfere with normal cellular proliferation and possibly with cellular differentiation as well, complex response curves seem likely on *a priori* basis. Micronuclei are not normally found in the circulating erythrocytes in blood because they are filtered out by the spleen.

Protocols

At least three different protocols for treatment and sampling have been used: (1) a single treatment followed by a series of samples at different intervals (Heddle, 1973); (2) two treatments, one at 30 hr and one at 6 hr prior to a single sample (Matter and Schmid, 1971); and (3) five daily treatments, the last one a few hours prior to sampling (Bruce and Heddle, 1979). The last two protocols include a treatment less than about 10 hr prior to sampling, which, as explained in the preceding section, cannot cause any significant production of micronuclei in this cell population and therefore should not be helpful. In an attempt to define the most efficient protocol, we have investigated the variation in micronucleus frequency as a function of time after single and multiple injections of one of three known clastogens.

The changes in the micronucleus frequency as a function of time after a single acute treatment correspond to the predicted pattern quite well. For example, as shown in Figure 21.1, no increase in the micronucleus frequency occurs within about 10 hr of treatment. This is the usual finding. The time at which the maximum frequency occurs is not uniform but rather varies with the agent used. This means that the time at which the assay is most sensitive varies and that no single time is best for all agents. We wondered, therefore, if two treatments would be better than one when appropriate sampling intervals were used. Figure 21.2 shows the result obtained for the same three agents with two injections 24 hr apart. A comparison of the results of Figure 21.2 with those of Figure 21.1 shows that the frequency of micronuclei found within 10 hr of the second treatment can be fully accounted for by the first treatment, again confirming that treatments within about 10 hr of sampling are wasted. As is evident, no single sampling time gives maximum sensitivity for all chemicals, even after two treatments. The data also indicate that the effect of the two treatments is nearly additive, which means that two treatments can increase the sensitivity of the assay over that of one treatment. If additivity is the rule, then multiple injections would further increase the sensitivity of the assay. Unfortunately, we have found that

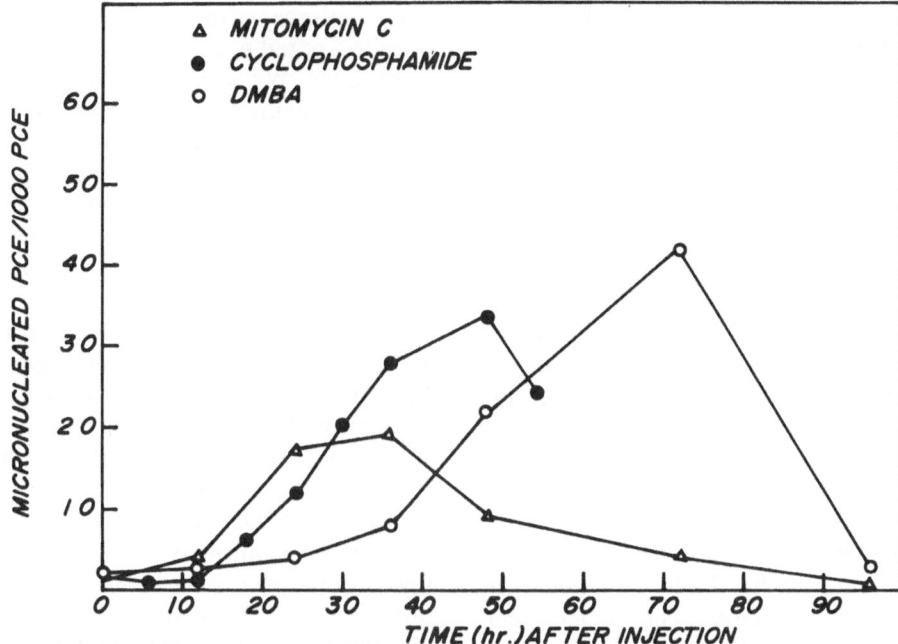

Figure 21.1.   Micronucleus formation in mouse bone marrow as a function of time after a single treatment (injection of drug) of MMC (1 μg/g in saline), CP (75 μg/g in saline), and DMBA (40 μg/g in DMSO). The $LD_{50}/7$'s for MMC, CP, and DMBA were 10.7, 585, and 84 mg/kg, respectively.

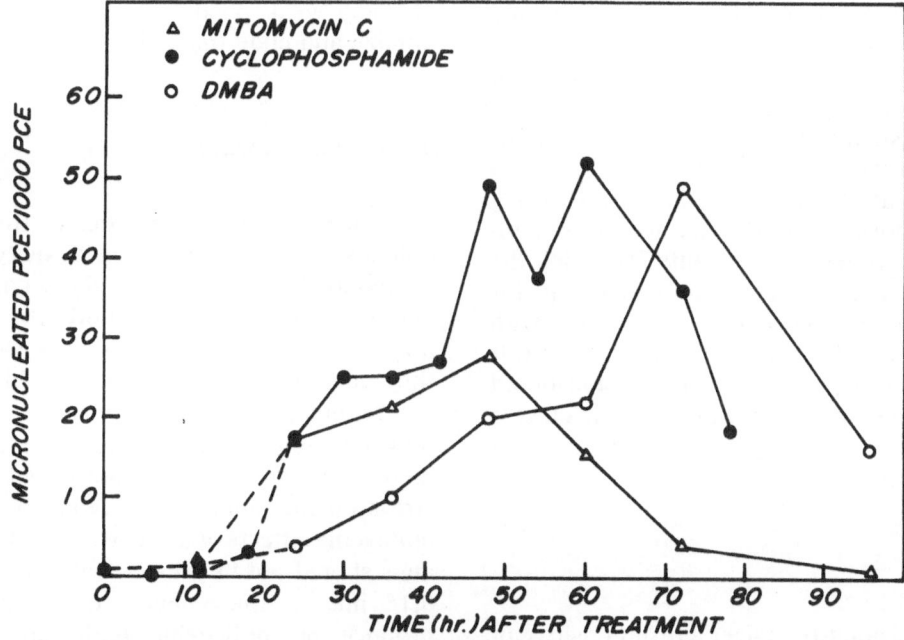

Figure 21.2.   Micronucleus formation in mouse bone marrow as a function of time after two treatments. Treatments were administered at 0 and 24 hours. The dashed portion of each curve represents micronucleus formation attributed to the initial treatment. After the 24-hour point the curves show the response attributed to the combined effects of both treatments.

multiple injections may give much less than additive increases. This raises the possibility that even a second treatment may reduce the assay's sensitivity in some cases.

These data show that none of the commonly used treatment and sampling protocols is best for all agents. Accordingly, we have begun to use a two-phase protocol involving multiple sampling times (Salamone et al., 1980; Heddle and Salamone, 1980). In the first phase, the treatment consists of two intraperitoneal injections of the agent 24 hr apart at as high a dose as possible, usually 80% of the LD_{50}, to maximize the response. Samples are taken at 48, 72, and 96 hr after the *first* injection. This means that the 48 hr sample is 48 hr after the first treatment but 24 hr after the second. If a positive response is obtained, then in the second phase the test is repeated at those times that proved positive, again with two injections 24 hr apart. If the repeat gives a positive response, then we categorize the agent as a confirmed positive. A dose response curve could then be obtained if desired. A negative second test calls for a third test, which is usually decisive, confirming either the original (positive) or the second (negative) test. An initial negative result is followed by a second phase of testing in which a single treatment at 50% of the LD_{50} is used with samples at 30, 48, and 72 hr. The lowering of the dose by about threefold is recommended to detect any agents for which the dose response curve bends over or for which the second injection interferes with the production of micronuclei by the first. If this test is negative, the agent is considered a confirmed negative, whereas if this test is positive, a repeat experiment is required for a decision.

Statistics

The decision to consider a result positive or negative is a statistical one. The statistics depend on the experimental design, includ-ing the number of animals used, the number of PCE scored per animal, and the number of samples. We presume that the vast majority of chemicals are genuinely inactive and that only a few, perhaps 1%, are genuinely active. In this situation, most positives will be false positives if the 5% level of significance is used. Our current practice is to score 500 PCE/mouse and 3–6 mice per sample. The frequency of mice, in our cumulative control samples, which have 0, 1, 2, 3, etc. spontaneous micronuclei per 500 PCE is adequately described by the Poisson distribution. It is easy, therefore, to choose our criteria for a positive response so that we get fewer than 10% false positives in each experiment. A false positive will be confirmed (falsely) 10% of the time so that we expect about 1% confirmed false positives. (Because of the possibility of a third experiment when the first two do not agree, it is necessary to set somewhat more stringent criteria to keep the false positives to 1%.) Fewer false positives can be obtained by setting more stringent criteria, but this will also reduce the sensitivity of the assay by leading to the rejection of some weak positives. We always randomize and code the slides before scoring.

Technical Details

Choice of solvent. In order to treat the animals, a solvent with a low toxicity must be found. We have used saline, corn oil, and dimethyl sulfoxide (DMSO). (It is possible to inject chemicals as a slurry if they cannot be dissolved at suitable concentrations.) The quantity of saline or corn oil used should be limited to less that 2–3 ml for a 20 g mouse so as not to put undue stress on the animals. To avoid toxic and mutagenic effects with DMSO, 20 g animals should not be treated with more than 0.15 ml. At these levels these solvents produce no noticeable toxic effect nor increases in the control levels of micronuclei. Other solvents such as benzene, dichloromethane, and alcohols are not as useful because of their toxicity.

Determination of the LD_{50}. To reduce the possibility of false negatives we always test compounds at the highest dose possible. Usually the upper limit that can be used is determined by the survival of the animals rather than the bone marrow toxicity. Accordingly, we estimate the $LD_{50/7}$, i.e., the dose that kills 50% of the animals within 7 days. We have chosen a 7-day LD_{50} because it is higher in some cases than the LD_{50} determined over longer times and, although 7 days is a slightly longer interval than our usual experiment, a week is a convenient interval for routine work. The $LD_{50/7}$ will vary with the mode of exposure (per os, intraperitoneal injection, etc.) and so should be determined for the mode to be used in testing. We have routinely used the intraperitoneal mode. The $LD_{50/7}$ also varies with the strain of mouse. We have used only B6C3F1/BR mice, which are F1 hybrids between C57BL/6 females and C3H males, so as to have a genetically uniform but outcrossed stock.

The actual determination of the $LD_{50/7}$ is done by the technique of Weil (1952) and Thompson and Weil (1952).

Preparation of Slides

The two methods currently used for the preparation of bone marrow slides with the micronucleus test are those of Heddle (1973) and Schmid (1973). The former method involves touch preparations, i.e., a portion of the marrow is touched to the slide in many places. This has the benefit of being a rapid procedure; however, it suffers in that the cells are sometimes clumped together, making it difficult to distinguish various cell types. The method developed by Schmid gets around this problem by using homogeneous smears. His procedure involves preparation of a centrifuged suspension of marrow from which

smears are made. The main drawback to this technique is the additional time required and costs incurred.

The method of slide preparation used here involves a slightly different technique. Mice are sacrificed by cervical dislocation and the femur removed and stripped clean of muscle. A pair of scissors is used to make a small opening in the iliac end of the femur and a pin the approximate size of the canal (we use a $2^1/_2$ cm safety pin) is introduced into the marrow canal at the epiphyseal end. The canal opening at the epiphyseal end is often obvious after the cap has been removed but, even when it is not, the pin can be easily introduced into the marrow canal. The pin is slowly pushed up and into the canal with a screwing motion. This causes the marrow to exude out the hole at the iliac end. This method removes most of the marrow from the bone cavity. The marrow is placed onto a slide to which a drop of fetal bovine serum from a Pasteur pipet is added. Using the edge of a clean slide, the marrow and serum are mixed until homogeneous and then spread as a smear. Additional slides of this marrow may be prepared by simply transferring some of the mixed preparation to other clean slides. This procedure gives a homogeneous preparation and yet is rapid. The interval from the time the femur is cleaned until the smear is prepared is less than a minute.

For staining purposes, any nucleophilic dye will be taken up by micronuclei. However, not all such dyes achieve good resolution between PCE and mature erythrocytes. To date, the best stain for both identifying micronuclei and distinguishing PCE is that developed by Giemsa. Both the Gurr R-166 and BDH (product code RO3055) commercial Giemsa stains have produced good results in this lab. Currently, the BDH product is used owing primarily to its greater accessibility and its better staining consistency.

After the smears are dry, they are fixed and stained with Giemsa using the following procedure:

1. Fix in absolute methanol for 5 minutes.
2. Air-dry for a few minutes in order to remove the methanol.
3. Stain in 5% Giemsa (the stain is dissolved in 0.01 M phosphate buffer (0.71 g Na_2HPO_4, 0.68 g KH_2PO_4, 1 liter distilled water), which is adjusted to pH 6.8).
4. Rinse in the phosphate buffer for about 30 seconds.
5. Rinse in water.
6. Air-dry.
7. Mount in DPX (BDH) or other neutral mountant.

Good results have been regularly obtained with this procedure. Generally, slides are stained within 24 hours but they have been left up to 3 days without any difference in the final quality of the preparation. A fresh stain solution is prepared weekly. After the first day, a sheen or film may develop on the surface of the solution. This should be cleared away with a tissue before reusing the stain. In general, the stain does not need to be filtered; however, certain commercial Giemsa preparations do form stain granules which should be removed by filtration.

Scoring

As mentioned earlier, we randomize and code all slides before scoring. We do this because there is a certain proportion of structures about which the observer may be in doubt. The coding relieves the observer of concern that the decisions in such cases are being subconsciously biased by knowledge of the treatment group being scored. Our scoring rule is that, with respect to any structure about which there is doubt, an experienced observer is to classify it by what he or she believes it to be rather than by any set of criteria. If the observer is unable to decide on the basis of his or her experience, then the rule is "when in doubt, leave it out." The criteria for a micronucleus is that is must resemble a nucleus in its texture, shape, and staining properties but be

smaller. In this way, the occasional strain granules or other artifacts are easily rejected.

The most difficult decisions are not whether structures are micronuclei but rather whether erythrocytes are polychromatic (von Ledebur and Schmid, 1973). This is a problem since there is a continual gradation from the polychromatic to the normochromatic staining characteristic as the cells develop. Good differentiation of the PCEs is therefore essential. It is to be expected that different observers may score slightly different PCE populations. The difference will lie at the distinction between the oldest PCE and the normochromatic erythrocytes. The fraction of PCE scored by one observer and not by another will be small. Only if this fraction contained a significant proportion of the micronuclei would the difference be an important influence on scoring. In practice, we find that two experienced observers can quickly agree upon the intensity of the bluish staining necessary to classify an erythrocyte as polychromatic, and that thereafter their scores are not distinguishable.

Concluding Remarks

The micronucleus assay should not be regarded as a fixed technique with no possibilities for improvement. There has been, as yet, no systematic search to determine whether some strains of mice are more suitable than others, although other species have been studied (Matter and Schmid, 1971, nor whether pretreatment with inducing agents increases the sensitivity of the assay. Other possibilities for improvement, including machine scoring, certainly exist. As improvements such as we have devised are incorporated into the assay, its rate of success should increase. In our view, the main advantages of the assay are the short time required to complete an *in vivo* assay and the fact that chromosomal breakage is a well-defined and important genetic event.

Acknowledgments

The authors wish to thank Earl Stuart for his technical assistance and Dr. Morris Katz for his support with the project. The authors also gratefully acknowledge the financial support, in part, of a grant from the Ontario Ministry of the Environment, Provincial Lottery Trust Fund, Project No. 78-010-33.

References

Bruce, W.R., Heddle, J.A. (1979): The mutagenic activity of 61 agents as determined by the micronucleus, *Salmonella*, and sperm abnormality assays. Can. J. Genet. Cytol., *21*: 319–334.

Heddle, J.A. (1973): A rapid *in vivo* test for chromosomal damage. Mutat. Res. *18*: 187–190.

Heddle, J.A., Bruce, W.R. (1977a): Comparison of tests for mutagenicity and carcinogenicity using assays for sperm abnormalities, formation of micronuclei and mutations in Salmonella. In: H.H. Hiatt, J.D. Watson, J.A. Winsten (eds.), Origins of Human Cancer, Book C. Cold Spring Harbor Laboratory, Cold Spring Harbor, pp. 1549–1557.

Heddle, J.A., Bruce, W.R. (1977b): On the use of multiple assays for mutagenicity, especially the micronucleus, *Salmonella* and sperm abnormality assays. In: D. Scott, B.A. Bridges, F.H. Sobels (eds.), Progress in Genetic Toxicology. Amsterdam: Elsevier/North-Holland Biomedical Press, pp. 265–274.

Heddle, J.A., Salamone, M.F. (1981): The micronucleus assay: a protocol based on multiple sample times and a statistical method to interpret the results (manuscript in preparation).

Jenssen, D., Ramel, C. (1978): Factors affecting the indication of micronuclei at low doses of x-rays, MMS, and dimethylnitrosamine in mouse erythroblasts. Mutat. Res. *58*:51–65.

Maier, P., Schmid, W. (1976): Ten model mutagens evaluated by the micronucleus test. Mutat. Res. *40*:325–328.

Matter, B., Grauwiler, J. (1975): The micronucleus test as a simple model, *in vivo*, for the evaluation of drug-induced chromosomal aberrations. Mutat. Res. *23*:239–249.

Matter, B., Schmid, W. (1971): Trenimon-induced chromosomal damage in bone marrow cells of six mammalian species, evaluated by the micronucleus test. Mutat. Res. *12*: 417–425.

Miller, R.C. (1973): The micronucleus test as an *in vivo* cytogenetic method. Environ. Health Perspect. *6*:167–170.

Salamone, M.F., Heddle, J.A., Stuart, E., Katz, M. (1980): Towards an improved micronucleus test: studies on 3 model agents, mitomycin C, cyclophosphamide and dimethylbenzanthracene. Mutat. Res.

Schmid, W. (1973): Chemical mutagen testing on *in vivo* somatic mammalian cells. Agents and Actions *3/2*:77–85.

Schmid, W. (1975): The micronucleus test. Mutat. Res. *31*:9–15.

Schmid, W. (1976): The micronucleus test for cytogenetic analysis. In: A. Hollaender (ed.), Chemical Mutagens: Principles and Methods for their Detection. Vol. 4. New York: Plenum Press, pp. 31–53.

Thompson, W.R., Weil, C. (1952): On the construction of table for moving average interpolation. Biometrics *8*:51–54.

Trzos, R.J., Petzold, G.L., Brunden, M.N., Swenberg, J.A. (1978): Evaluation of sixteen carcinogens in the rat using the micronucleus test. Mutat. Res. *58*:79–86.

von Ledebur, M., Schmid, W. (1973): The micronucleus test: Methodological aspects. Mutat. Res. *19*:109–117.

Weil, C. (1952): Tables for convenient calculation of medium effective dose (LD_{50} or ED_{50}) and instructions in their use. Biometrics *8*: 249–263.

22

The Micronucleus Assay. II. In Vitro*

JOHN A. HEDDLE, A. SUDHARSAN RAJ, AND ALENA B. KREPINSKY

Introduction

The genetic implications of chromosomal translocations, inversions, and deletions for human health are well known (Hamerton, 1971). In addition, the presence of chromosomal aberrations indicates that damage to DNA has occurred. Similarly, sister chromatid exchanges (SCE), whose genetic significance is not known, have been found to be useful as a sensitive assay for the presence of DNA damage. Normally, chromosomal aberrations are scored at the first mitosis after treatment whereas SCE are scored at the second mitosis. For this reason the assays are not usually combined. One of the reasons that aberrations are scored at the first mitosis is that most of the aberrations that are easily detected are cell lethal events (although they are almost always accompanied by their heritable counterparts). Many of the cell lethal aberra-

tions include an acentric fragment which, when not incorporated into the daughter nucleus, is detectable as a micronucleus. Micronuclei have been used as a measure of chromosomal damage in plant cells (Evans *et al.*, 1959; Ma *et al.*, 1978), mammalian cells *in vivo* (see preceding chapter), and mammalian cells *in vitro* (Countryman and Heddle, 1976; Adams *et al.*, 1977; Ashwood-Smith *et al.*, 1977; Heddle *et al.*, 1978a,b). Recently we have begun to measure micronuclei on cytological preparations made for the purpose of scoring SCE. The scoring of two effects at the same time has some practical advantages. Although limited, our experience with the combination of assays has been favorable (Krepinsky *et al.*, 1980).

The Underlying Biology

Chromosomal breakage in any population of dividing cells will lead to the loss of acentric fragments and thus to micronuclei. Since all fragments are not lost at the first

*This research has been supported by research grants from the Atkinson Charitable Foundation, the National Cancer Institute of Canada, and the National Science and Engineering Research Council. John Gingerich has been particularly helpful to us.

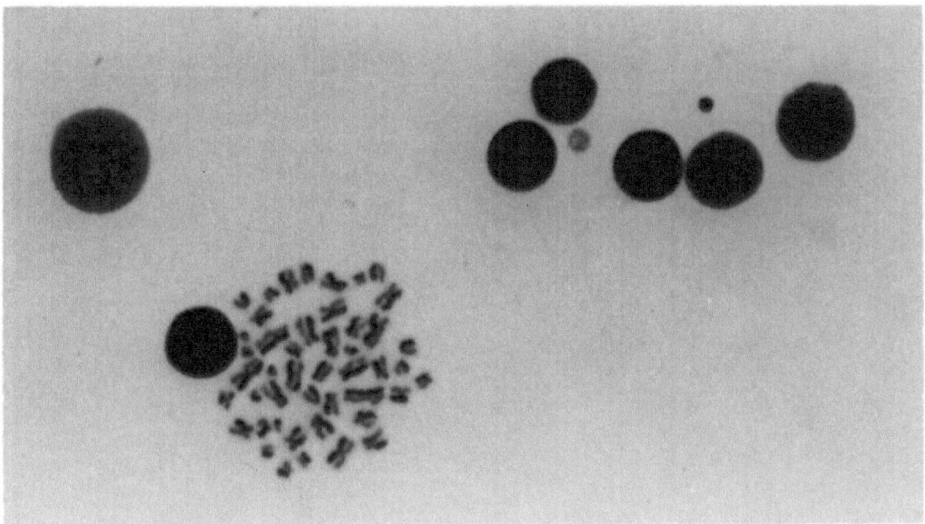

Figure 22.1.   Sister chromatid exchanges and micronuclei in human lymphoblasts.

division (Carrano and Heddle, 1973), the micronucleus frequency may increase for several cell divisions after treatment (Countryman and Heddle, 1976). As further divisions occur, cells containing micronuclei are diluted out of the population (Heddle, 1973). The maximum frequency of micronuclei, e.g., in X-irradiated human lymphocytes, occurs between the third and fourth mitoses (Heddle *et al.*, 1978a,b). Sister chromatid exchanges, in contrast, are usually scored at the second mitosis after treatment. Thus if micronuclei are scored on SCE preparations (Fig. 22.1), some loss in sensitivity must occur. This is compensated, however, by increased reliability of the micronucleus assay: the absence of micronuclei can be attributed only to a lack of chromosomal aberrations rather than to a failure of cells to divide. The combination of the two assays has some other advantages: agents such as X-rays and bleomycin that are potent inducers of aberrations but not of SCE (Solomon and Bobrow, 1975; Gebhart and Kappauf, 1978; Vig and Lewis, 1978) can be readily recognized (as could be agents that produce SCE but few aberrations, if any exist). In addition, agents that do not permit cells to proceed to second metaphase, such as actinomycin D, still give preparations on which micronuclei are detectable.

Some examples of the results that can be obtained by using the same preparations for both assays are shown for human fibroblasts in Fig. 22.2 and Chinese hamster ovary (CHO) cells in Figs. 22.3 and 22.4. It can be seen in the figures that increases in both SCE and micronuclei appear to have

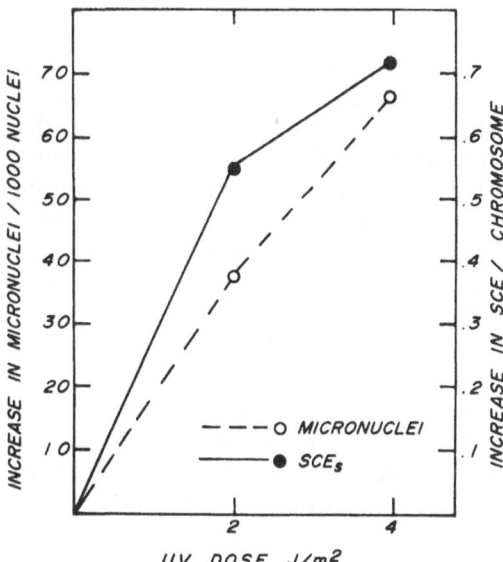

Figure 22.2.   Increase in frequency of SCE and micronuclei in Bloom's syndrome fibroblasts (strain GM 1492) after 254 nm UV irradiation (mean of three experiments). Solid lines and symbols (SCE); dashed lines and open symbols (micronuclei).

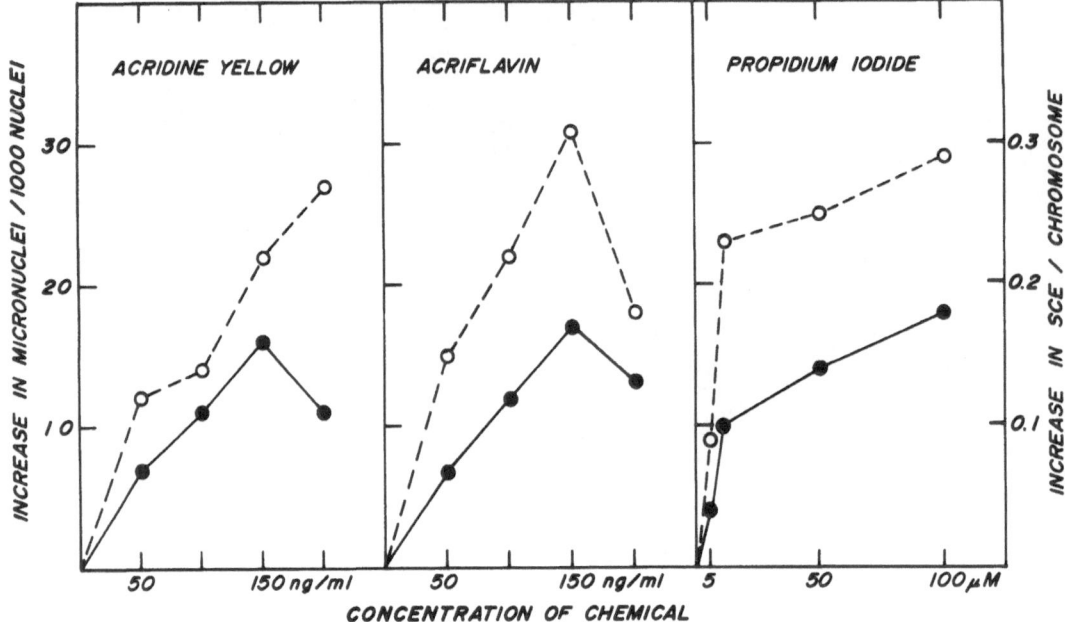

Figure 22.3. Frequencies of SCE (solid lines and symbols) and micronuclei (dashed lines and open symbols) in CHO cells treated with different intercalating agents.

no threshold and show similar, although not identical, dose response relations. Because the rate of cell division is vital for the micronucleus assay, there is no reason to expect the form of the curves to be similar at high doses. With the number of cells we normally sample, the assays have similar sensitivities. This is illustrated for a series of intercalating agents in Fig. 22.4. These three figures (Figs. 22.2 to 22.4) show that, in general, SCE are 200–500 times more frequent than micronuclei on a per cell basis. On the other hand, cells can be scored at least 200 times faster for micronuclei than for SCE. Although the SCE measurements take considerably longer, they are less variable. We must emphasize that SCE and micronuclei do *not* always occur together. We have never found an increase in SCE without an increase in micronuclei, but we have observed the reverse (Krepinsky *et al.*, 1980).

Technical Details

Cell culturing, slide preparation, and staining. There are no special techniques required for the combined assays beyond those required for SCE alone (this assay is described elsewhere in this volume by Wolff). The cell culture methods will vary for the particular cell line used. The slide preparation and staining methods used for SCE are also appropriate for micronuclei. The presence of stain granules is annoying when micronuclei are being scored but these can be eliminated by filtering the Giemsa stain through a 0.2 μm pore filter prior to use. The presence of colchicine for 2–4 hr to prevent anaphase has no appreciable effect upon the micronucleus frequency. A number of different treatment protocols have been used for the SCE assay. Those in which sampling occurs at the second or third mitosis after treatment are obviously preferable to those in which sampling occurs at the first mitosis since there will be more micronuclei at the second and third mitoses than at the first. We normally add the agent being tested and the 5-bromo-2'-deoxyuridine at the same time and sample at the second mitosis. The timing will vary with the cell type being used. For example, L5178Y mouse lymphoma cells can be sampled 12–16 hr after treatment begins;

the corresponding times for CHO cells, human fibroblasts, and human lymphocytes are 24–28 hr, 44–48 hr, and 68–72 hr respectively.

Testing protocol. The first requirement for testing is the selection of a suitable solvent because not all chemicals are soluble in culture medium. If the chemical is not soluble in culture medium, one should try to dissolve it in as little quantity as possible of another solvent and then dilute further either in water or cell culture medium to make up the necessary concentrations. In this way, one can minimize the effects of the solvent on the cultured cells. In our experiments with intercalating agents on CHO cells, all the chemicals were first dissolved in 0.2 ml of dimethyl sulfoxide (DMSO) and then made up to 1 ml with the growth medium. We found that CHO cells tolerate 2% DMSO, which is an excellent solvent for many agents that are difficult to dissolve in water.

The second requirement is to find a suitable dose range. To minimize the frequency of false negatives, one usually tests the highest concentrations possible. The upper limit in these assays is determined either by the degree of inhibition of cell progression or by the solubility of the chemical in medium. The limit can be estimated in trial runs, or by using a wide range of concentrations in the actual experiments. Once an estimate of the maximum concentration is available, a graded series of doses, including at least one that is expected to stop or greatly slow cell growth, should be used. This provides confirmation that no higher dose can be used. An arithmetic series of doses is preferable to a geometric series for weak mutagens, which are the ones that are most difficult to detect. Certain chemicals require metabolic activation. We have had no experience with this procedure for mammalian cells *in vitro* but see no reason why it should interfere with the combined assays.

It is vital to compare the results of the treated material with untreated material. When the chemical is dissolved in a solvent other than water the results obtained with the solvent alone constitute the negative control. Known mutagens like mitomycin C or ethyl methanesulfonate (EMS) may be used as positive controls. EMS is useful because it does not produce very large increases either in SCE or micronuclei. The usual standards for scoring SCE should be observed. For micronuclei the slides should be randomized and coded before being scored. For each experiment at least 500 interphase cells on each slide and two slides per treatment should be scored. Normally, two parallel experiments are carried out to give a total of 2000 scored cells per treatment. Micronuclei are readily visible as small bodies which are adjacent to and stain with less or equal intensity as the nucleus (Fig. 22.1). The diameter of a micronucleus

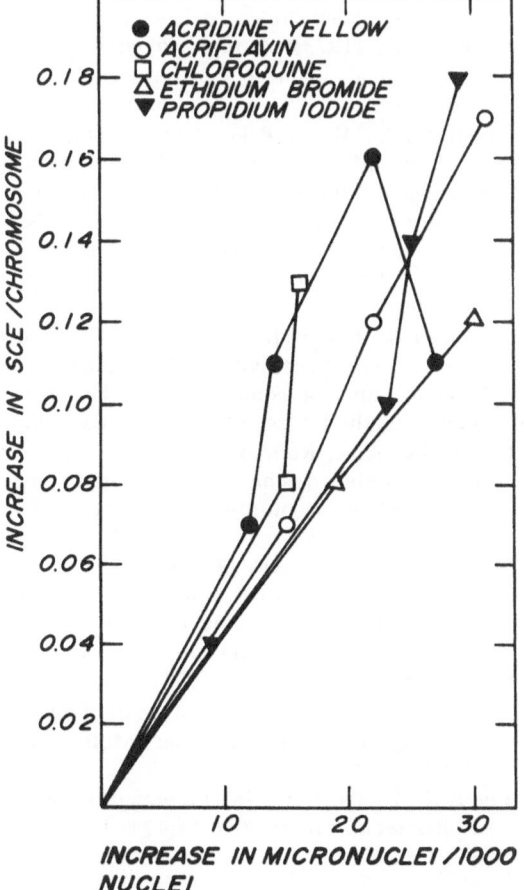

Figure 22.4. Relationship between the increase in SCE and the increase in micronuclei in CHO cells treated with different intercalating agents.

should be less than one-third that of the main nucleus, so that no confusion with binucleate cells occurs. The micronucleus should be located in the cytoplasm or, if this is not visible, within three or four nuclear diameters of the nucleus. In order to distinguish micronuclei from nuclear blebs, micronuclei touching the nucleus should not be scored. A cell with many micronuclei and no nucleus should not be scored: micronuclei in these cells probably arise through nuclear breakdown rather than from chromosomal fragments (Countryman and Heddle, 1976; Heddle et al., 1978a,b). Micronuclei are not necessarily uniformly distributed over the slide, so an effort should be made to sample several areas by moving some distance after each 100 cells scored. We have not investigated the statistics of micronucleus or SCE production in any of the cell systems that we have studied, but rather have depended upon the demonstration of reproducible dose response curves as a criterion of a positive response.

Concluding Remarks

Whenever an agent is tested using the SCE assay, the slides can also be scored for micronuclei. Chromosomal aberrations and SCE do not always go together, so the success rate of the combined assays should be superior to either alone. As the micronucleus scoring is much quicker, one can have two assays for virtually the price of one.

References

Adams, F. H., Norman, A., Mello, R.S., Bass, D. (1977): Effect of radiation and contrast media on chromosomes. Radiology *124*:823–826.

Ashwood-Smith, M.J., Grant, E.L., Heddle, J.A., Friedman, G.B. (1977): Chromosome damage in Chinese hamster cells sensitized to near-ultraviolet light by psoralen and angelicin. Mutat. Res. *43*:377–385.

Carrano, A.V., Heddle, J.A. (1973): The fate of chromosome aberrations. J. Theoret. Biol. *38*: 289–304.

Countryman, P.I., Heddle, J.A. (1976): The production of micronuclei from chromosome aberrations in irradiated cultures of human lymphocytes. Mutat. Res. *41*:321–332.

Evans, H.J., Neary, G.J., Williamson, F.S. (1959): The relative biological efficiency of single doses of fast neutrons and gamma rays on *Vicia faba* roots and the effect of oxygen. II. Chromosome damage: the production of micronuclei. Int. J. Radiat. Biol. *1*:216–229.

Gebhart, E., Kappauf, H. (1978): Bleomycin and sister chromatid exchanges in human lymphocyte chromosomes. Mutat. Res. *58*: 121–124.

Hamerton, J.L. (1971): Human Cytogenetics, Vol. I, 412 pp. and Vol. II, 545 pp. New York: Academic Press.

Heddle, J.A. (1973): A rapid *in vivo* test for chromosomal damage. Mutat. Res. *18*: 187–190.

Heddle, J.A., Benz, R.D., Countryman, P.I. (1978a): Measurement of chromosomal breakage in cultured cells by the micronucleus technique. In: H.J. Evans, D.C. Lloyd (eds.), Mutagen-Induced Chromosome Damage in Man. Edinburgh University Press, pp. 191–200.

Heddle, J.A., Lue, C.B., Saunders, E.F., Benz, R.D. (1978b): Sensitivity to five mutagens in Fanconi's anemia as measured by the micronucleus method. Cancer Res. *38*:2983–2988.

Krepinsky, A.B., Rainbow, A.J., Heddle, J.A. (1980): Studies on the ultraviolet light sensitivity of Bloom's syndrome fibroblasts. Mutat. Res. *69*:357–368.

Ma, T., Sparrow, A.H., Schairer, L.A., Nauman, A.F. (1978): Effect of 1, 2-dibromoethane on meiotic chromosomes of *Tradescantia*. Mutat. Res. *58*:251–258.

Solomon, E., Bobrow, M. (1975): Sister chromatid exchanges—a sensitive assay of agents damaging human chromosomes. Mutat. Res. *30*: 273–278.

Vig, B.K., Lewis, R. (1978): Genetic toxicology of bleomycin. Mutat. Res. *55*:121–145.

23

Automation in Cytogenetics*

John Melnyk, Kenneth R. Castleman, and Garland W. Persinger

Principle of Method

The standard laboratory procedures involved in cytogenetics for clinical evaluation or for screening potential clastogens and exposed personnel are time-consuming, costly, and by their nature, self-limiting in terms of volume production. The rapid developments in computer technology, especially in pattern recognition, provided a possible solution to the problems in processing and analyzing specimens. Several independent methodologies were developed during the period 1965–78 (Mendelsohn, 1976) for automated karyotyping. A recent workshop on automated human cytogenetics held in Copenhagen abstracted the projects in progress in Europe (Granum and Jakobsen, 1978).

The details of an automated karyotyping system (Castleman et al., 1976) and an automated system for processing blood and bone marrow cultures which follow represent the first computerized processes in cytogenetics to reach the field testing stage.

Test Organism

The cells which can be processed on the specimen preparation system are human or animal blood and bone marrow cells. Although testicular cells for meiotic studies have been processed with some success, further developmental work is necessary.

Testing Protocol

Manual Method

A brief review of the method used in most cytogenetic laboratories today will form a basis for understanding the automated system which will be described subsequently.

*Presented as a 16-mm motion picture sponsored by the National Institute of Child Health and Human Development, Mental Retardation and Developmental Disabilities Branch.

Cytogenetic methodology can be divided into two broad categories: (1) cell processing, which includes cell culturing and harvesting, and (2) karyotyping.

Cell culturing and harvesting. The manual blood culturing process used in most laboratories involves a test tube into which culture medium and the blood specimen are dispensed with a syringe. The cultures are incubated for 3 days and Colcemid is added during the last 1½ hours to arrest the cells at metaphase. After centrifugation of the cultures and aspiration of the supernatant with a Pasteur pipet, hypotonic fluid is added to expand the cells. Another centrifugation and aspiration are performed and an acetic-alcohol fixative is added to harden the cells before slide-making and staining.

A Pasteur pipet is used to dispense the fixed cells on dry or wet slides and a short oral blast of air is given to spread the cells. Following a drying period, the slides are stained individually or in groups.

Variations of this basic technique are used for bone marrow cells, which generally include a short culture period. Staining procedures vary considerably among laboratories and are not considered in detail here. Several automated staining systems are available commercially and some can be used for conventional, trypsin-Giemsa, and fluorescent staining of chromosomes.

Analysis and karyotyping. The manual procedure for analyzing metaphase cells involves a time-consuming search procedure for good quality metaphases under a microscope. A chromosome count is made and the number and kind of chromosomal breaks or sister chromatid exchanges are recorded visually. Other approaches involve photography of 1–200 cells or only abnormal ones on 35-mm film or on Polaroid.

In clinical usage, suitable metaphases are enlarged on 8 × 10 paper and individual chromosomes are cut out and aligned in a karyotype.

Automated Culturing and Harvesting

The first step in automating the cell processing system involved the design of a new culture tray. The trays are made of polystyrene and are 7.6 × 11.4 cm (Fig. 23.1). Each well has 5 ml capacity, although in practice 3 ml of culture medium is used to avoid spillage. The trays are produced commercially by Lab-Tek Products in sterile packages with or without caps. The trays have embossed numbers from 1 to 4 horizontally and the letters A, B, C vertically to identify individual wells.

The specimen preparation system consists of two major components, Console 1 and Console 2 (Figs. 23.2, 23.3). Console 1 functions in four modes to dispense medium, blood specimens, and Colcemid into trays, and fixed cells onto slides. Console 2 consists of an aspirator and a fluid dispenser for hypotonic fluid and fixative. Both consoles are operated automatically

Figure 23.1. Culture tray. Each well holds a maximum of 5 ml of fluid and has a rounded bottom for pelleting cells.

Figure 23.2. Console 1. This unit dispenses culture medium, specimens, and Colcemid into culture wells and fixed cells on microscope slides. Components of the system are (1) dispensing head, (2) specimen tray, (3) pipet tip rack, (4) tray magazine, (5) nebulizer, (6) slide housing, (7) bar-code scanners, (8) air manifold, and (9) slide magazines.

Figure 23.3. Console 2. Following the 3-day culture period, blood specimens are processed on this unit. The centrifuged specimens are aspirated (1), resuspended, and hypotonic or fixative solutions are added (2).

through a PDP 11/34 computer or manually through a control panel. The system also has a specimen identification subsystem, which is based on bar-code technology.

Specimen entry. Specimen entry is made in a book with pertinent patient and source information. A bar-code label is attached to the specimen tube and the duplicate is placed in the book opposite the entry. The specimen number (readable bar-code number) is entered at the PDP 11/34 keyboard and the label is scanned with a light pen for verification. A misentry or a misreading is marked by an intermittent beeper and the entry process is repeated. The specimen tube is placed in a carrier which is coupled to a photoelectric matrix signaling the location of the specimen to the computer. All information concerning the patient number and tube location are printed out at the terminal (a total of 48 specimens can be processed in one run).

The specimen carrier is placed in a sterile hood where the tube stoppers are removed. A sterile parafilm sheet is placed over the tubes and the carrier is then transferred to a mixer for 10 minutes.

Dispensing of medium. Under a sterile hood, four culture trays are inserted into a tray magazine (4 in Fig. 23.2), which is then placed in Console 1 over an elevator mechanism. A Filamatic syringe dispensing system operated by the computer transfers 3.5 ml of culture medium to the culture wells. The culture medium is drawn from a flask through sterile tubing to a special dispensing manifold mounted in one arm of the head. After the tray elevator moves and a tray is fed out under the dispensing head, four wells in a row are filled simultaneously. The table moves laterally and the second row is filled. As each tray is filled it is automatically retracted into the tray magazine and the process continues until all trays are filled.

Dispensing of specimens. In preparation for dispensing blood specimens into trays, the specimen carrier is transferred from the mixer to a table (2, Fig. 23.2) on Console 1 and the parafilm covering is removed. A rack of Oxford pipet tips (3, Fig. 23.2) is placed on an adjacent table and a pipet discard box is placed between positions 7 and 8 (Fig. 23.2).

A signal from the keyboard or central panel initiates the following sequence of events:

1. The four-armed head descends and a dispensing manifold picks up four pipet tips.
2. The head raises, rotates 45°, and descends again into the blood specimen tubes. A suction mechanism draws up a predetermined amount of blood. At the same time, the pipet rack advances and the second dispensing arm picks up another set of pipets.
3. The culture tray magazine lowers and a tray is fed out past a bar-code reader. The computer assigns the first four patient specimens or test substances to the first row of wells in this tray.
4. The dispensing head descends for the third time and inoculates the wells with the first four specimens. At the same time the second arm aspirates four more specimens and the third arm picks up four new pipets.
5. On the next descent, the first dispensing manifold is positioned over the pipet discard box where a pipet extractor locks in over the top flange of the pipets. As the head ascends, the pipets are pulled off and are dropped into the box.
6. As each tray is inoculated, it is retracted into the magazine, which lowers, and another tray is fed out.

The procedure continues automatically until all 48 specimens on the carrier are transferred into four trays in a magazine. The magazine is removed from the console and is inserted into a sterile plastic transfer box, which is placed in a CO_2 incubator for 72 hours.

Dispensing of Colcemid. Colcemid is added to the cultures at this stage with the Filamatic syringe system used previously for medium. An amount equal to a final concentration of 0.05 μg/ml is dispensed into each well and the trays are returned to the incubator for 1 hour. At this time a small bar magnet is placed in each well for resuspending the pellet after centrifugation.

Centrifugation. Following a 1-hour exposure to Colcemid, the tray magazines with four trays each are placed in an IEC centrifuge Model UV and are spun for 5 minutes at 1000 rpm.

Aspiration. The tray magazines are placed on an elevator mechanism in Console 2 (Fig. 23.3). A computer signal initiates the following series of events:

1. An arm drives a tray from the magazine to a position under an aspirator.
2. A bar-code label on the tray is read to monitor the stage of processing.
3. The aspirator head descends and suction applied by a vacuum pump removes the supernatants in 3 seconds.
4. The head ascends and the tray is withdrawn into the magazine.
5. The head descends and the aspirating needles are immersed in an ultrasonic cleaner where the combination of sonication and continuous suction decontaminates the needles.
6. The head ascends and the next cycle begins.

Dispensing of hypotonic fluid and fixative. The tray magazines are moved manually to the elevators in the fluid dispenser station (Fig. 23.3).

A single keyboard command initiates the following events:

1. As the trays are fed out, a bar-code reader monitors the stage of processing.
2. The bar magnets are activated by revolving magnets mounted beneath the working surface, resuspending the cells.

3. Hypotonic fluid is added by means of a peristaltic pump mounted in the chassis.
4. After 3 ml of fluid is added, the pump stops, the trays are retracted, and the tray magazines are removed manually for incubation.

Centrifugation and aspiration are repeated as before and acetic-alcohol fixative is dispensed as described for hypotonic fluid, using the same components. The cycle of fixative and centrifugation is repeated a total of three times.

Slide preparation. Following the last aspiration, the tray magazines are returned to Console 1 at the slide preparation station (Fig. 23.2). The console is prepared for dispensing of fixed cells onto microscope slides. The pipet tip rack is placed in the position previously held by the specimen carrier. The disposal tank is moved to the position that the pipet tip rack occupied during the specimen dispensing mode. Microscope slides are prelabeled manually and are stacked in the slide housing unit (6, Fig. 23.2).

A signal from the keyboard initiates the following events:

1. A tray is fed past a bar code scanner signaling to the computer the identity of the tray and the 12 specimens that are being dispensed.
2. The head descends and four pipet tips are picked up by the dispensing manifold.
3. Four slides are fed out simultaneously from the housing past bar-code scanners. The slide numbers are assigned to the first group of four specimens, which were stored in memory during specimen entry and the blood dispensing mode.
4. The head rises, rotates and descends, and specimens are picked up from four wells.
5. The head rises again, rotates, and descends over the microscope slides.
6. The equivalent of two drops is deposited on the slides.

7. As these are advanced to the next station where an air blast disperses the cells, four clean slides are fed out.
8. The blast of air is followed by a flow of moisturized air, which aids in chromosome spreading.
9. Slide warmers under the unit facilitate spreading and drying.
10. As the slides advance in the slide track, they are fed into slots in four slide magazines which are mounted on elevators at the end of the track.
11. The slide magazines are removed from the console at the completion of the run and are transferred to a staining machine.

Automated Karyotyping

The automated microscope is part of a separate system which was designed and constructed at Jet Propulsion Laboratories to perform chromosome analysis (Fig. 23.4). Although the system was designed for clinical studies, it has several features which are applicable to chromosome breakage studies. The following procedure describes the clinical usage which can be adapted to clastogen studies.

The analysis system is built around a PDP-11 minicomputer, which controls the motor-driven microscope stage and processes the chromosome images. A television camera mounted on the microscope supplies the video signal for a closed-circuit TV monitor, an automatic focus and chromosome spread detector circuit, and an image digitizer. The system also has a typewriter terminal and a special console to permit operator control of the analysis. A grayscale display system allows the computer to present processed images for operator inspection and a hardcopy image recorder produces permanent images for the patient's file.

The operation of the system may be divided into three phases: slide search, spread analysis, and picture generation. In the slide search phase, the computer program exercises the microscope stage in a search pattern while monitoring the spread detector output for an indication of the presence of a metaphase spread. The stage coordinates of positions containing metaphase spreads are recorded in memory. After a suitable area of the slide has been searched, the program returns the stage to each of the previously located cells for the operator's review. The operator then selects a suitable number of the best cells for analysis, rejecting poor-quality cells and nonchromosome material which may have confused the spread detector. The images of acceptable spreads are then fed into the computer for analysis. In the slide search phase, the initial cell location is done without operator assistance, but the final selection of cells is made by the operator.

In the analysis phase, the computer program locates each of the chromosomes in the cell, assigning each one a number. The numbered chromosomes are displayed to the operator who may reject nonchromosome material and separate touching chromosomes. Ordinarily, overlapping chromosomes are avoided in the cell selection phase.

After proper isolation the chromosomes are measured and classified into groups or, with banded preparations, into homologous pairs. The computer program then composes a karyotype image in which the individual chromosomes are arranged by group or autosomal type. This karyotype is displayed for operator verification. If the operator detects errors, he may indicate corrective action. This includes correcting centromere position and moving chromosomes about within the karyotype. After the karyotype is approved by the operator, it comes out as a photographic print on the hardcopy output device.

For any particular cell, the steps of slide search, analysis, and pictorial output are performed sequentially. However, in order to make most efficient use of the system components, the three functions are carried out continuously and simultaneously. For example, while one cell is being analyzed, the karyotype of the previous cell will be coming out of the hardcopy output device,

Figure 23.4. Automated light microscope. This system locates metaphases, focuses, counts chromosomes, and karyotypes cells automatically. Components of the system are (1) Dec writer, (2) Mag tape drive, (3) PDP 11/40 computer, (4) image digitizer and operation keyboard, (5) television camera, (6) automated microscope, and (7) line printer (hardcopy device not shown).

and simultaneously the microscope will be searching for another cell to analyze. The process then mimics an assembly line having three stations. The operator alternates between cell selection via the stage keyboard and karyotype verification at the gray-level display.

In production karyotyping the automated system averaged 7 min per cell on homogeneously stained specimens and 15 min per cell on trypsin-Giemsa-banded material. This includes all steps from cell location through production of the karyotype image. Chromosome counts, complete with cell images, averaged 90 sec per cell. In a study involving a 35% abnormal group of 435 patients, 1.6% of the karyotypes contained errors which led to an incorrect interpretation. These errors were due primarily to poor cell selection by the operator and to

improper microscope focus. Both of these error sources were subsequently reduced by procedural changes. With operator interaction to correct machine errors and a high-quality karyotype output the accuracy of the system should equal that of the manual technique.

System Performance

Chromosome Preparations

Time studies were conducted for the various functions of the specimen preparation system. These included specimen entry (recording, labeling, bar-code scanning, etc.) and dispensing of medium, specimens, Colcemid, hypotonic fluid, fixative,

Table 23.1.   *Processing Time Comparisons for 96 Specimens*

Processing mode	Time
Manual	22 hr, 48 min
Automated	3 hr, 13 min

and fixed cells onto slides. Actual times were obtained for processing 48 specimens and these were extrapolated to determine the time for 96 specimens, representing two runs. Actual times were also obtained for manual processing of 12 specimens which were extrapolated also. These results are shown in Table 23.1. The reduction in time achieved with the automated system is approximately sevenfold.

In order to evaluate the culture conditions with this system, cord blood specimens from 35 individuals were established in trays using RPMI 1640 medium and 15% fetal calf serum. The mean mitotic index obtained was 5.28. This compared favorably with the results of an earlier study in which the manual method using tube cultures and a prototype system using culture trays were evaluated under controlled conditions.

The advantages of the automated system for chromosome preparations over the manual process are:

1. Ability to process large numbers of specimens.
2. Uniform treatment of all specimens.
3. Ability to process controls and replicates in the tray.
4. A strict control over specimen identification.
5. Reduction in cost of processing.

The major advantage of the system is the control it provides over timing and treatment of groups of specimens. This has emerged as one of the most important aspects of consistent high-quality slide making in trypsin-Giemsa banding. The importance of uniform treatment of all specimens in clastogen studies cannot be overemphasized. If standardization of results between laboratories is to be achieved, a standardized procedure involving some form of automation will be required.

Clastogen Analysis

Although the karyotyping system described was developed for clinical cytogenetics, it has been tested in a chromosome aberration study which compared manual and automated methodologies (Table 23.2). In this study the operator instructed the automated microscope to find the metaphases and to display them for scoring, which was conducted directly on the TV monitor. Compared with the manual system, the preceding method was faster by a factor of two.

Although the automated search found slightly fewer abnormal cells, this difference is not statistically significant. The operator reported less fatigue with the automated system owing to the elimination of search time and objective changing, the more comfortable viewing position, and the periods of inactivity during the slide search. The rapid display of previously found cells made scoring easier. Specimen quality affected system performance since poor slides required longer search times. A modified system using two or more microscopes simultaneously could avoid the period of operator inactivity and significantly increase the cell scoring rate.

Table 23.2.   *Comparison of Chromosome Breakage Analyses*

Search mode	Cells examined	Abnormal cells found	Breakage rate (%)	Average time per cell (sec)
Manual	503	37	7.4	72.2
Automatic	541	31	5.7	31.8

Acknowledgments

This paper presents the results of one phase of research conducted by the Jet Propulsion Laboratory, California Institute of Technology, for the National Institutes of Health under Inter-Agency Agreement No. 1-Y01-HD-30001-00 between the National Institutes of Health and the National Aeronautics and Space Administration.

The paper also presents the results of a project conducted at the City of Hope Medical Center under Contract NO-1-HD-52483 for the National Institute of Child Health and Human Development.

References

Castleman, K.R., Melnyk, J., Frieden, H.J., Persinger, G.W., Wall, R.J. (1976): Karyotype analysis by computer and its application to mutagenicity testing of environmental chemicals. Mutat. Res. *41*:153–162.

Granum, E., Jakobsen, B. (eds.) (1978): Proceedings of the 1978 Workshop in Automated Human Cytogenetics. Copenhagen, The Technical University of Denmark.

Mendelsohn, M. (1976): Introduction. In: M. Mendelsohn (ed.), Automation of Cytogenetics: Asilomar Workshop. Springfield, Virginia, National Technical Information Service, pp. 1–2.

24

Mutagenesis Studies in Diploid Human Cells with Different DNA-Repair Capacities

J. Justin McCormick and Veronica M. Maher

Introduction

Human diploid fibroblasts in culture can be very useful for examining the cytotoxic and mutagenic effects of environmental agents (Albertini and DeMars, 1973; Buchwald, 1977; Cox and Masson 1976; Glover *et al.*, 1979; Jacobs and DeMars, 1978; Maher and McCormick, 1976, 1980; Maher and Wessel, 1975; Maher *et al.*, 1975, 1976a,b,c, 1977, 1979; Mankovitz *et al.*, 1974). Studies with these human cells provide a bridge between studies conducted with animal cells in culture and whole animals, on the one hand, and epidemiological studies of the accidental exposure of humans to various agents on the other. In addition, the use of human cells makes it possible to carry out comparative studies designed to elucidate the role of DNA repair in somatic cell mutagenesis (and carcinogenesis) because of the availability of repair-deficient strains derived from individuals with various inherited DNA-repair-deficiency diseases (Fujiwara *et al.*, 1977; German, 1972; Lehmann *et al.*, 1975; Robbins *et al.*, 1974; Taylor *et al.*, 1975). The DNA-repair-defective variants which have recently been isolated in certain animal cell lines, using various selection techniques, should also prove useful for this purpose when they have been fully characterized (Sato and Hieda, 1979; Schultz *et al.*, 1978; Thompson *et al.*, 1979).

Quantitative methods have been developed to assay the cytotoxic effect of environmental agents by measuring the cells' ability to survive and form a colony. Since cell replication requires the integration of multiple metabolic pathways so that perturbation of a large number of different steps will result in inactivation of colony formation, this method of determining cell survival has been shown to be more accurate and sensitive than other methods, such as trypan blue exclusion. Comparative cytotoxicity studies involving repair-deficient and repair-proficient cell strains can be used to determine whether an agent interacts with cellular DNA (see text) and can also shed light on the mode of action, e.g., formation of DNA crosslinks, induction of pyrimidine dimer-like lesions, etc. (Heflich et al., 1977; Maher et al., 1975).

Methods have also been developed for carrying out quantitative studies of the mutagenic effect of environmental agents in diploid human skin fibroblasts, both normally repairing and repair-deficient strains. The latter are especially useful for detecting mutation induction at very low concentrations of an agent (Yang et al., 1980).

Culture Conditions

Source of Cell Culture

Skin fibroblasts from normal individuals are derived from foreskins of newborn infants following circumcision. Skin biopsies from adults are also a useful source of such cells. DNA repair-deficient cells are best obtained from the American Type Culture Collection (Rockville, Maryland), the Institute for Medical Research (Camden, N.J.), or similar cell banks, since this assures that one is dealing with a well-characterized repair-deficient cell.

Human skin fibroblasts are very deficient in ability to metabolize such compounds as benzo(a)pyrene or 2-acetyl-aminofluorene. Consequently, these cells show little or no cytotoxic or mutagenic effect from exposure to large doses of such agents unless one provides an activating system, such as microsomal or S9 fractions from liver or a layer of metabolizing epithelial cells (Aust et al., 1980; Huberman, 1975; Newbold et al., 1977). This lack of ability to metabolize certain compounds can be advantageous if one seeks to quantitate a direct-acting mutagen in a complex mixture of polycyclic aromatic hydrocarbons or aromatic amine carcinogens (McCormick, 1979). However, it should be made clear that diploid human skin fibroblasts are not completely deficient in ability to metabolize carcinogens, as they have been shown to activate such compounds as 4-nitroquinoline-1-oxide (Ikenaga et al., 1977), as well as nitrofuran compounds (McCalla et al., 1978) and dihydrodiols of polycyclic hydrocarbons (Aust et al., unpublished observation).

Initiation of Cell Cultures

Tissue samples (skin biopsies or foreskins) are placed in a tube containing tissue culture medium and 250 μg/ml gentamycin (Schering Diagnostics, Union, N.J., or Sigma, St Louis, Missouri) and are placed in a refrigerator at 4° C. Samples are transported to the laboratory on ice and are transferred into a small dish containing a few drops of medium and divided into small pieces with a sterile iris scissor. Each fragment is transferred to a separate 35-mm-diameter glass culture dish containing a few drops of medium, minced finely with sterile scalpels, and made to attach to the dish with a drop of serum. After 24 hours at 37° C in a CO_2 incubator, additional medium can be added. Alternatively, individual cells of the small fragments can be dissociated in a 25 ml Erlenmeyer flask by 60 min incubation at 37° C in culture medium containing 0.5% collagenase (Sigma) on a magnetic stirrer with insulating styrofoam inserted between the flask and the stirring motor to prevent overheating. Cells are centrifuged in a 50 ml conical tube at 600 × g for 7 min. The supernatant is aspirated off, the cells are washed with culture medium, and after a second centrifugation are resuspended in culture medium and plated into a 250 ml flask. Foreskin cultures provide a very large initial population of cells which have great growth potential (> 60 doublings) and therefore are exceptionally well-suited for mutagenesis studies (Jacobs and DeMars, 1977). However, skin biopsies from older persons are also a satisfactory source of fibroblasts.

Culture Medium

The medium used for cell culture is Ham's F10 lacking hypoxanthine (HX) and supplemented with 15% fetal bovine serum and antibiotics. Medium lacking HX supports growth of cells as well as does medium containing this purine and is required for selection of cells resistant to purine analogues. Alternatively, Eagle's minimal essentials medium (MEM) with Earle's salts is substituted for Ham's F10, especially for culturing cells from patients with DNA repair defects which often do not grow as well as cells from normal individuals. Both media can be purchased in powdered form, e.g., from Grand Is. Biol. Co., Grand Is., N.Y., or K.C. Biologicals, Lenexa, Calif. Gentamycin is the antibiotic used for our stock cultures because of its stability at 37° pH 7.2. However, a mixture of penicillin (100 U/ml) and streptomycin (100 μg/ml) is substituted in the short-term experiments for economy.

Selective Media

To select cells resistant to 8-azaguanine (AG), Ham's F10 lacking HX and supplemented with 15% calf serum, antibiotics, and 20 μM AG is used; for selecting 6-thioguanine (TG) resistant cells, F10 lacking HX supplemented with 10% fetal bovine serum, antibiotics, and 40μM TG. Before choosing a suitable 100-liter lot of serum, a series of tests is carried out on a number of samples from different lots to find one which will support high cloning efficiency and growth of cells to high density, but allow the designated concentrations of purine analogues to prevent nonresistant cells from undergoing even a single population doubling. The concentrations of the selective agent, AG or TG, used in our experiments are 4 to 10 times higher than that required to reduce the colony-forming ability of the unirradiated population to background levels (cf. Fig. 1, Maher *et al.*, 1977, and Fig. 3, Maher and McCormick, 1980).

Trypsinization

In subculturing, great care is taken not to injure the cells either by overtrypsinization or by too vigorous pipetting. Cells are rinsed with phosphate-buffered saline (PBS), pH 7.2, and exposed to 0.25% trypsin at 37°C (1:250 trypsin, Grand Is. Biol. Co.) for as short a period of time as necessary to release their attachment (about 3 min). During the enzymatic digestion process the cells are viewed under an inverted microscope and the trypsin is removed at the first sign of cell detachment. Digestion by the remaining film of trypsin is allowed to continue for a short time and is then stopped by covering the cells with culture medium containing serum. Gentle pipetting yields a suspension of single cells. We routinely attain cloning efficiencies of greater than 50% with normal human fibroblasts and 15–35% for many of the DNA repair-deficient cell strains. Care is taken to maintain a pH of 7.2 to 7.4 in the medium by inspection and adjustment of the CO_2 level in the incubator at least once a week. We use a Fyrite Test Kit (Bacharach Instrument Co., Pittsburgh, Pa) routinely for this purpose since the incubator CO_2 gauge cannot be relied upon to give an accurate reading of the actual CO_2 levels.

Freezing Cell Cultures

Cells may be frozen in liquid nitrogen and stored until needed. For freezing, cells in exponential growth are trypsinized as described above, but the action of the film of trypsin is stopped by adding 2 ml of a freezing mixture (1 part DMSO, 9 parts media containing 15% serum and 50 μg/ml of gentamycin) per 250 ml flask. The cells are gently pipetted in this mixture to create a suspension of single cells and the mixture placed in glass or plastic freezing vials in an ice bath. The vials are sealed and transferred to a styrofoam box which is designed to permit a cooling rate of 1° per min when placed in a −80°C freezer. After 1.5–2 hours, the vials are transferred to a liquid nitrogen freezer.

Mutagenization of Cell Cultures

Ultraviolet Radiation

The medium is removed from cells attached to the surface of dishes and the cells rinsed with PBS and irradiated with the cover off, using a UVS-54 lamp (Ultraviolet Products, San Gabriel, California). To avoid a shadowing effect caused by the edge of the dish, the lamp is placed as far as possible above the dish and an incident dose of 0.1 or 0.2 J/m² per sec, as determined by a model IL 570 International Light radiometer (Newburyport, Massachusetts), is used. A wooden stand fitted with grooves is used to center each sized dish on a turntable directly under the lamp and dishes are rotated during irradiation to equalize exposure. If cells are to be released from the dish following irradiation and replated at lower densities, rather than being allowed to form clones *in situ*, the cells located around the circumference of the dish are removed using a sterile rubber policeman prior to irradiation to avoid the contribution of cells which, because of their location, did not actually receive the designated dose.

Chemicals

Carcinogens or other materials to be tested for mutagenicity are stored at −20°C and weighed as needed on a Cahn electrobalance (sensitivity and readability of 0.1 µg, Cahn Instruments, Cerritos, Calif.) under gold fluorescent light (500–700 nm). Just prior to use, the culture medium on attached cells is exchanged for serum-free medium and the chemical is dissolved in the appropriate solvent and introduced by mi-cropipet (final concentration of organic solvent < 0.5%). At the end of the exposure, medium containing the chemical is removed by aspiration into a special vacuum flask equipped with a trap and filled with an iodine solution (Wescodyne). Waste liquid containing carcinogen is treated as hazardous and disposed of by a commercial disposal company. The cells are refed with culture medium containing 15% fetal bovine serum. Since proteins possess strong nucleophilic centers, the serum can be expected to inactivate remaining traces of electrophilic reactants. All operations with chemical carcinogens are carried out in type 2, class B safety cabinet "hood" (Baker NCB, Baker Co., Sanford, Maine). Disposable gloves are worn and, in the case of volatile compounds, aspirating is carried out by hand with the aid of automatic pipetters to avoid formation of aerosols.

Comparative Cytotoxicity Studies

In situ *Technique*

Cells are plated at cloning densities (i.e., 100 to 6000 cells/60 mm diameter dish), irradiated or treated with chemicals as described, refed with freshly prepared culture medium, and allowed to form macroscopic colonies (about 14 days). Cells are refed once or twice during this period. When agents are to be compared for their cytotoxic effect on a series of cell strains, this is done simultaneously in a single experiment using a single stock concentration of carcinogen or a single setting of the source of UV radiation. This is especially important when comparing the effect of chemicals used at very low concentrations or which are unstable upon storage (Maher *et al.*, 1975). The cloning efficiency of the treated cells divided by the cloning efficiency of the untreated control cells (which receive solvent only) determines the cytotoxicity of the agent and is expressed as a percentage.

Replating Technique

A monolayer of exponentially growing cells or a confluent culture is exposed to radiation or chemicals, as described. Immediately following the treatment or at designated times, the cells are trypsinized, pooled, counted with an electronic counter (Coulter Corporation, Hialeah, Fl.) and a hemocytometer, diluted appropriately, plated at cloning densities, and allowed to form colonies as above.

Examples of Results with These Methods

The cytotoxic effects of a series of reactive derivatives of aromatic amides and polycyclic hydrocarbons in diploid human fibroblasts with different repair capabilities are compared in Figures 24.1 and 24.2. The XP2BE strain excises DNA lesions caused by these chemicals at a rate less than 50% that of the normal cells; XP12BE are virtually incapable of excision repair of such

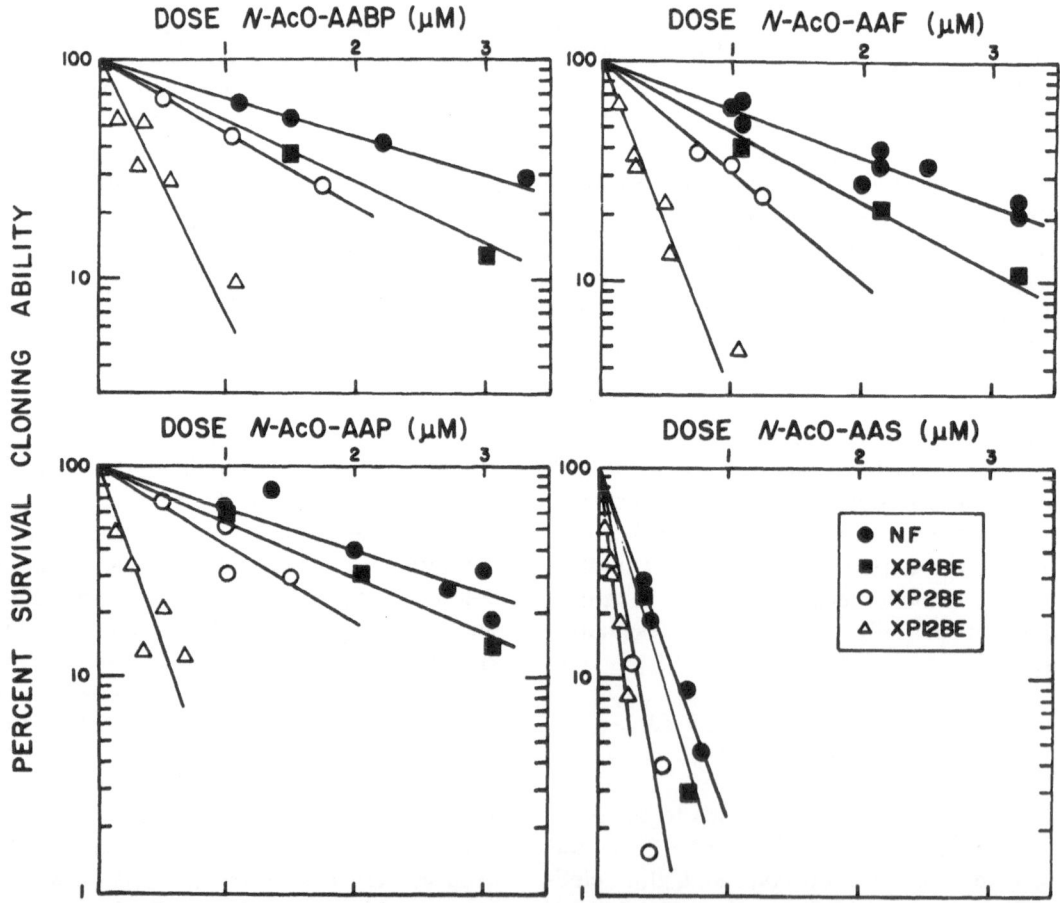

Figure 24.1. Comparison of the cytotoxic effect of reactive derivatives of four aromatic amine carcinogens, viz., N-acetoxy-4-acetylaminobiphenyl(N-AcO-AABP), N-acetoxy-2-acetylaminofluorene (N-AcO-AAF), N-acetoxy-2-acetylaminophenanthrene (N-AcO-AAP), and N-acetoxy-4-acetylaminostilbene (N-AcO-AAS), in normal (NF) and repair-deficient (XP) human fibroblasts. The excision capacity of XP2BE and XP12BE are described in the text. Cells from an XP-variant patient (XP4BE) excise DNA damage at a normal or near-normal rate, but are defective in replication of DNA from damaged templates (Lehmann *et al.*, 1975; Maher *et al.*, 1975, 1976a; McCaw *et al.*, 1978). (From Maher *et al.*, 1975.)

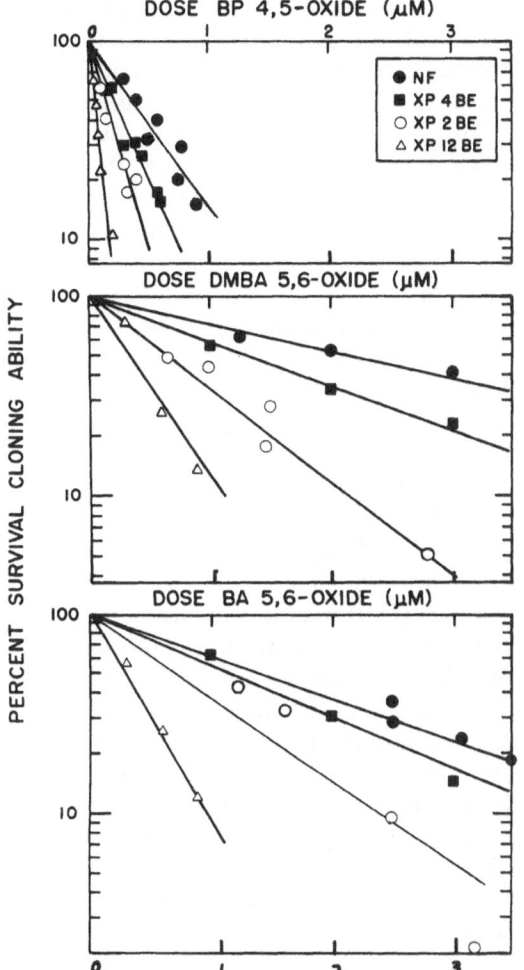

Figure 24.2. Comparison of the cytotoxic effect of reactive derivatives of three polycyclic aromatic hydrocarbons, viz. the K-region epoxides of benzo(a)pyrene, 7,12-dimethylbenz(a)-anthracene, and benz(a)anthracene in the four strains described in Figure 24.1.

al., 1975), or excision of thymine-containing dimers (Konze-Thomas *et al.*, 1979). The similarities in the ratios of the slopes of the respective survival curves indicate that the DNA lesions caused by exposure to these agents may be handled in a similar manner by these strains. Data of Heflich *et al.* (1977) show that one common feature is the degree of helix distortion which results from the binding of these DNA adducts.

Comparative Induced Mutation Frequency Studies

Mutagenicity Selection in situ

This term refers to procedures in which, for each experimental point, a large population of cells ($1-2.4 \times 10^6$) is plated into hundreds of individual dishes at low densities, exposed to an agent as described above, allowed time for the surviving cells to carry out 3.3 to 4 population doublings (5- to 8-day expression period) and then ex-

damage (Heflich *et al.*, 1980; Maher *et al.*, 1975; McCaw *et al.*, 1978; Yang *et al.* 1980). The cytotoxic effect of ultraviolet radiation is normal and XP strains, shown in Figure 24.3 for comparison, also reflects the excision repair capabilities of the three strains as measured by unscheduled DNA synthesis (Petinga *et al.*, 1977; Robbins *et al.*, 1974), repair replication (Heflich *et al.*, 1980; Konze-Thomas *et al.*, 1979), loss of endonuclease sensitive sites (Lehmann *et*

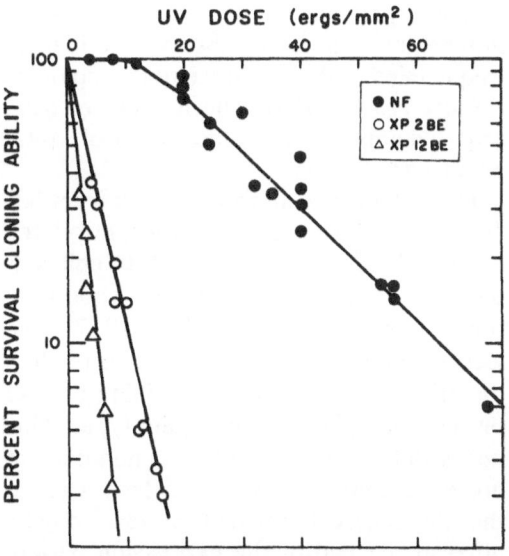

Figure 24.3. Comparison of the cytotoxic effect of UV radiation in normal and excision repair-deficient XP cells.

posed to a particular selective agent, e.g., 20 μM AG, *directly* in the original dishes. (An alternative to treating cells in individual dishes is to expose a large number of cells in a few dishes or flasks and then immediately plate them at lower densities for expression and then selection.) During the expression period the surviving cells *in situ* give rise to loose clusters of progeny cells. If a mutation inactivates the gene coding for hypoxanthine (guanine) phosphoribosyltransferase (HPRT, E.C. 2.4.2.8.), the number of HPRT molecules in the progeny cells will be diluted out with time as a result of cell division and half-life of the protein and its mRNA. Eventually the number will drop to a level where the progeny cells are unable to metabolize AG or TG and can be selected for by these guanine analogues. For *in situ* selection, AG is preferable to TG because under our experimental conditions it stops cell multiplication immediately, lessening the amount of metabolic cooperation (Subak-Sharpe *et al.*, 1969) and causes the nonresistant cells to lift off the dishes. This is a decided advantage for selection *in situ*, since loose clones of nonresistant cells have formed in the dishes at the beginning of selection. Our data indicate that at least 3.5 population doublings by the target cells must occur before selection begins in order to observe the highest frequencies of AG-resistant colonies (Maher *et al.*, 1979). The number of days required depends upon the extent of DNA damage (5–8 days).

Diploid human fibroblasts are very efficient at transferring the phosphorylated nucleoside of AG or TG by cell-to-cell contact into HPRT cells (metabolic cooperation) (Corsaro and Migeon, 1977; van Zeeland and Simons, 1976), which renders the HPRT cells sensitive to killing by AG or TG in spite of their genotype. This necessitates the use of large numbers of dishes to obtain sufficient surface area so that the cells will not be too close to each other at the end of the expression period. Thus, for the *in situ* technique, 500 to 1000 dishes of 60 mm diameter are commonly

used for a mutagenesis experiment involving a control and two or three experimental points. Selection with AG continues for about 21 days and requires refeeding every 2–3 days to circumvent enzymatic breakdown (van Zeeland and Simons, 1975). Although the *in situ* selection system with AG is laborious, it eliminates the need to trypsinize, pool, and replate the population of surviving target cells during the expression period or at the beginning of selection, and therefore avoids possible differences in cell doubling times between HPRT$^-$ cells and HPRT$^+$ cells in the population, which would affect the apparent mutation frequencies. This value is derived directly from the number of mutation events *per surviving cell originally plated*, determined from the number of dishes containing no colonies (P(0) method), using the Poisson distribution function to determine the chance of a mutant event per dish, and combining this with the number of clonable cells present per dish (Jacobs and DeMars, 1977). When combined with reconstruction studies (see below), this *in situ* selection technique scored by the P(0) method yields an accurate estimate of the frequency of mutations induced *per survivor*.

Mutagenicity Selection by Replating

A population of $1–5 \times 10^6$ exponentially growing cells is exposed to a concentration of mutagen and then plated into 150-mm-diameter dishes at a density calculated to give $2.5 – 4 \times 10^5$ *surviving* cells/dish (and a combined total of at least 10^6 survivors) and allowed to begin expression. Alternatively, cells are plated at these densities, allowed to attach, and then treated. The surviving cells are allowed to undergo two to three population doublings in these dishes and kept in rapid growth during expression by additional replating. On each of several days during expression, cells are pooled and $1–10 \times 10^6$ cells plated into dishes for selection in TG medium at a density of about 500 cells/cm². Since TG is stable in medium containing serum, only one or two refeedings are necessary during the 16–20 days of selection.

Reconstruction Studies

A known number of HPRT⁻ cells derived from patients with Lesch Nyhan (LN) syndrome (American Type Culture Collection or the Medical Research Institute, Camden, N.J.) are seeded into dishes containing cells at the same density as in the experimental dishes and exposed to the selective medium. The cloning efficiency of the LN cells in the presence of the HPRT⁺ cells, divided by their cloning efficiency alone, determines the efficiency of recovery of resistant colonies and is used to correct the observed mutation frequencies. Under the conditions described above, recovery ranges from 40 to 100% for AG and 80 to 100% for TG selection.

Examples of Results with These Methods

Figure 24.4 compares the frequency of mutations to AG resistance induced by benzo(a)pyrene 4,5-oxide (B(a)P 4,5-oxide) in normal cells and XP2BE cells as a function of dose. These data, obtained using the *in situ* selection technique, indicate that at low doses the normal cells which can excise the DNA damage much more rapidly are significantly less susceptible to mutations than are the excision-repair-defective cells derived from the XP patient. The cytotoxic effect of B(a)P 4,5-oxide in these two strains is also shown for comparative purposes. Low concentrations which cause killing in the repair-deficient cells but not in the normally repairing cells also cause mutations in the XP cells but not in the normal cells. The data in Figure 24.4 indicate that the excision repair process itself, which is being carried out by the normal cells, does

not introduce a measurable frequency of mutations. Instead, the process appears to protect cells from lesions which would lead to mutations if left unexcised.

The relationship seen in Figure 24.4 in which excision-repair-deficient XP cells are significantly more susceptible than normal human cells to the cytotoxic and mutagenic effect of B(a)P 4,5-oxide has also been observed for cells exposed to a series of agents, such as UV radiation (Maher *et al.*, 1979), N-acetoxy-2-acetylaminofluorene (Maher and McCormick, 1976; Maher *et al.*, 1975), 7,12-dimethylbenz(a)anthracene 5,6-oxide (Maher *et al.*, 1976a, 1977), dibenz(a,h)-anthracene 5,6-oxide (Maher *et al.*, 1976a, 1977), and (±)−7β,8α-dihydroxy-9α, 10α-epoxy-7,8,9,10-tetrahydrobenzo(a)pyrene (Yang *et al.*, 1980). This suggests that the excision repair processes which protect normal cells from potentially cytotoxic damage also act on potentially mutagenic damage and that

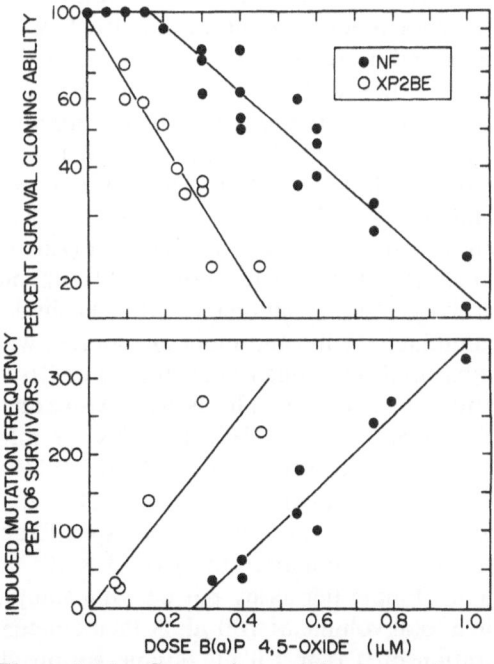

Figure 24.4. Comparison of the cytotoxic and mutagenic effect of B(a)P 4,5-oxide in normal NF and XP2BE cells using the *in situ* cytotoxicity and AG mutagenicity selection techniques described in the text.

the repair processes for these different agents may share one or more common steps which are defective or limiting in the XP cell strains. Biochemical data in support of this conclusion has recently been obtained (Levinson *et al.*, 1980; Maher *et al.*, 1980; Yang *et al.*, 1980).

Characterization of Drug-Resistant Colonies

Assay of Resistant Colonies for Levels of HPRT Activity

Resistant colonies randomly isolated at the end of selection period were dislodged with trypsin, suspended in culture medium, and seeded into 30-ml flasks and into 60-mm dishes, one of which contained a sterile cover slip. To avoid culturing a minor fraction of HPRT$^+$ cells which might be transferred along with the AG- or TG-resistant colony, cells in the flask were cultured in selective medium. Crude cell extracts were prepared by removing the medium, rinsing three times with 5 ml ice-cold PBS, and adding 1 ml of the following lysing mixture: 10 mM MgCl$_2$, 30 mM KCl, 0.1 mM dithiothreitol, 0.5% Triton X-100 (vol/vol), and 10 mM Tris-HCl at pH 7.4. The amount of protein was determined (Lowry *et al.*, 1951) and the level of HPRT enzyme activity, i.e., conversion of (^{3}H)-hypoxanthine into inosine monophosphate, was measured according to the method of Olsen and Milman (1974). The assay mixture contained 50 mM Tris HCl (pH 7.4), 9 mM MgCl$_2$, 2 mM dithiothreitol, 1 mM PRPP, 0.06 mM unlabeled HX (8-^{3}H)-hypoxanthine (New England Nuclear, specific activity 11 Ci/mmole) to give 3×10^4 to 3×10^5 cpm per assay per enzyme sample in a total volume of 100 μl. A final concentration of 0.1 or 1.0 ml bovine serum albumin was included to improve the stability of the enzyme fractions. The reaction mixture was incubated for 15 min at 37°C, and the incubation terminated by adding 5 ml of cold 50 mM Tris-HCl, pH 8, and placing

the mixture on ice. A 200 μl aliquot was removed and added to a Pasteur pipet column (3.5 × 0.5 cm) of DEAE-cellulose (Bio-Rad Cellex D, capacity 0.6 mEq per g) equilibrated with 50 mM Tris-HCl, pH 8, and washed with 5 ml of the same buffer to remove unreacted hypoxanthine. The nucleotide was eluted with 1.5 ml of 1 N HCl and counted in 10 ml of scintillation fluid and the activity compared with the activity of the HPRT in a crude extract of normal (HPRT)$^+$ human fibroblasts. (One unit of HPRT is defined as that amount of enzyme that catalyzes the conversion of 1 nmole HX into inosine monophosphate in 15 min.) The results are shown in Table 24.1.

Measurement of Hypoxanthine Incorporation into Macromolecules

Cells derived from the same resistant colonies used for the HPRT assay were grown on coverslips for several days in the presence of selective medium. The coverslips were rinsed several times in serum-free medium and transferred for 24 hr to F10 medium lacking HX but supplemented with 5 μCi/ml tritiated HX (New England Nuclear, generally labeled, 11 Ci/mmole). The coverslips were washed three times in buffered saline, fixed in Carnoy's, washed in 20% ethanol, then 10% ethanol, air-dried, and mounted on slides to be processed for autoradiography. None of the cells derived from 10 AG- or 4 TG-resistant colonies showed evidence of ability to incorporate HX (Maher and McCormick, 1980).

Determination of Maintenance of Resistant Phenotype in the Absence of Selection

Cells from these same resistant colonies plated into nonselective culture medium were allowed to grow for > 10 generations (i.e., increase 1000-fold) and then plated at cloning densities into (a) HAT medium, i.e., Ham's F10 which had been prepared

Table 24.1. *HPRT Activity of Randomly Isolated AG- or TG-Resistant Clones*

Code	Relative HPRT activity as percent of HPRT+NF strain	Code	Relative HPRT activity as percent of HPRT+ NF strain
NF-812	100.0	AG-8	5.8
AG-1	1.4	AG-9	13.9
AG-2	3.17	AG-10	7.7
AG-3	6.4	TG-1	4.7
AG-4	6.5	TG-2	4.7
AG-5	14.3	TG-3	0.0
AG-6	18.3	TG-4	0.0
AG-7	1.6	LN1112	0.0

without HX and thymidine but now supplemented with 30 μM HX, 0.1 μM aminopterin, and 30 μM thymidine and 10% FCS; (b) nonselective culture medium, i.e., F10 lacking HX but supplemented with 10% FCS, and (c) AG- or TG-selective medium. Growth in HAT medium requires a cell to possess HPRT activity (or alternatively, to be resistant to aminopterin). Growth in AG- or TG-selective medium requires a cell to lack HPRT activity or to have altered uptake of these purines. After three weeks, the dishes were stained and the number of clones in each set determined. The results showed that cells from the majority (9/11) of the AG- or TG-resistant colonies which grew in culture medium were able to grow in selective medium but not in HAT medium.

Conclusion

In summary, human skin fibroblasts in culture are valuable for studying the mechanisms of mutagenesis (and carcinogenesis) as well as for use in short-term assays for mutagenic potential of chemical carcinogens. Furthermore, the presence of DNA repair-deficient strains makes it possible to demonstrate the interaction of many agents with DNA since such strains show greater cytotoxicity and mutagenicity than cells from normal individuals when treated with comparable doses of agents which are repaired by the enzymatic repair

processes defective in XP cells. In the past, little research was carried out with diploid human fibroblasts because of the difficulties in culturing such cells. However, improved culture methods now give much higher cloning efficiencies with cells from normal individuals. Mutational studies using AG or TG resistance as the genetic marker are still quite complex, since the large size of human cells necessitates plating cells at a very low density to prevent loss of mutants through metabolic cooperation. Similarly, use of ouabain resistance as a marker presents some difficulty because the spontaneous and induced frequencies are 10- to 50-fold lower than those observed for AG resistance (Buchwald, 1977; Mankowitz *et al.*, 1974), and we found that diphtheria toxin can give irreproducible background frequencies (Chasin, 1979). Research is under way in a number of laboratories to improve methods used to quantitate mutagenesis and to develop mutational markers without these disadvantages. In the future, diploid human cells should prove as versatile for mutagenesis studies as are established animal cell lines.

Acknowledgments

We thank our colleagues, Drs. A.E. Aust, R.H. Heflich, B. Konze-Thomas, J.W. Levinson, and L.L.Yang, who carried out many of the experiments described in this review, as well as R. Corner, K.J. Falahee, R.M. Hazard, T. Kinney, and A.L. Mendrala, without whose excellent technical assistance this research would not have been possible. These investigations were supported by DHEW Grants CA-21247 and CA-21253 from the National Cancer Institute, and Contract ER-78-S-02-4659 from the Department of Energy.

References

Albertini, R.J., DeMars, R. (1973): Somatic cell mutation. Detection and quantification of x-ray induced mutation in cultured, diploid human fibroblasts. Mutat. Res. *18*:199–224.

Aust, A.E., Falahee, K.J., Maher, V.M., Mc-Cormick, J.J. (1980): Human cell-mediated benzo(a)pyrene cytotoxicity and mutagenicity in human diploid fibroblasts. Cancer Res. *40*:4070–4075.

Buchwald, M. (1977): Mutagenesis at the ouabain-resistance locus in human diploid fibroblasts. Mutat. Res. *44*:401–411.

Chasin, L.A. (Chairperson) (1979): Roundtable: Quantitative Mutational Systems—Evidence for Genetic Events. Banbury Report 2, Hsie, A.W., O'Neill, J.P., McElheny, V.K. (eds.), New York: Cold Spring Harbor Laboratory, pp. 99–121.

Corsaro, C.M., Migeon, B.R. (1977): Comparison of contact-mediated communication in normal and transformed human cells in culture. Proc. Natl. Acad. Sci. USA *74*: 4476–4480.

Cox, R., Masson, W.K. (1976): X-ray induced mutation to 6-thioguanine resistance in cultured human diploid fibroblasts. Mutat. Res. *37*:125–136.

Fujiwara, Y., Tatsumi, M., Sasaki, M.S. (1977): Cross-link repair in human cells and its possible defect in Fanconi's anemia cells. J. Mol. Biol. *113*:635–649.

German, J. (1972): Genes which increase chromosomal instability in somatic cells and predispose to cancer. Progr. Med. Genet. *8*: 61–101.

Glover, T.E., Chang, C.C., Trosko, J.E. (1979): Ultraviolet light induction of diphtheria toxin-resistant mutants in normal and xeroderma pigmentosum human fibroblasts. Proc. Natl. Acad. Sci. USA *76*:3982–3986.

Heflich, R.H., Dorney, D.J., Maher, V.M., Mc-Cormick, J.J. (1977): Reactive derivatives of benzo(a)pyrene and 7,12-dimethylbenz(a)anthracene cause S1 nuclease sensitive sites in DNA and 'UV-like repair. Biochem. Biophys. Res. Commun. *77*:634–641.

Heflich, R.H., Hazard, R.M., Lommel, L., Scribner, L.D., Maher, V.M., McCormick, J.J. (1980): A comparison of the DNA biding, cytotoxicity and repair synthesis induced in human fibroblasts by reactive derivatives of aromatic amide carcinogens. Chem.-Biol. Interact. *29*:43–56.

Huberman, E. (1975): Mammalian cell transformation and cell-mediated mutagenesis by carcinogenic polycyclic hydrocarbons. Mutat. Res. *29*:285–291.

Ikenaga, M., Takebe, H., Ishii, Y. (1977): Excision repair of DNA base damage in human cells treated with the chemical carcinogen 4-nitroquinoline 1-oxide. Mutat. Res. *43*: 415–427.

Jacobs, L., DeMars, R. (1977): Chemical mutagenesis with diploid human fibroblasts. In: B.J. Kilbey, M. Legator, W. Nichols, C. Ramel (eds.), Handbook of Mutagenicity Test Procedures. Amsterdam: Elsevier/North-Holland Biomedical Press, pp. 193–220.

Jacobs, L., DeMars, R. (1978): Quantification of chemical mutagenesis in diploid human fibroblasts: induction of azaguanine-resistant mutants by N-methyl-N'-nitro-N-nitrosoguanidine. Mutat. Res. *53*:29–53.

Konze-Thomas, B., Levinson, J.W., Maher, V.M., McCormick, J.J. (1979): Correlation among the rates of dimer excision, DNA repair replication and recovery of human cells from potentially lethal damage induced by ultraviolet radiation. Biophys. J. *28*:315–326.

Lehmann, A.R., Kirk-Bell, S., Arlett, C.F., Paterson, M.C., Lohmann, P.H.M., DeWeerd-Kastelein, E.A., Bootsma, D. (1975): Xeroderma pigmentosum cells with normal levels of excision repair have a defect in DNA synthesis after UV irradiation. Proc. Natl. Acad. Sci. USA *72*:219–223.

Lowry, O.H., Rosenbrough, N.J., Farr, A.L., Randall, R.J. (1951): Protein measurement with the folin phenol reagent. J. Biol. Chem. *193*:265–275.

Maher, V.M., McCormick, J.J. (1976): Effect of DNA repair on the cytotoxicity and mutagenicity of UV irradiation and of chemical carcinogens in normal and xeroderma pigmentosum cells. In: J.M. Yuhas, R.W. Tennant,

J.B. Regan (eds.), Biology of Radiation Carcinogenesis, New York: Raven Press, pp. 129–145.

Maher, V.M., McCormick, J.J. (1980): Comparison of the mutagenic effect of ultraviolet radiation and chemicals in normal and DNA repair deficient human cells. In: F.J. de Serres, A. Hollaender (eds.), Chemical Mutagens: Principles and Methods for Their Detection, Vol. 6, New York: Plenum Press, pp. 309–329.

Maher, V.M., Wessel, J.E. (1975): Mutations to azaguanine resistance induced in cultured diploid human fibroblasts by the carcinogen, N-acetoxy-2-acetylaminofluorene. Mutat. Res. 28:277–284.

Maher, V.M., Birch N., Otto, J.R., McCormick, J.J. (1975): Cytotoxicity of carcinogenic aromatic amides in normal and xeroderma pigmentosum fibroblasts with different DNA repair capabilities. J. Natl. Cancer Inst. 54:1287–1294.

Maher, V.M., Curren, R.D., Ouellette, L.M., McCormick, J.J. (1976a): Role of DNA repair in the cytotoxic and mutagenic action of physical and chemical carcinogens. In: F.J. de Serres, J.R. Fouts, J.R. Bend, R.N. Philpot (eds.), In Vitro Metabolic Activation in Mutagenesis Testing, Amsterdam: Elsevier/North-Holland Biomedical Press, pp. 313–336.

Maher, V.M., Ouellette, L.M., Curren R.D., McCormick, J.J. (1976b): Caffeine enhancement of the cytotoxic and mutagenic effect of ultraviolet irradiation in a xeroderma pigmentosum variant strain of human cells. Biochem. Biophys. Res. Commun. 71:228–234.

Maher, V.M., Ouellette, L.M., Curren, R.D., McCormick, J.J. (1976c): Frequency of ultraviolet light-induced mutations is higher in xeroderma pigmentosum variant cells than in normal human cells. Nature (Lond.) 261:593–595.

Maher, V.M., McCormick, J.J., Grover, P.L., Sims, P. (1977): Effect of DNA repair on the cytotoxicity and mutagenicity of polycyclic hydrocarbon derivatives in normal and xeroderma pigmentosum human fibroblasts. Mutat. Res. 43:117–138.

Maher, V.M., Dorney, D.J., Mendrala, A.L., Konze-Thomas, B., McCormick, J.J. (1979): DNA excision-repair processes in human cells can eliminate the cytotoxic and mutagenic consequences of ultraviolet irradiation. Mutat. Res. 62:311–323.

Maher, V.M., Aust, A.E., Yang, L.L., Zator, R.M., McCormick, J.J., Harvey, R.G. (1980): The anti-7,8-diol-9,10-epoxide of benzo(a)pyrene is not a more potent or efficient mutagen in human cells than the K-region epoxide. Carcinogenesis (submitted for publication).

Mankovitz, R., Buchwald, M., Baker, R.M. (1974): Isolation of ouabain-resistant human diploid fibroblasts. Cell 3:221–226.

McCalla, D.R., Arlett, C.F., Broughton, B. (1978): The action of AF2 on cultured hamster and human cells under aerobic and hypoxic conditions. Chem.-Biol. Interact. 21:89–102.

McCaw, B.A., Dipple, A., Young, S., Roberts, J.J. (1978): Excision of hydrocarbon-DNA adducts and consequent cell survival in normal and repair defective human cells. Chem.-Biol. Interact. 22:139–151.

McCormick, J.J. (1979): Studies of the effects of diesel particulate on normal and xeroderma pigmentosum cells. In: EPA Symposium on Health Effects of Diesel Engine Emissions, Cincinnati, Ohio, Sponsored by Health Effects Research Laboratory, U.S. Environmental Protection Agency.

Newbold, R.F., Wigley, C.B., Thompson, M.H., Brookes, P. (1977): Cell-mediated mutagenesis in cultured Chinese hamster cells by carcinogenic polycyclic hydrocarbons: nature and extent of the associated hydrocarbon-DNA reaction. Mutat. Res. 43:101–116.

Olsen, A.S., Milman, G. (1974): Chinese hamster hypoxanthine-guanine phosphoribosyltransferase. Purification, structural, and catalytic properties. J. Biol. Chem. 249:4030–4037.

Petinga, R.A., Andrews, A.D., Tarone, R.E., Robbins, J.H. (1977): Typical xeroderma pigmentosum complementation group A fibroblasts have detectable ultraviolet light-induced unscheduled DNA synthesis. Biochim. Biophys. Acta. 479:400–410.

Robbins, J.H., Kraemer, K.H., Lutzner, M.A., Festoff, B.W., Coon, H.G. (1974): Xeroderma pigmentosum. An inherited disease with sun sensitivity, multiple cutaneous neoplasms, and abnormal DNA repair. Ann. Intern. Med. 80:221–248.

Sato, K., Hieda, N. (1979): Isolation of a mammalian cell sensitive to 4-nitroquinoline-1-oxide. Int. J. Radiation Biol. 35:83–87.

Schultz, R., Trosko, J.E., Chang, C.C. (1980): Mutagen sensitive and DNA repair mutants of Chinese hamster fibroblasts. Environ. Mutagenesis.

Subak-Sharpe, H., Bürk, R.R., Pitts, J.D. (1969): Metabolic co-operation between biochemically marked mammalian cells in tissue culture. J. Cell Sci. *4*:353–367.

Taylor, A.M.R., Harnden, D.G., Arlett, C.F., Harcourt, S.A., Lehmann, A.R., Stevens, S., Bridges, B.A. (1975): Ataxia telangiectasia: a human mutation with abnormal radiation sensitivity. Nature (Lond.) *258*:427–429.

Thompson, L.H., Rubin, J.S., Cleaver J.E., Whitmore, G.F., Brookman K.B. (1980): A screening method for isolating DNA repair deficient mutants of CHO cells. Somatic Cell Genetics *6*:391–405.

Yang, L.L., Maher, V.M., McCormick, J.J. (1980): Error-free excision of the cytotoxic and mutagenic N^2-guanosine DNA adduct formed in human fibroblasts by (±)-7β, 8α-dihydroxy-9α, 10α-epoxy-7, 8, 9, 10-tetrahydrobenzo(a)pyrene. Proc. Natl. Acad. Sci. USA. *77*:5933–5937.

van Zeeland A.A., Simons, J.W.I.M. (1975): The effect of calf serum on the toxicity of 8-azaguanine. Mutat. Res. *27*:135–138.

van Zeeland A.A., Simons, J.W.I.M. (1976): The use of correction factors in the determination of mutant frequencies in populations of human skin fibroblasts. Mutat Res. *34*:149–158.

Abbreviations

NF, (normal fibroblasts) i.e., diploid fibroblasts derived from normal individuals; XP, xeroderma pigmentosum; AG, 8-azaguanine; TG, 6-thioguanine; HX, hypoxanthine; HPRT, hypoxanthine(guanine) phosphoribosyltransferase; PBS, phosphate buffered saline, pH 7.2.

25

Liver Culture Indicators for the Detection of Chemical Carcinogens

GARY M. WILLIAMS

Introduction

In any test system under consideration for inclusion in a battery of short-term tests for carcinogens, the two principal components that determine the usefulness of the test are the metabolizing system and the endpoint of the test (Williams, 1980a). Tests involving intact mammalian cells offer potential advantages in both components, by providing intracellular metabolism of chemicals within the target cell and reliable endpoints whose biological significance is clear. The development of liver-derived test systems has been pursued because liver possesses the broadest capability of any organ for metabolizing chemical carcinogens (Weisburger and Williams, 1975). We have suggested that the most useful endpoints in mammalian cells to supplement bacterial mutagenesis tests in a battery would be DNA repair, sister chromatid exchange, mutagenesis, and transformation (Weisburger and Williams, 1978; 1980b). Liver

culture systems utilizing three of these endpoints are now available.

The Hepatocyte Primary Culture/DNA Repair (HPC/DNA Repair) Test

Primary cultures of nonreplicating hepatocytes (Williams *et al.*, 1977) are used for the detection of carcinogens by their ability to induce DNA repair, measured autoradiographically as unscheduled DNA synthesis (Williams, 1976c), as recommended in other systems by Stich and associates (1977).

The approach used involves simultaneous exposure to the carcinogen and provision of ^{3}H-TdR for incorporation in regions of DNA repair that occur as a consequence of interaction of the test chemical or metabolites with DNA (Williams, 1977b). In addition, because of the evidence that metabolism was most active during the first 24 hr of culture (Williams, 1977b, Schmeltz *et al.*, 1978), exposure was commenced immediately after attachment of hepatocytes (Williams, 1978).

Thus, primary cultures of hepatocytes are allowed 1.5 hr for attachment and are then exposed to the test chemicals and 10 μCi/ml ^{3}H-TdR overnight for 18 hr. Repair synthesis is quantified autoradiographically as unscheduled DNA synthesis. The autoradiographic silver grains in the emulsion are counted electronically by a Count-All Model 600 colony counter with microscopic attachment (Williams, 1978). The grains overlying the nucleus are counted and then, for background, three counts are taken over the cytoplasm using the aperture setting for the nucleus of that cell. The highest cytoplasmic count is subtracted from the nuclear count to obtain the net nuclear count resulting from repair.

Hepatocytes have an advantage over other cell types used for repair studies (San and Stich, 1975; Martin *et al.*, 1978) by being capable of performing metabolism of test compounds without the addition of a metabolizing system. The metabolism of carcinogens in intact hepatocytes differs from that in liver S9 preparations (Schmeltz *et al.*, 1978) and is probably more representative of *in vivo* metabolism (Williams, 1979b). Hepatocytes also have an advantage because they have a low level of replicative DNA synthesis and thus the measurement of repair synthesis is relatively easy; furthermore, the few cells undergoing replicative DNA synthesis (Laishes and Williams, 1976) can be recognized by their heavy nuclear labeling in autoradiographs and can be completely excluded from the results. The procedure described is technically and logistically simple. The biggest difficulty with this approach has been the variable background (Williams, 1978). The background is higher over cells than in areas of the emulsion where no cells are present and therefore, seems to be due to radioactivity retained intracellularly, which probably represents nonspecifically bound ^{3}H-TdR (Williams, 1980a). Regardless, the background does not preclude the measurement of repair and since it is slightly higher over the cytoplasm than the nucleus (Williams, 1978), subtraction of the cytoplasmic background from the nuclear grains actually produces a slight underestimate of repair so that results are not biased in favor of a positive result. A shortcoming of the test may be that it requires 2–3 weeks to complete.

To establish the validity of this test, attention has been focused on carcinogens requiring metabolic activation; four activation-independent agents were readily positive in this test (Table 25.1), as in most. For activation-dependent carcinogens, related compounds of known biological activity representing structurally different types of chemicals requiring different pathways of metabolic activation have been examined (Williams, 1976c; 1977b; 1978; 1980b). Of 28 known carcinogens of 7 different activation-dependent types, 27 were positive, while only benz(a)anthracene was negative (Table 25.1). The positive carcinogens included, for most types, carcinogens that do not produce liver tumors. Among 13 related noncarcinogens, 12 were negative whereas N-4-fluorenylacetamide was positive. This compound is also positive in the Ames test (McCann *et al.*, 1975). Thus, the test has demonstrated a substantial ability to detect reliably activation-dependent carcinogens and a low or absent (depending upon the status of N-4-fluorenylacetamide) incidence of false positive responses. In addition, this test may be of particular value for certain types of compounds. For example, carcinogenic pyrrolizidine alkaloids were positive, whereas noncarcinogenic flavonoids that have been mutagenic in bacterial systems were negative. Recently, we have demonstrated that the HPC/DNA repair test can be performed with mouse and hamster hepatocytes. This finding opens the possibility to study the effect of chemicals with species selectivity.

The Adult Rat Liver Epithelial Cell Line/Hypoxanthine-Guanine Phosphoribosyl Transferase (ARL/HGPRT) Mutagenesis Assay

Because of the requirement for cell replication in mutagenesis studies, long-term rat liver epithelial lines (Williams, 1976a) are used. The mutation studied is in the hypoxanthine-guanine phosphoribosyl transferase (HGPRT) gene, which is on the X chromosome and, therefore, has only one functional locus in each cell, making the gene highly mutable. Since HGPRT is required for the utilization of exogenous purines, HGPRT$^-$ mutants can be isolated by their resistance to purine analogues such as 8-azaguanine and 6-thioguanine that are toxic to HGPRT$^+$ cells that utilize them. The analogue-resistant cells isolated from rat liver epithelial lines retained resistance for months in the absence of the selective agent and were HGPRT-deficient (Williams, 1978; Tong and Williams, 1978), thereby satisfying the major criteria required for identifying these mutants in mammalian cells.

In the test system that we have developed, exposure of proliferating cultures is employed because of our data showing that replicating cultures are more sensitive to mutagens than quiescent cultures (Williams, 1976b; Berman et al., 1977). Incidentally, this phenomenon dictates that, in comparing the potency of two mutagens, it is critical that the exposed cultures be in the same growth phase.

In order to obtain the maximum number of mutants from mutagen exposure, the interval between exposure and mutant selection must be optimized. This interval, called the expression time or phenotypic lag, allows for replication of DNA and fixation of the mutations, but probably more importantly, reduction of the level of HGPRT present in cells at the time of mutation. For the rat liver lines, the expression time is about 10 days (Tong and Williams,

1978), and thus, 14 days are allowed in the test.

For the selection of HGPRT$^-$ mutants of rat liver epithelial cells, 8-azaguanine is less stringent than 6-thioguanine (Tong and Williams, 1978). Furthermore, rat liver fibroblasts, as a result of high guanase and 5'-nucleotidase, are resistant to 8-azaguanine even though they have adequate levels of HGPRT (Berman et al., 1977). Therefore, 6-thioguanine is the preferred selective agent.

Thus, the protocol that has been defined for the mutagenesis assay (Tong and Williams, 1979, 1980) consists of exposure of proliferating cultures to a test chemical for 3 days. After this, the mutagen is removed and the cells are subcultured at confluency, and as required over the next 14 days. After the expression time, cells are subcultured and inoculated at the densities for colony-forming efficiency and mutant selection. Twenty-four hours after the final inoculation, 10 μg/ml 6-thioguanine is added to the one set of flasks for mutant selection. The flasks are maintained for 10-14 days with refeeding and then colonies are stained with Giemsa and counted.

Mammalian mutagenesis assays may be useful supplements to bacterial mutagenesis assays by providing an endpoint of closer relevance to the human situation. The advantages of the ARL/HGPRT mutagenesis assay over other mammalian mutagenesis assays is that it provides intrinsic metabolic activation and thereby extends this parameter in batteries containing bacterial mutagenesis assays dependent upon an added mammalian enzyme preparation for activation.

Using protocols of this general pattern, activation-independent and activation-dependent carcinogens were mutagenic to the rat liver epithelial lines (Table 25.2). The mutagenicity of activation-dependent carcinogens supports previous studies on transformation (Williams et al., 1973) and DNA damage (Williams 1975), indicating that these lines possess activation capabili-

Table 25.1. Results in the HPC/DNA Repair Test

Type and related chemicals	Carcinogenicity[a]	HPC/DNA repair test [b]		
		Results	Grain/nucleus[a]	Dose (M)
Alkylating agents, activation-independent				
Methyl Methanesulfonate	+	+[d]	14.0 ± 6.0	6×10^{-5}
N-Methyl-N'-Nitro-N-Nitrosoguanidine	+	+[e]	8.8 ± 1.0	5×10^{-5}
Methylazoxymethanol Acetate	+	+[g]	17.5	10^{-2}
Propylenemine	+	+[i]	20.4 ± 2.3	1.2×10^{-3}
Mycotoxins				
Aflatoxin B_1	+	+[d,e]	35.0 ± 14.5	10^{-4}
Aflatoxin B_2	W	+[e]	16.8 ± 4.1	10^{-4}
Aflatoxin G_1	+	+[e]	19.7 ± 11.7	10^{-5}
Aflatoxin G_2	−	−[e]	0.0	10^{-5}
Aminoazo dyes				
3'-Methyl-4-Dimethylaminoazobenzene	+	+[e]	22.1 ± 3.2	10^{-4}
4'-Methyl-4-Dimethylaminoazobenzene	+	+[e]	11.4 ± 4.2	10^{-4}
2-Methyl-4-Dimethylaminoazobenzene	+	+[e]	7.7 ± 2.5	10^{-4}
4-Dimethylaminoazobenzene	+	+[g]	12.5 ± 5.1	10^{-4}
4-Aminoazobenzene	W	+[e]	25.1 ± 9.4	10^{-4}
Polycyclic aromatic hydrocarbons				
3-Methylcholanthrene	+	+[g]	11.2 ± 7.0	10^{-4}
7,12-Dimethylbenz(a)anthracene	+	+[e]	37.8 ± 7.9	10^{-3}
Benz(a)anthracene	W	−[e]	0	10^{-3}
Anthracene	−	−[e]	0	10^{-3}
Benzo(a)pyrene	+	+[g]	19.0 ± 13.7	10^{-4}
Benzo(e)pyrene	−	−[g]	0	10^{-4}
Pyrene	−	−[g]	0.5 ± 0.6	10^{-4}
Nitrosamines				
Dimethylnitrosamine	+ +	+[e,h]	39.6 ± 11.4	10^{-2}
Diethylnitrosamine	+	+[h]	37.7 ± 9.4	10^{-2}
Dimethylformamide	−	−[e,h]	0.5 ± 0.9	10^{-2}
Diphenylnitrosamine	−	−[h]	2.1 ± 0.5	10^{-3}
Nitrosomorpholine	+ +	+[h]	41.1 ± 7.6	10^{-2}
Nitrosopyrrolidine	+	+[h]	33.2 ± 9.6	10^{-2}
Nitrosocarbazole	−	−[h]	0.6 ± 0.4	10^{-4}
Nitrosonornicotine	+	−[h]	24.0 ± 3.9	10^{-2}

Compound	Carcinogenicity[a]	Result	Mean ± SD[c]	Concentration
4-(N-methyl-N-nitrosamino)-1-(3-pyridyl)-1-butanone	+	+[h]	52.6 ± 10.0	10^{-2}
Aromatic amines				
N-2-Fluorenylacetamide	+	+[e]	42.4 ± 4.9	10^{-3}
N-4-Fluorenylacetamide	?	+[e]	12.3 ± 1.2	10^{-3}
2-Aminofluorene	+	+[g]	28.1 ± 3.6	10^{-4}
9-Aminofluorene	−	−[g]	0.5	10^{-4}
3-Methyl-2-Naphthylamine	+	+[g]	11.3	10^{-5}
Biphenyl-4,4'-Diamine (Benzidine)	+	+[f]	27.8 ± 3.7	10^{-5}
2',3-Dimethylbiphenyl-4-Amine	+	+[f]	54.1 ± 19.3	10^{-5}
Biphenyl-4-Amine	+	+[f]	71.8 ± 36.1	10^{-6}
Biphenyl-2-Amine	−	−[f]	2.0 ± 1.9	10^{-5}
Biphenyl	−	−[f]	0.2 ± 0.3	10^{-4}
4,4'-methylene-bis-2-chloroaniline	+	+[i]	32.6 ± 1.7	4×10^{-4}
Aniline	−	−[g]	0.1 ± 0	10^{-3}
Aza-aromatics				
Quinoline	+	+[i]	15.1 ± 1.6	10^{-3}
Isoquinoline	−	−[i]	2.6 ± 2.0	10^{-3}
Pyrrolizidine alkaloids				
Lasiocarpine	+	+[i]	21.3 ± 11.5	10^{-5}
Petasitenine	+	+[i]	46.7 ± 7.0	10^{-3}

[a] + = carcinogenic; W = weakly carcinogenic; − = noncarcinogenic.
[b] All results represent the highest previously reported results or unpublished results obtained under the same conditions of testing for each type.
[c] Mean ± standard deviation of triplicate coverslips.
[d] Williams (1976c).
[e] Williams (1977b).
[f] Williams (1978).
[g] Williams (1980b).
[h] Williams and Laspia (1979).
[i] Williams (unpublished).

Table 25.2. *Results in ARL/HGPRT Mutagenesis Assays*

Type and related chemicals	Carcino-genicity[a]	ARL/HGPRT Assay		Hepatocyte-mediated ARL/HGPRT assay	
		Results	Dose (M)	Results	Dose (M)
Alkylating agents, activation-independent					
Methyl Methanesulfonate	+	+[b]	3×10^{-3}		
N-Methyl-N'-Nitro-N-Nitrosoguanidine	+	+[c]	5×10^{-5}		
Mycotoxins					
Aflatoxin B_1	+	+[b]	1×10^{-6}		
Aflatoxin G_2	−	−[b]	5×10^{-6}		
Aminoazo dyes					
3'-Methyl-4-Dimethylaminoazobenzene	+	+[c]	1×10^{-5}		
4-Aminoazobenzene	w	−	1×10^{-4}		
Polycyclic aromatic hydrocarbons					
7,12-Dimethylbenz(a)anthracene	+	+	1×10^{-6}	+[d]	1×10^{-7}
Benz(a)anthracene	−	−	1×10^{-5}		
Benzo(a)pyrene	+	+	1×10^{-5}		
Aromatic amines					
N-2-Fluorenylacetamide	+	+[b]	1×10^{-6}	+[d]	1×10^{-5}
N-4-Fluorenylacetamide	−	−	1×10^{-4}		
Nitrosamines					
Dimethylnitrosamine	+	−[e]	1×10^{-3}	+[d]	1×10^{-3}
Diethylnitrosamine	+	−[f]	1×10^{-3}		
Nitrosopyrrolidine	+	+[c]	1×10^{-4}		

[a]+ = carcinogenic; W = weakly carcinogenic; − = noncarcinogenic.

[b]Tong and Williams (1978).

[c]Tong and Williams (1980).

[d]San and Williams (1977).

[e]Assayed in lines 6,14,19 only.

[f]Assayed in line 6 only.

ty. However, preliminary studies revealed that two of the rat liver epithelial lines, although derived from the inbred Fischer strain, differed significantly in their mutagenic response to specific carcinogens (Williams, 1976b) and this finding has been confirmed and extended in current studies (Tong and Williams, 1979). Such differences undoubtedly reflect differences in metabolic activation capability (Williams, 1980a).

Although the rat liver epithelial lines have significant carcinogen metabolizing capability, the activity is certainly less than that of the primary cultures of hepatocytes. Therefore, a variation of the test was developed (San and Williams, 1977) using the feeder system approach (Huberman, 1976) to mediate mutagenesis of cocultivated epithelial lines. In addition to the extensive activation capability of hepatocytes, they also have the advantage in such a system that they cannot be subcultured during the first few days after introduction into culture and, consequently, are eliminated from mixtures with epithelial lines by subculture following mutagen exposure. In using the epithelial lines as target cells, the potential for activation in both feeder (i.e., hepatocytes) and target cells is obtained. In addition, since transfer of metabolites is facilitated by cell contacts (Kuroki and Drevon, 1978), this process might be more efficient in mixtures of liver epithelial cells with hepatocytes since the epithelial cells may associate more closely with hepatocytes than might other potential target cells.

In the hepatocyte-mediated mutagenesis studies, freshly isolated hepatocytes were mixed in a ratio of about 2 to 1 with liver epithelial cells before inoculation into culture flasks (San and Williams, 1977). After attachment, the two cell types were found to be well mixed and in close contact. Exposure of mixed cultures to several activation-dependent carcinogens that produce DNA repair in hepatocytes also produced mutagenesis on the cocultivated rat liver epithelial lines under conditions of exposure in which the carcinogens were not mutagenic to the lines in the absence of cocultivated hepatocytes (Table 25.2). These results correspond closely to those of Green et al. (1977), who used whole liver cells for activation in a bacterial fluctuation test. Although the ability of primary cultures of hepatocytes to generate activated metabolites was confirmed in this assay, the measurement of repair in hepatocytes is simpler and more direct, and, therefore, is to be preferred for screening. In addition, hepatocyte-mediated mutagenicity may be limited in sensitivity owing to the necessity for transfer of the activated metabolite from the hepatocytes to the target cell.

Adult Rat Liver Epithelial Cell Line/Transformation (ARL/Transformation) Assays

In 1973, we reported that several activation-dependent carcinogens produced malignant conversion of rat liver epithelial lines derived from newborn rats (Williams et al., 1973). The tumors that arose from these transformed lines gave rise to carcinomas upon transplantation, indicating that the lines were truly epithelial (Williams, 1976b). These studies were confirmed by Montesano et al. (1973), but this approach had two major problems: the transformation could not be quantified and the interval of exposure or maintenance postexposure before tumorigenicity was acquired seemed excessively long. Thus, efforts have been exerted to find markers that could quantify the transformation process in its early stages (Montesano et al., 1977; San et al., 1979a,b). Using the same ARL lines derived from adult rat liver that were used in the mutagenicity assay, we have completed a survey of 11 different properties for transformation (San et al., 1979a,b). These assays were classified as population markers, which simply score the presence of transformed cells in the population, and cell markers, which quantify the fraction of transformed cells in the population. Among population markers, high 2-deoxyglucose uptake and biochemical assay of gamma-glutamyl transpeptidase (GGT) activity were found to be useful (Table 25.3). Growth characteristics were also considered to be of some value since they could be monitored routinely during maintenance of the lines. While the two former markers were specific in that only tumorigenic lines possessed high levels, they were not sensitive markers. Among cell markers, growth in soft agar, cytochemical GGT activity, and colony formation in low-calcium medium correlated well with tumorigenicity (Table 25.3). At this time, we believe that the combined use of these three cell markers will provide a reliable and sensitive means of detecting transformation in rat liver epithelial cultures.

The colony-forming efficiency in soft agar was measured according to the method developed by Macpherson and Montagnier (1964), as modified by Kakunaga and Kamahora (1968), and further modified by us as regards the nutrient medium (San et al., 1979b). The base layer of agar medium was prepared by mixing an equal volume of

Table 25.3. *Markers for Transformation in Adult Rat Liver Epithelial Cell Cultures*[a]

Cell line	Cell markers			Population markers			
	CFE in soft agar	Relative CFE in low and normal calcium	Cytochemical GGT	Growth characteristics		2-DG uptake (cpm $\times 10^3$/ 10^6 cells)	Biochemical GGT (nm/min/mg)
				Population increase[b]	Generation time (days)		
Normal liver (nontumorigenic)							
ARL 15	0	9	0	1.7	2.7	2.9	0.7
ARL 18	0	0	0	1.2	2.1	5.4	1.8
T 51	0	1	0	0.9	0.8	3.9	3.7
Normal liver (tumorigenic)							
ARL 6	8	70	Focal	1.1	0.7	2.1	2.6
ARL 14	9	7	Focal	1.8	0.9	2.8	2.2
ARL 16	22	55	3+	2.8	0.8	3.5	9.0
ARL 17	40	29	3+	3.8	0.9	3.5	14.8
Hepatocarcinoma							
HTC	100	97	4+	2.3	0.9	23.2	63.6
RH 35	100	16	3+	10.3	0.7	6.6	8.4
ND-HCl	100	94	3+	2.0	0.7	2.4	17.4
MH_1C_1	13	14	4+	1.0	2.4	3.5	55.7

[a]From San *et al.* (1979b).

[b]Ratio of number of cells at 3 days after confluency to that at confluency.

stock agar and double concentrated Williams' medium E (WME) supplemented with 11% FBS to give final concentrations of 0.5% Noble agar, 0.2% Bacto-Pepton, 0.05% NaCl, and 0.01% Na_2HPO_4. Five ml of base agar was poured into 21 cm^2 Petri dishes and allowed to solidify. Cells from log phase growth cultures were suspended in WME 10% FBS, and mixed with agar medium to obtain the final concentration of cells in 0.3% agar. One ml of agar-cell suspension was placed on top of the solidified base layer. After hardening the top layer at room temperature, the cultures were incubated at 37° C with 5% CO_2 and humidity. Each dish was immediately examined in 16 fields 0.4 cm^2 in area on a reticle, and the number of cell clusters containing 8 or more cells was recorded as the Day 0 colony count. On Day 14, reticle counts of the number of colonies were again made. The net colony count was obtained by subtracting the Day 0 count from the Day 14 count and was used in computing colony-forming efficiency (Williams, 1976b; San et al., 1979a).

For the cytochemical assay of GGT activity, cells were seeded at densities of 5×10^4 or 5×10^5 onto 25-mm sterile coverslips in 8 cm^2 tissue culture dishes. The cultures were fixed at 4 days by immersion in acetone at 4° C for 2 hr, air-dried at room temperature, and mounted on slides for staining. GGT activity was demonstrated cytochemically by the procedure of Rutenberg et al. (1969). Cells were incubated for 30 min at room temperature in a freshly prepared and filtered solution containing gamma-glutamyl-4-methoxy-2-naphthylamide as substrate, glycylglycine as accepter, and Fast Blue BBN (diazotized-4'-amino-2', 5'-diethyoxybenzanilide as the coupling agent for the liberated 4-methoxy-2-naphthylamine. The bright red GGT reaction was scored visually and graded from 0 to 4 +.

To measure colony-forming efficiency in low-calcium medium, cells were seeded at 200 per 25-cm^2 culture flask in WME with ± 10% FBS for attachment. After 24

hours of incubation, the cells were switched to low-calcium medium supplemented with 5% dialyzed FBS. Control cultures were fed with the low-calcium medium supplemented with 5% dialyzed FBS and calcium chloride to provide a normal calcium concentration. Low-calcium medium contained calcium pantothenate as the major source of calcium and, when supplemented with 5% dialyzed FBS, had a final concentration of 0.035 mM calcium. Medium for the control cultures supplemented with 5% dialyzed FBS and calcium chloride had a final calcium concentration of 1.8 mM as in regular WME. The cultures were fixed in formalin 8 days after inoculation and stained with Giemsa. All colonies containing 32 or more cells were scored to obtain the colony-forming efficiency.

Studies are now under way to determine the inducibility of these markers in non-tumorigenic lines. A major advantage of the ARL system is that mutagenesis and transformation can be examined in the same culture. The mutagenesis assay has already been demonstrated to be sensitive, and it must now be determined whether the transformation assays offer any advantage for the detection of carcinogens.

Discussion

The DNA repair and mutagenesis systems using rat liver cells have demonstrated considerable reliability as indicators for the detection of chemical carcinogens. Thus, they deserve consideration for inclusion as part of a battery of rapid in vitro screening tests. Of particular value for this purpose is their capacity to provide the metabolic capability of intact cells in a battery including bacterial mutagenesis tests that are dependent upon enzyme preparations for carcinogen metabolism.

An important application of the results from a battery of in vitro tests is envisioned to be the mechanistic classification of chemical carcinogens (Williams, 1977a;

1980b; Weisburger and Williams, 1978). Chemical carcinogens are generally defined operationally by their ability to induce neoplasms. To distinguish between different modes of action, we have proposed a mechanistic classification of chemical carcinogens (Williams, 1977a; 1979a; 1980b), in which the term genotoxic, recommended by Brookes *et al.* (1973) as a general expression to cover toxic, lethal, and heritable effects of carcinogens to genetic material, would be used to designate the category of carcinogens that damage nuclear DNA. This category would, therefore, include carcinogens that operate as electrophilic reactants as well as others that alter the genome through direct interactions. Carcinogens not displaying genotoxic properties would be provisionally categorized as epigenetic until their mechanism of action is elucidated. The major classes of carcinogens (Weisburger and Williams, 1980b) have been assigned to these two categories (Williams, 1979a, 1980b). Short-term tests measuring genetic effects of chemicals will be of considerable importance in identifying genotoxic carcinogens (Williams, 1977a, 1979a, 1980b). On the other hand, epigenetic carcinogens will not be active in tests measuring genotoxicity, i.e., they will be true negatives in short-term genetic tests.

The present liver-derived systems are anticipated to be of particular value in establishing the mechanism of action of carcinogens with an organotropism for liver (Williams, 1977a; Williams, 1979c). Of the 28 carcinogens studied in the DNA repair assay, 17 have liver as one of their major target organs and all of these were positive. If it can be documented that all genotoxic liver carcinogens are positive in this test, then negative results with a liver carcinogen would be strong evidence for an epigenetic mechanism. Such evidence could be of great importance in the classification of liver tumor-inducing agents such as certain drugs and organochlorine pesticides (Williams, 1979c), whose structure does not suggest that they give rise to an electrophilic reactant characteristic of genotoxic carcinogens.

The rat liver transformation assays are reliable markers for neoplastic conversion of the lines, but their sensitivity for the detection of carcinogens has not yet been demonstrated. The production of mutagenesis and transformation in the same cultures is presently being studied to determine which is the most sensitive for detection of genotoxic carcinogens. Even if transformation assays are less sensitive, they would still have a value in that their endpoint has the potential of demonstrating neoplastic conversion *in vitro*. Furthermore, one of the other reasons that transformation assays were included in the battery that we proposed (Weisburger and Williams, 1978) was for the possibility that these assays might detect epigenetic carcinogens. Thus, an important area to be explored with these assays is their response to hepatocarcinogens such as phenobarbital and certain drugs and organochlorine pesticides that have been suggested to be epigenetic carcinogens of the promoter class (Williams, 1977a; 1979b; 1979c; Weisburger and Williams, 1980b).

A positive response in a single *in vitro* test is not considered to be sufficient evidence to classify an agent of unknown *in vivo* activity as a carcinogen. The results from a properly constructed battery of tests including systems with diverse metabolic capability and reliable endpoints of different biological significance may eventually make such a conclusion possible. In particular, the Ames test and the HPC/DNA repair test, as described, may already provide a basis for such a decision. Carcinogen metabolism performed by intact hepatocytes is more similar to the *in vivo* situation than that provided by microsomes, as in the Ames test (Schmeltz *et al.*, 1978). Furthermore, DNA repair is not elicited by certain mutagens, such as pure intercalating agents, purine analogues, and flavonoids that are not reliably carcinogenic. Therefore, a positive result in the HPC/DNA repair test greatly strengthens a finding of bacterial mutagenesis. Indeed, current evidence has led us to suggest that positive results in these two tests will be a certain indication of carcinogenicity (Williams, 1980a,b). Regardless, it seems very likely that full-scale bioassay may be avoided for chemicals giving a positive response in a battery by following the "decision point approach" (Weisburger and Williams, 1978), which recommends further testing of such agents first in limited *in vivo* bioassays.

Acknowledgments

This work was supported by Contracts N01-CP-55705 from the U.S. National Cancer Institute and # 68-02-2483 from the U.S. Environmental Protection Agency.

I thank Bette Meyer and Karen Brummett for help with the preparation of the manuscript.

References

Berman, J., Tong, C., Williams, G.M. (1980): Differences between rat liver epithelial cells and fibroblast cells in sensitivity to 8-azaguanine. In Vitro *16*:661–668.

Berman, J.J., Tong, C., Williams, G.M. (1978): Enhancement of mutagenesis during cell replication of cultured liver epithelial cells. Cancer Lett. *4*:277–283.

Brookes, P., Druckrey, H., Ehrenberg, L., Lagerlof, B., Litwin, J., Williams, G.M. (1973): The relation of cancer induction and genetic damage. In: C. Ramel (ed.), Evaluation of Genetic Risks of Environmental Chemicals. Ambio Special Report #3, pp. 15–16.

Green, H.L., Bridges, B.A., Rogers, A.M., Horspool, G., Muriel, W.J. Briges, J.W., Fry, J.R. (1977): Mutagen screening by a simplified bacterial fluctuation test: use of microsomal preparations and whole liver cells for metabolic activation. Mutat. Res. *48*:287–294.

Huberman, E. (1976): Cell-mediated mutagenicity of different loci in mammalian cells by carcinogenic polycyclic hydrocarbons. In: R. Montesano, H. Bartsch, and L. Tomatis (eds.), Screening Tests in Chemical Carcinogenesis (IARC Scientific Publications No. 12). Lyon: International Agency for Research on Cancer, pp. 521–532.

Kakunaga, T., Kamahora, J. (1968): Properties of hamster embryonic cells transferred by 4-nitroquinoline-1-oxide *in vitro* and their correlations with the malignant properties of the cells. Biken J. *11*:313–332.

Kuroki, T., Drevon, C. (1978): Direct or proximate contact between cells and metabolic activation systems is required for mutagenesis. Nature (Lond.) *271*:368–370.

Laishes, B.A., Williams, G.M. (1976): Conditions affecting primary cell cultures of functional adult rat hepatocytes. II. Dexamethasone enhanced longevity and maintenance of morphology. In Vitro *12*:821–832.

MacPherson, I., Montagnier, L. (1964): Agar suspension culture for the selective assay of cells transformed by polyoma virus. Virology *23*:291–294.

Martin, C.N., McDermid, A.C., Garner, R.C. (1978): Testing of known carcinogens and noncarcinogens for their ability to induce unscheduled DNA synthesis in HeLa cells. Cancer Res. *38*:2621–2627.

McCann, J., Choi, E., Yamasaki, E., Ames, B.N. (1975): Detection of carcinogens as mutagens in the Salmonella/Microsome test: assay of 300 chemicals. Proc. Natl. Acad. Sci. USA *72*:5135–5139.

Montesano, R., Drevon, C., Kuroki, T., Saint Vincent, L., Handleman, S., Sandford, K.K., Defeo, D., Weinstein, I.B. (1977): Test for malignant transformation of rat liver cells in culture: cytology, growth in soft agar, and production of plasminogen activator. J. Natl. Cancer Inst. *59*:1651–1658.

Montesano, R., Saint Vincent, L., Tomatis, L. (1973): Malignant transformation in vitro of rat liver cells by dimethylnitrosamine and N-methyl-N'-nitroN-nitrosoguanidine. Br. J. Cancer *28*: 215–220.

Rutenberg, A.M., Kim, H., Fischbein, J.W., Hanker, J.S., Wasserdrug, H.L., Seligman, A.M. (1969): Histochemical and ultrastructural demonstration of gamma-glutamyl transpeptidase activity. J. Histochem. Cytochem. *17*:517–526.

San, R.H.C., Stich, H.F. (1975): DNA repair synthesis of cultured human cells as a

288   Gary M. Williams

rapid bioassay for chemical carcinogens, Int. J. Cancer *16*:284–291.

San, R.H.C., Williams, G.M. (1977): Rat hepatocyte primary cell culture-mediated mutagenesis of adult rat liver epithelial cells by procarcinogens. Proc. Soc. Exp. Biol. Med. *156*:534–538.

San, R.H.C., Laspia, M.F., Soiefer, A.I., Maslansky, C.J., Rice, J.M., Williams, G.M. (1979a): A survey of growth in soft agar and cell surface properties as markers for transformation in adult rat liver epithelial-like cell cultures. Cancer Res. *39*:1026–1034.

San, R.H.C., Shimada, T., Maslansky, C.J., Kreiser, D.M., Laspia, M.F., Rice, J.M., Williams, G.M. (1979b): Growth characteristics and enzyme activities in a survey of transformation markers in adult rat liver epithelial-like cell cultures. Cancer Res. *39*: 4441–4448.

Schmeltz, I., Tosk, J., Williams, G.M. (1979): Comparison of the metabolic profiles of benzo(a)pyrene obtained from primary cell cultures and subcellular fractions derived from normal and methylcholanthren-induced rat liver. Cancer Lett. *5*:81–89.

Stich, H.F., San, R.H.C., Lam, P., Koropatnick, J., Lo, L. (1977): Unscheduled DNA synthesis of human cells as a short-term assay for chemical carcinogens. In: H.H. Hiatt, J.D. Watson, J.A. Winsten (eds.), Origins of Human Cancer, Book C, Cold Spring Harbor Laboratory, New York, pp. 1499–1512.

Tong, C., Williams, G.M. (1978): Induction of purine analog-resistant mutants in adult rat liver epithelial lines by metabolic activation-dependent and -independent carcinogens. Mutat. Res. *58*:339–352.

Tong, C., Williams, G.M. (1979): Variation in response in the adult rat liver epithelial cell/hypoxanthine-guanine phosphoribosyl transferase (ARL/HGPRTase) mutation assay with different metabolic activation-dependent carcinogens. Environ. Mut. *1*:148.

Tong, C., Williams, G.M. (1980): Definition of conditions for the detection of genotoxic chemicals in the adult rat liver hypoxanthine-guanine phosphoribosyl transferase (ARL/HGPRT) mutagenesis assay. Mutat. Res. *74*:1–9.

Weisburger, J.H., Williams, G.M. (1975): Metabolism of chemical carcinogens. In: F.F. Becker (ed.), Cancer: A Comprehensive Treatis. New York: Plenum Press, pp. 185–234.

Weisburger, J.H., Williams, G.M. (1978): Decision point approach to carcinogen testing. In: I.M. Asher, C. Zervos (eds.), Proceedings FDA Symposium on Structure-Activity Correlations in Carcinogenesis. Office of Science, FDA, Rockville, Md., pp. 45–52.

Weisburger, J.H., Williams, G.M. (1980a): Chemical carcinogenesis. In: J.F. Holland, E. Frei, III, (eds.), Cancer Medicine. Philadelphia: Lea & Febiger.

Weisburger, J.H., Williams, G.M. (1980b): Chemical carcinogens. In: J. Doull, C. Klaasen, M. Amdur (eds.), Toxicology, The Basic Science of Poisons, 2nd Edition. New York: MacMillan Pub. Co., Inc., pp. 84–138.

Williams, G.M. (1975): The study of chemical carcinogenesis using cultured rat liver cells. In: L.E. Gerschenson, E.B. Thompson (eds.), Gene Expression and Carcinogenesis in Cultured Liver. New York: Academic Press, pp. 480–487.

Williams, G.M. (1976a): Primary and long-term culture of adult rat liver epithelial cells. In: D.M. Prescott (ed.), Methods in Cell Biology, Vol. XIV. New York: Academic Press, pp. 357–364.

Williams, G.M. (1976b): The use of liver epithelial cultures for the study of chemical carcinogenesis. Am. J. Pathol., *85*:739–753.

Williams, G.M. (1976c): Carcinogen-induced DNA repair in primary rat liver cell cultures; A possible screen for chemical carcinogens. Cancer Lett. *1*:231–236.

Williams, G.M. (1977a): The significance of rodent liver tumors in bioassay. Presented at the Toxicology Forum, Aspen, Colorado, July.

Williams, G.M. (1977b): The detection of chemical carcinogens by unscheduled DNA synthesis in rat liver primary cell cultures. Cancer Res. *37*:1845–1851.

Williams, G.M. (1978): Further improvements in the hepatocyte DNA repair test for carcinogens: Detection of carcinogenic biphenyl derivatives. Cancer Lett. *4*:69–75.

Williams, G.M. (1979a): Review of in vitro test systems using DNA damage and repair for screening of chemical carcinogens. J. Assoc. Off. Anal. Chem. *62*:857–863.

Williams, G.M. (1979b): A comparison of *in vivo* and *in vitro* metabolic activation systems. In: B. Butterworth (ed.), Critical Reviews in Toxicology: Strategies for Short-Term Testing for Mutagens/Carcinogens. West Palm Beach, Florida: CRC Press, pp. 96–97.

Williams, G.M. (1979c): Liver cell culture systems for the study of hepatocarcinogenesis. In: G.P. Margison (ed.), Advances in Medical Oncology, Research and Education, Proceedings of the XIIth International Cancer Congress, Vol. 1, Carcinogenesis. New York: Pergamon Press, pp. 273–280.

Williams, G.M. (1980a): DNA repair and mutagenesis in liver cultures as indicators in chemical carcinogen screening. In: N.K. Mishra, V.C. Dunkel, M.A. Mehlman (eds.), Mammalian Cell Transformation by Chemical Carcinogens: Advances in Environmental Toxicology. New York: Pathotex Publishers Inc.

Williams, G.M. (1980b): The detection of chemical mutagens/carcinogens by DNA repair and mutagenesis in liver cultures. In: F.J. de Serres, A. Hollaender (eds.), Chemical Mutagens: Principles and Methods for Their Detection, Vol. 6. New York: Plenum Press pp. 61–79.

Williams, G.M., Laspia, M.F. (1979): The detection of various nitrosamines in the hepatocyte primary culture/DNA repair test. Cancer. Lett. 6:199–206.

Williams, G.M., Elliott, J.M., Weisburger, J.H. (1973): Carcinoma after malignant conversion in vitro of epithelial-like cells from rat liver following exposure to chemical carcinogens. Cancer Res. 33:606–612.

Williams, G.M., Bermudez, E., Scaramuzzino, D. (1977): Rat hepatocyte primary cell cultures. III. Improved dissociation and attachment techniques and the enhancement of survival by culture medium. In Vitro 13: 809–817.

Williams, G.M., Tong, C., Berman, J.J. (1978): Characterization of analog resistance and purine metabolism of adult rat-liver epithelial cell 8-azaguanine-resistant mutants. Mutat. Res. 49:103–115.

26

An Escherichia coli *Differential Killing Test for Carcinogens Based on a* uvrA recA lexA *Triple Mutant*

M.H.L. GREEN AND D.J. TWEATS

Introduction

Virtually all short-term tests for mutagens and carcinogens detect the ability of an agent to damage DNA, either directly or following metabolic activation. A differential killing assay demonstrates this by increased killing of a strain unable to repair DNA damage, as compared to an isogenic repair-proficient strain. The best known of such tests are the *rec*-assay of Kada *et al.* (1972) and the *pol*-assay of Slater *et al.* (1971). We believe that the test described here may have two advantages over other assays.

Firstly, repair-deficient bacteria tend to grow poorly, in comparison with their repair-proficient counterparts, and this can complicate the interpretation of results. It was earlier observed by chance (Green *et al.*, 1973) that *E. coli* strains containing the repair-deficiency genes *recA* and *lexA* in combination retain the extreme repair deficiency of *recA* but show the nearly normal growth of *lexA*. We have therefore con-

structed *E. coli* strain CM871, a *uvrA recA lexA* triple mutant that grows far better than the equivalent *uvrA recA lexA*⁺ strain WP100. Our test also uses a *polA uvrA* strain, WP67, and as control the repair-proficient strain WP2.

Secondly, tests that involve measuring zones of inhibition from a spot of test agent have a number of limitations. We therefore determine viability following treatments using Miles and Misra plating (1938). This approach allows us to measure the actual surviving fraction in a population but with considerably less effort than by conventional plating.

Methods

Strains

Escherichia coli strains WP2 (UV-resistant) and WP67 *uvrA polA* were given to us by E.M. Witkin. The construction of strain

CM871 *uvrA recA lexA* is described elsewhere (Tweats *et al.*, in preparation). Stocks can be stored in soft agar stabs (Green and Muriel, 1976) at room temperature for periods up to several years. For storage up to 4 weeks, cultures are streaked onto Oxoid nutrient agar No. 2 plates, incubated at 37° C overnight, and kept at 4° C.

Test Agents

These are dissolved in water or dimethyl sulfoxide. Stock solutions are normally prepared at 50 mg/ml or 100 mg/ml.

Media etc.

For nutrient medium we use Oxoid Nutrient Broth No. 2 (Code CM67), and for buffer, a standard phosphate buffer (Boyle and Symonds, 1969).

The Test

Single colonies from streak plates are inoculated into 10 ml nutrient broth and incubated overnight at 37° C with shaking. Cultures are diluted in buffer to about 2×10^3 bacteria/ml. This can normally be achieved by using a 2.5×10^5-fold dilution for WP67 and CM871, and 5×10^5-fold for WP2. For experiments in which S9 is present, some growth of the strains occurs, and therefore dilutions of 7.5×10^5 and 1.5×10^6, respectively, are recommended. The diluted cultures are dispensed in small test tubes in 0.5 ml or 0.25 ml aliquots.

Either a series of dilutions of the stock solution of test agent are made and each strain treated with each dilution, or else the stock solution can be serially diluted in the bacterial suspensions. As dimethylsulfoxide at high concentration can interfere with the activity of the metabolizing system, it is advisable to restrict concentrations of this solvent to a maximum of 2% v/v. A routine test would consist of three dose levels of test agent plus untreated and positive controls.

The small test tubes are incubated for 2 hours at 37° C. Then, for each strain, beginning with the highest concentration of agent, 3×10 μl spots (Miles and Misra, 1938) are placed on a nutrient agar plate. Each concentration and the untreated control can be spotted using the same disposable tip on the same plate. The agar plates are incubated at 37° C overnight and counted early the next day.

S9 Activation

Standard Aroclor-induced rat liver S9 supernatant fraction (Ames *et al.*, 1975) is used. Our mixture contains 1 ml S9 (approximately 30 mg protein/ml), 1 ml 40 mM NADP, 1 ml 40 mM glucose-6-phosphate, 1 ml 70 mM $MgSO_4$, 5 ml 0.1 M phosphate buffer (pH 7.4), 1 ml H_2O (Mattern and Greim, 1978). We add 3 ml to 12 ml of diluted bacterial suspension, dispense and treat exactly as described above. We are obtaining evidence (R. Forster, personal communication) that optimal levels of S9 fraction tend to be lower with liquid incubation than in a plate incorporation assay such as the Ames test. Therefore, the killing test could be extended to include a treatment with one-fifth the amount of S9 in the mix or with S9 prepared by an alternative method of induction.

Evaluation

Any conclusions should be based on consistent results in independent experiments. Statistical significance can be determined by converting individual counts to percent survival and performing regression analysis. However, it is possible that some small effects not related to DNA damage will be significant by this criterion. A more conservative approach will be to accept agents as positive only if they give an effect at two or more dose levels at least × fourfold apart and give a greater than × fourfold dif-

Table 26.1. Typical Results from the E. coli Differential Killing Test

Test agent (solubility)	S9	Strain	Concentration (µg/ml)	Colony counts			Percent survival	Comments
Mitomycin C (soluble)	Absent	WP2 UV resistant	0	11	11	13	100	
			0.01	17	26	—	184.3	
			0.1	17	10	18	128.6	
			1	11	19	13	122.9	
		WP67 uvrA polA	0	21	20	15	100	
			0.01	13	23	16	92.9	
			0.1	9	9	10	50.0	
			1	0	0	0	0	Positive
		Cm811 uvrA recA lexA	0	11	14	18	100	
			0.01	8	7	0	34.9	
			0.1	2	0	0	4.7	
			1	0	0	0	0	Positive
Benzo(a)pyrene (insoluble at higher levels)	Present	WP2 UV resistant	0	20	25	25	100	
			1	24	31	23	111.4	
			10	20	19	23	88.6	
			100	14	19	19	74.3	
		WP67 uvrA polA	0	23	30	25	100	
			1	23	26	25	94.9	
			10	12	5	15	41.0	
			100	13	15	18	59.0	
		CM871 uvrA recA lexA	0	21	25	23	100	
			1	18	15	23	81.2	
			10	0	0	0	0	
			100	0	0	0	0	Positive
Ampicillin (soluble)	Absent	WP2 UV resistant	0	22	21	26	100	
			1	28	29	29	124.6	
			10	23	18	24	94.2	
			100	0	0	0	0	

	Strain	Conc.				%	
	WP67 *uvrA polA*	0	14	12	13	100	Actually more toxic to wild type
		1	13	21	15	125.6	
		10	5	8	12	64.1	
		100	5	8	3	41.0	
	Cm871 *uvrA recA lexA*	0	20	25	28	100	
		1	24	24	30	106.8	
		10	18	20	21	80.8	
		100	6	2	4	16.4	
Diethyl-stilbestrol (insoluble)	Absent						
	WP2 UV resistant	0	20	22	18	100	
		500	11	10	15	60	
		1000	7	15	19	68.3	
		2500	14	17	6	61.7	
	WP67 *uvrA polA*	0	19	19	22	100	Questionable because effect found only above limit of solubility
		500	19	17	16	86.7	
		1000	17	13	14	73.3	
		2500	4	5	3	20.0	
	CM871 *uvrA recA lexA*	0	18	17	17	100	
		500	7	5	4	30.8	
		1000	1	3	0	7.7	
		2500	0	0	1	1.9	

ference in survival at one or both doses. If the sensitive strain shows 0% survival, the wild type strain should show at least 25%. If the test agent is relatively insoluble, the test should be confined to concentrations below saturation, as false positive effects have been observed with poorly soluble compounds tested at concentrations where precipitation has occurred. Where there is an increased effect with activation, results above the apparent limit of solubility may be acceptable.

Examples of Typical Results

Table 26.1 shows results for separate experiments with mitomycin C as a clear positive not requiring activation, benzo(a)pyrene as a positive requiring activation, ampicillin as a clear negative, and diethylstilbestrol as a borderline positive of uncertain meaning.

These are typical single results from different experiments. In order to reach a conclusion it would be necessary to obtain effects such as those shown here consistently in independent experiments.

Discussion

The test described here is particularly easy, reproducible, rapid, and economical. It requires considerably less labor and materials than an Ames test and results are obtained overnight. It detects agents that are frameshift or base substitution mutagens, crosslinking agents, and agents requiring S9 activation. Agents such as mitomycin C, which are mutagenic to only particular tester strains of bacteria, are readily detected.

Disadvantages

There are two main problems. Firstly, apart from direct-acting agents, the test detects only those that are activated reasonably efficiently by S9, such as benzo(a)pyrene or 2-aminoanthracene. In general, unless enough active metabolite is generated to give an average of at least one or two lethal lesions per bacterial chromosome, an effect will not be seen. A considerably lower number of mutagenic lesions per chromosome can probably be detected in a test such as the Ames test. With direct-acting mutagens any lack of sensitivity may be overcome by simply adding more test agent (within the confines of solubility, etc.), but when S9 activation is required, it is likely that metabolism will limit the effective dose. Interestingly, the *uvrA recA lexA* triple mutant CM871 appears to be considerably better than the *uvrA polA* mutant WP67 at detecting agents requiring activation.

The second problem became apparent during the evaluation of the test in the International blind trial of 42 compounds, representing known classes of carcinogens and noncarcinogens (Anon., 1978). In a number of cases, weak but significant effects were obtained without S9 activation either for noncarcinogens or for carcinogens known to require activation to express mutagenicity. It is for this reason that we suggest that a more conservative criterion for a positive response may be desirable.

Future Developments

The most obvious possibility for improvement of the assay is to increase the permeability of the strains in order to increase their sensitivity, particularly to agents that require metabolic activation. If this is successful, it could be of considerable benefit.

Another area for investigation is the phenomenon described earlier of weak positive effects shown by particular noncarcinogens and indirect carcinogens, which cannot easily be related to DNA damage. The effects are real, but their relevance to carcinogenesis remains to be established.

Conclusions

Although differential killing tests such as the one described here are unlikely to be as sensitive as the best mutation assays, it must be stressed that they are exceptionally quick and economical. Thus, although they will not replace mutation assays, they can complement them by confirming positive effects and picking up certain "difficult" types of agent. We would suggest that they are ideally used before a mutation assay, in a battery of tests, since they can provide useful data on bacterial toxicity in addition to picking out potential carcinogens.

The relative merits of the various killing assays will be established only by testing a large number of agents, as in the International study. Nevertheless, our impression remains that the triple mutant CM871 *uvrA recA lexA* and our method of determining viability are likely to offer real advantages.

Acknowledgments

We thank Mr. R. Forster for permission to quote unpublished findings, Professor B.A. Bridges for advice and help and Dr. D.A.H. Pratt for helpful criticisms of the manuscript.

References

Ames, B.N., McCann, J., Yamasaki, E. (1975): Methods for detecting carcinogens and mutagens with the *Salmonella*/mammalian microsome mutagenicity test. Mutat. Res. *31*: 347–364.

Anon. (1978): International programme for the evaluation of short-term tests for carcinogenicity. Mutat. Res. *54*:203–206.

Boyle, J.M., Symonds, N. (1969): Radiation sensitive mutants of T_4D. 1. T_4y: a new radiation sensitive mutant: effect of the mutation on radiation survival, growth and recombination. Mutat. Res. *8*:431–439.

Green, M.H.L., Muriel, W.J. (1976): Mutagen testing using Trp^+ reversion in *Escherichia coli*. Mutat. Res. *38*:3–32.

Green, M.H.L., Gray, W.J.H., Sedgwick, S.G., Bridges, B.A. (1973): Repair of DNA damage produced by gamma-radiation in *Escherichia coli* K12 and a radiation sensitive *exrA* derivative during inhibition of protein synthesis and normal DNA replication by chloramphenicol. J. Gen. Microbiol. *77*:99–108.

Kada, T., Tutikawa, K., Sadaie, Y. (1972): *In vitro* and host-mediated "rec-assay" procedures for screening chemical mutagens; and phloxine, a mutagenic red dye detected. Mutat. Res. *16*:165–174.

Mattern, I.E., Greim, H. (1978): Report of a workshop on bacterial *in vitro* mutagenicity test systems. Mutat. Res. *53*:369–378.

Miles, A.A., Misra, S.S. (1938): The estimation of the bactericidal properties of blood. J. Hyg. (Camb.) *38*:732–749.

Slater, E.E., Anderson, M.D., Rosenkranz, H.S. (1971): Rapid detection of mutagens and carcinogens. Cancer Res. *31*:970–973.

27

Detection of Carcinogens Using the Fluctuation Test with S9 or with Hepatocyte Activation

S.A. Hubbard, M.H.L. Green, and J.W. Bridges

Introduction

The fluctuation test was originally developed by Luria and Delbruck (1943) to demonstrate the mutational origin of bacterial variants, but it is also a method of determining spontaneous mutation rates and has been adapted to measure small induced increases over the spontaneous mutation rate (Clarke and Wade, 1975; Green et al., 1976; Voogd et al., 1974). In particular, Green et al. (1976, 1977a) have used a modified form of the method, devised by Ryan (1955) for mutagen screening. It has been found in some cases to be a more sensitive method for measuring induced mutation than conventional plating tests (Green et al., 1977a, 1977b). A version of the test using Microtitre[R] trays has been developed by Gatehouse (1978).

The fluctuation test has been used to measure induced mutation in organisms other than bacteria, including *Saccharomyces cerevisiae* (Parry, 1977)

and L5178Y mouse lymphoma cells (Cole et al., 1976).

Many carcinogens require metabolic activation to exert their effect. Most metabolism in mutagenicity tests has been provided by using S9 fraction (9000 g supernatant) of rat liver homogenate (Ames et al., 1975; Garner et al., 1972; Malling, 1971). Microsomal or S9 fractions are somewhat artificial in nature, using fragmented endoplasmic reticulum and requiring very high nonphysiological cofactor levels, including the addition of NADP for optimal oxidation activity. Moreover, unless other cofactors are added, they cannot carry out conjugation as they lack conjugating systems (Bridges and Fry, 1977). An alternative to S9 is the use of isolated hepatocytes that are capable of various biochemical reactions, including microsomal mixed-function oxidase metabolism, and which also retain conjugating ability (Bridges and Fry, 1977). In drug metabolism studies they are generally considered to resemble more closely *in vivo* metabolic activation than metabolism by microsomal fractions (Bridges and Fry, 1977; Fry and Bridges, 1976). Hepatocytes tend to be nonviable when treated in agar overlay and so must be used in a protocol such as the fluctuation test which involves liquid medium.

The fluctuation test is performed by treating amino acid-requiring bacteria in suitable medium, either with or without a metabolizing system (S9 fraction or isolated hepatocytes), with the chemical to be tested and with a trace amount of the required amino acid (to allow a few generations of growth during which mutations to amino acid independence may be expressed). This mixture is immediately dispensed in small aliquots into the wells of disposable plastic trays and incubated for 16–18 hours. Selective indicator medium is then added to each well so that only revertants can continue to grow. The bacteria release acid on growth to full turbidity; therefore, a pH indicator, bromocresol purple, is added. At full turbidity the pH changes from 7 (purple) to 5.2–6.8 (yellow). The yellow wells are those in which one or more reversions to amino acid independence have occurred. The test has been modified since the original papers by Green *et al.* (1976, 1977a) to use disposable plastic trays, which avoid the problems of decontaminating and washing glass tubes and occupy less incubator space. The test with hepatocytes (Green *et al.*, 1977a) has been modified to use a mammalian culture medium, which is more suitable for hepatocytes but still allows the bacteria to grow. Tissue culture treated disposable trays have been utilized to permit hepatocyte attachment and extend metabolism over as long a period of time as possible.

Materials and Methods

Bacterial Strains and Maintenance of Stocks

Salmonella typhimurium strains TA98 and TA100 have been described by Ames *et al.*

(1975). Stocks containing a low background of His$^+$ revertants are obtained from Oxoid Nutrient Broth No. 2 cultures grown overnight at 37° C with shaking. 1 ml aliquots (with 10% DMSO) are stored frozen in a liquid nitrogen refrigerator. For an experiment, an aliquot is thawed and a series of 0.1 ml aliquots grown up in 10 ml nutrient broth overnight. The cultures are checked for sensitivity to crystal violet, resistance to ampicillin, background level of His$^+$ revertants, and spontaneous reversion rate, and the best cultures used for experiments. The cultures can be kept for up to a week at 4° C without significant loss of viability. Cultures can also be streaked on nutrient agar plates and kept for up to 4 weeks in a refrigerator.

Media and Equipment

All solutions used are sterile. DM salts: K_2HPO_4, 7 g; KH_2PO_4, 2 g; $(NH_4)_2SO_4$, 1 g; trisodium citrate, 0.25 g; $MgSO_4.7H_2O$, 0.1 g; distilled water, 1000 ml.

KCl solution: 330 mM; $MgCl_2.7H_2O$ solution, 80 mM; sodium phosphate buffer, 0.2 M, pH 7.4 (BDH Ampoules, BDH, Poole, England). PBS'A' (Phosphate buffered saline, Dulbecco A, Oxoid, London), after autoclaving 0.5 ml sterile glucose solution (20% w/v) is added to 100 ml PBS'A' to give a final concentration of 1 mg/ml. Hanks Mg^{2+}-free Balanced Salt Solution (BSS) without phenol red (Gibco-Biocult, Paisley, Scotland). $CaCl_2$ solution, 1 M; $NaHCO_3$ solution, 7.5% w/v (Flow Laboratories, Irvine, Scotland). Specially prepared Basal Medium Eagle's (BME) with Earle's Salts, 20 mM Hepes buffer, and without $NaHCO_3$, glutamine, phenol red, tryptophan and histidine (Flow Laboratories, Irvine, Scotland). BME was supplemented with 1% fetal bovine serum and 2 mM L-glutamine (Flow Laboratories); EGTA, 0.5 mM in PBS'A'; NADP (40 mM) and G-6-P (50 mM) are made up in sterile distilled water, filter sterilized, and stored frozen. Reference mutagens and test chemicals are made up in DMSO as concentrated stock solutions and added to the test at a maximum concentration of 10 μl/ml DMSO.

Linbro white vinyl trays are purchased from Flow Laboratories, repacked, and gamma-sterilized. Tissue culture grade trays are already sterile.

Animals

Sprague-Dawley male rats of 60–80 g in weight are used for isolation of hepatocytes. Rats of 200 g in weight are used for preparation of S9, except for direct comparison with hepatocytes. Animals are killed by cervical dislocation.

Preparation of S9 Fraction

The S9 or 9000 g supernatant fraction is prepared by the method of Ames et al., (1975) and stored in a liquid nitrogen refrigerator.

Preparation of Isolated Hepatocytes

Hepatocytes are prepared by the method of Fry et al. (1976). The liver lobes are removed under sterile conditions into cold PBS'A', dried between sterile filter papers, and sliced with a heavy duty blade (Swann-Morton PM40, Swann-Morton, Sheffield), so that slices 0.5–1.0 mm in thickness are obtained. The lobes are held on a filter paper in a glass Petri dish during this procedure. The slices are then incubated in 250-ml conical flasks (3 g liver per flask) in 10 ml PBS'A' at 37° C for 10 min in a shaking water bath (approximately 90 oscillations per min). The supernatant is discarded and the incubation repeated once more with 10 ml PBS'A', followed by two further incubations with 10 ml PBS'A' + 0.5 mM EGTA. The slices are then incubated for 60 min in the shaking water bath with enzyme solution consisting of 10 ml Hank's BSS containing 6 μg/ml phenol red, 5 mM $CaCl_2$ (50 μl

of 1 M solution) and collagenase (0.05% w/v Boehringer, Lewes) and hyaluronidase (0.1% w/v, Sigma, Poole, England). Approximately 75 μl of 7.5% sodium bicarbonate solution is added to make the solution just neutral as judged by the phenol red indicator in the Hank's BSS. Following incubation, the mixture is filtered through a layer of bolting cloth (150 μm pore size, Henry Simon Ltd., Cheadle Heath, Stockport) to remove undigested material and large cell clumps. The filtrate is centrifuged at approximately 100 g for 3 min and the cell pellet resuspended and recentrifuged once in 10 ml PBS'A' and twice in 10 ml BME. The pellet is finally resuspended in 10 ml BME, and the total cell yield and viability index (% of cells viable) calculated. Viability is judged by the ability to exclude the dye, trypan blue.

Preparations of viability below 80% are not used. Viability varies from 80–95%. The cell yield is usually 6–12 × 10⁶ viable cells/g wet tissue. The hepatocytes are stored on ice until required (not longer than 4 hours).

Fluctuation Tests

For a test, a bulk solution of 15 ml (see below) containing bacteria (approximately 2 × 10⁷ cells TA98 or 2 × 10⁶ cells TA100 total), test compound (at not more than 1% v/v DMSO or other solvent), with or without a metabolizing system, is distributed in 0.3 ml aliquots into 48 2-ml-capacity wells in Linbro white vinyl trays (tissue culture trays are used with hepatocytes). For experiments with TA100 and hepatocytes, 50 μl is added to tissue culture treated trays with a well capacity of 0.5 ml. The trays are incubated for 16–18 hours at 37° C, after which 1 ml/well of selective indicator is added (200 μl for TA100/hepatocytes). The indicator medium consists of DM salts supplemented with glucose (0.4%) and 6.5 μg/ml BCP. The trays are incubated for a further 3 days when yellow wells are scored positive and purple wells negative. A typical test consists of solvent control, positive control, and a series of concentrations of the chemicals to be tested.

Bulk Solutions for Fluctuation Tests

Without metabolic activation. 15 ml DM salts supplemented with 0.4 $\mu g/ml$ biotin, 0.4% w/v glucose, 1 $\mu g/ml$ L-histidine for a test with TA98 and 0.25 $\mu g/ml$ for a test with TA100, and the test chemical.

With metabolic activation by S9 fraction. The S9 is prepared as described by Ames *et al.* (1975). S9 mix contains in 10 ml: 1 ml S9 (about 30 mg protein per ml); 1 ml $MgCl_2$ solution; 1 ml KCl solution; 1 ml G-6-P solution; 1 ml NADP solution; 5 ml sodium phosphate solution (0.2 molar). The bulk test solution of 15 ml consists of: 3 ml S9 mix; 12 ml DM salts containing 0.4 $\mu g/ml$ biotin, 0.4% w/v glucose, and 0.5 $\mu g/ml$ L-histidine for TA98 (the S9 provides sufficient histidine for TA100)—plus, of course, the test chemical.

With metabolic activation by freshly isolated hepatocytes. The hepatocytes are prepared as described and diluted to 2×10^5 viable cells/ml in BME. To 15 ml are added bacteria and test chemical. The serum present in the medium provides sufficient histidine for the test. 1% serum has been found to be the highest level that can be used without increasing the number of spontaneous mutants excessively. Because of the high spontaneous mutation rate of TA100, the test has to be modified when it is used together with hepatocytes. In this case, 2.5 ml of diluted hepatocytes, together with about 2×10^6 bacteria and the test agent, are dispensed in 50-μl aliquots in tissue-culture-treated trays with a well capacity of 0.5 ml. 200 μl of selective indicator medium is later added and the experiment conducted as before.

Choice of Doses and Toxicity Tests

Doses of 100 $\mu g/ml$, 10 $\mu g/ml$ and 1 $\mu g/ml$ have been found suitable for an initial test for most chemicals. Doses are adjusted as necessary for subsequent tests. Most mutagenic chemicals that we have tested have shown both toxicity and mutagenicity within this range.

Toxicity is indicated by a reduction in the number of revertants compared to the control, and most often no positive wells are found when toxicity occurs. When no toxicity is seen, higher doses are used, and when there is toxicity or significant mutagenicity, doses are chosen so that the nonmutagenic and toxic limits of the dose range can be found. Solubility of the test chemical in DMSO or medium is another limiting factor.

Controls

A fluctuation test consists of one or two negative solvent controls, a positive reference control, and dose range of the chemicals to be tested. Results for some reference mutagens are shown in Table 27.1. The number of positive wells in the controls is characteristic for each strain and condition, but varies considerably between experiments. Levels of 3–16 and 16–32 positive wells out of 48 occur for TA98 and TA100, respectively, without metabolic activation and with hepatocytes. With S9 fraction values of 15–25 and 25–35, respectively, are usual.

Statistical Evaluation

The significance of an increase in the number of positive wells in a treatment over the corresponding control is determined by χ^2 with Liddell's correction.

	Wells negative	Wells positive	Total
Control	a	b	a + b
Treated	c	d	c + d
Total	a + c	b + d	a + b + c + d = n

$$\chi^2 = \frac{(ad - bc \pm \frac{1}{4}n)^2 \times n}{(a + b)(c + d)(a + c)(b + d)}$$

Table 27.1. Results Taken from Three Sample Experiments Using Reference Mutagens

Reference mutagen	Strain	Metabolizing system[a]	Concentration (μg/ml)	Number of wells positive[b]		
				Expt. 1	Expt. 2	Expt. 3
Furacin	TA 100	− S9	0	17	19	19
			0.1	30	25	34
4-nitro-ortho-phenyl-diamine	TA 98	− S9	0	7	9	6
			0.1	39	47	47
2AAF	TA 98	+ S9	0	15	22	9
			0.1	48	48	40
2-Aminoanthracene	TA 100	+ S9	0	32	26	30
			0.1	45	31	45
2AAF	TA 98	+ Cells	0	7	12	14
			1.0	25	43	31
2-Aminoanthracene	TA 100	+ Cells	0	24	26	27
			1.0	47	36	37

[a]Metabolizing systems − S9—without metabolic activation + S9—with metabolic activation by S9 fraction + Cells—with metabolic activation by isolated hepatocytes.

[b]48 wells treated per series.

χ^2 is estimated by summing together the total number of wells positive and negative in the treated and corresponding control and series in all the experiments done. Since χ^2 is determined only where there is an excess of positive wells in the treated series, this is a one-tailed test. The mutation frequency per well can be determined by taking the fraction of wells negative and applying the zero term of the Poisson distribution (e^{-m}). Paradoxically, the smallest increase over the spontaneous mutation rate can be detected when 50–80% of the wells in the control series are positive (Fisher, 1954), although results are satisfactory if 10–30% of the wells are positive in the controls.

Modifications

The test can be used with other *Salmonella* strains, such as TA1535 and TA1538, and with *E. coli* (Green *et al.*, 1977a, 1977b; Gatehouse, 1978; Gatehouse and Delow, 1979). Minor adjustments to amino acid levels may be required. S9 fraction from other tissues and S9 fraction from animals induced in various ways, for example, with phenobarbitone, 3-methylcholanthrene, or Aroclor 1254, can be used. Hepatocytes isolated from phenobarbitone and 3-methylcholanthrene-induced animals can be used, but the recovery of intact hepatocytes from Aroclor-induced animals is very poor. As has been stressed previously (Ames *et al.*, 1975), the amount of S9 used may be critical, and the levels present can be altered for specific chemicals. It is arguable that several different levels or types of S9 fraction should be used for each compound tested, since no one set of conditions will detect all carcinogens.

VB salts (Ames *et al.*, 1975) can, of course, be used instead of DM salts. Small test tubes may be used instead of disposable trays and are useful for chemicals that attack plastic. Gatehouse (1978) has used Microtitre[R] trays successfully, although the

number of bacteria tested is reduced and some sensitivity may be lost.

Evaluation of the Results

Figure 27.1 shows the type of response that is most often seen. 4NQO is a potent mutagen, showing toxicity to the bacteria at high levels and no mutagenicity at sufficiently low levels (below approximately 10 ng/ml). A dose response can be seen with most chemicals tested, but a disadvantage of the test is that often at high levels all the wells are positive and a dose response cannot be seen. This is observed with 4NQO and many other chemicals that are good mutagens.

Table 27.2 gives results for 2 AAF with metabolic activation by both S9 and hepatocytes. Hepatocytes are less sensitive as a metabolizing system in this instance.

Experiments are repeated on at least three separate occasions. Normally at this stage the pooled results are clearly positive or negative by χ^2. Rarely is a dose response not observed, but χ^2 for the highest nontoxic level after three experiments is between 2 and 3. In such cases, the experi-

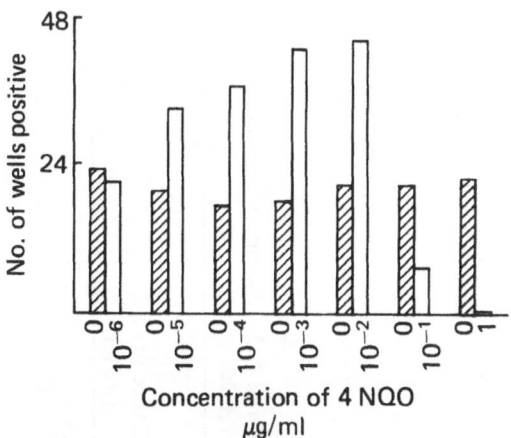

Figure 27.1. Results of experiments with 4NQO using TA100 and no metabolizing system (results average of two or three separate experiments at each dose level).

Table 27.2.   Summary of Results Obtained for 2AAF with Metabolic Activation by S9 and also by Isolated Hepatocytes Using TA 98

Concentration (μg/ml)	Number of experiments	Average number of wells positive[a]		χ^2	Probability	Induced mutants per well[b]
		Control	Treated			
Activation by S9						
10	3	21.7	48.0	113.28	<0.0005	—
1	3	22.0	48.0	107.09	<0.0005	—
0.1	3	15.3	45.3	113.58	<0.0005	2.49
0.05	2	18.5	44.0	60.80	<0.0005	1.99
0.01	3	15.3	21.0	2.99	n.s.	0.19
0.005	3	15.3	13.3	0.05	n.s.	—
Activation by isolated hepatocytes						
10	1	9.5	48.0	81.14	<0.0005	—
1	3	10.8	33.0	87.91	<0.0005	0.91
0.5	3	8.8	26.3	53.04	<0.0005	0.59
0.1	3	12.2	15.7	3.32	n.s.	0.10
0.01	3	7.7	8.0	1.57	n.s.	0.07

[a]48 wells per test are treated. 96 wells are occasionally used for the control series.
[b]Induced mutations determined by the zero-term of the Poisson distribution.

ment is repeated a further three times and χ^2 again determined. A chemical is considered positive if the result is significant with a probability of less than 0.05 (preferably 0.001) and the result is reproducible. When a dose response is seen, not all levels need to be tested three times, since the dose response is itself an indication of the validity of the test.

Experiments in which the positive control mutagen does not exhibit the usual response are not considered. Experiments occasionally occur in which the number of wells positive in the control is very high, and these are only considered if the reference mutagen shows adequate mutagenicity.

Chemicals Tested and Correlation with Carcinogenicity

The fluctuation test is being evaluated in the International paired compound trial (Anon., 1978). It appears that with hepatocytes there is generally less effect than with S9 fraction, but in one or two cases a greater mutagenic effect is observed. Until the paired compound trial has been evaluated it is impossible to predict the usefulness of hepatocytes for general screening.

Discussion

The fluctuation test, though slower to perform than the Ames test, is probably more amenable to automation. It has the advantage that it allows the bacteria to grow and mutate in the presence of a constant low concentration of test chemical. In a conventional Ames test, diffusion of chemical from the top agar overlay may make it necessary to use a higher initial concentration of chemical. Normally this will not matter, but in the case of toxic, weakly mutagenic agents, it may be critical. Another situation where the fluctuation test may be of advan-

tage arises when testing water samples for traces of mutagenic activity. By using the fluctuation test, up to 90% of the water in the treated sample can be replaced with the water sample under test.

As mentioned above, the test can be used with a variety of organisms. More importantly, it can be used to measure a variety of genetic endpoints in addition to reversion to amino acid dependence. Provided that during the first 18-hr incubation period, growth is limited in a nonselective way, in the second stage a suitable selective medium can be added and genetic endpoints such as mutation to bacteriophage or antibiotic resistance can be scored.

A third advantage of the fluctuation test is that it is a sensitive method of measuring mutation in liquid. Not only is there evidence that metabolism by an S9 fraction differs from metabolism *in vivo*, but also there is evidence that S9 fraction performs differently in liquid than in agar overlays (such as in the Ames test) (R. Forster, personal communication). Indeed, some agents such as dimethylnitrosamine are known to require liquid incubation before they can be detected. Although preincubation in liquid can be used in conjunction with the Ames test, quantities of S9 appropriate for the Ames test may well not be optimal for preincubation.

The final advantage of the fluctuation test is that it enables hepatocytes to be used as a metabolizing system. This does not necessarily make the test more sensitive. Many metabolites that *in vivo* would be conjugated and rendered harmless but *in vitro* are mutagenic after S9 metabolism are likely to be negative with hepatocytes. There are, of course, other reasons, in addition to detoxification by enzymes present in hepatocytes and absent in S9, that chemicals mutagenic with S9 may be less or nonmutagenic with hepatocytes. In particular, metabolites may not be formed in sufficient quantity in hepatocytes to cause a mutagenic effect, metabolites may be short-lived and not escape from the cell, or the metabolites formed in hepatocytes may be different from or may be found in different amounts from those found with S9.

Chemicals that give greater mutagenicity with hepatocytes are also found. When this is seen, care must be taken that the effect is not simply due to the chemical's being toxic to the hepatocytes, causing them to break up and release histidine into the medium.

While hepatocytes are unlikely to replace S9 in routine screening, they may provide a valuable adjunct to S9. They may be useful in determining species specificity with certain chemicals. Their closer relevance to *in vivo* makes them important in the study of ultimate carcinogenic metabolites, and confirmation of S9 results with hepatocytes makes it less likely that an effect is an artifact of S9 metabolism.

Conclusion

Since the fluctuation test requires more labor than the Ames test to generate a given amount of data, and since it normally gives the same answer, we obviously do not suggest its use as a general alternative. Nevertheless, in particular situations, the fluctuation test would appear to offer real advantages and we would hope that in these cases it will gain acceptance.

Acknowledgments

We would like to thank Professor B.A. Bridges for his interest in and advice with this work. We would also like to thank R. Forster for his personal communication. S.A. Hubbard was in receipt of an MRC research studentship.

Abbreviations

2AAF, 2-Acetamidofluorene; BCP, Bromocresol purple; BME, Basal Medium (Eagle); DMSO, Dimethyl sulfoxide; DM salts, Davis-Mingioli salts; EGTA, Ethylene glycol bis(B-amino-ethyl ether)-N-N-tetra-acetic acid; G-6-P, D-Glucose-6-phosphate; NADP, Nicotinamide adenine dinucleotide phosphate; NOPD, 2-Nitro-orthophenylenediamine; 4NQO, 4-Nitroquinoline-N-oxide; PBS'A', Dulbecco's phosphate buffered saline 'A' supplemented with glucose; VB, Vogel-Bonner medium.

References

Ames, B.N., McCann, J., Yamasaki, E. (1975): Methods for detecting carcinogens and mutagens with the *Salmonella*/mammalian microsome mutagenicity test. Mutat. Res. *31*: 347–364.

Anon. (1978): International programme for the evaluation of short-term tests for carcinogenicity. Mutat. Res. *54*:203–206.

Bridges, J.W., Fry, J.R. (1977): Drug metabolism in cell suspension and cultures. In: D.V. Parke and R.L. Smith (eds.), Drug Metabolism from Microbes to Man. London: Taylor and Francis, pp. 43–54.

Clarke, C.H., Wade, M.J. (1975): Evidence that caffeine, 8-methoxypsoralen and steroidal diamines are frameshift mutagens for *E. coli* K-12. Mutat. Res. *28*:123–125.

Cole, J., Arlett, C.F., Green, M.H.L. (1976): The fluctuation test as a more sensitive system for determining induced mutation in L5178Y mouse lymphoma cells. Mutat. Res., *41*:377–386.

Fisher, R.A. (1954): Statistical Methods for Research Workers. Edinburgh: Oliver and Boyd, pp. 61–67.

Fry, J.R., Bridges, J.W. (1976): The metabolism of xenobiotics in cell suspensions and cell cultures. In: J.W. Bridges and L.F. Chasseaud (eds.), Progress in Drug Metabolism, Vol. 2. New York: John Wiley and Sons, pp. 71–118.

Fry, J.R., Jones, C.A., Wiebkin, P., Belleman, H.P., Bridges, J.W. (1976): The enzymic isolation of adult rat hepatocytes in a functional and viable state. Anal. Biochem. *71*: 341–350.

Garner, R.C., Miller, E.C., Miller, J.A. (1972): Liver microsomal metabolism of aflatoxin B_1 to a reactive derivative toxic to *Salmonella typhimurium* TA1530. Cancer Res. *32*: 2058–2066.

Gatehouse, D.G. (1978): Detection of mutagenic derivatives of cyclophosphamide and a variety of other mutagens in a "Microtitre R" fluctuation test, without microsomal activation. Mutat. Res. *53*:289–296.

Gatehouse, D.G., Delow, G.F. (1979): The development of a "Microtitre R" fluctuation test for the detection of indirect mutagens and its use in the evaluation of a mixed enzyme induction of the liver. Mutat. Res. *60*:239–252.

Green, M.H.L., Muriel, W.J., Bridges, B.A. (1976): Use of a simplified fluctuation test to detect low levels of mutagens. Mutat. Res. *38*: 33–42.

Green, M.H.L., Bridges, B.A., Rogers, A.M., Horspool, G., Muriel, W.J., Bridges, J.W., Fry, J.R. (1977a): Mutagen screening by a simplified fluctuation test: Use of microsomal fractions and whole liver cells for metabolic activation. Mutat. Res. *48*:287–294.

Green, M.H.L., Rogers, A.M., Muriel, W.J.,

Ward, A.C., McCalla, D.R. (1977b): Use of a simplified fluctuation test to detect and characterize mutagenesis by nitrofurans. Mutat. Res. *44*:139–143.

Luria, S.E., Delbrück, M. (1943): Mutations of bacteria from virus sensitivity to virus resistance. Genetics *28*:491–511.

Malling, H.V. (1971): Dimethylnitrosamine: Formation of mutagenic compounds by interaction with mouse liver microsomes. Mutat. Res. *13*:425–429.

Parry, J.M. (1977): The use of yeast cultures for the detection of environmental mutagens using a fluctuation test. Mutat. Res. *46*: 165–176.

Ryan, F.J. (1955): Spontaneous mutation in nondividing bacteria. Genetics *40*:726–738.

Voogd, C.E., Van der Stel, J.J., Jacobs, J.J.J.A.A. (1974): The mutagenic action of nitroimidazoles. 1. Metronidazole, Nimorazole, Dimetridazole and Ronidazole. Mutat. Res. *26*:483–490.

28

In vitro *Mammalian Cell Transformation for Identification of Carcinogens, Cocarcinogens, and Anticarcinogens*

CHARLES H. EVANS AND JOSEPH A. DIPAOLO

Introduction

Short-term tests for carcinogen identification are an increasingly valuable means for rapid detection of the carcinogenic potential of a substance and for studying the mechanism of carcinogenic action removed from the myriad complex interactions that are present *in vivo*. A variety of mammalian cell culture systems exist in which morphologic transformation is a reliable indicator of carcinogenic potential of the test agent (DiPaolo and Casto, 1977; Heidelberger, 1977). These systems represent the closest short-term *in vitro* correlation to animal models for the study of carcinogenic action. Organ culture systems should present a means of identifying site-specific responses but are not yet suitably developed for the rapid detection and study of diverse carcinogens. Epithelial cell systems are not presently as versatile in terms of their speed or endpoints. *In vitro* tests with other than mammalian cells or that utilize subcellular organelles or fractions may be reliable assays for the identification of mutagens and in some cases carcinogens, but remain further removed from the intact animal and often do not permit the study of the modulation of carcinogenesis by cocarcinogens and anticarcinogens.

Freshly isolated diploid cell strains and long-term cultured cell lines are successfully being used for the identification and study of carcinogens. Thus far, models with a morphological endpoint employ fibroblast-like cells and the morphologically transformed cells will produce tumors in suitable hosts. As a result of our experiments with cell lines and strains, our preference is to use cell strains. Freshly isolated cell strains are diploid, possess higher and more diverse enzymatic activities, and are closer to the *in vivo* situation than long-term passaged cell lines.

Morphologic transformation of Syrian golden hamster embryo cells (HEC) is a sensitive, reproducible, and quantitative means to identify and study the carcinogenic potential of chemical (DiPaolo and Casto, 1979; DiPaolo *et al.*, 1969, 1972), physical (DiPaolo and Donovan, 1976), and biological (Casto *et al.*, 1973) agents individually or in combination. Morphological transformation can be studied in a 7-day colony (DiPaolo *et al.*, 1969) or 3-week focus assay (Casto *et al.*, 1977); freshly isolated as well as liquid nitrogen cryostored cells give equivalent results (DiPaolo, 1980). For compounds requiring metabolic activation, transformation can be obtained by administering the agents *in vivo* prior to excision of the embryos in a combination *in vivo/in vitro* host-mediated assay (DiPaolo *et al.*, 1973), or by addition of microsomal enzymes or hepatocytes as an activation system to the HEC in culture (Pienta, 1979). Morphological transformation is relevant to carcinogenesis as morphologically transformed HEC grow into progressively enlarging tumors when injected into animals (DiPaolo *et al.*, 1971d). The validity of the HEC transformation assay as a short-term test for carcinogens is shown by the high positive correlation between the *in vitro* transformation ability and *in vivo* carcinogenic potential of established carcinogenic and noncarcinogenic agents (Berwald and Sachs, 1965; DiPaolo and Donovan, 1967; Kuroki *et al.*, 1967; Pienta, 1979). More than 100 chemical, physical, and biological agents have been examined for their ability to induce transformation in the HEC system by this and other laboratories (DiPaolo and Casto, 1977; Pienta, 1979). No false positive results and only rare false negative results have occurred.

Freshly isolated or early passage guinea-pig (Evans and DiPaolo, 1975), human (Kakunaga, 1977; Milo and DiPaolo, 1978), and rat (Olinici and DiPaolo, 1974; Sekely *et al.*, 1973) fibroblast-like fetal cell strains can also be used to study transformation. Results with mouse cell strains can be difficult to interpret owing to the intrinsic high level of spontaneous transformation. Each species has innate responses to carcinogen insult and value in the detection and study of carcinogens. The rat is most

similar to the hamster. Guinea-pig and human cells, however, diverge dramatically from the temporal pattern of *in vitro* carcinogenesis exhibited by HEC. In the guinea-pig and human models, morphological transformation appears weeks or even months after carcinogen exposure and frequently precedes neoplastic transformation, the ability of morphologically transformed cells to produce a progressively growing tumor when injected into an appropriate animal.

The long latent period between carcinogen exposure and transformation of guinea-pig and human cells makes it difficult to quantify morphological and neoplastic transformation in relation to the concentration of carcinogen. The expanded latent period, however, provides unique opportunities to study the temporal acquisition of phenotypic changes, such as loss of density-dependent inhibition of growth, alteration of surface antigens, and other surface-associated molecules such as plasminogen activator activity, fibronectin, and susceptibility to lymphotoxin cytotoxicity (Evans *et al.*, 1977). The increased latent period also provides greater latitude to study cocarcinogenic action such as additive, synergistic, and promoting effects, and anticarcinogenic action. HEC transformation, however, with its short latent period and hence greater quantification *in vitro*, is a more useful short-term test for the identification of carcinogenic, cocarcinogenic, and anticarcinogenic action.

Hamster Embryo Cell Transformation

In the following paragraphs, the techniques of the HEC transformation assays are described. Examples are also presented of the quantitative dose response to diverse carcinogens obtained with fresh and cryopreserved cells and of cocarcinogenic and anticarcinogenic modulation of carcinogen-induced transformation. These demonstrate the versatility of the HEC transformation system as a short-term assay to identify and study the carcinogenic potential and modulating activity of chemical, physical, and biological agents.

Principle of the Hamster Embryo Cell Transformation Assays

The exposure of HEC to a carcinogen results in alterations in cell morphology in which a small percentage develop into a severe, persistent morphological derangement characterized by a loss of density-dependent regulation of growth, with a random crisscrossed piling up of cells defined as morphological transformation that is not seen in controls (DiPaolo et al., 1971d). Morphological transformation occurs in colonies when sparsely seeded cells are exposed to a carcinogen in the colony assay or in foci of densely piled up transformed cells on a background of oriented density-dependent growth-inhibited cells when heavily seeded cells are exposed to a carcinogen, the focus assay. In both assays, transformation is observed only after carcinogen treatment and the frequency of transformation varies directly with the concentration of the carcinogen. Transformation is a reliable indication of the carcinogenic potential of the test agent and relevant to malignancy, since isolation, propagation, and injection of the transformed cells into newborn hamsters or homozygous nude athymic NIH/Swiss mice is followed by growth of the cells as a progressively enlarging tumor (DiPaolo et al., 1971d; Evans and DiPaolo, 1976). The frequency of transformation, just as the incidence of tumors induced by a carcinogen in vivo, can be increased or decreased by other chemical, physical, and biological agents, allowing identification and study of cocarcinogenic and anticarcinogenic action in vitro (Casto and DiPaolo, 1973; Casto et al., 1977; DiPaolo and Donovan, 1976, 1978; DiPaolo et al., 1971b,c, 1974; Donovan and DiPaolo, 1974).

HEC Assay Conditions

Reproducible results depend on an ideal environment for cells and animals as well as on reagents of high quality. Broadly speaking, the procedures used in selecting animals, cell culturing, and the standard colony transformation assay are similar to those originally published in DiPaolo et al. (1969). The dispersed cells are grown in Dulbecco's modification of Eagle's minimal essential medium supplemented with 10% fetal bovine serum except that the phenol red is reduced to 5 mg/liter of medium. All cells are grown in Petri dishes in an incubator at 37° in a 10% CO_2 water-saturated air atmosphere.

HEC are obtained from pregnant Syrian golden hamsters at 13–14 days of gestation. Our laboratory currently uses NIH animals. Animals from a variety of commercial as well as other sources have yielded successful experiments, provided they come from a well-maintained, healthy, and pathogen-free colony.

Random-bred Syrian golden hamsters are maintained in a room with a regulated light cycle consisting of 12 hours light and 12 hours darkness. The room is automatically darkened at 4:00 P.M. and at 5:00 P.M. the planned matings are done. The modified light cycle for the animals makes it possible for animals to breed during the day rather than at night, since the onset of estrus is influenced by the lighting of the animal quarters. The temperature of the breeding room is 22–23° C. Temperature control appears essential for the maintenance of regular reproductive cycles in both males and females. The animals are examined regularly and frequently handled, since tame animals tend to have larger litters and frequent examination of animals makes it possible to eliminate runts or animals with obvious pathological conditions. Any animal that dies or is killed is given a complete autopsy. Mature males are housed individually and a female is placed in their cages. If the female turns on the male, she is removed from the cage and replaced with another female until one is found which assumes the primary lordotic position. The next morning the dam is provided with an individual cage containing sawdust bedding, mouse (NIH07) and guinea-pig pellet chow, and a crumpled up paper towel to be used in making a nest. Once the proven studs have been identified, they may be used for up to two years; females are usually bred the first time at 8–10 weeks of age and are used to provide fetal material usually after the second or third pregnancy.

The dam, at 13–14 days' gestation, when abdominal palpation indicates that the individual amnionic sacs are approximately 5/8-inch in diameter, just prior to elongation of the fetuses, is killed quickly by cervical dislocation. The abdomen is washed with 70% ethanol, aseptically opened, and the uterus aseptically excised and placed in a sterile Petri dish. A number of factors determine whether the embryos of any particular dam will be used for an experiment. Important conditions include pneumonia, change in weight of the liver, or changes in its color from mahogany to any other color, and size of the spleen or kidneys as well as the weight of the dam. Dams 13–14 days in gestation usually have fetuses 16–19 mm in length (crown-rump). If the fetuses are not evenly distributed within the uterine horns or between the horns, or if the number of fetuses is less than the average number of 12–20 expected at the second or third pregnancy, all embryos are discarded. If the uterus contains more than one dead fetus, or if the fetuses do not show a gradual, slight change in size with the smallest at the ovarian end and the largest at the vaginal end of the uterus, they are also all discarded.

The embryos are aseptically removed from the uterus and placed in a Petri dish with 37° C Dulbecco's formula for phosphate buffered saline (PBS), washed twice with PBS, the liver and head removed, the bodies transferred to a clean Petri dish, minced with iris scissors, transferred to a trypsinizing flask with additional PBS, and washed for 10–15 minutes to remove red blood cells. A magnetic stirrer is used with gentle stirring to facilitate the washing. After washing, the tissue pieces are allowed to settle for 3–5 minutes and the supernatant decanted. After the PBS is removed, 50 ml of PBS containing 0.0075% (Worthington $2 \times$ recrystallized) trypsin at 37° is added, stirred for 10 minutes, allowed to settle for 5 minutes and the supernatant discarded with any cells and debris that flow with it. To collect cells, fresh 0.002% trypsin in 50 ml PBS is added for 10 minutes for the first time and 7–10 minutes for the next cell collection. At each interval, the supernatant fluid containing cells is decanted and the trypsinizing action inhibited with 10% v/v fetal bovine serum. After collection, the cells are centrifuged at $400 \times g$ for 5 minutes and suspended in 50 ml complete medium with 10% fetal bovine serum. Further cell dilutions are made so that 10^7 cells may be transferred to and grown as a primary culture in 100-mm Petri dishes in complete medium with 10% serum or so that the trypsinate may be frozen. Antibiotics are no longer included in our cell cultures.

Cells are frozen for cryostorage with a Linde BF-4/6 Biological Freezing System. The cells of the trypsinate are diluted to contain 10^7 cells in 2 ml of medium with 10% serum and 10% DMSO in a 2-ml screw-cap glass dram vial or plastic 2 ml Nunc cryotube. The cooling rate associated with the survival of the greatest number of viable cells is 1°/min. Cooling at the controlled rate is continued to a temperature of -40 to $-50°$ before the preparation is transferred to storage in liquid nitrogen. The use of the controlled rate freezer protects the cells from rewarming when the heat of fusion is released at the time of the change from the liquid to the solid phase. Approximately 20 vials of cells are obtained from the fetuses of a single dam. When the frozen cells are to be used, the vials of cells are removed from the liquid nitrogen. The cells are quickly thawed at 37°, the contents of the test tube are transferred to a 100-mm plastic Petri dish, and 10 ml medium with 10% serum is added. The medium is replaced at 24 hours, and 24 hours later there are approximately 10^7 total cells per Petri dish. These primary cells can be used for an experiment or subcultured at the level of $2.5 \times 10^6/100$ mm Petri dish and used as secondary cultures 48 hours after seeding or reseeded for use as 2-day-old tertiary cultures. The same culture schedule is used for fresh cells except that primary cultures are subcultured at 3 days and 2-day secondary cultures are routinely used when seeding cells for the colony assay. Cells are harvested from the dishes by removal of the medium and addition of 10 ml 0.002% trypsin in PBS at 37°. After 3–4 minutes, when the cells begin to round up, they are pipetted from the plate, transferred to a 15-ml centrifuge tube containing 1 ml serum, centrifuged, resuspended in 10 ml medium-serum, counted, and recultured.

For the quantitative colony transformation assay, 300 previously unfrozen or cryostored cells from a primary, secondary, or tertiary hamster culture are seeded simultaneously in 60-mm plastic Petri dishes with 6×10^4 irradiated hamster cells in 4 ml complete medium with 10 or 15% serum. The serum concentration varies from lot to lot, with 10% being the usual concentration over the years. The irradiated cells serve as a feeder layer to increase the cloning efficiency (20–30%) of the early passage cells. Sister primary, secondary, or tertiary cultures to the cultures used for the transformation assay are used to obtain the feeder cells. The medium is removed from the dish, 3600 R of X-irradiation delivered at 100 Kvp and 5 ma from a Picker portable industrial X-ray machine (T55-433), and the cells trypsinized as described above. Test chemicals are usually added 24 hours later with the addition of 4 ml complete medium-serum. Chemicals are first dissolved and sterilized in acetone at 10

mg/ml. Polycyclic hydrocarbons are further diluted with warm complete medium to the desired concentration. Other chemicals are diluted in cold medium without serum until the final dilution when warm complete medium is used. All dilutions and additions to cell cultures are performed under yellow light to protect light-sensitive chemicals and decrease photo-induced reactions.

Transformation in the colony assay is scored 7 days after seeding of the cells. Cell growth is stopped by removing the medium, rinsing the dishes with PBS, and fixing the cells by adding 4–8 ml of methanol/dish for 30 minutes. The methanol is then removed, 4–8 ml 10% aqueous Giemsa stain (Fisher Scientific) added for 45 minutes, the Giemsa removed, and the dishes washed with distilled water and air-dried. Colonies greater than 2 mm in diameter are counted macroscopically and the morphology scored through the bottom of the transilluminated inverted dish, using a stereoscopic dissecting microscope at 50 to 500 × magnification. Ten to twelve dishes are analyzed for each test agent concentration. The criterion for transformation is altered morphology characterized by crisscrossing of cells not observed in controls (DiPaolo *et al.*, 1971d). The frequency of transformation is calculated from the average number of transformed colonies per dish or as the average number of transformed colonies per total colonies.

An additional method of exposing the cells to chemical agents is by the host-mediated *in vivo/in vitro* assay (DiPaolo *et al.*, 1973). This system differs from the standard *in vitro* method by including *in vivo* transplacental exposure of the fetuses to the chemical or to its metabolites. Pregnant hamsters at 11 days of gestation are given intraperitoneal injections of the chemical in 1 ml of saline, ethanol, or trioctanoin at a chemical concentration of 1–5 mg/100 g maternal weight. The injected animals are fed water and standard laboratory chow, and the embryos excised on day 13, 48–60 hours after maternal injection.

The preparation and culture of cells from these embryos and the identification of transformation are as described above.

In the focus assay, secondary HEC are plated at 5×10^4 cells/60-mm dish in 4 ml of BioLabs modified Dulbecco's medium containing 10% FBS and 0.11 g of $NaHCO_3$ per 100 ml (Casto *et al.*, 1977). After 24 hours, 4 ml of a chemical dilution (as $2 \times$ the desired final concentration) is added to each dish and it is incubated for 3 days at 37° in 5% CO_2. The culture fluid is replaced with fresh medium (0.22 g of $NaHCO_3$ per 100 ml) at 3 to 4 day intervals. The cells are fixed, stained, and scored for transformed foci at 21–25 days. Cell lethality is determined on dishes seeded with 1000 cells and treated as above. Surviving colonies are fixed, stained, and counted 8 days after seeding. The surviving fraction is determined by dividing the number of colonies on the test-treated dishes by the number on the solvent-treated dishes. The number of foci/dish may then be expressed as a transformation frequency or foci/10^5 surviving colonies. The focus assay is also used to study virus-induced transformation and virus-chemical interactions as described elsewhere (Casto *et al.*, 1973) and in this volume.

Identification of Carcinogenic Action and Its Modulation by Cocarcinogens and Anticarcinogens

The sensitivity and reproducibility of Syrian hamster embryo cells to transformation induced by a wide variety of organic and inorganic chemical carcinogens and UV irradiation is shown in Tables 28.1 to 28.8. A good correlation between the degree of metabolic activation, carcinogenicity, and transformation potential of a series of compounds can be obtained (DiPaolo *et al.*, 1969, 1972) and is shown for AAF, N-OH-AAF and N-Ac-AAF in Table 28.1. For compounds in which metabolic activation does not occur *in vitro*, such as urethan and DEN, transformation can be demonstrated by the host-mediated combination *in vivo/in vitro* assay as indicated in Table 28.1.

Table 28.1. *Transformation of Syrian Hamster Cells by Chemicals of Various Classes*

Compound	Dose (μg/ml)	Cloning efficiency, %	Transformed colonies, %
Aflatoxin B_1	0.5	5.2	7.0
11-Methylcyclopenta(a)-phenanthrene	5	4.6	5.8
N-2-acetylaminofluorene (AAF)	5	4.8	0.5
N-Hydroxy AAF	5	3.2	2.1
N-acetoxy AAF	5	4.4	15.4
Urethan	50	6.4	0
N-Hydroxyurethan	12.5	2.9	0
Diethylnitrosoamine (DEN)	100	8.9	0
N-Methyl-N-nitro-N-nitrosoguanidine (MNNG)	0.5	4.8	6.9
Methylazoxymethanol (MAM)	2.5	11	8.4
DEN transplacental	0.08[a]	11	14.3
Urethan transplacental	1[a]	18	17.9

[a]mg/g intraperitoneally on day 12 of gestation.

Quantitative dose responses are obtained with the focus (Table 28.2) or the colony (Tables 28.3 and 28.4) assay. Either freshly cultured cells or cells stored in liquid nitrogen can be utilized for quantitative identification of carcinogenic activity as indicated in Table 28.5. Cryostored cells are an advantage to the laboratory without a continuing supply of pregnant animals or facilities to house them. Equivalent results are obtained with cryostored cells when seeded in the transformation assay from primary, secondary, or tertiary cultures compared with noncryostored cells seeded from freshly established secondary cultures (Table 28.6). The variation in transformation frequency is the same among different pools of cryostored cells (Table 28.7) as

Table 28.2. *Focus Assay for Transformation of Syrian Hamster Cells by Various Chemical Carcinogens*

Chemical	Dose (μg/ml)	Surviving fraction[a]	Foci/ dish	Transformation frequency[b]
Ac-AAF	5.0	<0.003	3/19	>105.3
	2.5	0.11	10/20	9.1
	1.2	0.79	8/19	1.1
	0.6	1.08	7/20	0.6
	Acetone	0.84	0/20	0
Benzo(a)pyrene	5.0	0.66	5/8	1.9
	2.5	0.74	4/10	1.1
	1.2	0.60	1/9	0.4
	0.6	0.71	0/7	0
	Acetone	1.18	0/9	0
3-Methylcholanthrene	1.00	0.19	12/20	6.3
	0.50	0.28	16/20	5.7
	0.25	0.38	9/20	2.4
	0.12	0.56	6/20	1.1
	Acetone	0.84	0/20	0
Aflatoxin B_1	1.00	0.04	1/10	5.0
	0.50	0.15	1/10	1.3
	0.25	0.32	2/7	1.8
	0.12	0.47	0/8	0
	Acetone	1.07	0/9	0
β-PL	20.0	<0.01	0/10	0
	10.0	0.10	2/10	4.1
	5.0	0.31	1/10	0.6
	2.5	0.87	0/10	0
	Acetone	0.99	0/10	0
DB(a,h)A	5.0	1.20	1/10	0.7
	2.5	1.05	2/10	1.5
	1.2	1.07	1/10	0.7
	Acetone	0.93	0/10	0
EMS	400	0.014	3/10	42.8
	200	0.92	0/6	0
	100	0.99	0/10	0
	Acetone	1.19	0/10	0
MMS	50.0	0.0002	0/10	0
	25.0	0.11	3/10	2.7
	12.5	0.63	2/10	0.3
	Acetone	1.05	0/10	0

[a]The number of surviving colonies in treated dishes divided by the number of colonies in solvent control dishes, each seeded with 1000 cells compared to 50,000 cells in the dishes for foci formation.

[b]The number of foci per 10^5 surviving cells.

Table 28.3. *Transformation of Syrian Hamster Cells Treated with Diverse Inorganic Metal Salts*

Chemical	Dose (μg/ml)	Cloning efficiency, %	Transformed colonies/dish	Transformed colonies, %
$NiSO_4 \cdot 6H_2O$	0	28.1	0	0
	2.5	24.5	1.0	1.2
	5.0	20.1	1.5	1.4
	10.0	13.8	2.2	5.4
$(CH_3COO)_2Cd \cdot H_2O$	0	28.1	0	0
	0.1	22.3	1.0	1.5
	0.5	18.8	1.8	3.1
	1.0	9.3	1.1	4.0
$Na_2CrO_4 \cdot 4H_2O$	0	28.2	0	0
	0.1	27.8	0.6	0.7
	0.5	27.8	1.8	2.1
	1.0	17.3	1.8	3.5
Ni_3S_2	0	28.1	0	0
	1.0	14.5	0.7	1.5
	2.5	6.4	1.0	5.2
	5.0	5.1	1.8	11.5
$BeSO_4$	0	26.0	0	0
	2.5	16.4	0.9	1.8
	5.0	12.0	2.3	6.4
Na_2HAsO_4	0	29.1	0	0
	2.5	22.2	1.2	1.7
	5.0	17.5	2.1	4.1

among experiments with fresh cells (Table 28.8). Routinely, results are reproducible within a twofold range (DiPaolo, 1980; DiPaolo *et al.*, 1971a). Cryostored cells can be used for at least 2 years after placement in liquid nitrogen, as shown in Table 28.7 by the results from pool 30–76 stored for 2 years before being cultured in 1978 with cell pools 47–78 to 59–78. The ability to store the cells for extended periods per-

mits the use of one pool of cells for a series of assays. It further allows the exchange of cells with equivalent viability, sensitivity, and reproducibility between laboratories, facilitating comparison of test results.

In vitro transformation of HEC is appropriate for cocarcinogenesis studies. Tables 28.9 and 28.10 demonstrate the time-dependent synergistic induction of transformation by X-irradiation and chemical car-

Table 28.4. *Transformation of Syrian Hamster Cells by UV (254 nm)*[a]

Dose (J/m^2)	Cloning efficiency, %	Transformed colonies/dish	Transformed colonies, %
0	22.5	0	0
0.7	21.7	0.5	0.8
1.5	20.7	1.0	1.7
3.0	18.1	1.7	3.2
4.6	13.7	2.6	6.3
6.1	6.7	1.5	7.5

[a]For colony formation assay, 300 cells/60-mm dish were seeded; 24 hr later, the medium was replaced with phosphate buffered saline while the cells were treated with UV; after irradiation, the phosphate buffered saline was replaced with fresh medium; incubation was continued for an additional 7 days before the colonies were fixed for examination. Irradiation delivered from a single 15-watt germicidal GE 15T8 lamp through a Permanox top.

Table 28.5.   *Transformation of Cryostored Syrian Hamster Cells by Diverse Carcinogens*[a]

Chemical	Dose (μg/ml)		Cloning efficiency, %	Transformed colonies/dish	Transformed colonies, %
None	—		29	0	0
MNNG	0.15		23.6	0.5	0.7
	0.25		20.2	1.0	1.6
	0.50		10.1	1.7	5.6
AAF	1.25		29.2	0.1	0.1
	2.5		19.7	0.3	0.4
	5.0		26.9	0.7	0.8
N-Acetoxy AAF	1.25		27.4	1.2	1.6
	2.5		19.7	1.9	3.2
	5.0		11.7	5.2	14.0
Aflatoxin B$_1$	0.125		22.3	2.0	3.0
	0.25		19	2.8	5.0
	0.50		11.7	3.8	10.7
Cis Pt(II)	0.05		24.3	0.5	0.7
	0.1		21.5	1.5	2.4
	0.2		18.0	1.2	6.5
UV[b]	3	J/m^2	18.5	1.6	2.9
	4.5	J/m^2	12.9	2.4	6.2

[a]HEC for feeder and cells to be treated obtained from sister secondary cultures.

[b]See Table 28.4 for method of irradiation.

cinogens. X-irradiation doses up to and including 1000 R do not in themselves induce transformation but enhance chemical carcinogen induced transformation up to tenfold when administered 48 hours before the chemical carcinogen. If X-irradiation is administered 24 or 72 hours before the carcinogen (DiPaolo *et al.*, 1971b), the enhancement is less. A similar enhancement of transformation, maximal with pretreatment 48 hours before carcinogen addition, is seen with some weakly transforming alkylating agents (DiPaolo *et al.*, 1974), as shown for methyl methanesulfonate in Table 28.11. Caffeine, a noncarcinogen and nontransforming chemical, produces a different cocarcinogenic synergistic increase in transformation (Donovan and DiPaolo, 1974). The maximum enhancement of transformation occurs when the cells are exposed to caffeine 4 hours after treatment with carcinogen (Table 28.12).

The sequential addition of carcinogenic agents may also result in an additive cocarcinogenic increase in transformation rather than a synergistic enhancement, as indicated by the results in Table 28.13 when

UV was followed by N-Ac-AAF treatment. Another type of cocarcinogenic synergistic increase in transformation is shown in Table 28.14, where pretreatment of the cells with 7,8-benzoflavone or the weak carcinogen benz(a)anthracene enhanced the transformation induced by the carcinogen benzo(a)pyrene. The cocarcinogenic action in this instance presumably reflects a stimulation of the microsomal enzymes metabolizing the carcinogenic polycyclic hydrocarbon, as coincident with the enhanced transformation there is a decrease in the toxicity of the carcinogen (DiPaolo *et al.*, 1971c). Alteration in the sequence and/or timing of flavone or weak cocarcinogen exposure may result in a decrease in transformation, i.e., anticarcinogenic action.

Anticarcinogenic action, the inhibition of carcinogenesis *in vivo* or inhibition of *in vitro* transformation can be studied with the focus and colony assays. Table 28.15 shows the inhibition of foci formation resulting from a change in nutrient culture conditions to one selectively favoring the growth of tumor cells but which inhibit

Table 28.6. *Transformation by Benzo(a)pyrene of Liquid Nitrogen-Stored Syrian Hamster Cells Seeded from Primary. Secondary, or Tertiary Cultures and Fresh Secondary Cells Seeded on Fresh Feeder*

Dose (µg/ml)	Primary			Secondary			Tertiary			Fresh		
	C.E.[a]	T/d[b]	T/c[c]	C.E.	T/d	T/c	C.E.	T/d	T/c	C.E.	T/d	T/c
0	35.7	0	0	32.7	0	0	32	0	0	31.3	0	0
2.5	18.3	3.1	5.7	24.4	2.7	3.7	27.3	3.8	4.6	28.0	2.6	3.1
5.0	16.7	4.5	9.0	23.3	4.2	6.0	24.3	5.1	7.0	27.4	4.6	5.7
10.0	14.7	5.7	11.8	22.6	8.8	13.0	23.3	7.2	10.3	17.7	7.3	13.8

[a]C.E., cloning efficiency, determined by dividing the average number of colonies per plate by the number of cells seeded per plate multiplied by 100.

[b]Average number of transformed colonies per dish.

[c]Transformed colonies relative to the total colony count,%.

Table 28.7. *Variation in the Transformation Rate Resulting from 5 μg BP/ml Medium with Cells Obtained from Ten Different Cryostored Cell Pools*

Cell pool	Cloning efficiency,%	Transformed colonies/dish	Transformed colonies, %
30–76	23.8	3.2	4.8
90–77	12.4	3.0	8.0
3–78	20.6	4.2	6.9
45–78	23.3	4.2	6.0
46–78	19.7	3.2	5.4
47–78	17.9	2.2	4.0
49–78	23.3	4.2	6.0
50–78	21.3	4.0	6.2
59–78	17.6	3.8	7.2
6–79	23.8	4.1	5.7

Table 28.8. *Variation in Transformation Frequency of Syrian Hamster Cells Irradiated with UV (254 nm)[a]*

Experiment	Cloning efficiency, %	Transformed colonies/dish	Transformed colonies, %
1	18.0	1.7	3.2
2	11.2	1.9	5.6
3	19.1	1.7	3.0
4	21.8	2.1	3.2
5	15.4	1.5	3.2
6	14.0	1.7	4.0
7	16.5	1.6	3.3

[a]Irradiated with 3.0 J/m² through a Permanox top except for 6 and 7 which were irradiated without a top with 2.9 J/m² as described in Table 28.4.

Table 28.9. *Transformation of Syrian Hamster Cells X-irradiated 48 hr before Treatment with 10 μg BP/ml*

X-ray (R)	Cloning efficiency, %	Transformed colonies, %	Enhancement[a]
0	3.6	6.7	1.0
150	3.6	16.6	2.5
250	3.4	54.8	8.2
500	2.2	40.8	6.1
750	1.0	12.4	1.9
1000	0.4	11.4	1.7

[a]Ratio of frequency of transformed colonies from irradiated cells to unirradiated results defined as 1.0.

Table 28.10. *Transformation of Syrian Hamster Cells X-irradiated before Treatment with 254 nm UV[a]*

UV dose (J/m²)	Cloning efficiency, %	Transformed colonies/dish	Transformed colonies, %	Enhancement
No X-ray				
0	24.7	0	0	
1.5	23.4	0.9	1.2	1.0
3.0	17.9	1.5	2.7	1.0
X-ray				
0	17.6	0	0	
1.5	12.3	5.6	15.9	12.2
3.0	9.1	4.1	15.1	5.6

[a]For X-irradiation enhancement of transformation, hamster cultures were exposed to 250 R, cloned, incubated for 48 hr; medium was temporarily replaced with PBS during UV irradiation and cultures further incubated for 7 days. Enhancement-transformed colonies obtained with X-ray and UV to colonies obtained with UV only. The latter by definition is 1.

Table 28.11. Transformation of Syrian Hamster Cells after Sequential Treatment with Methylmethanesulfonate and Benzo(a)pyrene[a]

	0 hr			24 hr			48 hr			72 hr		
MMS µg/ml	0	11	27.5	0	11	27.5	0	11	27.5	0	11	27.5
Cloning efficiency, %	15.3	13.3	12.4	19.9	16.6	14.7	21.2	17.1	14.6	18.1	15.6	15.0
Total transformants	32	48	32	55	43	61	32	238	171	25	47	34
Transformed colonies, %	3.3	5.7	4.3	3.8	3.8	5.8	2.1	21.1	17.0	2.0	4.4	3.3
Enhancement	1.0	1.7	1.3	1.0	1.0	1.5	1.0	10	8.1	1.0	2.2	2.7

[a]Cells were seeded on a feeder layer and 24 hr later pretreated with MMS for 1 hr. BP at a final concentration of 2.5 µg/ml was added at the periods indicated: period 0 refers to first hour after MMS pretreatment. See Table 28.9 for definition of enhancement.

Table 28.12.   *Transformation of Hamster Cells after AcAAF or MNNG Treatment followed by Caffeine*

Hr Caffeine added after carcinogen[a]	Cloning efficiency, %	Transformed colonies/dish	Transformed colonies, %	Enhancement[b]
AcAAF				
—	23.0	0.8	1.1	1.0
1	14.9	1.9	4.3	3.8
3	8.1	2.2	9.1	8.2
4	7.7	3.6	15.6	13.9
6	9.0	1.1	4.2	3.8
14	20.2	1.8	3.0	2.1
24	26.2	0.9	1.2	1.2
48	24.7	1.0	1.3	0.8
MNNG				
—	24.1	0.7	1.0	1.0
1	18.7	3.4	6.1	6.1
4	17.4	5.2	9.9	9.9
12	19.0	2.2	3.9	2.7
24	17.4	5.0	9.6	7.4
48	16.3	4.2	8.5	6.1
72	18.4	2.7	4.8	2.9

[a] 300 cells (60-mm dish) were seeded for colony formation; 24 hr later the cells were treated with 1 μg AcAAF or 0.25 μg MNNG followed by 50 μg caffeine/ml medium at the times indicated; 48 hr after addition of caffeine, the medium was replaced by fresh medium without caffeine; corresponding dishes with AcAAF or MNNG only (——) were also refed.

[b] Ratio of frequency of transformed colonies obtained with AcAAF or MNNG and caffeine to colonies obtained with AcAAF or MNNG only; the latter by definition is 1.

Table 28.13.   *Transformation of Syrian Hamster Cells Sequentially Treated with UV Irradiation and with AcAAF[a]*

Treatment	Cloning efficiency, %	Transformed colonies/dish	Transformed colonies, %
UV	20.9	0.9	1.4
AcAAF	16.5	1.0	3.0
UV + AcAAF			
0.1 hr	9.6	1.1	3.7
1 hr	10.1	1.2	4.1
2 hr	12.4	1.5	4.1
3 hr	11.0	1.2	3.8
4 hr	10.6	0.7	2.1
Control	23.5	0	0

[a] After 24 hr growth of cells seeded for colony formation, the medium was removed and temporarily replaced with PBS during UV irradiation with 1.5 J/m². PBS was replaced by complete medium and at the times indicated, AcAAF at 0.5 μg/ml was added.

Table 28.14. *Transformation of Syrian Hamster Cells Sequentially Treated with a Flavone or Benz(a)anthracene 24 hr before Addition of Benzo(a)pyrene*

7,8-Benzoflavone (μg/ml)	BP (μg/ml)	Cloning efficiency, %	Transformed colonies, %	Enhancement[b]
0	0	13.7	0	
5	0	12.3	0	
0	1	8.5	2.5	1.0
0	5	7.2	5.3	1.0
1	1	12.6	9.2	3.7
5	1	13.6	13.1	5.3
1 × 3[a]	1	13.0	7.2	2.9
1	5	12.8	8.8	1.6
5	5	13.0	7.7	1.4
1 × 3[a]	5	13.0	5.3	1.0
Benz(a)anthracene				
0	0	13.7	0	
10	0	13.1	0.5	
0	1	8.5	2.5	1.0
0	5	7.2	5.3	1.0
1	1	11.6	5.0	2.0
5	1	13.4	10.8	4.4
1 × 3[a]	1	12.3	8.6	3.5
1	5	11.6	6.0	1.1
5	5	12.8	7.8	1.5
1 × 3[a]	5	12.3	7.7	4.3
10	5	12.8	23.1	2.7

[a]Three treatments at 48-hr intervals.

[b]See Table 28.9 for definition of enhancement.

Table 28.15. *Inhibition of Syrian Hamster Cell Focus Transformation by Change in Serum or Addition of Polysaccharide[a]*

Culture conditions	Foci/ dish	Inhibition of foci, %
10% Fetal bovine serum (FBS) throughout	0.5	—
FBS changed to 1%	0.05	90
FBS changed to 10% calf serum	0	100
Dextran sulfate added at 4 to 8 μg/ml medium	0	100
DEAE-dextran added at 2 μg/ml medium	0	100

[a]Serum was changed or polysaccharide added one day after removal of MCA. The MCA was used at 0.125 to 2.0 μg/ml medium and the inhibition was independent of MCA concentration.

transformation. Another example of anticarcinogenic activity is presented in Table 28.16 where the addition of lymphotoxin, one of the lymphokines produced by lymphocytes, inhibits the development of transformation.

The foregoing examples prove the versatility of HEC *in vitro* transformation assays for the identification of carcinogens, cocarcinogens, and anticarcinogens. The model system is also a valuable means to study the mechanism of carcinogenesis. Mechanistic investigations will help in selecting agents warranting a high priority for extensive evaluation and possible environmental control, particularly when consideration is given to methods to modulate their carcinogenic potential when elimination from the environment is not possible.

Table 28.16. *The Anticarcinogenic Action of Lymphotoxin during Transformation of Syrian Hamster Embryo Cells by Diverse Carcinogens*[a]

Carcinogen	Concentration	Lymphotoxin units/dish	Cloning efficiency, %	Transformed colonies, %	Inhibition of transformation, %
None	—	0–437	30.1–32.6	0	—
Benzo(a)pyrene	2.5 µg/ml	0	20.6	8.1	—
		61	12.1	1.4	83
		99	12.8	1.3	84
		437	6.6	0	100
Ultraviolet (254 nm)	3.0 J/m²	0	24.7	1.7	—
		61	28.9	0.4	76
		99	26.4	0	100
		437	22.9	0	100

[a]0.9 ml lymphotoxin was added/dish 48 hr after carcinogen. Lymphotoxin was prepared from Syrian hamster peritoneal leukocytes (Evans *et al.*, 1977). Lymphotoxin units were determined by measuring ³H release from HTdR labeled murine alpha-L929 cells incubated with lymphotoxin for 3 days (Zwilling *et al.*, 1975). One unit of lymphotoxin releases 50% of the ³H from 10⁴ L cells.

References

Berwald, Y., Sachs, L. (1965): *In vitro* transformation of normal cells to tumor cells by carcinogenic hydrocarbons. J. Natl. Cancer Inst. *35*:641–661.

Casto, B.C., DiPaolo, J.A. (1973): Virus, chemicals and cancer. Progr. Med. Virol. *16*:1–47.

Casto, B.C., Janosko, N., DiPaolo, J.A. (1977): Development of a focus assay model for transformation of hamster cells *in vitro* by chemical carcinogens. Cancer Res. *37*: 3508–3515.

Casto, B.C., Pieczynski, W.J., DiPaolo, J.A. (1973): Enhancement of adenovirus transformation by treatment of hamster cells with diverse chemical carcinogens. Cancer Res. *34*:72–78.

DiPaolo, J.A. (1980): Quantitative *in vitro* transformation of Syrian Golden hamster embryo cells with the use of frozen stored cells. J. Natl. Cancer Inst. *64*:1485–1489.

DiPaolo, J.A., Casto, B.C. (1977): Chemical carcinogens. *In*: Gallo, R.C. (ed.), Recent Advances in Cancer Research: Cell Biology, Molecular Biology and Tumor Virology, Vol. 1. Cleveland: CRC Press, pp. 17–48.

DiPaolo, J.A., Casto, B.C. (1979): Quantitative studies of *in vitro* morphological transformation of Syrian hamster cells by inorganic metal salts. Cancer Res. *39*:1008–1013.

DiPaolo, J.A., Donovan, P.J. (1967): Properties of Syrian hamster cells transformed in the presence of carcinogenic hydrocarbons. Exp. Cell Res. *48*:361–377.

DiPaolo, J.A., Donovan, P.J. (1976): *In vitro* morphologic transformation of Syrian hamster cells by U.V.-irradiation is enhanced by X-irradiation and unaffected by chemical carcinogens. Int. J. Radiat. Biol. *30*: 41–53.

DiPaolo, J.A., Donovan, P.J. (1978): Transformation frequency of Syrian golden hamster cells and its modulation by ultraviolet irradiation. *In*: Hanna, M.G. (ed.), International Conference on Ultraviolet Carcinogenesis. Natl. Cancer Inst. Monogr. *50*:75–80.

DiPaolo, J.A., Donovan, P.J., Casto, B.C. (1974): Enhancement by alkylating agents of chemical carcinogen transformation of hamster cells in culture. Chem.-Biol. Interact. *9*: 351–364.

DiPaolo, J.A., Donovan, P.J., Nelson, R.L. (1969): Quantitative studies of *in vitro* transformation by chemical carcinogens. J. Natl. Cancer Inst. *42*:867–876.

DiPaolo, J.A., Donovan, P.J., Nelson, R.L. (1971a): *In vitro* transformation of hamster cells by polycyclic hydrocarbons: Factors influencing the number of cells transformed. Nature (Lond.) *230*:240–242.

DiPaolo, J.A., Donovan, P.J., Nelson, R.L. (1971b): X-irradiation enhancement of transformation by benzo(a)pyrene in hamster embryo cells. Proc. Natl. Acad. Sci. USA *68*: 1734–1737.

DiPaolo, J.A., Donovan, P.J., Nelson, R.L. (1971c): Transformation of hamster cells *in vitro* by polycyclic hydrocarbons without cytotoxicity. Proc. Natl. Acad. Sci. USA *68*: 2958–2961.

DiPaolo, J.A., Nelson, R.L., Donovan, P.J. (1971d): Morphological, oncogenic, and karyological characteristics of Syrian hamster embryo cells transformed *in vitro* by carcinogenic polycyclic hydrocarbons. Cancer Res. *31*:1118–1127.

DiPaolo, J.A., Nelson, R.L., Donovan, P.J. (1972): *In vitro* transformation of Syrian hamster embryo cells by diverse chemical carcinogens. Nature (Lond.) *235*:278–280.

DiPaolo, J.A., Nelson, R.L., Donovan, P.J., Evans, C.H. (1973): Host-mediated *in vivo–in vitro* assay for chemical carcinogens. Arch. Pathol. *95*:380–385.

Donovan, P.J., DiPaolo, J.A. (1974): Caffeine enhancement of chemical carcinogen-induced transformation of cultured Syrian hamster cells. Cancer Res. *34*:2720–2727.

Evans, C.H., DiPaolo, J.A. (1975): Neoplastic transformation of guinea pig cells in culture induced by chemical carcinogen. Cancer Res. *35*:1035–1044.

Evans, C.H., DiPaolo, J.A. (1976): Comparison of nude mice with the host species for evaluation of the tumorigenicity of guinea pig and hamster cells transformed *in vitro* by chemical carcinogens. Cancer Res. *36*:128–131.

Evans, C.H., Rabin, E.S., DiPaolo, J.A. (1977): The susceptibility of guinea pig cells to the colony-inhibitory activity of lymphotoxin during carcinogenesis. Cancer Res. *37*:898–903.

Heidelberger, C. (1977): Oncogenic transformation of rodent cell lines by chemical carcinogens. *In*: Hiatt, H.H., Watson, J.D., and Winsten, J.A. (eds.), Origins of Human Cancer, Book C, Cold Spring Harbor Laboratory, New York, pp. 1513–1520.

Kakunaga, T. (1977): The transformation of human diploid cells by chemical carcinogens. *In*: Hiatt, H.H., Watson, J.D., and Winsten, J.A. (eds.), Origins of Human Cancer, Book C, Cold Spring Harbor Laboratory, New York, pp. 1537–1548.

Kuroki, T., Goto, M., Sato, H. (1967): Malignant transformation on hamster embryonic cells by 4-hydroxyanimoquinoline N-oxide in tissue culture. Tohuku J. Exp. Med. *91*: 109–118.

Milo, G.E., DiPaolo, J.A. (1978): Neoplastic transformation of human diploid cells *in vitro* after chemical carcinogenic treatment. Nature (Lond.) *275*:130–132.

Olinici, C.D., DiPaolo, J.A. (1974): Chromosome banding patterns of rat fibrosarcomas induced by *in vitro* transformation of embryo cells or *in vivo* injection of rats by 7, 12-dimethylbenz(a)anthracene. J. Natl. Cancer Inst. *52*:1627–1634.

Pienta, R.J. (1979): A hamster embryo model system for identifying carcinogens. *In*: Griffin, A.C., Shaw, C.R. (eds.), Carcinogens: Identification and Mechanisms of Action. New York: Raven Press, pp. 121–141.

Sekely, L.I., Malejka-Giganti, D., Gutman, H.R., Rydell, R.E. (1973): Malignant transformation of rat embryo fibroblasts by carcinogenic fluorenylhydroxamic acids *in vitro*. J. Natl. Cancer Inst. *50*:1337–1346.

Zwilling, B.S., Meltzer, M.S., Evans, C.H. (1975): Differential cytotoxicity of tumorigenic and nontumorigenic strain-2 guinea pig cells mediated by phytohemaggtutinin stimulated peritoneal exudate cells. J. Natl. Cancer Inst. *54*:743–747.

29

The Use of Cryopreserved Syrian Hamster Embryo Cells in a Transformation Test for Detecting Chemical Carcinogens*

R.J. PIENTA, W.B. LEBHERZ, III, AND R.F. SCHUMAN

Introduction

Syrian Golden hamster embryo cells have been employed extensively (Huberman and Sachs, 1966; DiPaolo et al., 1969, 1971, 1972, 1973; Casto et al., 1977; DiPaolo and Casto, 1979) to study the activity of diverse carcinogens in cultured cells since the original report by Berwald and Sachs (1963). In most of these studies target cells were derived from primary embryo cultures, which were prepared as needed for each experiment. Efforts to standardize procedures for the routine use of these cells in a carcinogenicity bioassay were hampered initially by the variability in response observed among different batches of cells (Pienta et al., 1977). We discuss here the methodology for using aliquot samples of cryopreserved hamster embryo cells as the source of target and feeder cells in a standardized in vitro carcinogenesis bioassay.

Materials and Procedures

Animals

Random-bred Syrian golden hamsters are used as the source of target or feeder cell cultures in the bioassay. Several strains of timed-pregnant animals (LVG/LAK, Charles River, Lakeview Hamster Colony, Newfield, NJ; ELA/ENG, Engle Laboratory Animals for Cancer Research, Farmersburg, IN; Graffi, Leo Goodwin Institute, Nova University, Fort Lauderdale, FL) have given satisfactory results. The animals must be healthy and free of concurrent viral or bacterial infections as well as ecto- and endoparasites. To prevent the introduction of unwanted agents, timed-preg-

*The work was supported by Contract NO1-CO-75380 with the National Cancer Institute, NIH, Bethesda, Maryland 20205.

nant animals are brought into a quarantine facility at approximately the 7th day of gestation. On the following day, blood samples are taken to monitor for antibodies to a spectrum of viruses* in addition, to monitor for bacterial or parasitic infections, fecal, hair, and anal samples are taken. On the thirteenth day of gestation, embryos from acceptable animals are collected aseptically and primary cultures are prepared.

Culture Medium and Serum

Dulbecco's Modified Eagle's Medium (DMEM), containing 1000 mg/l glucose and 110 mg/l sodium pyruvate and supplemented with 2 mM L-glutamine and 20% sterile, heat-inactivated (56° C for 30 minutes) fetal bovine serum (FBS), is used for all bioassay experiments. DMEM is purchased as sterile liquid medium or prepared from powdered stock with deionized double-distilled water and sterilized by pressure filtration through a 0.22-μm filter. No antibiotics are used in bioassay experiments. Individual batches of culture medium are pretested for the ability to support optimal growth of normal hamster embryo cells before being accepted for use in bioassay experiments. Petri dishes and flasks are similarly tested for the absence of cytotoxicity.

Not all batches of FBS are satisfactory for use in routine bioassays; therefore, it is necessary to select acceptable ones (Schuman et al., 1979). As part of the evaluation for routine use, aliquot samples of RBS, certified to be free of adventitious agents by the supplier, are retested by the Quality Control Laboratory at the Frederick Cancer Research Center to confirm the absence of mycoplasma and bacteriophage.

---

*Sendai, pneumonia virus of mice, minute virus of mice, mouse adenovirus, mouse hepatitis virus, lymphocytic choriomeningitis, ectromelia, lactic dehydrogenase virus, REO-3, GDVII, polyoma, and K virus.

Samples of trypsin are similarly tested for mycoplasma contamination.

Test portions of each candidate serum sample selected for further evaluation are compared to the standard serum currently in use for the ability to support the growth of transformed colonies following treatment of cells with benzo(a)pyrene or 3-methylcholanthrene. The general morphological appearance of the cells is also evaluated for any adverse effects due to the sera under test. Serum samples acceptable for use are procured in quantities large enough to ensure a supply for several months. Serum is stored either at −80° C or −20° C until used.

Target and Feeder Cell Cultures

This assay relies on the morphological alteration of colonies resulting from the treatment of individual cells in culture with carcinogenic chemicals. Therefore, target cells are planted under conditions that allow the formation of 50–100 discrete colonies per 50-mm Petri dish. The plating or colony-forming efficiency of early passage diploid hamster embryo cells is extremely low (1–3%). However, this problem is circumvented by the use of lethally X-irradiated feeder cells whose function is to enhance the plating efficiency to about 15–20%, thus obtaining the required number of colonies of target cells. Paradoxically, if dishes are seeded with more than 5000 cells, monolayer cultures rather than individual colonies are formed. Historically, both rat and hamster embryo cells have been used as feeder cells. However, in this laboratory, hamster embryo cells are used as the source of both feeder and target cells.

Although detailed procedures for the preparation of primary cultures used subsequently as target and feeder cells have been reported elsewhere (Pienta, 1979, 1980; Pienta et al., 1977, 1978), the salient points will be briefly reiterated.

To prepare a batch of primary cells, embryos are collected aseptically during the twelfth to thirteenth day of gestation. They are then decapitated and eviscerated, after which the carcasses are gently minced with scissors and washed with a calcium-, magnesium-free balanced salt solution (TCD)* to remove red blood cells and tissue debris. The minced tissue is gently disaggregated, at room temperature, to a suspension of single cells by several 20–30-minute cycles of trypsinization with 1X ENZAR-T (Reheis) solution containing 0.5% FBS. The serum serves to slow down the trypsinization process, thus minimizing damage to the cells. The cell suspensions, held on ice, are pooled, washed by centrifugation at $200 \times g$ for 10 minutes, and resuspended in a minimum measured amount of DMEM. Viable cells are counted by trypan blue exclusion in a hemocytometer and approximately 1×10^7 cells are planted per 75 cm² flask in 20 ml of DMEM containing 20% FBS and antibiotics (200 units of pencillin and 200 μg streptomycin per ml). The cells are incubated at 37° C in a humidified atmosphere of 10% CO_2 in air. The cultures are refed with antibiotic-free medium at 24 hours and daily until 90–100% confluence is reached. Cultures not reaching 90% confluence by three days are discarded. Acceptable cultures are then disaggregated with 1X ENZAR-T, washed once, counted and dispensed into glass ampules, usually at 2.5×10^6 cells/ml for target cells and 5×10^6 cells/ml for feeder cells, in DMEM supplemented with a final concentration of 7.5% dimethylsulfoxide (DMSO, Crown-Zellerbach, Camus, WA) and 20% FBS. The cell suspension is gassed with CO_2, if necessary, to pH 7.4. The ampules are flame-sealed and placed at $-80°$ C to $-75°$ C for 1.5 to 2 hours, then transferred and stored at $-195°$ C in liquid nitrogen until needed.

*A 20X stock solution is prepared by mixing 160.0 g NaCl, 8.0 g KCl, 0.71 g Na_2HPO_4 and 2.0 ml phenol red (1%) in 1 liter of deionized double-distilled H_2O. For use the stock solution is further diluted to 1X, sterilized by filtration, supplemented with 0.1% glucose and adjusted to pH 7.4 with 7.5% $NaHCO_3$.

Chemicals

Reference carcinogens should be obtained at the highest purity available. Whenever possible, all candidate chemicals should be analyzed by the supplier for the presence and amounts of impurities which may be carcinogenic themselves. Samples are stored at $-20°$ C or 4° C under desiccation as required. All bioassay experiments are performed under nonactinic yellow light (Sylvania F40GO lamps) to minimize photoinactivation of the chemicals.

To prevent exposure of personnel, chemicals are weighed in an enclosed glove box (Class III safety cabinet). Protective outer garments and disposable gloves are worn by personnel handling all bioassay materials. To prevent possible contamination glassware is not recycled. Instead, disposable plastic or glass labware is used. Chemically contaminated culture media or other liquid wastes are collected in plastic containers filled with absorbent paper, then disposed of by high temperature incineration.

It is necessary to test solvents used in the bioassay system to determine levels which are cytotoxic or have other adverse effects on cell growth and morphology. The highest nontoxic dose may be a limiting factor in the selection of doses at which candidate chemicals are tested. Stock solutions are prepared in laminar flow biological safety cabinets (Class II, Type B). Generally, for compounds not soluble in aqueous solution, DMSO or other appropriate solvents such as acetone or ethanol are used. Compounds soluble in aqueous solution are dissolved in DMEM or sterile distilled water. The stock solutions are further diluted in DMEM to obtain desired concentrations of the chemicals in 0.2% DMSO or nontoxic concentrations of other solvents.

Selection of Susceptible Target Cell Cultures

For unknown reasons, not all batches of primary hamster embryo cultures provide cells that are susceptible to transformation by known carcinogens (Pienta *et al.*, 1977; Pienta, 1979). However, responsive batches can be readily selected and these provide aliquot samples of the same population of cells, which can then be used routinely in repeated bioassay experiments. Initially, an aliquot sample from each of several batches of cryopreserved primary cultures is tested for its response to a standard carcinogen such as benzo(a)pyrene or 3-methylcholanthrene. Cultures that fail to respond are discarded. Cultures that respond initially are further tested against representative carcinogens from diverse chemical classes. Such a panel might include an aromatic amine, an aminoazo dye, a nitrosamine, a mycotoxin, a heterocyclic compound, and an inorganic metal salt. The batch of primary cultures giving the best overall response is then used as the source of target or feeder cells for routine bioassays.

Bioassay Procedure

The general procedures for transforming freshly prepared early passage Syrian golden hamster embryo cells have been described elsewhere (Berwald and Sachs, 1963, 1965; DiPaolo *et al.*, 1969, 1972). Although this cell system has been used extensively, procedures for using the cells were not standardized until recently (Pienta *et al.*, 1977). In the earlier studies, small numbers, i.e., 250–1000 of primary or secondary cells were seeded onto a lawn of X-irradiated feeder cells which were prepared from either hamster or rat embryo cultures. The cultures were then treated with chemicals and later examined for evidence of morphological transformation. Treatment of the cells with the carcinogens varied from 30 minutes to 8 days.

In the standardized system which relies upon the use of cryopreserved cells, careful attention is given to the logistics for performing the *in vitro* bioassay in a routine manner. Aliquot samples of cryopreserved cells from individual cell pools have been

used in bioassays for as long as 1.5 years. This procedure overcomes the variation in response exhibited among different pools of target cells derived from freshly prepared primary embryo cultures, and eliminates the need for pregnant hamsters for each bioassay experiment.

On Day 0, usually a Friday, a cryopreserved sample of feeder cells is thawed and planted (5×10^6 cells/150 cm² flask, i.e., 3.3×10^4/cm²) in DMEM supplemented with 20% FBS. The cells are refed later in the day to remove residual DMSO and any unattached cells. Two days later, the culture, generally 50–80% confluent, is subcultured into four to six 75 cm² flasks, depending upon the number of cells required. Cultures not having reached 50% confluence are discarded. After incubation for two days these cultures are trypsinized, pooled, and X-irradiated in suspension with 5000 R by means of a Picker Portable Industrial X-ray machine fitted with a 0.25-mm aluminum filter with the X-ray source 30 cm from the suspension of cells. The cells are immediately centrifuged at low speed, resuspended in complete DMEM, and planted at 8×10^4 cells per 50-mm dish in 2 ml DMEM. Twelve dishes are generally used for each dose of chemical and control groups. Early on the day that feeder cells are planted, an ampule of cryopreserved target cells is thawed and planted (2.5×10^6 cells/25 cm² flask, i.e., 10×10^4/cm²) in DMEM supplemented with 20% FBS. At this density the number of cell divisions is kept to a minimum. These cells are also refed later in the day to remove residual DMSO. On the next day, the degree of confluence as well as the general appearance of the culture is evaluated. If, at this time, the cells are not at least 40% confluent, the culture is discarded. Acceptable cultures are trypsinized and planted at 500 cells per dish in 2 ml DMEM onto the X-irradiated feeder cells seeded on the previous day. The volume of culture medium is now 4 ml per dish.

On the following day test chemicals, prepared at 2 × concentrations in DMEM, are added to the cultures in 4 ml amounts to obtain desired final concentrations of the chemicals in 8 ml medium. Each bioassay experiment includes several doses of a positive carcinogen control such as benzo(a)pyrene or 3-methylcholanthrene. Negative control cultures are treated with 0.2% DMSO or other appropriate solvents, as well as DMEM alone. All cultures are incubated for 8 days, without refeeding, at 37° C in a humidified atmosphere of 10% CO_2 in air. In a typical bioassay experiment, 4 or 5 doses of chemical are tested. All chemicals are tested under code. Code numbers are etched onto the plastic dishes with either a diamond- or carbide-tipped scriber. At the end of the experiment the medium is collected and the cultures are washed twice with TCD and once with methanol to remove residual unbound chemical in the dishes (Sansone *et al.*, 1976). The cultures are then fixed with methanol for 10 minutes and stained with 10% Giemsa for 20 minutes. Air-dried colonies are counted with an ARTEK model 880 electronic colony counter (New Brunswick Scientific Co., Edison, N.J.) or visually under low magnification. The dishes are further examined under $7-40\times$ magnification for the presence of morphologically transformed colonies.

Selection of Test Doses

In order to ascribe appropriate doses for testing candidate chemicals in the bioassay, a preliminary toxicity test is performed. For this purpose, multi-well cluster dishes are employed, each dish containing six 35-mm wells. One dish is used for each dose of chemical. As in the bioassay, each well is seeded with X-irradiated feeder cells, but at a cell density of 4×10^4 cells per well. On the following day 250 target cells are added and the cultures are incubated for 24 hr. Chemicals are then added at half-log incremental doses. The maximum dose is dependent upon the solubility of the compound in the appropriate solvent.

The treated cultures are incubated further for 8 days. The medium is then removed and the dishes are washed, fixed, and stained as in the standard procedure. Total numbers of colonies at each dose level are recorded and curves are drawn to graphically determine the dose at which the number of colonies is reduced by 50% (TD_{50}). In the standard bioassay the chemicals are then tested at twofold dose increments from $TD_{50}/4$ to $2\ TD_{50}$. If a TD_{50} is not attained in the toxicity test, that particular chemical is tested at twofold doses from 62.5 μg/ml to 1000 μg/ml. The transformation response is expressed as the number of transformed colonies per surviving colonies for each dose of chemical tested. Multiplication by 100 results in the percentage of transformation.

Criteria for Morphological Transformation

The target cells used in the bioassay are derived from primary cultures prepared from eviscerated torsos of hamster fetuses, taken at 12–13 days of gestation. These early passage cultures contain a variety of cell types; hence, the colonies derived from these cells exhibit morphological heterogeneity (Fig. 29.1a,b,c,d). Fibroblasts are predominant and include fusiform or spindle cells (long axis), as well as pyramidal (moderate axis) and polygonal (short or no axis) cells. Stellate, muscle, or epithelioid cells are observed less frequently. In addition to the morphological differences, colony density also varies. In some cases the colony, whether normal or transformed, is dense with three-dimensional growth in the center. In others the cells grow only to a monolayer (Fig. 29.2a,b,c,d) and are referred to as light or semicontiguous colonies (DiPaolo *et al.*, 1971). Normal or nontransformed colonies contain well-ordered arrays of cells with a high degree of contact inhibition and little or no crossing-over of cells (Fig. 29.3). The cellular environment, including pH, serum, medium,

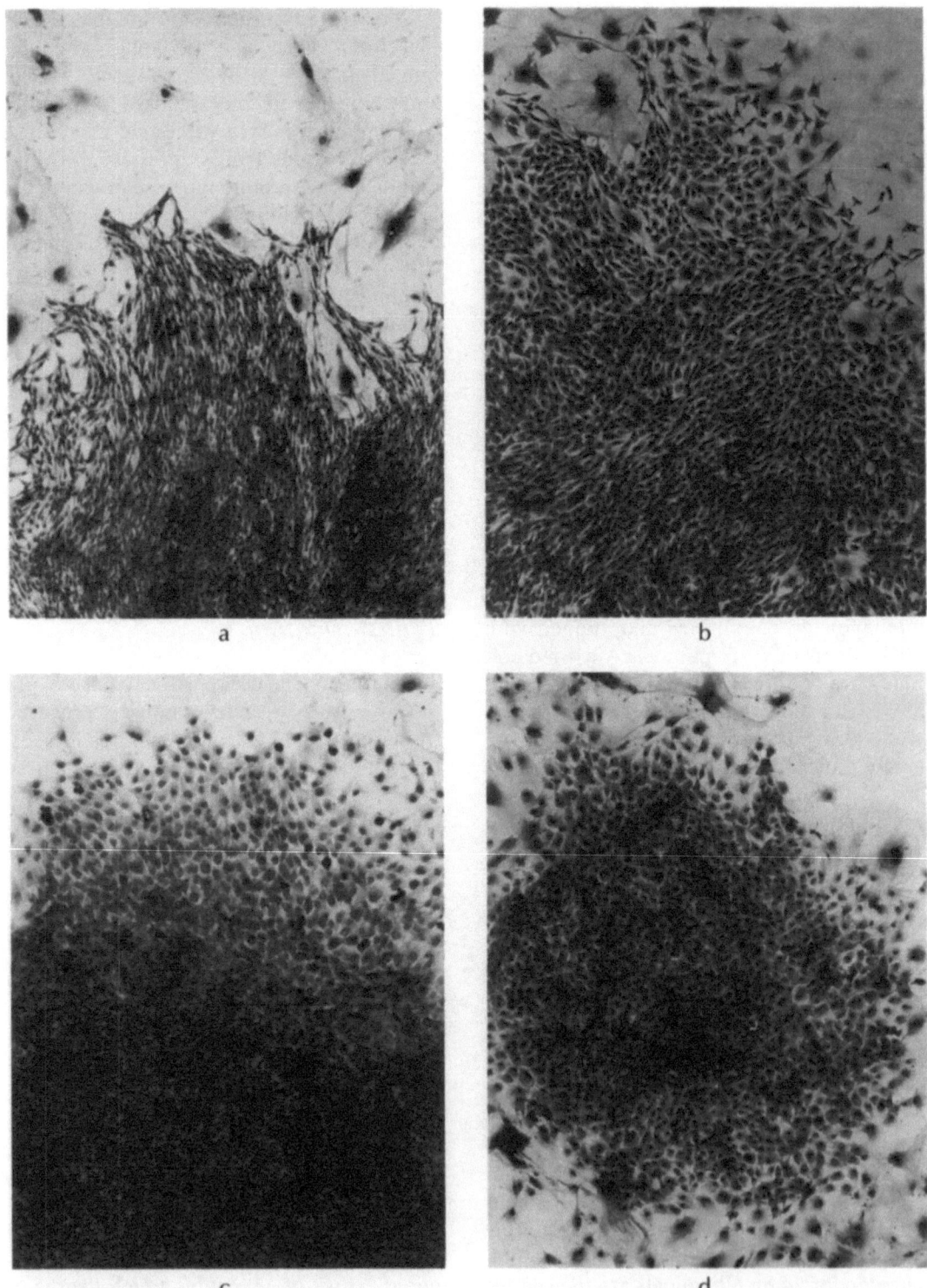

Figure 29.1.   (a) Periphery of a normal colony of untreated hamster embryo cells. Cells are fusiform. Giemsa, 40×. (b) Periphery of a normal colony of untreated hamster embryo cells. Cells are pyramidal. Geimsa, 40×. (c) Periphery of a normal colony of untreated hamster embryo cells. Cells are polygonal. Giemsa, 40×. (d) Whole colony of untreated hamster embryo cells. Cells are epithelioid. Giemsa, 40×.

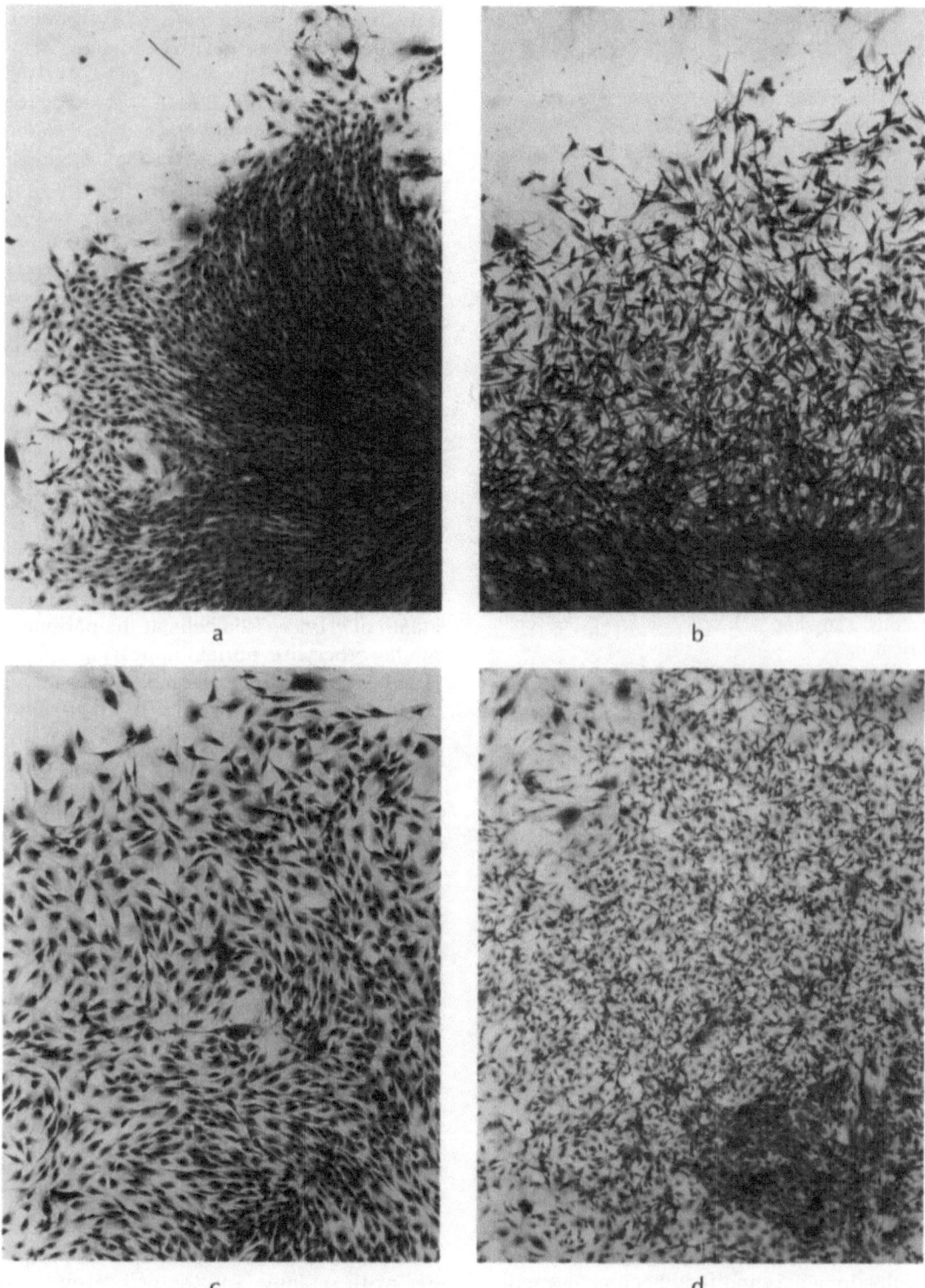

Figure 29.2.   (a) Portion of a normal, densely populated fibroblastic colony. Giemsa, 40 ×. (b) Portion of a densely populated transformed fibroblastic colony. Giemsa, 40×. (c) Portion of a semicontiguous colony of normal fibroblast calls. Note the intercellular space. Giemsa, 40×. (d) Portion of a transformed semicontiguous colony. Note intercellular space. Giemsa, 40×.

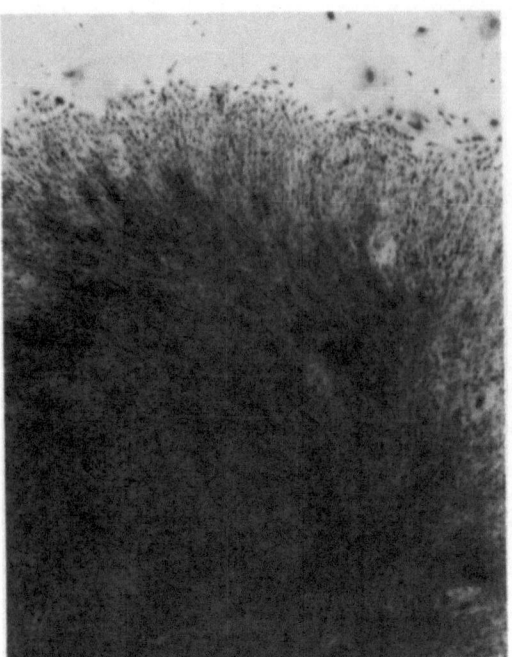

Figure 29.3. Portion of a normal fibroblast colony showing well-ordered arrays of cells. Giemsa, 25×.

and culture dishes can affect the overall appearance of the colonies. Therefore, reagents and conditions which result in a population of cells exhibiting optimal morphological characteristics for the evaluation of transformation are selected.

The morphological criteria for transformation of hamster cells have been described previously (DiPaolo *et al.*, 1971; Pienta *et al.*, 1978). Transformed colonies exhibit extensive random orientation and crossing-over of cells, particularly at the periphery. The cells are usually smaller but may also be variable in size and show increased basophilia, as well as an increase in the ratio of nucleus to cytoplasm (Sanford, 1974). The crossing-over and random orientation of cells are the major criteria for transformation used in this bioassay. These criteria are most easily recognized in cells with a long axis and they become more difficult to distinguish as the cellular axis shortens.

In this laboratory, a systematic approach has been taken to evaluate colonies for transformation. In the bioassay system cultured cells are treated with candidate chemicals and colonies are stained and examined after 8 days of incubation, thus showing the end results of transformation. Initial manifestations of transformation may have occurred much earlier. If the transforming event occurs early, when the colony is composed of few cells, randomly growing, crisscrossed cells may be observed over the entire colony. In a dense transformed colony many overlapping cells are observed (Fig. 29.4a,b), whereas in cultures of normal cells only peripheral extranuclear cellular processes may overlap. In a transformed colony which is semicontiguous and has not formed a monolayer, the individual cells may be random-oriented but exhibit little or no overlapping (Fig. 29.2d).

If several more population doublings take place before the initial transforming event is expressed, the colony may exhibit a halo of crisscrossed cells at the periphery of the otherwise normal-appearing colony (Fig. 29.5a,b). Occasionally, variations occur in which extensive crossing-over of cells is observed only in a sector of transformed colony (Fig. 29.6). This is due presumably to the transformation of only a portion of the early cell population within a colony.

To facilitate the evaluation of morphological transformation, each colony is arbitrarily subdivided into three zones, as shown in Figure 29.7a. Zone C is the central portion of the colony and is the most dense area. The cells are tightly packed and frequently exhibit three-dimensional growth. Cell boundaries are often unclear and judgments of morphological alteration are difficult to make. Observation of colonies from control and carcinogen-treated cells shows that this type of growth is found with both normal and transformed cells.

Zone B, the next outermost zone, extends from the edge of zone C to the periphery of the colony, and the cells grow to at least a monolayer. Normal cells grow in well-ordered arrays of contact-inhibited cells (Fig. 29.7a,b). Transformation is indicated by random growth or crossing-over

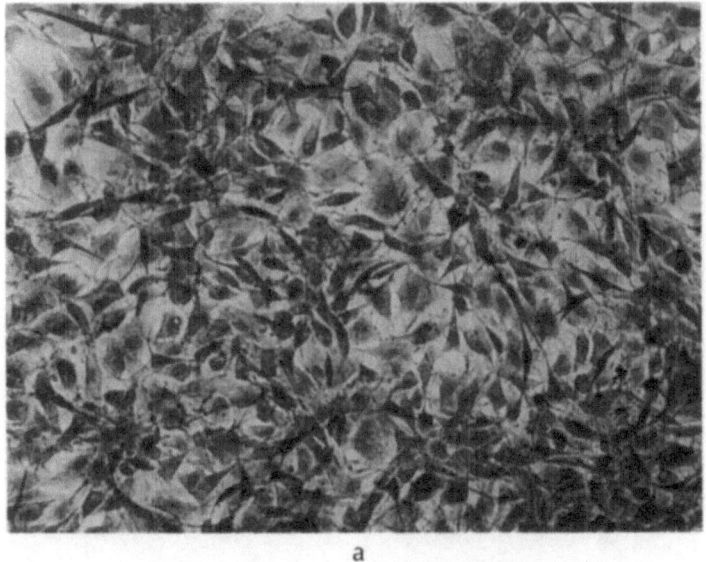

a

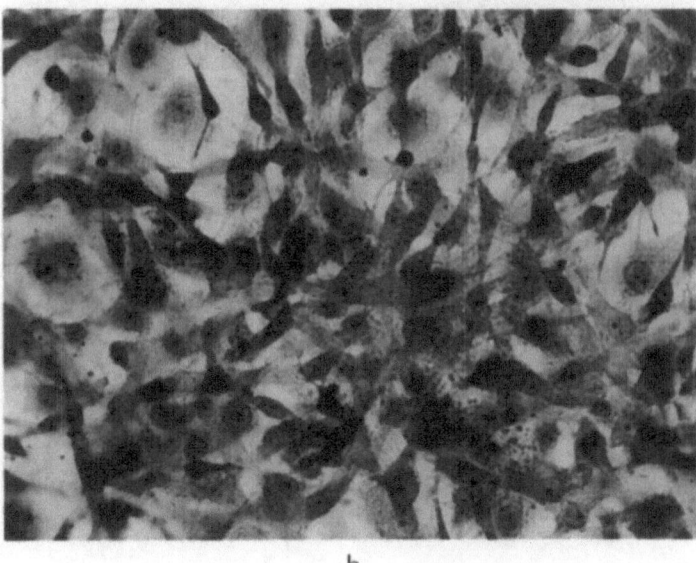

b

Figure 29.4. (a) Portion of a transformed colony showing the characteristic overlapping of cells. Giemsa, 100×. (b) Portion of a transformed colony showing overlapping cells. Giemsa, 200×.

of cells, primarily in this zone (Fig. 29.5a,b). For cells with a short axis, crossing-over is less distinct. Therefore, one should check for nuclear overlapping as an indication of crossing-over of cells.

Zone A is the sparsely populated area just beyond the periphery of the colony and occurs in only a very small number of colonies in any bioassay. This area contains the most newly formed cells and consequently is analogous to a freshly seeded population of cells not influenced by contact inhibition of growth. There is much intercellular space and little, if any, contact between

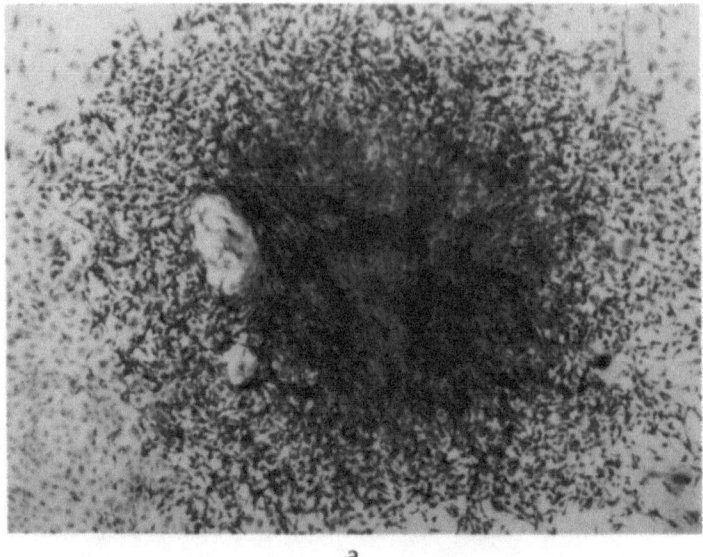

Figure 29.5.   (a) Colony showing a halo of transformed cells at the periphery. Giemsa, 25 ×. (b) Portion of a transformed colony of pyramidal fibroblast cells, showing the halo of disordered cells surrounding an ordered central area. Giemsa, 40×.

cells, although they may be randomly oriented (Fig. 29.7a). Occasionally, some overlapping of cellular processes is observed; overlapping of entire cells is rare. For these reasons, the morphological appearance of zone A is not definitive and is

not used in the evaluation of the colony for transformation.

In summary, a colony is judged to be morphologically transformed if randomness and crisscrossed, overlapping cells are observed in zone B, although these character-

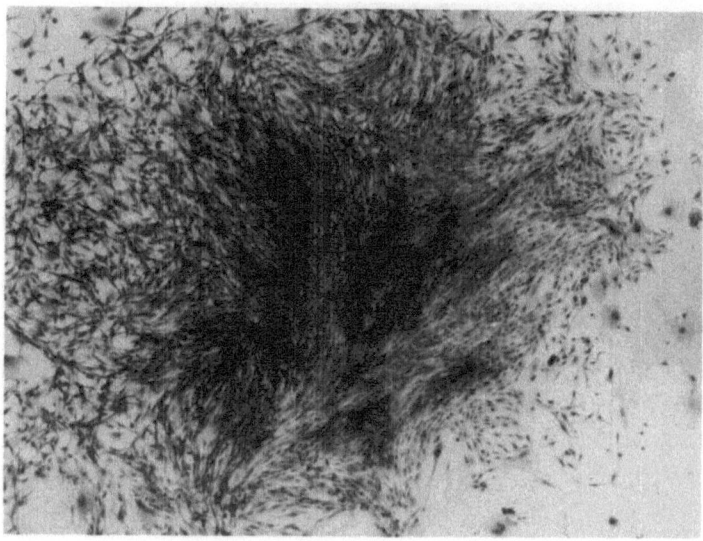

Figure 29.6.   Colony showing sectional transformation. The left half of the colony shows the typical transformed morphology. The right side is normal. Giemsa, 25×.

istics may also occur in other zones. None of these criteria for transformation have been observed in the standard bioassay when cells were treated with DMEM alone, DMEM and solvent, or when cells were treated with known noncarcinogens.

Metabolic Activation

Some procarcinogens may require further metabolic activation than that provided by the hamster fibroblast cells before transformation can occur. Exogenous metabolic activation systems have been provided by the addition of either hamster S9 homogenates (Pienta *et al.*, 1978; Pienta, 1979, 1980) or cultured hamster hepatocytes (Poiley *et al.*, 1978, 1979; Pienta, 1980). Hamster liver S9 homogenates are prepared according to Ames *et al.* (1975). Provisionally, S9 preparations are added concomitantly with, or immediately after, the candidate chemical. The medium is changed either 4–6 hours later or overnight, and the cultures are further incubated for a total of 8 days.

Primary cultures of hepatocytes are prepared according to a procedure modified from Leffert *et al.* (1977). The hepatocytes

are X-irradiated and added to the cell cultures immediately prior to the addition of candidate chemicals and remain for the duration of the bioassay. Standard procedures for the routine use of S9 or hepatocytes are under development. Exogenous metabolic activation is not routinely incorporated into every bioassay but is used only for those compounds that give negative responses in the standard bioassay.

Results and Discussions

Thus far, approximately 120 chemicals from diverse classes have been tested in the bioassay employing cryopreserved cells. These include representatives of polycyclic aromatic hydrocarbons, direct alkylating agents, nitrosamines and nitrosamides, aromatic amines, aminoazo dyes, inorganic metallic salts, and a number of miscellaneous compounds. Correlation of morphological transformation with reported carcinogenic activity of the chemicals was greater than 90% (Pienta *et al.*, 1977, 1978; Pienta, 1979, 1980). The incorporation of exogenous metabolic activation, provided

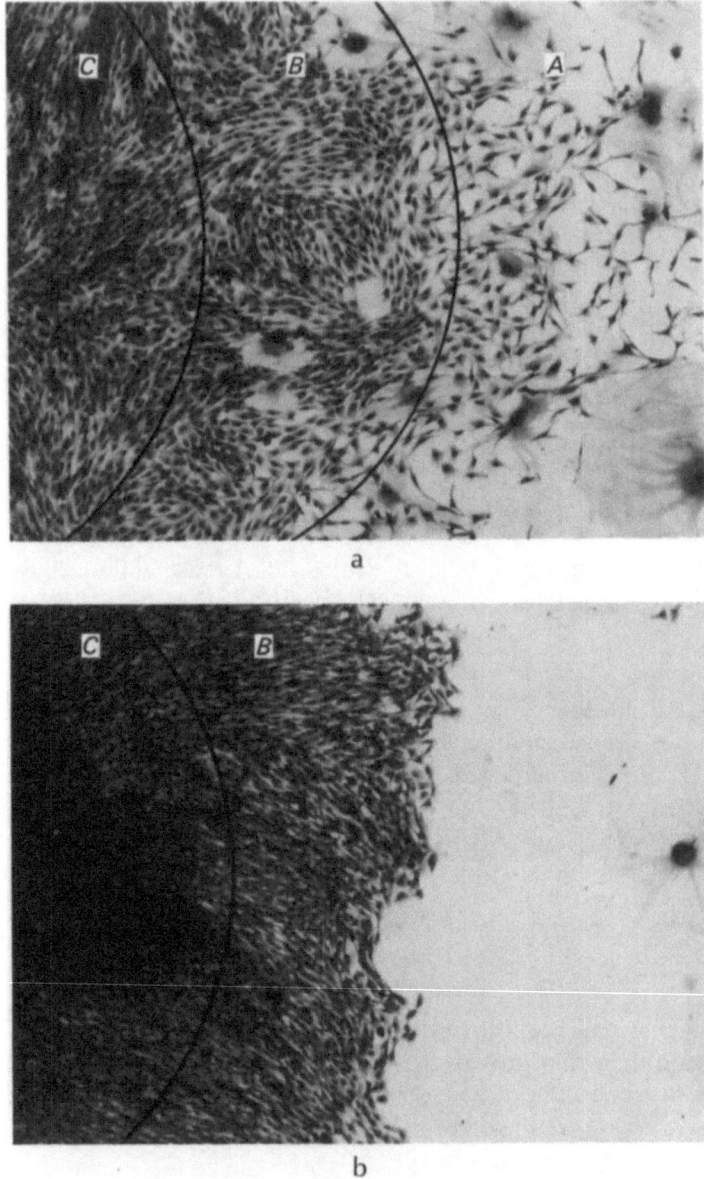

Figure 29.7.   (a) Portion of a normal fibroblast colony. The black lines show the approximate demarcations of growth zones A, B, and C. Giemsa, 40×. (b) Portion of a normal colony showing well-ordered arrays of cells in zone B. Zone A is not present in this colony. Giemsa, 40×.

by hamster liver S9 homogenates or cultured hamster hepatocytes, further extended the range of the system (Pienta, 1979, 1980; Poiley *et al.*, 1979) in that the correlation was increased and the incidence of apparently false negative results was reduced. Compounds that transformed cryopreserved hamster embryo cells are listed in Table 29.1. Those that gave negative responses are summarized in Table 29.2. Several of the carcinogens that gave negative results in the standard bioassay were retested in the presence of an exogenous metabolic activation system. The results with these are shown in Table 29.3.

Table 29.1. *Chemicals That Transformed Hamster Embryo Cells in the Standard Bioassay*

1. Acetamide
2. N-Acetoxy-2-acetylaminofluorene
3. N-2-Acetylaminofluorene
4. Acridine orange
5. Aflatoxin B_1
6. Aflatoxin B_2
7. 4-Aminobiphenyl
8. 3-Amino-1,2,4-triazole
9. 2-Anthramine
10. Azaserine
11. Benz(a)anthracene
12. Benzo(a)pyrene
13. Benzyl chloride
14. Beryllium sulfate tetrahydrate
15. Bis(p-dimethylamino)-diphenylmethane
16. 1,4-Butane sultone
17. Calcium chromate
18. Cholesterol epoxide
19. Chrysene
20. Cyclophosphamide
21. Dibenz(a,c)anthracene
22. Dibenz(a,h)anthracene
23. 1,2,3,4-Diepoxybutane
24. Diethylstilbestrol
25. 7,9-Dimethylbenz(c)acridine
26. N,N-Dimethyl-4-aminoazobenzene
27. 3,2'-Diemthyl-4-aminobiphenyl·HCl
28. 7,12-Dimethylbenz(a)anthracene
29. Dimethylcarbamyl chloride
30. 1,2-Dimethylhydrazine
31. 1,2-Epoxybutane
32. Ethionine
33. Ethyl-p-toluenesulfonate
34. 2-Fluorenamine
35. 2-(2-Furyl)-3-(5-nitro-2-furyl)-acrylamide
36. Glycidaldehyde
37. Glycidol
38. Hycanthone
39. Hydrazine sulfate
40. N-Hydroxy-2-acetylaminofluorene
41. 4-Hydroxylaminoquinoline-1-oxide
42. Lead acetate
43. Lithocholic acid
44. 3-Methylcholanthrene
45. Methyl iodide
46. Methylazoxymethanol acetate
47. 3'-Methyl-4-dimethylaminoazobenzene
48. N-Methyl-N'-nitro-N-nitrosoguanidine
49. 2-Naphthylamine
50. Nickel sulfate hexahydrate
51. 4-Nitrobiphenyl
52. N-[4-(5-Nitro-2-furyl)-thiazolyl]-formamide
53. 2-Nitronaphthalene
54. 2-Nitro-p-phenylenediamine
55. 4-Nitro-o-phenylenediamine
56. 4-Nitroquinoline-1-oxide
57. N-Nitrosodimethylamine
58. N-Nitrosodiphenylamine
59. N-Nitrosoethylurea
60. N-Nitrosopiperidine
61. m-Phenylenediamine
62. 1-Phenyl-3,3-dimethyltriazene
63. 1,3-Propane sultone
64. Propyleneimine
65. Safrole
66. Succinic anhydride
67. Thioacetamide
68. Thiourea
69. Titanocene dichloride
70. 2,4-Toluenediamine
71. 4-o-Tolylazo-o-toluidine
72. Uracil mustard

Table 29.2. *Chemicals That Did Not Transform Hamster Embryo Cells in the Standard Bioassay*

1. Acetone	25. 5-Iododeoxyuridine
2. N-4-Acetylaminofluorene	26. Limonene
3. 4-Aminoazobenzene	27. Lithocholic acid sulfate
4. 2-Aminobiphenyl	28. Methyl carbamate
5. Aniline	29. Methotrexate
6. Anthracene	30. 3-Methoxy-4-aminoazobenzene
7. 1-Anthramine	31. Methoxychlor
8. Aroclor 1254	32. 2-Methyl-4-dimethylaminoazobenzene
9. Auramine	33. 1-Naphthylamine
10. Benzo(e)pyrene	34. α-Naphthylisothiocyanate
11. Bromobenzene	35. Natulan · HCl
12. 5-Bromodeoxyuridine	36. 2-Nitrofluorene
13. Caffeine	37. N-Nitrosodiethylamine
14. ε-Caprolactone	38. Phenanthrene
15. o-Chloroaniline	39. Phenobarbital
16. p-Chloroaniline	40. Progesterone
17. Cholesterol	41. Propylene glycol
18. Dimethylformamide	42. Pyrene
19. Dimethyl sulfoxide	43. p-Rosaniline
20. Ethanol	44. Saccharin
21. Ethinylestradiol	45. Sodium nitrite
22. Ethyl carbamate	46. 12-0-tetradecanoyl-phorbol-13-acetate
23. 5-Fluorodeoxyuridine	47. 1,8,9-Trihydroxyanthracene
24. Hydroxylamine · HCl	

Table 29.3. *Chemicals That Transformed Hamster Embryo Cells When Tested in the Presence of an Exogenous Metabolic Activation System*

Chemical	Activation system
4-Aminoazobenzene	Hamster Hepatocytes
Auramine	Hamster Liver S9
Diethylnitrosamine	Hamster Hepatocytes, Hamster Liver S9
Ethyl Carbamate	Hamster Liver S9
3-Methoxy-4-aminoazobenzene	Hamster Liver S9
Natulan · HCl	Hamster Liver S9
2-Nitrofluorene	Hamster Hepatocytes, Hamster Liver S9
p-Rosaniline	Hamster Liver S9

References

Ames, B.N., McCann, J., Yamasaki, E. (1975): Methods for detecting carcinogens and mutagens with the *Salmonella*/mammalian microsome mutagenicity test. Mutat. Res. *31*: 347–364.

Berwald, Y., Sachs, L. (1963): *In vitro* transformation with chemical carcinogens. Nature (Lond.) *200*:1182–1184.

Berwald, Y., Sachs, L. (1965): *In vitro* transformation of normal cells to tumor cells by carcinogenic hydrocarbons. J. Natl. Cancer Inst. *35*:641–661.

Casto, B.C., Janosko, N., DiPaolo, J.A. (1977): Development of a focus assay model for transformation of hamster cells *in vitro* by chemical carcinogens. Cancer Res. *37*: 3508–3515.

DiPaolo, J.A., Casto, B.C. (1979): Quantitative studies of *in vitro* morphological transformation of Syrian hamster cells by inorganic metal salts. Cancer Res. *39*:1008–1013.

DiPaolo, J.A., Donovan, P.J., Nelson, R.L. (1969): Quantitative studies of *in vitro* transformation by chemical carcinogens. J. Natl. Cancer Inst. *42*:867–876.

DiPaolo, J.A., Nelson, R.L., Donovan, P.J. (1971): Morphological, oncogenic, and karyological characteristics of Syrian hamster embryo cells transformed *in vitro* by carcinogenic polycyclic hydrocarbons. Cancer Res. *31*:1118–1127.

DiPaolo, J.A., Nelson, R.L., Donovan, P.J. (1972): *In vitro* transformation of Syrian hamster embryo cells by diverse chemical carcinogens. Nature (Lond.) *235*:278–280.

DiPaolo, J.A., Nelson, R.L., Donovan, P.J., Evans, C.H. (1973): Host-mediated *in vivo–in vitro* assay for chemical carcinogenesis. Arch. Pathol. *95*:380–385.

Huberman, E., Sachs, L. (1966): Cell susceptibility to transformation and cytotoxicity by the carcinogenic hydrocarbon benzo(a)pyrene. Proc. Natl. Acad. Sci. USA *56*:1123–1129.

Leffert, H.L., Moran, T., Boorstein, R., Koch, K.S. (1977): Procarcinogen activation and hormonal control of cell proliferation in differentiated primary adult rat liver cell cultures. Nature (Lond.) *267*:58–61.

Pienta, R.J. (1979): A hamster cell model system for identifying carcinogens. In: A.C. Griffin and C.R. Shaw (eds.), Carcinogens: Identification and Mechanisms of Action. New York: Raven Press, pp. 121–141.

Pienta, R.J. (1980): A transformation bioassay employing cryopreserved hamster embryo cells. In: J.K. Mishra, V.C. Dunkel, and M.A. Mehlman (eds.), Mammalian Cell Transformation by Chemical Carcinogens: Advances in Environmental Toxicology. New York: Pathotex Publishers.

Pienta, R.J., Poiley, J.A., Lebherz, W.B., III (1977): Morphological transformation of early passage golden Syrian hamster embryo cells derived from cryopreserved priamry cultures as a reliable *in vitro* bioassay for identifying diverse carcinogens. Int. J. Cancer *19*:642–655.

Pienta, R.J., Poiley, J.A., Lebherz, W.B., III (1978): Further evaluation of a hamster embryo cell carcinogenesis bioassay. In: H.E. Nieburgs, V.E.O. Valli, and S.A. Kay (eds.), Cancer Prevention and Detection, Part I, Vol. 2. New York: Marcel Dekker, pp. 1993–2011.

Poiley, J.A., Raineri, R., Cavanaugh, D., Schuman, R.F., Pienta, R.J. (1978): Culture of hamster hepatocytes capable of high rates of carcinogen metabolism. In Vitro *14*:336–337.

Poiley, J.A., Raineri, R., Pienta, R.J. (1979): Use of hamster hepatocytes to metabolize carcinogens in an *in vitro* bioassay. J. Natl. Cancer Inst. *63*:519–524.

Sanford, K.K. (1974): Biologic manifestations of oncogenesis *in vitro*. J. Natl. Cancer Inst. *53*:1481–1485.

Sansone, E.B., Poiley, J.A., Pienta, R.J., Lebherz, W.B., III (1976): Potential hazard of tissue culture assays arising from carcinogenic compounds incompletely removed by washing. Cancer Res. *36*:2455–2458.

Schuman, R.F., Pienta, R.J., Poiley, J.A., Lebherz, W.B., III (1979): The effect of fetal bovine serum on 3-methylcholanthrene-induced transformation of hamster embryo cells *in vitro*. In Vitro *15*:730–735.

30

Assay of Chemically Induced Transformation of Human Cells

TAKEO KAKUNAGA

Principle of the System

Significance of the System

Many experimental systems have been developed as short-term tests for chemical carcinogens (Fig. 30.1). However, there have been three major problems for the extrapolation of the results obtained from these systems into human risk assessment. The first problem is species difference in the response to chemical carcinogens. It may ultimately be shown that the fundamental process of cancer development is common to all organisms. However, the rate-limiting factor can be quite different owing to small differences in the metabolism involved. For example, it is possible that point mutation is a major rate-limiting factor of carcinogenesis in the mouse whose control of chromosomes is unstable, but not in the human whose diploidy is quite stable. At least for quantitative assessment of human risk, the species dif-

ference is crucial. Secondly, the markers used in these test systems are widely different and their relation to induction of human cancer is unclear. The third problem is the genetic heterogeneity of the human population.

In order to solve these problems, there are only two possible experimental approaches. One is to develop a system of rapid assays for transformation using human tissues or cells. Another approach is to examine systematically the relation between these different test systems. It is particularly important to know the relation between the transformation of animal cells and human cells.

Markers of Transformation

Multistep process of transformation. Normal diploid fibroblast cells turn into malignant cells by acquiring different biological characters one by one (Kakunaga and Kamahora, 1968, 1969; Kuroki and

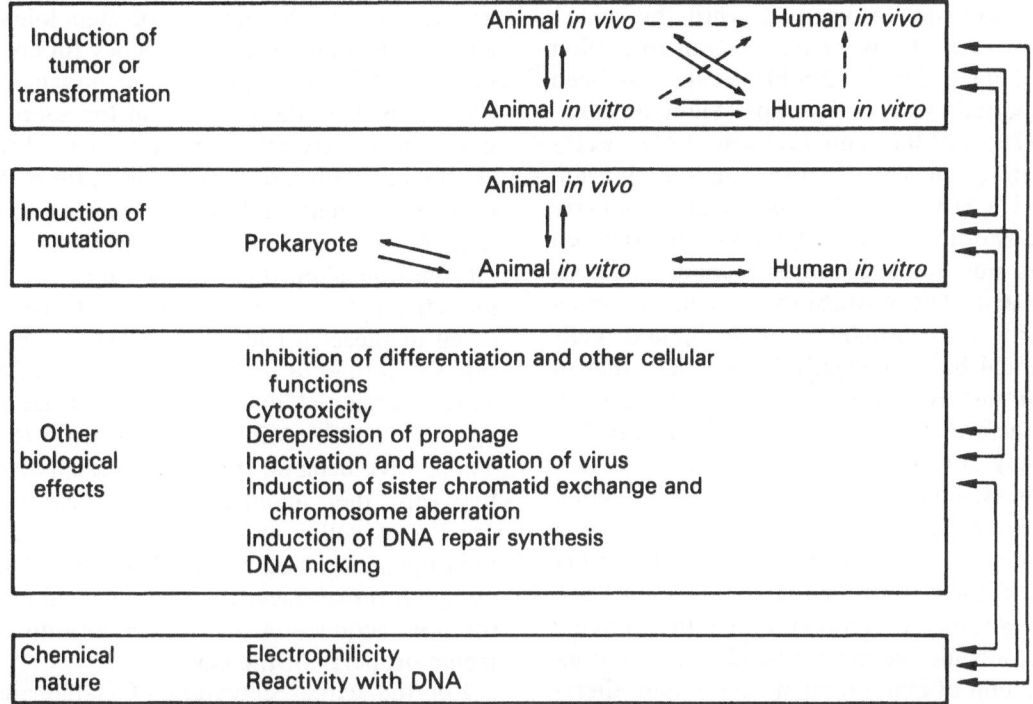

Figure 30.1. Experimental systems for assessment of human risk to environmental carcinogens. Solid arrow line indicates the relation that can be examined by experimental research. Dotted line indicates the extrapolation.

Sato, 1969; Kamahara and Kakunaga, 1970; Barrett and Ts'o 1978). There has been no report that diploid cells of normal tissue origin became transplant-tumorigenic in a one-step process regardless of causing agents. This stepwise development through qualitatively different stages is consistent with the concept of progression of tumor that was originally developed and generalized, based on the observation of carcinogenesis in animals and humans (Rous and Beard, 1935; Foulds, 1969, 1975). There is no evidence that the progressive development is always sequential. It is likely that the acquisition of each transformed phenotype of all those which compose the malignant properties of cells when they occur together is basically an independent event.

The mechanism of induction of phenotypic change (from one stage to the next stage) may be different between stages. It is unknown which step is the most crucial in the incidence of cancer in humans and what type of agent is effective for each step. Thus, in order to develop an optimum system for assay of potential carcinogens to humans, it is necessary to establish various transformation systems that are well defined for the steps involved.

Malignant transformation versus early transformation. Malignant transformation of normal diploid cells, i.e., the conversion of normal diploid cells to transplant-tumorigenic cells, involves the whole multistep process of transformation. This whole biological process is considered to directly represent the basic process of incidence of *in vivo* cancer. However, the entire process may be too complicated to utilize as a system for a short-term test of chemical carcinogens.

On the other hand, early transformation is considered to represent the earlier part of the multistep progress toward the malignant state. Early or partial transformation of human diploid fibroblast cells has been reported by several groups (Shimada *et al.*, 1976; DeMars and Jackson, 1977; Freeman *et al.*, 1977; Freedman and Shin, 1977). However, the clear relation of early transformation to malignant transformation has not yet been experimentally demonstrated. The probability of conversion of early transformant into malignant cells should be significantly higher than that of untransformed cells. A system using early transformation as a short-term test has many advantages, such as simplicity, rapidness, ease, and inexpensiveness, compared to the use of the whole process of malignant transformation. On the other hand, there is a remarkable drawback in an early transformation system in that it might not include the most crucial step in the induction of major human cancer and, therefore, might lead to inadequate assessment of human risk against environmental carcinogens. No system has been developed including only the latter part of the transformation process.

Markers for malignant transformation. The malignancy of cultured human cells is most convincingly determined by the tumorigenicity of the cells when they are transplanted into athymic nude mice. There is a possibility that the transplant-tumorigenicity of cultured cells in nude mice does not precisely reflect their malignant properties (Fogh and Hajdu, 1975; Kakunaga, 1980). However, at present, no other characteristics are more convincing than transplant-tumorigenicity as a criterion of malignancy of cultured human cells. To compensate for the uncertainty of transplant-tumorigenicity, other biological characteristics of cultured cells that are usually associated with the cells derived from tumor tissues or transplant-tumorigenic cells are examined *in vitro*. Included are the loss of anchorage-dependency for cell growth (determined by the ability to grow in gel), the escape from senescence (unlim-

ited growth of cell population), aneuploidy, and the loss of density-dependency for contact inhibition of cell growth and movement (determined by the increase in the saturation density, growth in medium with low serum concentration, irregular pattern of cell arrangement, and focus formation on a monolayer sheet of untransformed cells). All the transplant tumorigenic human cells including those of tumor origin showed part or all of these *in vitro* transformed phenotypes (Giard *et al.*, 1973; Fogh and Hadju, 1975; Zamacnick and Long, 1977; Kakunaga, 1977, 1978; Milo and DiPaolo, 1978). There have been no reports of dissociation between the acquisition of unlimited growth of cell population and transplant tumorigenicity of cultured human cells. The escape from senescence is an essential factor for acquisition of transplant-tumorigenic property of the cells.

For the actual detection of stable and progressively transformed cells, focus formation is utilized as an indicator of transformation, because in our results there has been good correlation between the focus-forming ability of the cells and their ability to become transplant-tumorigenic cells. In general, scoring transformed foci is an objective determination. To form a clear focus on a monolayer sheet of untransformed cells, the cells have to lose the sensitivity to the contact inhibition of cell growth and movement against untransformed cells as well as against transformed cells. On the other hand, alteration of clonal morphology (morphological transformation of colony level) requires the loss of sensitivity to the contact inhibition only between transformed daughter cells.

Markers for early transformation. It has been reported that the ability of the cells to grow in gel is one of the closest correlations with their transplant-tumorigenicity among the biological properties of cells in culture (Kakunaga and Kamahora, 1968; Shin *et al.*, 1975; Barrett & Ts'o, 1978). This parameter is also one of few properties that is quantitatively and objectively measurable. Thus, the ability of the cells to grow in gel is used as a marker of an early transformation of human fibroblastic cells.

Cell Culture

Cell

A human fibroblast cell strain, termed KD, is used as the indicator cell. This strain was initiated by Day (1974) from a skin biopsy sample taken from the lip of an adult female who had not shown evidence of any genetic diseases. The primary culture was made by plating the cells dissociated by stirring the biopsy material in trypsin/collagenase solution. The grown cells were subcultured at 1:2 dilution and stored in a liquid nitrogen freezer ($-100°$ C) at the fourth to eighth passage until use. The cells were used for experiments at least two passages after thawing of the frozen cell stock. Transformation was successful even when cloned KD cell population were used. The cells at any passage between the sixth and thirty-fourth passage (or approximately 6 to 35 cell generations) after isolation were usable. Mycoplasma tests were always negative with KD cultures. KD cells show the typical morphology characteristic of fibroblasts in culture although the original skin biopsy contained epidermal tissue as well. Cloning efficiency of KD cells is routinely about 50–80% under our culture conditions and they very stably maintain diploid karyotype until reaching near senescence.

Culture Medium

The regular culture medium consists of Eagle's minimum essential medium supplemented with 10% fetal calf serum, 0.4–0.5% human serum, and 2 mM L-glutamine. Fetal calf serum and human serum must be selected by testing serum samples of every lot for cloning efficiency and morphology of the cultured cells. The serum lot that gives the higher cloning efficiency usually gives less sign of abnormal morphology of cultured cells. Reservation of a sufficient quantity of selected serum in a $-60°$ C freezer is essential for reproducible results. Addition of human serum usually enhances the cloning efficiency and growth rate of transformed cells as well as untransformed cells. L-glutamine is added mainly for the purpose of precautionary protection against occasional disaster due to the defects in medium obtained from an outside source.

Maintenance of Culture

The cells are cultured in 100-mm plastic tissue culture dishes containing 10 ml of culture medium in a humidified CO_2 incubator at $37°$ C. Culture medium is changed two times a week. The subcultivation is made whenever the culture becomes confluent (approximately 3×10^6 cells per plate at confluency), by aspirating culture medium, adding 4 ml of 0.25% trypsin solution, and letting the culture stand at room temperature with trace amount of trypsin solution until the cells show a sign of digestion (minor change in morphology). Two ml of fresh culture medium is then added, and the cells are suspended and made into single cells by pipetting, and dispensed into new dishes as quickly as possible. The dilution ratio of the subculture is 1:2 before chemical treatment and 1:4 after the treatment.

There are remarkable variations in the quality of plastic dishes from different suppliers and even in different lots from the same supplier. It must be cautioned that a problem with the substrate is usually manifested slowly in culture.

Incubator

The function of the incubator should be checked by the yield of the cloning efficiency of human fibroblast cells. In addition to routine monitoring of temperature, checks must be made of variation of temperature in different areas within the chamber, CO_2 concentration, and humidity inside the chamber. (Gauges on the chamber are not always correct.) Contamination with toxic substances is not uncommon, such as a residual amount of detergent used for washing inside the chamber, adhesive used to secure the rubber gasket on the inside door, and the presence of trace amounts of bactericidal or fungicidal chemicals which are generally regarded to be unharmful to cultured mammalian cells.

Testing Protocol

Preliminary Determination of the Dose of Chemicals for Transformation Assay

Approximate toxicity of chemicals is determined by the following procedure. 200 KD cells are plated into 60-mm plastic dishes containing 5 ml of culture medium. At 18 hr after seeding of cells, the chemical to be tested is dissolved in dimethyl sulfoxide (DMSO) and directly added to cultures at tenfold dilution, ranging from 10^{-3} M to 10^{-7} M. Three days later, the medium containing the chemical is removed, and cultures are refed with fresh medium, and incubated for a further 10 days, at which time the cultures are fixed with methanol, stained with Giemsa, and the number of colonies scored which are composed of more than 50 cells. If necessary, toxicity of chemicals are retested at different concentrations until approximate survival curves are obtained. The three doses that will give approximately 10, 30, and 90% cell survivals are determined and used for transformation assay.

Malignant Transformation Assay

(Outline of protocol is shown in Figure 30.2.)

Preparation of the culture. The growing cultures at subconfluent state are refed with fresh culture medium. One day later the cultures are trypsinized, suspended in fresh culture medium, cell number counted, diluted with culture medium, and then plated at 3×10^6 cells per 10 ml medium per 100-mm plate for transformation assay and at 1×10^2 and 3×10^2 cells per 5 ml medium per 60-mm plate for plating efficiency. Minimum number of dishes per group is ten for transformation assay and five for plating efficiency assay. The cultures are immediately placed in a CO_2 incubator.

Treatment with chemicals. At 18 hr after seeding of cells, the cultures are exposed to test compounds. Test compounds are dissolved in DMSO (spectroanalyzed grade) at appropriate concentrations and, under protection from light, directly added to the culture medium in the dishes. The chemical-DMSO solution is prepared immediately before use. A control group is treated with DMSO only. Final concentration of DMSO in culture medium is 0.2–1.0%. As a positive control of transformation, cultures are treated with 4-nitro-quinoline-1-oxide at the concentration of 0.1 μg/ml for 30 min.

In cases which the test compounds are known to be short-acting in culture, the cells suspended in culture medium are treated with chemicals. The cells in growing conditions are suspended in culture medium at the concentration of 2×10^5 cells per ml, the chemicals added which were dissolved in DMSO and mixed well, incubated with gentle shaking at 37° C for 30 min, and centrifuged at 800 $\times$ g for 5 min; the chemical-containing medium decanted, the cells are resuspended in fresh culture medium, and then plated at 3×10^6 cells per 10 ml medium per 100-mm plate for transformation assay and 1×10^2 and 3×10^2 cells per 5 ml medium per 60-mm plate for toxicity assay.

Maintenance of culture and observation. The cultures for transformation assay are maintained with two medium changes a week, and successively subcultured at a 1:4 dilution ratio when they become confluent.

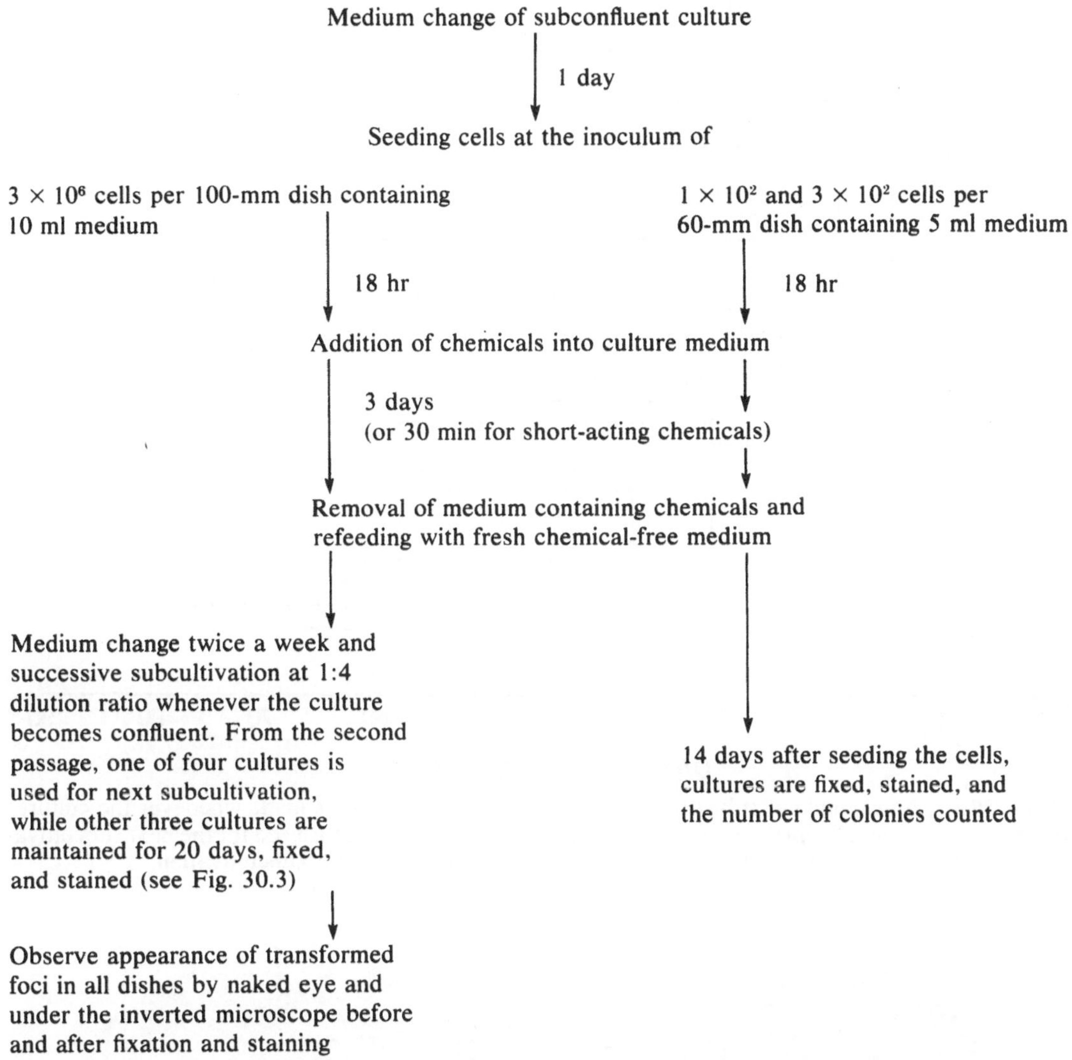

Medium change of subconfluent culture

| 1 day

Seeding cells at the inoculum of

3×10^6 cells per 100-mm dish containing 10 ml medium

1×10^2 and 3×10^2 cells per 60-mm dish containing 5 ml medium

18 hr   18 hr

Addition of chemicals into culture medium

3 days
(or 30 min for short-acting chemicals)

Removal of medium containing chemicals and refeeding with fresh chemical-free medium

Medium change twice a week and successive subcultivation at 1:4 dilution ratio whenever the culture becomes confluent. From the second passage, one of four cultures is used for next subcultivation, while other three cultures are maintained for 20 days, fixed, and stained (see Fig. 30.3)

14 days after seeding the cells, cultures are fixed, stained, and the number of colonies counted

Observe appearance of transformed foci in all dishes by naked eye and under the inverted microscope before and after fixation and staining

Figure 30.2.   Outline of assay procedure for malignant transformation using human fibroblastic cells, KD.

The first subcultivation is made with all dishes. From the second subcultivation, only one dish out of four is used for the subsequent subcultivation (see Fig. 30.3). Three other dishes are maintained without further subculturing, and fixed with methanol and stained with Giemsa 20 days after subcultivation.

All cultures are examined for the appearance of transformed foci at least once a week by naked eye and by using the inverted microscope.

Subcultivation is continued until clear transformed focus appears or as far as the cells grow and reach a confluent state.

Score of Transformation

Morphology of transformed foci. A typical transformed focus is the localized piling up of cells into multi-cell layer with mitotic figures and some disorder of cell arrangement (Figs. 30.4 and 30.5), and can be recognized by the naked eye as a white spot on a thin monolayer sheet of transformed cells. After Giemsa-staining, the focus area is distinctly densely stained. In most transformed foci, however, the disorganization and crisscross pattern of cell arrangement —the typical pattern of most transformed rodent cells—is not distinctive, and further

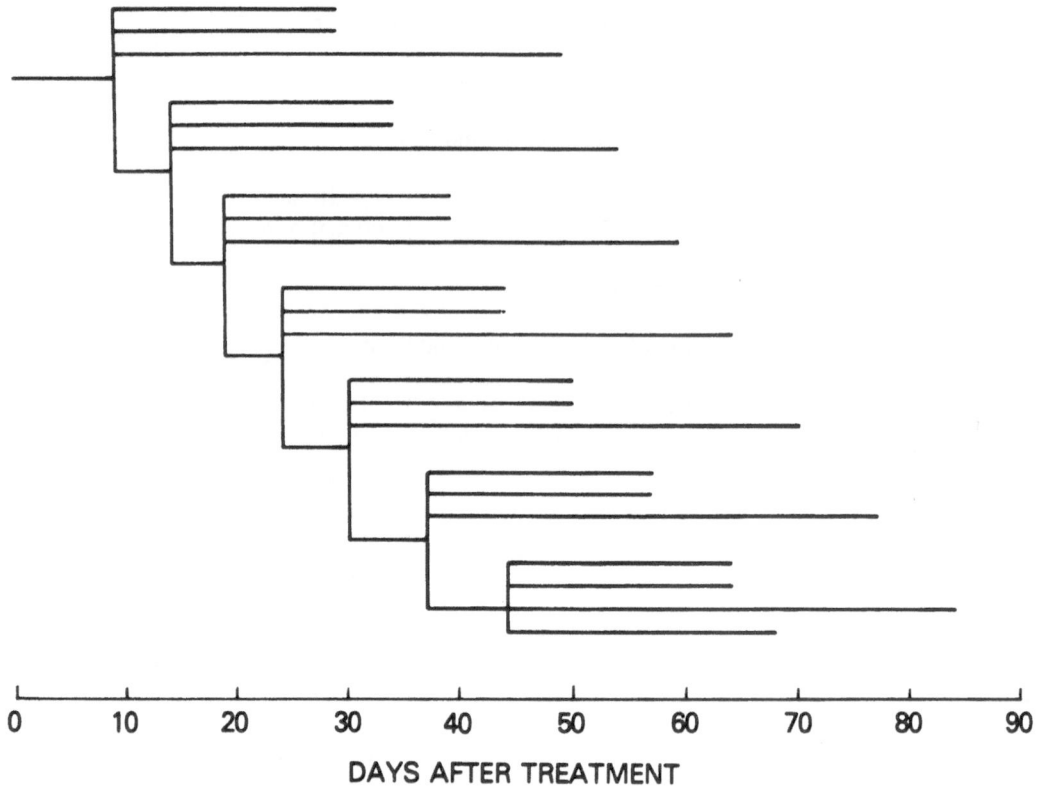

DAYS AFTER TREATMENT

Figure 30.3. An example of subcultivation schedule. Subcultivation is made whenever the cultures reach confluent state. From the second passage, only one of four dishes is used for the next subcultivation. Other three cultures are fixed and stained 20 days or longer after subcultivation.

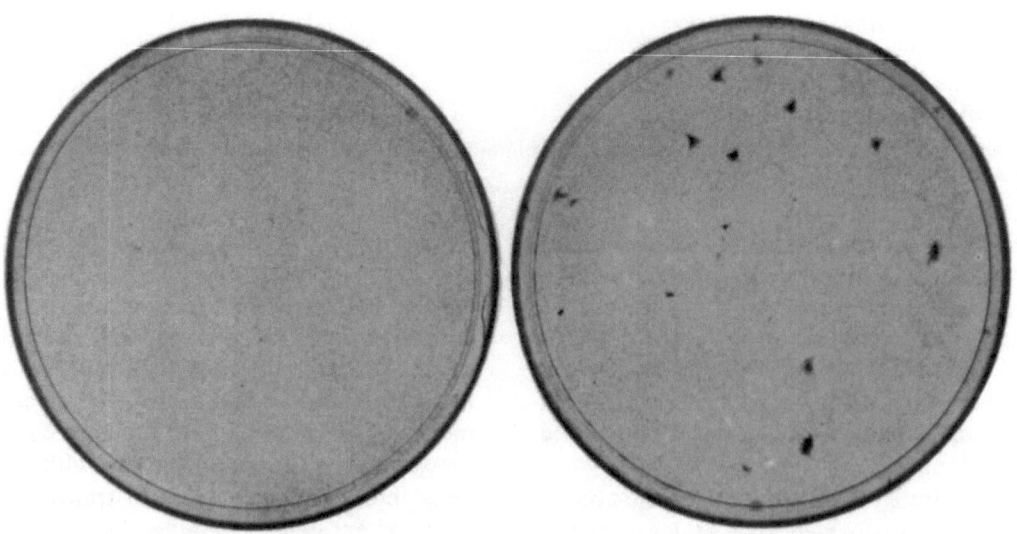

Figure 30.4. The focus formation by transformed cells. The dishes were derived from the cultures which were treated for 30 min with 0.3% DMSO (left) and 0.3 mg/ml of 4NQO (right), then subcultivated 11 times at a 1:4 dilution ratio whenever they became confluent, maintained for 20 days without further subculturing, and fixed and stained with Giemsa.

experiments are needed for the determination of transformation.

Confirmation of transformation. If only one suspicious focus is observed in the unfixed dish, the culture is subjected to subcultivation. The transformed cells should spread over the subcultured dish and form multiple numbers of foci. When more than one suspicious transformed foci is found in the unfixed dish, the cells from suspicious areas are picked up by sucking up into a fine glass pipet and are transferred into a new dish.

When the cells derived from a single focus grow near a confluent state in the new dish, they are tested for the ability to form foci on the monolayer of untransformed cells. For this test, a monolayer sheet of untransformed cells is previously prepared by plating 3×10^5 of KD cells into a 60-mm dish and incubating for one week without medium change. Then the cells to be tested are trypsinized, and suspended in culture medium at the concentration of 50 cells per 5 ml. Five ml of cell suspension are poured onto the previously prepared monolayer sheet of KD cells and incubated. After 10 days of incubation without medium change, the cultures are fixed, stained, and scored for focus formation. Formation of any number of clear foci is regarded as evidence that the tested cells are transformed.

Quantitation. Because of the extremely low frequency of transformation and the long time required for development of transformed foci, it is practically impossible to measure the frequency of transformation per treated or surviving cell. Rough estimation can be made by dividing the number of dishes which showed transformation after subcultivation by the number of dishes treated.

Early Transformation Assay

Preparation of the culture and treatment with chemicals are the same as those described in the preceding section on the assay of malignant transformation.

Maintenance of culture and observation. The cultures treated with chemicals are maintained with medium change (two times a week). When the cultures become confluent, the first subcultivation is made at a 1:4 dilution ratio with all dishes. From the second subcultivation only one of four dishes is used for the subsequent subcultivation, and the other three cultures are used for a soft agarose test. When the three cultures become 50–90% confluent, they are trypsinized, thoroughly made into single cells, suspended in culture medium, counted for cell number, and diluted with culture medium to 2×10^5 cells per ml. Aggregates or clumps should not exist in cell suspension. Cell suspension is warmed up at 37° C, mixed well at 1:1 ratio with 0.66% agarose (Sigma, Type V) solution which is prewarmed at 42° C, and 1 ml of mixture is immediately poured onto previously set 5 ml of a base layer, which is composed of culture medium and 0.66% agarose. After standing for 5–10 min to set top layer of cell-agarose mixture, the cultures are placed in a CO_2 incubator. Cells from every three subcultures are tested for the growth in agarose gel for 6 weeks after carcinogen treatment.

Scoring of colony formation in gel. Colony formation is examined after 3 weeks' incubation of agarose gel culture in a highly humidified CO_2 chamber. Colonies larger than 0.1 mm diameter are counted under an inverted microscope. Colony-forming efficiency is calculated by dividing the number of colonies per dish by 2×10^5.

Evaluation of Results

Malignant Transformation Assay

Reliability of results. It is almost impossible to obtain a dose-response curve for transformation in the present system. However, because no malignant transformation has occurred spontaneously in any human diploid cells of normal tissue origin, qualita-

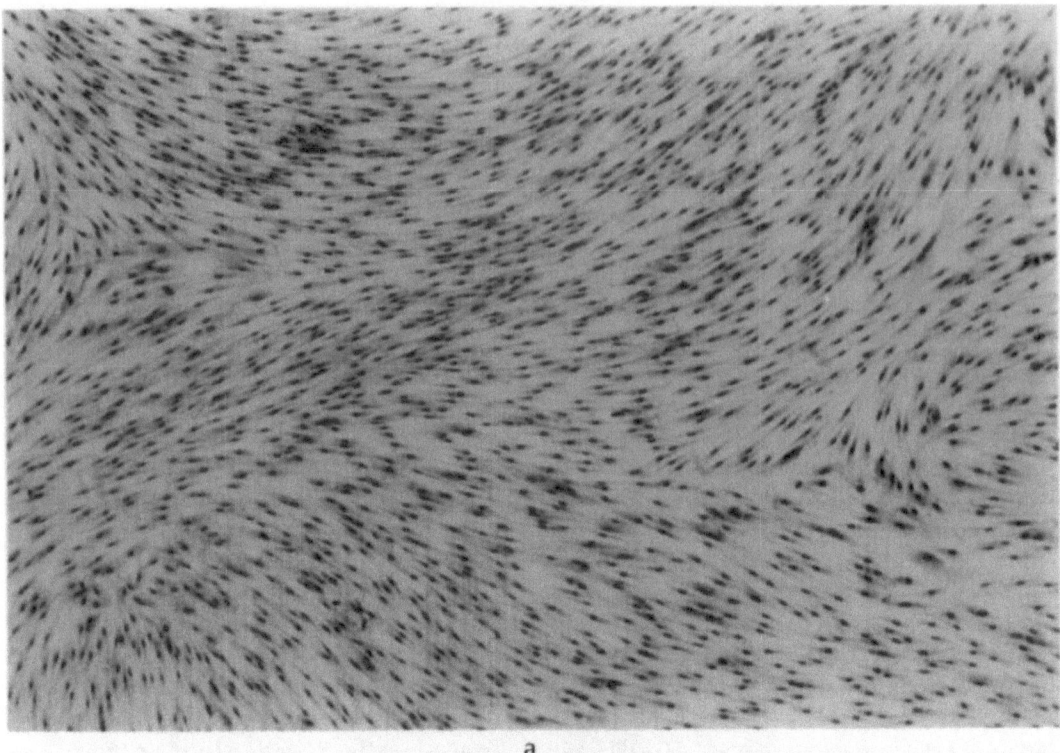

Figure 30.5. Photomicrographs after fixation and Giemsa staining of (a) untreated KD cells; (b) focus area of the morphologically altered cells in a culture treated with 4NQO (0.1 μg/ml); and (c) a higher magnification of a part of (b).

tive detection of induction by chemicals of malignant transformation is considered significant.

With respect to the malignance of the cells, the focus-forming cells may not yet possess transplant-tumorigenicity when they form a focus. However, it is definite that at least daughter cells of focus-forming cells are transplant-tumorigenic in nude mice. Thus the endpoint used in this assay is close to malignant transformation.

On the other hand, negative results need further investigation. Because the induction frequency of transformation has been very low even with potent carcinogens such as 4NQO and MNNG (Kakunaga, 1978), and because it is unknown why the spontaneous transformation is low, it is logically impossible to conclude with certainty that results are negative. Negative results from re-

peated or very large-scale experiments will give suggestive evidence that tested compounds are either inactive or very weakly active in transforming human fibroblastic cells.

Extrapolation of results into human risk assessment. Any compound which gives positive results in this assay is considered dangerous to humans even if that chemical has been negative in all other *in vivo* and *in vitro* tests for assay of carcinogens. On the other hand, it is impossible to draw any conclusive assessment from negative results unless two major weak points in this system are resolved: i.e., lack of quantitation and poor carcinogen-metabolizing activity of the target cell. KD cells showed very low activity of arylhydrocarbon hydroxylase and were not transformed by treatment with methylcholanthrene (Ka-

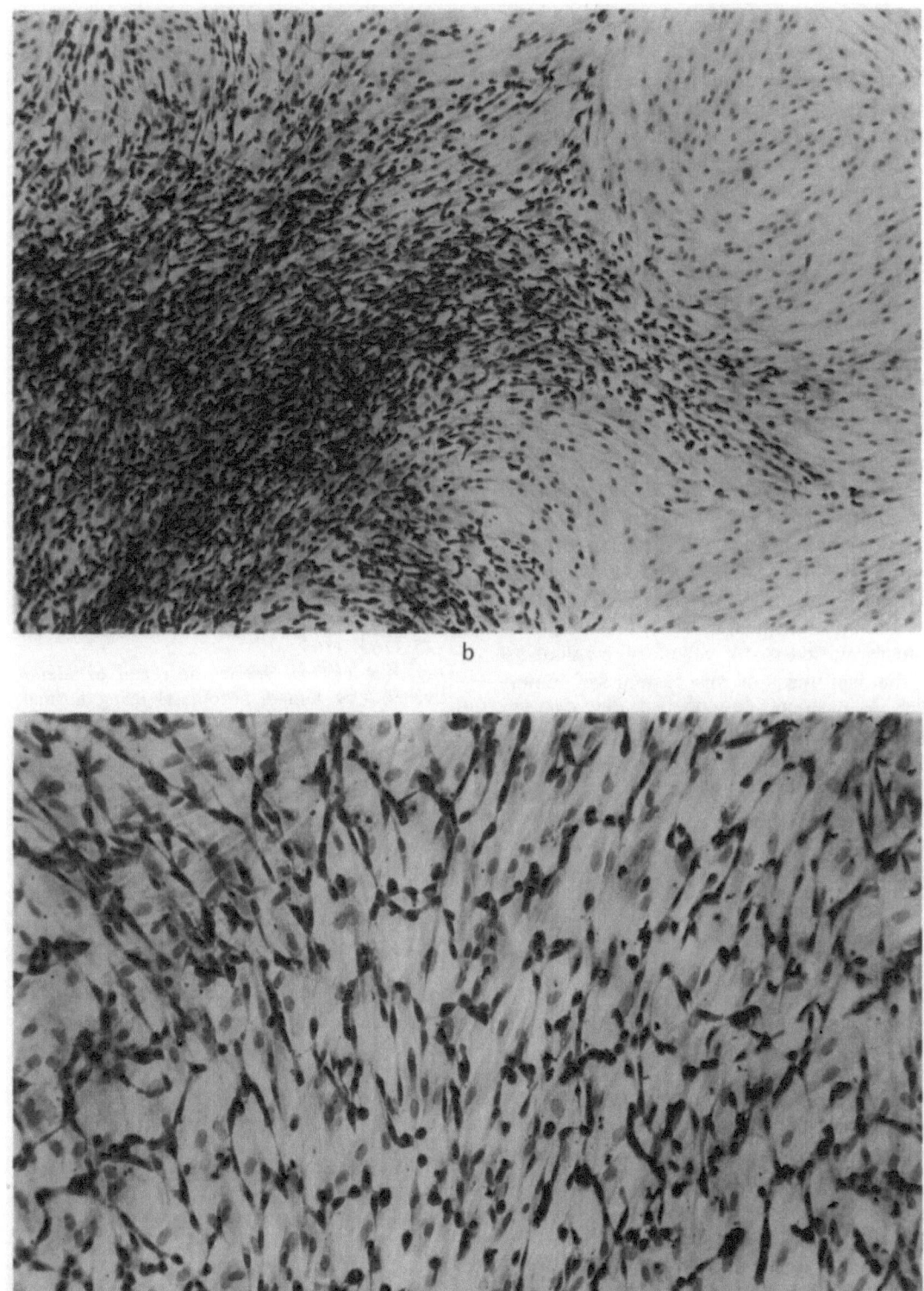

b

c

kunaga, 1978). Addition of liver microsome fraction or epithelial cells to carcinogen-target cell mixture will improve this shortcoming. Use of human epithelial cells of normal tissue origin as a target cell seems far away because of the extreme difficulty in culturing these cells.

Early Transformation Assay

Reliability of results. Occasionally an untreated control group may show low but significant colony-forming efficiency in gel for unknown reasons. It is possible and necessary to obtain a dose-response curve and statistical analysis of the results. Sensitivity of the assay system seems adequate for evaluation of negative results.

Extrapolation of results into human risk assessment. The relation of the temporarily increased ability of colony formation in gel of human fibroblastic cells to malignant transformation is not yet clear. Evaluation of the usefulness of this system for human risk assessment depends on the demonstration of this relationship. Blind tests using the chemicals which are etiologically known to be carcinogenic to humans may give a vague evaluation of this system. The poor carcinogen-metabolizing activity of target cells will be covered by modification of the system as described in the malignant transformation assay.

Conclusion

It is obvious that the transformation assay using human cells is not yet well developed. It has many disadvantages as a short-term test system, such as the complexity of the system, time-consuming nature, relatively high expense, low sensitivity, and the requirement for skillful technique. However, it is quite unknown whether the change in cellular genes is the only cause of human cancer by chemicals or what type of genetic change is the major source of chemical car-

cinogenesis in humans even if we assume a genetic cause for cancer incidence. Under such circumstances, a system that is the most directly analogous to in vivo human carcinogenesis, i.e., an assay for induction of transformation of human cells by chemicals, is of great importance. As a short-term test, there are many aspects that need to be modified and improved. The low frequency of transformation induction may well be the greatest problem. However, the distinct difference in the sensitivity to transformation between mouse, rat, or hamster and human cells suggests the importance of a transformation assay system using human cells.

References

Barrett, J.C., Ts'o, P.O.P. (1978): Evidence for the progressive nature of neoplastic transformation in vitro. Proc. Natl. Acad. Sci. USA 75:3761–3765.

Day, R.S. (1974): Studies on repair of adenovirus 2 by human fibroblasts using normal, xeroderma pigmentosum, and xeroderma pigmentosum heterozygous strains. Cancer Res. 34:1965–1970.

DeMars, R., Jackson, J.L. (1977): Mutagenicity detection with human cells. J. Environ. Pathol. Toxicol. 1:55–77.

Fogh, J., Hajdu, S. I. (1975): The nude mouse as a diagnostic tool in human tumor cell research. J. Cell Biol. 67:1172–1975.

Foulds, L. (1969): Neoplastic Development, Vol. 1. London, New York: Academic Press.

Foulds, L. (1975): Neoplastic Development, Vol. 2. London, New York: Academic Press.

Freedman, V.H., and Shin, S. (1977): Isolation of human diploid cell variants with enhanced colony-forming efficiency in semisolid medium after a single-step chemical mutagenesis. J. Natl. Cancer Inst. 58:1873–1875.

Freeman, A.E., Lake, R.S., Igel, H.J., Gernard, L., Pezzutti, M.R., Malone, J.M., Mark, C., Benedict, W.F. (1977): Heteroploid conversion of human skin cells by methylcholanthrene. Proc. Natl. Acad. Sci. USA 74: 2451–2455.

Giard, D.J., Aaronson, S.A., Todaro, G.J., Arnstien, P., Kersey, J.H., Dosik, H., Parks, W.P. (1973): In vitro cultivation of human tumors:

establishment of cell lines derived from a series of solid tumors. J. Natl. Cancer Inst. *51*:1417–1423.

Kakunaga, T. (1977): Transformation of human diploid cells by chemical carcinogens. In: Hiatt, H.H., Watson, J.D., and Winsten, J.A. (eds.), Origins of Human Cancer, Book C, Cold Spring Harbor Conferences on Cell Proliferation, Vol. 4, New York: Cold Spring Harbor Laboratory, pp. 1537–1548.

Kakunaga, T. (1978): Neoplastic transformation of human diploid cells by chemical carcinogens. Proc. Natl. Acad. Sci. USA *75*: 1334–1338.

Kakunaga, T. (1980): Approaches towards the development of human transformation assay system. In: Mishra, N.K., Dunkel, V., and Mehlman, M.A. (eds.), Advances in Environmental Toxicology: Mammalian Cell Transformation by Chemical Carcinogens. New York: Pathotex Publishers.

Kakunaga, T., Kamahora, J. (1968): Properties of hamster embryonic cells transformed by 4-nitroquinoline-1-oxide *in vitro* and their correlations with the malignant properties of the cells. Biken. J. *11*:313–332.

Kakunaga, T., Kamahora, J. (1969): Analytical studies on the process of malignant transformation of hamster embryonic cells in cultures with 4-nitroquinoline-1-oxide. Symposia Cell Chem. *20*:135–148.

Kamahora, J., Kakunaga, T. (1970): Significance of *in vitro* chemical carcinogenesis. Protein, Nucleic Acid Enzyme *15*:458–473.

Kuroki, T., Sato, H. (1969): Transformation and neoplastic development *in vitro* of hamster embryonic cells by 4-nitroquinoline-1-oxide and its derivatives. J. Natl. Cancer Inst. *41*: 53–71.

Milo, G.E., DiPaolo, J.A. (1978): Neoplastic transformation of human diploid cells *in vitro* after chemical carcinogen treatment. Nature (Lond.) *275*:130–132.

Rous, P., Beard, J.W. (1935): The progression to carcinoma of virus-induced rabbit papilloma. J. Exp. Med. *62*:523–548.

Shimada, H., Shibata, H., Yoshikawa, H. (1976): Transformation of tissue-cultured *Xeroderma pigmentosum* fibroblasts by treatment with N-methyl-N'-nitro-N-Nitrosoguanidine. Nature (Lond.) *264*:547–548.

Shin, S., Freedman, V.H., Risser, R., Pollack, R. (1975): Tumorigenicity of virus-transformed cells in nude mice is correlated specifically with anchorage independent growth in vitro. Proc. Natl. Acad. Sci. USA *72*: 4435–4439.

Zamacnick, P., Long, J.C. (1977): Growth of cultured cells from patients with Hodgkin's disease and transplantation into nude mice. Proc. Natl. Acad. Sci. USA *74*:754–758.

31

Chemical–Viral Interactions: Enhancement of Viral Transformation by Chemical Carcinogens

BRUCE C. CASTO

Introduction

A series of *in vitro* tests for the evaluation of potential carcinogenicity/mutagenicity of physical and chemical agents have been implemented in several laboratories. Of the various tests, those receiving most attention for prediction of carcinogenic activity are transformation of mammalian cells, mutation assays in microbial and mammalian cells, and the demonstration of damage to cell DNA by induction of repair synthesis or DNA fragmentation. Presently, none of these tests can be used alone to determine the carcinogenic potential of all chemical classes of suspect agents, although several of the above tests have been reported to have a 90% or greater correlation with known carcinogenic activity (McCann *et al.*, 1975; Casto *et al.*, 1978; Pienta *et al.*, 1977; Purchase *et al.*, 1976).

Another approach for the assay of carcinogens in mammalian cells has been described by Casto (1973, 1980) and Casto *et al.* (1973, 1974, 1976, 1979a). This assay quantitatively measures the capacity of physical or chemical agents to enhance the frequency of viral transformation. It is based upon the observations that treatment of cells with X-irradiation (Pollack and Todaro, 1968; Stoker, 1963; Coggin, 1969), DNA base analogues (Todaro and Green, 1964), or UV (Casto, 1973) increases the sensitivity of the treated cells to transformation by DNA viruses. Exposure of mammalian cells to the above agents often results in changes in cellular DNA. It has been postulated (Pollack and Todaro, 1968; Coggin, 1969; Casto and DiPaolo, 1973) that such agents interacting with cellular DNA may create additional sites for integration of viral DNA into cellular DNA during repair synthesis or in newly synthesized DNA during the subsequent period of scheduled DNA synthesis; the initiation of carcinogenesis by most chemical agents is also considered to involve changes in cell DNA by direct or indirect means. If the above suppositions are correct, discrete changes in cell DNA by chemicals might be detected by observing a quantitative increase in viral transformation among treated cells. Such increases in transformation have been demonstrated following treatment of cells with a variety of carcinogenic agents (Casto, 1973, 1980; Casto et al., 1973, 1974, 1976, 1979a; Diamond et al., 1974; Ledinko and Evans, 1973; Stich et al., 1972). Concurrently, Hirai et al. (1974) and Casto et al. (1979b) have shown that carcinogen treatment of cells results in an increased incorporation of viral DNA into cellular DNA. Subsequently, Casto (1980) has reported on over 150 chemicals (carcinogens and noncarcinogens) using enhancement of adenovirus transformation as an index of activity. The procedures used and the results of these studies are the subject of this presentation.

Test System

Cell cultures. Syrian hamster embryo cells (HEC) are prepared from fetuses after 13–14 days of gestation. The fetuses are decapitated, eviscerated, and minced with scissors. The mince is rinsed 3–5 min with 0.25% trypsin (Difco 1:250 prepared in Dulbecco's modified Eagle's medium [D-MEM] with 0.1 mM Ca^{++}), the trypsin solution decanted, and the tissue trypsinized for $1-1^1/_2$ hr at room temperature. The cells are separated from undigested material by filtration through cheesecloth, centrifuged, and resuspended in medium to yield approximately 1×10^6 total cells/ml. Five ml or 10 ml of cell suspension is plated into 60-mm or 100-mm plastic Petri dishes, respectively. The cells are cultured in a 5% CO_2 atmosphere in a modified DMEM (Casto, 1973) with 5% fetal bovine serum (FBS) and 0.22 g/100 ml of $NaHCO_3$. After 3 days at 37° C, cell counts range from $3.5-4.5 \times 10^6$ cells/60-mm dish.

Virus. Simian adenovirus, SA7, is cultured in a continuous line of monkey kidney cells (VERO) by inoculating 3–5 plaque-forming units/cell. After 2 hr adsorption, 5 ml of medium (Eagle's MEM with 5% FBS) is added to each 100-mm dish. Cytopathic effects are usually complete by 72 hr, at which time the cells are harvested by scraping with a rubber policeman, centrifuged, resuspended in $^1/_5$ of the supernatant fluid and frozen-thawed through four cycles. The virus pool is then checked for infectious virus by plaque formation in VERO cells and for transformation efficiency in HEC (Casto, 1969).

Transformation assay. For estimation of the transformation efficiency of a virus pool, 0.2 ml of twofold dilutions (1:4–1:64) of virus is added to each of two dishes in HEC. After adsorption for 3 hr, the cells are trypsinized, centrifuged, resuspended to 10^6/ml in DMEM with 10% FBS and 0.11 g/100 ml of $NaHCO_3$, and 2×10^5 cells (0.2 ml) plated into each of five dishes. Three ml of the latter medium is added to each dish; 48–72 hr later the medium is changed to DMEM (0.1 mM Ca^{++}), 10% FBS and 0.22 g/100 ml of $NaHCO_3$. After 5–6 days, the cultures are changed with 3 ml of the low-calcium medium containing 0.5 g/100 ml of Bacto-agar. At 4 to 5 day intervals thereafter, an additional 3 ml of agar medium is added to each dish. For fixation, 3 ml of 10% buffered formalin is added onto the agar overlay and the cells fixed overnight. The agar overlay is then removed and the cells stained with Giemsa (1:30 dilution of stock). The adenovirus-transformed foci appear as darkly stained areas of piled-up cells with a unique cytomorphology (Castro, 1969) against a monolayer of lightly stained background cells.

Checking procedures. A suitable pool of virus will have an infectious titer greater than 10^9 PFU/ml and a PFU/FFU ratio of $2-4 \times 10^5$. The VERO cells and virus stocks are monitored for mycoplasma contamination since the HEC are susceptible to these agents, often resulting in complete destruction of the cell sheet. Reagents used for preparation of the media (water, serum, basal media) are critical. The cloning efficiency of the HEC should not be below 10–12% when 500–600 cells are plated for survival. The media should allow the development of discrete (nondiffuse) foci within 10–14 days, which contrast sharply against background cells in stained or unstained cultures. The type and concentrations of agar used are necessary for the optimal development of transformed foci.

Viral Enhancement Assays

Determination of dose range. Following 3 days' incubation, the HEC medium is changed to the modified DMEM with 10% FBS. Each of two dishes are then treated with test 5–6 doses of chemical for 2 or 18 hr using fivefold dilutions with an initial concentration of 100 μg/ml. Following chemical exposure, the cells are trypsinized, centrifuged, diluted to 2.5×10^3 cells/ml, and plated into five dishes per dose at 500 cells/dish. Cloning medium is added (DMEM, 10% FBS, 0.11 g/100 ml of NaHCO$_3$) and changed after 48–72 hr with the above medium containing 0.22 g/100 ml of NaHCO$_3$. Cells are fixed in 10% buffered formalin and stained with crystal violet (0.02%) after 8 days' incubation. The highest dose to be used results in a 70–90% reduction in cloning efficiency and the two lowest doses cause no demonstrable cell lethality in clonal assays.

Enhancement assay. Chemicals are dissolved in acetone at concentrations of 1 or 10 mg/ml and further diluted in complete medium to give the desired concentrations. Primary HEC (3 to 4 days in culture) are treated with serial twofold dilutions of chemical for 2 or 18 hr using two dishes per dose and 5 to 7 doses per chemical. Following treatment, the treated and solvent control dishes are rinsed, inoculated with approximately 200 focus-forming units of SA7 in 0.2 ml medium (5% FBS), and incubated for 3 hr at 37° C. After adsorption, the cells are trypsinized, centrifuged, and plated as before for transformation and survival assays onto 10 or 5 dishes, respectively. Foci do not develop in nonvirus-treated cells, since the conditions favorable for virus transformation inhibit the development of chemically transformed HEC (Casto et al., 1977).

Enhancement is expressed as the increase in foci per 10^6 surviving cells. The fraction of surviving cells is determined by cloning assays from the same virus- and chemically treated cells as are used in the transformation assay. The total colonies in five dishes of treated cells is divided by the total number in five dishes of control cells to give the proportion of surviving chemically treated cells. The number of SA7 foci from virus-treated dishes is used as the control transformation frequency (number of foci per 10^6 SA7 inoculated cells). Among chemically treated cells, the frequency is determined by multiplying the actual number of foci by the reciprocal of the surviving fraction. The enhancement ratio is calculated by dividing the total foci in 10^6 surviving, treated cells by the number in 10^6 SA7-inoculated control cells.

For determination of statistical significance, a Table of Ratios (Casto et al., 1973) was constructed from the tables derived by Lorenz (1962) that are based upon the Poisson distribution. The total number of SA7 foci (control plus treated) for use with the table is taken from the actual number counted, not from the adjusted numbers. An increase in frequency is considered significant at the 1% or 5% confidence level if the increase exceeds the ratio between the treated and control value obtained from the Lorenz table.

Modifications. If compounds are not positive following the standard protocol, a repeat test is performed in which the SA7 is added prior to chemical treatment. Virus is adsorbed for 3 hr, the cells trypsinized and added to dishes in 3 ml of medium (200,000 for transformation and 600 for survival assays). Approximately 2 hr later, the test chemical (contained in an equal volume of medium) is added as a 2× concentration to each dish. With this procedure, the chemical remains in the culture medium for 48 hr at which time the cultures are changed to the low-calcium medium; the feeding schedule and agar overlays are as outlined previously.

The enhancement assay system using SA7 and hamster embyro cells has been extended to include human lung fibroblast cells and SV40 virus or rat embryo cells with SA7 virus. With the human lung fibroblast (HLF) assay, 5×10^4 cells are plated and treated with chemical after overnight incubation. After chemical treatment, SV40 is added, but unlike the SA7 assay, the cells are not transferred after virus adsorption. In two such assays, B(a)P showed no toxicity for HLF cells at 5 μg/ml but the number of SV40 foci increased from 28 in control cultures to 148 in B(a)P treated cells. At 50 μg/ml of MMS, the cloning of HLF was reduced by 22%, whereas the number of virus-transformed foci increased fourfold. Other chemicals shown to enhance SV40 transformation of human cells include: propane sultone, β-propiolactone, and 3-methylcholanthrene.

Rat embryo cells are transformed by human (Freeman et al., 1967) or simian (DiPaolo and Casto, 1976) adenoviruses. Using the same methodology as outlined for hamster embryo cells and SA7, a series of chemicals that were positive in hamster cells were also positive when tested in rat embryo cells for their capacity to enhance SA7 transformation (DiPaolo and Casto, 1976).

The demonstration that the virus enhancement assay system can be employed with other virus models (especially human cells) may lead to new enhancement assay systems that are more germane to the etiology of human cancer. However, in vitro tests for carcinogens in human cell strains or in many rodent cell lines have not been successful owing to the lack in these cells of adequate metabolizing enzymes that are necessary to convert the test chemical into an active compound. This problem is being circumvented in many laboratories by: (1) using tissue homogenates for activation or (2) coculturing metabolically competent cells with the target cells. The first of these two alternatives has been explored in the hamster embryo cell-SA7 system using a chemical (DMN) that is poorly activated by hamster cells in vitro. Liver homogenates (S9 fraction) were prepared from male hamsters inoculated i.p. with DMN on three occasions (2 mg per 100 g body weight). An S9 mix consisting of the S9 fraction and the necessary cofactors (Ames et al., 1973) was diluted 1:2, 1:4, and 1:8 in cell culture medium, added to an equal volume of $3 \times$ concentrations of DMN, and placed in 60-mm dishes containing cultures of hamster embryo cells. After an overnight incubation, SA7 was added and enhancement assays were performed as described previously. DMN without added S9 mix (up to 1000 μg/ml) did not enhance, but a 5.7-fold increase was induced by 40 μg/ml when added with the S9 activation system. Other experiments show that urethane can be similarly activated by urethane-primed hamster liver homogenates and enhance viral transformation.

Results

In Tables 31.1 to 31.3, data are presented from viral enhancement assays with six chemicals from several representative classes. The polycyclic hydrocarbons, including DMBA shown in Table 31.1, enhance when added prior to SA7 (Casto et al., 1973) but inhibit transformation when added after virus adsorption and cell transfer (Casto et al., 1973, 1976). The increase in transformation frequency was accompanied by a corresponding increase in the absolute numbers of SA7 foci per dish of treated cells; a twofold increase was observed where only 10% of the treated cells survived. Such data argue against the selection of transformation-sensitive, but chemically resistant cells, and suggest that the transforming potential of individual cells for SA7 is increased by direct interaction with the chemical carcinogens. The demonstration of an absolute increase in SA7 foci in the absence of a significant decrease in cell survival also shows that the increase in SA7 transformation frequency need not be related to cell lethality induced by chemical (Tables 31.1, 31.2, and 31.3).

Ethyl methanesulfonate, and other direct alkylating agents, will enhance SA7 transformation if added before or after virus inoculation and cell transfer (Casto et al., 1974, 1976). At high concentrations of chemical, where excessive lethality occurs, the enhancement ratio may be artificially high (Table 31.1). However, in many instances, in the presence or absence of inordinate cell killing at high doses of chemical, the frequency of transformation may be decreased (Casto et al., 1973, 1974) or may be similar to control values (Table 31.2).

A number of carcinogenic inorganic metal salts have been shown to enhance SA7 transformation (Casto et al., 1979a) and to induce an infidelity of DNA synthesis (Sirover and Loeb, 1976). Salts of antimony, arsenic, cadmium, chromium, and platinum were found to be most active in the viral enhancement assay, while those of beryllium, cobalt, copper, lead, manganese, nickel, and zinc were less active. An example of a combination of two enhancing metals, lead and chromium, is shown in Table 31.2. $PbCrO_4$ has a reduced level of cytotoxicity in contrast to K_2CrO_4 or Na_2CrO_4, but an absolute increase in foci greater than that observed with either lead or chromium salts alone.

The data from two compounds (propylene oxide and vinyl acetate) that are ineffective for enhancement when added prior to SA7, but do enhance when added 5 hr after virus, are shown in Table 31.3. Both cause a threefold to fourfold increase in the absolute number of virus-transformed foci at concentrations that reduce cell survival approximately 30% or less. Other chemicals that enhance more effectively when added after virus adsorption include: acry-

Table 31.1. *Enhancement of Viral Transformation by 7,12-dimethylbenz(a)anthracene and Ethyl Methanesulfonate following 18-hr Pretreatment*

Chemical[a]	μg/ml	Surviving fraction[b]	SA7 foci[c]	Enhancement ratio[d]
DMBA	0.05	0.09	180	22.9
	0.025	0.41	235	6.6
	0.012	0.070	171	2.9
	0.006	0.82	132	1.9
	0.0	1.00	86	1.0
EMS	400	0.006	28	128.0
	200	0.24	85	9.8
	100	0.71	60	2.4
	50	0.87	60	1.9
	0	1.00	36	1.0

[a]Chemical dilutions were added to mass cultures of HEC 18 hr before SA7. Virus was adsorbed for 3 hr and the cells transferred for survival (500–700 cells/dish) and for transformation assays (200,000 cells/dish).

[b]Determined from plates receiving 500–700 cells. The number of colonies from virus- and chemical-treated cells was divided by the number of colonies from virus-inoculated control cells to give the surviving fraction. Cloning efficiency of control cells was from 10 to 15%.

[c]Number of foci from 10^6 plated cells.

[d]Enhancement ratio was determined by dividing the transformation frequency (TF) per 10^6 treated cells (TF = SA7 foci × reciprocal of the surviving fraction) by that obtained from control cells. Underlined values are statistically significant at the 1% level.

Table 31.2. *Enhancement of Viral Transformation by Ethylene Dibromide and Lead Chromate following 18-hr Pretreatment*

Chemical[a]	$\mu g/ml$	Surviving fraction[b]	SA7 foci[c]	Enhancement ratio[d]
EDB	1000	0.39	89	<u>12.3</u>
	500	0.92	62	<u>3.7</u>
	250	0.99	22	1.2
	125	1.13	15	0.7
	0	1.00	18	1.0
PbCrO$_4$	100	0.79	35	1.2
	50	0.70	125	<u>4.7</u>
	25	1.02	127	<u>3.1</u>
	12	0.90	109	<u>3.2</u>
	0	1.00	37	1.0

[a]Chemical dilutions were added to mass cultures of HEC 18 hr before SA7. Virus was adsorbed for 3 hr and the cells transferred for survival (500–700 cells/dish) and for transformation assays (200,000 cells/dish).

[b]Determined from plates receiving 500–700 cells. The number of colonies from virus- and chemical-treated cells was divided by the number of colonies from virus-inoculated control cells to give the surviving fraction. Cloning efficiency of control cells was from 10 to 15%.

[c]Number of foci from 10^6 plated cells.

[d]Enhancement ratio was determined by dividing the transformation frequency (TF) per 10^6 treated cells (TF = SA7 foci × reciprocal of the surviving fraction) by that obtained from control cells. Underlined values are statistically significant at the 1% level.

lonitrile, cytosine arabinoside, 2-chloro-1, 3-butadiene, copper sulfide, and manganous chloride.

Although most of the carcinogenic compounds exhibit varying degrees of cytotoxicity, the potential for cell killing is not sufficient or necessary for the induction of viral transformation. As discussed pre-

Table 31.3. *Enhancement of Viral Transformation by Propylene Oxide and Vinyl Acetate: Addition of Chemical 5 hr after Virus Inoculation*

Chemical[a]	$\mu g/ml$	Surviving fraction[b]	SA7 foci[c]	Enhancement ratio[d]
PO	500	0.69	140	<u>6.0</u>
	250	0.88	111	<u>3.8</u>
	125	0.98	54	1.7
	62	0.93	41	1.4
	0	1.00	33	1.0
VA	1000	0.70	110	<u>4.7</u>
	500	0.87	120	<u>4.4</u>
	250	0.94	90	<u>3.0</u>
	125	1.02	64	<u>2.0</u>
	0	1.00	33	1.0

[a]Virus was adsorbed for 3 hr and the cells transferred for survival (500–700 cells/dish) and for transformation assays (200,000–300,000 cells/dish). Chemical dilutions were added approximately 5 hr after SA7.

[b]Determined from plates receiving 500–700 cells. The number of colonies from virus- and chemical-treated cells was divided by the number of colonies from virus-inoculated control cells to give the surviving fraction. Cloning efficiency of control cells was from 10 to 15%.

[c]Number of foci from 10^6 plated cells.

[d]Enhancement ratio was determined by dividing the transformation frequency (TF) per 10^6 treated cells (TF = SA7 foci × reciprocal of the surviving fraction) by that obtained from control cells. Underlined values are statistically significant at the 1% level.

viously, enhancement occurs at one or more dilutions of most carcinogens in the absence of cell killing. Noncarcinogenic, but toxic, chemicals do not cause enhancement when the data are judged by the criteria of a dose-response relationship or an absolute increase in SA7 foci per dish. Where only one dose of a chemical is positive, five to seven repeat tests are performed to confirm the reproducibility of the enhancement data in addition to testing the chemical at various time intervals in relation to virus inoculation. An example of two toxic, nonenhancing chemicals is presented in Table 31.4.

Discussion

The enhancement of viral transformation appears to be a sensitive indicator for chemical agents with the potential for damaging cell DNA either by direct or indirect methods and, therefore, may be useful as a screening tool to detect these chemi-

cals in the environment. In this context, the majority of agents examined have been: (1) those chemicals more commonly used in other short-term *in vitro* assays, (2) those used in various industrial applications, (3) chemicals annually produced in large volumes, and (4) inorganic metal salts. Table 31.5 summarizes the enhancement data for a large number of the compounds tested in the hamster embryo-SA7 assay system.

Data for many of the chemicals, including the polycyclic hydrocarbons and various alkylating agents used in mammalian cell transformation or mutagenesis experiments and in microbial assays, have previously been discussed (Casto *et al.*, 1973, 1974, 1976).

Forty-two of the top 50 production volume compounds have been tested at least twice in the enhancement assay. Among these compounds, ethylene dichloride, propylene oxide, and vinyl acetate have caused enhancement. Butadiene (1,3-)(the 27th ranked compound) has not been tested, but 2-chloro-1,3-butadiene enhances.

A total of 46 inorganic metals salts have

Table 31.4. *Effect of Ammonium Sulfate and Aluminum Sulfate on the Frequency of SA7 Transformation of HEC*

Chemical[a]	μg/ml	Surviving fraction[b]	SA7 foci[c]	Enhancement ratio[d]
$NH_4)_2SO_4$	1000	0.61	21	1.2
	500	0.67	31	1.6
	250	0.91	27	1.0
	125	0.96	25	0.9
	0	1.00	29	1.0
$Al_2(SO_4)_3$	1000	0.74	18	1.2
	500	0.83	16	1.0
	250	1.02	16	0.8
	125	1.04	15	0.8
	0	1.00	19	1.0

[a]Chemical dilutions were added to mass cultures of HEC 18 hr before SA7. Virus was adsorbed for 3 hr and the cells transferred for survival (500–700 cells/dish) and for transformation assays (200,000 cells/dish).

[b]Determined from plates receiving 500–700 cells. The number of colonies from virus- and chemical-treated cells was divided by the number of colonies from virus-inoculated control cells to give the surviving fraction. Cloning efficiency of control cells was from 10 to 15%.

[c]Number of foci from 10^6 plated cells.

[d]Enhancement ratio was determined by dividing the transformation frequency (TF) per 10^6 treated cells (TF = SA7 foci × reciprocal of the surviving fraction) by that obtained from control cells. Underlined values are statistically significant at the 1% level.

Table 31.5. *Enhancement of Viral Transformation by Various Chemical Carcinogens*

Chemical[a]	Response[b]	Concentration[c]
Acetic acid	0	1000
Acridine orange	+	2.5
Acrylonitrile	+	100
Adipic acid	0	1000
Aflatoxin B$_1$	+	0.1
alpha-Chloroacetamide	+	5
alpha-naphthylamine	+	100
Aluminum chloride	0	1000
Aluminum sulfate	0	400
Ammonium sulfate	0	1000
Antimony triacetate	+	1.2
Arsenic chloride	+	0.5
Barium chloride	0	1000
Benzene	0	5000
Benzidine	+	25
Benzidine 2HCl	+	12
Benzo(a)pyrene	+	0.08
Benzo(e)pyrene	+	5
beta-Naphthylamine	+	30
beta-Propiolactone	+	5
Cadmium acetate	+	0.25
Cadmium chloride	+	0.5
Caffeine	+	125
Calcium chloride	0	1000
Calcium chromate	+	1
Carbon black	0	1000
Chlordane (tech)	+	5
Chlorodimethyl ether	+	10
Chloroform	0	2000
cis-Diaminedichloroplatinum	+	0.06
Cobalt acetate	+	50
Cobalt molybdate	+	62
Cumene	0	1000
Cupric sulfate	+	12.5
Cuprous sulfide	+	62.5
Cyclohexane	0	1000
Cyclohexylamine	+	60
Cytosine arabinoside	+	0.06
Dibenz(a,c)anthracene	+	0.4
Dibenz(a,h)anthracene	+	0.4
Dichlorobenzidine	+	12.5
Dimethylterephthalate	0	1000
Ethanol	0	1000
Ethyl methanesulfonate	+	50
Ethylbenzene	0	1000
Ethylene dibromide	+	500
Ethylene glycol	0	1000
Ethylenethiourea	±	1000
Ferrous chloride	+	500
Ferrous sulfate	+	250
Fluocinolone acetonide	+	0.62
Heptachlor (tech)	+	10
Hydrazine sulfate	+	1000
Hydrochloric acid	0	1000
Hydroxylamine	+	200
Hydroxyurea	+	200

Table 31.5. *Enhancement of Viral Transformation by Various Chemical Carcinogens* (Continued)

Chemical[a]	Response[b]	Concentration[c]
Lead acetate	+	400
Lead chromate	+	12
Lead monoxide	+	100
Lithium chloride	0	1000
Magnesium acetate	0	1000
Maleimide	0	5
Manganous chloride	+	25
Mercuric chloride	+	12.5
Methanol	0	1000
Methyl methanesulfonate	+	5
Methylazoxymethanol acetate	+	2.5
Molybdenum chloride	+	500
Molybdenum sulfide	+	500
N-Acetoxy-acetylaminofluorene	+	0.2
N-Methyl-N-nitro-N-nitrosoguanidine	+	0.2
N-N-bis-2-Chloroethyl-2-naphthylamine	+	0.12
N-Nitrosodiethylamine	0	1000
N-Nitrosodimethylamine	0	1000
N-2-Acetylaminofluorene	0	20
Nickel sulfate	+	100
Nickel sulfide	+	1000
Nitric acid	0	1000
Pentachlorophenol	+	100
Perchloroethylene	0	1000
Perylene	0	20
Phenanthrene	0	20
Phenothiazine		
Phosphoric acid	0	1000
Phthalazinone	+	500
Platinum chloride	+	10
Potassium chromate	+	1
Propane sultone	+	3
Propylene oxide	+	250
Pyrene	0	20
Red dye #2	0	1000
Red dye #40	0	1000
Scarlet red	+	100
Selenium chloride	+	5
Silver nitrate	+	0.12
Sodium arsenite	+	0.3
Sodium carbonate	0	1000
Sodium hydroxide	0	1000
Sodium sulfate	0	1000
Strontium chloride	0	1000
Styrene	0	1000
Sulfuric acid	0	1000
Tetraethyl lead	+	12
Thallium acetate	+	12.5
Thallium chloride	+	25
Thioacetamide	+	12
Thiourea	±	1000
Titanic oxide	0	1000
Toluene	0	1000
Tolylazoxytoluidine	+	25
Trichloroethane	0	1000
Trichlorethylene	0	1000

Table 31.5.   *Enhancement of Viral Transformation by Various Chemical Carcinogens* (Continued)

Chemical[a]	Response[b]	Concentration[c]
Tungsten hexachloride	0	200
Tungsto-antimoniate	+	12.5
Unsym.-dimethylhydrazine	+	250
Urea	0	1000
Vanadium chloride	+	5
Vinyl acetate	+	62
Xylene	0	1000
Zinc chloride	0	200
Zinc chromate	+	0.3
Zinc sulfate	±	25
Zirconium tetrafluoride		
1,2-Dichloroethane	+	500
1,3-Diethyl-2-Thiourea	0	1000
1,4-Dioxane	0	1000
2-Chloro-1, 3-butadiene	+	125
3-Methylcholanthrene	+	0.08
4-Aminobiphenyl	+	100
4-Carboxybenzaldehyde	+	250
4-Nitrobiphenyl	+	60
4,4'-Methylenebis-(o-chloroaniline)	+	2.5
5-Fluorodeoxyuridine	0	100
6-Acetoxy-B(a)P	+	0.25
7,12-Dimethylbenz(a)anthracene	+	.004

[a]Chemicals are listed in alphabetical order and are not grouped according to class.

[b]Enhancement was considered + if statistically significant at the 5% level using the Lorenz tables (1962).

[c]Highest concentration tested that was negative or the lowest concentration giving positive enhancement.

been tested in the viral enhancement assay (Casto *et al.*, 1979a). Positive enhancement was found with salts of antimony, arsenic, beryllium, cadmium, cobalt, copper, iron, lead, manganese, molybdenum, nickel, platinum, thallium, vanadium, and zinc. Negative metals include the acetate, chloride, or sulfate salts of aluminum, barium, calcium, lithium, magnesium, potassium, strontium, titanium, and zirconium. Both the positives and negatives from above are in excellent agreement with the *in vitro* fidelity of DNA synthesis assay reported by Sirover and Loeb (1976), the only exception being FeCl$_2$, which was positive in the viral enhancement assay and negative for introducing copy error in the infidelity of DNA synthesis assay.

The enhancement data with metals also agrees with data obtained using rec assays with *B. subtilis* (Nishioka, 1975). In these studies of 56 metals, arsenic, cadmium, chromium, mercury, manganese, and molybdenum were considered positive. Three of the strongly positive metals, arsenic, chromium, and molybdenum, were also mutagenic in *E. coli*. The rec assays, however, failed to detect beryllium, copper, iron, lead, nickel, antimony, or zinc. There are several other reports on the carcinogenic and mutagenic activity of the metals shown to be positive in the viral enhancement assay (Sunderman, 1975; Furst and Haro, 1969).

The enhancement of viral transformation appears to be a sensitive assay for carcinogens in terms of both detection and the dose required to produce a positive response. Of 130 chemicals with known positive or negative activity, 94% agreed with their current classification; the only false positive was caffeine. False negatives (some with questionable carcinogenicity) include: diethylnitrosamine,   red   dye number 2, trichloroethylene, chloroform, dioxane, and acetylaminofluorene. These compounds have not been tested using an exogenous metabolic activation system that was sufficient to convert dimethylnitrosamine and urethane from negative to positive responses.

Because of the increasing evidence that many environmental chemicals are involved in the etiology of human cancer, it is necessary that reliable, sensitive assays for potential chemical carcinogens be developed. The strong positive correlation between carcinogenic activity and enhancement of viral transformation suggests that the viral enhancement assay may be a useful tool in screening for potential carcinogenic or mutagenic agents.

References

Ames, B.N., Durston, W.E., Yamasaki, E., Lee, R.D. (1973): Carcinogens are mutagens: A simple test system combining liver homogenates for activation and bacteria for detection. Proc. Natl. Acad. Sci. USA 70:2281–2285.

Casto, B.C. (1969): Transformation of hamster embryo cells and tumor induction in newborn hamsters by simian adenovirus SVII. J. Virol. 3:513–519.

Casto, B.C. (1973): Enhancement of adenovirus transformation by treatment of hamster cells with UV, DNA base analogs, and dibenz(a,h)anthracene. Cancer Res. 33: 402–407.

Casto, B.C. (1980): Detection of chemical carcinogens and mutagens in hamster cells by enhancement of adenovirus transformation. In: N.K. Mishra, V. Dunkel, and M.A. Mehlman (eds.), Advances in Environmental Toxicology: Mammalian Cell Transformation by Chemical Carcinogens, New York: Pathotex Publishers.

Casto, B.C., DiPaolo, J.A. (1973): Chemicals viruses, and cancer. Progr. Med. Virol. 16: 1–47.

Casto, B.C., Pieczynski, W.J., DiPaolo, J.A. (1973): Enhancement of Adenovirus transformation by pretreatment of hamster cells with carcinogenic polycyclic hydrocarbons. Cancer Res. 33: 819–824.

Casto, B.C., Pieczynski, W.J., DiPaolo, J.A. (1974): Enhancement of adenovirus transformation by treatment of hamster embryo cells with diverse chemical carcinogens. Cancer Res. 34:72–78.

Casto, B.C., Pieczynski, W.J., Janosko, N., DiPaolo, J.A. (1976): Significance of treatment interval and DNA repair in the enhancement of viral transformation by chemical carcinogens and mutagens. Chem.-Biol. Interact. 13:105–125.

Casto, B.C., Janosko, N., DiPaolo, J.A. (1977): Development of a focus assay model for transformation of hamster cells in vitro by chemical carcinogens. Cancer Res. 37: 3508–3515.

Casto, B.C., Janosko, N., Meyers, J., DiPaolo, J.A. (1978): Comparison of in vitro tests in Syrian hamster cells for the detection of carcinogens. Proc. Am. Assoc. Cancer Res. 19: 83.

Casto, B.C., Meyers, J., DiPaolo, J.A. (1979a): Enhancement of viral transformation for evaluation of the carcinogenic or mutagenic potential of inorganic metal salts. Cancer Res. 39: 193–198.

Casto, B.C., Miyagi, M., Meyers, J., DiPaolo, J.A. (1979b): Increased integration of viral genome following chemical and viral treatment of hamster embryo cells. Chem.-Biol. Interact. 25:255–269.

Coggin, J.H., Jr. (1969): Enhanced virus transformation of hamster embryo cells in vitro. J. Virol. 3:458–462.

Diamond, L., Knorr, R. Shimizu, Y. (1974): Enhancement of simian virus 40-induced transformation of Chinese hamster embryo cells by 4-nitroquinoline-1-oxide. Cancer Res. 34: 2599–2604.

DiPaolo, J.A., Casto, B.C. (1976): In vitro transformation: Interaction of chemical carcinogens with viruses and physical agents. In: R. Montesano, H. Bartsch, and L. Tomatis (eds.), Screening Tests in Chemical Carcinogenesis (IARC Scientific Publications No. 12), Lyon: International Agency for Research on Cancer, pp. 415–430.

Freeman, A.E., Black, P.H., Wolford, R., Huebner, R.J. (1967): Adenovirus type 12-rat embryo transformation system. J. Virol. 1: 362–367.

Furst, A., Haro, R.T. (1969): A survey of metal carcinogenesis. Progr. Exp. Tumor Res. 12: 102–133.

Hirai, K., Defendi, V., Diamond, L. (1974): Enhancement of simian virus 40 transformation and integration by 4-nitro-quinoline-1-oxide. Cancer Res. *34*:3497–3500.

Ledinko, N., Evans, M. (1973): Enhancement of adenovirus transformation of hamster cells by N-methyl-N'-nitro-N-nitrosoguanidine, caffeine, and hydroxylamine. Cancer Res. *33*: 2936–2938.

Lorenz, R.J. (1962): Zur Statistik des Plaque-Testes. Arch. Ges. Virusforsch. *12*:108–137.

McCann, J., Choi, E., Yamasaki, E., Ames, B.N. (1975): Detection of carcinogens as mutagens in the *Salmonella*/microsome test: assay of 300 chemicals. Proc. Natl. Acad. Sci. USA *72*:5135–5139.

Nishioka, H. (1975): Mutagenic activities of metal compounds in bacteria. Mutat. Res. *31*: 185–189.

Pienta, R.J., Poiley, J.A., Lebherz III, W.B. (1977): Morphological transformation of early passage golden Syrian hamster embryo cells derived from cryopreserved primary cultures as a reliable *in vitro* bioassay for identifying diverse carcinogens. Int. J. Cancer *19*: 642–655.

Pollack, E.J., Todaro, G.J. (1968): Radiation enhancement of SV40 transformation in 3T3 and human cells. Nature (London) *219*: 520–521.

Purchase, I.F.H., Longstaff, E., Ashby, J., Styles, J.A., Anderson, D., Lefevre, P.A., Westwood, F.R. (1976): Evaluation of six short-term tests for detecting organic chemical carcinogens and recommendations for their use. Nature (London) *264*:624–627.

Sirover, M.A., Loeb, L.A. (1976): Infidelity of DNA synthesis *in vitro*: screening for potential metal mutagens or carcinogens. Science *194*:1434–1436.

Stich, H.F., Hammerberg, O., Casto, B.C. (1972): The combined effect of chemical and virus on DNA repair, chromosome aberrations and neoplastic transformation. Can. J. Genet. Cytol. *14*:911–917.

Stoker, M. (1963): Effect of X-irradiation on susceptibility of cells to transformation by polyoma virus. Nature (London) *200*: 756–758.

Sunderman, F.W., Jr. (1975): A review of the carcinogenicities of nickel, chromium, and arsenic compounds in man and animals. Prev. Med. *5*:279–294.

Todaro, G.J., Green, H. (1964): Enhancement by thymidine analogs of susceptibility of cells to transformation by SV40. Virology *24*: 393–400.

32

The Calcium-Independence of Neoplastic Cell Proliferation: A Promising Tool for Carcinogen Detection*

A.L. BOYNTON, S.H.H. SWIERENGA, AND J.F. WHITFIELD

Introduction

At first glance, neoplastic transformation seems unquestionably to be caused by gene mutation, because cancer cells transmit their tumorigenic property to their progeny and many carcinogenic agents cause mutations in both prokaryotes and eukaryotes (Braun, 1974; Markert, 1978). However, this need not be the case, because changes in gene expression (e.g., those responsible for cell differentiation during embryonic development) can also be transmitted by cells to their progeny (Pierce and Cox, 1978; Pierce *et al.*, 1978). Because of this heritability, carcinogen-induced changes in gene expression responsible for neoplastic transformation would resemble mutations, although they would be completely reversible under appropriate circumstances. Support for this alternative is provided by the following facts: the frequency of neoplastic transformation is often very much higher

*Issued as N.R.C.C. No.: 17718.

than that of mutations (Braun, 1974); neoplastic transformation is often associated with changes in gene expression resulting in the reappearance of fetal antigens and inappropriate hormone production (Uriel, 1975); a large proportion of the progeny of the proliferating malignant cells in many tumors (e.g., squamous cell carcinomas) *in vivo* produce proliferatively inactivated, nontumorigenic, differentiated progeny (Pierce and Cox, 1978; Pierce *et al.*, 1978); and highly malignant mouse teratocarcinoma cells can cooperate with normal cells to produce a completely normal adult mouse after their insertion into a blastocyst (Mintz, 1978).

A proliferating differentiated cell replicates both its genes and the particular pattern of gene repression responsible for its characteristic phenotype. In biochemical terms, the cell replicates both its DNA and the protein-DNA complexes constituting higher order structures related to differential gene expression such as the nucleosomes (Li and Eckhardt, 1977). Some carcinogens might react only with DNA, thereby causing mutations, while at the same time altering the pattern of gene repression by interfering with DNA-protein interactions in the daughter chromosomes. Other carcinogens might react only with chromosomal proteins, thereby preventing the replication of the pattern of gene repression *without* causing mutations. If this is true, and neoplastic transformation is at least in some cases the result of altered gene expression rather than altered gene structure, mutagenesis and other indicators of DNA damage are at best indirect measures of the carcinogenicity of a particular group of compounds.

It follows from these considerations that no compound can reliably be labeled as carcinogenic or noncarcinogenic from the results of any *in vitro* test, the interpretation of which depends on neoplastic transformation being caused exclusively by mutation or altered gene expression. Clearly the most reliable *in vitro* test would be based on a property of mammalian cells which is easily and rapidly detectable *in vitro*, is as invariably associated with neoplastic transformation as tumorigenesis itself, and (like tumorigenesis) appears regardless of the cause of neoplastic transformation. In this communication, we will present some preliminary observations which suggest that the striking ability of neoplastic avian and mammalian cells to multiply rapidly in low-calcium media which do not support the multiplication of their nonneoplastic counterparts (Balk *et al.*, 1973; Boynton and Whitfield, 1976, 1978; Boynton *et al.*, 1976, 1977a, 1978; Parsons, 1978; Swierenga *et al.*, 1976, 1978; Whitfield *et al.*, 1979) may be just such a property upon which to base an *in vitro* test for carcinogenicity.

Materials and Methods

T51B adult rat liver epithelioid cells isolated by Swierenga *et al.* (1976) were maintained in T-75 flasks (LUX Scientific, Thousand Oaks, California) in a medium consisting of 90% (v/v) BME (Eagle's Basal Medium from Flow Laboratories, Rockville, Maryland) or Williams' Medium E (from Flow Laboratories), 10% (v/v) FBS (fetal bovine serum from Flow Laboratories) and 25 μg/ml gentamicin (Microbiological Associates, Bethesda, Maryland). Cells were subcultured at weekly intervals by first exposing them for 30 seconds to a 0.1% solution of crystalline trypsin (Sigma Chemical Co., St. Louis, Missouri) in Ca^{2+} - and Mg^{2+}-free PBS (phosphate-buffered saline; pH 7.2), removing the trypsin solution, allowing the flasks to stand at room temperature for 3 minutes, and finally suspending them in 10 ml of fresh medium. The cells were planted at one-tenth of the original density in 10 ml of fresh medium, and incubated at 37°C in a humidified atmosphere consisting of 95% air and 5% CO_2. Descriptions of the several experimental procedures are given in the appropriate figure and table legends.

The effects of calcium on the flow of T51B cells into the DNA-synthetic (S) phase of their cell cycle were determined autoradiographically according to Boynton *et al.* (1976).

Cellular cyclic AMP (adenosine 3',5'-monophosphate) contents were determined by radioimmunoassay according to Boynton *et al.* (1978).

Results and Discussion

The proliferative activity of a wide variety of nontumorigenic epithelial and mesenchymally derived avian and mammalian cells *in vitro* and *in vivo* depends on the extracellular ionic calcium concentration. These cells are unable to initiate DNA synthesis, multiply, and form colonies in low- (i.e., 0.001 to 0.1 mM) as opposed to high- (0.5 to 2.0 mM) calcium medium (Figs. 32.1 and 32.2) (Balk *et al.*, 1973; Boynton and Whitfield, 1976, 1978; Boynton *et al.*, 1976, 1977a, 1977b, 1978; Parsons, 1978;

364   A.L. Boynton, S.H.H. Swierenga, and J.F. Whitfield

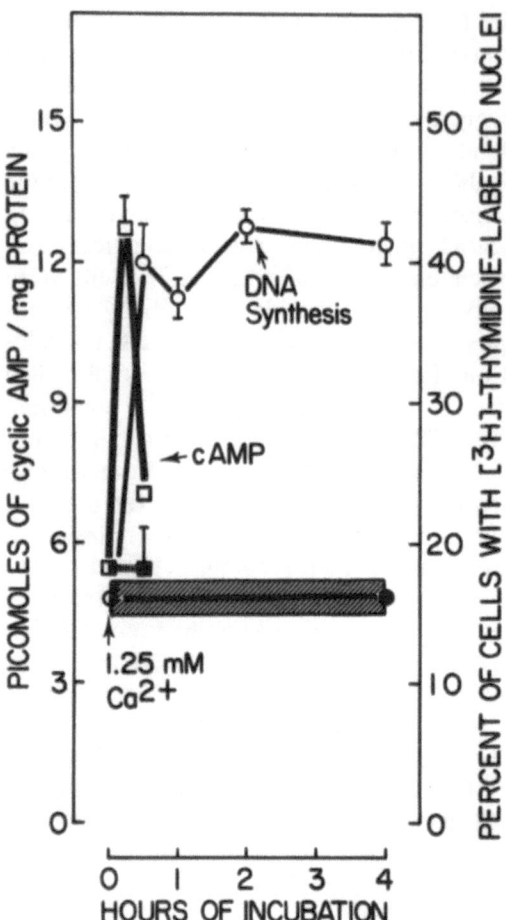

Figure 32.1. The cyclic AMP surge and initiation of DNA synthesis in calcium-deprived T51B rat liver cells caused by suddenly raising the extracellular calcium level. Cells were plated at a density of $0.75 \times 10^4/cm^2$ and incubated for 48 hours in high- (1.8 mM) calcium, 10%(v/v) FBS-90%(v/v) BME medium. The medium was then replaced with fresh, low- (0.01 mM) calcium medium. After 48 hours of calcium deprivation, the extracellular calcium level was raised to 1.25 mM by adding $CaCl_2$. The autoradiographic (for DNA-synthetic activity) and radioimmunoassay (for measuring cyclic AMP levels) procedures have been described fully elsewhere (Boynton and Whitfield, 1978; Boynton *et al.*, 1978). The points are the means ± S.E.M. of the values from four separate cultures.

Swierenga *et al.*, 1976, 1978; Whitfield *et al.*, 1979). The average ratio of these activities (for the 31 cell types so far tested) in low- and high-calcium media, the *neoplas-*

tic index (Swierenga *et al.*, 1978), is only 0.08 ± 0.01 (S.E.M.) (Table 32.1).

Lowering the extracellular calcium concentration from the usual 1.8 mM to 0.01 mM in cultures of nonneoplastic T51B liver epithelial cells (Swierenga *et al.*, 1978) reduces the proportion of DNA-synthesizing cells from 50% to 10–15% during the next 24–48 hours (Fig. 32.1) (Whitfield *et al.*, 1979). A large proportion of the calcium-deprived cells must be blocked at a late stage of prereplicative (G1) development, because suddenly raising the extracellular calcium concentration from 0.01 to 1.25 mM causes them to initiate DNA synthesis almost immediately (Fig. 32.1) and enter mitosis a few hours later (Whitfield *et al.*, 1979). This calcium-triggered initiation of DNA synthesis is probably mediated by cyclic AMP (adenosine 3',5'-monophosphate) because the increased DNA-synthetic activity is preceded by, and needs, a brief rise in the level of this cyclic nucleotide (Fig. 32.1) (Whitfield *et al.*, 1979), which by itself can stimulate calcium-deprived T51B cells rapidly (within 1 hour) to initiate DNA synthesis (Fig. 32.3) (Whitfield *et al.*, 1979).

Neoplastic transformation somehow, and seemingly invariably, disconnects the initiation of DNA synthesis from this calcium-cyclic AMP control system. Consequently, every type of epithelial and mesenchymally derived tumorigenic cell (produced by chemicals, viruses—both DNA- and RNA-containing—or spontaneously) so far examined makes DNA, multiplies, and forms colonies in low-calcium medium about as efficiently as it does in normal high-calcium medium. Thus, the average *neoplastic index* of neoplastic cells is more than 10 times greater (i.e., 1.14 ± 0.08 (S.E.M.); Table 32.2) than that of nontumorigenic cells (Table 32.1).

This common property of tumorigenic cells should be a direct *in vitro* indicator of neoplastic transformation, and may therefore serve as the basis of a test for carcinogenic activity. Obviously, exposure of populations of a standard test cell, such as

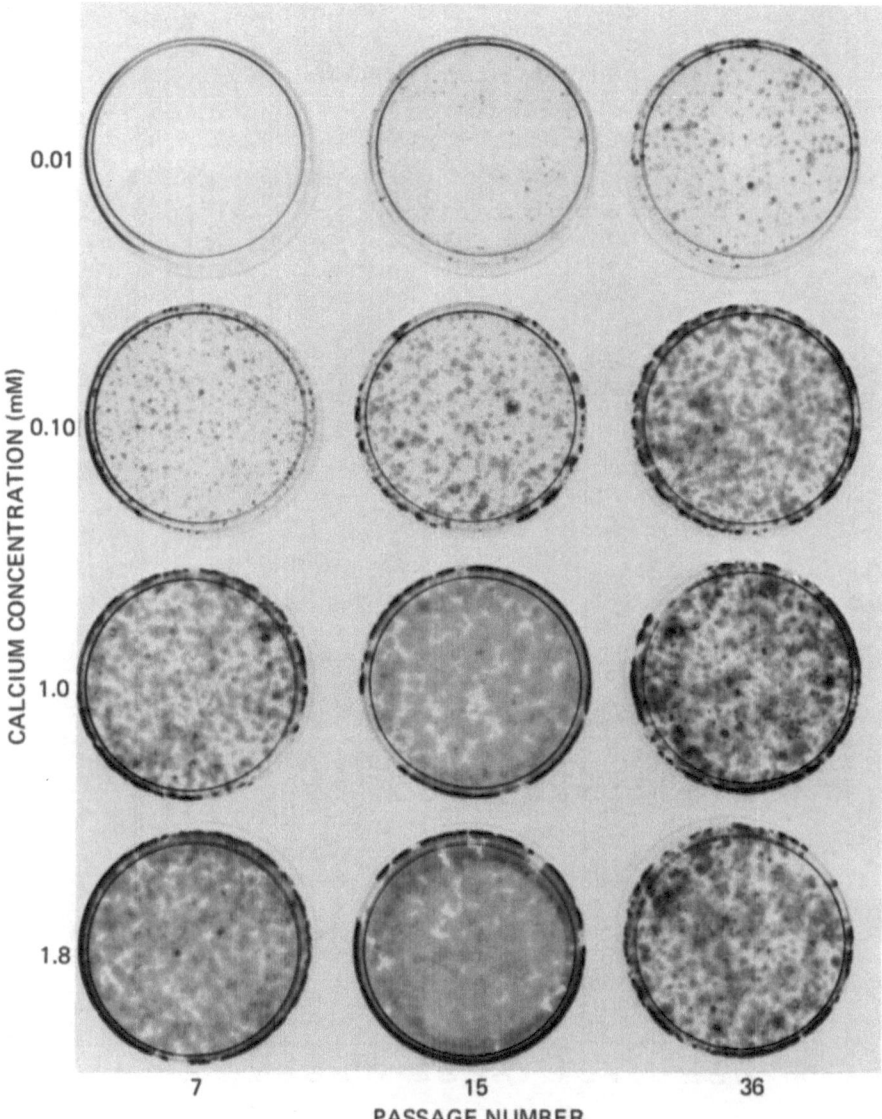

Figure 32.2. A demonstration of the passage-induced development of proliferative independence from the extracellular calcium concentration in cultures of T51B rat liver cells. After the appropriate number of passages, 2000 cells were plated in each 60-mm Petri dish in high- (1.8 mM) calcium, 10% (v/v) FBS-90% (v/v) BME medium and incubated for 24 hours. The medium was then discarded, and the cultures were washed twice with phosphate-buffered saline and exposed to test medium containing the appropriate calcium concentration. Fourteen days later, the cultures were fixed and stained.

6th to 12th passage T51B liver cells (which cannot proliferate to form colonies in low- (0.01 mM) calcium medium (Fig. 32.2), grow in soft agar, or produce tumors in athymic nude mice) to any carcinogenic agent (regardless of whether it is a mutagen or a modulator of gene expression) should cause the appearance of cells which can form colonies in low- (0.01 mM) calcium medium.

To begin the long process of testing this possibility (Table 32.3), T51B cell cultures were exposed to known carcinogens such as AFB_1, MCA or $MNNG_1$ (aflatoxin B_1,

Table 32.1. *The Neoplastic Indices of Various Types of Nonneoplastic Cell*

Cell type	Tumorigenicity	N.I.[e]	Reference
Endothelial Cells			
Mouse			
BALB/3T3-A31	—[c]	0.09	f
Epithelioid Cells			
Rate liver			
T51	NT[d]	0.00	g
RL34	—	0.07	g
18FR22	NT	0.23	g
T51A	NT	0.19	g
T51A	NT	0.15	g
T51B	—	<0.02	g
Bovine pancreatic			
Beta cells			
primary cultures	—	0.12	a
primary cultures	—	0.05	a
Mouse mammary cells	NT	0.03	b
		0.16	b
Skeletal Muscle Cells			
Fetal rat			
primary 1	NT	0.11	h
primary 2	NT	0.08	h
primary 3	NT	0.23	h
primary 4	NT	0.11	h

Cell type	Tumorigenicity	N.I.[e]	Reference
Fibroblasts			
Chicken			
primary culture	NT	0.04	i
Human			
BB	NT	<0.02	i
DJM	NT	<0.02	j
HA-1	NT	<0.02	j
HA-2	NT	<0.02	j
HFL	NT	<0.02	j
HFS	NT	<0.02	j
MM288-F	NT	<0.02	j
MM231-F	NT	<0.02	j
PGP	NT	<0.02	j
SGB	NT	0.11	j
WI-38	—	0.11	k
Mouse			
C3H1OT 1/2	—	0.09	l
MCA-C3H1OT 1/2 Type I		0.10	l
C3H fetal cells			
Pass 1	NT	0.17	l
Rat			
NRK	NT	0.09	m

MEAN NEOPLASTIC INDEX: 0.08 ± 0.01 (S.E.M.) $n = 31$

a A.L. Boynton, L.G. Healy, A. Sun, and J.F. Whitfield, unpublished observations.
b L. Zwierchowski, personal communication.
c —, nontumorigenic.
d NT, not tested.

e Neoplastic Index.
f Boynton and Whitfield (1976).
g Swierenga et al. (1978).
h Swierenga et al. (1976).
i Balk et al. (1973).

j Parsons (1978).
k Boynton et al. (1977a).
l Boynton et al. (1976).
m Boynton et al. (1977b).

Note: The neoplastic index is the ratio of DNA-synthetic, proliferative, or colony-forming activity in low-calcium medium to that in high-calcium medium. For the sake of brevity the particular activity and calcium concentrations used for each entry are not listed, but may be found in the appropriate references. In the unpublished references (a) and (b), the activity was cell multiplication (i.e., culture density after 5 to 7 days of incubation).

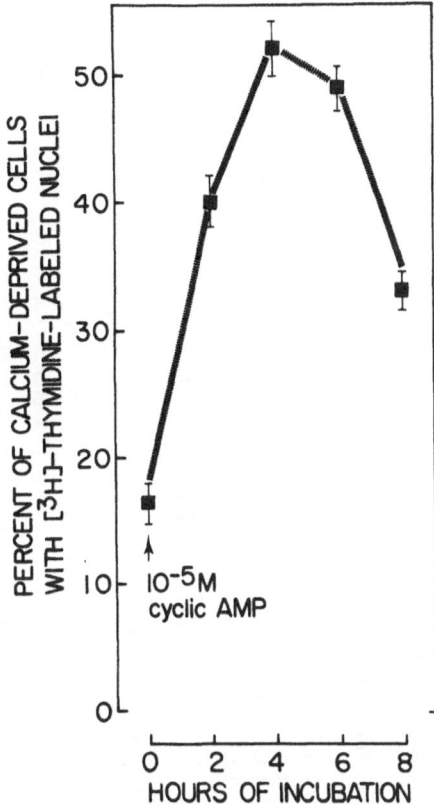

Figure 32.3. Demonstration of the ability of a maximally effective concentration of exogenous cyclic AMP to stimulate the initiation of DNA synthesis, as measured autoradiographically (Boynton *et al.*, 1976), by calcium-deprived, proliferatively quiescent T51B rat liver cells in 10% (v/v) FBS-90% (v/v) BME medium containing only 0.01 mM calcium. Cells were plated as in Fig. 32.1 before being exposed to 10^{-5} M cyclic AMP. The points are the means ± S.E.M. of the values from four separate cultures.

20-methylcholanthrene and N-methyl-N'-nitro-N-nitrosoguanidine, respectively) for 72 hours in high- (1.25 mM) calcium medium without supplementary activation systems. The exposure medium was then replaced with carcinogen-free, or solvent- (dimethyl sulfoxide (DMSO)) free medium and the cultures allowed to become confluent during the next 5–7 days. The cells in the confluent cultures were then suspended and plated in high- (1.8 mM) calcium medium for colony-forming assay. Twenty-four hours afterward this medium was replaced with low- (0.01 mM) calcium test medium,

and colonies allowed to develop during the next 10–14 days. In the next experiments (Table 32.4; Fig.32.3), the cells were exposed to MCA, MNNG, or NQO (4-nitroquinoline-1-oxide) and treated as previously, except that after the cultures became confluent they were incubated for 24–48 hours in low- (0.01 mM) calcium test medium before being finally suspended in fresh low- (0.01 mM) calcium test medium and plated for colony assay.

The results of these first efforts (Tables 32.3 and 32.4) showed that the four carcinogens indeed caused the appearance of many more calcium-insensitive cells than did their solvent DMSO. Moreover, they did so without the help of exogenous activation systems, probably because having descended from hepatocytes they still contained the necessary battery of activating enzymes. The yield of colonies, hence the sensitivity of the assay, was increased by passaging treated cultures (such as those in the experiments of Table 32.4) 2–6 times in high- (1.8 mM) calcium medium before finally plating them in low- (0.01 mM) calcium medium for the colony assay (Table 32.5). However, it should be noted that such additional passaging can be counterproductive because of the increase with passage number in the proportion of cells which spontaneously lose their proliferative calcium-dependence (Fig. 32.2).

In conclusion, these observations, preliminary and fragmentary though they are, clearly indicate that the property of proliferative calcium-independence (i.e., the ability to form colonies in low-calcium medium) may be a useful tool for the *in vitro* detection of carcinogens. Since it seems to be one of the two properties—growth in soft agar being the other (San *et al.*, 1979)—which invariably accompany neoplastic transformation and can be readily measured *in vitro*, its appearance in a cell culture is probably as reliable an indicator of neoplastic transformation as tumor production in the animal. In fact, it is this close linkage to neoplastic transformation which gives proliferative calcium-independence

Table 32.2. The Neoplastic Indices of Various Cell Types Neoplastically Transformed in vitro or Isolated from Tumors in vivo

Cell type	Tumorigenicity	N.I.[e]	Reference
Endothelial Cells			
Mouse			
BALB/3T3 Cl 2	+[c]	1.17	a
BALB/3T3 Cl 4	+	1.69	a
BALB/3T3 Cl 7	+	1.00	a
BALB/3T3 113(p.4)	+	0.99	a
BALB/3T3 113	+	0.90	a
K-BALB/3T3	+	1.02	f
MC5-S-BALB/3T3	+	1.03	f
SV$_{40}$-3T3 (F)	+	1.00	f
SV$_{40}$-3T3 (T)	+	0.92	f
Epithelioid Cells			
Bovine Pancreatic			
Beta Cells			
Passage 4	+	0.70	b
Passage 5	+	0.70	b
Passage 7	+	0.94	b
Passage 9	+	1.00	b
Passage 12	+	1.00	b
Passage 18	+	1.25	b
Human			
HeLa	+	1.20	g
KB (pharyngeal carcinoma)	+	3.20	g
Kidney T cells	+	1.00	g
Mammary tumor	+	0.91	g
Melanoma MM96	+	2.20	g
Melanoma MM138	+	1.10	g
Melanoma MM200	+	0.95	g
Melanoma MM253	+	1.90	g
Rat			
Liver CCL-144	+	0.75	h
Liver DAB4	+	1.19	h
Liver HNT	+	1.09	h
Liver HN4	+	1.25	h
Liver RLCC-1	+	1.18	h
Liver 5123tc	+	0.87	h
Fibroblasts			
Human			
SV$_{40}$-WI-38	+	1.00	i
Mouse			
MCA-C3H1OT 1/2 Type III	+	1.14	j
Rat			
B-77-NRK	NT	1.37	k
LA-23-NRK	NT	1.07	k
Skeletal Muscle Cells			
Rat			
Rhabdomyosarcoma 1	+	0.79	l
Rhabdomyosarcoma 2	+	0.86	l
Rhabdomyosarcoma 3	+	1.00	l
Rhabdomyosarcoma 4	+	0.90	l

MEAN NEOPLASTIC INDEX: 1.14 ± 0.08 (S.E.M.) n = 37

[a] A.L. Boynton and J.F. Whitfield, unpublished observations.

[b] A.L. Boynton, G. Healy, A. Sun and J.F. Whitfield, unpublished observations.

[c] +, tumorigenic.

[d] NT, not tested.

[e] Neoplastic Index.

[f] Boynton and Whitfield (1976).

[g] Parsons (1978).

[h] Swierenga et al. (1978).

[i] Boynton et al. (1977a).

[j] Boynton et al. (1976).

[k] Boynton et al. (1977b).

[l] Swierenga et al. (1976).

Note: Neoplastic index is defined in Table 32.1. The activity used in references ([a]) and ([b]) to determine the neoplastic index was culture density after 5 to 7 days of incubation.

Table 32.3. *Carcinogen-Induced Ability of T51B (passage 12) Rat Liver Epithelioid Cells To Form Colonies in a Low- (0.01 mM) Calcium Medium*[a]

Compound tested	Dose (μg/ml)	Total colonies in 0.01 mM Ca^{2+} medium[d]	Number of Dishes	Average colonies per dish
None	—	0	5	0
DMSO[b]	0.02(%)	0	4	0
DMSO[b]	0.05(%)	0	6	0
DMSO[c]	0.1 (%)	0	6	0
AFB$_1$[c]	1.0	12	4	3.0
MCA[b]	0.1	24	6	4.0
MNNG[b]	0.5	91	6	15.1

[a]Twenty-four hours after plating in 10 ml of medium in T-75 flasks, T51B cells were exposed for 72 hours to either the DMSO solvent (dimethyl sulfoxide) or carcinogen without supplementary activation systems. The medium was then removed, the cellular monolayers rinsed twice with PBS, and covered with fresh, carcinogen-free medium. The cultures were allowed to become confluent. The cells were then removed from the flask with trypsin and suspended in fresh culture medium for colony-forming assay. Two hundred cells were plated in 60-mm Petri dishes in high- (1.25 mM) calcium medium. Twenty-four hours later, this medium was replaced (after rinsing the cultures twice with PBS) with low- (0.01 mM) calcium test medium and the cultures incubated for an additional 10–14 days. At this time the resulting colonies were fixed in methanol and stained with a 1% crystal violet solution. Colonies were counted using a Leitz Diavert microscope fitted with 10X eyepieces and a 2.5X objective lens. Colonies were scored as positive if their size exceeded 6.25 mm^2 which represents about 30–50 cells.

[b]600 cells exposed to compound.

[c]100,000 cells exposed to compound.

[d]Low- (0.01 mM) calcium BME and Williams' Medium E were purchased from either Flow Laboratories or Grand Island Biological Co. (Grand Island, N.Y.). The calcium concentration in the FBS was reduced either by chelating the ionic calcium with the specific calcium chelator EGTA (ethylene glycol-bis-[β-amino ethyl ether]-N,N'-tetraacetic acid) according to the method of Borle and Briggs or by actually removing the calcium with Chelex ion exchange resin. The hydroxyl form of the Chelex 100 resin (Bio Rad), as a 50% slurry in distilled water, was adjusted to pH 7.0. One volume of settled (by centrifugation) resin and 2 volumes of serum were combined and stirred gently for 30 minutes, followed by centrifugation (to remove the resin and attached calcium), and sterilization by filtration through a 0.2 μm pore-size fitter (Nalgene). The final total calcium concentration of the FBS was 0.05 mM, which was determined by atomic absorption spectroscopy.

Note: Abbreviations: AFB$_1$, aflatoxin B$_1$; DMSO, dimethyl sulfoxide; MCA, 20-methylcholanthrene; MNNG, N-methyl-N'-nitro-N-nitrosoguanidine.

an advantage over other indicators such as bacterial mutagenesis and DNA repair, which may or may not be related directly to carcinogenesis.

Summary

Proliferative independence from extracellular calcium ions is a common property of neoplastic, but not nonneoplastic, cells. Therefore, its appearance in a cell population either *in vitro* or in vivo is as direct an indicator of neoplastic transformation as tumor formation. Since this property can be readily and fairly rapidly measured *in vitro* by colony formation in low-calcium medium, and since we have shown it to be induced in cultures of a strain of nonneoplastic liver cells by known carcinogens, it promises to become the basis of a reliable *in vitro* procedure for the rapid detection of carcinogens.

Acknowledgments

We thank R. Tremblay and R.J. Isaacs for excellent technical assistance and D.J. Gillan for preparing the illustrations.

Table 32.4. *Abilities of MCA, MNNG, and NQO To Cause the Appearance of T51B Liver Cells Which Are Able To Form Colonies in Low- (0.01 mM) Calcium Medium*[a]

Compound tested	Dose (μg/ml)	Total colonies in 0.01 mM Ca²⁺ medium	Number of dishes	Average colonies per dish
Experiment 1				
None	—	0	5	0
DMSO	0.1(%)	1	5	0.2
MNNG	0.5	27	5	5.4
MNNG	1.0	25	5	5.0
MCA	1.0	135	5	27.0
MCA	2.0	188	5	37.6
Experiment 2				
None	—	0	4	0
DMSO	0.1(%)	3	4	0.7
MNNG	1.0	5	4	1.2
MNNG	2.5	278	4	69.5
NQO	0.1	2	4	0.5
NQO	0.75	54	4	13.5

[a]250,000 T51B liver cells were plated in T-75 flasks. Twenty-four hours later they were exposed to the test compound for 48 hours without supplementary activation systems. The medium was then replaced, after rinsing the cultures twice with PBS, with a fresh DMSO- or carcinogen-free medium and incubated until the cultures became confluent (5–7 days later). They were then incubated for 24–28 hours in a low- (0.01 mM) calcium medium, after which they were trypsinized, suspended in fresh low-calcium (0.01 mM) medium and plated at a density of 2000 cells per 60-mm dish in 5 ml low- (0.01 mM) calcium medium. 10–14 days later, the resulting colonies were fixed in methanol, stained in a 1% crystal violet solution, and counted as described in the legend of Table 32.3.

Note: Abbreviations: DMSO, dimethyl sulfoxide; MCA, 20-methylcholanthrene; MNNG, N-methyl-N'-nitro-N-nitrosoguanidine.

Table 32.5. *Demonstration of the Effect of Additional Passages on the Yield of Colonies from Cells from Cultures of Carcinogen- and DMSO-treated T51B Rat Liver Cells*[a]

Compound tested	Dose (μg/ml)	Total colonies in 0.01 mM Ca²⁺ medium	Number of dishes	Average colonies per dish
Experiment 1[b]				
None	—	18	4	4.5
DMSO	0.1(%)	19	4	4.7
MNNG	0.5	40	4	10.0
MNNG	1.0	33	4	8.2
MNNG	2.0	296	4	74.0
MCA	0.5	136	4	34.0
MCA	2.0	373	4	93.2
Experiment 2[c]				
DMSO	0.1(%)	125	4	31.2
NQO	0.1	612	4	153.0
NQO	0.75	533	4	133.0

[a]Experimental details as described in Table 32.4 except that the cells were passaged 2 to 6 times in high-calcium medium before being plated for the colony-forming assay.

[b]Low-calcium colony assay performed (by A. Boynton, J.F. Whitfield at N.R.C.C.) after 6 passages each at a 1:4 split (12 population doublings).

[c]Low-calcium colony assay performed (by S.H.H. Swierenga at Health and Welfare Canada) after two passages each at a 1:10 split (about 10 population doublings).

Note: Abbreviations: DMSO, dimethyl sulfoxide; MCA, 20-methylcholanthrene; MNNG, N-methyl-N'-nitro-N-nitrosoguanidine; NQO, 4-nitroquinoline-1-oxide.

References

Balk, S.D., Whitfield, J.F., Youdale, T., Braun, A.C. (1973): Roles of calcium, serum, plasma, and folic acid in the control of proliferation of normal and Rous sarcoma virus-infected chicken fibroblasts. Proc. Natl. Acad. Sci. USA 70:675–679.

Borle, A.B., Briggs, F.N. (1968): Microdetermination of calcium in biological material by automatic fluorometric titration. Anal. Chem. 40:339–344.

Boynton, A.L., Whitfield, J.F. (1976): Different calcium requirements for proliferation of conditionally and unconditionally tumorigenic mouse cells. Proc. Natl. Acad. Sci. USA 73: 1651–1654.

Boynton, A.L., Whitfield, J.F. (1978): Calcium requirements for the proliferation of cells infected with a temperature-sensitive mutant of Rous sarcoma virus. Cancer Res. 38: 1237–1240.

Boynton, A.L., Whitfield, J.F., Isaacs, R.J. (1976): The different roles of serum and calcium in the control of proliferation of BALB/c 3T3 mouse cells. In Vitro 12:120–123.

Boynton, A.L., Whitfield, J.F., Isaacs, R.J., Tremblay, R.G. (1977a): Different extracellular calcium requirements for proliferation of nonneoplastic, preneoplastic, and noeplastic mouse cells. Cancer Res. 37: 2657–2661.

Boynton, A.L., Whitfield, J.F., Isaacs, R.J., Tremblay, R.G. (1977b): The control of human WI-38 cell proliferation by extracellular calcium and its elimination by SV-40 virus-induced proliferative transformation. J. Cell Physiol. 92:241–247.

Boynton, A.L., Whitfield, J.F., Isaacs, R.J., Tremblay, R.G. (1978): An examination of the roles of cyclic nucleotides in the initiation of cell proliferation. Life Sci. 22:703–710.

Braun, A.C. (1974): The Biology of Cancer. Reading, Massachusetts: Addison-Wesley.

Li, H.J., Eckhardt, R. (eds.) (1977): Chromatin and Chromosome Structure. New York: Academic Press.

Markert, C.L. (1978): Cancer: The survival of the fittest. In: G.F. Saunders (ed.), Cell Differentiation and Neoplasia. New York: Raven Press, pp. 9–22.

Mintz, B. (1978): Genetic mosaicism and in vivo analyses of neoplasia and differentiation. In: G.F. Saunders (ed.), Cell Differentiation and Neoplasia. New York: Raven Press, pp. 27–53.

Parsons, P.G. (1978): Selective proliferation of human tumour cells in calcium depleted medium. Austr. J. Exp. Biol. Med. Sci. 56: 297–300.

Pierce, G.B., Cox, W.F. (1978): Neoplasms as caricatures of tissue renewal. In: G.F. Saunders (ed.), Cell Differentiation and Neoplasia. New York: Raven Press, pp. 57–66.

Pierce, G.B. Shikes, R., Fink, L.M. (1978): A Problem of Developmental Biology. Englewood Cliffs, New Jersey: Prentice-Hall Inc.

San, R.H.C., Laspia, M.F., Soiefer, A.I., Maslansky, C.J., Rice, J.M., Williams, G.M. (1979): A survey of growth in soft agar and cell surface properties as markers for transformation in adult rat liver epithelial-like cell cultures. Cancer Res. 39:1026–1034.

Swierenga, S.H.H., Whitfield, J.F., Gillan, D.J. (1976): Alteration by malignant transformation of the calcium requirements for cell proliferation in vitro. J. Natl. Cancer Inst. 57: 125–129.

Swierenga, S.H.H., Whitfield, J.F., Karasaki, S. (1978): Loss of proliferative calcium dependence: simple in vitro indicator of tumorigenicity. Proc. Natl. Acad. Sci. USA 75: 6069–6072.

Uriel, J. (1975): Fetal characteristics of cancer. In: F.F. Becker (ed.), Cancer, Vol. 3, New York: Plenum Press, pp. 21–49.

Whitfield, J.F., Boynton, A.L., MacManus, J.P., Sikorska, M., Tsang, B.K. (1979): The regulation of cell proliferation by calcium and cyclic AMP. Mol. Cell. Biochem. 27: 155–179.

33

*Induction of a Resistant Preneoplastic Liver Cell as a New Principle for a Short-Term Assay in vivo for Carcinogens**

Emmanuel Farber and Hiroyuki Tsuda

Introduction

The past few years have seen the development of a large number (over 60) of new tests designed to detect and identify possible carcinogens in man's environment. The test organisms and the endpoints used vary widely in detail but in general the assays fall into two groups. The tests in the smallest group use a complex cellular response, cell transformation, as the indicator. This response resembles in several respects malignant neoplasia and is considered to be a valid *in vitro* analogue of cancer induction *in vivo*. The tests in the largest group use some aspect of DNA damage and/or repair or some presumed reflection of damage or repair of DNA, such as mutations, chromosome damage, chromatid exchanges, etc.

The chemical-biochemical basis of the transformation tests is unknown.

Thus, the majority of assays measure the ability of the test substance to effect some chemical or physicochemical change in DNA or in chromosomes, i.e., some genotoxic damage, while a few measure a late biological phenomenon. The majority of the genotoxic assays use rapidly proliferating prokaryotic or eukaryotic cells and generate a result quickly. Also, they are relatively simple and fairly easily carried out. Their advantages and ultimate utility in principle are obvious and generally accepted. Their usefulness in the screening for the majority of new genotoxic agents being introduced is evident.

However, the failure of the available assays to detect all known carcinogens, i.e., the presence of a significant number of false negatives, remains a disturbing aspect of this approach to screening. Also, since the genotoxic endpoints have no demonstrable relevance or relationship to cancer, there

*Supported by research grants from the National Cancer Institute of Canada, Medical Research Council of Canada (MA-5994) and the National Cancer Institute of the N.I.H. (CA-25094).

remains an obvious need for assays that utilize cancer or some response closely related to cancer. This is particularly so for those relatively few compounds that are either (a) of potential significant use and are negative in the screens, or (b) of great use and are positive.

Although the induction of malignant neoplasia is the best and only true index of carcinogenicity, its assay is slow and expensive. Acceptable valid assays in animals require large numbers of animals, with good animal care. Currently, 2-year protocols are becoming the only acceptable tests for carcinogenic activity in rodents by many regulatory agencies in some countries. This is not an unreasonable requirement for those chemicals or drugs that have been established to be of great benefit to man and yet may pose a carcinogenic risk when exposure is repeated and prolonged.

However, there is an evident need for a more screening type of assay in animals that has more obvious relevance to cancer. A short term *in vivo* assay could well bridge the current wide gap between the genotoxic assays and the long-term animal test. The role of different basic types of assays in the development of a rational screening tier has been discussed periodically over the past few years (e.g., Bridges, 1976). This is particularly important as we appreciate to an increasing degree the complexity of the carcinogenic process and the roles of different types of agents such as promoters in different steps of cancer induction.

Background

Two aspects of current knowledge relating to assays are impressive. These relate (a) to the need for metabolic activation for most carcinogens and the versatility of the liver in this respect and (b) to the multistep nature of this carcinogenic process.

(a) Most carcinogens are not active *per se* but require enzymatic conversion to one or more active derivatives which can interact with many cellular components including DNA. Cells in general and cells from different organs or tissues in any animal vary widely in their ability to effect the appropriate metabolic steps involved in activation. The liver is by far the most versatile in this activity. Because of these considerations, many *in vitro* assays using genotoxicity or cell transformation have an absolute requirement for suitable preparations of liver (S9) in order to generate an active derivative of the compound under test.

(b) There is increasing evidence that cancer induction occurs by a multistep process in which "new cell populations represent stages in the cellular evolution from normal, through initiated, preneoplastic and premalignant cells to highly malignant neoplasia" (Farber and Cameron, 1980; see also Foulds, 1975). Such new cell populations create the "material continuity of suspected sequential lesions" (Foulds, 1975) essential in cancer development. In this conceptual framework, the initiating carcinogen merely triggers a chain reaction and thus induces only a very early step in the sequence (Farber, 1973, 1980). Also, the number of very early carcinogen-induced focal lesions seen during the preneoplastic phase of the sequence exceeds by 10^3 to 10^4 the number of cancers seen ultimately in the liver (Farber and Cameron, 1980). Thus, the sensitivity of quantitation is far greater at the earlier than at the later times in the carcinogenic process.

Given (1) the widespread use of the activating liver preparations in the short-term assays for carcinogenicity, (2) the logical attractiveness of using the same tissue for both activation of a precarcinogen and detection of an active derivative, (3) the close relevance to carcinogenesis of a cellular alteration in a target organ induced by a carcinogen, and (4) the great amplification in sensitivity achieved with the use of an early cell change as compared to a late one, cancer, it seemed attractive to attempt to develop a short-term *in vivo* assay for carcinogenicity using the intact liver as both activator and indicator.

Principle

The principle underlying this new approach is based on the following hypothesis: it is proposed that an initiating dose of a carcinogen induces an alteration in a rare hepatocyte such that this hepatocyte is resistant to the "mito-inhibitory" effect (inhibition of cell proliferation) of carcinogens and thus can undergo proliferation under conditions which inhibit the surrounding hepatocytes (the majority of the cells, the "uninitiated" hepatocytes) from proliferating. Most carcinogens are potent inhibitors of cell proliferation. If a stimulus for cell proliferation, such as a partial hepatectomy, is applied in the presence of an inhibitory environment such as a low level of dietary 2-acetylaminofluorene (2-AAF), the liver generally does not proliferate. However, the rare carcinogen-induced resistant cell ("initiated") can respond and undergoes rapid cell proliferation (Solt and Farber, 1976). Focal collections of resistant cells show continuous proliferation for at least 10 to 12 cell cycles to form visible nodules of new hepatocytes. If the liver is exposed to an initiating carcinogen and no further treatment is performed, the rare resistant hepatocyte (1 cell per 10^5–10^6 original hepatocytes) persists for at least 44 weeks in an apparently unaltered form (Solt and Farber, 1977). If a suitable selection pressure is created—low level of dietary 2-AAF for 1 week plus partial hepatectomy (PH)—the resistant cells rapidly proliferate to produce nodules and these in turn undergo further change to act as precursors for the ultimate development of cancer (Solt et al., 1977; Ogawa et al., 1979).

In order to induce resistant hepatocytes, i.e., to initiate carcinogenesis, a round of cell proliferation is required. This cell cycle of hepatocytes must occur within 24 to 48 hours after exposure to the initiating dose of carcinogen for it to be effective. If cell proliferation is delayed beyond 48 hours, no resistant cells appear. This is presumably because of the presence of an effective repair (of damaged DNA?).

Cell proliferation can be induced by at least two mechanisms: (1) regenerative response following surgical removal of a portion of liver (PH), or (2) regenerative response to cell death (necrosis). PH is effective with many different carcinogens, either with classically nonliver carcinogens (Cayama et al., 1978; Tsuda and Farber, 1979; Ying and Sarma, 1979) or with subnecrogenic doses of liver carcinogens (Ying and Sarma, 1979). Virtually all liver carcinogens are necrogenic and thus induce cell proliferation at adequate dosage. In contrast, most carcinogens not active on liver do not induce necrosis and therefore do not initiate without an imposed round of cell proliferation.

Thus, with many different carcinogens, resistance can be induced in a rare hepatocyte if the liver is exposed at an appropriate time in relation to cell proliferation. Parenthetically, it is not known what mechanistic role cell proliferation plays in this initiation process, although it is presumed to be related to DNA replication in S phase.

The number of resistant cells can be counted by the use of a quantitative assay for initiation, based on foci of cell proliferation in an inhibited environment (Solt and Farber, 1976).

Assay Procedure

Male Fischer-344 rats (Charles River Breeding Laboratory, Detroit, Mich., U.S.A.), weighing 150 to 170 g are maintained on a moderately high protein (26%) semisynthetic diet (Bio-Serv. Inc. Frenchtown, New Jersey, U.S.A.—diet #101) and a daily cycle of alternating 12-hour periods of light and darkness for 1 week in order to acclimatize them before the start of the experiment.

At zero time, the rats are subjected to partial hepatectomy or to a sha.n operation (laparotomy plus handling of liver), using ether anesthesia. At 12 hours, the carcinogen is given in an appropriate vehicle (corn oil, dimethylsulfoxide, 0.9% NaCl solution, etc.) either by stomach tube (s.t.) or by intraperitoneal injection (i.p.). The largest dose given is one which just fails to produce death of any animal within 4 weeks of injection. The water-soluble compounds are generally given i.p. while the others are given s.t. Initially, each compound is given once at the maximum dosage.

After a 2-week recovery period all animals are placed on a basal diet (#101) containing 0.02% 2-AAF for 2 weeks. Halfway through this 2-week period, i.e., at the end of 1 week, the animals receive a single dose of carbon tetrachloride (CCl_4) at a dose of 2 ml/kg body wt. in corn oil by s.t. The purpose of this is to induce a stimulus for cell proliferation by an initial episode of cell death. At the end of the 2-week period on dietary 2-AAF, the rats are placed on the basal diet without the 2-AAF for 1 week and then killed.

The livers are examined grossly for grayish-white foci or tiny nodules, 1 to 3 ml in diameter. Pieces of liver are taken in a standardized fashion for routine histological examination and for histochemical analysis for γ-glutamyl transferase (transpeptidase) activity. This is done using low melting paraffin embedding and enzyme staining with γ-glutamyl-4-methoxy-2-naphthylamide (GMNA) as substrate and fast blue-BBN as coupling reagents (Rutenberg et al. 1969; Ogawa et al., 1980).

The foci of resistant hepatocytes stain intensely orange-red by this procedure and the number (as well as the diameter) is easily counted by low-power microscopy (or even with a good hand lens). The number is expressed as foci per cm^2 of section area. The area of the section is readily measured with a planimeter. Alternatively, an image analyzer can be used to determine number, area of each focus, and area of section examined.

The following points require some explanation:

1. Clearly, two episodes of cell proliferation are required—the first one in order to "fix" the initiation process and the second to stimulate the resistant initiated hepatocytes to proliferate. The second stimulus for cell proliferation can be produced by (a) PH (Solt and Farber, 1976), if the initiating carcinogen is necrogenic and therefore has no need for an exogenously induced round of cell replication, (b) use of CCl_4 or other necrogenic agent (Cayama et al., 1978), or (c) a chemical mitogen such as α-hexachlorocyclohexane (αHCH).

2. The selection pressure as used (2-AAF plus CCl_4 or PH or HCH) does not induce by itself any resistant hepatocytes. Although 2-AAF is a good liver carcinogen and can induce resistant hepatocytes ("initiate"), it does not do so under these conditions except very slowly (beginning after 4–5 weeks). This is probably because it inhibits cell proliferation so effectively that it prevents its own initiation. Choline-deficient diet or methotrexate in the drinking water seem to be possible substitutes for 2-AAF, but this is now under study.

3. The range of number of foci/cm^2 is from 3 or 4 to 60–70, depending on the carcinogen and its dosage. The background with selection alone, with carcinogen plus SH plus selection, or with saline plus PH in place of carcinogen is from 0 to 2 foci/cm^2.

4. It must be emphasized that the assay can be made more sensitive and even more selective when a substitute is found for 2-AAF and when the potential level of "scavengers" is reduced. For example, pretreatment of rats with naphthalene decreases the level of glutathione by at least 50% and increases very much the sensitivity with benzo(a)pyrene.

Compounds Tested

The compounds tested to date fall into several chemical groupings and are listed in Table 33.1. Included in the table is an evaluation as to whether any significant number of foci can be induced by omitting the initial

Table 33.1. *Efficacy in Induction of Resistant Hepatocytes by Various Carcinogens and Noncarcinogenic Analogues*

Compound	Resistant hepatocytes	Need for PH	Carcinogenicity
Polycyclic Aromatic Hydrocarbons			
Benzo(a)pyrene	+	+	+
7,12-dimethylbenz(a)anthracene	+	+	+
3-methylcholanthrene	+	+	+
Dibenzo(a,h)anthracene	+	+	+
Benzo(a)anthracene	+	+	+
Anthracene	−		− (?)
Naphthalene	−		−
Phenanthrene	−		−
Pyrene	−		−
Aromatic Amines			
α-naphthylamine	+	+	+
β-naphthylamine	+	+	+
2-acetylaminofluorene	+	−	+
4-aminobiphenyl	+	+	+
Auramine	+	+	+
Fluorene	−		−
N-Nitroso Compounds			
Diethylnitrosamine	+	−	+
Dimethylnitrosamine	+	−	+
N-methylnitrosourea	+	+	+
N-butyl-N-nitrosourea	+	+	+
N-methyl-N′-nitro-N-nitrosoguanidine	+	+	+
N-nitrosomorpholine	+	−	+
N-nitrosopiperidine	+	+	+
Analogues			
Morpholine	−		−
Piperidine	−		−
Miscellaneous Compounds			
Aflatoxin B$_1$	+	−	+
Methylazoxymethanol acetate	+	−	+
1,2-dimethylhydrazine	+	+	+
Ethyl carbamate (urethane)	+	+	+
Vinyl carbamate	+	+	+
Hycanthone	+	+	+
FANFT (N-[4-(5-nitro-2-furyl)-2-thiazolyl] formamide	+	+	+
Safrole	+	+	+
Dieldrin	+	+	+
Vehicles			
Corn oil	−		−
Dimethysulfoxide	−		−
0.9% NaCl	−		−

Score to-date:
  Resistant Hepatocytes − 26+   Carcinogenicity − 26+
                           7−   7−

  minus   PH − 5+ (all known liver carcinogens)

PH and by performing only a sham hepatectomy (SH). The number of foci seen with the vehicle alone (oil, saline, DMSO) or with almost all compounds without PH is very small and rarely exceeds 1.5 to 2 foci/cm². This shows that the selection pressure itself (2-AAF plus CCl₄) does not induce or select any appropriately resistant hepatocytes.

The correlation between carcinogenicity, as observed in long-term animal studies, and the results in this approach is so far very good (Table 33.1). However, the ease of induction of resistant hepatocytes is variable. Many carcinogens are effective with a single dose. Included among these are many that are normally not carcinogenic for the liver, such as benzo(a)pyrene (BP) and 1,2-dimethylhydrazine (DMH). Other less potent compounds, such as safrole and dieldrin, require at least three doses in order to induce a small but significant increase in the number of foci/cm². Still others, such as ethionine and 4-dimethylaminoazobenzene and derivatives, require a longer period of exposure (2 to 4 weeks) before a significant number of foci are induced.

Thus, based on the results to date with a reasonable but not a large number of compounds, it appears that the induction of resistant hepatocytes as putative preneoplastic foci is a valid index of carcinogenicity. With some components, such as DMH, safrole, and dieldrin, that are negative in most short-term *in vitro* assays, the test using liver *in vivo* is positive. One of these, DMH, is potent under these conditions even though hydrazines in general are difficult to detect in many *in vitro* assay systems.

It is noteworthy that all the compounds that show a positive result without PH are all known liver carcinogens and all induce acute liver cell injury with cell death. This reinforces the thesis that cell death and its subsequent stimulation of cell proliferation is important to the initiation of carcinogenesis in an organ such as liver that is composed of a quiescent cell population in the adult animal (Ying and Sarma, 1979).

General Considerations

It is clear from these early results with over 30 compounds that are known carcinogens, whether they induce liver cancer or not, that they can induce resistant hepatocytes and the measurement of these can be used as a short-term *in vivo* test for carcinogens. Since the induced resistant hepatocytes resemble very closely those seen with DEN and since the latter are precursors for liver cancer, it is very likely that the resistant cells seen with all the other carcinogens are also precursors for liver cancer. This is now being tested for many of the carcinogens listed in Table 33.1, with appropriate controls.

Also, the availability of a short-term carcinogen-induced biological endpoint that has an obvious relevance to cancer opens up many possibilities to relate mechanistically short-term effects of carcinogens in various unicellular organisms that are useful in carcinogenicity assays to a step in carcinogenesis. This has not been readily available heretofore except in the skin, which has only a limited capability to activate carcinogens.

The probable availability of a new valid short-term *in vivo* test for carcinogenicity makes the current spectrum of tests more versatile and offers a new opportunity to bridge the wide gap between effects of toxic agents in unicellular systems and those in mammals.

References

Bridges, B.A. (1976): Short-term screening tests for carcinogens. Nature (Lond.) *261*: 195–200.

Cayama, E., Tsuda, H., Sarma, D.S.R., Farber, E. (1978): Initiation of chemical carcinogenesis requires cell proliferation. Nature (Lond.) *275*:60–62.

Farber, E. (1973): Carcinogenesis—cellular evolution as a unifying thread: Presidential address. Cancer Res. *33*:2537–2550.

Farber, E. (1980): The sequential analysis of

liver cancer induction. Biochim. Biophys. Acta 605:149–166.

Farber, E., Cameron, R. (1980): The sequential analysis of cancer development. Adv. Cancer Res. 31:125–226.

Foulds, L. (1975): Neoplastic Development, Vol. 2. London: Academic Press.

Ogawa, H., Medline, A., Farber, E. (1979): Sequential analysis of hepatic carcinogenesis: A comparative study of the ultrastructure of preneoplastic, malignant, prenatal, postnatal and regenerating liver. Lab. Invest. 40: 22–35.

Ogawa, K., Solt, D., Farber, E. (1980): Phenotypic diversity as an early property of putative preneoplastic hepatocyte populations in liver carcinogenesis. Cancer Res. 40:725–730.

Rutenberg, A.M., Kim, H., Fischbein, J.W., Hanker, J.S., Wasserkrug, H.L., Seligman, A.M. (1969): Histochemical and ultrastructural demonstration of γ-glutamyl transpeptidase activity. J. Histochem. Cytochem. 17: 517–526.

Solt, D., Farber, E. (1976): New principle for the analysis of chemical carcinogenesis. Nature (Lond.) 263:701–703.

Solt, D., Farber, E. (1977): Persistence of carcinogen-induced initiated hepatocytes in liver carcinogenesis. Proc. Am Assoc. Cancer Res. 18:52.

Solt, D., Medline, A., Farber, E. (1977): Rapid emergence of carcinogen-induced hyperplastic lesions in a new model for the sequential analysis of liver carcinogenesis. Am. J. Pathol. 88:595–618.

Tsuda, H., Farber, E. (1979): Initiation of putative preneoplastic liver lesions by single doses of non-liver and liver carcinogens plus partial hepatectomy (PH). Proc. Am. Assoc. Cancer Res. 20:15.

Ying, T.S., Sarma, D.S.R. (1979): Role of cell necrosis in the induction of preneoplastic lesions. Proc. Am. Assoc. Cancer Res. 20:14.

34

Recent Achievements with Drosophila as an Assay System for Carcinogens

E. Vogel

Introduction

About 1911, T.H. Morgan began genetic studies with *Drosophila*. Since then the work with *Drosophila* has been greatly extended, because this test organism combines a eukaryotic organization with a unique range of test systems. *Drosophila* is the only *in vivo* test organism which permits the simultaneous and efficient testing of the various types of genetic lesions from the molecular to the chromosome level. Stocks are available or can be constructed to test in gonadal or somatic tissues for gene mutations, deletions, and for almost all possible types of chromosome rearrangements. Special test protocols have been devised to recover aneuploidy resulting from nondisjunctional events.

Choice of Genetic Endpoint

A point of central concern in the consideration of *Drosophila* as a detection system for carcinogens with mutagenic potential is the extent to which the various genetic endpoints are sensitive and reliable indicators of mutagenicity. Comparative investigation with a series of procarcinogens and directly acting compounds revealed that the identification of carcinogens by mutagen testing in *Drosophila* is largely dependent on the type of genetic endpoint used as the diagnostic criterion. The major finding of these studies was that considerably lower mutagen concentrations are required to raise significantly the number of point mutations, scored as recessive lethals in the F_2, relative to those needed for the production of chromosomal damage, such as dominant lethals, translocations, and chromosome losses. The list of mutagenic agents that showed this behavior is very extensive and includes agents with differing structure, mode of action, functionality, and activation requirements. The concept of "Two Effective Concentration Levels" for point mutations vs. damage resulting from chromosomal breakage was deduced from those experimental findings with a variety of reference mutagens (Vogel, 1975a; Vogel and Leigh, 1975; Vogel and Sobels, 1976).

The explanation of these phenomenological observations is the following. A comparison of the chromosome-breaking ability of a series of AA (alkylating agents) with differing s revealed that compounds with low selectivity (low s), like ENU, show low chromosome-breaking ability, while highest effectiveness to produce chromosomal damage was attributed to AA with high s, MMS and DMS (Vogel and Natarajan, 1979a; 1979b). The conclusion was that there are two variables determining the type as well as the frequency of genetic damage induced by monofunctional AA in *Drosophila*. These two parameters are dose (intensity of alkylation) and reaction pattern (site of alkylation). For each AA a "dose" could be defined, as measured by the yield of recessive lethals which served as biological dosimeter, below which detection of chromosomal aberrations was hardly possible (Table 34.1). Thus, for MMS-type mutagens, the starting point for the production of chromosomal damage, i.e., for translocations, is relatively low; translocations occur already at MMS-doses producing a fivefold increase in the frequency of recessive lethals (Table 34.1). However, Table 34.1 also shows that with an ENU-type mutagen, the recovery of chromosomal aberrations can be expected only at an extremely high degree of DNA-alkylation, i.e., at a dose leading to about a 100–300-fold increase in the frequency of recessive lethal mutations. If this dose level is not reached, as was the case with the more toxic DEN, chromosomal aberrations are not observed. The essence of all this experimental evidence is that the recessive lethal test is by far the most reliable assay in *Drosophila* to screen compounds for mutagenic activity. Thus, in this chapter, a brief outline of the performance and the most essential points of the test will be given. For a more detailed description and consideration of the method, the reader is referred to the literature (Auerbach, 1962a; Muller and Oster, 1961; Spencer and Stern,

1948; Abrahamson and Lewis, 1971; Vogel and Sobels, 1976; Würgler *et al.*, 1977).

Principle of Method

The recessive lethal test can be readily designed to detect the induction of genetic lesions in a large part of the *Drosophila* genome. Recessive lethals are a heterogeneous class. They comprise point mutations (forward mutations and deletions), and both small and large rearrangements (Auerbach, 1962a). Two generations are required for the detection of recessive lethals on the X-chromosome, which represents about 20% of the entire genome. It is estimated that about 700–800 of the 1000 loci on the X-chromosome can mutate to give rise to recessive lethal mutations (Schalet, personal communication).

The advantages of the test are manifold:

1. The criterion used to decide whether a mutation is present or not is very objective. The decision is based on the observation whether in the F_2-generation one entire class of males is absent or not (Fig. 34.1). Therefore, personal bias is reduced to a minimum.
2. Lethals are much more frequent than other types of genetic lesions, i.e., viable visible mutations (Spencer and Stern, 1948).
3. A representative part of the *Drosophila* genome is covered by this multi-locus test.

Test Organism

Life Cycle

Drosophila melanogaster undergoes complete metamorphosis. Depending on the temperature, this fly requires 9 to 20 days to complete one generation. At 25° C,

Table 34.1. Relationship between Reaction Pattern of Monofunctional AA and Chromosome-Breaking Efficiency

Compound	s^a	Doubling concentrations recessive lethals[b] (mM)	LC_{50} (mM)	$DC:LC_{50}$ recessive lethals	Critical "dose level" for induction of[c]		
					Translocations	Ring-X loss	Y-rearrangements
MMS	0.86	0.10	7.8	~1:100	5	~ 5	> 25
DMS	0.86	0.10	7.7	~1:100	5–10	5–10	> 25[d]
EMS	0.67	0.50	197	~1:400	25–50	~ 25	> 100
MNU	0.42	0.18	26	~1:150	10–30	10–20	> 35
iPS	0.31	0.50	72	~1:150	50–100	~ 50	> 100
ENU	0.26	0.06	28	~1:450	>100–300[d]	~100	> 100

[a]Swain-Scott *s* factor.

[b]Extrapolated from CM_4 values (concentration producing 4% point mutations after injection).

[c]"Dose" expressed as *n*-fold increase in the frequency of point mutations (recessive lethals) above spontaneous background.

[d]Close to LC_{50}; DC, Doubling Concentration; LC_{50}, Lethal Concentration − 50 (50% survival).

Data from Vogel and Natarajan (1979a).

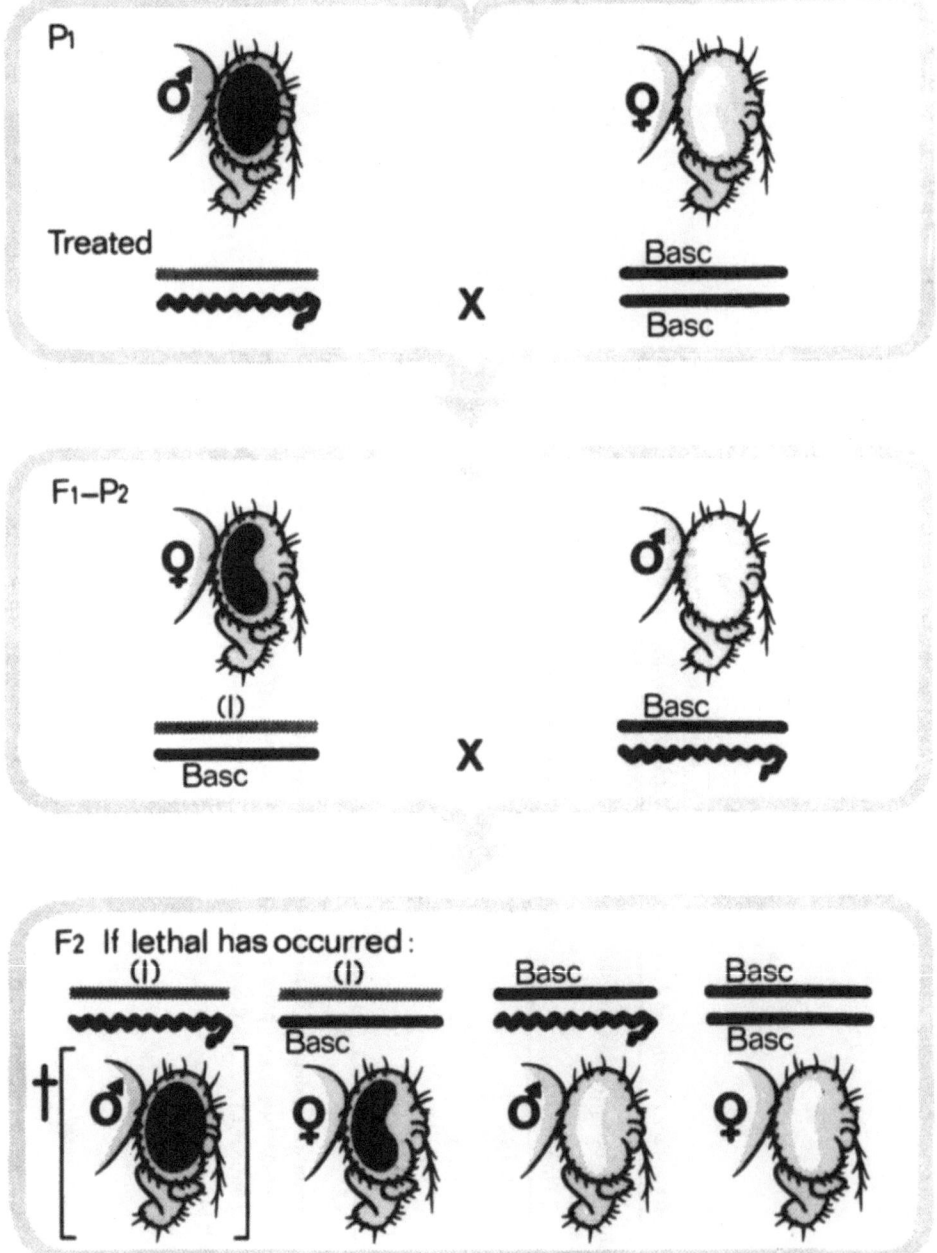

Figure 34.1.   The Basc test for the detection of X-chromosome recessive lethal mutations.

the culture temperature preferred in most laboratories, the major stages in the life cycle are: embryonic development, 1 day; first larval instar, 1 day; second larval instar, 1 day; third larval instar, 2 days; prepupa, 4 hours; pupa, 4.5 days (Würgler et al., 1977). Thus at 25° C one generation lasts only 9–10 days.

List and Nomenclature

The book by Lindsley and Grell (1968) entitled "Genetic Variations of *Drosophila melanogaster*" represents the last exhaustive compilation of the mutants of *Drosophila*. This book gives the nomenclature used by *Drosophila* geneticists, together with a detailed description of mutants, chromosome aberrations, special balancer chromosomes, cytological markers, and wild-type stocks. This guide is indispensable when working with *Drosophila*.

Equipment and Laboratory Techniques

There exist several detailed descriptions of mutation work on *Drosophila*, including culture medium, equipment, stock culturing, and handling of flies. These are the "*Drosophila* Guide" by Demerec and Kaufmann (1973), the review article by Abrahamson and Lewis (1971) and Würgler *et al.* (1977).

Testing Protocol

Performance

The most relevant features of the X-chromosomal recessive lethal test are shown in Fig. 34.1. Males from a wild-type strain are treated (or kept untreated as controls) and are then mated (P_1) to *virgin* females which are homozygous for the X-markers B (Bar, dominant) and w^a (white-apricot, recessive), affecting the shape and color of the eyes (Lindsley and Grell, 1968). The "*Basc*" X-chromosome further carries an inversion to prevent in the heterozygotes (F_2-P_2) crossing-over of a lethal from the treated (paternal) chromosome to its homologue. Thus the marker genes serve to distinguish "treated" (paternal) from "untreated" (maternal) chromosomes. In the F_2, which splits into four genotypes that can be identified by their different phenotypes, it is possible to distinguish the two classes of flies carrying copies from a treated chromosome (left side) from those which do not (right side). If a complete recessive lethal mutation is induced in an X-bearing germ cell of the treated P_1 male, all the somatic cells of the resulting F_1 female will be heterozygous for the muta-

tion, and also 50% of its eggs will carry it. If a lethal is induced, half of the F_2 males will be hemizygous for it and will therefore die. But this can be seen only when pair-mating is done in the F_1, which is an absolute prerequisite for the proper performance of the test.

Female treatment is not recommended in routine testing procedures, because the females may contain pre-existing lethals which have to be crossed out before starting an experiment.

Treatment Procedures

The procedures by which chemicals are most commonly administered to *Drosophila* consist of either injection into the body cavity or feeding, at adult or larval stages (Würgler *et al.*, 1977). Other methods to treat flies are inhalation (Magnusson and Ramel, 1976, 1978a; Verburgt and Vogel, 1977) and aerosol treatment (Sega and Lee, 1970).

For screening purposes, it is recommended to treat primarily adult males (Vogel and Lüers, 1974) since females are more readily sterilized by chemicals and have, so far, proved more refractory to the induction of heritable genetic changes (Clark and Sobels, unpublished; Vogel, 1971).

In spite of the fact that injection has widely been used in mutagenesis with *Drosophila*, there are several limitations to using this technique for testing purposes. One major problem becomes obvious when comparing the concentrations required to produce, after either feeding or injection, similar rates of mutations. When injecting a chemical, it is often observed that five- to tenfold higher concentrations are needed relative to 24-hr or 48-hr feeding to produce equal rates of recessive lethals. MMS, EMS, ENU, MNU, DMN, and DEN can be cited as examples of this observation (Hotchkiss and Lim, 1968; Kortselius, 1979; Vogel and Natarajan, 1979a). With hydrophilic compounds, preparation and injection of high doses causes no problems at all, but lipophilic mutagens may be registered as negative in the test, simply because the concentration which can be administered does not amount to a level sufficient to produce a positive effect. Thus the alkaryltriazene 2,4,6-CL$_3$-PDMT, although a very potent mutagen when fed, failed to produce recessive lethals when injected (Vogel and Lüers, 1974). On the other hand, injection is more reliable for the detection of mutagens that undergo rapid decomposition such as the unstable β-propiolactone and chloroethylene oxide (Kortselius, 1979). These instances may serve to illustrate that the choice of the application technique can be crucial to the outcome of the genetic lethal test. In chemical screening, preference should be given to adult feeding procedures, while injection may be indicated in those cases where the compound under test is very short-lived.

Toxicity Tests

Concentration-mortality relationships are considered useful to express the general biological reactivity of chemicals. For practical purposes, the greatest significance of such relations is the possibility of adequate planning of further experiments and, if the compound under test is a mutagen, of predicting the condition that will produce the maximum yield of mutations without killing the animal as a result of lethal overdose. Thus, pilot experiments should give an approximate idea of the possible toxicity of the test compound. Technical aspects of toxicity tests are described by Würgler *et al.* (1977). Actual results of toxicity tests with a series of monofunctional AA are reported by Vogel and Natarajan (1979a).

Brooding

It is well known that chemical mutagens often exhibit stage specificity, i.e., show more or less pronounced mutagenic effects in different stages of germ-cell develop-

ment. It is, therefore, very necessary to analyze the progeny from treated spermatozoa, late and early spermatids, and spermatocytes. Analysis of offspring from treated spermatids (early stage) is considered of particular importance, since this is the stage at which procarcinogens (nitrosamines, alkaryltriazenes, halo-olefins) exhibit peak mutagenic activity. This phenomenon has been explained as resulting from activation in situ in metabolically active germ cells, i.e., spermatid and spermatocyte stages (Vogel, 1975a). On the other hand, there seems no need to include in the test the analysis of spermatogonia, because there are only few cases of mutagens which affect spermatogonia but are not active in meiotic or postmeiotic stages (Auerbach, 1962b; Herskowitz, 1947).

With the brooding technique, the spatial pattern of spermatogenesis is translated into a temporal pattern of successive broods. Treated males are therefore remated at regular intervals of 2–3 days to fresh *virgin* females. An excess of 3–5 females per males serves to sample all germ cells which are in the mature stage. A total sampling period of 7–9 days (3 to 4 broods) is considered sufficient for mutagen screening.

Control and Replica Experiments. Sample Size

Würgler *et al.* (1975) prepared sample size tables which are very helpful for adequately planning recessive lethal experiments. The most important points in this respect are (Würgler *et al.*, 1975):

1. The dependence of the outcome of the genetic test on the number of the chromosomes tested in the treated group
2. The dependence of the result on the frequency of spontaneous mutations
3. How a given number of tests (chromosomes) should be divided into the treated and the control group

4. On statistical grounds, the optimal number of tests to be performed

It is highly recommended that, before starting an experiment, particular attention is drawn to these statistical problems. To give an example, with a spontaneous frequency of 0.2% lethals (10 lethals in 5000 progeny) and a sample size of 4500 in the treated group, 0.47% is the lowest value to prove statistically that a mutagenic effect was observed (Fig. 34.2). What is also seen from Fig. 34.2 is that an increase in the number of tested chromosomes above 5000 does not really help to improve the resolving power of the experiment. Such considerations help decide how large an experiment has to be planned.

From our experience with the recessive lethal test, two experiments, consisting of three successive broods each, can easily be handled with the aid of one technician per week. If 600 to 800 cultures are set up per each brood, then a testing capacity of 1800–2400 chromosomes results per experiment.

The next aspect to be considered is the general strategy to be followed when utilizing *Drosophila*. One might intend to start at the highest possible concentrations which permit the testing for recessive lethal mutations. Acute toxicity, reduced fertility, and solubility problems may then be the limiting factors. In the beginning, control experiments should be run concurrently. At a more experienced stage, the testing may be started with 1–2 concentrations, but without running a control. One replication experiment should be conducted in any case. This procedure will be sufficient to come to a firm conclusion about the genetic activity of the test compound, if the recessive lethal test is either positive (> 1–2% lethals) or negative (lethal frequency ≤ his-

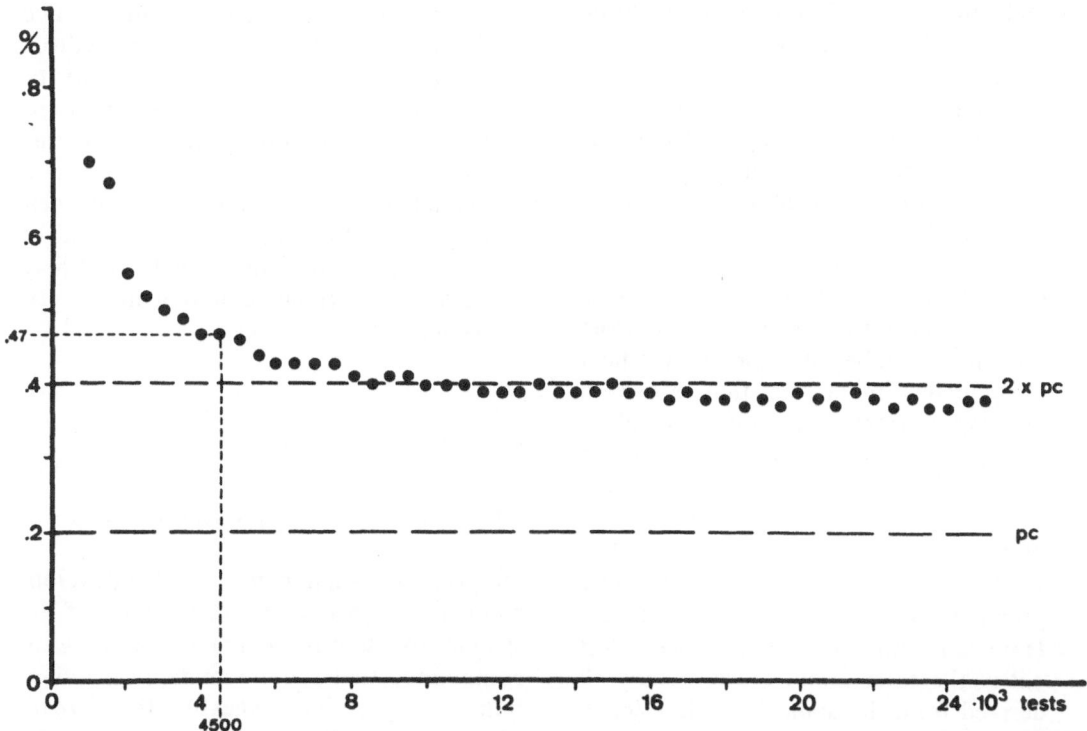

Figure 34.2. Lowest mutation frequencies significantly different from a control with 10 mutations in 5000 tests (=0.2%) for increasing number of tests in the treated group (from Würgler *et al.*, 1975).

torical control rate). Complicated "borderline cases," however, need far more experimental effort, and only in these cases are large control runs considered obligatory.

Pitfalls

The scoring of induced recessive lethal mutations is a highly objective method for exploring the mutagenic potential of a compound. Nevertheless, there exist few cases of misclassification due to incorrect performance of the test. Although several authors (Auerbach, 1962a; Abrahamson and Lewis, 1971; Spencer and Stern, 1948; Würgler *et al.*, 1977) discuss in detail precautions to be taken in carrying out recessive lethal tests, it may be profitable to summarize some of the obvious problems which can arise in the design of such experiments in the beginning.

1. As a standard rule, *pair-mating* (one male and 3–5 virgin females) should be applied to identify the very rare cases of spontaneous clusters, i.e., mutants of common origin. It is then possible to keep track of the F_1 family of cultures derived from each P_1 culture. Clusters will tend to appear in families. If, in postmeiotic broods, large clusters of lethals are observed among F_2 progeny derived from the same P_1 culture, it is recommended to eliminate these from the final score, because they may reflect spontaneous mutations that arose in dividing spermatogonia during development of that particular P_1 male (Würgler *et al.*, 1977). In a recent experiment with the procarcinogen 2-acetylaminofluorene, a recessive lethal frequency of 0.39% (7 lethals in 1807 progeny) was obtained, for instance. However, another 24 lethals were left out in the calculation, because they all derived from the same P_1 male (Vogel, unpublished).

2. Great care must be exercised in the scheme to ensure the usage of *virgin* females in the P-generation. Thus all F_1 females must be heterozygous for the treated X-chromosome and the Ins B w^a (Basc) balancer chromosome, and at least three different phenotypes must be present in the F_2 generation (see Fig. 34.1).

3. The presence of an extra Y-chromosome in a female of the P_1 generation will result in the appearance of non-Bar males in the F_1 generation. In such a case, the F_2 generation could have appreciable numbers of non-Bar males, even when there is a lethal present in the treated X-chromosome. Thus, care must be taken to exclude from further testing any F_1 culture in which non-Bar males are present.

4. Chemical mutagens which predominantly generate delayed mutations mosaically expressed in gonadal tissue may be considered nonmutagenic, if lethals are scored only in F_2, since an F_3 generation must be set up for determining these effects. Recently, Shukla and Auerbach (1979) reported that hydroxylamine produced only delayed effects, scored in the F_3 generation. The existence of such apparently exceptional cases, in our opinion, is not common enough to recommend the regular setting-up of three generations instead of two, because the only actual result may be a considerable reduction in test capacity.

Evaluation of Results

Mutagenicity of Procarcinogens

Of principal significance in the development of test protocols is the ability of the system to identify mutagenic activity of compounds that require metabolic activation. An early clue suggesting that *Drosophila* can activate procarcinogens into genetically active metabolites was the demonstration of the mutagenic activity of some nitrosamines (Pasternak, 1962; 1963; 1964) and pyrrolizidine alkaloids (Clark, 1959; 1960; 1963) in the recessive lethal test. Similar observations on the mutagenicity in *Drosophila* of other classes of procarcinogens soon followed in other laboratories.

Table 34.2 summarizes results of testing in the recessive lethal test some 60 to 70 procarcinogens from several classes for mutation induction. Most of these versatile carcinogens are clearly mutagenic in *Drosophila* males, and their detection in the recessive lethal test does not involve any problem. This is particularly true with the nitrosamines which, in contrast to the findings in other systems, belong to the strongest mutagens known in *Drosophila*.

With most procarcinogens, maximal mutagenic activity occurs in progeny from treated spermatids, whereas spermatocytes are particularly susceptible to killing. The explanation of this uniform stage response is that activation of carcinogens takes place *in vivo* within gonadal tissue (Vogel, 1975a; Vogel and Sobels, 1976). This observation is considered of particular significance for the identification of carcinogens such as the nitrosamines and azoxyalkanes, which form short-lived metabolites.

One problem which has yet to be solved in *Drosophila* is the weak mutagenic effectiveness of carcinogenic polycyclic hydro-

Table 34.2.   *Mutagenic Activity of Procarcinogens in* Drosophila

Compound	Mutagenicity in Drosophila[a]	References
1. Nitrosamines		
DMN; N-nitrosodimethylamine	+	Pasternak, 1962; Fahmy and Fahmy, 1975; Vogel and Natarajan, 1979a
DEN; N-nitrosodiethylamine	+	Pasternak, 1963; Fahmy *et al.*, 1966; Vogel and Leigh, 1975; Vogel and Natarajan, 1979a
N-nitrosobenzylamine	+	Pasternak, 1964
N-nitrosovinylamine	+	Pasternak, 1964
N-nitrosomethylphenylamine	+	Vogel, unpublished
N-nitrosomorpholine	+	Henke *et al.*, 1964; Vogel, 1976
N-nitroso-N'-methylpiperazine	+	Pasternak, 1964
N-nitrosopyrrolidine	+	Vogel, unpublished
N-nitrosopiperidine; NP	+	Nix *et al.*, 1979
3-Chloro NP	+	Nix *et al.*, 1979
4-Chloro NP	+	Nix *et al.*, 1979
3,4-Dichloro NP	+	Nix *et al.*, 1979
3,4-Dibromo NP	+	Nix *et al.*, 1979
3,5-Dimethyl NP	+	Nix *et al.*, 1979
2-Methyl NP	+	Nix *et al.*, 1979
3-Methyl NP	+	Nix *et al.*, 1979
4-Methyl NP	+	Nix *et al.*, 1979
N-Nitroso-4-piperidinol	0	Nix *et al.*, 1979
2. Hydrazo- and Azoxy alkanes		
Azoxymethane	+	Vogel, 1976b
1,2-Dimethylhydrazine	+	Vogel and Sobels, 1976
Procarbazine	+	Blijleven and Vogel, 1977

Table 34.2. *Mutagenic Activity of Procarcinogens in* Drosophila (Continued)

Compound	Mutagenicity in Drosophila[a]	References
3. Alkaryltriazenes		
a. 1-Phenyl-3,3-dimethyltriazenes (PDMT)[a]	+	Vogel, 1971; Kolar et al., 1974
2,4,6-Cl$_3$-PDMT	+	Kolar et al., 1974
2,4,6-Br$_3$-PDMT	+	Kolar et al., 1974
4-Br-PDMT	+	Kolar et al., 1974
4-Cl-PDMT	+	Kolar et al., 1974
4-C$_2$H$_5$-PDMT	+	Kolar et al., 1974
4-CH$_3$-CONH-PDMT	+	Kolar et al., 1974
4-OH-PDMT[a]	+	Kolar et al., 1974
3-OH-PDMT[a]	+	Kolar et al., 1974
2,4,6-F$_3$-PDMT	+	Vogel, unpublished
4-F-PDMT	+	Vogel, unpublished
2,4,6-Cl$_3$-PMMT (monomethyl-)[a]	+	Vogel, 1977, Vogel unpublished
b. Miscellaneous triazenes		
3,3-dimethyl-1-N-oxide (3-pyridyl)-triazene	+	Vogel, 1971; Vogel et al., 1973
3,3-dimethyl-1-(3-pyridyl)-triazene	+	Vogel, 1971; 1976b
3,3-diethyl-1-(3-pyridyl)-triazene	+	Vogel, unpublished
5-(3,3-dimethyl-1-triazeno)-imidazole-4-carboxamide	+	Vogel, 1977
4. Oxazaphosphorines		
Cyclophosphamide	+	Bertram and Höhne, 1959; Röhrborn, 1968; Vogel, 1975
Trofosfamide	+	Vogel, 1975
Ifosfamide	+	Vogel, 1975
5. Pyrrolizidine Alkaloids		
Heliotrine	+	Brink, 1963, 1966, 1969; Clark, 1959; 1960; 1963
Echinatine	+	Clark, 1960
Echimidine	+	Clark, 1960
Lasiocarpine	+	Clark, 1960
Senecionine	+	Clark, 1960
Supinine	+	Clark, 1960
Jacobine	+	Clark, 1960
Platyphylline	+	Clark, 1960
Monocrotaline	+	Clark, 1960; Cook and Holt, 1966
Fulvine	+	Cook and Holt, 1966
Retrorsine	+	Cook and Holt, 1966
6. Halo-olefins and Halo-alkanes		
Vinyl chloride	+	Magnusson and Ramel, 1976; 1978a; Magnusson et al., 1979; Verburgt and Vogel, 1977
Vinyl bromide	+	Vogel, unpublished
1,2-Dibromoethane[a]	+	Vogel and Chandler, 1974

Table 34.2. *Mutagenic Activity of Procarcinogens in* Drosophila (Continued)

Compound	Mutagenicity in *Drosophila*[a]	References
6. Halo-olefins and Halo-alkanes		
1,2-Dibromopropane[a]	+	Vogel and Chandler, 1974
1,2-Dichloroethane[a]	+	King et al., 1979
Tris-(2,3-dibromopropyl)phosphate	+	Berkowitz, 1978
2-Chloro-1,3-butadine[a]	+	Vogel, 1979
1-Chloro-1,3-butadiene[a]	+	Vogel, 1979
1,4-Dichlorobutene-2[a]	+	Vogel, 1979
7. Miscellaneous		
Aflatoxin B_1	+	Lamb and Lilly, 1971
HEMPA	+	Srám, 1972
Thioacetamide	+	Magnusson and Ramel, 1978b
Safrole	0	Vogel, unpublished
1-OH'-safrole	0	Vogel, unpublished
Estragole	+	Vogel, unpublished
1-OH-estragole	+	Vogel, unpublished
4-Nitroquinoline-N-oxide	+	Kramers, unpublished

[a]No activation required.

carbons and aromatic amines. These compounds were ineffective or produced only marginal effects when assayed for the induction of recessive lethals (Table 34.3).

The MFO System in Adult Flies and in Larvae

One experimental approach to improving the detection capacity of *Drosophila* for polyclics is an analysis of the enzymes involved in their activation, and the usage of enzyme inducers. Such attempts resulted in the identification of several types of cytochrome P-450 in whole body homogenates of numerous *Drosophila* strains (Baars et al., 1977). Microsomal fractions also showed aryl hydrocarbon hydroxylase (AHH) activity and hydratase action, while the postmicrosomal supernatant contained glutathione-S-transferase (Baars *et al.*, 1979).

Table 34.4 shows that the spectral and enzymic features of the larval cytochrome differ considerably from that present in adult flies, i.e., in the peaks of the cytochrome P-450 and in its content per gram

body weight. A difference between larvae and adult flies was also observed in paranitroanisole demethylation, whereas this was not the case for benzo(a)pyrene hydroxylation (Table 34.4).

From these data a differential response to procarcinogens would be expected after larvae or adult treatment. This was found, indeed. BP and 2-AAF, which are either inactive or produce only marginal effects when fed adult males, have a slight positive effect after larvae feeding (Table 34.5). What is of relevance here is that lower concentrations of test substances can be applied to produce a positive effect in larvae.

It also follows from Table 34.4 that there is quite a variation in the enzymic spectrum within the species *Drosophila*, as shown for strains Berlin K and Hikone R. In adult flies, this can be seen most clearly in the differences in benzo(a)pyrene hydroxylation and paranitroanisole demethylation. When microsomal preparations from larvae are analyzed, a large deviation is observed between the two strains in the peaks of their cytochromes, and again in paranitroanisole demethylation and benzo(a)pyrene hydroxylation. Moreover, distinct strain differences

Table 34.3. *Mutagenic Activity of Aromatic Amines and Polycyclic Hydrocarbons in the Recessive Lethal Test*

Chemical	Recessive lethals	References
N-Acetyl-2-aminofluorene	0	Demerec *et al.,* 1949; Fahmy and Fahmy, 1970b, 1972a; 1972b; Nix *et al.,* 1979
N-Hydroxy-N-acetyl-2-aminofluorene (N-HO-AAF)	0	Fahmy and Fahmy, 1970b
N-Acetoxy-N-acetyl-2-aminofluorene (N-AcO-AA)	0	Fahmy and Fahmy, 1970b;
	+	Fahmy and Fahmy, 1972b
N-Acetyl-aminofluorene-N-sulfate (AAF-N-SO$_4$)	0	Fahmy and Fahmy, 1972b
Benzidine	(+)	Fahmy and Fahmy, 1977
2-Naphthylamine	0	Demerec *et al.,* 1949; Fahmy and Fahmy, 1970b
N,N-Dimethyl-4-aminoazobenzene (DAB)	0	Demerec *et al.,* 1949; Fahmy and Fahmy, 1972a, 1972c
4-Dimethylamino-trans-stilbene	0	Vogel, 1976b
Benzo(a)pyrene (BP)	0	Demerec *et al.,* 1949; Nix *et al.,* 1979;
	(+)	Fahmy and Fahmy, 19773a
3-Methylbenzo(a)pyrene (3-MBP)	+	Fahmy and Fahmy, 1973a
6-Hydroxymethyl-benzo(a)pyrene (6-HMPB)	+	Fahmy and Fahmy, 1973a
1,6-Dimethylbenzo(a)pyrene (1,6-DMBP)	(0)	Fahmy and Fahmy, 1973a
3,6-Dimethylbenzo(a)pyrene (3,6-DMBP)	+	Fahmy and Fahmy, 1973a
7-Methylbenz(a)anthracene (MBA)	0	Fahmy and Fahmy, 1973b
7,12-Dimethylbenz(a)-anthracene (DMBA)	0	Fahmy and Fahmy, 1969; 1970c; 1972a; 1970a; 1973b
7-Bromomethyl-12-methylbenz(a)anthracene (7-BrM, 12-MBA)	0	Fahmy and Fahmy, 1970a
7-Bromomethylbenz(a)anthracene	(+)	Fahmy and Fahmy, 1973b
3-Methylcholanthrene (3-MCH)	0	Burdette, 1952; Demerec *et al.,* 1949; Fahmy and Fahmy, 1970b; 1972a; 1973b

Table 34.4. *Some Spectral and Enzymic Features of the MFO System in* Drosophila

Parameters	Flies		Larvae	
	Berlin-K	Hikone-R	Berlin-K	Hikone-R
Microsomal protein per gram body weight	10.5 ± 0.5	10.5 ± 0.5	5.3 ± 0.3	5.5 ± 0.3
Peak cytochrome P-450 (NADPH-reduced CO difference spectr.)	451.8 ± 0.2	451.6 ± 0.3	450.2 ± 0.1	448.1 ± 0.1
nmole cyt. P-450 per gr. body weight	1.25 ± 0.05	1.45 ± 0.05	0.19 ± 0.02	0.39 ± 0.02
Benzo(a)pyrene hydroxylation nmole 3-OH B.P. per nmole P-450 per minute (25°C)	0.55 ± 0.05	0.10 ± 0.03	0.58 ± 0.05	0.15 ± 0.03
Paranitroanisole demethylation nmole per nmole P-450 per minute (25°C)	1.0 ± 0.3	7.8 ± 0.7	0.6 ± 0.2	29 ± 1

Unpublished data from J.A. Zijlstra.

in dosage-mortality characteristics and mutation induction were observed when males from stocks Berlin K and Hikone R were exposed to numerous alkaryltriazenes, nitrosamines, and to azoxymethane (Vogel, 1976b; 1979). The triazene carcinogens revealed higher mutagenic activity and cytotoxicity in Berlin K males, while the opposite picture is found after exposure to DMN, DEN, or AM.

Since the analysis of the components of the MFO system in *Drosophila* is currently in its inventory phase, in the present status of knowledge we are unable to develop a standardized protocol as to what strain may be the most profitable for mutagen testing.

Correlation between Test Results and Carcinogenicity

More than 600 chemicals have been investigated for their ability to induce recessive lethal mutations in *Drosophila*. Chemicals which gave positive effects cover all classes of procarcinogens and a wide array of directly acting agents. The carcinogens mitomycin C, procarbazine, methylphenylnitrosamine, DEN, DMN, urethane,

thioacetamide, griseofulvin, 1,2-dimethylhydrazine, carcinogens causing problems in the *Salmonella* test of Ames (Hollstein *et al.*, 1979), are readily detectable in the recessive lethal test (Table 34.2; Magnusson and Ramel, 1978b). However, the results of testing 600 chemicals in *Drosophila* have not systematically been compared with carcinogenicity data, and we are, therefore, at present unable to comment on the relationship between a positive response in *Drosophila* and carcinogenicity in mammals.

Direction of Research

Enzyme Induction

For the present, it seems necessary to continue studies on the nature and the properties of the components of the MFO system in *Drosophila*. One approach to this problem is the usage of the enzyme inducers phenobarbital, 3-methylcholanthrene or arochlor. Zijlstra (personal communication) showed that the effect of treating larvae from strain Berlin K

Table 34.5. Mutagenicity of 4-DAS, BP, and 2-AAF in Strain Berlin-K

Chemical	Treatment	Concentration (mM)	Exposure	Experiment[a]	Chromosomes tested	Lethals	% L.	Probability
4-DAS	Adult feeding	1.5–2.0	48–72 h	9	12,469	44	0.35 ± 0.05	<0.01
		Control	48–72 h	4	5,511	6	0.11 ± 0.04	
BP	Adult feeding	1.3	48 h	1	1,630	0		
		2.6	48 h	1	3,268	5	0.15 ± 0.07	
		7.5	48 h	1	1,848	8	0.43 ± 0.15	≤0.01
	Larvae feeding[b]	0.1	9–10 days	3	3,556	20	0.56 ± 0.13	≤0.01
		Control		3	3,521	7	0.20 ± 0.08	
2-AAF	Adult feeding	2.0	3 days	1	1,205	2	0.17 ± 0.05	
		3.8	48 h	2	4,862	8	0.16 ± 0.06	
		Control	2–3 days	2	3,087	6	0.19 ± 0.08	
	Larvae feeding[b]	0.9–1.0	9–11 days	4	4,557	18	0.39 ± 0.09	0.07
		Control		3	3,521	7	0.20 ± 0.08	

[a] Number experiments/concentration (control) 4-DAS, 4-dimethylamino-trans-stilbene; BP, benzo(a)pyrene; 2-AAF, 2-acetylaminofluorene.

[b] W.G.H. Blijleven, unpublished.

with phenobarbital was a ninefold increase in the content of P-450; this effect was paralleled by a sixfold enhancement in the specific benzo(a)pyrene hydroxylating capacity, and this resulted in a 54-fold increase in this activity per animal. The result of exposing adult flies to PB was a doubling of the amount of cytochrome P-450, and also an increased benzo(a)pyrene hydroxylation. Substantial changes in the level of P-450 were also observed after treatment with 3-MC.

Magnusson and Ramel (1978a) reported that pretreatment with phenobarbiturate caused an increase in the mutagenic effect of vinyl chloride. Distinct differences in the enzyme induction and activation of vinyl chloride were noticed between different *Drosophila* strains (Magnusson *et al.*, 1979; Hällström and Ramel, personal communication).

Mutagen-sensitive Strains

The finding that cytochrome P-450 and substantial AHH activity are present in adult and larval flies offers a new approach to the problems associated with the detection of aromatic amines and polycyclic hydrocarbons. An alternative possibility is the development and usage of special tester strains, i.e., of mutants with an increased sensitivity to mutagens. Several of the recently isolated mutagen-sensitive strains (Baker *et al.*, 1976; Smith, 1973; 1976; Boyd *et al.*, 1976a; Nguyen *et al.*, 1978) have been shown to be deficient in DNA repair (Boyd and Setlow, 1976; Boyd *et al.*, 1976b; Nguyen and Boyd, 1977). Thus males bearing the *mei-9ª mei-41D5* mutants when treated with benzo(a)pyrene produced significantly more recessive lethals than did control (yellow) males (Nguyen *et al.*, 1979).

Experimental work conducted in our laboratory with 2-acetylaminofluorene, N-hydroxy-2-acetylaminofluorene, 7,12-dimethylbenzanthracene, and 4-dimethylamino-transstilbene revealed a some-

what improved detection capacity of strains *mei-9L1* and *mei-9ª mei-41D5* relative to the wild-type tester strain Berlin K (Table 34.6). All these carcinogens could be identified as weak mutagens following carcinogen exposure of male larvae or adult males. These results are encouraging, since in the past most attempts failed to detect these mutagenic carcinogens in the recessive lethal test. However, more research must be done to determine the response and the sensitivity of the various mutants to several classes of procarcinogens before firm conclusions and recommendations are reached for their usage in mutagen testing.

Conclusions and Recommendations

The discussion has been concerned mainly with the description and validation of the test for X-chromosomal recessive lethal mutations. Much attention has been drawn also to the ability of *Drosophila* to activate procarcinogens. The positive response of some 70 to 80 procarcinogens in the recessive lethal test clearly expresses the substantially activating potency of *Drosophila*. Bioactivation of procarcinogenes takes place in numerous tissues including metabolically active material in the gonads. As a consequence, carcinogens which release short-lived metabolites, such as the dialkylnitrosamines, can readily be detected as mutagens in *Drosophila*.

For testing protocols the following procedure is suggested:

1. Absolute priority should be given to the conventional assay on X-chromosomal recessive lethals. The test constitutes the only *Drosophila* assay that can be recommended for testing without reservations. The fact that lowest adverse effect levels were recorded for recessive X-linked lethals clearly indicates its superior discriminating power. The argument in favor of recessive lethal tests is further supported by the fact that (i)

Table 34.6. Mutagenicity of 4-DAS, 2-AAF, N-OH-2-AAF, and DMBA in Strains mei-9^{L1} and y mei-9^a mei-41^{D5}

Chemical	Strain	Treatment	Concentration (mM)	Exposure time	Experiment[a]	Chromosomes	Lethals	% L.	Probability
4-DAS	mei-9^{L1}	Adult feeding	1.5–2.0	48–72 h	5	9,037	90	1.0 ± 0.10	≤0.01
		Adult feeding	0 (Control)	48–72 h	3	4,964	28	0.56 ± 0.11	
2-AAF	y mei-9^a mei-41^{D5}	Adult feeding	1.0	72 h	1	1,125	3	0.27 ± 0.15	>0.05
			2.0	72 h	1	1,119	8	0.71 ± 0.25	
			0 (Control)	72 h	1	1,546	8	0.56 ± 0.19	
2-AAF	y mei-9^a mei-41^{D5}	Larvae feeding	0.1	9–11 d	2	2,929	20	0.68 ± 0.15	<0.05
			0 (Control)	9–11 d	2	3,371	10	0.30 ± 0.09	
(N-OH)-2-AAF	mei-9^{L1}	Adult feeding	2.0–3.0	72 h	4	6,725	41	0.61 ± 0.09	≤0.01
			0 (Control)	72 h	3	4,358	10	0.23 ± 0.07	
DMBA	mei-9^{L1}	Adult feeding	4.0	72 h	2	3,598	21	0.58 ± 0.13	≤0.01
			0 (Control)	72 h	2	3,867	7	0.18 ± 0.07	

[a]Number experiments/concentration (control).

recessive lethals include forward mutations, deletions and gross structural rearrangements, and (ii) a sizeable part of the entire genome (700–800 loci or 20% when dealing with the X-chromosome) is covered by recessive lethal tests.

2. Tests for the induction of translocations or ring-X losses are considered suitable to gain additional information regarding the genetic properties of carcinogens. This allows correlation of the various endpoints under the same test conditions.

3. The dominant lethal test in *Drosophila* is not recommended. The major shortcomings of the assay are its low sensitivity and the fact that "dominant lethality" can result from nongenetic damage.

4. Low resolving power is also the major disadvantage of the test for the induction of rod-X-loss, Y-loss, and Y-rearrangements, which makes the assay of very little value for screening purposes.

Acknowledgment

This work was supported by the National Institute of Environmental Health Sciences (U.S.A.), Contract ESO 1027 - 05/06, and the "Stichting Koningin Wilhelmina Fonds" (The Netherlands), Contract SG 77-42. Part of this investigation also received support from the Association Contract 139-77-1 ENV N between the European Communities (Environmental Research Programme) and the University of Leiden. I am very grateful to W.G.H. Blijleven and J.A. Zijlstra for permitting me to quote their unpublished data.

References

Abrahamson, S., Lewis, E.B. (1971): The detection of mutations in *Drosophila melanogaster*. In: A. Hollaender (ed.), Chemical Mutagens: Principles and Methods for Their Detection. New York: Plenum Press, pp. 461–487.

Auerbach, C. (1962a): Mutation. An Introduction to Research on Mutagenesis. Part I: Methods, Edinburgh, London: Oliver and Boyd, 176 pp.

Auerbach, C. (1962b): The production of visible mutations in *Drosophila* by chloroethyl methylsulfonate (CB 1506). Genet. Res. (Camb.) *3*:461–466.

Baars, A.J., Zijlstra, J.A., Vogel, E., Breimer, D.D. (1977): The occurrence of cytochrome P-450 and aryl hydrocarbon hydroxylase activity in *Drosophila melanogaster* microsomes, and the importance of this metabolizing capacity for the screening of carcinogenic and mutagenic properties of foreign compounds. Mutat. Res. *44*:257–268.

Baars, A.J., Zijlstra, J.A., Jansen, M., Vogel, E., Breimer, D.D. (1979): Xenobiotica-metabolizing enzymes in *Drosophila melanogaster*. In: 21st Congress of the European Society of Toxicology Dresden, June (abstract).

Baker, B.S., Boyd, J.B., Carpenter, A.T.C., Green, M.M., Nguyen, T.D., Ripoll, P., Smith, P.D. (1976): Genetic controls of meiotic recombination and somatic DNA metabolism in *Drosophila melanogaster*. Proc. Natl. Acad. Sci. USA *73*:4140–4144.

Berkowitz, S. (1978): The induction of II-III translocations by tris (2,3-dibromopropyl) phosphate in *Drosophila*. Mutat. Res. *57*:385–387.

Bertram, C., Höhne, G. (1959): Über die radiomimetische Wirkung einiger Zytostatika im Mutationsversuch an *Drosophila*. Strahlentherapie *43*:388–391.

Blijleven, W.G.H., Vogel, E. (1977): The mutational spectrum of procarbazine in *Drosophila melanogaster*. Mutat. Res. *45*:47–59.

Boyd, J.B., Setlow, R.B. (1976): Characterization of postreplication repair in mutagensensitive strains of *Drosophila melanogaster*. Genetics *84*:507–526.

Boyd, J.B., Golino, M.D., Nguyen, T.D., Green, M.M. (1976a): Isolation and characterization of X-linked mutants of *Drosophila melanogaster* which are sensitive to mutagens. Genetics *84*:485–506.

Boyd, J.B., Golino, M.D., Setlow, R.B. (1976b): The mei-9[a] mutant of *Drosophila melanogaster* increases mutagen sensitivity and decreases excision repair. Genetics *84*: 527–544.

Brink, N.G. (1963): The effect of cyanide and azide on the mutagenic activity of the pyrrolizidine alkaloid heliotrine in *Drosophila melanogaster*. Z. Vererbungsl. *94*:331–335.

Brink, N.G. (1966): The mutagenic activity of heliotrine in *Drosophila*. I. Complete and mosaic sex-linked lethals. Mutat. Res. *3*:66–72.

Brink, N.G. (1969): The mutagenic activity of the pyrrolizidine alkaloid heliotrine in *Drosophila melanogaster*. II. Chromosome rearrangements. Mutat. Res. *8*:138–146.

Burdette, W.J. (1952): Tumor incidence and lethal mutation rate in *Drosophila* treated with 20-methylcholanthrene. Cancer Res. *12*:201–205.

Clark, A.M. (1959): Mutagenic activity of the alkaloid heliotrine in *Drosophila*. Nature (Lond.) *183*:731–732.

Clark, A.M. (1960): The mutagenic activity of some pyrrolizidine alkaloids in *Drosophila*. Z. Vererbungsl. *91*:74–80.

Clark, A.M. (1963): The brood pattern of sensitivity of the *Drosophila* testis to the mutagenic action of heliotrine. Z. Vererbungsl. *94*: 115–120.

Cook, L.M., Holt, A.C.E. (1966): Mutagenic activity in *Drosophila* of two pyrrolizidine alkaloids. J. Genet. *59*:273–274.

Demerec, M., Kaufmann, B.P. (1973): *Drosophila* Guide: Introduction to the Genetics and Cytology of *Drosophila melanogaster*. Washington DC, Carnegie Institute, 45 pp.

Demerec, M., Wallace, B., Witkin, E.M., Bertani, G. (1949): The gene. Carnegie Institute of Washington Yearbook, *48*:156–186.

Fahmy, M.J., Fahmy, O.G. (1977): Mutagenicity of hair dye components relative to the carcinogenic benzidine in *Drosophila melanogaster*. Mutat. Res. *56*:31–38.

Fahmy, O.G., Fahmy, M.J. (1969): Specific genetic deletions by a carcinogenic hydrocarbon in *Drosophila*. Nature (Lond.) *224*:1328–1329.

Fahmy, O.G., Fahmy, M.J. (1970a): Genetic deletions at specific loci by polycyclic hydrocarbons in relation to carcinogenesis. Int. J. Cancer 6:250–260.

Fahmy, O.G., Fahmy, M.J. (1970b): Gene elimination in carcinogenesis: reinterpretation of the somatic mutation theory. Cancer Res. *30*:195–205.

Fahmy, O.G., Fahmy, M.J. (1970c): Induction of bobbed (bb) mutations by polycyclic aromatic carcinogens in *Drosophila*. Mutat. Res. *9*:239–243.

Fahmy, O.G., Fahmy, M.J. (1972a): Mutagenic selectivity for the RNA-forming genes in relation to the carcinogenicity of alkylating agents and polycyclic aromatics. Cancer Res. *32*:550–557.

Fahmy, O.G., Fahmy, M.J. (1972b): Mutagenic properties of N-acetyl-2-aminofluorene and its metabolites in relation to the molecular mechanisms of carcinogenesis. Int. J. Cancer *9*:285–298.

Fahmy, O.G., Fahmy, M.J. (1972c): Genetic properties of substituted derivatives of N-methyl-4-aminobenzene in relation to azo-dye carcinogenesis. Int. J. Cancer *10*:194–206.

Fahmy, O.G., Fahmy, M.J. (1973a): Mutagenic properties of benzo(a)pyrene and its methylated derivatives in relation to the molecular mechanisms of hydrocarbon carcinogenesis. Cancer Res. *33*:302–309

Fahmy, O.G., Fahmy, M.J. (1973b): Oxidative activation of benz(a)anthracene and methylated derivatives in mutagenesis and carcinogenesis. Cancer Res. *33*:2354–2361.

Fahmy, O.G., Fahmy, M.J. (1975): Mutagenic selectivity of carcinogenic nitroso compounds. II. N,N-Dimethylnitrosamine. Chem.-Biol. Interact. *11*:395–412.

Fahmy, O.G., Fahmy, M.J., Massasso, J., Ondrej, M. (1966): Differential mutagenicity of the amine and amide derivatives of nitroso compounds in *Drosophila melanogaster*. Mutat. Res. *3*:201–217.

Henke, H., Höhne, G., Künkel, H.A. (1964): Über die mutagene Wirkung von Röntgenstrahlen, N-nitroso-N-methyl-urethan und N-nitroso-morpholin bei *Drosophila melanogaster*. Biophysik *1*:418–421.

Herskowitz, I.H. (1947): A new method for treating *Drosophila* gametes with chemicals. Evolution *1*:111–112.

Hollstein, M., McCann, J., Angelosanto, F.A., Nichols, W.W. (1979): Short-term tests for carcinogens and mutagens. Mutat. Res. *65*:133–226.

Hotchkiss, S.K., Lim, J.K. (1968): Mutagenic specificity of ethyl methanesulfonate affected by treatment method. Dros. Inf. Serv. *43*:116.

King, M.-T., Beikirch, H., Eckhardt, K., Gocke, E., Wild, D. (1979): Mutagenicity studies with X-ray-contrast media, analgetics, antipyretics, antirheumatics and some other pharmaceutical drugs in bacterial, *Drosophila* and mammalian test systems. Mutat. Res. *66*:33–43.

Kolar, G.F., Fahrig, R., Vogel, E. (1974): Structure-activity dependence in some novel ring-substituted 3,3-dimethyl-1-phenyltriazenes. Genetic effects in *Drosophila melanogaster* and in *Saccharomyces cerevisiae* by a direct and a host-mediated assay. Chem.-Biol. Interact. *9*:365–378.

Kortselius, M.J.H. (1979): Induction of sex-linked recessive lethals and autosomal translocations by beta-propiolactone in *Drosophila*: influence of the route of administration on mutagenic activity. Mutat. Res. *66*:55–63.

Lamb, M.J., Lilly, L.J. (1971): Induction of recessive lethals in *Drosophila melanogaster* by aflatoxin B₁. Mutat. Res. *11*:430–433.

Lindsley, D.L., Grell, E.H. (1968): Genetic Variations of *Drosophila melanogaster*, Washington, DC, Carnegie Institute Publication No. 627, 472 pp.

Magnusson, J., Ramel, C. (1976): Mutagenic effects of vinyl chloride in *Drosophila melanogaster*. Mutat. Res. *38*:115.

Magnusson, J., Ramel, C. (1978a): Mutagenic effects of vinyl chloride on *Drosophila melanogaster* with and without pretreatment with sodium phenobarbiturate. Mutat. Res. *57*:307–312.

Magnusson, J., Ramel, C. (1978b): Mutagenic effects of thioacetamide in *Drosophila melanogaster*. Mutat. Res. *58*:253–262.

Magnusson, J., Hällström, I., Ramel, C. (1979): Studies on metabolic activation of vinyl chloride in *Drosophila melanogaster* after pretreatment with phenobarbital and polychlorinated biphenyls. Chem.-Biol. Interact. *24*:287–298.

Muller, H.J., Oster, J.J. (1961): Some mutational techniques in *Drosophila*. In: W.J. Burdette (ed.), Symposium on Methodology in Basic Genetics. San Francisco: Holden Day Inc., pp. 249–267.

Nguyen, T.D., Boyd, J.B. (1977): The meiotic-9 (mei-9) mutants of *Drosophila melanogaster* are deficient in repair replication of DNA. Mol. Gen. Genet. *158*:141–147.

Nguyen, T.D., Green, M.M., Boyd, J.B. (1978): Isolation of two X-linked mutants in *Drosophila melanogaster* which are sensitive to gamma-rays. Mutat. Res. *49*:139–143.

Nguyen, T.D., Boyd, J.B., Green, M.M. (1979): Sensitivity of *Drosophila* mutants to chemical carcinogens. Mutat. Res. *63*:67–77.

Nix, C.E., Brewen, B., Wilkerson, R., Lijinsky, W., Epler, J.L. (1979): Effects of N-ni-

trosopiperidine substitutions on mutagenicity in *Drosophila melanogaster*. Mutat. Res. *67*:27–38.

Pasternak, L. (1962): Mutagene Wirkung von Dimethylnitrosamin bei *Drosophila melanogaster*. Naturwissenschaften. *49*:81.

Pasternak, L. (1963): Untersuchungen über die mutagene Wirkung verschiedener Nitrosamine und Nitrosomethylharnstoff. Acta. Biol. Med. Ger. *10*:436–438.

Pasternak, L. (1964): Untersuchungen über die mutagene Wirkung verschiedener Nitrosamin- und Nitrosamid-Verbindungen. Arzneimittelforsch. *14*:802–804.

Röhrborn, G. (1968): Chemische Konstitution und mutagene Wirkung. IV. Zyklische N-Lostderivate. Mol. Gen. Genet. *102*:50–68.

Sega, G.A., Lee, W.R. (1970): A vacuum injection technique for obtaining uniform dosages in *Drosophila melanogaster*. Dros. Inf. Serv. *45*:179.

Shukla, P.T., Auerbach, C. (1979): The delayed mutagenic action of hydroxylamine in *Drosophila*. Mutat. Res. *61*:399–400.

Smith, P.D. (1973): Mutagen-sensitivity of *Drosophila melanogaster*. I. Isolation and preliminary characterization of a MMS-sensitive strain. Mutat. Res. *20*:215–220.

Smith, P.D. (1976): Mutagen-sensitivity of *Drosophila melanogaster*. III. X-Linked loci governing sensitivity to methyl methanesulfonate. Mol. Gen. Genet. *149*:73–85.

Spencer, W.P., Stern, C. (1948): Experiments to test the validity of the linear r-dose/mutation frequency relation in *Drosophila* at low dosage. Genetics *33*:43–74.

Srám, R.J. (1972): The differences in the spectra of genetic changes in *Drosophila melanogaster* induced by chemosterilants TEPA and HEMPA. Folia Biol. (Praha) *18*:139–148.

Verburgt, F.G., Vogel, E. (1977): Vinyl chloride mutagenesis in *Drosophila melanogaster*. Mutat. Res. *48*:327–336.

Vogel, E. (1971): Chemische Konstitution und mutagene Wirkung. VI. Induktion dominanter und rezessiv-geschlechtsgebundener Letalmutationen durch Aryldialkyltriazene bei *Drosophila melanogaster*. Mutat. Res. *11*: 397–410.

Vogel, E. (1975a): Some aspects of the detection of potential mutagenic agents in *Drosophila*. Mutat. Res. *29*:241–250.

Vogel, E. (1975b): Mutagenic activity of cyclophosphamide, trofosfamide, and ifosfamide in *Drosophila melanogaster*. Specific induction of recessive lethals in the absence of detectable chromosome breakage. Mutat. Res. *33*: 221–228.

Vogel, E. (1976a): The relation between mutational pattern and concentration by chemical mutagens in *Drosophila*. In: R. Montesano, H. Bartsch and L. Tomatis (eds.), Screening Tests in Chemical Carcinogenesis (IARC Scientific Publications No. 12). Lyon: International Agency for Research on Cancer, pp. 117–132.

Vogel, E. (1976b): Mutagenicity of carcinogens in *Drosophila* as a function of genotype-controlled metabolism. In: F.J. de Serres, J.R. Fouts, J.R. Bend, and R.M. Philpot (eds.), In Vitro Metabolic Activation in Mutagenesis Testing. Amsterdam: Elsevier/North-Holland Biomedical Press, pp. 63–79.

Vogel, E. (1977): Identification of carcinogens by mutagen testing in *Drosophila*: the relative reliability for the kinds of genetic damage measured. In: H.H. Hiatt, J.D. Watson, and J.A. Winsten (eds.), Origins of Human Cancer, Book C. New York, Cold Spring Harbor Laboratory, pp. 1483–1497.

Vogel, E. (1979): Mutagenicity of chloroprene, 1-chloro-1,3-trans-butadiene, 1,4-dichlorobutene-2 and 1,4-dichloro-2,3-epoxybutane in *Drosophila melanogaster*. Mutat. Res. *67*: 377–381.

Vogel, E., Chandler, J.L.R. (1974): Mutagenicity testing of cyclamate and some pesticides in *Drosophila melanogaster*. Experientia *30*:621–623.

Vogel, E., Leigh, B. (1975): Concentration-effect studies with MMS, TEB, 2,4,6-TriCl-PDMT, and DEN on the induction of dominant and recessive lethals, chromosome loss and translocations in *Drosophila* sperm. Mutat. Res. *29*:383–396.

Vogel, E., Lüers, H. (1974): A comparison of adult feeding to injection in *Drosophila melanogaster*. Dros. Inf. Serv. *51*:113–114.

Vogel, E., Natarajan, A.T. (1979a): The relation between reaction kinetics and mutagenic action of monofunctional alkylating agents in higher eukaryotic systems. I. Recessive lethal mutations and translocations in *Drosophila*. Mutat. Res. *62*:51–100.

Vogel, E., Natarajan, A.T. (1979b): The relation between reaction kinetics and mutagenic action of monofunctional alkylating agents in higher eukaryotic systems. II. Total and partial sex-chromosome loss in *Drosophila*. Mutat. Res. *62*:101–123.

Vogel, E., Sobels, F.H. (1976): The function of *Drosophila* in genetic toxicology testing. In: A. Hollaender (ed.), Chemical Mutagens: Principles and Methods for Their Detection, Vol. 4. New York: Plenum Press, pp. 93–142.

Vogel, E., Fahrig, R., Obe, G. (1973): Triazenes, a new group of indirect mutagens; comparative investigations of the genetic effects of different aryldialkyl triazenes using *Saccharomyces cerevisiae*, the host-mediated assay, *Drosophila melanogaster*, and human chromosomes *in vitro*. Mutat. Res. *21*: 123–136.

Würgler, F.E., Graf, U., Berchtold, W. (1975): Statistical problems connected with the sex-linked recessive lethal test in *Drosophila melanogaster*. I. The use of the Kastenbaum-Bowman test. Arch. Genet. *48*:158–178.

Würgler, F.E., Sobels, F.H., Vogel, E. (1977): *Drosophila* as assay system for detecting genetic changes. In: B.J. Kilbey, M. Legator, W. Nichols, and C. Ramel (eds.), Handbook of Mutagenicity Test Procedures. Amsterdam: Elsevier/North-Holland Biomedical Press, pp. 335–373.

35

Strategy for Breeding Test Animals of High Susceptibility to Carcinogens*

F. Anders, M. Schwab, and E. Scholl

Introduction

Many microbial (Ames *et al.*, 1973; Nagao and Sugimura, 1972) as well as other sub-mammalian (e.g., *Neurospora*: Ong and de Serres, 1972; Yeast: Koske and Stich, 1973) and mammalian (Stich and San, 1970; Stich *et al.*, 1971, 1976) cell systems for the detection of potential carcinogens have been developed and successfully applied. They are economic and time-saving as compared to the systems consisting of entire test animals such as laboratory mice, rats, and others. However, one should keep in mind that the feasibility of these cell systems suffers from a certain ambiguity in relating the *in vitro* effects (e.g., mutation, numerical and structural chromosome aberrations, sister chromatid exchanges, focus-forming capacity, etc.) to the *in vivo* event of neoplastic transformation in the entire

animal. *In vitro* cell test systems, therefore, cannot satisfactorily replace *in vivo* test systems. On the other hand, one should also keep in mind that little is known about the factors governing the susceptibility of a particular animal to the agent. Many of the wild ancestors of the laboratory animals are almost insusceptible to carcinogens and are, therefore, unsuitable for the test. In contrast, laboratory animals selected as test organisms frequently show a high rate of "spontaneously" developing neoplasms. The high rate of spontaneous neoplasia makes it difficult to relate neoplasia to the agent. The application of these test systems may also lead to erroneous conclusions as to whether a certain agent behaves as a carcinogen or cocarcinogen.

In order to overcome the dilemma of our ignorance on the sensitivity of higher test organisms to carcinogens, we are going to construct genotypes that stably confer to the animals a high susceptibility to carcinogens and at the same time a low rate of spontaneous neoplasia.

*This paper contains parts of the habilitation thesis of M. Schwab and parts of the dissertation of E. Scholl.

As the experimental system we are using neoplasia in *Xiphophorus*, which has been studied extensively for five decades (Gordon, 1927, 1941, 1959; Kosswig, 1927; Häussler, 1928; Anders, 1967; Anders and Anders, 1978; Anders *et al.*, 1973a,b).

The present article shall be restricted to the melanoma, mainly because this type of tumor, owing to the natural pigmentation of the melanoma cells, is most easily detectable and, therefore, highly suitable for screening tests aiming to detect potential carcinogens. However, the principles presented here for the melanoma are also valid for several other types of neoplasms of epithelial, mesenchymal, and neurogenic origin (Kollinger *et al.*, 1979a; Schwab *et al.*, 1979a; Schwab and Anders, 1980; Schwab *et al.*, 1978a,b, 1979b).

The Melanophore System of *Xiphophorus*

Differentiation of Normal Pigment Cells

See Fig. 35.1, left side. On the fourth day of life of the embryo some thousand neural crest cells start migrating. Those entering their definite places, including the corium of the skin, the peritoneum, the brain membrane, etc., are committed to become pigment cells and therefore are considered to be chromatoblasts. These cells give rise to the stem-melanoblasts (S-melanoblasts), which may reproduce identically throughout the life of the fish but may also differentiate to the intermediate melanoblasts (I-melanoblasts), which irreversibly continue differentiation to the dopa-positive advanced melanoblasts (A-melanoblasts) (Anders *et al.*, 1979, 1980). The A-melanoblasts differentiate to melanocytes and these finally to melanophores. These cells, after reaching a certain age, are removed by macrophages. The whole system maintains a homeostasis between the different stages of melanophore differentia-

tion, which is apparently controlled by distance regulation.

Pigment Cells Competent for Neoplastic Transformation

See Fig. 35.1, hatching. A certain mutant (golden, *gg*), in which the melanophore differentiation is almost completely blocked at the stage of the S-melanoblasts, fails to develop melanomas, indicating that this stage as well as those of the preceding chromatoblasts and neural crest cells are noncompetent for neoplastic transformation.

On the other hand, A-melanoblasts, melanocytes and melanophores, which can easily be recognized by their shape and content of pigment, have never been found to undergo neoplastic transformation, indicating that they are too far advanced to become transformed.

The competent cells for the transforming activity of *Tu* are therefore the I-melanoblasts.

Differentiation of Transformed Pigment Cells

See Fig. 35.1, right side. The I-melanoblasts, after being transformed (TI-melanoblasts; all transformed cells are designated T-cells), differentiate to the easily recognizable, large, proliferating dopa-positive TA-melanoblasts (Anders *et al.*, 1979, 1980). These T-cells differentiate to the heavily pigmented, obviously endopolyploid T-melanocytes, which represent the predominant cells of the malignant melanoma. T-Melanocytes may differentiate to the final stage, represented by the giant T-melanophores, which are unable to divide any more and which are removed by macrophages. The radically different feature of the T-melanophores as compared to the normal melanophores is, however, that they are not subjected to distance-dependent regulation. Their lobules and dendrites interlace to each other and to the T-melanocytes and TA-melanoblasts, thus forming compact three-dimensional accumulations of the neoplastically transformed cells.

The differentiation of the neoplastically transformed cell is controlled by diffusible factors of the surrounding tissue (Schartl, 1979).

The Genetic Information for Neoplastic Transformation

As evidenced by interpopulational and interspecific chromosome substitutions as well as by the analyses of structural changes of the chromosomes, the genetic information coding for neoplastic transformation is represented by a certain chromosomal gene which is located terminally. This gene, which represents the genetic basis for the susceptibility to the development of neoplasia, was designated "tumor gene" (Tu). Tu is normally repressed by regulating genes (R) which suppress the development of melanoma. Certain R-genes are linked to Tu and operate in cis-position only, while others are nonlinked. If Tu becomes derepressed, it may mediate neoplastic transformation.

Induced and "Spontaneously" Developing Melanoma

With regard to the genetic makeup of a test system for the detection of carcinogens, it is important to distinguish between (a) really carcinogen-induced, and (b) "spontaneously" developing neoplasia. The different kinds of biology of neoplasia are summarized in Figure 35.1 using the melanoma as the example.

Derepression, for instance, may be induced by mutation of R-genes in an A-melanoblast (Fig. 35.1, Aa). This cell is no longer competent to neoplastic transforma-

tion and therefore Tu activity remains undetectable.

Derepression of Tu may also be induced by mutation of R-genes in an I-melanoblast (Fig. 35.1, Ab). This cell is competent and becomes neoplastically transformed. Following the processes involving cell division and cell differentiation, the T-cell gives rise to an easily detectable cell clone. The smallest clones observed so far consist of 8 T-melanocytes, indicating that there were 3 cell divisions between the mutation event and the occurrence of the T-melanocytes. These T-cells may continue to divide, but finally differentiate to T-melanophores. The origin of such a somatic mutation-conditioned melanoma is unicellular, and the melanoma grows exclusively by proliferation.

Furthermore, derepression of Tu may also be induced by mutation in an S-melanoblast (Fig. 35.1, Ac). This cell is not yet competent. It remains still nontransformed and may multiply over a long period ("latent period") as a normal stem cell. Later on, those descendants reaching the stage of competence by differentiation are transformed simultaneously. After some cell divisions, paralleled by cell differentiation, they become visible as a large cell clone consisting of hundreds or thousands of dividing TA-melanoblasts and T-melanocytes, which give rise to the melanoma. Those S-melanoblasts which do not further differentiate may reproduce identically throughout the further life of the fish and may serve as a permanent source of I-melanoblasts, which then become neoplastically transformed. The origin of such a melanoma is multicellular, although it can be traced back to a single mutational event in a somatic cell. The melamona grows by both permanent transformation and proliferation of the descendants of the mutated cell.

Besides the derepression of Tu in somatic cells, this gene may also be derepressed by mutation of R-genes (Fig. 35.1, B) or by crossing-conditioned elimination of R-genes (Fig. 35.1, C; those which are

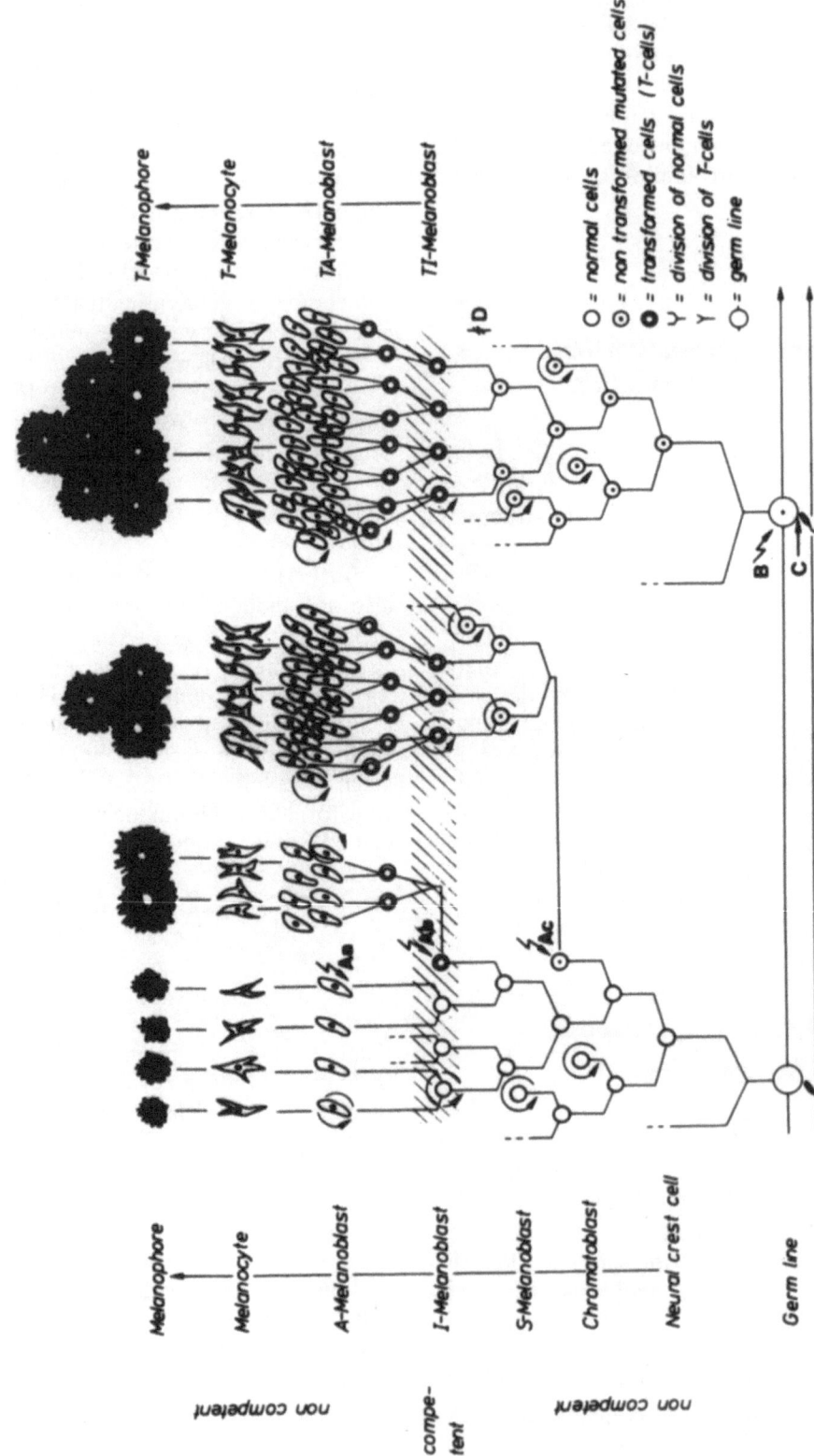

Figure 35.1. Differentiation of normal and of neoplastically transformed pigment cells. Consequences of the derepression of *Tu* in germline and in somatic cells. For details, see text.

nonlinked) in the germ line. As a consequence, hereditable melanomas develop in the progeny "spontaneously" as soon as the noncompetent S-melanoblasts differentiate to the competent I-melanoblasts. The origin of such melanomas is highly multicellular, and the melanoma grows by both permanent transformation and proliferation.

Finally, conditions have been analyzed in which Tu, although already derepressed by mutation or elimination of R-genes, cannot mediate neoplastic transformation because pigment cell differentiation is delayed in the stage of the incompetent S-melanoblasts. Agents such as cyclic AMP, corticotropin, methyltestosterone, as well as general environmental changes such as the decrease of the temperature and the increase of the salinity of the water in the tank, promote almost simultaneously the differentiation of large amounts of the noncompetent cells to competent ones, which subsequently become neoplastically transformed by Tu and give rise to an "induced" melanoma (Fig. 35.1, D). Epigenetic factors related to sexual maturity and sexual propagation, probably steroid sex hormones, may trigger the same type of melanoma, which in this case appears as a "spontaneously" developing neoplasm. From the point of view of our experiments, it seems to be also clear that these agents, although capable of triggering neoplasia, are not real carcinogens.

Desired Features of the Genetic System Suitable for the Detection of Carcinogens

The ideal test stock would consist of animals which are all homozygous for the tumor gene Tu, and all homozygous for one highly potent pigment cell-specific R, which is (a) the only one in the system, which is (b) linked to Tu, and (c) operates exclusively in cis-position. Mutation of the R in a competent or not yet competent cell would give rise to the melanoma in such a system, indicating that the agent in question acts as a carcinogen.

The presence of only one R is very important for the feasibility of the system, as shown by the following simple calculation (Table 35.1).

In the pigment cell system of a young fish, the average of cells competent and not yet competent to neoplastic transformation is about 10^6. The mutation rate induced, for instance, by 1000 roentgen (X-rays) is estimated to be about 10^{-6} which is a reasonable value for a vertebrate cell. In this case the tumor incidence is 1 (on the average the treated animals develop one tumor each). If, however, Tu is regulated by two R-genes, the rate of simultaneous mutations of both of these R-genes in one cell is 10^{-12}, and the tumor incidence is 10^{-6}. The calculation shows that it is extremely unlikely to induce a somatic mutation-conditioned neoplasm, if Tu is controlled by more than one R-gene. Furthermore, "spontaneously" developing melanomas induced by the "spontaneous" event of R-mutation is expected to occur very exceptionally.

Table 35.1. *Relationship between (a) the Number of Regulating Genes* (R) *for the Tumor Gene* (Tu), *(b) the Simultaneous Mutations in a Single Pigment Cell, and (c) the Incidence of Somatic Mutation Conditioned Melanoma*[a]

(a) R-genes	(b) Rate of simultaneous mutation	(c) Melanoma incidence
1 R	10^{-6}	10^0
2 R	10^{-12}	10^{-6}
3 R	10^{-18}	10^{-12}

[a]Premises: Melanoblasts competent for transformation in an
  individual: 10^6
  mutation rate: 10^{-6}

Working on the Genetic Makeup for the Test System

The first step for the composition of the genetic makeup of the test system is based on the classic crossings of a spotted platyfish (*Xiphophorus maculatus*) and a swordtail (*Xiphophorus helleri*) by Gordon and by Kosswig about 50 years ago. Using the appropriate parental animals and applicating the appropriate crossing procedures, the offspring develop predictably hereditary conditioned melanomas without any treatment (Fig. 35.2).

The following is only one example in which six genes of the platyfish are taken into consideration: first, the tumor gene Tu; second, R', a regulating gene for Tu that is impaired by mutation; third, R_{Df}, a regulating gene for Tu that has lost the capacity to repress Tu in the compartment of the dorsal fin (Df); fourth, $R_{Pp'}$, a regulating gene for Tu that has lost the capacity to repress Tu in the compartment of the posterior part of the trunk (P_p); fifth, R_{Diff}, a regulating gene that controls cell differentiation posttransformationally; sixth, $Est\text{-}1$, a marker gene coding for the esterase-1 (Ahuja *et al.*, 1977). Tu, R', R_{Df}, and $R_{Pp'}$ (abbreviated $Tu\text{-}R'$-complex in the following) are linked on the X-chromosome* while R_{Diff} and $Est\text{-}1$ represent an autosomal linkage group. The platyfish used in the crossings is homozygous for all these genes.

Following the crossing of the spotted platyfish with the nonspotted swordtail, the F_1-progeny, which is hemizygous for both the $Tu\text{-}R'$-complex and the $R_{Diff}Est\text{-}1$ develops benign melanomas instead of spots.

Following backcrossing of the F_1 with the swordtail, the animals of the BC_1-generation segregate as expected into: (a) 50%, which lack the T-cells because of the lack of the $Tu\text{-}R'$-complex, (b) 25%, which develop benign melanomas because of the presence of both the $Tu\text{-}R'$-complex and the $R_{Diff}Est\text{-}1$, and (c) 25%, which develop malignant melanomas because of the presence of the $Tu\text{-}R'$-complex and the lack of the $R_{Diff}Est\text{-}1$.

Further backcrossings (not shown in Fig. 35.2) of the fish carrying benign melanoma with the swordtail results in a BC_2 that exhibits the same segregation as the BC_1. The same applies for further backcrosses of this type.

Backcrossing of the fish carrying a malignant melanoma with the swordtail results in a BC_2, in which 50% of the animals do not develop melanomas because of the lack of the $Tu\text{-}R'$-complex, while the remaining animals develop malignant melanomas because of the $Tu\text{-}R'$-complex, but $R_{Diff}Est\text{-}1$ is lacking. The same applies for further backcrosses.

The second step for the composition of the genetic makeup of the test system consists of the replacement of the $Tu\text{-}R'$ complex by a $Tu\text{-}R$-complex, the R of which is highly potent to repress Tu (Fig. 35.3) (the whole $Tu\text{-}R$-complex is originally located on the Y chromosome and became associated to a $Tu\text{-}R'$-lacking X-chromosome. $R_{Bs'}$ represents a regulating gene that has lost the capacity to repress Tu in the compartment of the side of the body). Since the potent R is inherited along with Tu, the $Tu\text{-}R$ animals do not develop spontaneous melanomas. However, following the treatment of the fish with carcinogens such as X-rays or N-methyl-N-nitrosourea, depending on the presence or absence of the R_{Diff}, which can be determined by a chemical test for esterase-1 (see arrows), the $Tu\text{-}R$ hybrids develop somatic mutation conditioned melanomas according to type Ab and Ac in Figure 35.1. The $Tu\text{-}R$ backcross hybrids lacking the $R_{Diff}Est\text{-}1$ develop large malignant melanomas, while

---

*In addition to the Tu on the X-chromosome there are several copies of Tu on different chromosomes of the platyfish and the swordtail that are completely repressed by linked regulating genes each. These cryptic Tu-copies do not influence the activity of the Tu considered in this experiment, and they are therefore not shown in Fig. 35.2. They may, however, become active following mutagen-carcinogen-induced germ line mutations and somatic mutation of the R-genes. There are indications that Tu might be related to an endogenous virus (Kollinger *et al.*, 1979b).

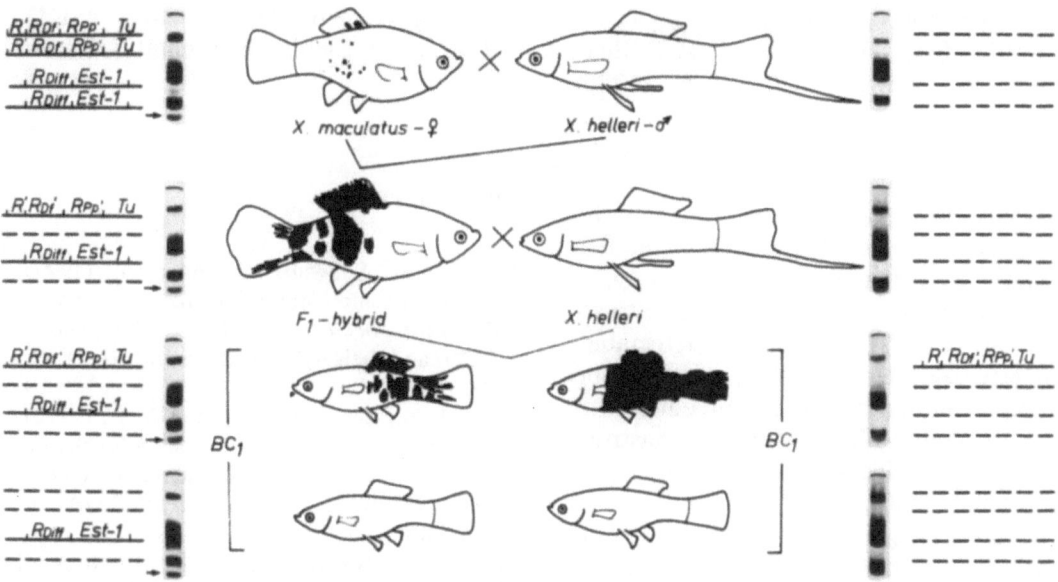

Figure 35.2.   Crossing scheme displaying the genetic conditions for the lack of T-cells and for the development of spots, benign melanomas and malignant melanomas. _____, chromosomes of *X. maculatus*; --------, chromosomes of *X. helleri*; *Tu*, tumor gene; $R_{Pp'}$ and $R_{Df'}$, impaired regulating genes controlling *Tu* in the compartments of the posterior part of the body (P_p) and of the dorsal fin (Df); *R'*, impaired compartment nonspecific regulating gene; R_{Diff}, regulating gene controlling differentiation of neoplastically transformed cells; *Est-1*, locus for esterase-1 of *X. maculatus* (see arrows; polyacrylamid-gel electrophoresis from homogenates of the eye).

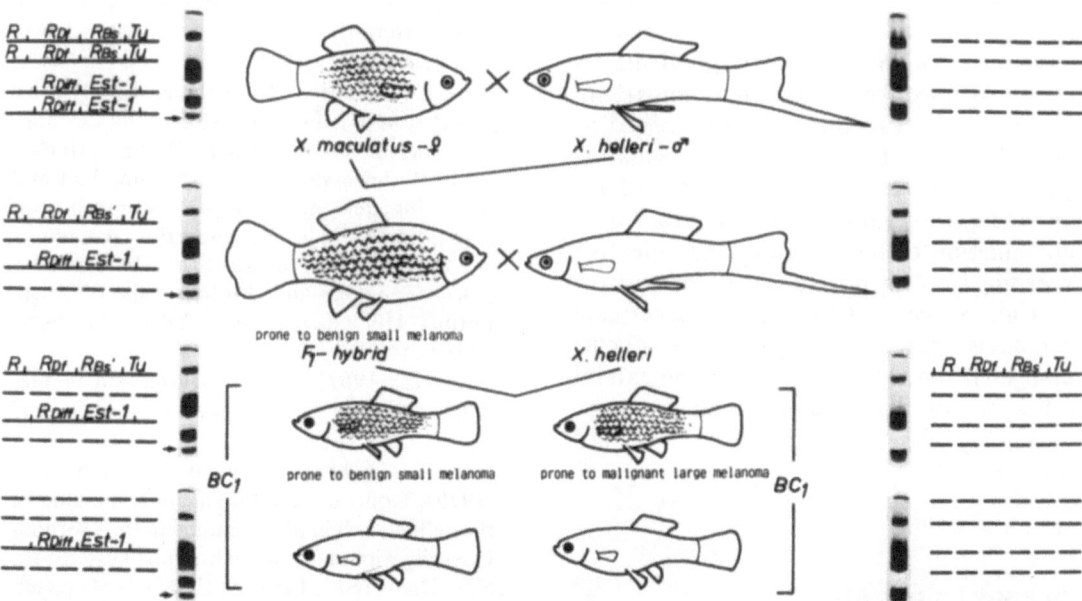

Figure 35.3.   Crossing scheme displaying the procedure for obtaining fish of genotypes prone to benign melanoma or to malignant melanoma following carcinogen treatment. Abbreviations as in Fig. 35.2. $R_{Bs'}$, impaired regulating gene controlling *Tu* in the compartment of the body side. *Est-1* analyzed from small pieces of fin tissue can be used for monitoring the breeding program designed to obtain fish exclusively prone to develop malignant melanoma.

those carrying this linkage group develop small melanomas, consisting in many cases of only a few cells and tending, furthermore, to regression (Schwab, 1980; Schwab and Scholl, 1980) (Fig. 35.3). Consequently, in a program in which large numbers of fish are employed for carcinogen screening, in order to keep the costs for the personal watching of the fish low, it is inevitable to have fish genotypes in which the effects of the carcinogen treatment are clearly and rapidly detectable. The esterase-1 test, therefore, seems to be important in monitoring our breeding program in *Xiphophorus* aiming to construct genotypes particularly suitable for carcinogenicity tests.

We are now going to breed a test stock consisting exclusively of animals homozygous for the highly sensitive *Tu-R*-complex, and lacking the $R_{Diff}Est-1$.

Conclusions

Genetic analysis performed during the last five decades has led to a stage at which we have in *Xiphophorus* a fairly good understanding about the regulation of gene(s) involved in the susceptibility to neoplasia as well as in the function of the regulating genes themselves. We are making use of this knowledge now in order to construct by a combination of mutagenesis and selective breeding stable genotypes that are sensitive enough to be suitable for a large-scale screening for carcinogens and cocarcinogens, and that at the same time exhibit a low rate of spontaneous neoplasia. Some promising progress has been made toward this aim.

Acknowledgments

Supported by the Deutsche Forschungsgemeinschaft through Sonderforschungsbereich 103, "Zellenergetik und Zelldifferenzierung," Marburg, by Deutsche Forschungsgemeinschaft (Schw 251/1), and by Justus-Liebig-Universität, Giessen.

References

Ahuja, M.R., Schwab, M, Anders, F. (1977): Tissue specific esterase in the xiphophorine fish *Platypoecilus maculatus*, *Xiphophorus helleri* and their hybrids. Biochem. Genet. *15*: 601–609.

Ames, B.N., Lee, F.D., Durston, W.E. (1973): An improved bacterial test system for the detection and classification of mutagens and carcinogens. Proc. Natl. Acad. Sci USA *70*: 782–786.

Anders, A., Anders, F. (1978): Etiology of cancer as studied in the platyfish-swordtail system. Biochim. Biophys. Acta *516*:61–95.

Anders, A., Anders, F., Klinke, K. (1973a): Regulation of gene expression in the Gordon-Kosswig melanoma system. I. The distribution of the controlling genes in the genome of the xiphophorine fish, *Platypoecilus maculatus* and *Platypoecilus variatus*. In: J.H. Schröder (ed.), Genetics and Mutagenesis of Fish. Berlin, Heidelberg, New York: Springer-Verlag, pp. 33–52.

Anders, A., Anders, F., Klinke, K. (1973b): Regulation of gene expression in the Gordon-Kosswig melanoma system. II. The arrangement of chromatophore-determining loci and regulating elements in the sex chromosome of xiphophorine fish, *Platypoecilus maculatus* and *Platypoecilus variatus*. In: J.H. Schröder (ed.), Genetics and Mutagenesis of Fish. Berlin, Heidelberg, New York: Springer-Verlag, pp. 53–63.

Anders, F. (1967): Tumour formation in platyfish-swordtail hybrids as a problem of gene regulation. Experientia *23*:1–10.

Anders, F., Diehl, H., Schwab, M., Anders, A. (1979): Contributions to an understanding of the cellular origin of melanoma in the Gordon-Kosswig xiphophorine fish tumor system. In: S.N. Klaus (ed.), Pigment Cell, Vol. 4. Basel: Karger, pp. 142–149.

Anders, F., Diehl, H., Scholl, E. (1980): Differentiation of normal melanophores and of neoplastically transformed melanophores in the skin of *Xiphophorus*. In: R.I.C. Spearman (ed.), Zoological Society of London Symposia. New York: Academic Press.

Gordon, M. (1927): The genetics of a viviparous topminnow *Platypoecilus*—the inheritance of two kinds of melanophores. Genetics *12*: 253–283.

Gordon, M. (1941): Genetics of melanomas in fishes. Cancer Res. *1*:656–659.

Gordon, M. (1959): The melanoma cell as an imcompletely differentiated pigment cell. In: M. Gordon (ed.), Pigment Cell Biology. New York: Academic Press, pp. 215–239.

Häussler, G. (1928): Über Melanombildung bei Bastarden von *Xiphophorus helleri* und *Platypoecilus maculatus* var. Rubra. Klin. Wochenschr. *7*:1561–1562.

Kollinger, G., Schwab, M., Abdo, S. (1979a): Structure and ultrastructure of the mutagen/carcinogen-induced melanoma and neuroblastoma in *Xiphophorus*. Verh. Dtsch. Zool. Ges. *251*.

Kollinger, G., Schwab, M., Anders, F. (1979b): Virus-like particles induced by bromodeoxyuridine in melanoma and neuroblastoma of *Xiphophorus*. J. Cancer Res. Clin. Oncol. *95*: 239–246.

Koske, R.E., Stich, H.F. (1973): A sensitive yeast assay system for carcinogenic nitroquinoline oxides. Mutat. Res. *19*:265–298.

Kosswig, C. (1927): Über Bastarde der Teleostier *Platypoecilus* und *Xiphophorus*. Z. indukt. Abstamm. Vererbungsl. *44*:253.

Nagao, M., Sugimura, T. (1972): Sensitivity of repair-deficient mutants and similar mutants to 4-nitroquinoline 1-oxide and their derivatives. Cancer Res. *32*:2369–2374.

Ong, T., de Serres, F.J. (1972): Mutagenicity of chemical carcinogens in *Neurospora crassa*. Cancer Res. *32*:1890–1893.

Schartl, M. (1979): Studies on cell autonomy in *Xiphophorus*—transplantation of prospective and definitive melanoma tissue in embryos. Verh. Dtsch. Zool. Ges. *249*.

Schwab, M. (1980): Sensitivität für Induktion von Neoplasmen durch N-Methyl-N-nitrosohaarstoff als Folge des Kombination von Artgenomen bei *Xiphophorus*. Habilitationsschrift, Giessen.

Schwab, M., Anders, A. (1980): Carcinogenesis in *Xiphophorus* and the role of the genotype in tumor susceptibility. In: H.E. Kaiser (ed.), Neoplasms—Comparative Pathology of Growth in Animals, Plants, and

Man. Baltimore: Williams & Wilkins Book Co.

Schwab, M., Scholl, E. (1980): Neoplastic pigment cells induced by N-methyl-N-nitrosourea (MNU) in *Xiphophorus* and genetic control of their terminal differentiation. Differentiation.

Schwab, M., Abdo, S., Ahuja, M.R., Kollinger, G., Anders, A., Anders, F., Frese, K. (1978a): Genetics of susceptibility of xiphorine fish to develop fibrosarcoma and rhabdomyosarcoma following treatment with N-methyl-N-nitrosourea (MNU). Z. Krebsforsch. *91*:301–315.

Schwab, M., Haas, J., Abdo, S., Ahuja, M.R., Kollinger, G., Anders, A., Anders, F. (1978b): Genetic basis of the susceptibility for the induction of neoplasms by N-methyl-N-nitrosourea (MNU) and X-rays in the platyfish-swordtail tumor system. Experientia *34*: 780–782.

Schwab, M., Abdo, S., Kollinger, G. (1979a): Relation between the inducibility of neoplasms and the genotype of the individual in *Xiphophorus (Poeciliidae)*. Verh. Dtsch. Zool. Ges. *250*.

Schwab, M., Kollinger, G., Haas, J., Ahuja, M.R., Abdo, S., Anders, A., Anders, F. (1979b): Genetic basis of the susceptibility for development of neuroblastoma following treatment with N-methyl-N-nitrosourea (MNU) and X-rays in *Xiphophorus*. Cancer Res. *39*:519–526.

Stich, H.F., San, R.H.C. (1970): DNA repair and chromatid anomalies in mammalian cell exposed to 4-nitroquinoline 1-oxide. Mutat. Res. *10*:389–404.

Stich, H.F., San, R.H.C., Kawazoe, Y. (1971): DNA repair synthesis in mammalian cells exposed to a series of oncogenic and non-oncogenic derivatives of 4-nitroquinoline 1-oxide. Nature (Lond.) *229*:416–419.

Stich, H.F., San, R.H.C., Lam, P.P.S., Koropatnick, D.J., Lo, L.W., Laishes, B.A. (1976): DNA fragmentation and DNA repair as an *in vitro* and *in vivo* assay for chemical precarcinogens, carcinogens and carcinogenic nitrosation products. In: R. Montesano, H. Bartsch and L. Tomatis (eds.), Screening Tests in Chemical Carcinogenesis (IARC Scientific Publications No. 12). Lyon: International Agency for Research on Cancer, pp. 617–636.

36

Methods for Human and Murine Sperm Assays*

A.J. WYROBEK

Introduction

Recent human studies with dibromochloro-propane (DBCP) illustrate that human exposure to a chemical toxin can have profound antispermatogenic effects, even at doses that show no other clinical signs of toxicity (e. g., Whorton *et al.*, 1977; Sandifer *et al.*, 1979). Studies in domestic animals, rodents, and humans show that sperm anomalies can be used as indicators and, in certain cases, dosimeters of induced antispermatogenic effects. Although induced sperm anomalies are clearly linked to testicular damage, their relationship to induced, heritable, genetic defects is not yet clear and is an active research area. Since sperm are such easy cells to obtain, can represent damage to the gonads, and can be studied in humans and model animals, sperm anomaly assays have received much attention in the fields of chemical mutagenesis and carcinogenesis screening.

*This work was supported by the United States Department of Energy under contract #W-7045-ENG-48 and by the U.S. EPA under the Pass-Through Agreement.

The following is an overview of the methodology of human sperm studies including descriptions and applications of the various sperm assays, their relative sensitivities, guidelines for new human sperm studies, the development of new assays, and the predictive value of induced human sperm changes. The mouse sperm abnormality assay is also described, including methods for testing an unknown agent, the value of the assay in screening for testicular toxins of relevance to man, its relationship to other mouse sperm assays, and its role as an animal model to help understand the heritable consequences of induced sperm anomalies.

Methods for Human Sperm Assays

Description of Assays and Examples of Their Applications

Visual studies of semen and microscopic analyses of sperm have a long history in fertility diagnosis in many species, especially in domestic animals and man (Lagerlöf, 1934; Farris, 1950). As a result, most of the human studies assessing testicular function in occupational and therapeutic settings have used the parameters developed in these initial fertility studies, namely sperm count, sperm motility, and sperm morphology.

One of the best examples of the use of semen assays to assess testicular function in a workplace setting is the study of lead workers in a storage-battery plant in Romania (Lancranjan *et al.*, 1975). One hundred men with a mean occupational exposure of 8.5 years and 50 men who worked in annex workrooms of the plant were studied and compared to 50 unexposed controls. As shown in Fig. 36.1, men were assigned to dosage groups by the amounts of lead detected in their peripheral blood and urine. The proportion of men with asthenospermia (reduced sperm motility), hypospermia (reduced sperm numbers), or teratospermia (increased

frequencies of abnormally shaped sperm) was dose-related. Sperm-shape abnormalities showed the largest lead-related effect.

The study of DBCP workers is another example of the use of semen assays to assess testicular function in the workplace (Whorton *et al.*, 1977). Semen samples of approximately 110 men working in a DBCP area of a chemical plant were compared to those of workers from other areas of the plant. Figure 36.2 shows the reduced sperm counts in the DBCP workers; sperm motility and morphology were not assessed. The levels of blood FSH, LH, and testosterone were also measured in the exposed men. Only in cases of azoospermia were there any appreciable changes in blood gonadotropin level, suggesting that these blood indices are not as responsive to changes in testicular function as is sperm count. In another group of DBCP workers, Kapp *et al.* (1979) found an increased frequency of sperm with two fluorescent bodies, which are thought to be two Y chro-

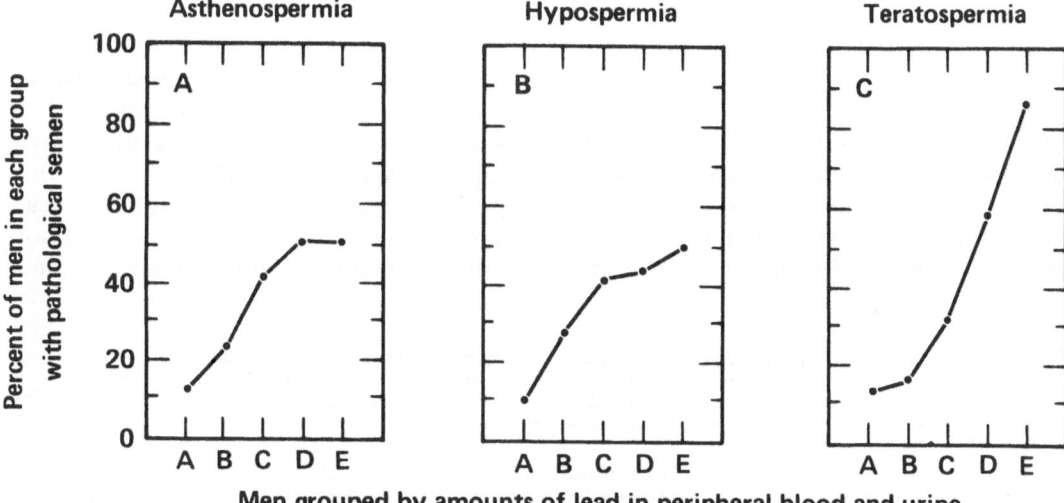

Figure 36.1. Semen abnormalities in lead workers. A total of 200 men were divided into five groups according to their exposure to lead (A to E). Group A contained the 50 men who were not exposed to occupationally toxic materials (controls). Groups B through E consisted of the 150 men who worked in a lead storage-battery plant. The 50 men in B were technicians and office workers of the plant. Groups C, D, and E included 35, 42, and 23 men, respectively, who worked in the plant for an average of 8.5 years. The average concentrations of lead in blood (μg/100 ml) and urine (μg/liter) respectively in each group were: 23 and 92 for Group B; 41 and 101 for C; 52.8 and 251 for D; and 74.5 and 385 for E. Each data point represents the percent of men in each group that showed asthenospermia (Panel A), hypospermia (Panel B), and teratospermia (Panel C). Redrawn from the data of Lancranjan *et al.* (1975).

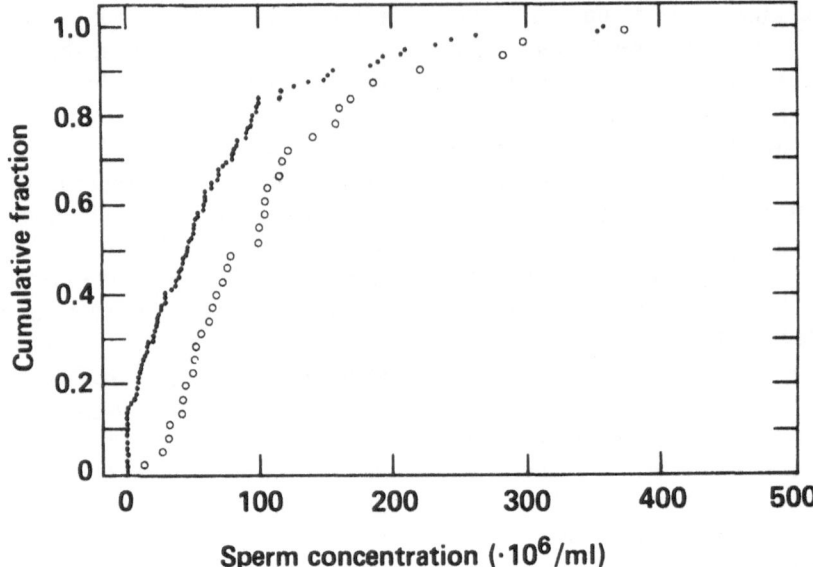

Figure 36.2. Sperm count depression in DBCP workers. The semen of 110 men working in a DBCP area of a chemical plant (closed circles) was compared with that of 35 men working in non-DBCP areas of the plant (open circles). Each data point represents the sperm count in a single semen collection from an individual male. Graph is redrawn from the data of Whorton *et al.* (1977).

mosomes in the same sperm resulting from meiotic errors in disjunction.

Relative Statistical and Biological Sensitivities

In Table 36.1, the relative statistical sensitivities of the sperm assays for counts, morphology, and two F bodies are compared. These data were obtained from the control group in one of our occupational studies (Wyrobek *et al.*, 1980b). Of the three assays, the morphology assay requires the smallest sample size to detect with confidence a 25% increase above the mean control values. However, it is important to realize that the statistical sensitivities may be independent of the relative biological sensitivities of these assays. For example, Lancranjan *et al.* (1975) found that sperm morphology was as sensitive to lead or even more sensitive than were sperm counts or motility. On the other hand, in a

Table 36.1. *The Relative Statistical Sensitivities of Three Assays of Human Sperm*

	Counts	Morphology	2F bodies
Assumed Distribution	Log-normal	Normal	Normal[a]
Mean[b]	$132 \cdot 10^6$/ml	41.9%	0.8%
Standard Deviation	$160 \cdot 10^6$/ml	12.4%	0.7%
25% change in mean (Std. Dev. Units)	0.2	0.8	0.3
Sample size for 5% level test with 80% power[c]	214	26	41

[a]Square-root transformation (Snedecor and Cochran, 1978).

[b]Based on 35 unexposed, new hires in a pesticide plant.

[c]Method of Owen (1962).

study of DBCP workers, the analyses of sperm morphology showed no DBCP-related differences, while counts showed major differences (Sandifer *et al.*, 1979). We have recently completed a human study in which sperm morphology was the only assay that yielded positive findings in a group of pesticide production workers scored for changes in sperm counts, morphology and two F bodies (Wyrobek *et al.*, 1980b). Since the data are still insufficient to predict which assay would be most sensitive to an agent, the conservative approach to assess changes in human testicular function should include assays for sperm count, morphology, two F-bodies, and, whenever practical, motility.

Guidelines for New Human Sperm Studies

These and other published studies, as well as ongoing projects in our laboratory of men exposed to mercury, DBCP, and anesthetic gases, suggest the following guidelines for an effective study of human testicular function in the workplace.

Identifying populations of men at risk of exposure to testicular toxins. Groups of men may be considered at risk if they are exposed to any agent or analogue with positive mutagenicity, carcinogenicity, teratogenicity, or testicular toxicity in any mammal. However, groups may be identified even more directly, as shown in the study of men exposed to DBCP, which was initiated by workers, complaints of unintentional childlessness (Whorton *et al.*, 1977).

The number of men exposed to the agent as well as the likelihood of high exposures are important for study design. Since between-male variability in semen characteristics among men is high, even among fertile and presumably healthy men, rather large numbers of subjects are required to establish differences between control and exposed groups in cross-sectional studies in which each individual is usually sampled only once (see Table 36.1). When samples

are not available because of vasectomies and personal beliefs, even larger numbers of men must be examined. Cross-sectional studies are, therefore, typically large, e.g., 170 in the study of smoking effects (Viczian, 1969) and 200 in the study of lead effects (Lancranjan *et al.*, 1975).

Longitudinal study designs may be more appropriate when fewer men are available for sampling. In this study design, repeated semen samples from an individual are compared to assess chemically induced sperm defects. Since variation of sperm morphology within an individual is considerably less than variation among individuals (MacLeod 1965, 1974; Sherins *et al.*, 1977), in principle, fewer people are required to detect induced changes. These studies, however, have several major constraints: (a) no precedent exists for such studies in the workplace although studies have been successfully conducted on men exposed to X-rays and men receiving drugs (MacLeod, 1965; Heller *et al.*, 1965; Rowley *et al.*, 1974); (b) repeated samplings during a period of months and perhaps years are required; (c) samples before exposure are needed (or within days of an acute exposure, which is probably insufficient time for induced semen defect to be seen); and (d) the number of men needed for an effective study is unknown.

Identifying a control group. All control groups must be selected with care. Control groups typically include (a) new hires, (b) administrative persons, (c) workers from other areas of the plant, (d) workers in another plant, and (e) historical controls. Each of these groups may have disadvantages, such as differences in age, socioeconomic factors, other chemical exposures, etc. The conservative approach seems to be to use two controls: the best available concurrent control group and a historical control.

Recruiting volunteers. Numerous methods for recruiting volunteers are possible (e.g., an announcement at general meetings, the circulation of a form letter, etc.). We have found that, regardless of the method of introduction, the best procedure includes a session with each prospective donor on matters of information, clarification, and motivation. Recruiting local, concurrent controls is usually more difficult than recruiting exposed men.

Collecting exposure, medical, and personal data from volunteers. Detailed questionnaires are needed to obtain dose information on work exposure (dates of employment, types of jobs held, types of agents used, etc.) with special emphasis on the year before semen collection. Detailed medical histories are needed because ill health and associated drug consumption may affect seminal quality (MacLeod, 1965). Personal data usually collected includes smoking and drinking habits, as well as number and dates of children, miscarriages, abortions, and birth defects.

Assigning exposed males to dosage groups. Because the dose-dependent response is an important criterion in identifying a testicular toxin, every effort should be made to group men according to dose. In general, chemical dosimetry in the workplace is inadequate, and other methods of dose-grouping are necessary. For example, Lancranjan *et al.* (1975) grouped men according to the lead concentrations in urine and peripheral blood. Men can also be grouped by the type of job held with some understanding of which jobs result in greater exposure.

Collecting semen samples. In a cross-sectional study, one semen sample per individual is required. Longitudinal studies require several samples from each man. Before collection day, each donor is given an instruction form on the importance of obtaining a complete and fresh sample as well as the accepted methods of collection (i.e., glass bottle or Mylar sheath used with either coitus interruptus or masturbation). All samples are coded at the laboratory, and all samples are analyzed as blind studies.

Analyses of semen samples. Conventional methods for semen analyses: sperm count, motility (Amelar, 1966), and mor-phology (MacLeod, 1965; Wyrobek and Bruce, 1978; Wyrobek and Gledhill, 1980) are generally simple and rapid assays, but require fresh semen. Sperm counts can be determined with a hemocytometer or Coulter counter, motility by visual observation under a light microscope (numerous automated methods are also available), and morphology from air-dried smears prepared shortly after collection. Air-dried sperm smears can be shipped and stored for extended periods of time (we have kept slides for months), fixed, stained by a Papanicolaou method (Humason, 1972), and scored for sperm morphology at a later date. We typically score 500 sperm per individual and classify each sperm into one of the ten categories shown in Figure 36.3. The criteria for morphological analyses must be carefully controlled. To ensure validity, the slides of the concurrent controls should be analyzed together with slides of exposed people. We find that uniformity of scoring criteria for morphology is best controlled by repeated comparisons to a set of standard slides at regular intervals throughout the analysis.

For the analyses of sperm with double fluorescent bodies, air-dried smears were fixed and stained with quinacrine dihydrochloride by the method of Pearson *et al.* (1970). Five hundred sperm are scored under a fluorescent microscope as OF (sperm containing no fluorescent body and presumably those with no Y chromosome), IF (those sperm presumably with one Y chromosome), and two F (sperm presumably containing two Y chromosomes).

Statistical treatment of data. The effects of age, smoking, illness, medication, and other confounding factors must be considered. Distributions of sperm data from control and exposed men can be analyzed by the t-test or by the Kolmogorov-Smirnov test as in the DBCP study of Whorton *et al.* (see Fig. 36.2) or by analyses of the proportion of men with abnormal semen as in the lead study (Lancranjan *et al.*, 1975) (see Fig. 36.1). Dose-response data should be evaluated whenever possible.

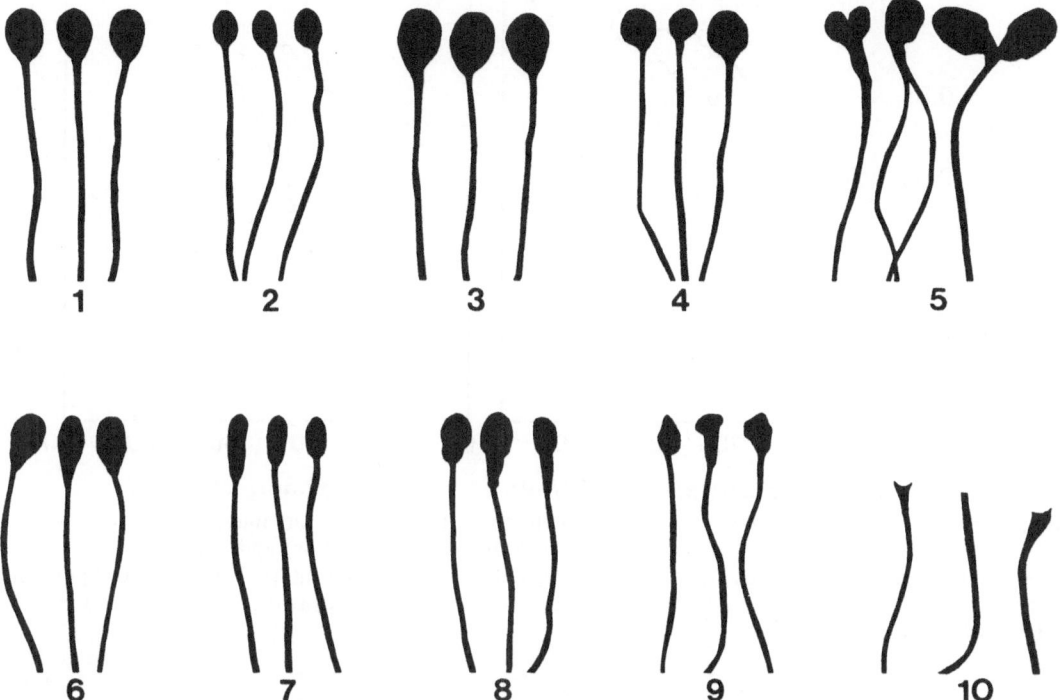

Figure 36.3.   Variations in shape of human sperm. The head shapes of the sperm in category 1 are oval and considered by us as normal. Those in categories 2 to 10 are scored as being abnormal. The sperm in category 2 are small; 3, large; 4, rounded; 5, doubles; 6, narrow at the base of the head; 7, narrow; 8, pear-shaped; 9, amorphous; and 10, ghost-like.

The Predictive Value of Induced Sperm Changes

Although it is generally agreed that major reductions in sperm counts and motility are linked to reduced fertility, it remains unclear as to which sperm parameters are indicative of embryonic failure or heritable genetic abnormalities. Human data is very limited. In one study (see Fig. 36.4), fathers in 201 spontaneous abortions showed significantly higher sperm abnormalities and lower sperm counts than 116 fathers in normal pregnancies (Furuhjelm *et al.*, 1962). Although other studies generally support this finding (Czeizel *et al.*, 1967; Joel, 1966; Takala, 1957) some studies found no correlation (e.g., Kneer, 1957). Clearly, more human studies are needed to compare exposure of the male parent, induced sperm defects, and reproductive outcome.

Most of the work on genetic validation has been done in the mouse. The mouse assay for chemically induced sperm-head defects provides three lines of evidence that link induction of abnormal and heritable genetic abnormalities. First, considerable evidence exists that sperm shaping and the production of abnormal sperm is polygenically controlled by autosomal as well as sex-linked genes (Beatty, 1972; Krzanowska, 1972; Brozek, 1970; Hugenholtz and Bruce, 1979; Wyrobek, 1979). Therefore, any agent that affects sperm shaping interferes with a system that has strict genetic controls. Second, in three independent studies with numerous mutagens and nonmutagens, it was found that germ-cell mutagens generally induced sperm abnormalities, while nonmutagens generally had no effect (Wyrobek and Bruce, 1975, 1978; Bruce and Heddle, 1979; Topham, 1979). Third, in several compelling studies

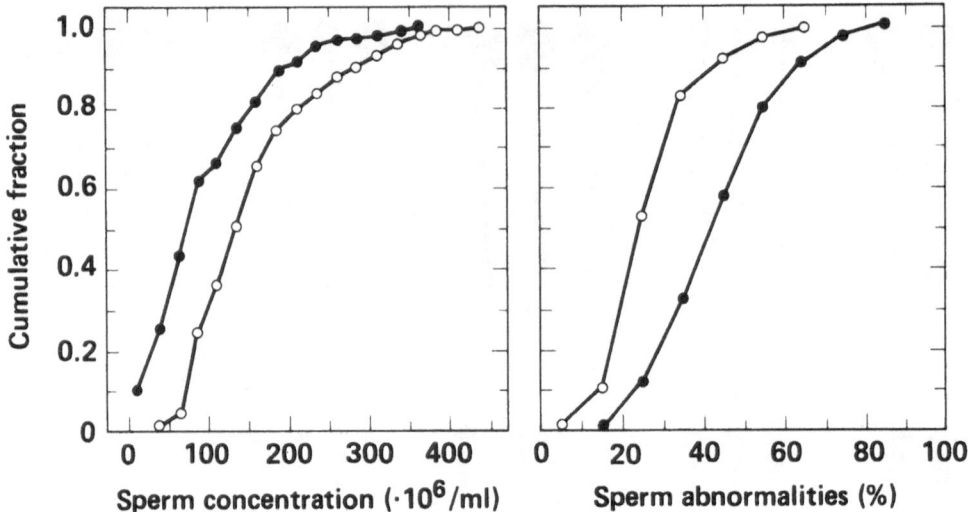

Figure 36.4. Sperm counts and sperm abnormalities in fathers in spontaneous abortions. Semen was collected from 201 fathers in spontaneous abortions (closed circles) and 116 fathers in normal pregnancies. Each point represents a grouping of males in increasing intervals of $25 \cdot 10^6$ sperm/ml (left panel) and 10% sperm abnormalities (right panel). Graphs are drawn from the data of Furuhjelm *et al.* (1962).

with agents that induce sperm abnormalities in exposed males, sperm abnormalities were transmitted to the male offspring of the exposed mice (Wyrobek and Bruce, 1978; Hugenholtz and Bruce, 1979; Topham, 1979; Staub and Matter, 1976/1977; Sotomayor, 1979). Further murine studies are needed to better understand the quantitative relationships among (a) dosage regime, (b) appearance of abnormal sperm shapes in the semen, (c) time between exposure and conception, (d) the frequency of offspring with sperm defects, and (e) the fertility of the abnormal offspring.

Improvement and Development of Sperm Assays

Two human sperm assays are realizing rapid advances in automation: motility (e.g., Overstreet *et al.*, 1979) and morphology (e.g., Schoevaert-Brossault, 1980). In principle, automation not only adds speed, but improves objectivity, provides more accurate quantitation and when used in conjunction with slide-based scanning systems

often yields increased sensitivity to small change. Assays for changes in DNA fluorescence using cell-orienting flow systems are already effective for mouse sperm (Van Dilla *et al.*, 1980). Their application to man is a topic of much current research interest (Wyrobek and Gledhill, 1980). Other new sperm assays are needed, especially those that can quantitate mutational changes in individual sperm.

Methods for Sperm Assays in Mice

Methods for the Sperm Morphology Assay

Of the sperm parameters used in the human studies, sperm morphology is the one best supported by a murine model system. The sperm morphology assay in mice has been used in a number of laboratories to screen agents for their ability to induce testicular damage (Wyrobek and London, 1973; Wyrobek and Bruce, 1975, 1978; Topham, 1979; Bruce and Heddle, 1979; Soares *et al.*, 1979; Matter *et al.*, 1979; Land *et al*, 1979). The methodology of the assay is simple (see Fig. 36.5) and has been well described elsewhere (Wyrobek and Bruce, 1975, 1978). The following is a protocol that can be used when testing unknown solids or liquids.

1. The solubility limits of each compound are determined for various solvents: usually in water or dimethylsulfoxide and occasionally in saline or corn oil. The solvent with the highest solubility limit is usually the one best suited for the study.

2. The lethal dose-50 (LD_{50}) is determined by single intraperitoneal injection. At least five mice are exposed at each of six dosage levels ranging from a 5-day total dose of 4 times LD_{50} with each subsequent dosage level lower by a factor of 2.

3. Mice are injected daily for 5 days with the agents diluted in small volumes of vehicle (typically 0.1 ml). Unless solvent-specific stability data is available, dilutions should be prepared fresh daily. Mice exposed to vehicle only are used as negative controls and methylmethanesulfonate-treated mice as positive controls.

4. At 5 weeks from the first injection, mice are sacrificed, epididymides excised, sperm suspensions prepared, and the average number of sperm per epididymis determined. We typically also measure the testicular wet weight for each mouse.

5. Sperm aliquots are stained with eosin-Y (0.1% aqueous), smears prepared, and 500 sperm per mouse scored for morphological abnormalities according to the criteria of Wyrobek and Bruce (1978).

6. Agents are scored as positive inducers of abnormal sperm if two consecutive dosages show statistically significant elevations in sperm abnormalities and as negatives if no increase in sperm abnormalities is seen up to dosages that cause

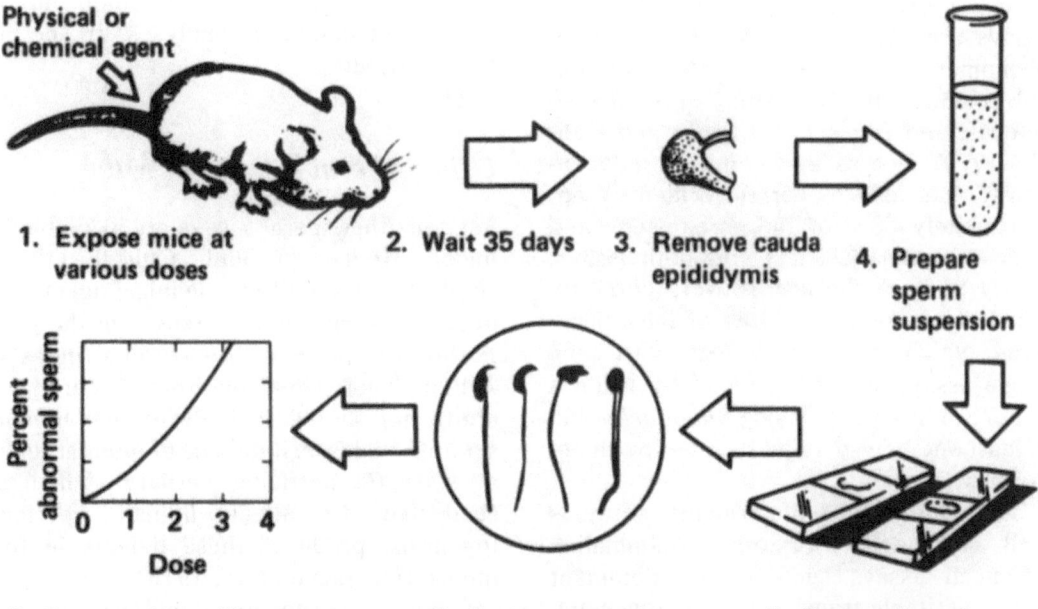

Figure 36.5. Methodology for the mouse sperm abnormality assay. Key steps in the assay are noted. Refer to text and cited references for details.

whole-animal lethality. Results not fulfilling these criteria demand retesting. Small adaptations in this protocol would make it suitable for inhalation studies and studies with different routes of exposure.

Correlations of Sperm Abnormality Results with Other Short-Term Assays for Carcinogenesis and Mutagenesis

As part of the international program for the evaluation of short-term tests for carcinogenesis, six carcinogen/noncarcinogen pairs and five unpaired carcinogens were surveyed as unknowns on the mouse sperm morphology assay. No false positive responses were found, although a high rate of false negative results was suggestive that not all carcinogens induce sperm abnormalities (Wyrobek et al., 1980). A comparison of the potency (dose required to double background frequencies) of some 30 agents studied in both the Salmonella/microsome and sperm abnormality assays shows no apparent correlation, suggesting that the assays are measuring different biological phenomena (Wyrobek and Holstein, unpublished results). In a study of 61 agents, (Heddle and Bruce 1977; Bruce and Heddle, 1979) each assay when used independently was able to correctly identify approximately 60% of the carcinogens and noncarcinogens tested. Using both assays, nearly 90% of the agents were correctly identified. Thus the number of false negatives, which is relatively high with each single assay, may be reduced by using a battery of assays including Salmonella and at least one in vivo assay such as sperm abnormality.

The sperm-shape abnormality assay is well correlated with other mammalian germ-cell assays such as the dominant lethal, heritable translocation, cytogenetic, and specific locus tests (Wyrobek and Bruce, 1975; Topham, 1979). Since this sperm assay is much easier to use than any of the other mammalian germ-cell assays, it is best suited as a primary screen for agents that affect mammalian testicular function and are potential in vivo germ-cell mutagens.

Improving the Sensitivity of the Morphology Assay

The human eye is very effective in detecting subtle shape differences but is very poor in discriminating sizes. For example, size changes induced by ionizing radiation were readily detected photographically yet were imperceptible to the scorer (Wyrobek et al., 1980b). By making four specific measurements of sperm-head dimensions we were able to assign to each sperm a statistical difference (Mahalanobis's distance) from the control mean (Moore et al., 1980). This method requires that only 50 sperm be scored, yielding a doubling dose of approximately 10 rads in comparison to the doubling dose of approximately 60 rads when 500 sperm per mouse were visually scored. Automated scoring can be easily applied to this method, using a scanning light microscope and computer analyses of sperm-head contour.

Other Sperm Assays in Mice

Several other sperm assays are available in mice. Moutschen and Colizzi (1975) showed that numerous chemical agents induce dose-dependent increases in the proportion of sperm with acrosomal defects. Although acrosomal abnormalities are generally not scored in conventional human seminal analyses, it may be of interest to do so since the pesticide, carbaryl, which is suspected to affect human spermatogenesis, produces these defects in the mouse (Degraeve et al., 1976).

Chemical exposure can also reduce sperm count in mice. For example, Lu and Meistrich (1979) have used sperm-head counts in the testes to quantify germ-cell toxicity and stem-cell survival in mice exposed to cancer chemotherapeutic drugs.

F_1 sperm assays are the most direct way to assess the heritable consequences of chemically induced sperm defects. Briefly, male mice are exposed to an agent, mated during their presterile period to unexposed females, and, when mature, the F_1 male offspring scored for sperm abnormalities. The final score obtained with this assay is the fraction of affected descendents, i.e., the fraction of male progeny with elevated sperm abnormalities. Studies with X-rays (Wyrobek and Bruce, 1978; Hugenholtz and Bruce, 1979), lead, methylmethanesulfonate (Wyrobek and Bruce, 1978), triethylenemelamine (Staub and Matter, 1976/1977), cyclophosphamide (Sotomayor, 1979), and ethylmethanesulfonate (Sotomayor, 1979; Topham, 1979) show that when mice are exposed to these agents, and then mated to normal females, increased proportions of offspring with defective sperm shaping are seen. The F_1 sperm-abnormality assay is not limited to studying sons of exposed males. Topham (1979) used METEPA, to show heritable abnormalities in the F_1 male progeny of treated female mice.

The availability of F_0 and F_1 mouse sperm assays provides a means of assessing the time and dose relationships between exposure of the adult male, the appearance of sperm defects in the semen, and the likelihood of heritable consequences. The combined use of the different assays in mice also provides an approach to assess the fertility changes and heritable consequences that may be associated with chemically induced changes in human sperm.

Conclusions

Human sperm assays in the context of an epidemiological study provide an effective means to identify agents that affect testicular function. Four assays for human sperm anomaly (counts, motility, morphology, and double-fluorescence bodies) have been successfully used for a variety of human occupational and drug-related exposures. However, more research is needed to validate the genetic implications of sperm defects induced by exposure to chemicals *in vivo*. The F_0 and F_1 mouse-sperm assay, especially for induced morphological defects, provides a method of screening agents that may be potentially harmful to man and provides a means to assess the antifertility effects and consequences of induced sperm defect.

Acknowledgments

Special thanks to G. Watchmaker and L. Gordon whose technical accomplishments made this review possible. The author is very grateful for the incisive and helpful comments by M. Mendelsohn, B. Ishida, and B. Gledhill in the preparation of this manuscript.

References

Amelar, R.D. (1966): Infertility in Men: Diagnosis and Treatment. Chapter II: The semen analysis. Philadelphia: F.A. Davis Co., pp. 30–53.

Beatty, R.A. (1972): The genetics of size and shape of spermatozoon organelles. In: R.A. Beatty and S. Gluecksohn-Waelsch (eds.), The Genetics of the Spermatozoon: Proceedings of an International Symposium, Department of Genetics, University of Edinburgh, pp. 97–115.

Brozek, C. (1970): Proportion of morphologically abnormal spermatozoa in two inbred strains of mice, their reciprocal F_1 and F_2 crosses and backcrosses. Acta Biol. Cracov. (Ser. Zool.) *13*:189–198.

Bruce, W.R., Heddle, J.A. (1979): The mutagenic activity of 61 agents as determined by the micronucleus, Salmonella, and sperm abnormality assays. Can. J. Genet. Cytol. *21*:319–334.

Czeizel, E., Hancsok, M., Viczian, M. (1967):

418   A.J. Wyrobek

Examination of the semen of husbands of habitually aborting women. Orvosi Hetilap, *108*:1591–1595.

Degraeve, N., Moutschen-Damen, J., Moutschen-Damen, M., Houbrechts, N., Colizzi, A. (1976): A propos des risquen d'un insecticide: le carbaryl utilisé seul et en combination avec les nitrites. Bull. Soc. Roy. Scs. Liège *45*:46–57.

Farris, E. J. (1959): Human Fertility and Problems of the Male. White Plains, New York: Author's Press.

Furuhjelm, M., Jonson, B., Lagergren, C.G. (1962): The quality of human semen in spontaneous abortion. Int. J. Fertil. *7*:17–21.

Heddle, J.A., Bruce, W.R. (1977): Comparison of tests for mutagenicity or carcinogenicity using assays for sperm abnormalities, formation of micronuclei and mutations in *Salmonella*. In: H.H. Hyatt, J.D. Watson and J.A. Winsten (eds.), Origins of Human Cancer, Book C, Cold Spring Harbor Conferences on Cell Proliferation, Vol. 4. New York: Cold Spring Harbor Laboratory, pp. 1549–1557.

Heller, C.G., Wootton, P., Rowley, M.J., Lalli, M.F., Brusca, D.R. (1965): Action of radiation upon human spermatogenesis. Proc. 6th Pan-American Congress of Endocrinology: Excerpta Med. Int. Congr. Ser. No. 112, pp. 408–410.

Hugenholtz, A.P., Bruce, W.R. (1979): Radiation-induced heritable sperm abnormalities in mice. Environ. Mutag. *1*:127–128.

Humason, G.L. (1972): Animal Tissue Techniques. San Francisco: W.H. Freeman and Co., pp. 456–457.

Joel, C.A. (1966): New etiologic aspects of habitual abortion and infertility, with special reference to the male factor. Fertil. Steril. *3*:374–380.

Kapp, R.W., Picciano, D., Jacobson, C.B. (1979): Y-Chromosomal nondisjunction in dibromochloropropane-exposed workmen. Mutat. Res. *64*:47–51.

Kneer, M. (1957): Der habituelle Abort. Dtsch, Med. Wochenschr., *82*:1059–1060.

Krzanowska, H. (1972): Influence of Y chromosome on fertility in mice. In: R.A. Beatty and S. Gluecksohn-Waelsch (eds.), The Genetics of the Spermatozoon: Proceedings of an International Symposium, Department of Genetics, University of Edinburgh, pp. 370–386.

Lagerlöf, N. (1934): Morphologische Untersuchungen über Veränderungen im Spermabild und in den Hoden bei Bullen mit ver-minderter oder aufgehobener Fertilität. Acta: Path. Microbiol. Scand. Suppl. *19*:1–254.

Lancranjan, I., Popescu, H.I., Gavanescu, O., Klepsch, I. Serbanescu, M. (1975): Reproductive ability of workmen occupationally exposed to lead. Arch. Environ. Health *30*:396–401.

Land, P.C., Owen, E.L., Linde, H.W. (1979): Mouse sperm morphology following exposure to anesthetics during early spermatogenesis. Anesthesiology *50*:S259.

Lu, C.C., Meistrich, M.L. (1979): Cytotoxic effect of chemotherapeutic drugs on mouse testes cells. Cancer Res. *39*:3575–3582.

MacLeod, J. (1965): Human seminal cytology following the administration of certain anti-spermatogenic compounds. In: C.R. Austin and J.S. Perry (eds.), A Symposium on Agents Affecting Fertility. Boston: Little Brown and Co., pp. 93–123.

MacLeod, J. (1974): Effects of environmental factors and of antispermatogenic compounds on the human testis as reflected in seminal cytology. In: R.E. Mancini and L. Martini (eds.), Male Fertility and Sterility: Proceedings of the Sereno Symposia, Vol. 5. New York: Academic Press, pp. 123–248.

Matter, B.E., Jaeger, I., Suter, W., Tsuchimoto, T., Deyssenroth, H. (1979): Action of an anti-spermatogenic, but non-mutagenic indenopyridine derivative in mice and *Salmonella typhimurium*. Mutat. Res. *66*:113–127.

Moore, D.H., II, Bennett, D.E., Wyrobek, A.J., Kranzler, D. (1980): Mahalanobis, distance and variable selection to optimize dose response. Biometrics (submitted for publication).

Moutschen, J., Colizzi, A. (1975): Absence of acrosome: an efficient tool in mammalian mutation research. Mutat. Res. *30*:267–272.

Overstreet, J., Katz, D., Hanson, F., Fouseca, J. (1979): A simple, inexpensive method for objective assessment of human sperm movement characteristics. Fertil. Steril. *31*:162–172.

Owen, D.B. (1962): Handbook of Statistical Tables. Palo Alto, Ca: Addison-Wesley Publishing Co., p. 42.

Pearson, P.L., Bobrow, M., Vosa, C.G. (1970): Techniques for identifying Y chromosomes in human interphase nucleus. Nature (Lond.) *226*:78–80.

Rowley, M.J., Leach, D.R. Warner, G.A.,

Heller, C.G. (1947): Effects of graded doses of ionizing radiation on the human testis. Radiat. Res. *59*:665–678.

Sandifer, S.H., Wilkens, R.T., Loadholt, C.B., Lane, L.G., Eldridge, J.C. (1979): Spermatogenesis in agricultural workers exposed to dibromochloropropane (DBCP). Bull. Environ. Contam. Toxicol. *423*:703–710.

Schoevaert-Brossault, D. (1980): Analyse morphologique et classification automatique de spermatoziodes humains.

Sherins, R.J., Brightwell, D., Sternthal, P.H. (1977): Longitudinal analysis of semen of fertile and infertile men. In: P. Troen and H.R. Nankin (eds.), The Testis in Normal and Infertile Men. New York: Raven Press, pp. 473–488.

Snedecor, G.W., Cochran, W.G. (1978): Statistical Methods. Ames, Iowa: Iowa State University Press, pp. 325–327.

Soares, E.R., Sheridan, W., Haseman, J.K., Segall, M. (1979): Increased frequencies of aberrant sperm as indicators of mutagenic damage in mice. Mutat. Res. *64*:27–35.

Sotomayor, R.E. (1979): Spermatid head abnormalities in translocation heterozygotes from EMS- or CPA-treated sires. Environ. Mutag. *1*:129.

Staub, J.E., Matter, B.E. (1976/1977): Heritable reciprocal translocations and sperm abnormalities in the F_1 offspring male mice treated with triethylenemelamine (TEM). Arch. Genet. *49*/150:29–41.

Takala, M.E. (1957): Studies on the seminal fluid of fathers of congenitally malformed children (199 sperm analyses). Acta Obstet. Gynec. Scand. *36*:29–41.

Topham, J. (1979): A sensitive hybrid for the detection of sperm head abnormalities and its potential for the detection of transmissible mutations in mice. Environ. Mutag. *1*:126–127.

Van Dilla, M.A., Pinkel, D., Gledhill, B.L.,

Lake, S., Watchmaker, G., Wyrobek, A.J. (1980): Flow cytometry of mammalian sperm: progress report. Acta Path. Microbiol. Scand.

Viczian , M. (1969): Ergebnisse von Spermauntersuchungen bei Zigarettenrauchern. Z. Haut Geschlechtskr. *44*:183–187.

Whorton, D., Krauss, R.M., Marshall, S., Milby, T.H. (1977): Infertility in male pesticide workers. Lancet *2*:1259–1261.

Wyrobek, A.J. (1979): Changes in mammalian sperm morphology after X-ray and chemical exposures. Genetics (Suppl.) *92*:s105–s119.

Wyrobek, A.J., Bruce, W.R. (1975): Chemical induction of sperm abnormalities in mice. Proc. Natl. Acad. Sci. USA *72*:4425–4429.

Wyrobek, A.J., Bruce, W.R. (1978): The induction of sperm-shape abnormalities in mice and humans. In: A. Hollaender, F.J. de Serres (eds.), Chemical Mutagens: Principles and Methods for Their Detection, Vol. 5. New York: Plenum Press, pp. 257–285.

Wyrobek, A.J., Gledhill, B.L. (1980): Human semen assays for workplace monitoring. In: P. Infante, M. Legator (eds.), Proceedings of the Workshop on Methodology for Assessing Reproductive Hazards in the Workplace, Washington D. C.: National Institute for Occupational Safety and Health.

Wyrobek, A.J., London, S.A. (1973): Effect of hydrazines on mouse sperm cells. In: Proc. 4th Annual Conference on Environmental Toxicology, Springfield, VA, National Technical Information Service, pp. 417–432.

Wyrobek, A.J., Gordon, L., Watchmaker, G. (1980a): The effects of 17 chemical agents including 6 carcinogen/non-carcinogen pairs on sperm shape abnormalities in mice. In: International Programme for the Evaluation of Short-Term Tests for Carcinogenicity; Amsterdam: Elsevier.

Wyrobek, A.J. *et al.* (1980b): Environ. Health Perspect. (submitted for publication).

37

In vitro *Assay for Tumor Promoters**

James E. Trosko, Larry P. Yotti, Betty Dawson, and
Chia-Cheng Chang

Principle of Method

Any screening assay system designed to
detect potential human mutagens, carcinogens, and teratogens is based on some
fundamental assumptions concerning the
mechanisms of mutagenesis, carcinogenesis, or teratogenesis (Chang *et al.*,
1978). For example, it is assumed by those
who use the Ames assay to detect mutagens that not only will the detected mutagens in a bacterial system be relevant to
human mutagenesis but also that mutations
contribute to genetic or heritable diseases,
as well as to carcinogenesis. Evidence that
mutations play a role in carcinogenesis
seems to be overwhelming (see review,
Trosko and Chang, 1978).

However, it does seem fair to state that,
where mutagenesis does play a role in car-

cinogenesis, it is a necessary, but only partial, step in the complex carcinogenic process. There is growing evidence that carcinogenesis in many organ systems in several species, including *Homo sapiens*, involves several distinguishable stages,
namely *initiation* and *promotion* (see Slaga
et al., 1978). Initiation seems to be the
result of an irreversible event which happens to a cell after its exposure to physical,
chemical, or viral agents that are known to
damage or change the DNA molecules
(Trosko and Chang, 1980a). It has been
postulated that mutagenesis is the consequence of this DNA damage and that
mutations in a few critical genes are responsible for the initiation phase (Boutwell,
1974; Trosko *et al.*, 1977; O'Brien and
Diamond, 1978).

On the other hand, promotion appears to
be a "reversible" process (up to a point),
which depends on repeated treatment of the
initiated cell by physical or chemical agents
that are, by themselves, non- or weak carcinogenic initiators (Boutwell, 1974; Frei,

*Research was supported by grants from the Na-
tional Cancer Institute to J.E. Trosko (CA21104) and
from the National Institute of Environmental Health
Sciences to C-C. Chang (Young Environmental Sci-
entist Award, ES01809).

1976; Slaga *et al.*, 1978; Diamond *et al.*, 1978). Moreover, the manner by which an "initiated" cell can be promoted seems to include exogenous promoters, wounding, cytotoxicity of an initiating agent, hepatectomy, and growth stimuli (Frei, 1976; Trosko and Chang, 1980b) (Fig. 37.1). The list of chemicals that are known or suspected promoters is expanding—e.g., phorbol esters, bile salts, butylated hydroxytoluene, phenobarbital, and saccharin (Sivak, 1978; Slaga *et al.*, 1978).

Consequently, it is now evident that environmental monitoring of carcinogenic promoters will be as important as it is for that of carcinogenic initiators. At present, it is suspected that initiators are, in fact, mutagens (Trosko and Chang, 1978) and tumor promoters are nonmutagenic gene modulators (Chang *et al.*, 1978). Since it is very clear that all carcinogenic initiators do not act the same way in all tissues in all organisms (e.g., X-rays, UV light, direct or metabolized chemicals), the practical task of identifying initiators with *one* screening procedure seems to be an impossible task.

The common unifying feature of initiators seems to be their ability to induce a variety of DNA damages which can act as substrates for *mutation fixation*, even though the initiator's mode of damaging DNA and their fixation into a mutation may be very different. It would not be unreasonable, *a priori*, to suspect that carcinogenic promoters would act, *epigenetically,* via a variety of ways to induce changes in gene expression (i.e., gene modulators).

A general property of tumor promoters seems to be the fact that they are nonmutagens and they do not inhibit the repair of DNA damaged by other agents (see review, Trosko and Chang, 1978). They do seem to alter the phenotype of cells, block terminal differentiation in some cells, but induce terminal differentiation in other cells, stimulate protein, RNA, DNA, phospholipid, and prostaglandin synthesis in some systems, phosphorylate histones, affect cell surface glycoproteins, modify membrane transport of 2-deoxyglucose, induce plasminogen activator, ornithine decarboxylase, and latent viruses, and enhance the *in vitro* transformation of cells previously treated with physical or chemical carcinogens (see reviews, Trosko and Chang, 1980a,b).

Since the specific unifying molecular basis (if there is one) is unknown, the design of screening systems for detecting tumor promoters will be a task that may be more difficult than the design of mutation or transformation screening systems.

Several years ago, we reported that 12-0-tetradecanoyl-phorbol-13-acetate (TPA), a powerful mouse skin tumor promoter, enhanced the recovery of UV-induced hypoxanthine guanine-phosphoribosyl-transferase–negative mutants (HG-PRT$^-$) and ouabain-resistant mutants (oua^r) in Chinese hamster cells (Trosko *et al.*, 1977). This was later confirmed by Lankas *et al.* (1977, 1978) after the cells were initiated by several carcinogens. In order to test that interpretation, we uncovered an interesting observation which confirmed our original observation (i.e., TPA enhanced the recovery of mutated HG-PRT$^-$ and oua^r cells), but which negated the specific form of our original interpretation.

The observation to which we now refer is the fact that TPA seems to inhibit "metabolic cooperation," known to play an important role in the recovery of mutant HG-PRT$^-$ cells in the 6-thioguanine or 8-azaguanine selection system (van Zeeland *et al.*, 1972). In brief, when a HG-PRT$^-$ cell is induced in a population of wild-type HG-PRT$^+$ cells, the wild-type cells can transport the toxic metabolite (phosphorylated 6-thioguanine) formed in the presence of the HG-PRT$^+$ enzyme and 6-thioguanine to the HG-PRT$^-$ cells, consequently killing some mutant cells. We have recently demonstrated that TPA blocks this "metabolic cooperation," thereby allowing the "rescue" of the HG-PRT$^-$ cells (Yotti *et al.*, 1979). Therefore, based on the assumption that TPA-like promoters might act by blocking "metabolic coopera-

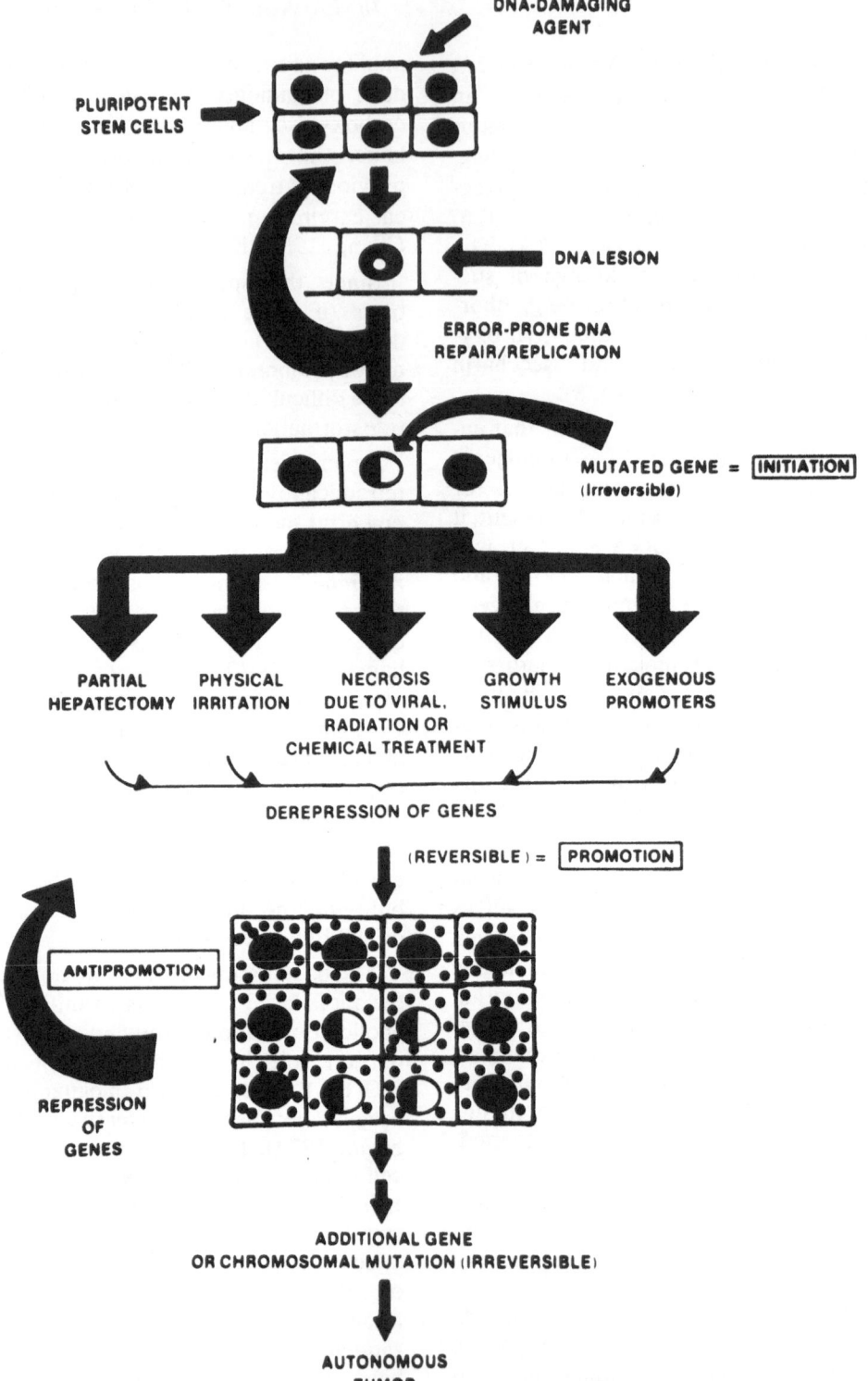

Figure 37.1. A diagrammatic heuristic scheme to depict the postulated mechanisms of the initiation and promotion phases of carcinogenesis. DNA damages, induced by physical or chemical mutagens, are substrates which can be fixed if they are not removed in an error-free manner prior to DNA replication. Promotion includes those conditions (wounding, cytotoxicity) in which a pluripotent but surviving initiated cell can escape the nonproliferative state. The build-up of initiated cells allows them to "resist" the antimitotic influence of neighboring noninitiated cells. This, together with a second mutation, might allow a given cell the autonomous, invasive properties of a malignant cell.

tion," we have designed a simple test system to pick up tumor promoters.

Test Organism

List and nomenclature of organism. This assay utilizes the Chinese hamster cell line (V79). It is an aneuploid cell line, derived originally from the lung of a male Chinese hamster (*Cricetulus griseus*, 2n=22) (Ford and Yerganian, 1958). HG-PRT⁻ mutants were isolated and cloned from spontaneous mutations.

Culture techniques. Chinese hamster V79 cells, (HG-PRT⁺ and HG-PRT⁻) are grown in modified Eagle's medium (Eagle, 1959) (Earle's balanced salt solution with a 50% increase in all vitamins and essential amino acids except glutamine), supplemented with nonessential amino acids (100% increase), 1mM sodium pyruvate, and 5% fetal calf serum. Under the incubation condition with 5% CO_2 in humidified air at 37° C, cells not contact-inhibited have a generation time of about 12 hours. Cells needed for assay are grown in large plastic flasks and removed from flasks using 0.01% crystalline trypsin (Sigma Chemical Company, St. Louis, Mo.) in phosphate-buffered saline without calcium and magnesium ions.

Equipment and chemicals. Standard and routine tissue culture facilities and supplies are needed for this assay. Several bench-top, water-jacketed CO_2 incubators are needed to house several assays. A biohazard hood is recommended for handling cultures with known or suspected tumor promoters. Adequate disposable procedures and facilities also should be available for discarding media and plates exposed to the promoters. 6-Thioguanine is prepared by dissolving 2-amino-6-mercaptopurine (Sigma, No. A-4882) in double-distilled water at a final concentration of 10 μg/ml. Known or suspected tumor promoters are dissolved in appropriate solvents and delivered to the tissue culture plates at noncytotoxic concentrations. All solutions are delivered in microliter quantities utilizing Eppendorf pipets.

Testing Protocol

Protocol. All cells used in the assay were originally derived from V79 Chinese hamster lung cells. Wild-type cells (6-thioguanine-sensitive, presumptive HG-PRT⁺) are seeded in 100-mm tissue culture dishes at a density of 9×10^5 cells per dish (Fig. 37.2). Immediately after plating of wild-type cells, 100 6-thioguanine–resistant cells (HG-PRT⁻) are seeded in the same dishes. Both cell lines are given four hours for attachment, at which time the tumor promoter is added directly into each individual dish. 6-Thioguanine is added immediately after treatment with the tumor promoter. The cells are grown for three to four days without interruption, at which time the medium is changed. The tumor promoter is removed and growth is continued in the selective medium. Four more days of growth result in colonies of a size sufficient to score visually. The medium is decanted, each dish is rinsed with 0.85% saline; the plates are air-dried, fixed with 95% ethanol, and stained with Giemsa. The resulting colonies are scored visually.

Controls. For each treatment six control plates are used to determine the colony-forming ability of 100 6-thioguanine–resistant cells grown in the presence of the drug under consideration and 6-thioguanine. The control plates are handled identically to the treatment plates with the exclusion of the 6-thioguanine–sensitive cells. The data is calculated upon the percentage of recovery of 6-thioguanine–resistant cells, i.e., the total number of colonies scored per number of colonies expected, as determined by the control plates. Moreover, when testing unknown but presumptive tumor promoters, both TPA (a positive control) and phorbol (a negative control) ought to be included as an internal check of the experiment.

Statistical considerations. Statistical significance is determined by the Student-Newman-Keuls' test, a multiple-range test for judging the significance of a set of differences (see Steel and Torrie, 1960). This is calculated as follows: each individual

In Vitro Tumor Promotion Assay

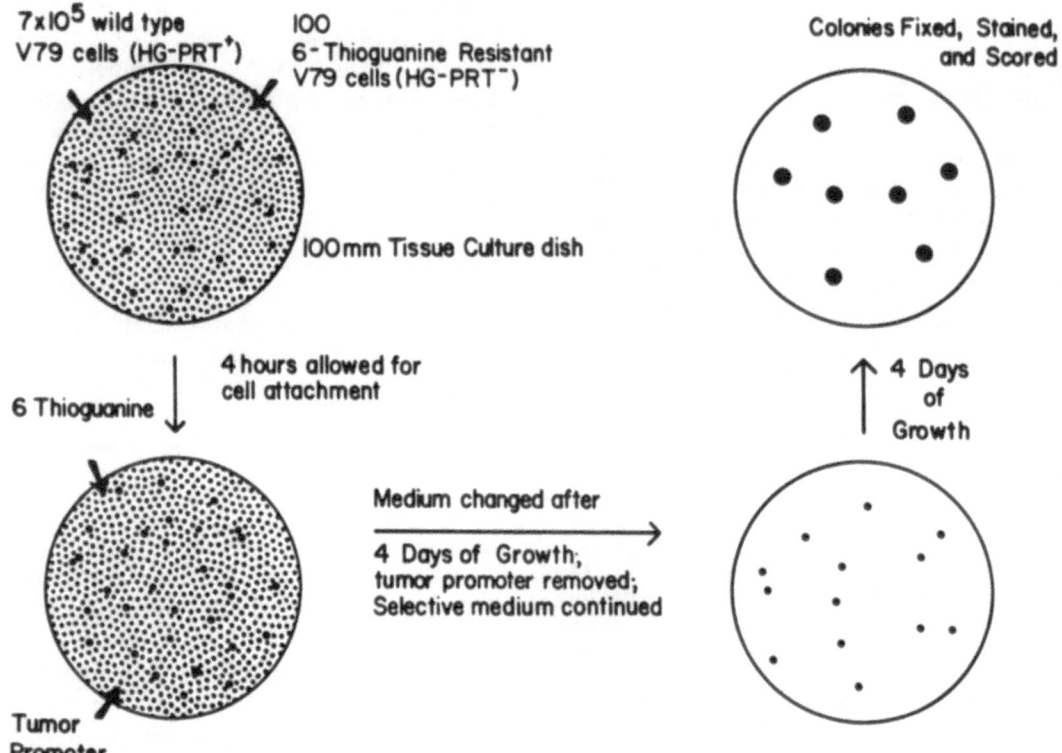

Figure 37.2. A diagrammatic representation of the *in vitro* promotion assay protocol (see text for details).

plate is converted to a percentage based upon the number of colonies expected in that plate in the absence of 6-thioguanine–sensitive cells. These values are determined for all plates in all treatment groups (usually 15–21 plates per treatment). An analysis of variance is performed to determine the error mean square for the experiment. Using the value for the error mean square, the appropriate SNK value is generated. Multiplication of this value by the appropriate studentized range value at the 0.01 level of significance results in a series of values that can then be rank-ordered from lowest to highest. If the difference between the percentage or recovery of 6-thioguanine–resistant cells for any two treatment groups exceeds the calculated value, statistical significance is assumed.

Correlation between Test Results and Carcinogenicity

Although this assay has been developed only recently, we have already tested several major classes of tumor promoters. The chemicals tested have been shown to promote initiated cells of a wide range of tissues in several organisms—i.e., saccharin as bladder promoter in rats (Cohen *et al.*, 1979); phenobarbital, DDT, and butylated hydroxytoluene as promoters of rat liver tumors (Peraino *et al.*, 1975, 1977); TPA (Boutwell, 1974), bile salts as promoters of colon tumors (Reddy *et al.*, 1978), and Tween 60 as a skin tumor promoter in mouse skin (Boutwell, 1978). Those chemicals which have been shown not to be promoters—phorbol in mice skin, cholesterol in germ-free mice (Reddy and Wa-

Table 37.1. List of Chemicals Tested in the "Metabolic Cooperation" in vitro Chinese Hamster V79 Assay

Compound	Relative *in vitro* response	Reported *in vivo* response as a promoter
TPA	+++	+++
Phorbol 12,13-Didecanoate (PDD)	++	++
4α-Phorbol 12, 13-Didecanoate (4α-PDD)	−	−
4-0-Methyl Phorbol-12-Myristate-13-Acetate (MPMA)	+	+
Phorbol 12,13-Diacetate (PDA)	+	+
Phorbol 12,13-Dibutyrate (PDB)	++	++
Phorbol	−	−
Phenobarbital	++	++
Butylated Hydroxytoluene	++	++
Mezerein	+++	(See Kensler and Mueller, 1978.)
Melittin	+	?
DDT	++	+
Lindane	+	?
Chlordane	−	?
Anthralin	+	+
Saccharin (Pure)	+	+
Saccharin (Impure)	+	+
Tween 60	++	+
Tween 80	+	?
Cholesterol	−	(In germ-free animals, see Reddy and Watanabe, 1979.)
Deoxycholic Acid	++	+
Lithocholic Acid	++	+
Taurodeoxycholic Acid	−	?
Cytocalasin B	+	?
Cytocalasin D	−	?

tanabe, 1979)—proved to be negative in this assay. Chemicals such as retinoic acid, butyric acid, dibutyryl-cAMP, and DMSO, which have been shown to be antipromoters or which block the promoting activity of TPA (Chang *et al.*, 1978), do not seem to affect the basic phenomenon on which this assay depends (i.e., the mechanism affecting metabolic cooperation between cells). Several chemicals which showed up as positives have either been shown to negate TPA promoting activity *in vivo*—i.e., antipain (Troll *et al.*, 1970)—or have not yet been shown to be a promoter in any system (Melittin: bee's venom). Antipain in our system acts as though it could be a promoter. There is

some *in vitro* (Kuroki and Drevon, 1979; Goetz *et al.*, 1972) evidence that antipain is working at the membrane level and, therefore, could be a promoter. Melittin has also been shown in other *in vitro* systems to have general properties of other promoters (Mufson *et al.*, 1979).

The major question at this point is, "Are these positive results in an *in vitro* system, using Chinese hamster cells, relevant to the human, *in vivo*, situation?" Since there is no "experimental" human data, little epidemiological data, and no comparable data from an *in vitro* human test system, the extrapolation to human beings at this time must, of course, be done with a great deal of caution. Moreover, since some of these chemical promoters seem to be organ-specific in animal systems (i.e., DDT, phenobarbital, butylated hydroxytoluene, saccharin) or cell-type–specific in some *in vitro* systems (O'Brien and Diamond, 1978), either the distribution within the organism, tissue-specific metabolism or cell membrane structure ("receptors") might determine the promoting or nonpromoting characteristics of a chemical.

Development of a human *in vitro* system, utilizing Lesch-Nyhan (HG-PRT⁻) cells as the test cells, might help to determine if there is or is not a difference in response to promoters by species. Moreover, development of an epithelial *in vitro* cell assay system should be accomplished to determine if there might be a differential response to promoters based on cell type (i.e., fibroblast versus epithelial).

Chemicals Tested

At the time of this writing, Table 37.1 includes those chemicals known or suspected as promoters in other test systems. Not shown in the table are several antipromoters and solvent carriers for the promoters which were negative in this assay [dimethylsulfoxide; ethanol; dibutyryl-cyclic AMP; butyric acid; and retinoic acid].

References

Boutwell, R.K. (1974): The function and mechanism of promoters of carcinogenesis. CRC Crit. Rev. Toxicol. *2*:419–443.

Boutwell, R.K. (1978): Biochemical mechanism of tumor promotion. In: T.J. Slaga, A. Sivak and R.K. Boutwell (eds.), Carcinogenesis: Mechanisms of Tumor Promotion and Cocarcinogenesis, Vol. 2. New York: Raven Press, pp. 49–58.

Chang, C.C., Trosko, J.E., Warren, S.T. (1978): *In vitro* assay for tumor promoters and antipromoters. J. Environ. Pathol. Toxicol. *2*: 43–64.

Cohen, S.M. Arai, M., Jacobs, J.B., Friedell, G.H. (1979): Promoting effect of saccharin and DL-tryptophan in urinary bladder carcinogenesis. Cancer Res. *39*:1207–1217.

Diamond, L., O'Brien, T.G., Rovera, G. (1978): Tumor promoters; Effects on proliferation and differentiation of cells in culture. Life Sci. *23*:1979–1988.

Eagle, H. (1959): Amino acid metabolism in mammalian cell cultures. Science *130*: 432–437.

Ford, D.K., Yerganian, G. (1958): Observation on the chromosomes of Chinese hamster cells in tissue cultures. J. Natl. Cancer Inst. *21*: 393–425.

Frei, J.V. (1976): Some mechanisms operative in carcinogenesis. A review. Chem. Biol. Interact. *13*:1–25.

Goetz, I.E., Weinstein, C., Roberts, E. (1972): Effects of protease inhibitors on growth of hamster tumor cells in culture. Cancer Res. *32*:2469–2474.

Kensler, T.W., Mueller, G.C. (1978): Retinoic acid inhibition of the comitogenic action of mezerein and phorbol esters in bovine lymphocytes. Cancer Res. *38*:771–775.

Kuroki, T., Drevon, C. (1979): Inhibition of chemical transformation in C3H/10T¹/₂ cells by protease inhibitors. Cancer Res. *39*: 2755–2761.

Lankas, G.R., Baxter, C.S., Christian, R.T. (1977): Effect of tumor promoting agents on mutation frequencies in cultured V79 Chinese hamster cells. Mutat. Res. *45*:153–156.

Lankas, G.R., Baxter, C.S., Christian, R.T. (1978): Effect of alkane tumor-promoting agents on chemically induced mutagenesis in cultured V79 Chinese hamster cells. J. Toxicol. Environ. Health *4*:37–41.

Mufson, R.A., Laskin, J.D., Fisher, P.B., Weinstein, I.B. (1979): Melittin shares certain cellular effects with phorbol ester tumor promoters. Nature (Lond.) *280*:72–74.

O'Brien, T.G., Diamond, L. (1978): Metabolism of tritium-labeled 12-0-tetradecanoyl phorbol-13-acetate by cells in culture. Cancer Res. *38*: 2562–2566.

Peraino, C., Fry, R.J.M., Staffeldt, E., Christopher, J.P. (1975): Comparative enhancing effects of phenobarbital, amobarbital, diphenylhydantoin, and dichloro-diphenyltrichlorethane on 2-acetylaminofluorene-induced hepatic tumorigenesis in the rat. Cancer Res. *35*:2884–2890.

Peraino, C., Fry, R.J.M., Staffeldt, E., Christopher, J.P. (1977): Enhancing effects of phenobarbital and butylated hydroxytoluene on 2-acetylamino-fluorene-induced hepatic tumorigenesis in the rat. Food Cosmet. Toxicol. *15*:93–96.

Reddy, B.S., Watanabe, K. (1979): Promoting effect of cholesterol metabolites and bile acids on colon carcinogenesis in germfree and conventional rats. Proc. Am. Assoc. Cancer Res. *20*:124.

Reddy, B.S., Weisburger, J.H., Wynder, E.L. (1978): Colon cancer: bile salts as tumor promoters. In: T.J. Slaga, A. Sivak, and R.K. Boutwell (eds.), Carcinogenesis: Mechanisms of Tumor Promotion and Cocarcinogenesis, Vol. 2. New York: Raven Press, pp. 453–464.

Sivak, A. (1978): Cocarcinogenesis. Biochem. Biophys. Acta *560*: 67–89.

Slaga, T.J., Sivak, A., Boutwell, R.K. (eds.) (1978): Carcinogenesis: Mechanisms of Tumor Promotion and Cocarcinogenesis, Vol. 2. New York: Raven Press.

Steel, R.G.D., Torrie, J.H. (1960): Principles and Procedures of Statistics. New York: McGraw-Hill Book Company.

Troll, W., Klassen, A., Janoof, A. (1970): Tumorigenesis in mouse skin: inhibition by synthetic inhibitors of proteases. Science *169*:1211–1218.

Trosko, J.E., Chang, C.C. (1978): Environmental carcinogenesis: an integrative model. Quart. Rev. Biol. *53*:115–141.

Trosko, J.E., Chang, C.C. (1980a): Role of mutations and epigenetic changes in carcinogenesis: correlations between chemical and radiation-induced carcinogenesis. In: J.Lett and H. Adler (eds.), Advances in Radiobiology, Vol. 9. New York: Academic Press.

Trosko, J.E., Chang, C.C. (1980b): The role of DNA repair capacity and somatic mutations in carcinogens and aging. In: H.T. Blumenthal (ed.), Handbook of the Diseases of Aging: A Pathogenic Perspective. New York: Van Nostrand Reinhold.

Trosko, J.E., Chang, C.C., Yotti, L.P., Chu, E.H.Y. (1977): Effect of phorbol myristate acetate on the recovery of spontaneous and ultraviolet light-induced 6-thioguanine and ouabain-resistant Chinese hamster cells. Cancer Res. *37*:188–193.

van Zeeland, A.A., van Diggelen, M.C.E., Simons, J.W.I.M. (1972): The role of metabolic cooperation in selection of hypoxanthine-guanine-phosphoribosyl (HGPRT)-deficient mutants from diploid mammalian cell strains. Mutat. Res. *14*:355–363.

Yotti, L.P., Chang, C.C., Trosko, J.E. (1979): Elimination of metabolic cooperation in Chinese hamster cells by tumor promoters. Science *206*:1089–1091.

38

Inhibition of Chemical Mutagenesis: An Application of Chromosome Aberration And DNA Synthesis Assays Using Cultured Mammalian Cells

Lan Wei, Robert F. Whiting, and Hans F. Stich

Introduction

In this paper the application of short-term tests using mammalian cell cultures to examine inhibition of chemical carcinogenesis/mutagenesis will be discussed. The choice of the mammalian cell system was made on the assumption that quantitative data obtained with cultured cells could be reliably extrapolated to the whole animal.

The two cell types employed in this study are (1) human skin fibroblasts with normal or deficient capacity of repair and (2) Chinese hamster ovary (CHO) cells. Of the two short-term assays successfully used, the first assay is the induction of *chromosome aberrations* (Evans and O'Riordan, 1977; German, 1974), a direct measure of DNA damage caused by chemical carcinogen. The second, *DNA repair synthesis* or unscheduled DNA synthesis, represents an indirect assessment of DNA damage via the repair process (Rasmussen and Painter, 1966; Stich and Laishes, 1973). A radioactive DNA base analogue (tritiated thymidine) is incorporated into the damaged DNA. The normal replicative synthesis of DNA is blocked so that the amount of base incorporation reflects only the repair synthesis. A sound correlation between the carcinogenicity/mutagenicity of a chemical and its capacity to induce chromosome aberrations or unscheduled DNA synthesis has been established (Stich *et al.*, 1974, 1976a; Hollstein *et al.*, 1979).

For purely conceptual clarity and ease of presentation, we have divided the chemical carcinogen/mutagens into three categories: (1) direct-acting (MNNG), (2) metabolically activated (aflatoxin B_1, DMN), and (3) chemically activated (N-nitrosation of methylguanidine, metal-catalyzed autoxidation of dimethylhydrazine, isoniazid, etc.). Following a brief description of methodology, model studies representing all three categories on inhibition of carcinogen-induced DNA damage will be presented.

Methodology

Cell cultures. Fibroblasts were obtained from skin biopsies of 18 to 25 years old xeroderma pigmentosum patients and normal females. Cells having gone through two to nine passages were used throughout the studies. Both human skin fibroblasts and Chinese hamster ovary (CHO) cells were cultured in minimum essential medium (MEM) supplemented with 15% fetal calf serum and antibiotics at 37° C in a water-saturated CO_2 incubator.

Chromosome aberration experiment. Cells were seeded on coverslips in plastic culture dishes and grown in MEM with 15% serum for 2–3 days. When 60–80% confluency was reached, the cells were treated with chemicals in MEM supplemented with 2.5% serum. The chemicals (carcinogens, inhibitors, and cofactors) were mixed directly in the culture dishes. The order of addition was: cofactors, inhibitors, and finally carcinogen. An example of such a complex system is described in the legend of Figure 38.2. The chemical treatment time was 1 to 3 hours. Cells were washed and allowed to recover in MEM with 15% serum for 20 hours. Colchicine (10 μg/ml) was added 4 hours before harvesting to arrest the dividing cells at metaphase. Following hypotonic treatment, fixation, and staining, the cells were scored for aberrations.

DNA repair synthesis experiment. In order to obtain unambiguous measurements of DNA repair synthesis, cells were blocked from the normal DNA replicative synthesis. This was achieved by placing the cells, which were seeded and grown for 3 days in MEM with 15% serum, in an arginine-deficient and low-serum (2.5%) medium for 2–3 days. Cells were then treated for 1 to 3 hours with chemicals. Methyl-^{3}H-thymidine (5–10 μCi/ml) was present either during or immediately following the chemical treatment. After harvesting and staining, the cells were processed for autoradiography (Stein and Yanishevsky, 1979).

Analytical Methods

Chromosome aberrations. For chromosome aberration analysis, only distinct exchanges and breaks were scored. This includes both the chromatid- and chromosome-type of breaks and exchanges (intra- as well as interchanges). For each sample, 50 to 200 well-spread metaphase cells were analyzed. The frequency of aberrations was expressed in two ways: (1) the total percentage of cells having at least one break or exchange, and (2) the average number of exchanges per metaphase cell. "Toxicity" and "mitotic inhibition" were used as a semi-quantitative measure of the effect of chemicals at high concentrations. "Toxicity" designates cells with no detectable mitosis and variable degree of cell loss. "Mitotic inhibition" indicates that there is less than 1 mitotic cell among 5000 cells.

DNA repair synthesis. Blockage of cell proliferation in low serum and arginine-deficient medium resulted in a nondividing cell population with less than 0.1% of cells in S-phase.

Scoring was done by manually counting the silver grains over individual nuclei. As a rule, at least 30 diploid nuclei, which were evenly stained and of similar size, were counted. Wherever possible, experiments were designed so that the highest grain counts were 30–100 grains per nucleus. Background and control cells should have 1–2 grains per nucleus. The average net grain counts (subtracting the background grain counts around the nucleus from the grain counts over the nucleus) were reported along with the 95% confidence limits of the mean.

Results and Discussion

Inhibition of DNA Damage Induced by a Direct-acting Carcinogen

One of the most extensively studied carcinogens, MNNG, is used here as a model compound (Schulz and McCalla, 1969; Lawley and Thatcher, 1970). The interactions of various amino acids with MNNG was determined by two independent parameters. The first is a biological index (frequency of chromosome aberrations), and the second is a chemical one (optical absorbance at 402 nm). Results showing the direct chemical interaction between MNNG and cysteine, a common amino acid with a reductive sulfhydryl group, are summarized in Figure 38.1. MNNG alone at 5×10^{-5} M induced a significant level of aberrations (of a total 109 metaphase cells analyzed, 21 showed breaks and exchanges, 19.3%) with high rate of chromosome and chromatid exchanges (of a total 109 cells scored there are 34 distinct chromatid- and chromosome-type exchanges, 31.2%). Cysteine alone did not induce any detectable adverse effects over three logs of concentrations (10^{-5}–10^{-2}M). When CHO cells were treated with a mixture of MNNG (5×10^{-5} M) and cysteine, the level of chromosome aberration showed an inverse dependence on the concentration of cysteine. Chromosome abnormalities were not detected when the cysteine concentration in the mixture exceeded 3×10^{-4} M. The level of chromosome breaks was reduced to that of the nontreated control cells, and no exchanges were observed. However, when cells were treated with MNNG and equimolar or lower concentrations of cysteine (5×10^{-5} M), no inhibition was detected. Cysteine also reduced the cytostatic and clastogenic effects of higher or lower concentrations of MNNG. Inhibitory effects of cysteine were also observed when MNNG and cysteine were mixed half an hour prior to addition to cells.

The interaction of MNNG and cysteine was also detected by following the optical absorption pattern of MNNG and cysteine mixtures between 350-450 nm (Fig. 38.1b). MNNG alone showed λmax at 402nm with ϵ equal to 165 (Lawley and Thatcher, 1970) while cysteine showed no absorption in the visible range. When MNNG (10^{-2} M) and

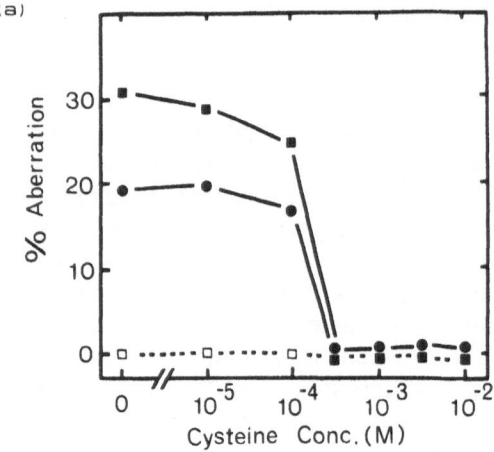

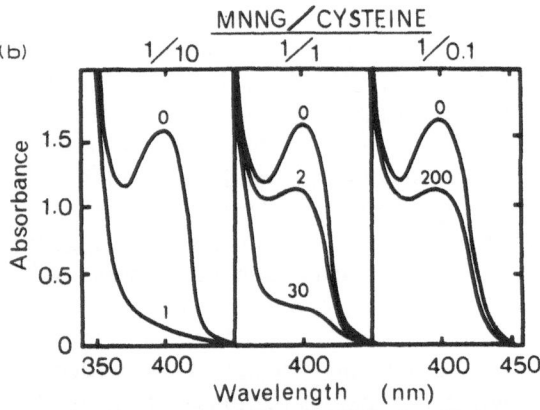

Figure 38.1. Interaction of MNNG with cysteine. (a) Inhibitory effect of cysteine on the induction of chromosome aberrations by MNNG. Chinese hamster ovary (CHO) cells were incubated for 2 hours with MNNG (5×10^{-5} M) and cysteine (0, 10^{-5}–10^{-2} M) in MEM supplemented with 2.5% calf serum. ●——● , per cent of metaphase cells containing chromosome aberrations; ■——■ , average number (%) of chromatid- and chromosome-type *exchanges* per metaphase cell; □——□ , aberrations in CHO cells incubated with cysteine alone. (b) Effect of cysteine on the decomposition of MNNG. Optical absorption curves of mixtures of MNNG (10^{-2} M) and cysteine (10^{-1}, 10^{-2} and 10^{-3} M) were recorded between 350–500 nm at various time intervals following mixing. The numbers on the absorption curves designate *minutes* after mixing MNNG with cysteine. Serial dilutions were made from stock solutions of MNNG (0.1 M in ethanol) and cysteine (0.2 M in phosphate buffer saline, PBS) in PBS (containing 7.9 mM of phosphate).

cysteine (10^{-1} M) were mixed at pH 7.0 in a phosphate-buffered saline solution (PBS, 7.9 mM phosphate), within 1 minute of mixing the absorbance dropped to near the baseline level. The decomposition rate of MNNG decreased as the cysteine concentration decreased (MNNG/cysteine = 1:10, 1:1, and 1:0.1, Fig. 38.1b).

Based on the result of chromosome aberration studies and the spectrophotometric measurements, it is clear that direct chemical interaction between MNNG and cysteine leads to the inhibition of DNA damage. This mechanism is further supported also by experiments using analogues of cysteine (R—SH, R=CH_2CHNH_2COOH). For example, alanine (R—CH_3), lacking the sulfhydryl group, did not reduce the cytostatic and clastogenic activity of MNNG. As well, the spectrophotometric measurements showed that only 50% of the MNNG had decomposed two hours after the mixing of MNNG (10^{-2} M) and alanine (10^{-1}M). Other amino acids, serine (R—OH) and methionine (R—CH_2SCH_3), showed weak inhibitory activity. Glutathione, a tripeptide containing cysteine, was found to have similar activity to that of cysteine.

Inhibition of DNA Damage Induced by Metabolically Activated Carcinogens

Dimethylnitrosamine (DMN) and aflatoxin B_1 were used as models for the inhibition study of metabolically activated carcinogens. Both compounds have been extensively studied *in vivo* and *in vitro* (Magee and Farber, 1962; Swann and Magee, 1971; Laishes and Stich, 1973; Wogan, 1973; Malling, 1974; Garner and Wright, 1975; Stich and Laishes, 1975; Lin *et al.*, 1977; Sarasin *et al.*, 1977; Autrup *et al.*, 1979). Figure 38.2 illustrates the use of both DNA-repair synthesis and chromosome aberration assays to examine the inhibitory action of a phenolic antioxidant. Propyl gallate (10^{-5}–5×10^{-4} M) reduced

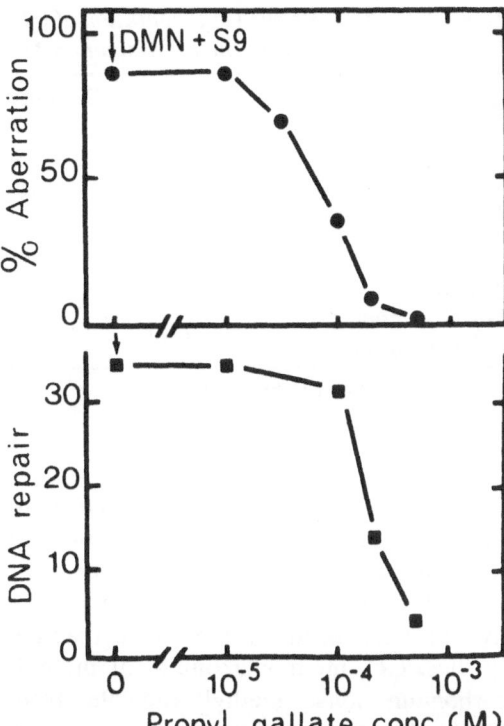

Figure 38.2. Inhibition of the mutagenic effects of DMN by propyl gallate. CHO cells (●——●) and human skin fibroblasts (■——■) were incubated for 1 hour with DMN (10^{-2} M and 5×10^{-2} M respectively), liver microsomes with cofactors (NADPH, $MgCl_2$ and glucose-6-phosphate), and propyl gallate (0, 10^{-5}–5×10^{-4} M). ●——● , per cent of metaphase cells with chromosome aberrations. ■——■ , DNA repair synthesis was detected as average number of silver grains per nucleus; cells were incubated in ADM with 2.5% calf serum and ^{3}H-TdR (10 μCi/ml) for 3 hours after the chemical treatment.

the frequency of chromosome aberrations induced by activated DMN from 85.7% to 1.4%. Similarly, propyl gallate decreased the incorporation of ^{3}H-thymidine (from 36 to 4 grains per nucleus) during the repair of DMN-induced damage of DNA. Addition of other reducing agents such as ascorbate, cysteine, and cysteamine also led to reduced DNA damage and repair (Table 38.1).

In this category, where *in vitro* detection of a carcinogen requires metabolic activa-

Table 38.1.   *Unscheduled Incorporation of [³H] TdR (Grains Per Nucleus) into Cultured Human Cells Exposed to DMN, the S9 Activation Mixture, and Different Reducing Agents*[a]

Concentration of reducing agent (M)	Dimethylnitrosamine plus S9 plus reducing agent			
	Ascorbate	Cysteine	Cysteamine	Propyl gallate
1×10^{-2}	3.9	9.7	4.4	2.8
3×10^{-3}	6.5	14.4	9.6	4.7
1×10^{-3}	7.9	15.7	10.9	7.8
3×10^{-4}	23.7	25.7	22.6	16.4
DMN without reducing agent	23.0	26.1	24.6	27.2

[a]L. Wei Lo and H.F. Stich (1978).

tion, the observed inhibition of DNA damage could arise from the interaction of a reducing agent with (1) the carcinogen prior to enzyme activation, (2) the metabolic activation system, and (3) the final reactive species, e.g., methyldiazonium or methylcarbonium ions, methyl radicals from DMN. The first mechanism, direct chemical interaction between an inhibitor and a carcinogen, could be estimated quantitatively by chemical measurements similar to examples given in the previous section (spectrophotometric, chromatographic, etc.). Evaluation of the second mechanism relies on biochemical analysis of metabolites of the carcinogen and inhibitor (Blumberg, 1978; Benson *et al.*, 1979). The third mechanism, interaction of inhibitor with the electrophilic reactive species (Miller, 1978), has been examined by electron spin resonance (ESR) spectrometry (Ts'o *et al.*, 1977). One example is the carcinogen benzo(a)pyrene. A characteristic ESR signal of benzo(a)pyrene radical (Fig. 38.3a) was detected when the benzene solution of benzo(a)pyrene was oxidized by a strong oxidant ($CF_3COOH + H_2O_2$). When the phenolic antioxidant BHT was added, the signal of benzo(a)pyrene radical disappeared and a stable signal of BHT radical (Fig. 38.3b) was detected.

Figure 38.4 illustrates complex interactions in which the above mechanisms are not clearly differentiated. Both reducing agents, sodium ascorbate and butylated

hydroxytoluene (BHT), reduced the DNA-damage and repair synthesis induced by activated DMN (Fig. 38.4b). A different pattern of inhibition was observed with aflatoxin B_1 (Fig. 38.4a). BHT reduced the DNA-damaging activity of aflatoxin B_1 in human skin fibroblasts, while sodium ascorbate exerted no effect.

Inhibition of DNA Damage Induced by Chemically Activated Carcinogens

This category includes chemicals which, upon reacting with an organic or inorganic compound, yield DNA-damaging agents. To illustrate the inhibition aspects, two examples will be given: Firstly, inhibition of the formation of N-nitrosomethylguanidine for substrates methylguanidine and nitrite (Lo and Stich, 1978). The second case deals with inhibition of the clastogenic activity of hydrazine compounds and transition metals (Whiting *et al.*, 1979, 1980).

Neither methylguanidine nor nitrite alone induced DNA repair synthesis in human skin fibroblasts. The nitrosation products of methylguanidine increased the incorporation of ³H-thymidine to 110 grains per nucleus (Fig. 38.5). Inhibition of this induced DNA repair process was detected when ascorbic acid was present during nitrosation. Ascorbic acid exhibited

(a)

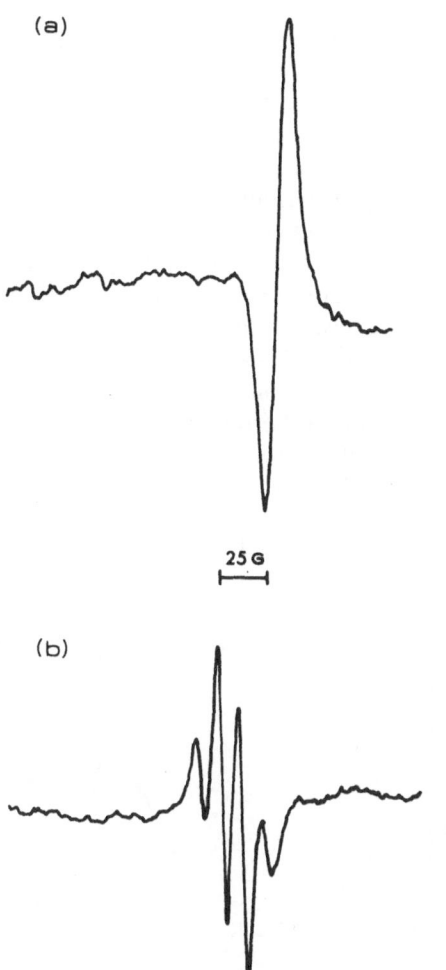

25 G

(b)

Figure 38.3. Electron spin resonance spectra measured at 22° C in benzene solution of (a) reaction product of benzo(a)pyrene and oxidants ($CF_3COOH + H_2O_2$) and (b) reaction product of benzo(a)pyrene, BHT, and oxidants ($CF_3COOH + H_2O_2$). The reaction mixture contained 20 mg/ml of benzo(a)pyrene or BHT, and 20 μl/ml of each of CF_3COOH and 30% H_2O_2.

Isoniazid, a tuberculostatic drug, and dimethylhydrazine, a colon carcinogen, represent a group of reducing agents which are auto-oxidized in the presence of oxygen. This auto-oxidation is catalyzed by essential metals such as copper, manganese, and iron (Skov and Vonderschmitt, 1975; Stich *et al.*, 1976 a,b, 1979; Whiting *et al.*, 1979, 1980). When cells were incubated with low concentrations of metal complex and isoniazid, high frequencies of chromosome aberrations were induced (Table 38.2). In this case the addition of catalase or glutathione reduced the cytostatic and

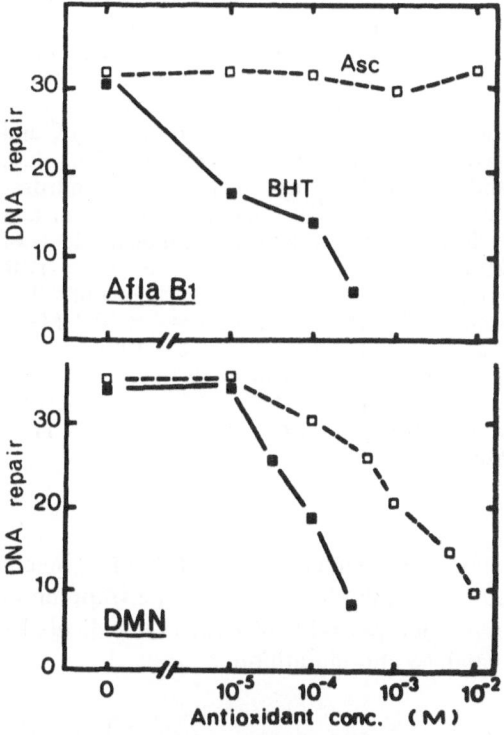

Figure 38.4. Effect of ascorbate and BHT on the mutagenicity of activated aflatoxin B_1 and DMN. Human skin fibroblasts were incubated for 1 hour with carcinogen [aflatoxin B_1 (2×10^{-5} M) or DMN (5×10^{-2} M)], liver microsomes with cofactors, and reducing agent [ascorbate (0, 10^{-5}–10^{-2} M) or BHT (0, 10^{-5}–3×10^{-4} M)]. Cells were then incubated in ADM with 2.5% calf serum and H^3-TdR (10 μCi/ml) for 3 hours. DNA repair synthesis was measured as average number of silver grains per nucleus ($\square$---$\square$, ascorbate; $\blacksquare$——$\blacksquare$, BHT).

complete inhibition when its molar ratio exceeded that of nitrite. Using chromosome aberration assay a similar response was observed: reduction of aberration frequencies when ascorbic acid was present (Lo and Stich, 1978). The inhibition was mainly due to competitive reaction between ascorbic acid and methylguanidine for nitrite, leading to decreased N-nitrosation.

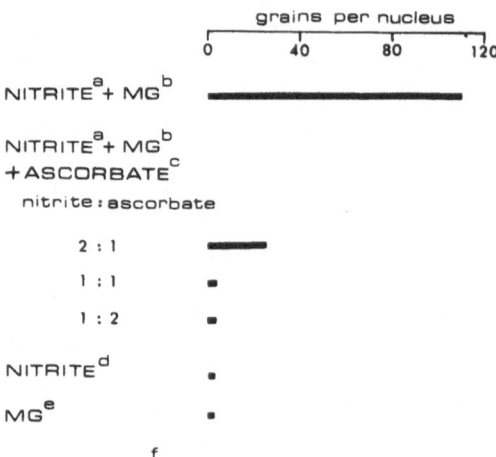

Figure 38.5.   Inhibitory effect of ascorbate on nitrosation of methyl guanidine (MG). DNA repair synthesis in human skin fibroblasts was used to measure the mutagenicity of nitrosation products of MG. Nitrosation was carried out at 0° C, pH 0.1 for 4 hours; the nitrosation mixtures contained 0.2 M MG, 0.6 M NaNO$_2$ and varied concentrations of ascorbate (0, 0.3, 0.6, and 1.2 M). After neutralization of the nitrosation mixtures, serial dilutions were made, and cells were exposed to the nitrosation products of MG for 3 hours, followed by 3 hours of ^{3}H-TdR (10 μCi/ml) treatment. Concentrations of a. Nitrite = 1.2×10^{-3} M, b. MG = 4×10^{-4} M, c. Ascorbate = 6×10^{-4}, 1.2×10^{-3} and 2.4×10^{-3} M, d. Nitrite = $10^{-6} - 2 \times 10^{-3}$ M, e. MG = $10^{-5} - 10^{-3}$ M, f. Ascorbate = $10^{-5} - 3 \times 10^{-3}$ M (data from H.F. Stich *et al.*, 1976a).

clastogenic effects (Table 38.2). The mechanism of inhibition is likely the trapping of hydrogen peroxide or hydroxyl radicals by catalase and glutathione.

Conclusion

Having presented various model studies which illustrate the application of chromosome aberration and DNA repair synthesis assays to investigate the inhibition of DNA damage, this discussion would not be complete without mentioning some limitations. Since the chromosome aberration assay detects the damage rendered by chemicals,

there is less ambiguity in interpreting the data of inhibition experiments. Decreased DNA damage as reflected by a decreased level of cytostatic and clastogenic activity is a straightforward measure of the inhibitory activity of various chemicals. However, the DNA repair synthesis assay is a more complex one. The damage is estimated indirectly by the repair process. It is therefore crucial to ascertain that the reduced level of DNA repair observed is actually due to inhibition of damage rather than inhibition of the repair process (Painter, 1981; Seiler, 1981). Results shown in Figure 38.6 illustrate this point. When a complete concentration range of carcinogen is tested, usually the dose-response

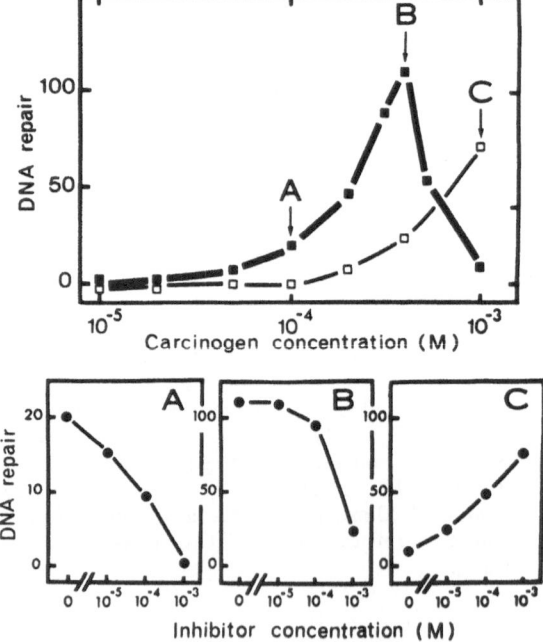

Figure 38.6.   The varied effects of an inhibitor on DNA repair synthesis induced by a carcinogen. ■——■ , levels of DNA repair synthesis (grains per nucleus) induced by a carcinogen alone. □——□ , DNA repair synthesis induced in cells treated with carcinogen plus inhibitor (10^{-3} M). ●——● , levels of DNA repair synthesis induced by carcinogen at three concentrations: A (10^{-4} M, sublethal), B (4×10^{-4} M, maximum DNA repair), and C (10^{-3} M, toxic); the inhibitors were present at 0 or 10^{-5}–10^{-3} M.

Table 38.2. *Effect of Metal Complexes, Glutathione, and Catalase on Chromosome Aberrations Induced by Isoniazid*

| Isoniazid conc. (M) | Isoniazid alone | Isoniazid plus | | Isoniazid plus 10^{-5}M Mn(II)[f] | Isoniazid/Mn(II) plus | | Isoniazid plus 10^{-4}M Cu(II)[f] | Isoniazid/Cu(II) plus catalase | Isoniazid plus 10^{-4}M Fe(III)[g] |
		Catalase (0.1 mg/ml)	Glutathione (10^{-2}M)		Catalase (0.1 mg/ml)	Glutathione (10^{-2}M)			
1×10^{-1}	30.0[b](0.71)[c]	5.1(0.22)	3.4(0.14)	Toxic[d]	6.3(0.28)	2.7(0.13)	M.I.	M.I.	2.5(0.07)
5×10^{-2}	41.0(0.95)	0.8(0.00)	0.5(0.00)	Toxic[d]	2.9(0.18)	1.5(0.01)	M.I.	23.1(0.25)	1.6(0.03)
2×10^{-2}	42.1(2.10)	0.5(0.00)	0.7(0.00)	Toxic[d]	1.2(0.00)	0.5(0.00)	M.I.	3.4(0.00)	
1×10^{-2}	27.0(0.65)	0.8(0.00)	0.7(0.00)	Toxic[d]			27.5(0.50)	1.0(0.00)	
5×10^{-3}	9.6(0.20)	0.5(0.00)	0.8(0.00)	Toxic[d]			19.3(0.47)	1.2(0.00)	
2×10^{-3}	2.8(0.06)	0.7(0.01)	0.5(0.00)	Toxic[d]	2.0(0.02)	0.7(0.01)	2.6(0.03)	1.0(0.00)	0.6(0.00)
1×10^{-3}	0.6(0.00)			Toxic[d]			1.3(0.01)	1.5(0.00)	
5×10^{-4}	1.3(0.00)	0.7(0.00)	0.5(0.00)	M.I.[e]	1.5(0.00)	0.4(0.00)	2.4(0.01)	0.8(0.00)	1.4(0.00)
2×10^{-4}	1.5(0.01)			53.1(2.53)	0.8(0.00)	0.7(0.00)			
1.5×10^{-4}	0.8(0.00)	0.5(0.01)	0.5(0.01)	31.2(1.75)	0.6(0.00)	0.5(0.01)	0.6(0.00)	0.9(0.00)	0.5(0.00)
1×10^{-4}	1.2(0.00)			8.6(0.41)					
5×10^{-5}				1.8(0.06)					
O[h]	0.6(0.00)	0.6(0.00)	1.1(0.00)	0.8(0.00)	0.5(0.00)	1.0(0.00)	1.0(0.00)	1.3(0.00)	1.0(0.00)

Frequency of chromosome aberrations (%) in CHO cells[a]

[a] For each treatment at least 100 metaphase plates were scored for breaks and exchanges.

[b] Frequency (percent) of metaphase figures having at least one aberration.

[c] Average number of exchanges per metaphase plate.

[d] Toxic: no detectable mitosis and variable cell loss.

[e] Mitotic inhibition: less than 1 metaphase among 5000 cells.

[f] Cu(II) and Mn(II) were added as glycine complexes at a molar ratio of metal:glycine = 1:10.

[g] Fe(III) was added as an EDTA complex (Fe:EDTA = 1:2).

[h] Controls: cells treated with chemical mixtures lacking isoniazid (e.g., 0.8(0.00) is the frequency of aberrations induced by 10^{-5}M Mn complex).

From R.F. Whiting et al. (1980).

curve rises with increasing concentrations of carcinogen until a maximum is reached, followed by a decreased capacity of repair. With the addition of a compound which reduces DNA damage (an inhibitor) the corresponding DNA repair synthesis shows reduced DNA damage and repair only when the carcinogen concentrations are at sublethal level (Fig. 38.6,A and B). However, at the region where carcinogen alone induced cytotoxicity and inhibition of repair, addition of an inhibitor led to a continuous rise in DNA repair synthesis (Fig. 38.6C) which might be misinterpreted as increased DNA damage. The net result of the addition of an inhibitor to a carcinogen at cytotoxic concentrations could be a decrease in cytotoxicity, yet an increase in mutagenicity. Therefore, it is essential to carry out inhibition experiments with a complete concentration range of a carcinogen rather than just one or two concentrations. This ensures the disclosure of a thorough pattern of inhibition.

Abbreviations

Afla B_1, aflatoxin B_1; BHT, butylated hydroxytoluene; DMN, dimethylnitrosamine; MG, methylguanidine; MNNG, N-methyl-N'-nitro-N-nitrosoguanidine.

References

Autrup, H., Essigmann, J.M., Croy, R.G., Trump, B.F., Wogan, G.N., Harris, C.C. (1979): Metabolism of aflatoxin B_1 and identification of the major aflatoxin B_1-DNA adducts formed in cultured human bronchus and colon. Cancer Res. 39:694–698.

Benson, A.M., Cha, Y.-N., Bueding, E., Heine, H.S., Talalay, P. (1979): Elevation of extrahepatic glutathione S-transferase and epoxide hydratase activities by 2(3)-tert-butyl-4-hydroxyanisole. Cancer Res. 39:2971–2977.

Blumberg, W.E. (1978): Enzymic modification of environmental intoxicants: the role of cytochrome P-450. Quart. Rev. Biophys. II 4: 481–542.

Evans, H.J., O'Riordan, M.L. (1977): Human peripheral blood lymphocytes for the analysis of chromosome aberrations in mutagen tests. In: B.J. Kilbey, M. Legator, W. Nichols, and C. Ramel (eds.), Handbook of Mutagenicity Test Procedures. Amsterdam: Elsevier/North-Holland Biomedical Press, pp. 261–274.

Freese, E.B., Gerson, J., Taber, H., Rhaese, H.J., Freese, E. (1967): Inactivating DNA alterations induced by peroxides and peroxide-producing agents. Mutat. Res. 4:517–531.

Freese, E., Sklarow, S., Freese, E. (1968): DNA damage caused by antidepressant hydrazines and related drugs. Mutat. Res. 5: 343–348.

Garner, R.C., Wright, C.M. (1975): Binding of [C^{14}] aflatoxin B_1 to cellular macromolecules in the rat and the hamster. Chem.-Biol. Interact. 11:123–131.

German, J. (ed.) (1974): Chromosomes and Cancer. New York: J. Wiley & Sons.

Hollstein, M., McCann, J., Angelosanto, F., Nichols, W.W. (1979): Short-term tests for carcinogens and mutagens. Mutat. Res. 65: 133–226.

Laishes, B.A., Stich, H.F. (1973): Repair synthesis and sedimentation analysis of DNA of human cells exposed to dimethylnitrosamine and activated dimethylnitrosamine. Biochem. Biophys. Res. Commun. 52:827–833.

Lawley, P.D., Thatcher, C.J. (1970): Methylation of DNA in cultured mammalian cells by MNNG: influence of cellular thiol concentration on extent of methylation and O-6 guanine as a site of methylation. Biochem. J. 116: 693–707.

Lin, J.K., Miller, J.A., Miller, E.C. (1977): 2,3-Dihydro-2-(guan-7-yl)-3-hydroxy-aflatoxin B_1, a major acid hydrolysis product of aflatoxin B_1-DNA or ribosomal RNA adducts formed in hepatic microsome-mediated reactions and in rat liver in vivo. Cancer Res. 37: 4430–4438.

Lo, L. Wei, Stich, H.F. (1978): The use of short-term tests to measure the preventive action of reducing agents on formation and activation of carcinogenic nitroso compounds. Mutat. Res. 57:57–67.

Magee, P.N., Farber, E. (1962): Toxic liver injury and carcinogenesis; methylation of rat liver nucleic acids by dimethylnitrosamine *in vivo*. Biochem. J. *83*:114–124.

Malling, H.V. (1974): Mutagenic activation of dimethylnitrosamine and diethylnitrosamine in the host-mediated assay and the microsomal system. Mutat. Res. *27*:465–472.

Miller, E.C. (1978): Some current perspectives on chemical carcinogenesis in human and experimental animals. Cancer Res. *38*:1479–1496.

Painter, R.B. (1981): DNA synthesis inhibition in mammalian cells as a test for mutagenic carcinogens. In: H.F. Stich and R.H.C. San (eds.), Short-Term Tests for Chemical Carcinogens. New York: Springer-Verlag, pp. 59–64.

Rasmussen, R.E., Painter, R.B. (1966): Radiation-stimulated DNA synthesis in cultured mammalian cells. J. Cell Biol. *29*:11–19.

Sarasin, A.R., Smith, C.A., Hanawalt, P.C. (1977): Repair of DNA in human cells after treatment with activated aflatoxin B_1. Cancer Res. *37*:1786–1793.

Schulz, U., McCalla, D.R. (1969): Reaction of cysteine with N-methyl-N-nitroso-p-toluene-sulphonamide and MNNG. Can. J. Chem. *47*:2021–2027.

Seiler, J.P. (1981): The testicular DNA-synthesis inhibition test (DSI-test). In: H.F. Stich and R.H.C. San (eds.), Short-Term Tests for Chemical Carcinogens. New York: Springer-Verlag, pp. 94–107.

Skov, K.A., Vonderschmitt, D.J. (1975): Kinetics of iron and copper catalysis of ascorbate oxidation. Bioinorg. Chem. *4*:199–213.

Stein, G.H., Yanishevsky, R. (1979): Autoradiography. In: W.B. Jakohy and I.H. Pastan (eds.), Methods in Enzymology, Vol. 58. New York: Academic Press, pp. 279–292.

Stich, H.F. Laishes, B.A. (1973): DNA repair and chemical carcinogens. In: H.L. Ioachim (ed.), Pathobiology Annual. New York: Appleton-Century-Crofts, pp. 341–376.

Stich, H.F., Kieser, D., Laishes, B.A., San, R.H.C. (1974): The use of DNA repair in the identification of carcinogens, precarcinogens, and target tissue. In: P.G. Scholefield (ed.), Proceedings of the Tenth Canadian Cancer Conference. University of Toronto Press, Toronto, pp. 83–110.

Stich, H.F., Laishes, B.A. (1975): The response of xeroderma pigmentosum cells and controls to the activated mycotoxins, aflatoxins and sterigmatocystin. Int. J. Cancer *16*:266–274.

Stich, H.F., San, R.H.C., Lam, P.P.S., Koropatnick, D.J., Lo, L.W., Laishes, B.A. (1976a): DNA fragmentation and DNA repair as an *in vitro* and *in vivo* assay for chemical precarcinogens, carcinogens and carcinogenic nitrosation products. In: R. Montesano, H. Bartsch, and L. Tomatis (eds.), Screening Tests in Chemical Carcinogenesis (IARC Scientific Publications No. 12). Lyon: International Agency for Research on Cancer, pp. 617–636.

Stich, H.F., Karim, J., Koropatnick, J., Lo, L.W. (1976b): Mutagenic action of ascorbic acid. Nature (Lond.) *260*:722–724.

Stich, H.F., Wei, L., Whiting, R.F. (1979): The enhancing effect of transition metals on the chromosome damaging action of ascorbate. Cancer Res. *39*:4145–4151.

Swann, P.F., Magee, P.N. (1971): Nitrosamine-induced carcinogenesis. The alkylation of N-7 of guanine of nucleic acids of the rat by diethylnitrosamine, N-ethyl-N-nitrosourea and ethyl methanesulfonate. Biochem. J. *125*:841–847.

Ts'o, P.O.P., Caspary, W.T., Lorentzen, R.J. (1977): Involvement of free radicals in chemical carcinogenesis. In: W.A. Pryor (ed.), Free Radicals in Biology, Vol. III. New York: Academic Press, pp. 251–303.

Whiting, R.F., Wei, L., Stich, H.F. (1979): Enhancement by transition metals of unscheduled DNA synthesis induced by isoniazid and related hydrazines in cultured normal and xeroderma pigmentosum human cells. Mutat. Res. *62*:505–515.

Whiting, R.F., Wei, L., Stich, H.F. (1980): Enhancement by transition metals of chromosome aberrations induced by isoniazid. Biochem. Pharmacol. *29*:842–845.

Wogan, G.N. (1973): Aflatoxin carcinogenesis. In: H. Busch (ed.), Methods in Cancer Research, Vol. I. New York: Academic Press, pp. 309–344.

39

Detection of Cocarcinogens and Anticarcinogens with Microbial Mutagenicity Assays

D.R. STOLTZ

Introduction

A meaningful discussion of cocarcinogens and anticarcinogens with reference to microbial assays is difficult. Although carcinogenesis is a whole-animal phenomenon, the apparent steps in the process, i.e., initiation, promotion, and progression, may be mimicked to some degree *in vitro* with mammalian cells and, to a much lesser degree, with microbial cells. In this paper we will examine the interactive effects of pairs of compounds and mixtures in cancer tests, attempt to relate these effects to specific mechanisms, and discuss the ability of microbial mutagenicity assays to predict co- and anticarcinogenic activity.

It is presently fashionable to imagine three major steps in carcinogenesis. The first, initiation, appears to be mutational; the second, promotion, may involve recombination and/or gene derepression; and the third, tumor progression, is characterized by cell proliferation, clonal evolution, and escape from homeostatic controls. Modifi-

ers of carcinogenesis may act at any of these steps or on processes as yet not recognized.

In descriptions of mechanisms of cocarcinogenesis, Berenblum (1978) has distinguished between permissive influences and actual involvement in the carcinogenic process (Table 39.1). Although clearly illustrated in some experimental systems, the role of permissive influences is impossible to discern in most carcinogenicity tests not specifically designed for the purpose. Initiators, of course, may act as cocarcinogens but they do not always do so. Promoters are a very special subset of cocarcinogens, defined procedurally in that the promoting agent must be administered after completion of initiating action, whereas permissive cocarcinogens can operate before, during, or after initiation of the carcinogenic process. Anticarcinogens come in just as many shapes and colors as cocarcinogens.

Although we can describe some discrete mechanisms of cocarcinogenesis, chemi-

Table 39.1. *Some Mechanisms of Cocarcinogenesis*

1. *Permissive Influences*
   (a) on the carcinogen (affecting metabolic activation, detoxification)
   (b) on the responding tissue (hormones prime mammary tissue)
   (c) systematically, on the body as a whole (immunosuppression)
2. *Actual Involvement in the Carcinogenic Process*
   (a) initiation
   (b) promotion

Adapted from: I. Berenblum (1978). Historical perspective. In: T.J. Slaga, A. Sivak and R.K. Boutwell (eds.), Mechanisms of Tumor Promotion and Cocarcinogenesis. New York: Raven Press, pp. 1–10.

Table 39.2. *Influences on the Carcinogen*

1. *Cocarcinogenic*
   (a) nitrosation (nitrite + amine)
   (b) transnitrosation (N-nitrosodiphenylamine + amine)
   (c) enzyme induction/inhibition (phenobarbital)
2. *Anticarcinogenic*
   (a) enzyme induction/inhibition (coumarin)
   (b) trapping agents

cals often show more than one type of activity. For example, cyclophosphamide is metabolized into a series of alkylating derivatives mutagenic to microbes and presumed to initiate carcinogenesis. Furthermore, reversible immunosuppression by cyclophosphamide appears to be a consequence of repairable alkylation of lymphocyte DNA (Shand and Howard, 1979). The fact that cyclophosphamide shows no specific organ carcinogenicity suggests that immunosuppression may be as important as initiation in carcinogenesis induced by cyclophosphamide.

Permissive Influences Acting on the Carcinogen

Examples of such influences (Table 39.2) include nitrosation, transnitrosation, en-zyme induction and inhibition, and the effects of trapping agents (which are discussed in the following chapter).

Nitrosation and transnitrosation reactions illustrate specific cocarcinogenic interactions. Nitrite plus amine or amide are cocarcinogenic in certain combinations, producing the same tumor type and site as the corresponding N-nitroso derivative believed to be formed from the nitrite and amine compound (World Health Organization, 1977). This type of cocarcinogenic interaction can be predicted for combinations of nitrite and certain amino compounds using microbial plate assays. Table 39.3 illustrates the ease of formation of direct acting mutagenicity from nitrite plus ethylenethiourea, simply added together in the plate incorporation assay. Although this facile comutagenesis assay would miss many amines requiring the more acidic gastric environment for nitrosation, it would serve for rapid screening of food extracts and hydrolysates for readily nitrosatable amino compounds (Neurath *et al.*, 1977), which might be cocarcinogenic with nitrite.

Transnitrosation involves the transfer of a nitroso group to an amine or other recep-

Table 39.3. *Nitrosation of Ethylenethiourea[a]*

μm/plate	His+/plate TA1535-S9M			
	ETU	NaNo₂	ETU + NaNO₂	Nitroso-ETU
5	25.3	38.0	153.3	1078.0
1	22.7	19.3	23.0	1157.5
0.5	19.0	16.7	17.6	1200.0
0.1	ND	ND	ND	238.0

[a]Results include 16.5 revertants/plate background for solvent (H₂O) control.

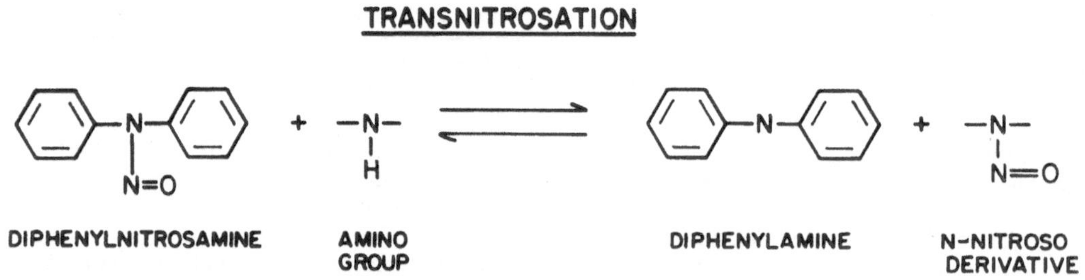

TRANSNITROSATION

DIPHENYLNITROSAMINE   AMINO GROUP   DIPHENYLAMINE   N–NITROSO DERIVATIVE

Figure 39.1.   Example of a transnitrosation reaction.

tor, a reaction also favored in acidic gastric conditions (Figure 39.1). Transnitrosation may account for the recent demonstration of carcinogenicity of N,N-diphenylnitrosamine, a compound considered noncarcinogenic on theoretical grounds, i.e., absence of α-oxidizable hydrogens, several previous negative cancer tests, and negative mutagenicity assays (Cardy *et al.*, 1979; Yahagi *et al.*, 1977).

The literature holds one example of possible transnitrosation detected in a microbial mutagenicity assay. In this instance, transfer of the nitroso group from dimethylnitrosamine to acetylaminofluorene in the presence of a cell-free *E. coli* extract resulted in direct-acting frameshift mutagenic activity (Mandel *et al.*, 1977).

It is conceivable that further demonstration of transnitrosation could be achieved in microbial assays by testing nitroso compounds in combination with a suitable nitroso receptor. In particular, the correlation between carcinogenicity and mutagenicity within this class of chemicals (Elespuru and Lijinsky, 1976; Lee *et al.*, 1977; Montesano and Bartsch, 1976) might improve if carcinogenic but nonmutagenic nitroso compounds (e.g., diphenylnitrosamine, N-nitroso-N-methylphenylamine, nitrosohydantoic acid, 2,6-dimethyl1-1,4-dinitroso-piperazine) were tested in combination with appropriate nitroso receptors.

Few generalizations can be made about the effects of enzyme inducers and inhibitors on carcinogenesis. An exceedingly complex situation may arise since many classic carcinogens, such as polycyclic aromatic hydrocarbons, are effective inducers of their own metabolism, and the most po-

tent microsomal enzyme inducers, e.g., chlorinated insecticides, are weak carcinogens. Thus, although many examples of co- and anticarcinogenesis by chemicals altering enzyme activity are available, the cause-effect relationship is not obvious.

Compounds that alter enzyme activity may be either co- or anticarcinogenic, depending upon the sequence of administration. For example, phenobarbital exerts a protective effect against hepatic tumorigenesis when administered simultaneously with 2-acetylaminofluorene, *para*-dimethylaminoazobenzene or dimethylnitrosamine, but enhances hepatocarcinogenesis initiated by prior treatment with these carcinogens (Peraino *et al.*, 1978). Therefore, it seems unlikely that microbial assays could be applied in a manner that would distinguish co- from anticarcinogenic influences mediated by enzyme activities. Nevertheless, microbial assays can certainly detect chemicals capable of altering metabolism of carcinogens and therefore potentially capable of co- and/or anticarcinogenic effects.

Potential co- and anticarcinogens may be detected by employing the suspect compound as the *in vivo* inducer of the liver metabolic activation system incorporated into a microbial mutagenicity test with several select carcinogens. Pitfalls to this approach are plentiful and may derive from choice of species, sex, organ, cofactors, and carcinogen. However, a number of chemicals have demonstrated the following triad of activities: (A) they alter enzyme activity upon *in vivo* treatment; (B) they show co- or anticarcinogenic effects; and (C) they have been responsible for altered *in vitro* microbial mutagenicity response, when used to pretreat the animal source of the *in vitro* metabolic activation system (Table 39.4). Compounds exhibiting A + B, and for which C should be determined, include tryptophan (Evarts and Brown, 1977), indoles (Matsumoto *et al.*, 1977a; Wattenberg and Loub, 1978), coumarin (Wattenberg *et al.*, 1979), Sudan III and IV (Huggins *et al.*, 1978), butylated hydroxytoluene (Peraino *et al.*, 1978) and saccharin (Chowaniec and Hicks, 1979). This type of *in vivo/in vitro* assay for modifying agents should be distinguished from the *in vitro* protocols more commonly associated with comutagenesis.

Two types of *in vitro* microbial comutagenesis assays have been reported. The first, represented by the work of Andrews with tryptophan metabolites (Bowden *et al.*, 1976) and bile acids (Silverman and Andrews, 1977) involves simultaneous testing of a threshold dose of mutagen plus suspect cocarcinogen. The other comutagenesis assay stems from the Japanese discovery that norharmane enhances mutagenicity of a number of carcinogens (Nagao *et al.*, 1977c). We will deal with these assays separately because they really are quite different.

1. Low-dose mutagen/carcinogen + suspect cocarcinogen.
2. Enhancer of mutagenicity (norharmane) + suspect carcinogen.

In the first case, the biological activity of one component, the mutagen/carcinogen, is known, but in the second, the activity of norharmane vis-á-vis carcinogenicity and cocarcinogenicity is still under investigation. Thus, the second assay cannot be deemed a corollary of the first. Nevertheless, some interesting results have come from both types of assays even though most of these results are uninterpretable in terms of co- or anticarcinogenesis.

The cocarcinogenicity of certain tryptophan metabolites and bile acids is well documented. However, when tested for comutagenicity in the presence of 2 μg 2-aminoanthracene, most of the suspect cocarcinogens failed to enhance the mutagenic response. It is possible that the high background of several thousand colonies per plate obscured any interactions (Bowden *et al.*, 1976; Silverman and Andrews, 1977). In a similar comutagenesis assay (Fig. 39.2) employing a low dose of benzo(a)pyrene, we and others (Matsumoto *et al.*, 1977b) were able to demonstrate comutagenic activity for skatole but negligible or marginal activity for tryptophan and indole. Since tryptophan as well as the two metabolites shown here have all shown co- or anticarcinogenic effects, it appears that this simple comutagenesis assay has limited predictability for cocarcinogenicity.

Other compounds or mixtures which enhance or inhibit activity of a mutagen in a comutagenesis assay include: ellipticines

Table 39.4.   *Co- and Anticarcinogenic Enzyme Inducers/Inhibitors*

	References[a]		
Compound	A	B	C
1. phenobarbital	52	52	1
2. 2,3,7,8-tetrachlorodibenzo-p-dioxin	17	17	21
3. 3-methylcholanthrene	49	39	1
4. ethanol	38	38	38
5. DDT	4	52	4
6. Aroclor 1254	70	70	1
7. aminoacetonitrile	4	4	4
8. B-naphthoflavone	18	70	18
9. dietary fat	11	26	11
10. tobacco smoke	6	69	6

[a]A, enzyme activity; B, carcinogenicity; C, mutagenicity (see text for elaboration).

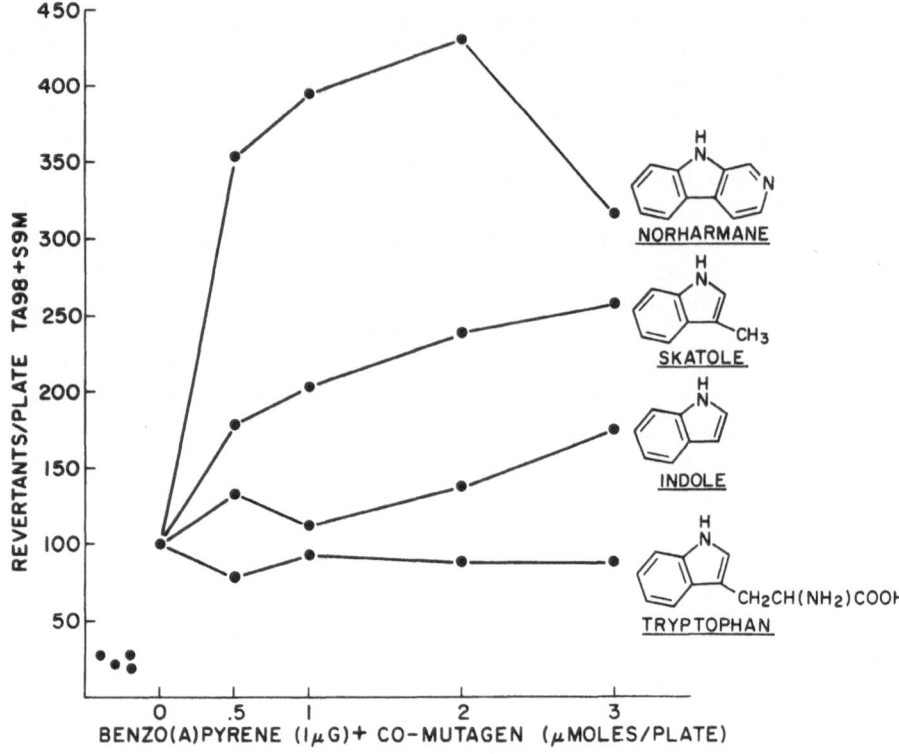

Figure 39.2. Comutagenic activity of norharmane, skatole, indole, and tryptophan.

(Lesca *et al.*, 1979), riboflavin (Sugimura *et al.*, 1977), paraoxon (Okuno *et al.*, 1979), epoxide hydrase inhibitors (Oesch *et al.*, 1977), vegetable extracts (Dierickx, 1979; Kada *et al.*, 1978), urine (Minnich *et al.*, 1976), eosin (Nagai, 1976), neohesperidin dihydrochalone (MacGregor, 1979), and cumene hydroperoxide (Callen *et al.*, 1978).

Although norharmane has been reported to enhance the mutagenicity of a number of mutagen/carcinogens, the activity of skatole is more specific (Table 39.5). Interestingly, two nonmutagens, aniline and *ortho*-toluidine, showed potent mutagenic activity only when tested in the presence of norharmane (Nagao *et al.*, 1977a). This finding came under fire since the evidence for carcinogenicity of aniline was largely negative, although *ortho*-toluidine is accepted as a bladder carcinogen. This criticism should be re-examined in view of a recent cancer study in which tumors were induced by aniline (Anon., 1978). Norharmane and

some analogues, possibly including tryptophan metabolites, may exhibit a variety of actions, including DNA binding (Hayashi *et al.*, 1977) and inhibition of repair (Remsen and Cerutti, 1979), but the critical effect seems to be on the balance of activating and deactivating enzyme activities in the liver homogenate (Ashby and Styles, 1978; Fujino *et al.*, 1978; Wright *et al.*, 1979).

If the comutagenic action of norharmane is due to enzyme inhibition/activation, we could predict that a number of drugs affecting microsomal mixed function oxidases (Cinti, 1978; Ivanetich *et al.*, 1978), as well as chemicals acting as competitive enzyme substrates (Ashby *et al.*, 1978), would inhibit or enhance the *in vitro* mutagenicity of carcinogens requiring metabolic activation. In this regard we have observed that two modifiers, metyrapone (Leibman and Ortiz, 1973) and pargyline (Demaster and Nagasawa, 1978), enhance the mutagenicity of benzo (a) pyrene but not aniline in this type

Table 39.5. *Revertants per Plate (TA98 + S9M) for Carcinogens Alone and in Presence of Norharmane or Skatole (2 μm/plate)*

	Carcinogen	+ Norharmane	+ Skatole
B(a)P	102	398	234
DAB	38	376	40
o-tol	21	287	18
anil	21	184	17

B(a)P, benzo(a)pyrene 1 μg; DAB, dimethylaminoazobenzene 50 μg; o-tol, o-toluidine 50 μg; anil, aniline 50 μg.

of comutagenesis assay. Whether these and similar results have any bearing on *in vivo* co- or anticarcinogenesis is not known, but would likely be signaled by *in vivo* pharmacological activity.

It has been suggested that routine incorporation of norharmane in microbial mutagenicity screening assays should be resisted until the results of such tests can be interpreted. On the other hand, it can be argued that the widely used S9M is such a poor representation of normal *in vivo* metabolism (Selkirk, 1977), that further distortion by norharmane will not harm the assay but might even improve it. Although that issue remains unresolved, both types of *in vitro* comutagenesis assays are recommended without reserve for compounds such as saccharin and amaranth where any clue regarding mechanism of action would be welcome. Unfortunately, saccharin is neither converted into a mutagen in the presence of norharmane (Nagao *et al.*, 1977b) nor does it act as a comutagen for several carcinogens (Rao *et al.*, 1979).

Amaranth (red number 2) is another food additive of equivocal carcinogenicity that has been examined for *in vitro* comutagenicity. Again, norharmane failed to cause conversion of amaranth into a mutagen, but in the other type of comutagenesis assay, an interaction was observed: amaranth suppressed the mutagenicity of α- and β-naphthylamines (Stoltz *et al.*, 1979). Whether this observation is related to the ability of Sudan III and IV, other azo dyes, to prevent 7,8,12-trimethylbenz(a)anthracene-induced leuke-

mia in rats (Huggins *et al.*, 1978) is unknown.

In the examples given above, attempts have been made to relate the results of microbial mutagenicity testing of pairs of compounds to co-/anticarcinogenesis. Although it was implied that enzyme activation/inhibition was responsible for any interactions reported, alternate ideas such as saturation of specific DNA repair pathways have not been excluded (Clarke and Shankel, 1975). The foregoing discussion inspires little confidence in the ability of *in vitro* microbial mutagenicity assays to detect modifiers of carcinogenesis, unless the modifier was employed to pretreat the animal source of the metabolic activation system.

Furthermore, it is apparent that results of testing mixtures in the standard *Salmonella* assay with an Aroclor-induced A9 must be interpreted cautiously with respect to whether or not a mutagen was present in the mixture, irrespective of the issue of co-/anticarcinogenesis. The inhibition of naphthylamine mutagenicity by amaranth indicates that when screening for a known mutagenic contaminant, one must test spiked samples to ensure that mutagenicity is indeed expressed in the mixture. Similarly, complex mixtures which show no mutagenicity should be spiked with at least one mutagen to rule out inhibitory effects as demonstrated in Athabasca tar sand fractions by Shahin and Fournier (1978). Inhibitors of mutagenesis have also been demonstrated in extracts of meat (Pariza *et al.*, 1979) and feces (Bruce *et al.*, 1977). On the

Table 39.6. Comutagenicity of Organic Solvent-Soluble Impurities of Sodium Saccharins

Saccharin	Revertants/plate (TA1538 + S9M)		
	I^a	BP^b	I + BP
1648/38	24.5	40.7	188
1648/38 fraction A	19.3	36.8	148.7
fractions B-D	14.3–20.3	36.8	37.0–49.0
subfraction E[c]	22.3	35.4	79.0
subfractions F-I	10.3–15.7	35.4	26.3–37.0
1648/41	26.5	32.3	120.0
1648/41	23.0	41.5	199.5
1648/41 fraction 1	10.0	40.3	145.3
fractions 2-14	12.0–24.7	40.3	30.3–44.7

[a] Impurities.

[b] Benzo(a)pyrene 1 μg/plate.

[c] E to I are subfractions of A.

other hand, when attempting to isolate a mutagen from a complex mixture, one should bear in mind the norharmane experience. If mutagenicity disappears upon fractionation of the mixture, a comutagenicity assay may be essential to isolation of the component of interest. Such an approach was adapted in our attempts to isolate mutagens from saccharin impurities (Table 39.6). In this instance, although the results suggest the existance of a comutagen among the impurities, subsequent studies indicated the presence of trace amounts of several mutagenic substances.

Permissive Influences Acting on the Responding Tissue

Microbial mutagenicity assays would not be expected to respond to cocarcinogens, such as hormones which appear to promote specific organ carcinogenesis.

Permissive Influences Acting Systemically

Although many chemical carcinogens can alter immune functions, few co- or anticarcinogens are known to act specifically by this mechanism. One would not expect microbial mutagenicity assays to detect such chemicals unless the effects on immune status were due to DNA damage in a susceptible cell population.

Involvement in the Carcinogenic Process: Initiation

The outcome of concurrent treatment of animals with two initiating agents may range from synergistic to antagonistic effects (National Health and Welfare, 1973; Schmähl, 1976). Similarly, nonadditive results may be encountered in microbial mutagenicity assays with pairs of mutagens, particularly when metabolic activation is considered (Guttenplan, 1979; Saffiotti *et al.*, 1979; Salamone *et al.*, 1979; Schubert *et al.*, 1978). Lessons learned from mutagenicity testing of combinations of mutagens have relevance primarily to the mutagenicity testing of complex mixtures.

Involvement in the Carcinogenic Process: Promotion

Recent research on the biochemical effects of a potent tumor promoter, 12-0-tetradecanoylphorbol-13-acetate (TPA), inactive TPA analogues, and antipromoter actions of anti-inflammatory agents permits speculation on the mechanism of action of promoters. Steps in promotion by TPA may include (a) liberation of lipids from cell membranes as substrate for (b) prostaglandin synthesis resulting in (c) perturbation of cyclic nucleotide levels, which permits (d) gene segregation leading ultimately to (e) recessive homozygosis and expression of carcinogen-induced recessive mutations. Pertinent intermediary events may include gene derepression, polyamine metabolism, and alterations in deoxyribonucleotide pools. In this scheme, most of the reported activities of promoters are incorporated into one linear relationship. It remains to be seen whether this concept is correct or whether several distinct pathways can effect promotion. If the latter is true, the super-promoter TPA may activate multiple pathways, while less effective promoters may stimulate single chains of events.

Although reports suggest that TPA may be comutagenic (Soper and Evans 1977) and not mutagenic in bacterial assays (Fresen and Zur Hausen, 1976), the proposed steps in promotion, if they bear any resemblance to fact, indicate that tumor promotion should be addressed with mammalian cells.

Protease inhibitors may provide an example of a class of anticarcinogens acting at the level of promotion by preventing the induction of error-prone or recombination repair. This activity may be reflected by antipain inhibition of SOS functions in *E. coli* (Troll *et al.*, 1978), antimutagenic effects of elastatinal in *S. typhimurium* TA1535 (Umezawa *et al.*, 1979), and inhibition of sister chromatid exchanges by antipain and leupeptin (Kinsella and Radman, 1978; Nagasawa and Little, 1979).

Conclusion

Cocarcinogens and particularly promoters may play the critical role in the pathogen-

esis of human cancer. Thus it is imperative that more effort be directed toward development of short-term detection assays for co- and anticarcinogens, validation with *in vivo* systems, elucidation of mechanisms, and development of regulatory strategies. The contribution of microbial mutagenicity assays in detecting co- and anticarcinogens is not yet clearly defined.

References

Anon. (1978): Clearinghouse finds seven compounds are carcinogenic, threat to humans. The Cancer Letter 4(35): 5.

Ashby, J., Styles, J.A. (1978): Comutagenicity, competitive enzyme substrates, and *in vitro* carcinogenicity assays. Mutat. Res. 54:105–112.

Ashby, J., Styles, J.A., Paton, D. (1978): *In vitro* evaluation of some derivatives of the carcinogen butter yellow: implications for environmental screening. Br. J. Cancer 38:34–50.

Berenblum, I. (1978): Historical perspective. In: T.J. Slaga, A. Sivak, and R.K. Boutwell (eds.), Mechanisms of Tumor Promotion and Cocarcinogenesis. New York: Raven Press, pp. 1–10.

Bowden, J.P., Chung, K.-T., Andrews, A.W. (1976): Mutagenic activity of tryptophan metabolites produced by rat intestinal microflora. J. Natl. Cancer Inst. 57:921–924.

Bruce, W.R., Varges, A.J., Furrer, R., Land, P.C. (1977): A mutagen in the feces of normal humans. In: H.M. Hiatt, J.D. Watson, and J.A. Winsten (eds.), Origins of Human Cancer, Book C, Cold Spring Harbor Conferences on Cell Proliferation, Vol. 4. Cold Spring Harbor Laboratory, New York, pp. 1641–1646.

Callen, D.F., Wolf, C.R., Philpot, R.M. (1978): Cumene hydroperoxide and yeast cytochrome P-450: spectral interactions and effect on the genetic activity of promutagens. Biochem. Biophys. Res. Commun. 83:14–20.

Cardy, R.H., Lijinsky, W., Hilderbrandt, P.K. (1979): Neoplastic and non-neoplastic urinary bladder lesions induced in Fischer 344 rats and B6C3F$_1$ hybrid mice by N-nitrosodiphenylamine. Ecotoxicol. Environ. Safety 3:29–35.

Chowaniec, J., Hicks, R.M. (1979): Response of the rat to saccharin with particular reference to the urinary bladder. Br. J. Cancer *39*:355–374.

Cinti, D.L. (1978): Agents activating the liver microsomal mixed function oxidase system. Pharmacol. Ther. *A2*:727–749.

Clarke, C.H., Shankel, D.M. (1975): Antimutagenesis in microbial systems. Bacteriol. Rev. *39*:33–53.

Demaster, E.G., Nagasawa, H.T. (1978): Inhibition of aldehyde dehydrogenase by propiolaldehyde, a possible metabolite of pargyline. Res. Commun. Chem. Path. Pharmacol. *21*:497–505.

Dierickx, P.J. (1979): Deactivation of sodium azide in the Salmonella/microsome test. Bull. Environ. Contam. Toxicol. *22*:660–665.

Elespuru, R.K., Lijinsky, W. (1976): Mutagenicity of cyclic nitrosamines in *Escherichia coli* following activation with rat liver microsomes. Cancer Res. *36*:4099–4101.

Evarts, R.P., Brown, C.A. (1977): Effect of L-tryptophan on diethylnitrosamine and 3′-methyl-4-N-dimethylaminoazobenzene hepatocarcinogenesis. Food Cosmet. Toxicol. *15*:431–435.

Fresen, K. O., Zur Hausen, H. (1976): Establishment of EBNA-expressing cell lines by infection of Epstein-Barr virus (EBV)-genome-negative human lymphoma cells with different EBV strains. Int. J. Cancer *17*:161–166.

Fujino, T., Fujiki, H., Nagao, M., Yahagi, T., Seino, Y., Sugimura, T. (1978): The effect of norharman on the metabolism of benzo(a)pyrene by rat liver microsomes *in vitro* in relation to its enhancement of the mutagenicity of benzo(a)pyrene. Mutat. Res. *58*:151–158.

Guttenplan, J.B. (1979): Comutagenic effects exerted by N-nitroso compounds. Mutat. Res. *66*:25–32.

Hayashi, K., Nagao, M., Sugimura, T. (1977): Interactions of norharman and harman with DNA. Nucleic Acids Res. *4*:3679–3685.

Huggins, C.B., Ueda, N., Russo, A. (1978): Azo dyes prevent hydrocarbon-induced leukemia in the rat. Proc. Natl. Acad. Sci. USA *75*:4524–4527.

Ivanetich, K.M., Lucas, S., Marsh, J.A., Ziman, M.R., Katz, I.D., Bradshaw, J.J. (1978): Organic compounds: their interaction with and degradation of hepatic microsomal drug-metabolizing enzymes *in vitro*. Drug Metab. Disp. *6*:218–225.

Kada, T., Morita, K., Inoue, T. (1978): Antimutagenic action of vegetable factor(s) on the mutagenic principle of tryptophan pyrolysate. Mutat. Res. *53*:351–353.

Kinsella, A.R., Radman, M. (1978): Tumor promoter induces sister chromatid exchanges: relevance to mechanisms of carcinogenesis. Proc. Natl. Acad. Sci. USA *75*:6149–6153.

Lee, K., Gold, B., Mirvish, S.S. (1977): Mutagenicity of 22 N-nitrosamides and related compounds for *Salmonella typhimurium*. Mutat. Res. *48*:131–138.

Leibman K.C., Ortiz, E. (1973): Metyrapone and other modifiers of microsomal drug metabolism. Drug Metab. Disp. *1*:184–189.

Lesca, P., Lecointe, P., Paoletti, C., Mansuy, D. (1979): Ellipticines as potent inhibitors of microsomes-dependent chemical mutagenesis. Chem.-Biol. Interact. *25*:279–287.

MacGregor, J.T. (1979): Mutagenicity study of flavonoids in vivo and *in vitro*. Toxicol. Appl. Pharmacol. *48*:A47.

Mandel, M., Ichinotsubo, D., Mower, H. (1977): Nitroso group exchange as a way of activation of nitrosamines by bacteria. Nature (Lond.) *267*:248–149.

Matsumoto, M., Oyasu, R., Hopp, M.L., Kitajima, T. (1977a): Suppression of dibutylnitrosamine-induced bladder carcinomas in hamsters by dietary indole. J. Natl. Cancer Inst. *58*:1825–1829.

Matsumoto, T., Yoshida, D., Mizusaki, S. (1977b): Enhancing effect of harman on mutagenicity in *Salmonella*. Mutat. Res. *56*:85–88.

Minnich, V., Smith, M.E., Thompson, D., Kornfeld, S. (1976): Detection of mutagenic activity in human urine using mutant strains of *Salmonella typhimurium*. Cancer *38*:1253–1258.

Montesano, R., Bartsch, H. (1976): Mutagenic and carcinogenic N-nitroso compounds: possible environmental hazards. Mutat. Res. *32*:179–228.

Nagai, S. (1976): Counteracting effect of eosin and related dyestuffs on the production of respiration-deficient mutants in yeast by 4-nitroquinoline 1-oxide. Mutat. Res. *34*:187–194.

Nagao, M., Yahagi, T., Honda, M., Seino, Y., Matsushima, T., Sugimura, T. (1977a): Demonstration of mutagenicity of aniline and o-toluidine by norharman. Proc. Japan. Acad. *53*:34–37.

Nagao, M., Yahagi, T., Kawachi, T., Seino, Y., Honda, M., Matsukura, N., Sugimura, T.,

Wakabayashi, K., Tsuji, K., Kosuge, T. (1977b): Mutagens in foods, and especially pyrolysis products of protein. In: D. Scott, B.A. Bridges and F.H. Sobels (eds.), Progress in Genetic Toxicology. Amsterdam: Elsevier/North-Holland Biomedical Press, pp. 259–264.

Nagao, Yahagi, T., Kawachi, T., Sugimura, T., Kosuge, T., Wakabayashi, K., Mizusaki, S., Matsumoto, T. (1977c): Comutagenic action of norharman and harman. Proc. Japan Acad. 53:95–98.

Nagasawa, H., Little, J.B. (1979): Effect of tumor promoters, protease inhibitors, and repair processes on X-ray-induced sister chromatid exchanges in mouse cells. Proc. Natl. Acad. Sci. USA 76:1943–1947.

National Health and Welfare (1973): The Testing of Chemicals for Carcinogenicity, Mutagenicity and Teratogenicity. Ottawa, Canada, pp. 44–45.

Neurath, G.B., Dunger, M., Pein, F.G., Ambrosius, D., Schreiber, O. (1977): Primary and secondary amines in the human environment. Food Cosmet. Toxicol. 15:275–282.

Oesch, F., Raphael, D., Schwind, H., Glatt, H.R. (1977): Species differences in activating and inactivating enzymes related to the control of mutagenic metabolites. Arch. Toxicol. 39:97–108.

Okuno, S., Takeishi, K., Seno, T. (1979): Differential effect of a microsomal deacetylase inhibitor on the mutagenicity in Salmonella typhimurium of 2-acetylaminofluorene by liver homogenates of guinea pigs, mice and rats. Cancer Lett. 6:1–5.

Pariza, M.W., Ashoor, S.H., Chu, F.S., Lund, D.B. (1979): Mutagens and inhibitors of mutagenesis in pan-fried hamburger. Proc. Am. Assoc. Cancer Res. 20:39.

Peraino, C., Fry, R.J.M., Grube, D.D. (1978): Drug induced enhancement of hepatic tumorigenesis. In: T.J. Slaga, A. Sivak and R.K. Boutwell (eds.), Mechanisms of Tumor Promotion and Cocarcinogenesis. New York: Raven Press, pp. 421–432.

Rao, T.K., Stoltz, D.R., Epler, J.L. (1979): Lack of enhancement of chemical mutagenesis by saccharin in the Salmonella assay. Arch. Toxicol. 43:141–145.

Remsen, J.F., Cerutti, P.A. (1979): Inhibition of DNA repair and DNA synthesis by harman in human alveolar tumor cells. Biochem. Biophys. Res. Comm. 86:124–129.

Saffiotti, U., Donovan, P.J., Rice, J.M. (1979): Interactions of multiple carcinogens in the Salmonella mutagenesis assay (Ames). Proc. Am. Assoc. Cancer Res. 20:191.

Salamone, M.F., Heddle, H.A., Katz, M. (1979): The use of the Salmonella/microsomal assay to determine mutagenicity in paired chemical mixtures. Can. J. Genet. Cytol. 21:101–107.

Schmähl, D. (1976): Combination effects in chemical carcinogenesis (experimental results). Oncology 33:73–76.

Schubert, J., Riley, E.J., Tyler, S.A. (1978): Combined effects in toxicology— a rapid systematic testing procedure: cadmium, mercury and lead. J. Toxicol. Environ. Health 4:763–776.

Selkirk, J.K. (1977): Divergence of metabolic activation systems for short-term mutagenesis assays. Nature (Lond.) 270:604–607.

Shahin, M.M., Fournier, F. (1978): Suppression of mutation induction and failure to detect mutagenic activity with Athabasca tar sand fractions. Mutat. Res. 58:29–34.

Shand, F.L., Howard, J.G. (1979): Induction in vitro of reversible immunosuppression and inhibition of B cell receptor regeneration by defined metabolites of cyclophosphamide. Europ. J. Immunol. 9:17–21.

Silverman, S.J., Andrews, A.W. (1977): Bile acids: co-mutagenic activity in the Salmonella-mammalian-microsome mutagenicity test. J. Natl. Cancer Inst. 59:1557–1559.

Soper, C.J., Evans, F.J. (1977): Investigations into the mode of action of the cocarcinogen 12-o-tetradecanoyl-phorbol-13-acetate using auxotrophic bacteria. Cancer Res. 37:2487–2491.

Stoltz, D.R., Stavric, B., Iverson, F., Bendall, R., Klassen, R. (1979): Suppression of naphthylamine mutagenicity by amaranth. Mutat. Res. 60:391–393.

Sugimura, T., Nagao, M., Kawachi, T., Honda, M., Yahagi, T., Seino, Y., Sato, S., Matsukura, N., Matsushima, T., Shirai, A., Sawamura, M., Matsumoto, H. (1977): Mutagen-carcinogens in food, with special reference to highly mutagenic pyrolytic products in broiled foods. In: H.H. Hiatt, J.D. Watson and J.A. Winsten (eds.), Origins of Human Cancer, Book C, Cold Spring Harbor Conferences on Cell Proliferation, Vol. 4. Cold Spring Harbor Laboratory, New York, pp. 1561–1577.

Troll, W., Meyn, M.S., Rossman, T.G. (1978):

Mechanisms of protease action in carcinogenesis. In: T.J. Slaga, A. Sivak and R.K. Boutwell (eds.), Mechanisms of Tumor Promotion and Cocarcinogenesis. New York: Raven Press, pp. 301–312.

Umezawa, K., Matsushima, T., Sugimura, T. (1979): Antimutagenic effect of elastatinal, a protease inhibitor from Actinomycetes. Proc. Japan Acad. 53:30–33.

Wattenberg, L.W., Loub, W.D. (1978): Inhibition of polycyclic aromatic hydrocarbon-induced neoplasia by naturally occurring indoles. Cancer Res. 38:1410–1413.

Wattenberg, L.W., Lam, L.K.T., Fladmoe, A.V. (1979): Inhibition of chemical carcinogen-induced neoplasia by coumarins and α-angelicalactone. Cancer Res. 39:1651–1654.

World Health Organization (1977): Environmental Health Criteria. 5. Nitrates, Nitrates and N-Nitroso Compounds. Geneva, pp. 63–65.

Wright, E.E., Bird, J.L., Feldman, J.M. (1979): The effect of harmane and other monoamine oxidase inhibitors on N-acetyltransferase activity. Res. Commun. Chem. Path. Pharmacol. 24:259–272.

Yahagi, T., Nagao, M., Seino, Y., Matsushima, T., Sugimura, T., Okada, M. (1977): Mutagenicities of N-nitrosamines on Salmonella. Mutat. Res. 48:121–130.

40

The Use of a Bacterial Assay to Identify Which Agents Modify Carcinogen-Induced Mutagenesis

Miriam P. Rosin

Introduction

Carcinogens are present in the environment and in man himself as components of complex mixtures. Hence, the overall activity of a given carcinogen is dependent on the presence of a variety of modifying agents, some of which enhance while others suppress the action of the carcinogen. Thus, a realistic appraisal of a carcinogen's hazard to man requires the identification of such modifying agents in his environment.

One efficient method of identifying such agents involves an analysis of an agent's ability to alter carcinogen-induced mutagenesis in bacteria assays. This approach has been used frequently during the last 3 years (Oesch *et al.*, 1976; Guttenplan, 1977; Jacobs *et al.*, 1977; Nagao *et al.*, 1977; Rosin and Stich, 1978a, 1978b, 1979, 1980; Rannug *et al.*, 1978; Kada *et al.*, 1978; Moriya *et al.*, 1978; Shamberger *et al.*, 1979; Lesca *et al.*, 1979). Our laboratory uses a modification of the *Salmonella* mutagenesis assay (Ames *et al.*, 1975) to obtain a "quantitative" approach to the study of the influence of various antioxidants, vitamins, and food additives on the mutagenic activity of model carcinogens. This modified procedure involves a measurement of the influence of test agents on the frequency of reversion to histidine prototrophy in *S. typhimurium* cultures which had been suspended in treatment mixtures of carcinogens and agents prior to plating on minimal glucose agar plates. In this report, we will attempt to summarize the advantages and limitations of this assay, as well as to indicate the type of data on carcinogen-modifying agent interactions which can be obtained with this experimental approach.

Methods

Material

Nutrient broth, nutrient agar plates, minimal glucose agar plates, top agar, and S9 mix were prepared according to Ames and co-workers (1975).

Growth of Bacterial Cultures

Cultures of *Salmonella typhimurium* strains TA 100, TA 1535 and TA 98 were obtained from Ames and co-workers (Berkeley, California, U.S.A.). Permanent stocks were prepared by mixing overnight broth cultures of the bacteria with dimethylsulfoxide (DMSO) to a final concentration of 8% DMSO and freezing the stocks in liquid nitrogen ($-80°$ C). Working stocks were prepared in a similar fashion but kept in a freezer at $-20°$ C.

Aliquots of frozen working stock were transferred with sterile wooden sticks to nutrient broth (5 ml) in sterile polypropylene tubes. These cultures were grown overnight on a rotary wheel (31 rpm) in an incubator at 37° C. After 16 hours, 0.1 ml of this growth was reinoculated into fresh nutrient broth (5 ml) and returned to the rotary wheel for 4 hours. Cultures at this time were in late logarithmic stage growth with $7-10 \times 10^8$ cells/ml.

Preparation of Treatment Mixtures

All chemical solutions were prepared immediately before the experiment began. Water-soluble chemicals were dissolved directly into phosphate-buffered saline (PBS : components/liter, 8 g sodium chloride, 0.2 g potassium chloride, 1.15 g dibasic sodium phosphate, and 0.2 g monobasic potassium phosphate, pH 7.4). Chemicals with lower water solubility (N-acetoxy-2-acetylaminofluorene, N-hydroxy-2-acetylaminofluorene, propyl gallate, disulfiram, α-tocopherol succinate) were dissolved in DMSO followed by dilution with PBS. The lowest possible DMSO concentration was employed. Serial dilutions of these stock solutions were made with a DMSO-PBS mixture so that all final DMSO concentrations would be the same. Experiments in which a comparison was

made between mutagenic activities of carcinogens in the presence or absence of modifying agents all contained appropriate controls for the influence of solvent concentration on the carcinogen itself. These precautions were taken because of the large influence DMSO appeared to have on both carcinogen and modifying agent activities in this assay (Rosin and Stich, 1979). For example, the mutagenic activity of N-methyl-N'-nitro-N-nitrosoguanidine increased significantly with higher DMSO concentrations in the testing mixture (Fig. 40.1). The activity of another carcinogen, N-hydroxy-2-acetylaminofluorene, could be completely abolished by adding sufficient DMSO (10% by total volume) to treatment mixtures (data not shown). The highest DMSO concentration employed was 10% DMSO. This solvent concentration was nontoxic to bacteria for a 20-minute exposure period. All solutions were adjusted to pH 7.4 before use.

S9 mix (Ames *et al.*, 1975) was incorporated into treatment mixtures involving carcinogens which required metabolic activation. The quantity of liver homogenate used in the S9 mix was adjusted (60–210 μl/ml

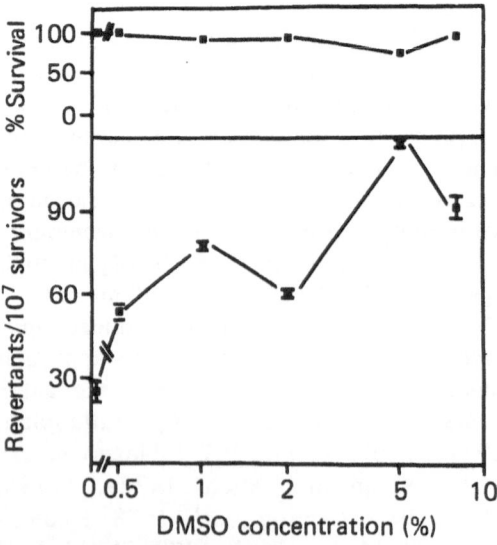

Figure 40.1. The influence of solvent concentration of DMSO on MNNG-induced mutagenesis. Bacteria strain was TA 1535. Values are mean ± S.D. (n = 3).

S9 mix) to give maximum carcinogen activity. Liver homogenate obtained from arochlor 1254-treated rats was supplied by Litton Bionetics, Kensington, Maryland.

Chemicals were prepared at two times (to assay modifying effect on direct-acting carcinogen) or three times (effect on carcinogen requiring S9 activation) final desired concentrations and mixed together (in the order of modifying agent, S9 mix, carcinogen) just prior to suspending the bacteria in them. For treatments involving S9 mix, 1 ml of a mixture of 0.5 ml modifying agent (or solvent), 0.5 ml S9 mix, and 0.5 ml carcinogen was added to the bacteria. Direct-acting carcinogens were mixed at a 1:1 ratio with the modifying agent to be tested and the bacteria were suspended in this mixture.

Experimental Procedure

Logarithmically growing bacteria cultures were vortexed and pooled in a sterile flask. Aliquots (1 ml) of this culture were distributed into sterile 5 ml centrifuge tubes. The bacteria were pelleted with a GLC-2B centrifuge (Dupont Instruments, Newtown, CT, U.S.) (1600 g, 5 min). The nutrient broth was siphoned off the bacterial pellets and the bacteria were resuspended in the treatment mixtures (total volume 1 ml/centrifuge tube) with a vortex. Centrifuge tubes were capped with parafilm and returned to the incubator (37° C) for 20 minutes. At the end of this time, the bacteria were pelleted by centrifugation. The treatment mixtures were siphoned off the pellets, and the bacteria were resuspended in PBS. After centrifugation, this PBS wash was removed and the bacteria were again resuspended in PBS (total volume 1 ml/centrifuge tube).

Assay for Mutagenicity

Aliquots of the treated bacteria (0.1 ml, equivalent to 10^8 cells) were added to 2.0 ml molten top agar (Ames et al., 1975) in sterile polypropylene tubes. The tubes were gently agitated and the mixtures were poured onto minimal glucose agar plates. A minimum of three replica plates were prepared per treatment. Plates were incubated at 37° C for 48 hours.

Assay for Toxicity

Aliquots of the treated bacteria suspension were diluted by 10^7 with a 0.85% NaCl solution and 0.5 ml of this dilution was spread onto nutrient agar plates (produces about 50 colonies per plate). A minimum of three replica plates were used per treatment. Plates were incubated for 48 hours at 37° C.

Scoring Results

Colonies were counted with an Artek Counter, model 880 (Artek Systems Corporation, Farmingdale, N.Y. 11735). Mutagenic activity was expressed as the number of histidine revertant colonies per 10^7 surviving bacteria cells (R/S). All values for mutagenic activity were corrected for the extent of spontaneous mutation in the presence of the appropriate solvent concentration. Average values for the various strains used were: TA1535, 1.2; TA 98, 1.6; and TA 100, 4.8 revertants per 10^7 survivors.

The calculation of % inhibition of carcinogen-induced mutagenesis was:

$$\frac{R/S \text{ carcinogen alone} - R/S \text{ carcinogen + agent}}{R/S \text{ carcinogen alone}} \times 100$$

Values of % inhibition of mutagenic activity which were 20% or less were considered as not significant since such values were only slightly above the variation observed between replica treatments (bacteria resuspended in same treatment mixture in more than 1 tube, ± 10% variation in revertants/10^7 survivors or R/S).

Results and Discussion

Our laboratory has used this microbial mutagenesis assay to study the influence of a variety of chemicals on the mutagenic activity of carcinogens from different molecular groups. The data obtained with such studies characterizes such interactions in the following ways:

1. *The assay may be used to study the influence of modifying agents on the toxic as well as mutagenic activities of carcinogens.* It has been reported that the antioxidant cysteamine inhibits 7,12-dimethyl-benz-(a)anthracene-induced transformation of mouse fibroblast cultures without affecting the toxic effect of this carcinogen (Marquardt *et al.*, 1974). This work suggests that the toxic and oncogenic effects of a carcinogen such as 7,12-dimethyl-benz(s)anthracene are a result of at least two different metabolites, with the modifying agent cysteamine interacting with only the oncogenic form(s). The *Salmonella* mu-

tagenesis assay provides a quick method of examining carcinogen-agent interactions for the presence of such effects. For ex-

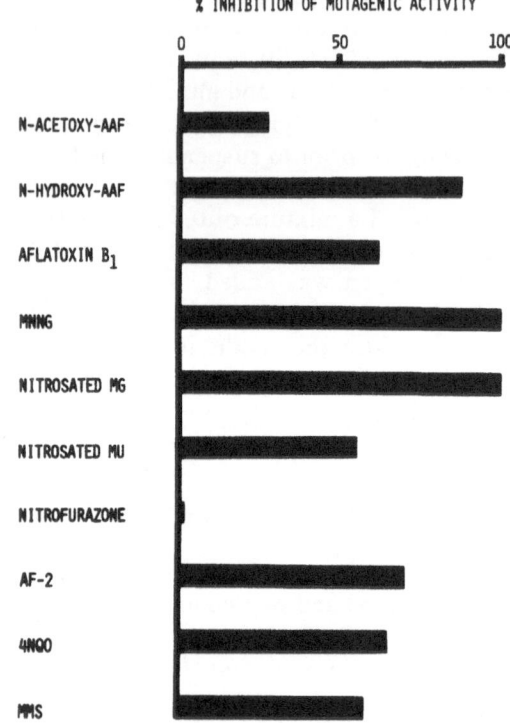

Figure 40.3. The inhibition of mutagenic activities of several carcinogens in the presence of cysteine. Cysteine (2.5 mM) was added to treatment mixtures consisting of the following carcinogens: N-acetoxy-AAF, N-acetoxy-2-acetylaminofluorene, 5×10^{-5} M; N-hydroxy-AAF, N-hydroxy-2-acetylaminofluorene, 5×10^{-4} M; aflatoxin B_1, 4×10^{-6} M; MNNG, N-methyl-N'-nitro-N-nitrosoguanidine, 10^{-5} M; nitrosated methylguanidine (MG), and nitrosated methylurea (MU) with starting concentrations of MG and MU of 5×10^{-5} M and 1.2×10^{-3} M, respectively (Rosin and Stich, 1978b); nitrofurazone, 5-nitro-2-furaldehyde semicarbazone, 10^{-5} M; AF-2, 2-(2-furyl)-3-(5-nitro-2-furyl)-acrylamide, 10^{-7}M; 4NQO, 4-nitroquinolone-1-oxide, 2×10^{-6} M; and MMS, methyl methanesulfonate, 2×10^{-2} M. Bacteria strain was TA1535. Carcinogen concentrations were those which resulted in the induction of a similar mutagenic frequency (40–70 revertants/10^7 survivors) during the treatment time. S9 mix was employed with aflatoxin B_1 treatments. Cysteine at higher doses (20–300 mM) also reduced nitrofurazone-induced mutagenesis.

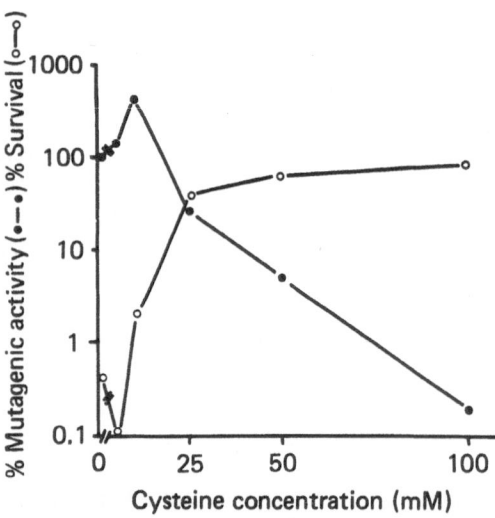

Figure 40.2. The influence of cysteine on the toxic and mutagenic actions of 4-nitroquinoline-1-oxide. Bacteria strain was TA 100. The percent mutagenic activity was activity compared with sample containing 4NQO (10^{-4} M) alone (312 R/S, 100%). The percent survival was survival of indicated treatment compared with survival in 4NQO alone.

ample, as the cysteine concentration decreases in treatment mixtures of 4-nitroquinoline-1-oxide (10^{-4}M), the inhibitory influence of this antioxidant on both carcinogen-induced mutagenesis and toxicity decreases (Fig. 40.2). However, at lower doses ($5 \times 10^{-3} - 10^{-2}$M) cysteine appears to enhance the mutagenic activity of this concentration of 4NQO while having a minimal effect on cell survival.

2. *The assay can indicate the specificity of carcinogen-modifying agent interactions.* Data from these studies may be used to determine which agents influence the *in vitro* mutagenic activity of particular carcinogens (Rosin and Stich, 1979). For example, cysteine inhibits the mutagenic activity of N-methyl-N-nitro-N-nitroso-guanidine (MNNG) as well as a range of other carcinogens (Fig. 40.3; Rosin and Stich, 1978b). In contrast, MNNG-induced mutagenesis is not affected by the presence of butylated hydroxyanisole, α tocopherol-succinate, tryptophan or methionine (Table 40.1). This specificity of inhibition of carcinogen-induced mutagenesis may be related to the different

mechanisms of action of the modifying agents. For example, agents such as ellipticines, which are felt to inhibit microsome-dependent chemical mutagenesis by interacting with cytochromes P-450, may be inactive with carcinogens which do not require such activation (Lesca *et al.*, 1979).

3. *It is possible to predict which modifying agent is the more potent inhibitor of the mutagenic activity of a carcinogen* (Rosin and Stich, 1979). For example, sodium bisulfite and cysteamine are very potent inhibitors of MNNG-induced mutagenesis, whereas ascorbate and serine are less effective (Table 40.1). By assaying a range of concentrations of the test agent in combination with a given carcinogen, an estimation can be made of the concentration of the agent necessary to cause an inhibition of 50% of the mutagenic action of the carcinogen (termed IN_{50}, Rosin and Stich, 1979). By employing such data, comparisons can be made among various agents for efficiencies of inhibition of carcinogen-induced mutagenesis. Such values can also be used to compare the degree of inhibition observed with the same carcinogen-agent

Table 40.1.   *Effect of Various Agents on Mutagenic Activity of MNNG*[a]

| Modifying agent | % Inhibition of MNNG-induced mutagenesis, concentration of agent (M) | | | |
	10^{-2}	10^{-3}	10^{-4}	10^{-5}
Cysteine	100	100	NT[3]	NT
Glutathione	100	100	NT	NT
Cysteamine	100	100	65	20
Sodium selenite	93	54	47	27
Ascorbate	87	12	15	19
Methionine	0	23	NT	NT
Serine	37	39	NT	NT
Propyl gallate	98	36	36	33
3,3'-Thiodiproprionic acid	22	17	NT	NT
Tryptophan	26	11	NT	NT
Sodium bisulfite	100	100	90	26
α-Tocopherol succinate	—[c]	—	12	4
Disulfiram[b]	—	—	31	22
Butylated hydroxyanisole	—	—	1	0

[a]Strain TA 1535, MNNG concentration was 5×10^{-6} M, mutagenic activity in absence of modifying agent was 503 R/S.

[b]Tetraethylthiuram disulfide.

[c]NT, not tested; —, unable to test due to toxicity or water-insolubility of agent.

Table 40.2. *Effect of Propyl Gallate on Carcinogen-Induced Mutagenesis*[a]

Carcinogen	Carcinogen concentration (M)	Mutagenic activity, propyl gallate concentration used (mM)				
		0	0.1	0.5	1	5
4NQO	6×10^{-6}	12.4 ± 1.9	42.1 ± 2.6	49.0 ± 4.5	74.3 ± 0.2	17.5 ± 0.8
N-hydroxy-AAF	2.5×10^{-4}	7.6 ± 1.0	19.3 ± 3.0	44.8 ± 2.8	44.4 ± 3.6	54.4 ± 2.7
N-acetoxy-AAF	1.5×10^{-4}	37.7 ± 8.2	30.8 ± 4.7	16.2 ± 1.5	15.9 ± 4.2	14.6 ± 1.6
MNNG	5×10^{-6}	55.7 ± 8.1	39.1 ± 5.1	25.1 ± 5.5	25.8 ± 2.8	4.0 ± 3.0
Aflatoxin B_1	1×10^{-5}	80.2 ± 8.1	68.2 ± 2.8	42.5 ± 2.0	22.3 ± 2.7	8.4 ± 0.9

[a]Mutagenic activity is expressed as number of histidine revertants per 10^7 survivors $\pm$ standard deviation (n = 3). All values are corrected for spontaneous frequency. Bacteria strains used were: TA 1535 (MNNG), TA 100 (4NQO), TA 98 (N-hydroxy-AAF, N-acetoxy-AAF and aflatoxin B_1).

combinations under different experimental conditions, in different laboratories, and with different experimental endpoints (e.g., carcinogen-induced chromosome aberration, unscheduled DNA synthesis, etc.).

4. *Data from this assay also indicates agents which enhance carcinogen-induced mutagenesis.* The food additive, propyl gallate, has been shown to enhance the mutagenic activities of some carcinogens while suppressing at equimolar doses the mutagenic activities of others (Rosin and Stich, 1980; Table 40.2). For example, N-hydroxy- 2-acetyl -aminofluorene –induced mutagenesis is enhanced by the presence of propyl gallate in treatment mixes. However, the ultimate form of this carcinogen, N-acetoxy-2-acetylaminofluorene, has a reduced mutagenic action in the presence of the same food additive. These results indicate the necessity for studying a large number of interactions of particular modifying agents with various carcinogens before drawing any generalizations. Since the *Salmonella* assay is both quick and inexpensive, it appears to be a valuable technique for such studies.

5. *Other uses.* Since this assay involves a limited exposure of bacteria to a test mixture in solution, it is possible to utilize the procedure to obtain a variety of data. The test agent can be applied as a pretreatment or as a posttreatment to bacteria in order to determine if concurrent exposure of modifying agent and carcinogen is essential

for the carcinogen-induced mutagenesis to be affected (Rosin and Stich, 1978a). Agents which act on the cytochrome P-450 metabolizing system may be effective when given to the bacteria prior to carcinogen exposure. This possibility is currently under study in our laboratory. The use of a liquid suspension test also allows the investigator to follow some agent-carcinogen interactions with a spectrophotometer and to use such data in studies of mechanisms of inhibition of carcinogen-induced mutagenesis (Guttenplan, 1978). Finally, this procedure could also be done in combination with enzyme kinetic assays to follow the effect of agents on specific carcinogen activation-deactivation systems both biochemically and genetically. This would indicate whether a modification in carcinogen metabolite profile was linked to a reduction or enhancement in microbial mutagenesis.

Limitations of the Assay

There are several limitations involved in the use of this assay to study carcinogen-modifying agent interactions. First, there is a physical limitation in studies employing agents which are very toxic and/or very water-insoluble, e.g., α-tocopherol-succinate, butylated hydroxyanisole, disulfiram. It may be difficult to obtain solutions of such agents high enough in concentration to be active without the solvent concentration being prohibitive. Another disadvantage of this assay is that the bacteria appear to be insensitive to the action of some agents, such as peroxide radical-producing compounds. These agents fail to induce reverse mutations in *Salmonella* assays although the same compounds will induce chromosomal aberrations and DNA repair synthesis in Chinese hamster ovary cells, particularly in the presence of trace metals (Stich *et al.*, 1979). Unfortunately, several of these compounds (cysteamine, ascorbate, glutathione, and selenite) appear to be very effective inhibitors of carcinogen-induced microbial mutagenesis (Rosin and Stich, 1978a, 1979). Finally, the assay as yet can not be used to identify agents which affect promotion phases of the carcinogenic process.

As with other short-term *in vitro* assays, it is essential to employ caution when using results from the *in vitro* situation to predict results in *in vivo* long-term animal studies or in man himself. However, a large number of studies employing carcinogens and antioxidants have already demonstrated that inhibitory effects of these agents on microbial mutagenesis parallel reductions in carcinogen-induced neoplasia in animals (Wattenberg, 1976). Short-term *in vitro* tests such as the one described in this paper may contribute much to our understanding of the modulation of carcinogen activity in animals.

References

Ames, B.N., McCann, J., Yamasaki, E. (1975): Methods for detecting carcinogens and mutagens with the *Salmonella*/mammalian-microsome mutagenicity test. Mutat. Res. *31*:347–364.

Guttenplan, J.B. (1977): Inhibition by L-ascorbate of bacterial mutagenesis induced by two N-nitroso compounds. Nature (Lond.) *268*:368–370.

Guttenplan, J.B. (1978): Mechanisms of inhibition by ascorbate of microbial mutagenesis induced by N-nitroso compounds. Cancer Res. *38*:2018–2022.

Jacobs, M.M., Matney, T.S., Griffin, C.A. (1977): Inhibitory effects of selenium on the mutagenicity of 2-acetylaminofluorene (AAF) and AAF derivatives. Cancer Lett. *2*:319–322.

Kada, T., Morita, K., Inoue, T. (1978): Antimutagenic action of vegetable factor(s) on the mutagenic principle of tryptophan pyrolysate. Mutat. Res. *53*:351–353.

Lesca, P., Lecointe, P., Paoletti, C., Mansuy, D. (1979): Ellipticines as potent inhibitors of microsomes-dependent chemical mutagenesis. Chem.-Biol. Interact. *25*:279–289.

Marquardt, H., Sapozink, M.D., Zedeck, M.S. (1974): Inhibition by cysteamine-HCl of oncogenesis induced by 7,12-dimethylbenz(a)anthracene without affecting toxicity. Cancer Res. *34*:3387–3390.

Moriya, M., Kato, K., Shirasu, Y. (1978): Effects of cysteine and a liver metabolic activation system on the activities of mutagenic pesticides. Mutat. Res. *57*:259–263.

Nagao, M., Yahagi, T., Kawachi, T., Sugimura, T., Kosuge, T., Tsuji, K., Wakabayashi, K., Mizusaki, S., Matsumoto, T. (1977): Comutagenic action of norharman and harman. Proc. Japan Acad. *53*:95–98.

Oesch, F., Bentley, P., Glatt, H.R. (1976): Prevention of benzo(a)pyrene-induced mutagenicity by homogenous epoxide hydratase. Int. J. Cancer *18*:448–452.

Rannug, U., Sundvall, A., Ramel, C. (1978): The mutagenic effect of 1,2-dichloroethane on *Salmonella typhimurium* 1. Activation through conjugation with glutathione *in vitro*. Chem.-Biol. Interact. *20*:1–16.

Rosin, M.P., Stich, H.F. (1978a): Inhibitory effect of reducing agents on N-acetoxy- and N-hydroxy-2-acetylaminofluorene-induced mutagenesis. Cancer Res. *38*:1307–1310.

Rosin, M.P., Stich, H.F. (1978b): The inhibitory effect of cysteine on the mutagenic activities of several carcinogens. Mutat. Res. *54*:73–81.

Rosin, M.P., Stich, H.F. (1979): Assessment of the use of the *Salmonella* mutagenesis assay to determine the influence of antioxidants on carcinogen-induced mutagenesis. Int. J. Cancer *23*:722–727.

Rosin, M.P., Stich, H.F. (1980): Enhancing and inhibiting effects of propyl gallate on carcinogen-induced mutagenesis. J. Environ. Pathol. Toxicol. *4*:159–167.

456   Miriam P. Rosin

Shamberger, R.J., Corlett, C.L., Beaman, K.D., Kasten, B.L. (1979): Antioxidants reduce the mutagenic effect of malonaldehyde and β-propiolactone. Part IX. Antioxidants and cancer. Mutat. Res. 66:349–355.

Stich, H.F., Wei, L., Whiting, R.F. (1979): The enhancing effect of transition metals on the chromosome-damaging action of ascorbate.

Cancer Res. 39:4145–4151.

Wattenberg, L.W. (1976): Inhibition of chemical carcinogenesis by antioxidants and some additional compounds. In: P.N. Magee, S. Takayama, T. Sugimura and T. Matsushima (eds.), Fundamentals in Cancer Prevention. Tokyo, Tokyo University Press/Baltimore: University Park Press, pp. 153–166.

41

Quantitative Measures of Induced Mutagenesis

FRIEDERIKE ECKARDT AND ROBERT H. HAYNES

Introduction

A surprisingly large number of substances have been found recently to act both as mutagens and as carcinogens (McCann *et al.*, 1975). Unfortunately, it is not enough merely to know whether or not a chemical *is* a mutagen. It is essential also to determine quantitatively *how mutagenic* various agents are in a given assay system, and also to determine *how mutable* various biological systems are when treated with a given mutagen. Knowledge of the relative mutagenicity of substances is a critical element in the assessment of genetic risks associated with environmental chemicals; and knowledge of the relative mutability of organisms is important in assessing the merits of short-term test systems.

In this paper we show how measurement of the *mutant yield*, that is, mutant clones induced per cell initially treated, can be used to compare the mutagenicity of different agents as well as the mutability of different test systems. We denote mutant

yield by the function $Y(x)$ where x is the mutagen exposure dose. Two other basic quantities which will concern us here are *mutant frequency*, $M(x)$, that is, induced mutants per survivor from the initially treated cell population; and the surviving fraction of cells $S(x)$ (Haynes and Eckardt, 1979).

It is convenient to distinguish between experiments in which the mutagen exposure is *acute* or *chronic*; between test systems in which the DNA is *nonreplicating* or *replicating* during treatment with the mutagen; and, in the case of selective assays, between protocols in which the selective conditions are applied *immediately* after treatment or *delayed* in order to allow full expression of the mutant phenotype. Furthermore, the exposed cells may be *synchronized* or *randomly distributed* throughout the cell cycle. For purposes of mathematical analysis, the simplest protocols are those which involve acute exposures of synchronized, nonreplicating cells requiring no expression time; or chronic exposure of replicating cells in fluctuation tests (Green et al., 1976; Parry, 1977) or in continuous culture (Kubitschek, 1970). There is no point in drawing a distinction between "acute" and "chronic" exposures for nonreplicating systems, except perhaps for studies on the dose-rate effects of radiation, or on tests of time-concentration reciprocity for chemical mutagens.

Mammalian cell systems often involve acute exposures to replicating cells followed by subculturing before the frequency of induced mutants is measured (see, for example, Hsie et al., 1978; Thacker, 1979). In such growing cultures, mutant frequency is measured as the fraction of the total viable cell population which exhibits a particular mutant phenotype at a given time after exposure to the mutagen (Thompson and Baker, 1973). Thus, mutant frequencies obtained from experiments on growing mammalian cell cultures are not necessarily equivalent as quantities to the mutant frequencies measured in stationary phase microbial systems. Data from mammalian cell culture experiments require a more complex mathematical treatment than that presented here, especially if the mutant and nonmutant cells should multiply at different rates during subculturing, or if the number of generations required to express the mutant phenotype should be dose-dependent.

It is important to distinguish mutant frequency from *mutation rate* in growing cultures. Mutation rate is the (spontaneous) probability of mutation at a given locus *per generation* (Luria and Delbrück, 1943; Drake, 1970; Kondo, 1972; Kubitschek, 1970; von Borstel, 1978). The mutant frequency in a growing culture obviously depends upon the mutation rate in that culture and the number of generations of culture growth. In establishing a mathematical relation between mutant frequency and mutation rate it is necessary to make a number of nontrivial assumptions regarding the clonal growth patterns of the mutant and nonmutant individuals, and the extent of mutational and/or lethal sectoring which may exist in the system employed (Drake, 1970). The phrase "mutation rate" also has

been used in mammalian cell experiments to describe the mutant frequency *per unit dose* of mutagen (see, for example, Albertini and DeMars, 1973). If the mutant frequency increases *linearly* with dose, then this latter quantity is a measure of the (constant) slope of the dose-response curve, and should not be confused with measures of mutation rates per cell generation. If the dose-response curve is nonlinear, then the ratio of mutant frequency to dose will, of course, be dose-dependent.

Clearly, the various quantitative measures of mutagenesis have confusingly similar names and their precise definitions depend on subtleties of experimental protocols which vary from one biological system to another. There is a real need in mutation research to standardize the basic nomenclature and definitions in a way that takes account of the different protocols associated with various short-term test systems. However, it would appear that the so-called "*in situ* method" for measuring mutant frequencies in mammalian cells provides quantitative data which can be compared with those obtained in microbial systems (Maher and McCormick, 1976).

The Nonreplicating Yeast System

In this paper we use data primarily from our experiments on UV- (2537A) induced mutagenesis in the simple eukaryote *Saccharomyces cerevisiae* (baker's yeast) to illustrate the mathematical analysis. Haploid yeast provides an unusually simple system, simpler even than *Escherichia coli*, for such analysis. Highly uniform, uninucleate G1 cell populations can be harvested from stationary phase batch cultures and resuspended for mutagen treatment in non-nutritive buffer (or distilled water). Under these conditions macromolecular synthesis comes virtually to a halt, and, although the cells do not grow, they do remain viable for several days (Patrick et al., 1964). These G1 cells can be exposed in suspension (or on nonnutritive agar plates) to acute doses of various mutagens, and subsequently tested in appropriate plating assays for viability and induced forward or reverse mutations (see Eckardt and Haynes, 1977a,b, for further experimental details). Usually it is possible to carry out these plating assays at cell densities well below those at which crowding effects might bias the data. For many mutagens and mutant phenotypes—in particular, UV-induced reversions of nutritional auxotrophies—yeast does not exhibit any phenotypic lag in mutation expression (Lemontt, 1977). Thus, normally it is not necessary to take expression time and the occurrence of plate mutants into account in the calculation of mutant yields and frequencies as it is for *E. coli* (Green and Muriel, 1976). It is an equally simple matter to measure mitotic crossing-over and gene conversion in appropriately constructed diploid strains (Zimmermann, 1975), and to study the induction of mutations defined at the level of amino acid and nucleotide sequences (Sherman and Stewart, 1978).

Basic Definitions

Consider a homogeneous suspension of N_o (per unit volume) nonreplicating, but equally viable, uninucleate haploid G1 cells that is uniformly treated with various acute exposure doses, x, of some mutagen. As indicated above, such cell suspensions can be obtained readily with *S. cerevisiae*. After each dose the number of induced mutants, $N_m(x)$, and of surviving cells, $N_s(x)$, are scored in suitable assay systems. We define three basic quantities as follows:

Surviving fraction of cells:

$$S(x) = N_s(x)/N_o \qquad (1)$$

Induced mutant yield:

$$Y(x) = N_m(x)/N_o \qquad (2)$$

Induced mutant frequency:

$$M(x) = N_m(x)/N_s(x) \qquad (3)$$

Note that $M(x) = Y(x)/S(x)$. Thus, to report frequency data alone is to suppress information since $Y(x)$ and $S(x)$ are independent observables. It is important to realize that mutant frequency and mutant yield are fundamentally different quantities. The toxic effects of a mutagen are "cancelled out" in calculating induced mutant frequencies, and so simple inspection of log-log plots of frequency versus dose provides a convenient way to determine the kinetics of the mutation induction process (Eckardt and Haynes, 1977a).

It is important to remember that the number of *induced* mutants is obtained by suitably correcting the number of mutants actually counted N_{mc}, for plate mutants N_{mp} (if they occur), and for the number of surviving spontaneous (or preexisting) mutants N_{mo}, from the original population; that is

$$N_m(x) = N_{mc} - N_{mp} - N_{mo}S(x) \qquad (4)$$

Normally $N_{mp} = 0$ for yeast. For plate tests with bacteria which exhibit phenotypic lag a formula equivalent to (4), but with different notation, has been given by Green and Muriel (1976).

Stochastic Description of Mutation and Killing

A formal description of induced mutagenesis and killing, based on Poisson statistics, can be derived from the following assumptions. All of these assumptions can be relaxed, if necessary, for further generalization of the mathematics.

1. The initially untreated cell population is homogeneous with respect to mutability and sensitivity to killing by the agent in question.
2. The all-or-none character of the biological endpoints (mutant versus nonmutant; survivor versus nonsurvivor) allows the application of single-event Poisson statistics.

3. Mutation and killing are stochastically independent processes in the sense that clone formation after a mutagen dose x is the same for both mutated and nonmutated cells.

It is important to distinguish between "physical hits" and "biological hits" (Haynes and Eckardt, 1979). To avoid ambiguity we refer to "physical hits" as potentially lethal or premutational *lesions* in whatever macromolecular targets are relevant to the endpoints assayed. We use the word "hit" only in the *biological* context of lethal or mutational hits. Complex biochemical processes are involved in the "conversion" of physical lesions to biological hits. We denote the expected, or average number of biological hits per cell in the treated population by the function $H(x)$. On the basis of single-event Poisson statistics, $H(x)$ biological hits are said to have occurred at a dose x which leaves a fraction $exp[-H(x)]$ of cells unaffected with respect to the biological endpoint measured. If the relevant physical lesions are formed in direct proportion to dose, and if no dose-dependent processes are involved in the conversion of initial lesions to biological hits, then the hit function will be linear. More generally, $H(x)$ could be some more complex function which, nonetheless, can always be represented by an infinite power series in x with no constant term (since there can be no induced hits for zero dose). We denote the expected number of lethal and mutational hits at dose x by $H_k(x)$ and $H_m(x)$ respectively, Thus, on the basis of the assumptions stated above we can write

$$S(x) = exp\ [-H_k(x)] \qquad (5)$$

and,

$$M(x) = 1 - exp[-H_m(x)] \sim H_m(x) \qquad (6)$$

whence,

$$Y(x) = M(x) \cdot S(x) = H_m(x) \cdot exp[-H_k(x)] \qquad (7)$$

It is evident from equations (5) and (6) that surviving fraction and mutant frequency are the probabilities that a cell has not suffered a lethal hit, and has sustained a mutational hit, respectively. The approximation in equation (6) is valid to high accuracy because mutant frequencies normally are small numbers, very much less than one. The mutant yield is the joint probability that a cell has sustained a mutational hit but not a lethal hit. In general, mutagenesis and killing are competing processes in that the probability of mutation increases with dose, but survival declines with dose (equation 7). Thus yield curves always possess a maximum whose position and magnitude depends on the kinetics of mutation and killing, that is, on the nature of the terms in $H_m(x)$ and $H_k(x)$ for each particular mutagenic system. Analysis of yield curves provides an integrated view of the action of mutagens on cells because most mutagens are toxic as well as mutagenic, and the interplay of these two effects can be determined from plots of mutant yield versus dose. In addition, if mutant yield is plotted against lethal hits $[H_k(x)]$, then it becomes possible to compare the mutagenic efficiencies of various mutagens whose physical exposure doses otherwise would be incommensurable. Finally, it is worth remembering that it is mutant yield, and not frequency, that actually is measured in the Ames test (Ames *et al.*, 1975).

If we relax the third assumption above then we must allow for the possibility that mutation and killing could be stochastically *dependent* processes (Eckardt and Haynes, 1977a). Under these circumstances the probability of clone formation for nonmutant cells, $S(x)$, would be different from that for mutant cells. For the latter we would write therefore

$$S(x) = exp[-H_{km}(x)] \qquad (8)$$

and the yield would become

$$Y(x) = H_m(x) \cdot exp[-\delta(x)H_k(x)] \qquad (9)$$

where we write for convenience

$$\delta(x) = H_{km}(x)/H_k(x) \qquad (10)$$

Stochastic *independence* of mutation and killing thus implies that $\delta(x) = 1$.

The mutational and lethal hit functions introduced above can be represented by infinite power series in x as follows

$$H_m(x) = m_1 x + m_2 x^2 + \cdots \qquad (11)$$
$$H_k(x) = k_1 x + k_2 x^2 + \cdots \qquad (12)$$

where the constants m_i and k_i can be regarded as empirical coefficients of formal one-lesion, two-lesion, etc., processes of mutagenesis and killing respectively.* [It is important to note that these coefficients are both agent and test system specific. A similar power series could be written for $H_{km}(x)$.] Theoretically any combination of the k_i's and m_i's could exist, but in practice we need concern ourselves only with a few of the simplest linear and quadratic cases. Here, by way of illustration, we will consider only the case of linear killing (exponential survival curves) coupled with linear mutation induction (denoted Lk,Lm); we have described the more complex kinetic response patterns in detail elsewhere (Haynes and Eckardt, 1979). Here all terms except the first in equations (11) and (12) are zero, so that the surviving fraction of cells (equation 5) becomes simply $S(x) = exp(-k_1 x)$, mutant frequency is $M(x) = m_1 x$, and mutant yield is

$$Y(x) = m_1 x \cdot exp(-k_1 x) \qquad (13)$$

Properties of Mutant Yield Curves

It is evident from equation (13) that $Y(x)$ is a bell-shaped curve which rises linearly from the origin, because of the factor $m_1 x$, but ultimately reaches a maximum and then declines to zero as the exponential killing factor comes to dominate the expression at high doses. By differentiation of equation (13) it is easy to establish the following properties of this yield curve (Fig. 41.1, upper panel curve L).

(a) The initial slope of $Y(x)$ is equal to the cross section m_1 [which is also the slope of the mutant frequency curve $M(x)$].

(b) The position of the maximum yield occurs at a dose, x, corresponding to one lethal hit [$x = 1/k_1 \simeq LD_{37}$].

(c) The magnitude of the maximum yield is directly proportional to the mutagenicity (m_1) and inversely proportional to the toxicity (k_1) of the mutagen [$Y_{max} = m_1/k_1 e$, where e is the base of the natural logarithms].

(d) At sufficiently high doses the yield declines at the same rate as does the survival curve. Thus, if yield and survival are plotted together on semilogarithmic paper, the final slopes of the two curves are equal. This is *not* the case if mutation and killing are stochastically dependent processes (equation 9). In this situation the final slope of the yield curve gives the sensitivity of the mutant subpopulation and not that of the nonmutant majority.

(e) The area under the yield curve (the "integral yield," I_x) is proportional to the maximum yield such that $I_x = eY_{max}/k_1$.

For more complex kinetic response pat-

---

*For purists, the coefficients m_i and k_i also are sources of terminological ambiguity. The active terms "mutagenicity" and "toxicity" should be used when referring to properties of the *agent* tested; the passive terms "mutability" and "sensitivity to killing" should be used when referring to properties of the *cells* tested. (It would be convenient in this context to use "toxibility" as the passive form of toxicity rather than the phrase "sensitivity, or susceptibility, to killing.") The magnitude of the lethal and mutational responses is jointly dependent on properties of the agent tested and the cells exposed. Thus, from a mechanistic standpoint, the coefficients (or better, *cross-sections*) m_1 and K_1 reflect simultaneously the mutagenicity (and toxicity) of the agent and the mutability (and "toxibility") of the cells. In situations where one wishes to make reference to the phenomena or processes of mutagenesis and/or killing, it is perhaps best to use the terms "mutational cross-section" and "cross-section for killing" (or "lethal cross-section") to describe the m_1 and k_1. However, in *comparative* studies of the response of a given biological test system to various agents, the m_1 and k_1 can be used to compare the intrinsic mutagenicity and toxicity of the agents; conversely, in studies of the responses of various cellular systems to a given agent, they can be used to compare the intrinsic mutability and toxibility of the cells.

terns the yield curves remain bell-shaped and their final slope (on semilog paper) is equal to the corresponding survival curve slope if $\delta(x) = 1$. However the initial slope of the yield curve becomes zero for quadratic or higher powers of mutation induction; indeed, the initial slope of the yield curve is non-zero, and equal to the cross-section m_1, *only* if there is a single-lesion component in mutagenesis (cf. Figs. 41.1 and 41.2). Furthermore, the position $(\hat{x})$

and the magnitude (Y_{max}) of the maximum yield differs from the values stated above for the purely linear response pattern. For example, for linear killing and purely quadratic mutation induction (Lk, Qm), the

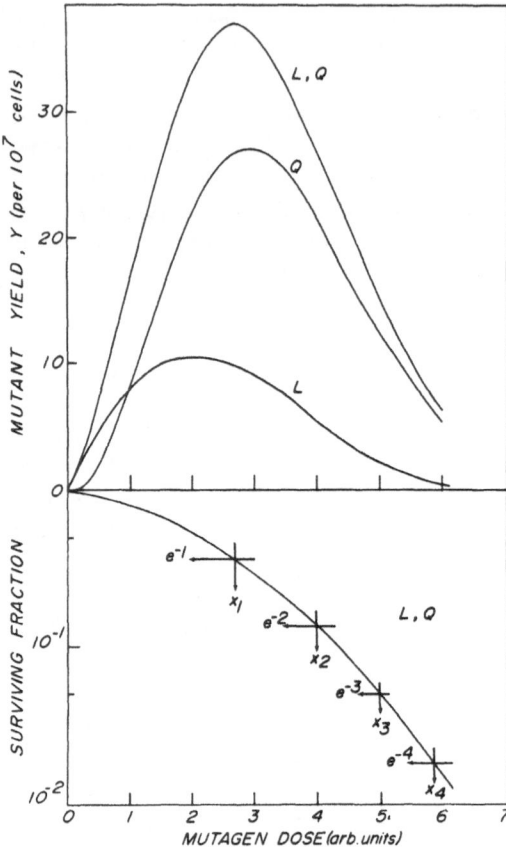

Figure 41.2. Calculated mutant yield and survival curves for biphasic, linear-quadratic killing (LQk) coupled with linear (L), quadratic (Q) or biphasic (LQ) mutation induction over dose in arbitrary units (these units were chosen so that the LD_{37} value would be the same in Figures 41.1 and 41.2). The doses corresponding to 1, 2, 3 and 4 lethal hits (x_1, x_2, x_3, x_4) are shown at the appropriate survival levels $(e^{-1}, e^{-2}, e^{-3}, e^{-4})$. Note that for shouldered survival curves the doses for successive lethal hit units (1, 2, 3,...) do not increase by constant intervals as they do for exponential survival curves. The values of the mutational and lethal cross-sections used in the calculation were: $m_1 = 10^{-6}(units)^{-1}$; $k_1 = 10^{-1} (units)^{-1}$; $k_2 = 10^{-1} (units)^{-2}$. Note that the initial slope of $Y(x)$ is zero for purely quadratic mutation induction; it is non-zero whenever there is a linear component in mutagenesis. For (LQk, Lm) $\hat{x} = \frac{1}{2}x_2$; for (LQk, Qm) $\hat{x} = \frac{1}{2}x_4$; for (LQk, LQm) $\frac{1}{2} x_2 < \hat{x} < \frac{1}{2}x_4$.

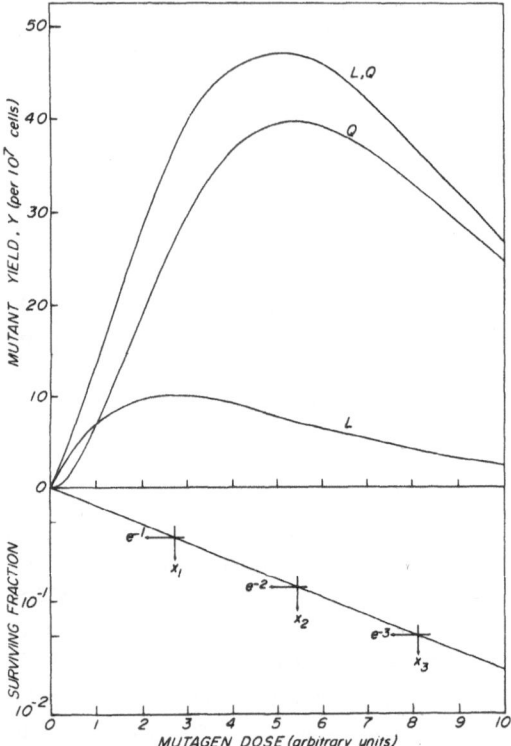

Figure 41.1. Calculated mutant yield and survival curves for linear killing (i.e., exponential survival) coupled with linear (L), quadratic (Q) or biphasic, linear-quadratic (LQ) mutation induction over dose (in arbitrary units). The doses corresponding to 1, 2 and 3 lethal hits (x_1, x_2, x_3) are shown at the appropriate survival levels (e^{-1}, e^{-2}, e^{-3}). The values of the mutational and lethal cross-sections used in the calculations were: $m_1 = 10^{-6}(units)^{-1}$; $m_2 = 10^{-6} (units)^{-2}$; $k_1 = 3.7 \times 10^{-1} (units)^{-1}$. Note that the initial slope of $Y(x)$ is zero for purely quadratic mutation induction while it is non-zero if there is a linear component in mutagenesis. For (Lk, Lm), $\hat{x} = x_1$; for (Lk, Qm), $\hat{x} = x_2$; for (Lk, LQm), $x_1 < \hat{x} < x_2$ (see text for further explanation).

maximum yield occurs at the dose corresponding to two lethal hits $[\hat{x} = 2/k^1 \simeq LD_{14}]$; for linear killing and biphasic, linear-quadratic mutation induction (Lk, LQm), the maximum yield occurs somewhere between the doses for one and two lethal hits (Fig. 41.1, upper panel curves Q, and L, Q respectively). For shouldered survival curves described by the linear-quadratic lethal hit function $H_k(x) = k_1x + k_2x^2$, coupled with linear mutation induction (LQk, Lm), the maximum yield occurs at one-half the dose for two lethal hits ($\simeq \frac{1}{2}LD_{14}$); for purely quadratic mutation induction (LQk, Qm) the maximum yield is at one-half the dose for four lethal hits ($\simeq \frac{1}{2}LD_2$); and for (LQk, LQm) the maximum yield occurs between the two positions just described (Fig. 41.2). Finally, it should be noted that for $\delta(x) > 0$ the maximum yield shifts to lower doses for all kinetic patterns, whereas a $\delta(x) < 0$ produces an opposite shift toward higher doses. Experimental demonstrations of some of these shifts in $\hat{x}$ are given elsewhere (Haynes and Eckardt, 1979).

Mutant yield (and frequency) also can be expressed as functions of lethal hits, z. By definition $z = H_k(x)$ and the solutions of this equation for the various types of survival curves provide formulae for x as a function of z. Thus for linear killing we have simply $x = z/k_1$ and by direct substitution in equation (13) we obtain

$$Y(z) = (m_1/k_1)z \cdot exp(-z) \qquad (14)$$

The maximum yield occurs at $\hat{z} = 1$ and the initial slope of $Y(z)$ is m_1/k_1 [which is also the slope of the frequency curve $M(z)$]. The final slope of $Y(z)$ on semilog paper is -1. The integral yield I_z also equals $m_1/k_1 (= eY_{max})$. The initial slope of $Y(z)$ is zero for linear killing and purely quadratic mutagenesis; in fact, this initial slope is non-zero only if both mutation and killing increase at the same power of dose (Haynes and Eckardt, 1979). It is algebraically cumbersome to write down explicit

formulae for $Y(z)$ for shouldered survival curves, because in this case the equation $z = H_k(x)$ becomes quadratic (or even higher powered).

Comparative Mutant Yields

The relative mutagenicities of various agents, and the relative mutabilities of different test systems, can be assessed from plots of mutant yield over lethal hits. In practice the simplest way to construct such curves is to plot mutant yield on a linear scale versus survival on a logarithmic scale since equation (5) tells us that the number of lethal hits is equal to the negative logarithm of survival [i.e., $H_k(x) = -\ln S(x)$]. The merits of such a plot are that first, lethal hits provide a common biological measure of "dose" for agents whose physical exposure doses are incommensurable; second, for similar kinetic response patterns the maximum yield occurs at a fixed position on a lethal hit scale; and third, the initial slopes of $Y(z)$ curves reflect the interplay of mutation and killing in generating viable mutant cells.

In Figure 41.3 we have plotted, as functions of mutagen dose in arbitrary units, two examples (A and B) of induced mutant yields, the corresponding mutant frequency, and cell survival for the purely linear response pattern (Lk, Lm) of equation (13). The mutational cross-section m_1 is the same in both examples, but in B the cross-section for killing k_1 is threefold greater than in A. The curves in Figure 41.3 could refer to situations in which either a single mutagen was applied to two different cellular systems (A and B), or two different mutagens were applied to the same system. Of course, the latter case could be realized experimentally only in those very few instances where mutagen doses can be measured in physically comparable units, for example, energy absorption for different types of ionizing radiation. However, the theoretical outcome of the subsequent discussion is the same in either situation and so let us imagine first that we are dealing with two mutagens A and B with comparable dosimetry in the same assay system.

Relative Mutagenic Efficiency (RME)*

In Figure 41.3 the induced mutant frequency $M(x)$ is the same straight line (upper panel, dashed line) for both A and B because m_1 has the same value in both cases. However, the yield curves rise to different maxima even though their initial slopes are the same, Y_{max} occurs at a higher dose, and more viable mutants survive in A as compared with B for all except very low doses. In Figure 41.4 we have plotted the same yield and frequency versus lethal hits. Again note that $Y_A(z) > Y_B(z)$, but in this plot the initial slopes of the yield curves, and hence also the slopes of the linear frequency curves $M_A(z)$ and $M_B(z)$, are different. This is because of the way that the ratio of mutagenicity to toxicity (m_1/k_1) of the two agents enters into the calculation of yield and frequency as functions of z (equation **14**). Note also that the position of the maximum yield occurs at the same value of z for both A and B, whereas in Figure 41.3 Y_{max} occurs at different values of dose x.

Suppose now we wish to compare the efficiencies of these two mutagens. On the basis of a dose-frequency plot we might say they were equally efficient because the slope of $M(x)$ is the same for both A and B (Fig. 41.3, dashed line, upper panel). However, even though their inital slopes are equal, the areas under the curves are very different. Agent A is less toxic than B, and so for A more mutants remain viable, after all except very low doses; furthermore, a greater dose is required to produce the

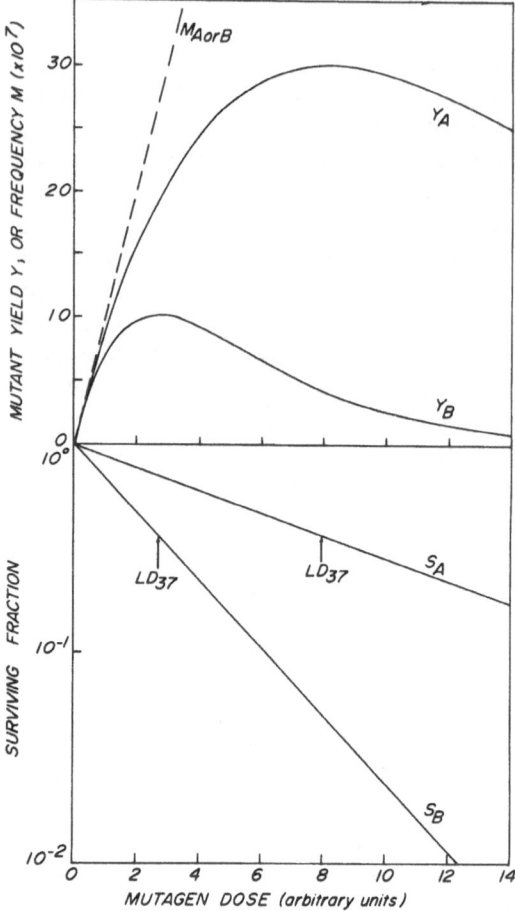

Figure 41.3. Mutant yields $Y(x)$, frequencies $M(x)$ and survivals $S(x)$ plotted over dose in arbitrary units for the purely linear kinetic response pattern (Lk, Lm). For both cases A and B m_1 10^{-6} (units)$^{-1}$. In A, $k_1 = 1.23 \times 10^{-1}$ (units)$^{-1}$; in B, $k_1 = 3.7 \times 10^{-1}$ (units)$^{-1}$; that is, the cross-section for killing in B is threefold greater than in A. Note that the frequency curve (dashed line) is the same for both A and B whilst the areas under the yield curves are very different. In each case Y_{max} is at the dose for one lethal hit ($\simeq$ LD$_{37}$).

maximum yield in A than in B. Clearly, a biologically meaningful definition of mutagenic efficiency cannot be framed unless mutagenicity, toxicity, and the position of the maximum mutant yield along the abscissa are taken into account. Kinetic con-

---

*It is useful to distinguish between the words mutagenic *effectiveness* and mutagenic *efficiency*. The word effectiveness should be used when comparing mutant yields with respect to the *physical dose* of the mutagen; the word efficiency should be used when comparing mutant yields with respect to some other *biological effect* of the mutagen, such as killing, chromosome breakage, recombination, etc.

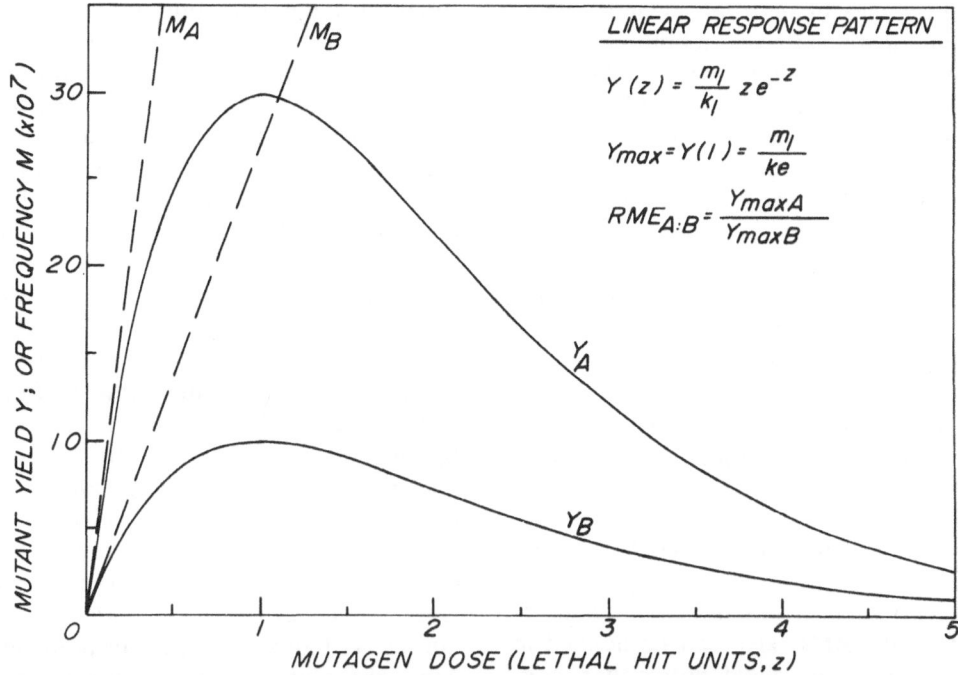

Figure 41.4. Mutant yields $Y(z)$ and frequencies $M(z)$ plotted over lethal hits for the purely linear kinetic response pattern (Lk, Lm). The numerical values of m_1 and k_1 are the same as in Figure 41.3 for both A and B. Survival is not shown since on a semilog plot all such curves are merely straight lines passing through surviving fraction e^{-1} for $z = 1$. Note that the frequencies (dashed lines) are the tangents to $Y(z)$ at the origin; the slope of these lines is (m_1/k_1). In A, $m_1/k_1 = 8.1 \times 10^{-6}$, while in B, $m_1/k_1 = 2.7 \times 10^{-6}$. The maxima of both curves occur at $z = 1$. The dose x in Figure 41.3 is converted to lethal hits by the relation $z = k_1 x$.

siderations aside, intuitively it would seem that mutagenic efficiency should be proportional directly to mutagenicity and inversely to toxicity. Thus a yield versus lethal hit plot for different mutagens which are governed by the *same kinetic response pattern* satisfies our requirements for comparing mutagenicities: the lethal hit scale provides a common biological measure of mutagen "dose"; the maximum yields occur at the same position along the abcissa; and the initial slopes of such curves are proportional directly to m_1 and inversely to k_1. Because of the mathematical properties of $Y(z)$ functions described above, it is possible to define the relative mutagenic efficiency (RME) of two agents (A:B) which have the same kinetic response pattern (Lk, Lm) in a given assay system as any of the following dimensionless ratios, *all of which are equal* (Haynes and Eckardt, 1979):

(a) The ratio of the (linear) slopes of mutant frequency versus lethal hits $[M'_A(z)/M'_B(z)$, where the prime symbols denote differentiation]

(b) The ratio of the initial slopes of mutant yield versus lethal hits $[Y'_A(z \to 0)/Y'_A(z \to 0)]$

(c) The ratio of the areas under the $Y(z)$ curves, that is, the ratio of the integral yields $[I_z(A)/I_z(B)]$

(d) The ratio of the maximum mutant yields $[Y_{max}(A)/Y_{max}(B)]$

These various mathematically explicit definitions of RME also hold for nonlinear kinetic patterns; however, definition (b) holds only if there is a linear [m_1 term in $H_m(x)$].

It is more hazardous to suggest mathematically simple definitions of RME for kinetic patterns having biphasic components of killing and/or mutation; or to compare the efficiencies of mutagens which follow different kinetic patterns. In both situations the problem is associated with the fact that the position of the maximum yields even on a lethal hit axis need not be the same. However, we argue that if two mutagens have the same maximum and integral yields, but one maximum occurs at a lower "lethal hit dose" than the other, then the former is a more efficient mutagen than the latter. Thus, we suggest that mutagenic efficiency be defined as being proportional to the integral yield over lethal hits (I_z) and inversely proportional to the number of lethal hits at which the maximum yield occurs $(\hat{z})$ viz.,

$$RME(A:B) = \hat{z}_B I_z(A)/\hat{z}_A I_z(B) \qquad (15)$$

Since it is tedious in practice to calculate integral yields, RME also can be defined in terms of maximum yield rather than integral yields, viz.,

$$RME(A:B) = \hat{z}_B Y_{max}(A)/\hat{z}_A Y_{max}(B) \qquad (16)$$

Equations (15) and (16) differ only by a constant factor since the integral yield is proportional to the maximum yield. However, it must be emphasized that both of these definitions are *arbitrary* and must be used with caution, especially in cases where the maximum yields occur at significantly different positions on the lethal hit axis, or where there is a strong δ-effect for one agent but not for the other. In such cases it may not be wise to try to calculate a single numerical value for RME, but rather to represent RME as a *function* of z calculated as the ratio of mutant yields for A to B at each value of z.

An Illustrative Calculation of RME

Even though the amount of work on mutagenesis has increased dramatically in recent years it is surprisingly hard to find data in the literature which can be used to calculate RME's of various mutagens. Thus, recently we measured, under similar conditions, the induction of *lys*2-1 locus reversion by UV and 4-nitroquinoline N-oxide (4NQO), in an excision proficient (RAD wild-type) strain (HT4-20A) of haploid yeast. We used our standard procedures for measuring the UV-induced reversions (Eckardt and Haynes, 1977a). To obtain the chemical data, stationary phase cells suspended in buffer were incubated with various concentrations of 4NQO (Fluka, Switzerland) at 30° C for 30 minutes, after which time the reaction was stopped with 5% sodium thiosulfate; the cells were washed twice with ice-cold buffer prior to plating. The survival curves and the mutant frequency curves from these two experiments are shown in Figures 41.5 and 41.6, respectively. It is evident that the survival curves have a shoulder (LQk) and the mutant frequency curves are biphasic (LQm) for both agents. Thus the kinetic response pattern is the same (LQk, LQm) in each case, and on a dose axis the maximum yield should lie on the range $^1/_2 LD_{14} < \hat{x} < LD_2$, its precise

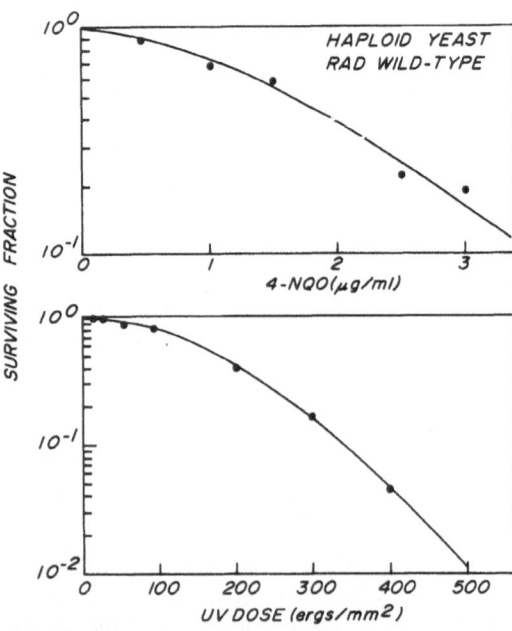

Figure 41.5. Comparative survival curves for UV and 4NQO in the same strain of stationary phase, haploid RAD wild-type yeast. In each case the data are best fit by a biphasic, linear-quadratic survival curve (LQk).

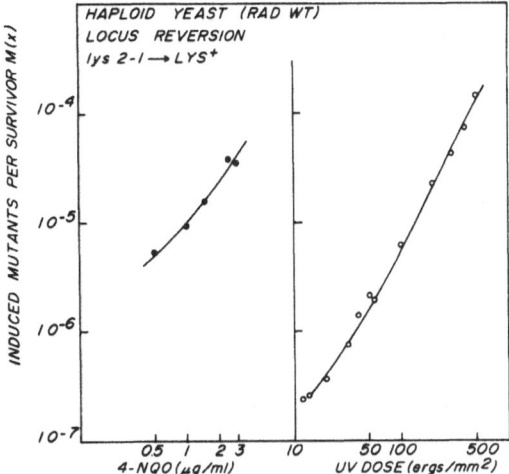

Figure 41.6. Comparative mutant frequency curves (same experiments as in Figure 41.5) for UV and 4NQO induced revertants at the same locus and in the same strain under the same physiological conditions. In each case the data are best fit by a biphasic, linear-quadratic (LQm) function.

position depending on the values of k_1 and k_2 for each survival curve. For both UV and 4NQO $\hat{x}$, lies roughly midway between $\frac{1}{2}LD_{14}$ and $\frac{1}{2}LD_2$ which happens to be close to the dose for one lethal hit. When the data of Figures 41.5 and 41.6 are plotted in the form of mutant yield versus lethal hits (Fig. 41.7), it can be seen that the maximum mutant yields for both agents indeed

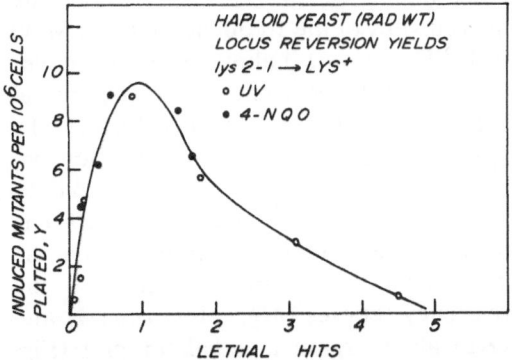

Figure 41.7. Induced mutant yield curves for UV and 4NQO plotted over lethal hits $Y(z)$ [same experiments as shown in Figures 41.5 and 41.6]. Note that the maximum yield in each case is near one lethal hit. The data for the two agents are superimposable, within experimental error, and so we conclude that RME (4NQO:UV) $\simeq$ 1.

lie close to $z = 1$; furthermore, the $Y(z)$ curves appear to be superimposable within experimental error over the measured range of lethal hits. Thus, according to our definition of relative mutagenic efficiency (equation 15) we conclude that RME(4NQO:UV) $\simeq$ 1. This is a satisfying result in view of the fact that these two agents appear to act in an equivalent way at both the macromolecular and genetic levels (Ikenaga *et al.*, 1975) in *E. coli*, even though the effects of the various *rad* genes on UV- and 4NQO-induced reversions in yeast do not always parallel one another (Prakash, 1976). The data shown in Figure 41.7 also imply that it is possible to construct a linear "calibration" relation between UV and 4NQO exposure doses which is meaningful with respect to both mutation and killing in the yeast system employed here.

It would be interesting to obtain a rank ordering of the RME's of various mutagens (relative to some arbitrarily chosen "standard mutagen"). Although the mutagenic efficiency of 4NQO is equal to that of UV in this yeast system, we suspect that X-rays would prove to be much less efficient than UV because of the very low mutant yields normally obtained by X-irradiation of yeast. Such a rank ordering of mutagens based on RME could then be compared with the rank ordering of their "mutagenic potencies" measured as reciprocal exposure doses. If these two rank orderings were equivalent then one could conclude that the mutagens on the list acted according to biologically similar mechanisms, and that the effective absorbed doses were proportional to the exposure doses for each agent. Obviously such a project would involve a massive amount of painstaking work.

Relative Mutability of Excision-Deficient Yeast

It has been known for some years that premutational lesions produced in DNA by UV (and other mutagens) can be eliminated prior to mutation fixation by error-free modes of excision repair. Large numbers of induced mutations at various loci are produced in excision-deficient strains of *E. coli* at UV doses far below those at which such mutations can be detected in excision-proficient parental stains (Hill, 1965; Witkin, 1966). Thus, the mutability of microbial strains used in short-term test systems generally can be enhanced by the inclusion of appropriate excision repair deficiencies (Ames *et al.*, 1973a). However, from a quantitative standpoint, comparison of the mutability of various strains by a given mutagen is not as straightforward as it might seem.

Figure 41.8 shows UV survival curves for an excision-proficient RAD wild-type strain of *S. cerevisiae* and two radiation-sensitive derivatives. The strain *rad*1 is deficient in the excision of pyrimidine dimers (Unrau *et al.*, 1971); the mutant r_1^s (Eckardt *et al.*, 1975) belongs to the same epistatic group of loci as *rad*1-1 but it has a more complex phenotype than other members of this group (F. Eckardt, unpublished observations). The RAD and r_1^s strains have shouldered survival curves (LQk), whereas $S(x)$ for *rad*1-1 is exponential (Lk). Figure 41.9 shows conventional log-log plots of locus reversion frequencies

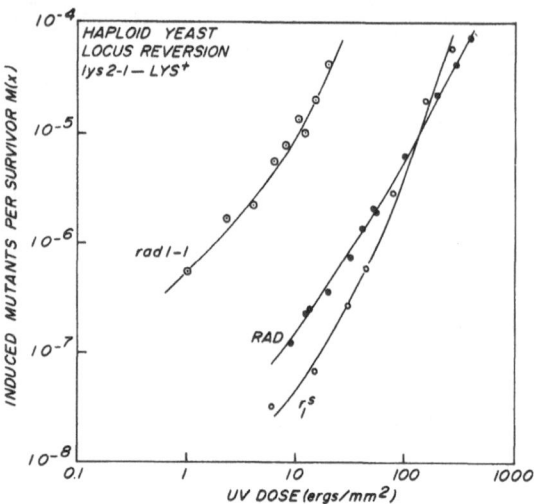

Figure 41.9. Comparative mutant frequency data (log-log plot) for UV induced lysine reversions in the strains whose survival curves are shown in Figure 41.8. Note that the curve for *rad*1-1 lies well *above* that for RAD wild-type.

versus UV dose for these three strains. It would appear from this plot that *rad*1-1 is more "mutable" than r_1^s or RAD wild-type; however, if mutant frequency is plotted versus lethal hits (here on semilog paper) the converse appears to be true—the points for RAD wild-type fall substantially *above* those for the two radiation-sensitive strains (Fig. 41.10). The situation is clarified by plotting the mutant yields versus dose and lethal hits (Figs. 41.11 and 41.12). The upward shift of the frequency curves $M(x)$ for RAD and *rad*1-1 above r_1^s on a dose plot (Fig. 41.9) reflects the increasing initial slopes of the yield curves $Y(x)$ (Fig. 41.11), that is, the mutational cross-section m_1 increases as one moves from r_1^s to RAD to *rad*1-1. From this point of view *rad*1-1 is indeed more "mutable" than RAD. On the other hand, the upward shift of the frequency curves $M(z)$ for *rad*1-1 and RAD above r_1^s on a lethal hit plot (Fig. 41.10) reflects the increasing initial slopes of the corresponding yield curves $Y(z)$ (Fig. 41.12), that is, the ratio of the mutational to lethal cross-section (m_1/k_1) increases as one moves from r_1^s to *rad*1-1 to RAD. Thus, the excision deficiency in *rad*1-1 has the ef-

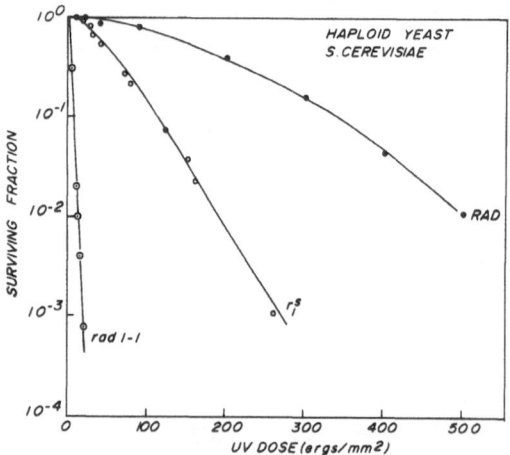

Figure 41.8. Comparative survival curves for three strains of yeast. Strain *rad*1-1 is deficient in pyrimidine dimer excision.

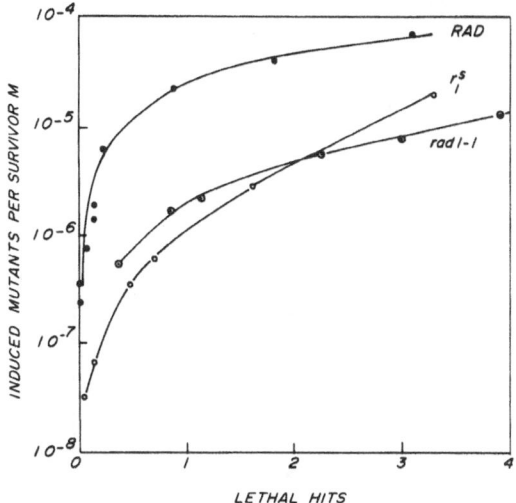

Figure 41.10. Comparative induced mutant frequency data $M(z)$ (semi-log plot) plotted versus lethal hits; same data as in Figure 41.9 except that here the frequencies are plotted against lethal hits. Note that in this plot the curve for *rad*1-1 lies well *below* that for RAD wild-type (cf. Fig. 41.9).

fect of increasing the value of m_1 over wild-type but the accompanying increase in k_1 is even greater. Thus, although the mutational response is increased in *rad*1-1 at very low doses, this increase is accompanied by a substantial loss in mutant yield.

It is evident from Figures 41.9 to 41.12 and the foregoing discussion that one must consider three factors in assessing the relative merits of excision-deficient and excision-proficient short-term test systems. We call these factors the relative *mutability* (Rmt), *mutational sensitivity* (Rms), and *mutational resolution* (Rmr). By analogy with the concept of RME defined in equation **(16)** the net mutability of a system may be measured by the ratio of the maximum yield to the number of lethal hits at which the maximum yield occurs. On the basis of this definition the relative mutabilities of the strains RAD:*rad*1-1:r_i^s are in the ratio 26:1.7:1 respectively. Thus, we assert that

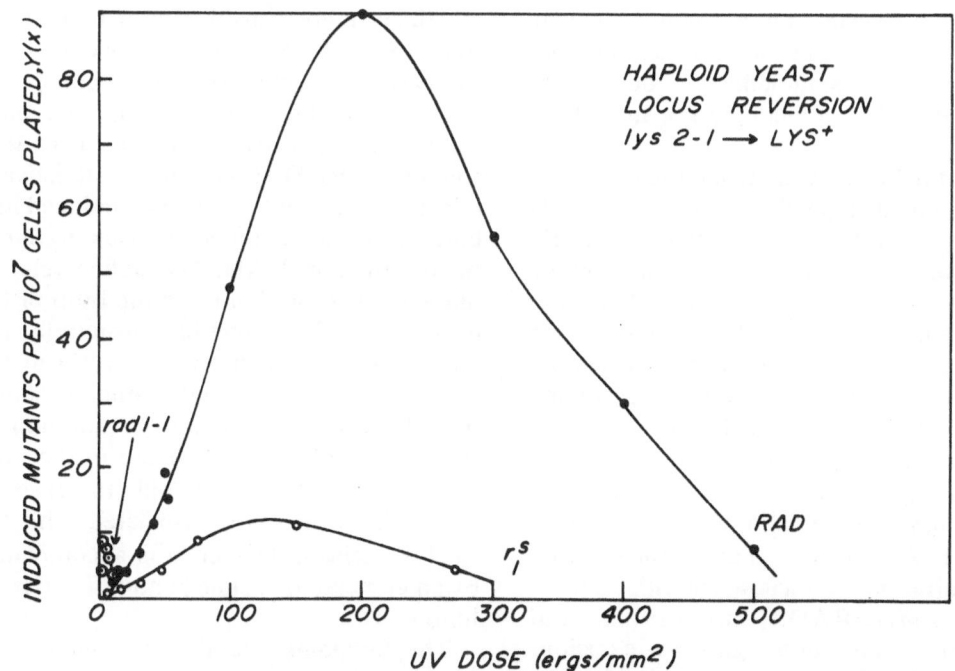

Figure 41.11. Comparative mutant yield curves $Y(x)$ for UV-induced reversions in the strains whose survival curves are shown in Figure 41.8; the corresponding frequency curves are plotted *versus* dose in Figure 41.9 and lethal hits in Figure 41.10. See text for explanation.

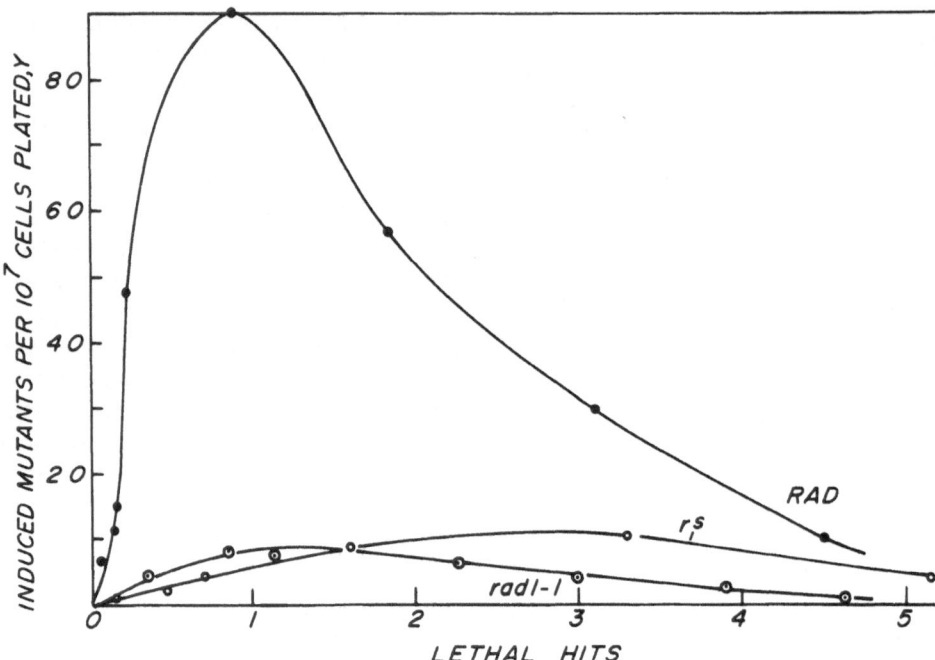

Figure 41.12. Comparative mutant yield curves $Y(z)$. Same data as that shown in Figure 41.11, but here plotted against lethal hits. It is evident from the areas under these curves that the excision-proficient RAD wild-type strain is more highly mutable than either of the radiation-sensitive derivatives. See text for further explanation.

in this yeast system the excision-proficient strain actually is 15 times more mutable than the excision-deficient derivative. However, this is obviously not the whole story.

It is further evident from Figure 41.11 that the initial slope, that is, the value m_1, is much greater for $rad1$-1 than for RAD. Thus the strain $rad1$-1 is a better detector of mutants at very low doses of UV than RAD, and much better than r_1^s. Hence, we say that of the three strains $rad1$-1 has the greatest mutational sensitivity. We define relative mutational sensitivity to be the ratio of Y_{max} to x for the strains compared. (For the purely linear case this ratio is proportional to the ratio of the two values of m_1.) A rough estimate shows that the relative mutational sensitivities of the strains $rad1$-1:RAD:r_1^s are in the ratio 33:5.5:1. Thus, although the excision-proficient RAD wild-type strain is 15 times more mutable than the excision-deficient strain $rad1$-1, the latter strain is about

six times more sensitive as a detector of mutants at low doses. However, this increase in detection sensitivity comes at a cost in mutational resolution. The maximum yield of $rad1$-1 is much lower than that of the RAD strain, and so statistically it is more difficult to "resolve" induced mutants from the spontaneous background in $rad1$-1 than in RAD. We define relative mutational resolution to be the ratio of the maximum yields above background for the strains compared, that is, the ratio of the values of Y_{max}. A rough estimate shows that the relative mutational resolution of the strains RAD:r_1^s:$rad1$-1 are in the ratio 11:1.3:1. Thus the sixfold increase in mutational sensitivity associated with the $rad1$-1 excision deficiency is accompanied by an elevenfold decline in mutational resolution.

The foregoing quantitative comparisons of these three yeast strains as short-term test systems can be visualized most readily in a plot of induced mutant yield versus log-

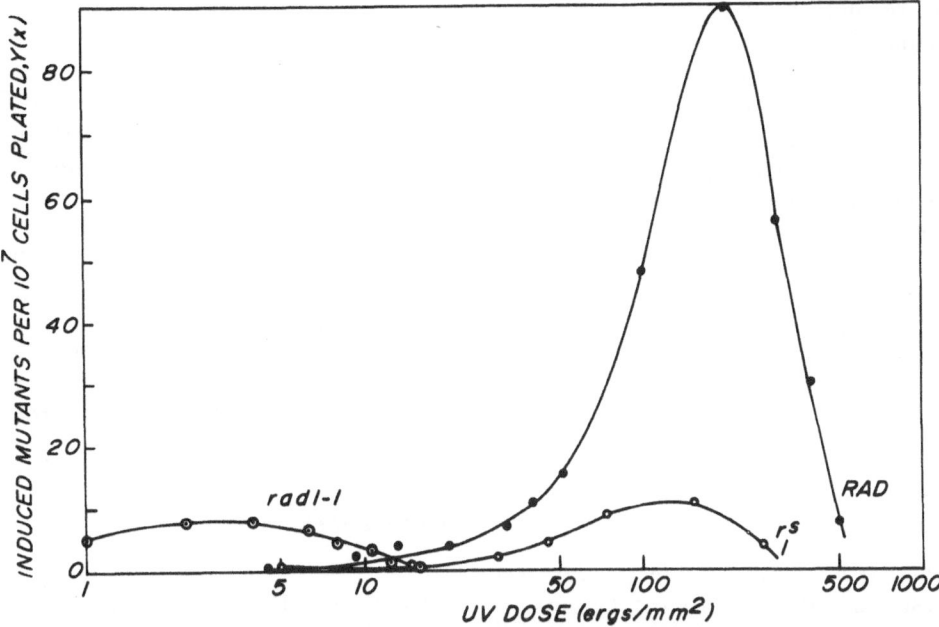

Figure 41.13. Comparative mutant yield curves plotted against log dose (in order to expand the abscissa at low doses). It is evident that the RAD wild-type strain is more mutable, and has a higher capacity for mutational resolution than either of the radiation sensitive derivatives; however, the excision-deficient strain *rad*1-1 has greater mutational sensitivity at very low doses, but this greater sensitivity is accompanied by a reduced capacity to "resolve" induced mutants from the spontaneous background (see text for further details).

dose (Fig. 41.13). This semilog plot "expands" the dose scale so that the yield data for the excision-defective strain are not so extremely compressed toward the origin as they are in Figure 41.11. From Figure 41.13 it is evident immediately that the integral yield and the value of Y_{max} is greatest for RAD. Thus, this strain has the greatest overall mutability and mutational resolution of the three. It would be the preferred strain for the detection of weak mutagens (all other biological factors being assumed equal for the strains). However, *rad*1-1 has the greatest yield at very low doses. Hence, this strain would be best of the three for the detection of very low mutagen doses. Finally, despite the fact that r_1^s is UV-sensitive, and presumably carries certain repair deficiencies, it is poorly mutable, and has neither the mutational sensitivity of *rad*1-1 nor the mutational resolution of RAD wild-type.

Concluding Remarks

The type of yield analysis described above can be carried out for genetic endpoints other than mutation. Two that come immediately to mind are mitotic recombination (Kunz and Haynes, 1978) and neoplastic transformation (Han and Elkind, 1979). Very roughly, it would appear, for some yeast loci at least, that the mutagenic effectiveness of UV is tenfold greater than X-rays. However, for mitotic recombination the relative efficiency of UV to X-rays is less, perhaps because of the importance of DNA single-strand breaks in recombination. On the other hand, fission-spectrum neutrons from the Argonne Janus reactor are about four times more efficient than UV for the neoplastic transformation of C3H mouse embryo cells (M.M. Elkind, personal communication). Thus it is conceivable that recombination-like events in

DNA are more proximally related to the formation of neoplasms than are induced somatic mutations. Indeed, is it possible that the reason for the correlation between mutagens and carcinogens (Ames *et al.*, 1973b) lies in the fact that most mutagens are also recombinagens? This urgent question might be answered by detailed comparative studies of the relative mutagenic and recombinagenic efficiencies of various mutagens in microbial and mammalian systems.

Summary

We feel that analysis of mutant yield data is the best way to compare the actions of genetically active agents in simple cellular systems. The quantitative features of yield curves reflect the interplay of lethal and genetic effects and so provide a broad view of the overall biological effect of mutagens on cells. The initial slope $[Y'(x)]$ and the position of the maximum yield $(\hat{x})$ is diagnostic of the kinetic response pattern of the system. The final slope of the $\ln Y(x)$ curve is parallel to the final slope of the $\ln S(x)$ curve if mutation and killing are stochastically independent processes; if these final slopes are not parallel then there is a δ-effect in the system. Yields, not frequencies, are the quantities normally measured, and rectilinear plots of such data baldly reveal the statistical error within the experiments and the amount of induction above background. Plots of yield versus lethal hits $[Y(-\ln S)]$ provide a simple way of assessing quantitatively the relative mutagenic efficiency of agents whose physical exposure doses would be otherwise incommensurable. Plots of yield versus log dose $[Y(\log x)]$ provide a simple way of visualizing the relative mutability and the mutational sensitivity and resolution of different test systems, particularly excision-proficient and excision-deficient strains of the same organism.

Acknowledgments

We thank Ingeborg Vranesic for the cheerful way in which she typed this manuscript despite the lunatic schedule imposed upon her. We also thank Veronica M. Maher, R.M. Baker, M.A. Hannan, J.A.M. Heddle, and B.A. Kunz for helpful discussions. The work reported here was supported by grants from the National Research Council of Canada, the Natural Sciences and Engineering Research Council of Canada, and the Deutsche Forschungsgemeinschaft.

References

Albertini, R.J. DeMars, R. (1973): Somatic cell mutation: detection and quantification of X-ray-induced mutation in cultured, diploid human fibroblasts. Mutat. Res. *18*:199–224.

Ames, B.N., Lee, F.D., Durston, W.E. (1973a): An improved bacterial test system for the detection and classification of mutagens and carcinogens. Proc. Natl. Acad. Sci. USA *70*:782–786.

Ames, B.N., Durston, W.E., Yamasaki, E., Lee, F.D. (1973b): Carcinogens are mutagens: a simple test system combining liver homogenates for activation of bacteria for detection. Proc. Natl. Acad. Sci. USA *70*:2281–2285.

Ames, B.N., McCann, J., Yamasaki, E. (1975): Methods for detecting carcinogens and mutagens with the *Salmonella*/mammalian-microsome mutagenicity test. Mutat. Res. *31*:347–364.

Drake, J.W. (1970): The Molecular Basis of Mutation. San Francisco: Holden-Day, pp. 40–50.

Eckardt, F., Haynes, R.H. (1977a): Kinetics of mutation induction by ultraviolet light in excision-deficient yeast. Genetics *85*:225–247.

Eckardt, F., Haynes, R.H. (1977b): Induction of pure and sectored mutant clones in excision proficient and deficient strains of yeast. Mutat. Res. *43*:327–338.

Eckardt, F., Kowalski, S., Laskowski, W. (1975): The effects of three *rad* genes on UV induced mutation rates in haploid and diploid *Saccharomyces* cells. Mol. Gen. Genet. *136*:261–272.

Green, M.H.L., Muriel, W.J., (1976): Mutagen testing using TRP+ reversion in *Escherichia coli*. Mutat. Res. *38*:3–32.

Green, M.H.L., Muriel, W.J., Bridges, B.A. (1976): Use of a simplified fluctuation test to detect low levels of mutagens. Mutat. Res. *38*:33–42.

Han, A., Elkind, M.M. (1979): Transformation of mouse C3H/10T¹/₂ cells by single and fractionated doses of x-rays and fission spectrum neutrons. Cancer Res. *39*:123–130.

Haynes, R.H., Eckardt, F. (1979): Analysis of dose-response patterns in mutation research. Can. J. Genet. Cytol. *21*:277–302.

Hill, R.F. (1965): Ultraviolet-induced lethality and reversions to prototrophy in *Escherichia coli* strains with normal and reduced dark repair ability. Photochem. Photobiol. *4*:563–568.

Hsie, A.W., O'Neill, J.P., Couch, D.B., San Sebastian, J.R., Brimer, P.A., Machanoff, R., Fuscoe, J.C., Riddle, J.C., Li, A.P., Forbes, N.L., Hsie, M.H. (1978): Quantitative analysis of radiation- and chemical-induced lethality and mutagenesis in Chinese hamster ovary cells. Radiat. Res. *76*:471–492.

Ikenaga, M., Ichikawa-Ryo, H., Kondo, S. (1975): The major cause of inactivation and mutation by 4NQO in *Escherichia coli*: excisable 4NQO-purine adducts. J. Mol. Biol. *92*:341–356.

Kondo, S. (1972): A theoretical study of spontaneous mutation rate. Mutat. Res. *14*:365–374.

Kubitschek, H.E. (1970): An Introduction to Research with Continuous Cultures. Englewood Cliffs, N.J.: Prentice-Hall, Inc.

Kunz, B.A., Haynes, R.H. (1978): Analysis of UV-induced mitotic recombination in yeast. In: Abstracts. 9th International Conference on Yeast Genetics and Molecular Biology, University of Rochester, Rochester, N.Y., p. 50.

Lemontt, J.F. (1977): Mutagenesis of yeast by hydrazine: dependence upon post-treatment cell division. Mutat. Res. *43*:165–178.

Luria, S.E., Delbrück, M. (1943): Mutations of bacteria from virus sensitivity to virus resistance. Genetics *28*:491–511.

Maher, V.M., McCormick, J.J. (1976): Effect of DNA repair on the cytotoxicity and mutagenicity of UV irradiation and of chemical carcinogens in normal and xeroderma pigmentosum cells. In: J.M. Yuhas, R.W. Tennant, and J.D. Regan (eds.) Biology of Radiation Carcinogenesis. New York: Raven Press, pp. 129–145.

McCann, J., Choi, E., Yamasaki, E., Ames, B.N. (1975): Detection of carcinogens as mutagens in the *Salmonella*/microsome test: assay of 300 chemicals. Proc. Natl. Acad. Sci. USA *72*:5135–5139.

Parry, J.M. (1977): The use of yeast cultures for the detection of environmental mutagens using a fluctuation test. Mutat. Res. *46*:165–176.

Patrick, M.H., Haynes, R.H., Uretz, R.B. (1964): Dark recovery phenomena in yeast. I. Comparative effects with various inactivating agents. Radiat. Res. *21*:144–163.

Prakash, L. (1976): Effect of genes controlling radiation sensitivity on chemically induced mutations in *Saccharomyces cerevisiae*. Genetics *83*:285–301.

Sherman, F., Stewart, J.W. (1978): The genetic control of yeast iso-1 and iso-2-cytochrome c after 15 years. In: M. Bacila, B.L. Horecker and A.O.M. Stoppani (eds.), Biochemistry and Genetics of Yeast: Pure and Applied Aspects. New York: Academic Press, pp. 273–316.

Thacker, J. (1979): The involvement of repair processes in radiation-induced mutation of cultured mammalian cells. In: S. Okada, M. Imamura, T. Terasima, and H. Yamaguchi (eds.), Radiation Research, Proceedings of the VIth International Congress on Radiation Research. Tokyo, Maruzen Co. Ltd., pp. 621–630.

Thompson, L.H., Baker, R.M. (1973): Isolation of mutants of cultured mammalian cells. In: D.M. Prescott (ed.), Methods in Cell Biology. Vol. IV. New York: Academic Press, pp. 209–281.

Unrau, P., Wheatcroft, R., Cox, B.S. (1971): The excision of pyrimidine dimers from DNA of ultraviolet irradiated yeast. Mol. Gen. Genet. *113*:359–362.

von Borstel, R.C. (1978): Measuring spontaneous mutation rates in yeast. In: D.M. Prescott (ed), Methods in Cell Biology, Vol. 20. New York: Academic Press, pp. 1–24.

Witkin, E.M. (1966): Radiation induced mutations and their repair. Science *152*:1345–1352.

Zimmermann, F.K. (1975): Procedures used in the induction of mitotic recombination and mutation in the yeast *Saccharomyces cerevisiae*. Mutat. Res. *31*:71–86.

42

Tests for Potential Carcinogens: Unresolved Problems

John Ashby

Introduction

At least 30 short-term tests for chemical carcinogens have been described during the past few years (Anon., 1978). They range from bacterial DNA repair assays *in vitro* through to several mammalian tests conducted *in vivo*. Several of these tests, such as the *Salmonella* reverse mutation assay developed by Ames *et al.* (1975), have been used extensively and with a wide range of chemical carcinogens (McCann *et al.*, 1975; Purchase *et al.*, 1978; Sugimura *et al.*, 1976). In contrast, the majority of tests have been evaluated using only a limited range of chemicals and in few laboratories. Some of these tests will make a unique contribution to the rapid detection of new potential carcinogens, while others may be either repetitive of other assays or of doubtful value *per se*. Several technical and philosophical problems must be resolved before any of these tests can achieve their true potential (Fig. 42.1), and even so there exists the possibility that their inadvertent

misuse will damage the credibility of the whole science (see the "Definition of Aims" at the end of this chapter).

The problems listed in Figure 42.1 are as follows:

The Predictive Accuracy of Short-Term Tests

The acceptance by most people that short-term tests have an important role to play in environmental screening programs is due largely to the good correlation observed, with some assays, between activity *in vitro* and the animal carcinogenicity of the test chemical. Most of these correlative studies have been conducted on the *Salmonella* assay, and predictivity figures of between 70–90% have been confirmed in several different laboratories (McCann *et al.*, 1975; Purchase *et al.*, 1978; Sugimura *et al.*, 1976). Many subsequent tests have relied implicitly on "validations" conducted on

UNRESOLVED PROBLEMS.

1. **REFINE EXISTING CORRELATIONS - CENTRAL INDICATOR ?**

2. **IS 'AROCLOR S-9 MIX' THE BEST ACTIVATION SYSTEM ?**

3. **DEFINITION OF A POSITIVE RESPONSE-ERRORS-REPRODUCIBILITY**

4. **'MANIPULATION' OF TEST RESPONSE - SIGNIFICANCE ?**

5. **DO MUTAGENIC NON-CARCINOGENS EXIST ?**

6. **POTENCY CORRELATIONS**

7. **TEST BATTERIES-WHY ARE THEY NEEDED-EVIDENCE ?**

8. **CONSIDERATION OF DISTRIBUTION/EXCRETION _IN_ _VIVO_**

9. **SHORT TERM TESTS _IN_ _VIVO_ ?**

10. **SYNERGISM AND PROTECTION, _IN_ _VIVO_ AND _IN_ _VITRO_**

11. **THRESHOLDS AND RAD-EQUIVALENTS ?**

12. **THE CHEMISTRY OF ORGANIC CARCINOGENS - RELEVANCE ?**

13. **INTEGRATION OF 'PRIVATE PATCHES'**

14. **INTERNATIONAL DEFINITION OF AIMS**

Figure 42.1.   Unresolved problems of short-term testing.

the *Salmonella* assay, although some have been individually evaluated in similar detail. Percentage-predictivity figures can be misleading; in particular, they are sensitive to the class of chemical being evaluated (Purchase *et al.*, 1978); nonetheless, they can act as a useful indication of the overall utility of a test. Most of these correlative studies were made using carcinogens defined before 1970 and none have been adjusted to take account of subsequent findings. The increasing number of good quality animal carcinogenicity bioassays being reported in the literature present an opportunity to continually update the overall predictivity of short-term assays. This would require that each bioassay report should be accompanied by a set of responses given by several short-term tests. These extra data could be collected from the published literature or be commissioned during the course of the bioassay. A central repository of

such data could provide a valuable indicator of the strengths and weaknesses of each test for chemicals of different classes. The fact that the very detailed bioassay reports of the National Cancer Institute fail to record even the response given by the test chemical in the *Salmonella* assay underlines this need.

A problem probably unique to the *Salmonella* assay relates to the development of new and more sensitive tester strains for the detection of carcinogens previously found to be negative. There is a tendency to add these new strains to the core battery of four or five tester strains and thereby assume that the test has been improved. The disadvantages of this practice, in terms of new false positive predictions, could well outweigh the apparent advantages, and this can only be determined if each new tester strain is validated prior to general use with a wide selection of those noncarcinogens found negative by the initial core of tester strains.

Metabolic Activation *in vitro*

Most *in vitro* assays rely upon a homogenate of rat liver (the S9 mix) to activate the test chemical into a DNA-reactive species. From the outset (Ames *et al.*, 1975), and based on very limited evidence, it has been usually accepted that the liver from which the S9 mix is made should be preinduced *in vivo* with chemicals. Most workers use either Aroclor 1254 or phenobarbital for this purpose. The implied rationale for induction is that the "activation" enzymes will be enhanced at the expense of the "deactivation" enzymes, and this is unlikely to be invariably true. The need for chemical induction needs to be reviewed; while it remains it presents a wall of unreality between effects observed *in vitro* and those predicted to occur *in vivo*.

Definition of a Positive Response: System Errors: Reproducibility

The generally accepted definition of a positive test response is based on two criteria. The first is that at least a doubling of the background level of change should occur, and the second, that any such effects should be dose-related. This definition is arbitrary and could encompass routine biological variations. As the results generated by such tests now have the power to seriously impede, if not curtail, the development of a new chemical, an attempt should be made to place the endpoint of each assay on a firm and internationally agreed basis. Likewise, operator errors will sometimes occur and these should not act as a permanent blight on the unlucky chemical in question. Related to these concerns is the need to ensure that any tests incorporated into a battery of tests (see "Test Batteries" below) should be shown to give a response for a given chemical that is capable of being reproduced in several laboratories. This is often assumed to be so, but is rarely established.

Manipulation of Test Responses

There are many ways of reproducibly increasing or decreasing the magnitude of a test response *in vitro* (Figs. 42.2 and 42.3) (Ashby, 1979; Ashby and Styles, 1980). These effects are usually consistent with the technical aspects of enzyme-mediated activation or deactivation, DNA-reactivity or test protocol, but they are not consistent with attempts to interpret a given test response within a rigid framework, such as with the implication of potency correlations

INCREASE RESPONSE IN VITRO

1. ADD PURIFIED ACTIVATION ENZYME OR CO-FACTOR

2. ADD REACTIVE CHEMICAL

3. ADD NORHARMAN

4. ADD COMPETITIVE SUBSTRATE

5. ANAEROBIC INCUBATION

6. INDUCTION ?

Figure 42.2. Factors which increase the response of *in vitro* short-term tests.

DECREASE RESPONSE IN VITRO

1. ADD PURIFIED DEACTIVATION ENZYME

2. ADD COMPETITIVE NUCLEOPHILE

3. CHANGE SOURCE OF S-9 MIX

4. INDUCTION ?

Figure 42.3. Factors which decrease the response of *in vitro* short-term tests.

(Ames and Hooper, 1978; Ashby and Styles, 1978).

Carcinogenic Nonmutagens and Mutagenic Noncarcinogens

The incorrect predictions of short-term tests, in particular, of those having a mutagenic endpoint, fall into one or other of the above two categories. An assessment of the reality of these categories can follow only from a detailed evaluation of all of the data available in each. It is clearly important to understand why such noncorrelations occur, and this will increase rather than decrease the value of these tests. Probably the best studied carcinogen that is nonmutagenic to bacteria is saccharin (Ashby *et al*., 1978a), and styrene seems to represent a noncarcinogenic bacterial mutagen (Ashby and Styles, 1980; DeMeester *et al.*, 1977; Anon., 1979a,b), but both of these chemicals have yet to be fully evaluated for carcinogenicity in animals. An example of a noncarcinogenic mammalian germ-cell mutagen has yet to be identified, and such a profile of activity is unlikely to be capable of being foretold by tests conducted *in vitro*.

Potency Correlations

A short-term test that is capable of predicting the carcinogenic potency of a potential

mammalian carcinogen would be extremely useful, and an extensive study to determine if the *Salmonella* assay fulfills this role is underway (Ames and Hooper, 1978). The greatest obstacles to the development of such an assay are the absence of an agreed definition of carcinogenic potency and the intrinsic variability, and arbitrary nature of the "Aroclor S9 mix" that is usually employed *in vitro*.

Test Batteries

Short-term tests such as the *Salmonella* assay achieved the position of importance that they currently enjoy because of their *established* ability to detect as positive the majority of the known mammalian carcinogens. Proposals are now being made to use these and other assays for the additional purpose of defining potential human (germ-cell) mutagens, and although no such chemicals are known, their probable existence justifies the undertaking of such a search. This duality of purpose has led to the confusing situation of the same test being employed to predict two separate activities, the first being real, the second putative.

There is an *established* need (Purchase *et al.*, 1978) to employ more than one assay for the detection of potential human carcinogens, and this is being confused with the *theoretical* need to use several tests to detect potential human mutagens. An attempt is made below to separate these endpoints; the need for and the credibility of the former should not automatically support the latter—each should be capable of individual justification.

Carcinogen test batteries. The need for a battery of tests to detect potential mammalian carcinogens can be supported by examples (Purchase *et al.*, 1978). Figure 42.4 outlines an exercise that could be conducted on the available literature; the questions posed as a result (Fig. 42.4) are self-explanatory. In particular, the need for a

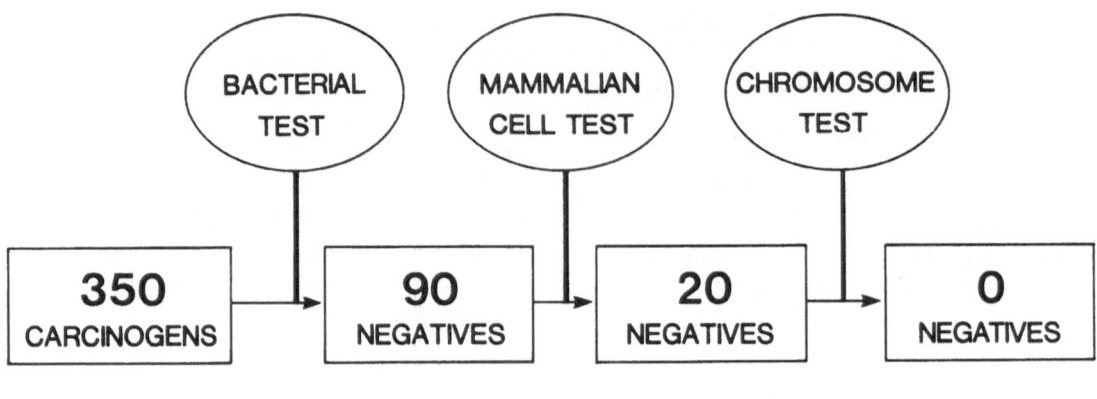

Figure 42.4.   Predictive accuracy of short-term tests.

third test would have to be supported by a list of established carcinogens that are *uniquely* detected by it. Furthermore, the *general* reliability of the third (or subsequent) test would have to be clearly established. For example, a test that successfully detected the carcinogens benzene, diethylstilbestrol, and chloroform as positive would also have to be capable of confirming the true negatives discerned by the first two tests (i.e., the later tests would have to possess high *specificity* and appropriate *sensitivity*).

Mutation test batteries. In the absence of any established human mutagens a different approach to that adopted for the construction of the carcinogen test battery must be adopted. The approach that is generally adopted is to list all of the *possible* types of genetic damage that a chemical *might* produce in man and then collect tests to detect each of these endpoints. The tests usually considered are selected from the following: bacterial or mammalian cell point mutation or DNA repair assays, chromosome aberration tests which measure endpoints ranging from breaks to aneuploidy and which use hosts that can

range from yeast and *Drosophila* to whole mammals, and finally, whole animal studies such as the dominant lethal assay.

The questions raised in connection with the carcinogen test battery (Purchase *et al.*, 1978) (Fig. 42.4) will almost certainly apply to a mutagen test battery, but the general paucity of mammalian data will preclude them from being answered. The fear exists, therefore, that cumulative "system-errors" or "false predictions" will severely reduce the effectiveness of such a battery. [I believe that a fundamental error was made with the decision to derive two separate test batteries. It should be possible to detect the vast majority of any potential human carcinogens or mutagens using a single battery of three or four carefully chosen assays. For example, the combination of a bacterial-cell and a mammalian-cell point mutation assay, a cell-transformation assay, and a chromosome aberration assay would detect the vast majority, if not all, of the known mammalian carcinogens and mutagens. Should *confirmation* of potential mammalian carcinogenic or heritable mutagenic effects be required, a carcinogenicity bioassay or an appropriate mammalian germ-cell assay could be undertaken.]

Distribution and Excretion of Chemicals *in vivo*

In most short-term test systems, the test chemical is confined in close contact with "induced" liver enzymes and nuclear DNA. This is not representative of the situation that exists during exposure of a mammal to a chemical. Attempts to discern overriding factors unique to the situation *in vivo* (such as the rapid detoxification or nonabsorption of a particular chemical) will aid the extrapolation of effects observed *in vitro* to the situation *in vivo*.

Short-Term Tests *in vivo*

Many of the problems discussed above would disappear if a simple and reliable *in vivo* assay were to be developed. Efforts being made in this direction are discussed elsewhere in this volume. Whatever problems remain to be solved, such tests are clearly worthy of continued development.

Synergistic and Protective Effects

Carcinogenic synergism and carcinogenic protection are both well-defined phenomena, and apparently similar effects have been observed *in vitro* (Ashby and Styles, 1980; Ashby *et al.*, 1978b; Mondal *et al.*, 1978; Rosin and Stich, 1978). Human exposure to a single, pure chemical is rare; thus the need to study cooperative effects seems to be great. As would be expected, synergistic or protective effects observed *in vivo* are not always faithfully reflected *in vitro* (Ashby and Styles, 1980), and vice versa; consequently, the possible significance to

man of cooperative effects observed *in vitro* should be assessed with caution.

Thresholds and Rad-equivalents

The debate as to the existence or otherwise of threshold dose levels for individual chemical carcinogens or individual mammals will continue and only be resolved when the appropriate experimental data become available. However, the equivalence that is often assumed to exist between mutagenic effects produced by either radiation or chemicals *can* be questioned (the so-called rad-equivalent dose level of a chemical carcinogen). Figure 42.5 shows the effect produced upon the mammalian cell transforming properties of the bacterial mutagen magic methyl (Ashby *et al.*, 1978c) by increasing doses of the simple amine piperidine (a decreasing biological response is observed when increasing levels of this amine are incorporated within the test medium). This illustrates that chemical electrophiles are subject to diversionary reaction with nucleophiles other than DNA, a restraint absent with radiation. It would therefore be expected that piperidine would not effect the DNA-modifying effects of an apparently equivalent dose level of radiation. Although piperidine is not a biological amine, many do exist, and these together with hydroxy and sulfhydryl groups (e.g., cysteine) (Rosin and Stich, 1978) will modulate the genetic activity of chemical carcinogens. Such factors, together with the effect of detoxification enzymes (which again do not apply directly to radiation), present the prospect of individual carcinogenic thresholds.

The Chemistry of Organic Carcinogens

The initial lesion produced within a cell by a chemical carcinogen, whether it be on

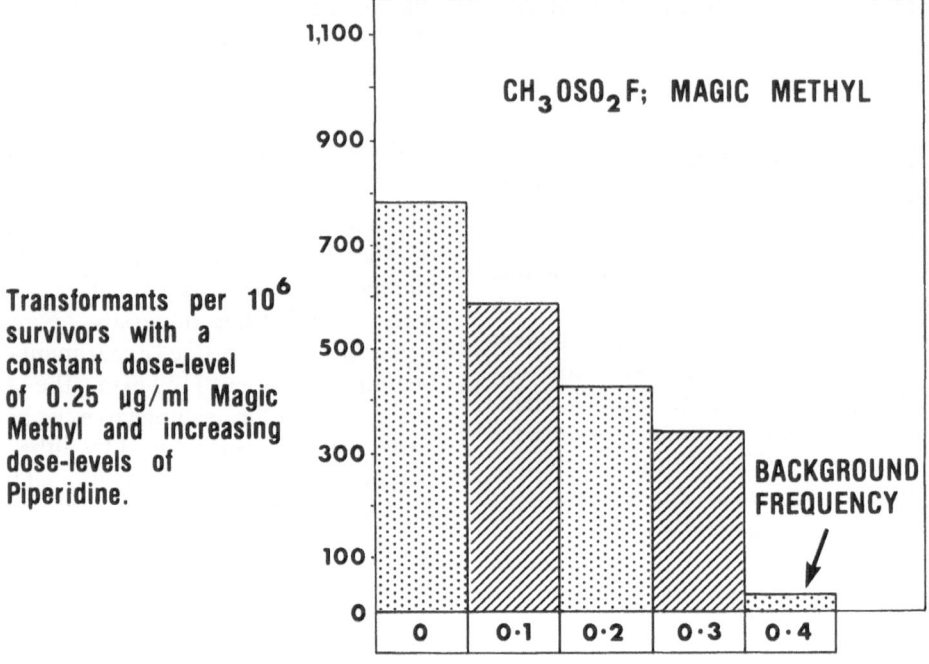

Figure 42.5.   The effect of piperidine on the transformation capacity of magic methyl.

DNA, RNA, or proteins, is the product of a simple (as they all are at the last moment) chemical reaction. Studies to discern the underlying chemical reactivity of chemical carcinogens are therefore bound to prove useful to any attempt to discern new chemical carcinogens. Based on the facts gleaned to date (Ashby, 1978), it is usually possible to predict if a new chemical is likely to possess electrophilic properties of biological significance, and this can aid the sorting of chemicals before any tests *in vitro* are conducted. From a chemical viewpoint, there are very few surprises as each new organic carcinogen is announced; the majority of surprises emanate from newly defined noncarcinogens. This indicates that this "screen" is over-sensitive, and as such is useful for discerning chemicals worthy of evaluation *in vitro*. General chemical awareness and studies to discern structure activity relationships (SAR) are the Cinderella of the environmental sciences. Most of the workers in this discipline confine their

attention to "why" and "how" benzo-(a)pyrene, aflatoxin B_1, dimethylnitrosamine, or simple alkylating agents produce the biological effects that they do. While this is a necessary pursuit, those involved could do much to help the current situation; they themselves are one of the oldest and most neglected of short-term tests.

Integration of "Private Patches"

A new science has recently been created aimed at rapidly and reliably detecting those chemicals that will either cause cancer in man or affect his progeny. This has necessitated that scientists from many disciplines combine their efforts, and this must continue. In particular, we must guard against any particular aspects of this cognate science being regarded as preeminent.

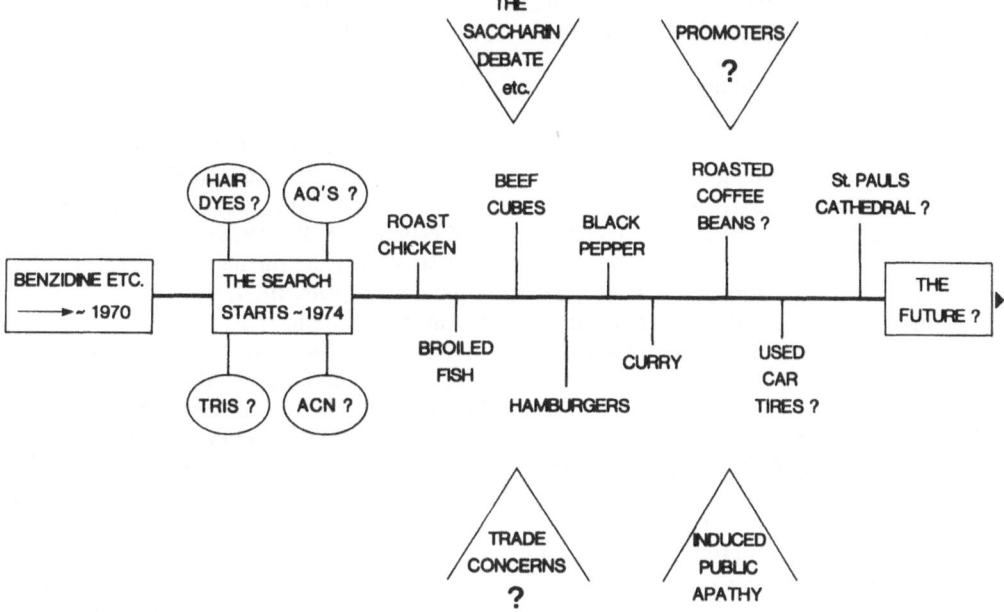

Figure 42.6. Issues of concern. In particular, it is suggested that "used" car tires and the Victorian patina of St. Paul's Cathedral will be mutagenic to bacteria.

Definition of Aims

Few people would doubt the need to remove from the environment those chemicals that will either cause cancer in exposed humans or affect adversely their progeny; however, this issue is becoming seriously confused by parallel attempts to investigate the possible genetic hazard presented to man by natural environmental chemicals and foodstuffs. While some of these "natural" agents may, in fact, contribute to the incidence of human cancer, their confusion with commodity chemicals will not help either cause. Some of these concerns and some related dilemmas are shown diagrammatically in Figure 42.6, and it is suggested that these must be addressed by those involved with short-term tests. If such attempts are not made, induced public apathy (Weinberg, 1979) ("does exposure to chemical 'X' present a greater or lesser hazard to man than the eating of a hamburger?" etc.) may overtake our efforts to create a safer environment.

References

Ames, B.N., Hooper, N.K. (1978): Does carcinogenic potency correlate with mutagenic potency in the Ames assay? Nature (Lond.) *274*:19–20.

Ames, B.N., McCann, J., Yamasaki, E. (1975): Methods for detecting carcinogens and mutagens with the *Salmonella*/mammalian-microsome mutagenicity test. Mutat. Res. *31*:347–364.

Anon. (1978): International Programme for the evaluation of short-term tests for carcinogenicity. Mutat. Res. *54*:203–206.

Anon. (1979a): NCI clearinghouse subgroup recommends retesting of styrene. Food Chem. News *20*:45–46.

Anon. (1979b): Maltoni's work on styrene. Plastics in Retail Packaging Bull. *2*:6.

Ashby, J. (1978): Structural analysis as a means of predicting carcinogenic potential. Br. J. Cancer *37*:904–923.

Ashby, J. (1979): Implications of carcinogenicity. In: G.E. Paget (ed.), Mutagenesis in Sub-Mammalian Systems 165, Lancaster England: MTP Press, pp. 165–184.

Ashby, J., Styles, J.A. (1978): Does car-

482   John Ashby

cinogenic potency correlate with mutagenic potency in the Ames assay? Nature (Lond.) *271*:452–455.

Ashby J., Styles, J.A. (1980): Carcinogenic synergism and its reflection *in vitro*. Br. Med. Bull. *36*:63–70.

Ashby, J., Styles, J.A., Anderson, D., Paton, D. (1978a): Saccharin: an epigenetic carcinogen/mutagen? Food Cosmet. Toxicol. *16*:95–103.

Ashby, J., Styles, J.A., Paton, D. (1978b): *In vitro* evaluation of some derivatives of the carcinogen butter yellow: implications for environmental screening. Br. J. Cancer *38*:34–50.

Ashby, J., Anderson, D., Styles, J.A. (1978c): The potential carcinogenicity of methyl fluorosulphonate (CH_3OSO_2F; Magic Methyl). Mutat. Res. *51*:285–287.

De Meester, C., Poncelet, F., Roberfroid, M., Rondelet, J, Mercier, M. (1977): Mutagenic activity of styrene and styrene oxide. A preliminary study. Arch. Int. Physiol. Biochim. *85*:398–399.

McCann, J., Choi, E., Yamasaki, E., Ames, B.N. (1975): Detection of carcinogens as mutagens in the *Salmonella*/microsome test: assay of 300 chemicals. Proc. Natl. Acad. Sci. USA *72*:5135–5139.

Mondal, S., Brankow, D.W., Heidelberger, C. (1978): Enhancement of oncogenesis in C3H/10T1/2 mouse embryo cell cultures by saccharin. Science *201*:1141–1142.

Purchase, I.F.H., Longstaff, E., Ashby, J., Styles, J.A., Anderson, D., Lefevre, P.A., Westwood, F.R. (1978): An evaluation of 6 short-term tests for detecting organic chemical carcinogens. Br. J. Cancer *37*:873–903.

Rosin, M.P., Stich, H.F. (1978): The inhibitory effect of cysteine on the mutagenic activities of several carcinogens. Mutat. Res. *54*:73–81.

Sugimura, T., Sato, S., Nagao, M., Yahagi, T., Matsushima, T., Seino, Y., Takeuchi, M., Kawachi, T. (1976): Overlapping of carcinogens and mutagens. In: P.N. Magee, S. Takayama, T. Sugimura and T. Matsushima (eds.), Fundamentals in Cancer Prevention. Tokyo: Tokyo University Press/Baltimore: University Park Press, pp. 191–215.

Weinberg, A.M. (1979): Another cancer scare . . or is it hypochondria? Nature (Lond.) *279*:286.

43

Mutagenicity Testing: Problems in Application*

Marvin S. Legator and Stephen J. Rinkus

Introduction

In this decade, genetic toxicology has emerged as an interdisciplinary science that aims to understand the mechanisms of action of agents that induce mutations and to appraise the human risks posed by exposure to such agents. Our research into chemical mutagenesis has shown us that some chemical types in themselves or after metabolic activation are capable of altering the informational content encoded in the base sequence of DNA. It now becomes understandable that adverse health effects in man such as carcinogenesis, mutagenesis, and teratogenesis may result from overexposure to such chemicals that alter genetic material. Thus it behooves us as scientists to use the many mutagenicity testing methods that have been developed —and are being developed—to study these

genotoxic substances. In addition, it behooves us as a society to be ready to institute measures that will minimize human exposure to such agents.

The interpretation of results of mutagenicity testing in terms of possible human experience is an exceedingly difficult one. The problem is made no easier by the eagerness of many genetic toxicologists to rely heavily, if not exclusively, on *in vitro* and nonmammalian systems in their evaluations. The attractiveness of the relatively simple and inexpensive short-term tests should not be a deterrent to the development of more relevant procedures in whole mammals and to the study of chemical mutagenesis in high-risk groups such as certain clinical and industrial populations. With this theme in mind, this paper will review the arguments for the necessity of combined *in vitro* and *in vivo* testing in mutagenicity screening and discuss the importance of cytogenetic studies in mutagenicity testing.

*This manuscript has been supported by EPA Grant R804621020 (Project Officer: Dr. J.F. Stara).

Mutagenicity Screening: The Need For *in vitro* and *in vivo* Systems

The evaluation of a chemical for mutagenic activity is essentially a two-phase process. First there is the *qualitative* identification of the mutagenic activity. The testing in this phase is designed primarily to answer the question of whether the chemical is mutagenic. Subsequently, qualitative mutagens are assessed *quantitatively,* the ultimate goal being to understand the potential risk to user populations. Ideally, quantitative testing answers the question of what is the dose response in a mammalian system. This general principle of qualitative followed by quantitative testing has been the topic of several papers (Bochkov *et al.,* 1976; Bora, 1976; Bridges, 1973, 1974, 1976; Dean, 1976; Flamm, 1974; Legator and Zimmering, 1975; Mayer and Flamm, 1975; Schöneich, 1976) and is popularly known as the tier approach to mutagenicity testing. Keeping in mind that no matter how many tiers are proposed the evaluation process still remains biphasic and sequential, the crucial importance of the initial phase of testing dictates that several criteria should be satisfied (Committee 17 of the Environmental Mutagen Society, 1975; Flamm, 1974; de Serres, 1974; de Serres *et al.*, 1973). The qualitative testing should be capable of detecting the entire spectrum of mutagenic events; it should account for the importance of metabolism in activating and deactivating mutagens; it should produce no false negatives unnecessarily and only a nominal amount of false positives; and it should provide reproducible results within any given laboratory and among different laboratories. Obviously, this last criterion necessitates the use of standardized protocols in order to judge reproducibility. Also, from a regulatory agency's viewpoint, standardized protocols for testing would be essential to the evaluation process. Finally, it would be desirable, if not essential, that the qualitative testing make use of short-term test procedures as opposed to long-term animal studies.

It is now generally accepted that no single test procedure can satisfy these criteria, and consequently a battery of test procedures will be necessary. However, there are still differences of opinion over the exact prescription of test procedures that should make up the qualitative testing. The differences in opinion essentially concern whether approval for conditional use of a chemical should require satisfactory (negative) testing results with both *in vitro* and *in vivo* systems or just with *in vitro* systems alone without any testing *in vivo*.

That the qualitative testing for mutagens of the sophistication previously described cannot be accomplished by any battery of *in vitro* test procedures is best illustrated by reviewing the complexity of chemical mutagenesis in a mammalian system. This complexity is depicted schematically in Figure 43.1 as having three levels of consideration: the mammalian chromosomal organization, mammalian metabolism, and the mammalian cell membrane. The importance of each of these levels as it concerns mutagenicity screening is discussed below.

Chromosomal Organization

It can be assumed that not all chemical mutagens have the same mode of action in causing mutagenesis. An appreciation of these distinctions should therefore be reflected in the selection of tests that are to be used for screening chemicals for mutagenicity. Several recognized or presumed modes of actions of chemicals known to cause gene mutations or chromosome damage are illustrated on the chromosome in Figure 43.1. Chemicals that provide a precedence for these distinctions are also indicated in this figure.

The best-studied modes of action of chemical mutagens are *intercalation* and adduct formation with DNA, i.e., *alkylation, arylamination, arylation, acylation,* and *strand linking.* In the case of intercalation, some planar compounds consisting of fused, aromatic rings appear to become wedged electrostatically into the rouleau of base pairs of DNA (e.g., actinomycins, acridines, daunomycin). External positioning of some planar compounds in the narrow groove of the DNA helix has also been proposed and is depicted in Figure 43.1. These planar chemicals cause gene mutations that are characterized by addition or deletion of base pairs during DNA synthesis. The altered DNA template thus codes for a mRNA with a shifted reading frame (Brenner et al., 1961; Lerman, 1964; Terzaghi et al., 1966). The potency of some intercalating agents as frameshift mutagens is increased many fold by the addition of a reactive side group like nitrogen mustard (Ames and Whitfield, 1966). Some intercalating agents act potently in causing chromosome damage (e.g., Benedict et al., 1977a; Hsu et al., 1977; Vig et al., 1969). Presumably, this clastogenicity is related to intercalation, although other mechanisms, for example, one that is not limited to just extensive intercalation of the DNA and that involves the suprastructure of the chromosome, may also be envisaged.

In the case of adduct formation, any chemical modification of DNA has the potential to interfere with the template function of DNA and thereby lead to miscoding during replication. For example, alkylation at 0–6 of guanine or 0–4 of thymine in DNA by N-methyl-N-nitrosourea (Lawley et al., 1973) prevents these oxygen atoms from participating in hydrogen bonding and consequently could result in the anomalous base pairing of guanine with thymine when the DNA is replicated. However, not all adduct formations with DNA produce solely base-substitution mutations. Frameshift mutations are also observed with some electrophilic derivatives of aromatic amines (Ames et al., 1972a) and aryl compounds (Ames et al., 1972b). The mechanism for this frameshift mutation may be analogous to that intercalation that is observed with acridine mustards (Ames and Whitfield, 1966). Displacement of the modified base out of the interior of the double helix, so called base displacement, has also been proposed (Levine et al., 1974).

Acylation of DNA has not received much attention as a type of adduct formation. It had been suggested that the cellular damage done by acylation of intracellular nucleophiles could be repaired by various proteolytic enzymes (Jones and Young 1968). However, given the known mutagenicity and carcinogenicity of some acylating agents, this concept—at least as it relates to these two toxicities—must be revised. Unlike alkylated or arylated DNA, acylated DNA is difficult to isolate owing to the instability of acylamine groups during the acidic conditions used for degrading DNA into its constituents. Acylation may be inferred if deamination resulting from acid hydrolysis alters the base composition of DNA (e.g., deaminated cytosine is uracil) (Nery, 1969). The number of acylating agents that are known to be mutagenic or carcinogenic is relatively small. Dimethylcarbamyl chloride is the best example of an acylating agent that is both mutagenic (McCann et al., 1975) and carcinogenic (Van Duuren et al., 1972). The strong carcinogenicity of N-acetylimidazole (Stoner et al., 1975) deserves serious recognition. Other known or possible acylating agents that may prove to be genotoxic include maleic anhydride (Dickens and Jones, 1963), succinic anhydride (Dickens and Jones, 1965), cantharidin (Laerum and Iversen, 1972), and some unsaturated γ-lactones (Jones and Young, 1968). Interestingly, one of the many mechanisms of action proposed for the powerful carcinogen urethane involves ethoxycarbonylation by the N-hydroxy metabolite (Nery, 1969). Hence, it may be the case that some carbamyl compounds are converted metabolically to acylating agents in vivo.

Strand linking can be considered another type of adduct formation whereby a polyfunctional agent covalently reacts with both strands of the DNA helix. This crosslinking of complementary strands prevents the DNA helix from denaturing; it therefore prevents DNA synthesis and probably leads to cell death. In the attempted removal of the crosslinks, mutations may arise from error-prone excision repair or repair that is unfinished at the critical time of replication (Kondo *et al.*, 1970). Mitomycin C and some of the pyrrolizidine alkaloids (White and Mattocks, 1972) are examples of chemicals known to crosslink DNA strands.

Several mycotoxins inhibit nucleic acid synthesis by *DNA binding which is apparently not covalent*. However, in many cases, the exact mechanism of this binding and the relationship of the binding to the antibiotic and antitumor properties of the chemicals remains to be elucidated. For instance, luteoskyrin is a hepatotoxin and a hepatocarcinogen in rodents. The compound inhibits DNA synthesis and causes chromosomal aberrations in Ehrlich ascites tumor cells presumably by its binding to DNA (Schachtschabel *et al.*, 1969). Studies of this binding to isolated DNA indicate the formation of two different types of strong complexes with DNA when magnesium ion is present (reviewed in Goldberg and Friedman, 1971). Similarly, anthramycin binds to DNA, albeit slowly, and exhibits potent antitumor, antimicrobial, and chemosterilant properties (in insects) which are attributable to its interference with DNA functioning (reviewed in Horwitz, 1975). The compound also induces mutations and crossing-over in the eukaryote *Saccharomyces cerevisiae* but is essentially too toxic to test in the *Salmonella*/S9 system (Hannan *et al.*, 1978). Other mycotoxins that are known to interact with DNA and to be cytogenetically active include chromomycin A (Sentein, 1974) and netropsin (Wobus *et al.*, 1977).

Another mode of mutagenic action is suggested by two other mycotoxins that have antitumor activity, bleomycin and streptonigrin. These compounds, although very different in structure from each other, both cause single *strand scissions* in DNA. Bleomycin is a complicated glycopeptide that upon binding to DNA will break the phosphodiester backbone generally on the 3' side of the deoxyribose. Preferential splitting of thymine bases from DNA is another characteristic property of bleomycin. These peculiar specificities have been interpreted by some as evidence for an enzyme-like mechanism for the degradative action of bleomycin upon DNA (reviewed in Remers, 1979). Not unexpectedly then, chromosomal aberrations are evident in hamster cells treated *in vitro*; a dose-dependent increase in the morphological transformation frequency of murine cells is also observed (Benedict *et al.*, 1977a). Similarly, streptonigrin can introduce single-strand breaks upon incubation with isolated DNA. This activity may require the reduced form of streptonigrin and the presence of oxygen. Notably, an *E. coli* mutant which was resistant to X-irradiation was also resistant to the antimicrobial effects of streptonigrin (reviewed in Kremer and Laszilo, 1975). The potent clastogenicity of streptonigrin is well known (Cohen *et al.*, 1963).

Unique to eukaryotes are the packaging of DNA into several chromosomes and the segregating of these chromosomes during cell division by an intracellular system of microtubules. Therefore, also unique to eukaryotes are mutations that arise from disfunctioning of this mitotic apparatus. The *mitotic poisons* colchicine and podophyllotoxin bind to microtubule protein at a site that is different from the binding site of the vinca alkaloids (Bryan, 1972). However, griseofulvin inhibits mitosis without affecting microtubule assembly (Grisham *et al.*, 1973). The halogenated anesthetics halothane, enflurane, and methoxyflurane also induce segregational errors but not chromosomal breakage in mammalian cells treated *in vitro* and in chick embryos treated *in vivo*; the nature of this antimitotic activity remains to be investigated (Kusyk and Hsu, 1976).

Finally, there are the mutations that arise from *antireplication* effects. Theoretically, agents exist that can decrease the accuracy of DNA synthesis by affecting the DNA polymerases (designated "R" in Figure 43.1) involved in replication or repair. This premise is the basis of the fidelity assay which measures the accuracy of a DNA polymerase to copy an artificial DNA template (Loeb *et al.*, 1978). Likewise, antimetabolites that are so structurally related to nucleic acid or its precursors as to be anabolized and incorporated into DNA, e.g., 6-mercaptopurine, 5-bromodeoxyuridine (BUdR), show a propensity to mispair and introduce base substitutions during DNA synthesis. Nucleic acid deficiency during replication may also lead to mutation. Thymidine starvation of thymine auxotrophs of *Salmonella* induced what appear to be deletions (Holmes and Eisenstark, 1968). Clastogenicity is produced in mammalian cells *in vitro* that are presumably depleted of their thymidine reserves by treatment with the antifolate methotrexate; morphological transformation of murine cells was also noted in that study (Benedict *et al.*, 1977a).

Metabolism

Figure 43.1 also depicts the involvement of metabolism in chemical mutagenesis. First, one can consider the metabolic activation of proximate mutagens to their ultimate, reactive species. This activation can take place in the same cell that is later mutated or it can occur at sites distant to the target cell (e.g., activation by the microflora of the intestines). In the case of the latter, transport of the ultimate species to the target cell occurs. Similarly, the actual activation of the proximate mutagen may proceed in several steps, the final one(s) being completed in the target cell. The activation of many proximate mutagens which covalently react with DNA is generally attributed to the mixed-function oxidases of the endoplasmic reticulum. Recent work also suggests that the mixed-function oxidases present in the nucleus may play a heretofore unrecognized role in mutagenesis (Bresnick, 1976). For instance, these nuclear enzymes may provide a nuclear location for the production of radicals from reductive dehalogenation and electrophilic sulfur from oxidative desulfurization, both of which are species that are presumably too short-lived to migrate from the endoplasmic reticulum to the nucleus to react with the DNA.

Not shown in Figure 43.1 is the possibility of mitochondrial activation. This is suggested by studies on 1-(o-chlorophenyl)-1-(p-chlorophenyl)-2,2-dichloroethane.This compound, called Mitotane, is used in the treatment of adrenal cortex carcinoma, since it selectively necrotizes this tissue. Mitotane is an analogue of the pesticide DDT and some of its metabolites. The adrenocorticolytic action of Mitotane has been related to its enzymatic conversion to a covalently binding species by mitochondria but not microsomes or the soluble fraction isolated from canine adrenals. This activation requires NADPH, is inhibited by carbon monoxide, and may therefore be mediated by mitochondrial P-450 (Martz and Straw, 1977).

A second aspect of metabolism to be considered in chemical mutagenesis is that of hormones. The biological action of steroidal hormones is mediated by their binding to a cytoplasmic protein in target cells and the translocation of this steroid-protein complex into the nucleus for association with a nuclear receptor (Leclercq *et al.*, 1973). In contrast, thyroxine, another lipophilic hormone, simply diffuses to its chromosome receptor without any intervening protein carrier although one apparently exists (Oppenheimer and Surks, 1975). Nonlipophilic hormones bind to a receptor in the plasma membrane of target cells. This binding triggers the production of cAMP which in turn complexes with a cytoplasmic protein. In bacteria, a cAMP-protein complex acts directly at the gene level (Pastan and Adhya, 1976), and conceivably this activity is descriptive of what occurs in mammalian cells, assuming the activity has been conserved throughout evolution.

Given that the chromosome is the site of action for both hormones and mutagens, there is at least some basis to suspect that hormones, antihormones, and chemicals which induce oversecretion of endogenous

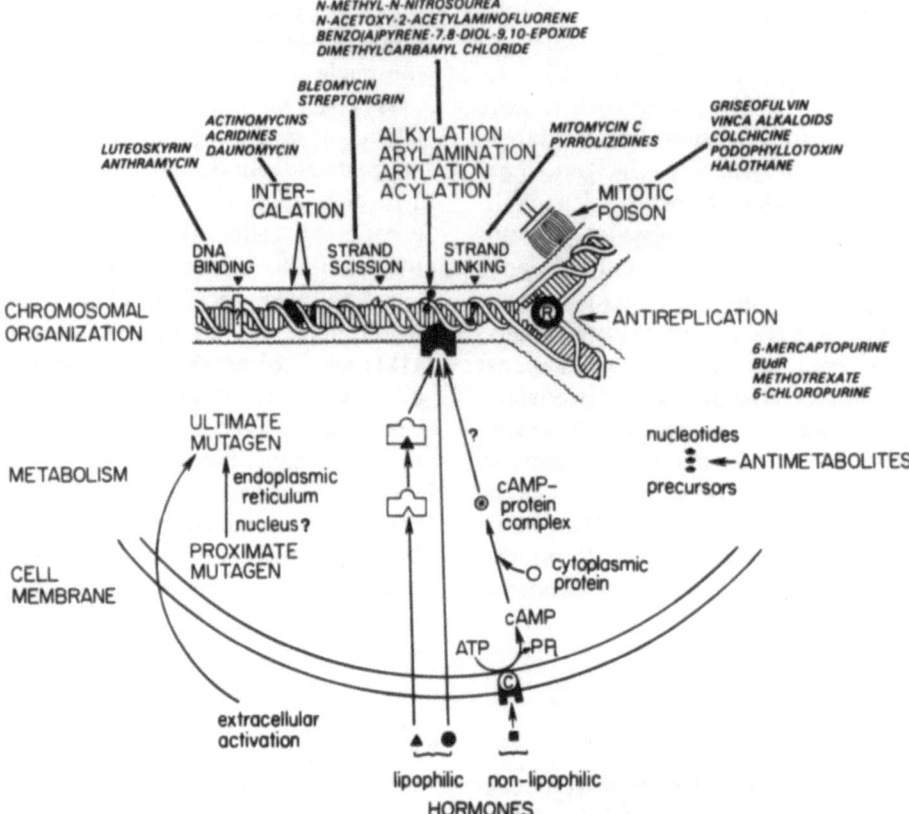

Figure 43.1.   Complexity of chemical mutagenesis in a mammalian system. Chromosomal organization, intracellular and extracellular metabolism, and the cell membrane are three levels of complexity in chemical mutagenesis. Known or assumed modes of action by which mutagens can be classified are illustrated in the diagram of a replicating chromosome ("R," DNA-replicating enzyme). Examples of chemicals that are presumably active in these manners are also shown. Three aspects of mammalian metabolism that may be important in chemical mutagenesis are the activation of proximate mutagens by mixed-function oxidases embedded in the endoplasmic reticulum and the nucleus, excessive stimulation by lipophilic and possibly nonlipophilic hormones or their analogues ("C," adenyl cyclase), and interference with the proper production of nucleotide pools or competition by nucleic acid analogues during replication (antireplication). The cell membrane becomes a consideration when test systems that are nonmammalian are used to predict chemical mutagenesis in a mammalian setting. Modified after Figure 5.1 in Gale *et al.* (1972) with permission.

hormones may act directly or indirectly as mutagens. Several investigators have reported that some steroids are cytogenetically active *in vivo* during meiosis in both female mice (Badr and Badr, 1974; Röhrborn and Hansmann, 1974) and male dogs (Williams *et al.*, 1971), and during mitosis in bone marrow cells of rats (Shimazu *et al.*, 1976). L-Thyroxine has been reported to increase the incidence of aneuploidy in bone marrow cells of mice treated *in vivo*

(Duca-Marinescu and Negoescu, 1972). The antiestrogen DES behaves like the mitotic poison colchicine in that it causes polyploidy in cells treated *in vitro* (Sawada and Ishidate, 1978). Reports on the mutagenicity testing of nonlipophilic hormones are scarce and seemingly all negative. This may be indicative of a general lack of interest in the testing of these hormones. There is also the possibility of organotropism in which the membrane receptors of target cells confer specificity to the biological as well as mutagenic action of these hormones. Hence, demonstration of mutagenicity (e.g., antimitotic effects) would depend on a proper pairing of hormone and cell type. This type of specificity of mutagenic action could also be associated with lipophilic hormones. However, the ease with which the lipophilic hormones can cross any cell membrane and the possibility that at high enough concentrations these hormones can saturate the intracellular environment may mask this specificity with a nonspecific mutagenic effect. Presumably this masking would be more likely to occur with *in vitro* testing rather than with *in vivo* testing.

Thirdly, one can consider metabolism as it relates to antimetabolites and DNA synthesis. As mentioned previously, depletion of nucleic acid reserves during replication can cause mutations. Hence, all the enzymes involved in the anabolism of purines and pyrimidines which support DNA synthesis may be potential targets for inhibition by chemicals that are generally (though not necessarily) marked similar to natural substrates. One can also consider the "activation" of antimetabolites as illustrated by the case of 6-chloropurine. This antipurine causes base-substitution mutations in the *Salmonella* strain G46 only in the host-mediated assay but not in the direct plate test (Ellis *et al.*, 1974); with mammalian cells it causes chromosome breakage *in vivo* but not *in vitro* (Holden *et al.*, 1974). The biological activity of 6-chloropurine is ascribed to its ribonucleoside or ribonucleotide (Hampton and Paterson, 1966). Hence, to borrow a term from the other side of Figure 43.1, the "ultimate" form of this mutagen may be 6-chloropurine ribonucleoside, ribonucleotide, or their deoxy derivatives. The important point is that one can speak of activation of a promutagen that does not involve the enzymes found in S9. Furthermore, this activation does not appear to be carried out by *Salmonella* enzymes that are involved in DNA synthesis.

Of course, this statement assumes that 6-chloropurine was assimilated by *Salmonella* in the first place. The importance of this consideration is discussed below.

Cell Membrane

The cell membrane is presented as a separate level of complexity. In a sense, the cell membrane is another aspect of metabolism in that what determines a chemical's entrance into a cell is the chemical's lipophilicity or the presence of an enzyme system in the membrane that allows for transport into the cell. Like the other aspects of metabolism, membrane specificity which is mediated by the so-called permeases will vary with different organisms. Consequently, when test systems that are nonmammalian are used to detect mutagens active in mammalian systems, membrane specificity may cause erroneous results (e.g., false negatives). For example, non-folic acid–requiring bacteria are relatively insensitive to the antifolates as measured by growth inhibition, because these bacteria are unable to assimilate them. However, these bacteria's dihydrofolate reductases, the target enzyme of the antifolates, are just as sensitive to these drugs as those enzymes of folic acid–requiring bacteria (Wood *et al.*, 1961). This may explain why methotrexate which is clastogenic to hamster cells treated *in vitro* (Benedict *et al.*, 1977a), presumably by causing thymidine deficiency, is not mutagenic in the *Salmonella* tester strains (Benedict *et al.*, 1977b), which do not require folic acid in their medium in order to grow. On the other hand, actinomycin D is apparently too large to pass through the outer membrane of the *Salmonella* tester strains to be bactericidal or mutagenic (Benedict *et al.*, 1977b). This contrasts with its clastogenicity in hamster cells treated *in vitro*; actinomycin D also morphologically transformed cells in that study (Benedict *et al.*, 1977a).

Conclusions and Further Comments

From the relationship of chromosomal organization to mechanisms of action of known mutagens, it is clear that eukaryotic organisms are needed to detect mutagens that behave as clastogens or mitotic poisons, whereas prokaryotic systems would suffice theoretically for detecting those mutagens that covalently modify or intercalate with DNA. It can also be argued that since some chemicals may cause mutations by more than one mode of action, more information about mutagens, and therefore the process of chemical mutagenesis, can be gained by developing a profile of testing in multiple systems.

From the relationships of mammalian metabolism and cell membrane to chemical mutagenesis, the need for mammalian systems in mutagenicity screening is once again clearly indicated. It was recognition of the importance of mammalian metabolism in the activation of promutagens that originally led to the innovation of S9 for *in vitro* testing (Malling, 1971). However, such metabolic complexity cannot be adequately, if at all in some respects, simulated by just using the solution of liver enzymes popularly known as S9. One must be aware of the inability of *in vitro* procedures to detect proximate mutagens activated by routes other than the hepatic mixed-function oxidases isolated from the endoplasmic reticulum. For instance, chemicals that are metabolized by the intestinal bacteria include: carcinogens, e.g., 1,2-dimethylhydrazine (Reddy et al., 1975); drugs, e.g., 4-isothiocyanato-4'-nitrodiphenylamine (Batzinger et al., 1978); food additives, e.g., cyclamate (Draser et al., 1972); food dyes, e.g., citrus red no. 2 (Radomski, 1961); and endogenous chemicals, e.g., bile acid (Draser and Hill, 1974). As previously suggested, some mutagens may be activated to highly reactive species, e.g., radicals, by the mixed-function oxidases in the nucleus. Presumably a nuclear location for the actual activation would allow these short-lived species to reach the DNA and thereby cause mutations and induce cancer. If this is the case, the use of liver microsomes to provide this metabolism may produce the ultimate species, but the mutagenicity may not be detected with the indicator organism owing to competing reactions with other constituents in the test system.

Recently we (Rinkus and Legator, 1979) have reviewed the *Salmonella* correlation studies. For the purpose of this paper, these correlation studies provide a data base for illustrating some points about *in vitro* testing. Table 43.1 lists the carcinogens that are demonstrably mutagenic in the *Salmonella*/S9 system. It is the detection of these mutagens that has made this system

Table 43.1. *Carcinogens Mutagenic in the* Salmonella/S9 *System*

Ultimate mutagens	Proximate mutagens
Diazoalkanes	Alkyltriazenes
Alkylnitrosamides, alkylnitrosoureas	Azoxyalkanes
Dimethylcarbamyl chloride	Alkylnitrosamines
Diaryl alkynyl carbamates	Aromatic amines
Aziridines	Polyaromatics
Oxiranes	Heteroaromatics
β-propiolactone, β-butyrolactone	Oxazaphosphorine mustards
Monohalomethyls	Haloethylenes
N-, O- and S-mustards	Monoalkyl hydrazines[a]
Alkylsulfates, alkylsulfonates, sultone	Nitroaromatics[b]
Trimethylphosphate	

[a]N-(2-Hydroxyethyl)hydrazine is also mutagenic (albeit less so) without S9 activation (Rinkus and Legator, 1979).

[b]Bacterial nitroreductases and not the S9 are involved in the activation of nitroaromatic compounds in the standard *Salmonella* test (Rosenkranz and Speck, 1975).

Table 43.2. *Enzymatic Activation of Promutagens in the* Salmonella/S9 *System:*
Few Enzymatic Steps to an Electrophile

Enzymatic reaction	Chemical category	Electrophile
C-Hydroxylation of aliphatic carbon adjacent nitrogen	Alkyltriazene Azoxyalkane Alkylnitrosamine	Alkyl carbonium ion
	Oxazaphosphorine mustard	Acrolein, phosphoramide mustard, bis(2-chloroethyl)-amine
Epoxidation	Aromatic, heteroaromatic[a] Haloethylene[b]	Epoxide
Formation of N-hydroxy compound	Monoalkyl hydrazine[c]	Alkyl carbonium ion
	Aromatic amine Nitroaromatic[d]	Aromatic hydroxylamine and corresponding esters

[a]See Selkirk (1977) for a discussion of the difference in the metabolite profiles of microsomes versus whole cells for benzo(a)pyrene.

[b]In the case of the haloethylene pesticide diallate (S-2,3-dichloroallyl diisopropylthiocarbamate), Schuphan *et al.* (1979) have proposed an activation mechanism that does not involve the formation of an epoxide.

[c]N-(2-Hydroxyethyl)hydrazine is also mutagenic (albeit less so) without S9 activation (Rinkus and Legator, 1979).

[d]Bacterial nitroreductases and not the S9 are involved in the activation of nitroaromatic compounds in the standard *Salmonella* test (Rosenkranz and Speck, 1975).

the premier testing system that it is today. The most salient commonality among these chemicals is the electrophilicity that is intrinsic to these chemicals or is introduced by enzymatic modification (S9 or nitroreductase in *Salmonella*). The former can thus be considered ultimate mutagens and the latter proximate mutagens. There is a shared metabolic feature among the proximate mutagens listed in Table 43.1. Generally, these compounds have a major metabolic pathway which in one or two enzymatic steps results in the production of an electrophile (Table 43.2). Presumably, it is this simplicity that allows these chemicals to be activated and detected as mutagens in the *Salmonella*/S9 system.

Examination of those carcinogens that are not demonstrably mutagenic in this system and that are known to be metabolized to electrophiles *in vivo* suggests the following three trends (Table 43.3). Firstly, promutagens metabolized to highly reactive electrophiles may not be detected owing to their short-lived nature. Secondly, S9 may

not be capable of activating promutagens that require several enzymatic steps to reach an electrophile. Finally, promutagens may not be detected if they have several major metabolic directions in which to go.

One must also appreciate the variability associated with *in vitro* activation. The fact that this variability exists argues against standardized testing, which was presented earlier as a criterion for qualitative testing. Rather, test conditions—e.g., amount of liver enzymes (Ames *et al.*, 1975), composition of S9 (Yahagi *et al.*, 1978), the liver donor's sex, species (Brusick, 1977), and strain (Felton and Nebert, 1975), and other variables (Ashby and Styles, 1978) —should be optimized on an individual basis for each chemical before *ultimately* concluding that a chemical is not mutagenic. However, when a chemical is shown to be mutagenic *in vitro*, the qualitative nature of this finding is illustrated by the experience with the trichomonacide metronidazole. A highly mutagenic metabolite isolated from the urine of patients

Table 43.3. *Examples of Proximate Carcinogens Not Detected in the* Salmonella/*S9 System*

Trend	Chemical category	Possible ultimate species
One enzymatic step to a short-lived electrophile	Phenyl compounds	Benzene-oxide
	Thiocarbamates	Electrophilic sulfur
	Polyhalogenated compounds	Radicals
Several enzymatic steps to an electrophile	Urethane	N-Hydroxy esters
	Symmetrically substituted dialkyl hydrazines	Alkyl carbonium ion
	Safrole	Esters of 3-hydroxy propenyl group
Several major metabolic pathways, some of which lead to an electrophile	Maneb	Ethylenethiourea, CS_2
	Phenacetin ⎱ Monuron ⎰	Aromatic hydroxylamine and corresponding esters, epoxide

and mice administered the drug apparently is produced by oxidative metabolism in the liver, since mice pretreated with hepatotoxic carbon tetrachloride do not excrete this metabolite in their urine. This same metabolite is not produced in either the phenobarbital-induced or Aroclor-induced Salmonella/S9 system, although the drug is mutagenic with or without S9 activation (Connor *et al.*, 1977). In this case, the tester organism itself carries out the *in vitro* activation through the action of its nitroreductases (Rosenkranz and Speck, 1975). Thus the S9, though hepatic in origin, still lacks the enzymes, cofactors, or conditions necessary to perform that metabolism that occurs *in vivo*. More importantly, what mutagenic response that is observed is not entirely representative of the mutagenic potential of this drug *in vivo*.

Thus the problem with qualitative testing that utilizes only *in vitro* systems is the false negatives that will surely be incurred. Such inadequacies are intrinsic to *in vitro* systems, be they mammalian, microbials, or batteries thereof. By necessity, the qualitative identification of mutagens must rely on a battery of *in vitro* and *in vivo* systems.

Mammalian Cytogenetics

Among some genetic toxicologists there is a tendency to view mammalian cytogenetic assays as tests of less significance and applicability than, for instance, the Salmonella/S9 system, in a mutagenicity screening program. This idea is exemplified by the following statement that appeared in the introductory section of seven volumes of the internationally read IARC Monographs on chemical carcinogenicity (IARC, 1975-1978):

"Although a correlation has often been observed between the ability of a chemical to cause chromosomal breakage and its ability to induce gene mutations, data on chromosomal breakage alone do not provide adequate evidence for mutagenicity, and therefore less weight should be given to prescreening that is based on the use of peripheral leukocyte cultures." Happily, this wording does not appear in the more recent issues of this fine publication, although the sentiment undoubtedly continues to exist in some circles.

The observations that implicate a relationship between instability in the structure and number of chromosomes and carcinogenesis have been discussed in detail elsewhere (Harnden and Taylor, 1979; Knudson, 1977; Mulvihill, 1975) and are quickly summarized here as follows.

1. Marker chromosomes which are derived from normal chromosomes through breakage and rearrangement can be identified in some clones of cancer cells. In fact, the consistency of marker chromosomes in cancer cells of a given host provides one of the main bodies of evidence supporting a clonal origin of cancer. While most marker chromosomes are unique to that one clone of cancer cells, at least two human cancers consistently have the same chromosome abnormality: the Philadelphia chromosome in as many as 90% of the cases of chronic granulocytic leukemia (Ezdinli et al., 1970); and an extra band on the long arm of chromosome 14 in cells from Burkitt's tumors and in cell lines derived from these tumors (Manolov and Manolova, 1972).

2. In general, progression from simple chromosomal anomalies to more complex ones is associated with increasing malignancy of a cancer (e.g., Lamb, 1967).

3. Aneuploidy is another common finding in cancer cells. By comparison, certain aneuploid states in man have higher incidences of certain cancers: leukemia in Down's syndrome (Miller, 1970); breast cancer in Klinefelter's syndrome (Harnden et al., 1971); cancers of neural crest origin in Turner's syndrome (Wertelecki et al., 1970); and gonadoblastoma in the dysgenetic gonad having a Y-chromosome cell line (Schellas, 1974). Furthermore, in vitro irradiation with X-rays induces a greater frequency of chromosomal aberrations in lymphocytes cultured from individuals with Down's syndrome (Countryman et al., 1977) as well as other trisomic (but not monosomic) disorders than in cells cultured from normal diploid donors (Sasaki et al., 1970).

4. De novo chromosomal aberrations are observed in lymphocytes cultured from individuals at increased risk of developing cancer because of their inheritance of certain genetic diseases: ataxia telangiectasia, Bloom's syndrome, and Fanconi's anemia. Lymphocytes cultured from these afflicted individuals also show a heightened radiosensitivity as measured by an increased frequency of γ-ray-induced chromosomal aberrations (Higurashi and Conen, 1973; Taylor, 1978). In like manner, fibroblasts cultured from individuals with the genetic disease xeroderma pigmentosum, which predisposes these individuals to develop skin cancer, are more sensitive to the cell-killing and chromosome-breaking effects of UV-irradiation (Cleaver and Bootsma, 1975; Parrington et al., 1971) and some chemical agents (Stich and Laishes, 1975, and references therein).

5. De novo chromosomal aberrations are observed in cells cultured from individuals at increased risk of developing cancer because of their exposure to ionizing radiation (Bloom et al., 1970) and certain chemical carcinogens [e.g., benzene (Forni et al., 1971), cyclophosphamide (Musilova et al., 1979) and vinyl chloride (Purchase et al., 1976)].

Of course, many of these observations can have different interpretations. For example, marker chromosomes and worsening karyotype could just be features that are temporally but not causally associated with cancer induction; or they may, indeed, contribute causally to the progression of a cancer that was initiated by some other preceding event(s). Regarding patients with aneuploidy and rare genes for chromosome instability, these individuals suffer from many afflictions, including immunological disorders, which could also account for their predisposition to cancer development.

However, the argument that some known human carcinogens cause chromosome damage is not as easily rebuked. Similarly, many known or suspected animal carcinogens break mammalian chromosomes *in vitro* and *in vivo*. The counter arguments have asserted that a seemingly high percentage of false negatives and false positives are observed in mammalian-cytogenetic studies of carcinogenic and noncarcinogenic chemicals. Let us consider the problems of false negatives and false positives separately.

False Negatives

Stoltz *et al.* (1974) and more recently Bridges and Fry (1978) have reviewed the feasibility of using *in vitro* mammalian cytogenetics as a test for chemical carcinogens. Both reviews highlighted the fact that without metabolic activation this assay would be expected to incur, and indeed has incurred, high rates of false negatives with promutagens. However, an entirely analogous situation with *in vitro* gene assays had existed earlier and this deficiency was overcome to a large extent by the development of metabolic activation procedures. With the increasing use of microsomes (Bimboes and Greim, 1976) and S9 (Matsuoka *et al.*, 1979) in *in vitro* cytogenetics, it now seems that many promutagens are detected as clastogens. A more recently devised approach detects chromosome damage in cultured rat-liver cells that can metabolically activate some carcinogens (Dean and Hodson-Walker, 1979). Of course, *in vivo* cytogenetic studies have already confirmed the clastogenic potential of many promutagens that reach the target cells in the blood or bone marrow (Shimazu *et al.*, 1976; Wild, 1978). Consequently, one can reasonably state that the sensitivity of mammalian cytogenetics to detect carcinogens is sufficiently high to warrant its use in a mutagenicity screening program. However, false negatives may still be expected to occur from inadequate activation when testing *in vitro* or organotropism when testing *in vivo*.

What is argued then is that the occurrence of mutagens with a true specificity of action toward genes but not chromosomes is rare, if not undocumented. The distinction between this point and the work of Vogel and Sobels (1976) with *Drosophila* should be appreciated. In *Drosophila,* some mutagens do not cause chromosome breaks (dominant lethals) or, for those that do, the aberrations occur at concentrations substantially higher than those causing gene mutations (recessive lethals) (Table 43.4). These findings clearly testify to the greater sensitivity of the recessive lethal test versus the dominant lethal test in the *Drosophila* system. As Vogel (1977) has speculated, the difference in sensitivity may reflect the fact that a rate-limiting factor like high toxicity or inadequate activation prevents a sufficient amount of ultimate mutagen from reaching the target cells and causing chromosomal damage; but this restricted amount is sufficient to induce gene mutations. However, the occurrence of these two-effect levels in *Drosophila* does not mean that these mutagens will not break chromosomes in other systems. For example, the anticancer agents procarbazine and cyclophosphamide will break chromosomes in both somatic and gametic cells when metabolically activated. Indeed, these findings may indicate significant differences between the genetic lesions induced by the same chemical in *Drosophila* and in mammalian systems. The alternative explanation, that the rate-limiting factor in *Drosophila* is not present in the mammalian system, is also plausible.

False Positives

Stoltz *et al.* (1974) have described what they contend is a serious problem of false positives in mammalian cytogenetic testing of presumably noncarcinogenic compounds. As proof of high rates of false positives, the reviews of Franz (1970) and Gebhart (1970) which list mainly nucleic acid antimetabolites known to cause chromosomal damage in various systems were cited. The authors have apparently as-

Table 43.4. *Comparison of Mutagenic Activity in Drosophila with Mammalian Cytogenetics*[a]

Chemical	Drosophila[b]		Mammalian cytogenetics		
	Recessive lethals	Dominant lethals	In vivo		In vitro
			Somatic	Gametic	
Procarbazine	+	(+)	+,MNT (Wild, 1978)	+,DL (Ehling, 1974)	+,MA (with S9 activation) (Matsuoka et al., 1979)
Diethylnitrosamine	+	0	0,MNT[c] (Wild, 1978)	0,DL (Propping et al., 1972)	
1-(2,4,6-Trichlorophenyl)-3,3-dimethyltriazene	+	(+)	+,MNT (Wild, 1978)		
1-(3-Pyridyl)-3,3-dimethyltriazene	+	0	+,MNT (Wild, 1978)		
Cyclophosphamide	+	0	+,MNT (Wild, 1978)	+,DL (Propping et al., 1972)	+,MA (with S9 activation) (Benedict et al., 1978)
			+,MA, human (Musilova et al., 1979)		+,MA (rat-liver cells) (Dean and Hodson-Walker, 1979)
Trofosfamide	+	0	+,MNT (Wild, 1978)		
Ifosfamide	+	0	+,MNT (Wild, 1978)		
Vinyl chloride	+	0	+,MA, human (Purchase et al., 1976)	0,DL[d] (Purchase et al., 1976)	
Hycanthone	+	0	+,MNT (Weber et al., 1975)	(+),DL (Russell, 1975)	+,MA (Benedict et al., 1977a)

[a] Symbols: +, positive response; 0, negative response; (+), marginal increase over control values. Abbreviations: DL, dominant lethal test; MA, metaphase spread analysis; MNT, micronucleus test.

[b] From Vogel and Sobels, 1976; Vogel, 1977.

[c] Karyotypic changes in liver cells isolated from rats treated with diethylnitrosamine has been reported (Grover and Fischer, 1971).

[d] Excessive miscarriages have been noted in wives of male workers exposed to vinyl chloride in the workplace (Infante et al., 1976).

sumed that such nonelectrophilic compounds are categorically not carcinogenic. However, this assumption may have its exceptions. Firstly, as Miller and Miller (1971) had noted earlier, there has been only limited testing of antimetabolites like base analogues for carcinogenicity. Secondly, there are now a few reports in the literature indicating that some antimetabolites are carcinogenic in animal testing. For example, the antifolate pyrimethamine and the base analogues 5-azacytidine and β-deoxythioguanosine (β-TGdR) induced lung tumors in mice after repeated i.p. injections (Stoner et al., 1973). β-TGdR also produced carcinomas of the ear canal in female rats by the same route of administration (National Cancer Institute, 1978). 6-Mercaptopurine (6-MP) has been reported to induce thymic lymphomas in neonatal mice (Doell et al., 1967) and hematopoietic tumors in mice and rats (Prejean et al., 1972). Azathioprine, a derivative of 6-MP, also caused thymic lymphomas in a strain of mice with autoimmune disease (Casey, 1968).

Whether the carcinogenicity reported in these various studies can be attributed to genotoxic versus immunosuppressive effects of these antimetabolites has to be considered. With regard to the former, the antifolate methotrexate (but not 6-MP) (Benedict et al., 1977a), and the base analogues 1-β-D-arabinofuranosylcytosine (Kouri et al., 1975), 5-fluorouracil, and 5-fluorodeoxyuridine (FUdR) (but not the bromo or iodo analogues or trifluorothymidine) morphologically transform cells in vitro (Jones et al., 1976). Given the lack of understanding of the mechanism of transformation, these findings offer only the minimal evidence for a genotoxic effect in the carcinogenicity of such antimetabolites. Interestingly, Jones et al. (1976) have speculated that chromosome breakage is possibly responsible for the transformation of cells by FUdR.

Of course, the idea that mutagens like antimetabolites which do not depend on their electrophilicity for their mutagenicity are not carcinogenic raises a difficult question. By a mutational theory of carcinogenesis, how can a somatic mutation induced by an electrophile (e.g., N-methyl-N-nitrosourea) be different from that induced by an antimetabolite (e.g., 6-MP), such that the former leads to carcinogenesis but the latter does not? One could envision that repair of chemically modified bases involves the removal and error-prone replacement of sizable lengths of DNA while no such repair process is invoked when a base analogue has substituted for the natural substrate in DNA. However, it would seem that in either case informational changes will be expressed in some protein whose prime importance is reflected in its disfunctioning, which initiates the neoplastic transformation. Another reconciliation would be to abandon the mutational theory. This possibility was stressed by Miller and Miller (1971). From this brief exercise, one can appreciate the lack of understanding of the biology of cancer—a fact that is not always reflected in our vigorous attempts to use mutagenicity assay systems to identify potential carcinogens.

Apart from the biology side of the matter raised by Stoltz et al. (1974), two practical comments can also be made. Firstly, if antimetabolites are a problem for cytogenetic testing for carcinogens, they are no more so than for other assay systems. The ability of some antimetabolites to transform cells in vitro previously was discussed. Similarly, some antimetabolites are mutagenic in mammalian cells (Huberman and Heidelberger, 1972) and in the popular microbial systems (Janion, 1978). Secondly, regardless of whether antimetabolites can cause somatic mutations which can lead to cancer, the possibility of mutagenic effects in gametic cells deserves proper attention. Mutagenesis at the gametic level is as much a social concern as carcinogenesis and teratogenesis with regard to its study and prevention. The ability of clinically used antimetabolites like 6-MP (Generoso et al., 1975), azathioprine (Clark, 1975) and methotrexate (Propping et al., 1972) to cause dominant lethals in rodent testing testifies to the need for concern. Consequently, in many, if not most instances, the testing of chemicals in the various *in vitro* and *in vivo* mutagenicity assays will be done for the dual purpose of identifying both potential carcinogens and (gametic) mutagens.

Bridges and Fry (1978) have also tried to document examples of false positives by analyzing the results of *in vitro* chromosomal testing of 134 chemicals (Ishidate and Odashima, 1977). By their analysis, there is an association between animal carcinogenicity, mutagenicity in bacteria, and chromosomal damage in Chinese hamster cells; but there are so many exceptions that it disqualifies this cytogenetic test as a routine component in mutagenicity screening of carcinogens. To substantiate this conclusion, these authors presented a table showing (among other things) that as many as 30 "noncarcinogens" break chromosomes *in vitro*. However, parts of their analysis appear to be in error. For reasons not explained, these authors have assumed that chemicals designated in the list as *untested for carcinogenicity* are noncarcinogenic. Consequently, chemicals like the following were counted as noncarcinogens that break chromosomes: 4-aminoquinoline 1-oxide, *N*-amyl-*N'*-nitro-N-nitrosoguanidine, seven N-acyl-N-alkylnitrosamines, four N-alkyl-N-(1-acetoxy or 1-methoxy)alkylnitrosamines, two nitrophenylenediamines, saccharin, and sodium nitrite. The mistake of this assumption is seen in the fact that at least these 17 chemicals are analogues of known carcinogens; or, as is definitely the case with saccharin (Arnold *et al.*, 1977) and sodium nitrite (Newberne, 1979), their carcinogenicity has been reported more recently. Furthermore, of these 17 chemicals, 16 were listed as having been tested in a bacterial system (identity not stated); 15 were mutagenic without enzymic activation, the only exception being saccharin.

The assumed noncarcinogenicity of many of the remaining 13 compounds is also not that obvious. If they are indeed noncarcinogenic, they exhibit curious structure-activity trends, as noted by Ishidate and Odashima (1977). Ethyl carbamate (urethane) and aryl N-methylcarbamates were clastogenic, but ethyl N-alkylcarbamates were not. Urea and N-methyl-N'-acetylurea were clastogenic but N-alkylureas were not. Nitroguanidine and N-(methyl or ethyl)-N'-nitroguanidines were clastogenic, but N-(n-butyl, i-butyl, or amyl)-N'-nitroguanidines were not.

However, as Bridges and Fry (1978) noted, the clastogenicity of several compounds like potassium sorbate, sodium benzoate, and urea belies their reasonably assumed nongenotoxicity. Thus, from the testing of 134 chemicals, only a few —not as many as 30—false positives appear to have been documented. Since the frequency of false positives is apparently as low as that encountered with other assay systems, cytogenetics qualifies as an appropriate endpoint in a screening program. As Bridges and Fry also noted, doubts about the validity of a cytogenetic effect can be partially allayed by having accurate blind scoring of the kinds and incidences of chromosomal damage and by demonstrating a dose-response. Such measures should help to distinguish a true cytogenetic effect from chromosomal fragmentation associated with dead or dying cells (Fishbein *et al.*, 1970).

Conclusions

Mammalian cytogenetics can be viewed as a gross measure of effects on the genome of cells. This makes this procedure a logical choice for an assay to be used in a screening program for genotoxic agents. Given our present understanding of carcinogenesis, it is therefore no less sound in theory than the *in vitro* gene and transformation assays as a method to detect potential carcinogens. Practical experience with cytogenetics also confirms its usefulness. In fact, several carcinogens which have proved negative in several other assays are active cytogenetically, e.g., asbestos (Sincock, 1977), benzene (Hite *et al.*, 1980), hexachlorocyclohexane, ethinylestradiol, testosterone propionate (Shimazu *et al.*, 1976), diethylstilbestrol (Sawada and Ishidate, 1978), and urethane (Wild, 1978). Regarding false negatives, *in vitro* cytogenetics, like any *in vitro* testing, cannot be considered complete unless metabolic activation procedures have been included. Similarly, *in vivo* testing needs to be complemented with a proper *in vitro* analysis any time organotropism may be masking a cytogenetic effect (e.g., diethylnitrosamine, Table 43.4). False positives should be partially avoidable by quality control during the procedure or otherwise suggested by their slight slopes on dose-response curves *in vitro* and lack of activity *in vivo*.

Cytogenetics also offers the decided advantages of being able to use human cells in the testing and to apply the procedure to exposed human populations. The latter allows us to identify and to monitor high-risk groups such as certain clinical and industrial populations. Therefore, this approach can serve as an advance-warning system for the detection of genetic agents long before cancer or ill effects on reproduction manifest themselves in these populations.

Epilogue

Let us look back over the last two decades in genetic toxicology to sense the direction of the field for the upcoming decade. Based on the work in the mid-1960s and early 1970s with dimethylnitrosamine and polycyclic hydrocarbons, the concept of proximate carcinogens was developed. This realization that mammalian cells can metabolize chemically unreactive compounds to electrophiles, which could react with intracellular nucleophiles, including DNA, provided a logical nexus for the long-sought relationship between carcinogenesis and mutagenesis. The twin concept of prox-

imate mutagens naturally followed and led to the development of microbial and mammalian cell mutagenicity assays, which employ cellular homogenates of mammalian livers, or fractions thereof, to simulate *in vitro* the metabolic activation of the test chemical. The early success of this experimental approach is clearly epitomized by the near-90% correlation between chemical carcinogenesis and mutagenesis that was soon being quoted by many by the mid-1970s. These results were so impressive that they generated a high level of optimism that these *in vitro* assays would suffice in themselves as an efficient means to identify potential genotoxins.

However, by the late 1970s, it became apparent that *in vitro* systems have critical shortcomings that detract from their efficiency as general screening assays. Firstly, the understood activation of some chemicals to electrophiles is more complex than just one or two P-450-mediated oxidations, and this complexity does not lend itself easily to simulation *in vitro*. Secondly, with regard to the popular microbial systems, their phylogenetic *un*similarity to mammalian organisms poses at least two other problems: their inability to detect the entire spectrum of mutational events; and differences in passage of some chemicals across microbial *versus* mammalian cell membranes. Consequently, the 1970s closed with the growing consensus that a battery of *in vitro* and *in vivo* systems would be the best strategy in screening for genotoxic chemicals.

In the 1980s, one can expect that this renewed emphasis on *in vivo* studies will continue and will result in new levels of sophistication in testing. Increasing emphasis will be placed on two areas: animal studies that measure both specific genetic alterations and their accompanying phenotypic manifestations, e.g., histocompatibility changes, behavioral effects; and new procedures for monitoring certain genetic as well as clinical and industrial populations for chemical mutagenesis. Ideologically, there will also be a shift in the perception of the magnitude of the problem of gametic mutagenesis in the human community as a whole. This will be in response to further and further identifications of such human mutagens, e.g., as evidenced by dominant lethal effects in exposed males and expressed as increased spontaneous abortions in their unexposed wives. By the end of the 1980s, it may very well be that there will be international committees verifying examples of chemically induced genetic diseases and thereby developing an operational listing of human mutagens. In such an atmosphere, chemical mutagenesis will assume an importance equaling, if not excelling, that of chemical carcinogenesis and chemical teratogenesis with regard to the study and prevention of these toxicities.

References

Ames, B.N., Whitfield, H.J., Jr. (1966): Frameshift mutagenesis in *Salmonella*. Cold Spring Harbor Symp. Quant. Biol. *31*:221–225.

Ames, B.N., Gurney, E.G., Miller, J.A., Bartsch, H. (1972a): Carcinogens as frameshift mutagens: metabolites and derivatives of 2-acetylaminofluorene and other aromatic amine carcinogens. Proc. Natl. Acad. Sci. USA *69*:3128–3132.

Ames, B.N., Sims, P., Grover, P.L. (1972b): Epoxides of carcinogenic polycyclic hydrocarbons are frameshift mutagens. Science *176*:47–49.

Ames, B.N., McCann, J., Yamasaki, E. (1975): Methods for detecting carcinogens and mutagens with the *Salmonella*/mammalian microsome mutagenicity test. Mutat. Res. *31*:347–363.

Arnold, D.L., Charbonneau, S.M., Moodie, C.A., Munro, I.C. (1977): Long-term toxicity study with *ortho*-toluenesulfonamide and saccharin. Sixteenth Annual Meeting of the Society of Toxicology, Toronto, Abstract 78.

Ashby, J., Styles, J.A. (1978): Factors influencing mutagenic potency *in vitro*. Nature (Lond.) *274*:20–22.

Badr, F.M., Badr, R.S. (1974): Studies on the mutagenic effect of contraceptive drugs. I. Introduction of dominant lethal mutations in female mice. Mutat. Res. *26*:529–534.

Batzinger, R.P., Bueding, E., Reddy, B.S., Weisburger, J.H. (1978): Formation of a mutagenic drug metabolite by intestinal microorganisms. Cancer Res. *38*:608–612.

Benedict, W.F., Banerjee, A., Gardner, A., Jones, P.A. (1977a): Induction of morphological transformation in mouse C3H/10T¹/₂ clone 8 cells and chromosomal damage in hamster A(T₁)C̄l-3 cells by cancer chemotherapeutic agents. Cancer Res. *37*:2202–2208.

Benedict, W.F., Baker, M.S., Haroun, L., Choi, E., Ames, B.N. (1977b): Mutagenicity of cancer chemotherapeutic agents in the *Salmonella*/microsome test. Cancer Res. *37*:2209–2213.

Benedict, W.F., Banerjee, A., Venkatesan, N. (1978): Cyclophosphamide-induced oncogenic transformation, chromosomal breakage, and sister chromatid exchange following microsomal activation. Cancer Res. *38*:2922–2924.

Bimboes, D., Greim, H. (1976): Human lymphocytes as target cells in a metabolizing test system *in vitro* for detecting potential mutagens. Mutat. Res. *35*:155–160.

Bloom, A.D., Nakagoma, Y., Awa, A.A., Neriiski, S. (1970): Chromosome aberrations and malignant disease among A-bomb survivors. Am. J. Publ. Health *60*:641–644.

Bochkov, N.B., Sram, R.J., Kuleshov, N.P., Zhurkov, V.S. (1976): System for the evaluation of the risk from chemical mutagens for man: basic principles and practical recommendations. Mutat. Res. *38*:191–202.

Bora, K.C. (1976): A hierarchical approach to mutagenicity testing and regulatory control of environmental chemicals. Mutat. Res. *41*: 73–82.

Brenner, S., Barnett, L., Crick, F.H.C., Orgel, A. (1961): The theory of mutagenesis. J. Mol. Biol. *3*:121–124.

Bresnick, E. (1976): Activation and inactivation of polycyclic hydrocarbons and their interaction with macromolecular components. In: F.J. de Serres, J.R. Fouts, J.R. Bend, and R.M. Philpot (eds.), In Vitro Metabolic Activation in Mutagenesis Testing. Amsterdam: Elsevier/North Holland Biomedical Press, pp. 91–104.

Bridges, B.A. (1973): Some general principles of mutagenicity screening and a possible framework for testing procedures. Environ. Health Perspect. 6:221–227.

Bridges, B.A. (1974): The three-tier approach to mutagenicity screening and the concept of radiation-equivalent dose. Mutat. Res. *26*: 335–340.

Bridges, B.A. (1976): Use of a three-tier protocol for evaluation of long-term toxic hazards, particularly mutagenicity and carcinogenicity. In: R. Montesano, H. Bartsch, and L. Tomatis (eds.), Screening Tests in Chemical Carcinogenesis (IARC Scientific Publications No. 12), Lyon, International Agency for Research on Cancer, pp. 549–568.

Bridges, J.W., Fry, J.R. (1978): Mammalian short-term tests for carcinogens. In: A.D. Dayan and R.W. Brimblecombe (eds.), Carcinogenicity Testing: Principles and Problems. Baltimore: University Park Press, pp. 29–52.

Brusick, D.J. (1977): *In vitro* mutagenesis assays as predictors of chemical carcinogenesis in mammals. Clin. Toxicol. *10*:79–109.

Bryan, J. (1972): Definition of three classes of binding sites in isolated microtubule crystals. Biochemistry *11*:2611–2616.

Casey, T.P. (1968): The development of lymphomas in mice with autoimmune disorders treated with azathioprine. Blood *31*:396–399.

Clark, J.M. (1975): The mutagenicity of azathioprine in mice, *Drosophila melanogaster* and *Neurospora crassa*. Mutat. Res. *28*:87–99.

Cleaver, J.E., Bootsma, D. (1975): Xeroderma pigmentosum: biochemical and genetic characteristics. Ann. Rev. Genet. *9*:19–38.

Cohen, M.M., Shaw, M.W., Craig, A.P. (1963): The effects of streptonigrin on cultured human leukocytes. Proc. Natl. Acad. Sci. USA *50*:16–24.

Committee 17 of the Environmental Mutagen Society (1975): Environmental mutagenic hazards. Science *187*:503–514.

Connor, T.H., Stoeckel, M., Evrard, J., Legator, M.S. (1977): The contribution of metronidazole and two metabolites to the mutagenic activity detected in the urine of treated humans and mice. Cancer Res. *37*:629–633.

Countryman, P.I., Heddle, J.A., Crawford, E. (1977): The repair of X-ray-induced chromosomal damage in trisomy 21 and normal diploid lymphocytes. Cancer Res. *37*:52–58.

Dean, B.J. (1976): A predictive testing scheme for carcinogenicity and mutagenicity of industrial chemicals. Mutat. Res. *41*:83–88.

Dean, B.J., Hodson-Walker, G. (1979): An *in vitro* chromosome assay using cultured rat-liver cells. Mutat. Res. *64*:329–337.

de Serres, F.J. (1974): Report of discussion group no. 12: mutagenic assessment. In: L. Goldberg (ed.), Carcinogenesis Testing of Chemicals. Cleveland, Ohio: CRC Press, pp. 101–107.

de Serres, F.J., Ehling, U., Kihlman, B., Kilbey, B.J., Loprieno, N., Marquardt, H., Ramel, C., Zetterberg, G. (1973): Report of group 5: genetic tests for evaluating environmental mutagens. In: C. Ramel (ed.), Evaluation of Genetic Risks of Environmental Chemicals, Ambio Special Report No. 3, pp. 23–27.

Dickens, F., Jónes, H.E.H. (1963): Further studies on the carcinogenic and growth inhibitory activity of lactones and related substances. Br. J. Cancer *17*:100–108.

Dickens, F., Jones, H.E.H. (1965): Further studies on carcinogenesis by lactones and related substances in the rat and mouse. Br. J. Cancer *19*:392–403.

Doell, R.G., St. Cyr, C. de V., Grabar, P. (1967): Immune reactivity prior to development of thymic lymphoma in C57BL mice. Int. J. Cancer *2*:103–108.

Draser, B.S., Hill, M.J. (1974): Human Intestinal Flora. New York: Academic Press.

Draser, B.S., Renwick, A.G., Williams, R.T. (1972): The role of the gut flora in the metabolism of cyclamate. Biochem. J. *134*:881–890.

Duca-Marinescu, D., Negoescu, I. (1972): The mutagenic effects of thyroxin. Rev. Roum. Endocrinol. *10*:149–152.

Ehling, U.H. (1974): Differential spermatogenic response of mice to the induction of mutations by antineoplastic drugs. Mutat. Res. *26*:285–295.

Ellis, J.H., Jr, Ray, V.A., Holden, H.E. (1974): Comparative studies of 6-chloropurine in host-mediated and *in vitro* bacterial assays. Mutat. Res. *26*:455.

Ezdinli, E.Z., Sokal, J.E., Crosswhite, L., Sandberg, A.A. (1970): Philadelphia-chromosome-positive and -negative chronic myelocytic leukemia. Ann. Intern. Med. *72*:175–182.

Felton, J.S., Nebert, D.W. (1975): Mutagenesis of certain activated carcinogens *in vitro* associated with genetically mediated increases in monooxygenase activity and cytochrome P_1-450. J. Biol. Chem. *250*:6769–6778.

Fishbein, L., Flamm, W.G., Falk, H.L. (1970): Chemical Mutagens, New York: Academic Press, p. 90.

Flamm, W.G. (1974): A tier system approach to mutagen testing. Mutat. Res. *26*:329–333.

Forni, A., Capellini, A., Pacifico, E., Vigiliani, E.C. (1971): Chromosome changes and their evolution in subjects with past exposure to benzene. Arch. Environ. Health 22:373–378.

Franz, J. (1970): Chromosomal aberrations induced by antimetabolites. In: F. Vogel and G. Röhrborn (eds.), Chemical Mutagenesis in Mammals and Man. New York: Springer-Verlag, pp. 342–349.

Gale, E.F., Candliffe, E., Reynolds, P.E., Richmond, M.H., Waring, M.J. (1972): The Molecular Basis of Antibiotic Action. New York: John Wiley and Sons, p. 174.

Gebhart, E. (1970): The treatment of human chromosomes in vitro: results. In: F. Vogel and G. Röhrborn (eds.), Chemical Mutagenesis in Mammals and Man. New York: Springer-Verlag, pp. 367–382.

Generoso, W.M., Preston, R.J., Brewen, J.G. (1975): 6-Mercaptopurine, an inducer of cytogenetic and dominant-lethal effects in premeiotic and early meiotic germ cells of male mice. Mutat. Res. 28:437–447.

Goldberg, I.H., Friedman, P.A. (1971): Antibiotics and nucleic acids. Ann. Rev. Biochem. 40:775–810.

Grisham, L.M., Wilson L., Bensch, K.G. (1973): Antimitotic action of griseofulvin does not involve disruption of microtubules. Nature (Lond.) 244:294–296.

Grover, S., Fischer, P. (1971): Cytogenetic studies in Sprague-Dawley rats during the administration of a carcinogenic nitroso compound—diethylnitrosamine. Europ. J. Cancer 7:77–82.

Hampton, A., Paterson, A.R.P. (1966): Conversion of 6-chloropurine to its ribonucleotide by ascites tumor cells. Biochim. Biophys. Acta 114:185–187.

Hannan, M.A., Hurley, L.H., Gairola, C. (1978): Mutagenic and recombinogenic effects of the antitumor antibiotic anthramycin. Cancer Res. 38:2795–2799.

Harnden, D.G., Taylor, A.M.R. (1979): Chromosomes and neoplasia. Adv. Human Genet. 9:1–70.

Harnden, D.G., Maclean, N., Langlands, A.O. (1971): Carcinoma of the breast and Klinefelter's syndrome. J. Med. Genet. 8:460–461.

Higurashi, M., Conen, P.E. (1973): In vitro chromosomal radiosensitivity in "chromosomal breakage syndromes." Cancer 32:380–383.

Hite, M., Pecharo, M., Smith, I., Thornton, S. (1980): The effect of benzene in the micronucleus test. Mutat. Res. 77:149–155.

Holden, H.E., Ray, V.A., Florio, J., Wahren-burg, M.G. (1974): In vivo and in vitro cytogenetic studies with 6-chloropurine. Mutat. Res. 26:455–456.

Holmes, A.J., Eisenstark, A. (1968): The mutagenic effect of thymine-starvation on Salmonella typhimurium. Mutat. Res. 5:15–21.

Horwitz, S.B. (1975): Anthramycin. In: A.C. Sartorelli and D.G. Johns (eds.), Handbook of Experimental Pharmacology, Vol. 38, Part 2. New York: Springer-Verlag, pp. 642–648.

Hsu, T.C., Collie, C.J., Lusby, A.F., Johnston, D.A. (1977): Cytogenetic assays of chemical clastogens using mammalian cells in culture. Mutat. Res. 45:233–247.

Huberman, E., Heidelberger, C. (1972): The mutagenicity to mammalian cells of pyrimidine nucleoside analogs. Mutat. Res. 14:130–132.

IARC (1975-1978): IARC Monographs on the Evaluation of Carcinogenic Risk of Chemicals to Man. Vols. 10-13, 15, 16, Lyon, International Agency for Research on Cancer.

Infante, P.F., Wagoner, J.K., Waxweiler, R.J. (1976): Carcinogenic, mutagenic and teratogenic risks associated with vinyl chloride. Mutat. Res. 41:131–142.

Ishidate, M., Jr., Odashima, S. (1977): Chromosome tests with 134 compounds on Chinese hamster cells in vitro—a screening for chemical carcinogens. Mutat. Res. 48:337–354.

Janion, C. (1978): The efficiency and extent of mutagenic activity of some new mutagens of base-analogue type. Mutat. Res. 56:225–234.

Jones, J.B., Young, J.M. (1968): Carcinogenicity of lactones. III. The reactions of unsaturated γ-lactones with L-cysteine. J. Med. Chem. 11:1176–1182.

Jones, P.A., Benedict, W.F., Baker, M.S., Mondal, S., Rapp, U., Heidelberger, C. (1976): Oncogenic transformation of C3H/10T¹/₂ clone 8 mouse embryo cells by halogenated pyrimidine nucleosides. Cancer Res. 36:101–107.

Knudson, A.G., Jr. (1977): Genetics and etiology of human cancer. Adv. Hum. Genet. 8:1–66.

Kondo, S., Ichikawa, H., Iwo, K., Kato, T. (1970): Base-change mutagenesis and prophage induction in strains of Escherichia coli with different DNA repair capacities. Genetics 66:187–217.

Kouri, R.E., Kurtz, S.A., Price, P.J., Benedict, W.F. (1975): 1-β-D-Arabinofuranosylcytosine-induced malignant transformation of hamster and rat cells in culture. Cancer Res. 35:2413–2419.

Kremer, W.B., Laszilo, J. (1975): Streptonigrin. In: A.C. Sartorelli (ed.), Handbook of Experimental Pharmacology, Vol. 38, Part 2. New York: Springer-Verlag, pp. 633–641.

Kusyk, C.J., Hsu, T.C. (1976): Mitotic anomalies induced by three inhalation halogenated anesthetics. Environ. Res. 12:366–370.

Laerum, O.D., Iversen, O.H. (1972): Reticuloses and epidermal tumors in hairless mice after topical skin applications of cantharidin and asiaticoside. Cancer Res. 32:1463–1469.

Lamb, D. (1967): Correlation of chromosome counts with histological appearances and prognosis in transitional-cell carcinoma of bladder. Br. Med. J. 1:263–277.

Lawley, P.D., Orr, D.J., Shah, S.A., Farmer, P.B., Jarman, M. (1973): Reaction products of N-methyl-N-nitrosourea and deoxyribonucleic acid containing thymidine residues. Biochem. J. 135:193–201.

Leclercq, G., Hulin, N., Heuson, J.C. (1973): Interaction of activated estradiol-receptor complex and chromatin in isolated uterine nuclei. Europ. J. Cancer 9:681–685.

Legator, M.S., Zimmering, S. (1975): Integration of mammalian, microbial, and Drosophila procedures for evaluating chemical mutagens. Mutat. Res. 29:181–188.

Lerman, L.S. (1964): Acridine mutagens and DNA structure. J. Cell. Comp. Physiol. 64 (Suppl. 1):1–18.

Levine, A.F., Fink, L.M., Weinstein, I.B., Grunberger, D. (1974): Effect of N-2-acetylaminofluorene modification on the conformation of nucleic acids. Cancer Res. 34:319–327.

Loeb, L.A., Sirover, M.A., Agarwal, S.S. (1978): Infidelity of DNA synthesis as related to mutagenesis and carcinogenesis. Adv. Exp. Med. Biol. 91:103–115.

Malling, H.V. (1971): Dimethylnitrosamine: formation of mutagenic compounds by interaction with mouse liver microsomes. Mutat. Res. 13:425–429.

Manolov, G., Manolova, Y. (1972): Marker band in one chromosome 14 from Burkitt lymphomas. Nature (Lond.) 237:33–34.

Martz, F., Straw, J.A. (1977): The in vitro metabolism of 1-(o-chlorophenyl)-1-(p-chlorophenyl)-2,2-dichloroethane (o-p′-DDD) by dog adrenal mitochondria and metabolite covalent binding to mitochondrial macromolecules. Drug. Metab. Disp. 5:482–486.

Matsuoka, A., Hayashi, M., Ishidate, M., Jr. (1979): Chromosomal aberration tests on 29 chemicals combined with S-9 mix in vitro. Mutat. Res. 66:277–290.

Mayer, V.W., Flamm, W.G. (1975): Legislative and technical aspects of mutagenicity testing. Mutat. Res. 29:295–300.

McCann, J., Spingarn, N.E., Kobori, J., Ames, B.N. (1975): Detection of carcinogens as mutagens: bacterial tester strains with R factor plasmids. Proc. Natl. Acad. Sci. USA 72:979–983.

Miller, E.C., Miller, J.A. (1971): The mutagenicity of chemical carcinogens: correlations, problems, and interpretations. In: A. Hollaender (ed.), Chemical Mutagens: Principles and Methods for Their Detection. New York: Plenum Press, pp. 83–119.

Miller, R.W. (1970): Neoplasia and Down's syndrome. Ann. N.Y. Acad. Sci. 171:637–644.

Mulvihill, J.J. (1975): Congenital and genetic diseases. In: J.F. Fraumeni, Jr. (ed.), Persons at High Risk of Cancer. New York: Academic Press, pp. 3–37.

Musilova, J., Michalova, K., Urban, J. (1979): Sister-chromatid exchanges and chromosomal breakage in patients treated with cytostatics. Mutat. Res. 67:289–294.

National Cancer Institute (1978): Bioassay of β-TGdR for Possible Carcinogenicity, Carcinogenesis Technical Report Series No. 57, DHEW Publication No. (NIH) 78-1363. Washington, DC, US Government Printing Office.

Nery, R. (1969): Acylation of cytosine by ethyl N-hydroxycarbamate and its acyl derivatives and the binding of these agents to nucleic acids and proteins. J. Chem. Soc. (C), 1860-1865.

Newberne, P.M. (1979): Nitrite promotes lymphoma incidence in rats. Science 204:1079–1081.

Oppenheimer, J.H., Surks, M.I., (1975): Biochemical basis of thyroid hormone action. In: G. Litwack (ed.), Biochemical Actions of Hormones. New York: Academic Press, pp. 119–157.

Parrington, J.M., Delhanty, J.D.A., Baden, H.P. (1971): Unscheduled DNA synthesis, UV-induced chromosome aberrations and SV-40 transformation in cultured cells from xeroderma pigmentosum. Ann. Hum. Genet. *35*:149–160.

Pastan, I., Adhya, S. (1976): Cyclic adenosine 5'-monophosphate in *Escherichia coli*. Bacteriol. Rev. *40*:527–551.

Prejean, J.D., Griswold, A.E., Casey, A.E., Peckham, J.C., Weisburger, J.H., Weisburger, E.K., Wood, H.B., Jr. (1972): Carcinogenicity studies of clinically used anticancer agents. Proc. Am. Assoc. Cancer Res. *13*:112.

Propping, P., Röhrborn, G., Buselmaier, W. (1972): Comparative investigations on the chemical induction of point mutations and dominant lethal mutations in mice. Mol. Gen. Genet. *117*:197–209.

Purchase, I.F.H., Richardson, C., Anderson, D. (1976): Chromosomal and dominant lethal effects of vinyl chloride. Lancet *2*:410–411.

Radomski, J.L. (1961): The absorption, fate and excretion of citrus red no. 2 (2,5-dimethoxyphenyl-azo-2-naphthol) and ext. D and C red no. 14 (1-xylylazo-2-naphthol). J. Pharmacol. Exp. Ther. *134*:100–109.

Reddy, B.S., Narisawa, T., Wright, P., Vukusich, D., Weisburger, J.H., Wynder, E.L. (1975): Colon carcinogenesis with azoxymethane and dimethylhydrazine in germ-free rats. Cancer Res. *35*:287–290.

Remers, W.A. (1979): The Chemistry of Antitumor Antibiotics, Vol. 1. New York: John Wiley and Sons, pp. 176–220.

Rinkus, S.J., Legator, M.S. (1979): Chemical characterization of 465 known or suspected carcinogens and their correlation with mutagenic activity in the *Salmonella typhimurium* system. Cancer Res. *39*:3289–3318.

Röhrborn, G., Hansmann, I. (1974): Oral contraceptives and chromosome segregation in oocytes of mice. Mutat. Res. *26*:535–544.

Rosenkranz, H.S., Speck, W.T. (1975): Mutagenicity of metronidazole: activation by mammalian liver microsomes. Biochem. Biophys. Res. Commun. *66*:520–525.

Russell, W.L. (1975): Results of tests for possible transmitted genetic effects of hycanthone in mammals. J. Toxicol. Environ. Health. *1*:301–304.

Sasaki, M.S., Tonomura, A., Matsubara, S. (1970): Chromosome constitution and its bearing on the chromosomal radiosensitivity in man. Mutat. Res. *10*:617–633.

Sawada, M., Ishidate, M., Jr. (1978): Colchicine-like effect of diethylstilbestrol (DES) on

mammalian cells *in vitro*. Mutat. Res. *57*:175–182.

Schachtschabel, D.O., Zilliken, F., Saito, M., Foley, G.E. (1969): Inhibition of DNA synthesis and chromosome aberrations in cultured Ehrlich ascites tumor cells following treatment with luteoskyrin. Exp. Cell Res. *57*:19–28.

Schellas, H.F. (1974): Malignant potential of the dysgenetic gonad. Obstet. Gynecol. *44*: 298–309, 455–462.

Schöneich, J. (1976): Safety evaluation based on microbial assay procedures. Mutat. Res. *41*:89–94.

Schuphan, I., Rosen, J.D., Casida, J.E. (1979): Novel activation mechanism for the promutagenic herbicide diallate. Science *205*:1013–1015.

Selkirk, J. (1977): Divergence of metabolic activation systems for short-term mutagenesis assays. Nature (Lond.) *270*:604–607.

Sentein, P. (1974): Action de la chromomycine A3 sur les chromosomes dans les blastomères en segmentation de l'oeuf de pleurodèle, Comparaison avec d'autres substances chromatoclasiques. C.R. Hebd. Séances Acad. Sci. Ser. D. *278*, 295–298.

Shimazu, H., Shiraishi, N., Akematsu, T., Ueda, N., Sugiyama, T. (1976): Carcinogenicity screening tests on induction of chromosomal aberrations in rat bone marrow cells *in vivo*. Mutat. Res. *38*:347.

Sincock, A.M. (1977): Preliminary studies of the *in vitro* cellular effects of asbestos and fine glass dusts. In: H.H. Hiatt, J.D. Watson, and J.A. Winsten (eds.), Origins of Human Cancer, Book A, New York: Cold Spring Harbor Laboratory, pp. 941–954.

Stich, H.F., Laishes, B.A. (1975): The response of xeroderma pigmentosum cells and controls to the activated mycotoxins, aflatoxins and sterigmatocystin. Int. J. Cancer *16*:266–274.

Stoltz, D.R., Poirier, L.A., Irving, C.C., Stich, H.F., Weisburger, J.H., Grice, H.C. (1974): Evaluation of short-term tests for carcinogenicity. Toxicol. Appl. Pharmacol. *29*:157–180.

Stoner, G.D., Shimkin, M.B., Kniazeff, A.J., Weisburger, J.H., Weisburger, E.K., Gori, G.B. (1973): Test for carcinogenicity of food additives and chemotherapeutic agents by the pulmonary tumor response in strain A mice. Cancer Res. *33*:3069–3085.

Stoner, G.D., Weisburger, E.K., Shimkin, M.B.

(1975): Tumor response in strain A mice exposed to silylating compounds used for gasliquid chromatography. J. Natl Cancer Inst. 54:495–497.

Taylor, A.M.R. (1978): Unrepaired DNA strand breaks in irradiated ataxia telangiectasia lymphocytes suggested from cytogenetic observations. Mutat. Res. 50:407–418.

Terzaghi, E., Okada, Y., Streisinger, G., Emrich, J., Inouye, M., Tsugita, A. (1966): Change of a sequence of amino acids in phage t4 lysozyme by acridine-induced mutations. Proc. Natl. Acad. Sci. USA 56:500–507.

Van Duuren, B.L., Goldschmidt, B.M., Katz, C., Seidman, I. (1972): Dimethylcarbamyl chloride, a multipotential carcinogen. J. Natl. Cancer Inst. 48:1539–1541.

Vig, B.K., Kontras, S.B., Aubele, A. (1969): Sensitivity of G_1 phase of mitotic cell cycle to chromosome aberrations induced by daunomycin. Mutat. Res. 7:91–97.

Vogel, E. (1977): Identification of carcinogens by mutagen testing in Drosophila: The relative reliability for the kinds of genetic damage measured. In: H.H. Hiatt, J.D. Watson, and J.A. Winsten (eds.), Origins of Human Cancer, Book C. New York: Cold Spring Harbor Laboratory, pp. 1483–1497.

Vogel, E., Sobels, F.H. (1976): The function of Drosophila in genetic toxicology testing. In: A. Hollaender (ed.), Chemical Mutagens: Principles and Methods for Their Detection. New York: Plenum Press, pp. 93–142.

Weber, E., Bidwell, K., Legator, M.S. (1975): An evaluation of the micronucleus test using triethylenemelamine, trimethylphosphate, hycanthone and niridazole. Mutat. Res. 28:101–106.

Wertelecki, W., Fraumeni, J.F., Jr., Mulvihill, J.J. (1970): Nongonadal neoplasia in Turner's syndrome. Cancer 26:485–488.

White, I.N.H., Mattocks, A.R. (1972): Reaction of dihydropyrrolizines with deoxyribonucleic acids in vitro. Biochem. J. 128:291–297.

Wild, D. (1978): Cytogenetic effects in the mouse of 17 chemical mutagens and carcinogens evaluated by the micronucleus test. Mutat. Res. 56:319–327.

Williams, D.L., Hagen, A.A. and Runyan, J.W., Jr. (1971): Chromosome alterations produced in germ cells of dogs by progesterone. J. Lab. Clin. Med. 77:417–429.

Wobus, A.M., Wobus, U., Schöneich, J. (1977): Effects of the antibiotic neotropsin on mouse ascites tumour chromosomes in vitro. Experientia 33:1212–1213.

Wood, R.C., Ferone, R., Hitchings, G.H. (1961): The relationship of cellular permeability to the degree of inhibition by amethopterin and pyrimethamine in several species of bacteria. Biochem. Pharmacol. 6:113–124.

Yahagi, T., Nagao, M., Matsushima, T., Seino, Y., Sawamura, N., Shirai, A., Kawachi, T., Sugimura, T. (1978): An improved method for detecting mutagens. Mutat. Res. 53:285.

44

Short-Term Genetic Tests Extended to the Human*

Mortimer L. Mendelsohn

This meeting has given clear demonstration of the diversity and power of short-term laboratory tests to assess mutagenicity and carcinogenicity. It has also laid bare the enormous impediments to the extrapolation of such test results to humans. Testing specific chemicals for toxicity is not enough; we must have reliable information to estimate the human exposure to, load of, and consequences from specific or general environmental mutagens and carcinogens. Thus, it is imperative that we find some way to bridge the gap between the laboratory and the clinic.

The obvious and immediate strategy is to extend to the human the laboratory tests now being done in mammals—i.e., to use easily sampled human cells to examine the DNA, chromosomes, and gene products from individuals. Many laboratory tests will have no such reasonable counterpart in the clinic, but a surprisingly large number can and are being extended to people; these constitute a broad, potentially powerful tool for addressing clinical genetic toxicology.

The Tests

Table 44.1 summarizes the short-term genetic tests that are already in clinical use or could be so extended. The tests are familiar; they include the measurement of repair synthesis, DNA adducts, aberrations, micronuclei, sister chromatid exchanges (SCEs), and specific locus mutations. Each test is described by eight characteristics:

1. Feasibility refers to the readiness of the test for application:
   0   not feasible at present
   ?   should be possible but yet to be done
   +   done on a very limited scale
   ++   fully applicable

*Work performed under the auspices of the U.S. Department of Energy by the Lawrence Livermore Laboratory under contract no. W-7405-ENG-48.

Table 44.1. *Characteristics of Short-Term Genetic Tests*

Test		Feasibility	Specificity	Directness	Heritability	Size of lesion	Sensitivity	Latency	Persistence
Repair Synthesis {	Soma	+	no	−	−	?	?	0	Hours
	Sperm	0							
DNA Adducts {	Soma	+	yes	+	−	?	?	0	Hours to days
	Sperm	?	yes	+	+	?	?	0	Weeks
Sperm Morphology		++	no	−	−	?	++	1 mo.	Months to years
Chromosome Aberration {	Soma	++	yes	+	−	macro	+	0	Years
	Germ	+	yes	+	+	macro	+	0	Weeks to years
Micronucleus		++	yes	+	−	macro	++	2 days	5 days
SCE		++	yes	−	−	mixed	++	0	Days to months
Specific Locus									
RBC-Sickle hemoglobin		+	yes	+	−	micro	++	2–3 mo's	? decades
Lymphocytes-HGPRT		++	yes	+	−	mixed	++	? days	? years
Sperm-LDH-X		?	yes	+	+	mixed	++	months	? decades

2. Specificity rates the genetic exclusiveness of the damage being measured.

    no   includes other than genetic damage
    yes is strictly genetic

3. Directness distinguishes lesions or mutations from the responses to lesions and is scored:

    +   a positive measure of genetic damage
    −   an indirect measure such as repair or SCE

4. Heritability describes the relation to Mendelian inheritance and is scored:

    +   a germinal, potentially transmissible lesion
    −   a somatic lesion or a germinal nontransmissible lesion

5. Size of lesion is scored:

    micro small compared to a gene
    macro large compared to a gene
    mixed both of above
    ?   indeterminate or unknown

6. Sensitivity refers to detection power in routine use and is scored:

    +   can resolve ten times background
  ++ can resolve two times background
   ?   unknown

7. Latency is the time from exposure to reasonable expression of the effect being measured.

8. Persistence is the time from exposure to loss of reasonable expression of the effect being measured.

Two types of tests have not been included. One is the detection of mutagens or their by-products in urine, blood, stool, or other human body components. Such tests relate to exposure but not to genetic damage. The second type is the scoring of F_1 heritable damage, such as inborn errors of metabolism, chromosomal abnormalities, and electrophoretic or other biochemical evidence of variant gene products in the offspring of exposed parents. These are not short-term tests and they have yet to develop the sensitivity to be useful in monitoring human populations.

Repair synthesis is comparable in human and animal tissues as are the methods for its measurement, with the important proviso that for general human application the incorporation of radioactive precursors of DNA must be carried out *in vitro* after the tissues have been sampled. As reported in these proceedings, Stich *et al.* (1981) have applied the method clinically to small bowel samples obtained at surgery. The method is necessarily indirect, uptake of radioactivity need not be specific for repair, and there can be false negatives due to toxic inhibition of repair. Following an acute exposure to a mutagen, repair synthesis is likely to be largely complete in a day or so. The corresponding method for sperm would be difficult to extend from the mouse to man because accessible mature sperm can no longer repair and systemic radioactivity is required to demonstrate repair at an earlier stage of spermiogenesis.

DNA adducts have been measured in human cells and tissues exposed in culture (see for example McCormick and Maher, 1981), and are readily measured in tissues and sperm of laboratory animals (Ehrenberg *et al.*, 1974). Sensitivity in such experiments currently requires radioactive adducts, an asset that would not be available for general human use. Feasibility in the human is, therefore, uncertain but should improve, perhaps to the point of applicability as methods improve for chemical analysis of specific adducts (see, for example, Harris *et al.*, 1979). Sperm would be prime test targets because of their genetic relevance, their availability, and their accessibility to adduct formation during their final preejaculatory weeks when repair is no longer present. Adducts to macromolecules other than DNA may be useful as well (Ehrenberg *et al.*, 1977).

An increase in the frequency of abnormally shaped sperm in response to mutagenic exposure is common to mammals and man, and is described here by Wyrobek (1981). Obtaining the cooperation of subjects to get samples has not been a problem, and field trials are in the literature. Sperm morphology is sensitive to many factors, some of which may not be genetic. Expression could well be an indirect effect on the spermiogenic differentiation process; heritability of abnormal sperm, although demonstrated with several mutagens, is probably a correlated effect that is mechanistically separate from abnormal sperm formation in the parent. The test is highly sensitive and easy to use. A latency of one month is expected in man and persistence could be months to years.

Scoring for chromosome aberrations is the best established, most widely practiced clinical test for genetic damage. The test is usually done with mitotically stimulated blood cultures and is specific and direct for clastogenesis. Unstable aberrations such as dicentrics decay exponentially with a half-life of a year or two, while stable aberrations such as translocations may persist for decades in long-lived lymphocytes. Meiotic analyses are well documented but involve rarely available testicular samples. It was recently shown that human sperm penetrate hamster oocyte nuclei *in vitro* and may allow analysis of the male haploid chromosome set as fertilization-like processes initiate (Yanagimachi *et al.*, 1976). Chromosomal analyses for aberrations are expensive, and in routine use the relatively small number of detected aberrations make the test insensitive.

The micronucleus test as described herein by Heddle and Salamone (1981) and Heddle *et al.* (1981) is well documented as a sensitive, relatively inexpensive, direct and specific test for clastogenesis. It is best known in its application to polychromatic red cells, which have a latency of a few days and a persistence of perhaps five days. Application to other cells and tissues is possible and would have different sensitivity, latency, and persistence.

Sister chromatid exchanges (SCE's), as described here by Wolff (1981), are an indirect, very sensitive measure of mutagenic effect. The test in humans requires two rounds of cell division *in vitro* in the presence of BUdR. It is widely used but is still controversial as to mechanism, specificity, and persistence. Calibration of SCE against specific locus mutations indicates widely varying yields from one mutagen to the next, although any one compound gives consistently proportional results with the two types of endpoint (Carrano *et al.*, 1978).

Human-oriented specific locus tests are a relative newcomer and seem very promising in terms of sensitivity, specificity, directness, and diversity of genetic targets. The furthest developed of these tests is the HGPRT assay based on 6-thioguanine–resistance of peripheral blood lymphocytes (Strauss and Albertini, 1979). Significantly elevated levels of resistant cells have been found in psoriatic patients with and without psoralen/UV treatment, in breast cancer patients with and without chemotherapy and radiotherapy, and in vitiligo patients but only those who have received therapy with psoralen and UV light. The available data indicate selection against the resistant lymphocytes; persistence may be reduced to months for this test.

A second human-oriented, specific locus test uses fluorescent antibodies against sickle hemoglobin to detect the presence of sickle-trait red cells in normal non-sickle-trait individuals. Manual and flow cytometric measurements indicate a relative frequency of around one in 10^7 for such cells in normal individuals (Papayannopolou *et al.*, 1976; Mendelsohn *et al.*, 1980). This test is presumably measuring transition of a specific single base, the ultimate microlesion. Persistence could well be for decades. We are extending the capability to other variant hemoglobins and to markers on the red cell membrane. Related approaches use radioimmunoassay (Doherty, personal communication) and the appearance of isoleucine (Popp *et al.*, 1976) to detect variant hemoglobins; since these last two methods search for the variants among hemoglobin molecules (rather than among red cells), they include errors of translation and transcription.

One can expect the list of specific locus methods to increase rapidly and to extend to many other organs and cell types. Eventually tests with a wide range of persistence and response to various classes of mutagenic lesion should be available to assess somatic mutation in particular organs. Some fraction of these tests should permit sampling of people by innocuous, relatively noninvasive means; others will require needle biopsy, surgical procedures, or autopsy material.

A recent development also raises the exciting possibility of extending the method to mature sperm and thus of measuring potentially transmissible, specific locus mutation as a short-term test in man. Ansari and Malling (personal communication) have succeeded in producing species-specific antibodies to LDH-X, the lactate dehydrogenase enzyme that appears only in testis and sperm. The mouse and rat enzymes differ by several amino acids and elicit antibodies which can be selected to be non-cross-reacting between the two species. Using fluorescently labeled antibody to rat LDH-X, Ansari and Malling find that one in 4×10^7 mouse sperm becomes fluorescent. They have found a clear-cut dose-response curve to procarbazine in DBA/2 mice; at the highest dose the frequency of fluorescent sperm rises 20-fold to one in 9×10^6. A similar strategy using human and perhaps monkey LDH-X should be feasible.

Discussion

The existence and capacity for rapid growth of this array of human short-term genetic tests is reassuring, but many challenging questions remain to be addressed before the tests can be used to full advantage. Validation and calibration of new tests in the human are particularly difficult. Cancer therapy with radiation and chemicals is perhaps the only situation where carefully controlled human exposure to known mutagens is permissible. Interpretation of such results is clouded, however, by the underlying disease process, by

the complexity of current therapy schedules, and by effects of massive cell killing, selection, and persistence. The alternate approach is to use accident victims and occupationally or environmentally exposed suspect populations, in which case doses are generally obscure and the offending agent may not even be certain. Compounding all human studies will be the background effects due to individual genetic variation, medical history, specific factors such as drugs or tobacco, and nonspecific effects of life style or environment in the broadest sense.

The value of a cheap, readily available, easily repeatable, sensitive and reliable battery of human genetic toxicologic assays would be enormous. With such a tool one could address such pressing problems as:

The identification, analysis and management of DNA-repair-defective individuals

The nature and diversity of human activation systems

The existence and control of genetically hazardous occupations or lifestyles or environments

The estimation of cumulative genetic load and its correlation with longevity, specific health effects, and well-being

The theme of this conference suggests yet another question, namely what bearing such genetic tests will have on the problem of human carcinogenesis? Two types of prediction can be made. One has been implicit in much of the conference and is that the biology of carcinogenesis coupled to the strong positive correlation between mutagens and carcinogens makes it extremely likely that organ-specific mutagenesis will be predictive of a large fraction of human cancer risk. The second prediction is that the technology and biological insight which made possible these genetic tests can also be applied to short-term tests of human carcinogenicity. The development of sensitive markers for carcinogenesis may someday allow us to detect rare or subtle precancerous cellular changes with sufficient facility to identify the people at risk, to define causation, and to interrupt the carcinogenic process before it becomes irreversible.

References

Carrano, A.V., Thompson, L.H., Lindl, P.A., Minkler, J.L. (1978): Sister chromatid exchange as an indicator of mutagenesis. Nature 271:551–553.

Ehrenberg, L., Hiesche, K.D., Osterman-Golkar, S. (1974): Evaluation of genetic risks of alkylating agents: tissue doses in the mouse from air contaminated with ethylene oxide. Mutat. Res. 24:83–103.

Ehrenberg, L., Osterman-Golkar, S., Segerback, D., Svensson, K., Calleman, C.J. (1977): Evaluation of genetic risks of alkylating agents. III. Alkylation of haemoglobin after metabolic conversion of ethene to ethene oxide in vivo. Mutat. Res. 45:175–184.

Harris, C.C., Yolken, R.H., Krokan, J., Change Hsu, I. (1979): Ultrasensitive enzymatic radioimmunoassay: application to detection of cholera toxin and rotavirus. Proc. Natl. Acad. Sci. USA 76:5336–5339.

Heddle, J.A., Salamone, M.F. (1981): The micronucleus assay. I. In vivo. In: H.F. Stich and R.H.C. San (eds.), Short-Term Tests for Chemical Carcinogens. New York: Springer-Verlag, pp. 243–249.

Heddle, J.A., Raj, A.S., Krepinsky, A.B. (1981): The micronucleus assay. II. In vitro. In: H.F. Stich and R.H.C. San (eds.), Short-Term Tests for Chemical Carcinogens. New York: Springer-Verlag, pp. 250–254.

McCormick, J.J., Maher, V.M. (1981): Mutagenesis studies in diploid human cells with different DNA-repair capacities. In: H.F. Stich and R.H.C. San (eds.), Short-Term Tests for Chemical Carcinogens. New York: Springer-Verlag, pp. 264–276.

Mendelsohn, M.L., Bigbee, W.L., Branscomb, E.W., Stamatoyannopoulos, G. (1980): The detection and sorting of rare sickle-hemoglobin containing cells in normal human blood. In: Flow Cytometry IV. Norway: Universitetsforlaget.

Papayannopoulou, T., McGuire, T.C., Lin, G., Gaszel, E., Nute, P.E., Stamatoyannopoulos, G. (1976): Identification of haemoglobin S in red cells and normoblasts, using fluorescent anti-Hb S antibodies. Br. J. Haematol. 66:25–31.

Popp, R.A., Baliff, E.G., Hirsch, G.P., Conrad, R.A. (1976): Errors in human hemoglobin as a function of age. Interdiscipl. Top. Geront. 9:209–218.

Stich, H.F., San, R.H.C., Freeman, H.J. (1981): DNA repair synthesis (UDS) as an in vitro and in vivo bioassay to detect precarcinogens, ultimate carcinogens and organotropic carcinogens. In: H.F. Stich and R.H.C. San (eds.), Short-Term Tests for Chemical Carcinogens. New York: Springer-Verlag, pp. 65–82.

Strauss, G.H., Albertini, R.J. (1979): Enumeration of 6-thioguanine-resistant peripheral blood lymphocytes in man as a potential test for somatic cell mutation arising in vivo. Mutat. Res. 61:353–379.

Wolff, S. (1981): The sister chromatid exchange test. In: H.F. Stich and R.H.C. San (eds.), Short-Term Tests for Chemical Carcinogens. New York: Springer-Verlag, pp. 236–242.

Wyrobek, A.J. (1981): Methods for human and murine sperm assays. In: H.F. Stich and R.H.C. San (eds.), Short-Term Tests for Chemical Carcinogens. New York: Springer-Verlag, pp. 408–419.

Yanagimachi, R., Yanagimachi, H., Rogers, B.J. (1976): The use of zone-free animal ova as a test system for the assessment of the fertilizing capacity of human spermatozoa. Biol. Reprod. 15:471–476.

Index

N-Acetoxy acetylaminofluorene (N-AAAF) 16
  DNA repair polymerization induced by 150
  DNA viruses reactivation and 23–24, 31
Acylating agents 138
Adenovirus
  irradiated, V antigen formation by 22–23, 25, 30
  reactivation of
    chemical and physical agents affecting 31
    host-cell 21–27
  simian, enhancement of viral transformation
    and 350–360
Adult rat liver epithelial cell line (ARL) 279–285
Aflatoxin(s)
  biochemical phage induction assay for 3
  DNA repair polymerization induced by 150
  DNA repair synthesis assay for 75, 77
  DNA viruses reactivation and 31
  hepatocyte primary culture/DNA repair test for 280
  mutagenesis assay for 282
  Salmonella mutagenicity test for 110
  transformation of Syrian hamster cells by 314
Alkaline elution of DNA assay 48–57
  carcinogenicity and mutagenicity relation to 52
  compounds evaluated with 53–56
  control procedures for 51
  modifications of 51–52
  principles of 48–49
  protocol for 50–52
  results, evaluation of 52
  statistical applications of 51
  test systems of 49–50
  in vitro 50–51
  in vivo 51
Alkaloids, pyrrolizidine. See Pyrrolizidine alkaloids
Alkanes, hydrazo- and azoxy- 387
Alkyl halides 121–124
Alkylating agents
  alkaline elution of DNA assay for 54
  DNA-modifying activity of 138
  DNA repair polymerization induced by 149
  hepatocyte primary culture/DNA repair test for 280
  mutagenesis assay for 282
  testicular DNA synthesis inhibition test for 98
Allium cepa 201–203

Aluminum sulfate 356
Amides 138
Amines, aromatic
  alkaline elution of DNA assay for 54
  cytotoxic effects of 268
  DNA-modifying activity of 138
  hepatocyte primary culture/DNA repair test for
    281
  mutagenesis assay for 282
  resistant preneoplastic liver cells induced by 376
  testicular DNA synthesis inhibition test for 98
Ammonium sulfate 356
Aneuploidy
  defined 187
  environmentally induced, in Neurospora 187–197
    agents tested 193–194
    gamma rays and 194
    other systems used 190, 196
    protocol for 193
  in human diseases 188–189, 197, 493
  mitotic vs. meiotic 196
  in tumors 188–189, 197
Angelicin 31
Anticarcinogens
  enzyme inducers/inhibitors and 441
  mammalian cell transformation and 306–315
  microbial mutagenicity assays and 438–445
Antimetabolites 136
Antimicrobial agents, nitroheterocyclics as 42–44
Antineoplastic agents
  with DNA excision repair 140–157
  DNA repair polymerization induced by 155
Arabinopsis thaliana 209
Aromatic amines. See Amines, aromatic
Aromatic polycyclic hydrocarbons. See Hydrocarbons,
  polycyclic, aromatic
Asthenospermia 409
Ataxia telangiectasia 20, 23, 217
Automation, in cytogenetics 255–262
  cell culturing and harvesting 256–260
  clastogen analysis 262
  karyotyping 260–261
  system performance 261–262
8-Azaguanine 266, 270–273

Bacteria. See also Antibacterial agents; Microbial assays
　　nitroheterocyclics activation by 36−45
　　nitroreductases of 37−40
Bacteriophage
　　biochemical induction assay 1−10
　　lambda promoters 1−2
　　PM2, DNA lesions in 12−19
Basc test 382
Benzo(a)pyrene(BP)
　　Escherichia coli differential killing test for 292
　　transformation of Syrian hamster cells by 315
　　unscheduled DNA synthesis and 86
Bioautography 5, 8−9
Biochemical phage induction assay (BIA) 1−10
　　bacterial strains for 2−3
　　compounds tested with 3
　　β-galactosidase and 2, 6
　　one-tube quantitiative 3−6
　　spot test 4−5, 7
　　　bioautography and 5, 8−9
Bleomycin 141, 155
Bloom's syndrome 20, 23, 217
Breeding test animals 399−406
Brooding technique, in Drosophila assay 384

Caffeine
　　aneuploidy induced by 196
　　DNA damage, UV-induced, effect on 25
Calcium, neoplastic proliferation and 362−369
Cancer. See also Transformation, malignant
　　aneuploidy in 188−189, 197
　　family syndromes of 217, 226−228
Carcinogen(s)
　　-activation dependent 278, 431−434
　　breeding test animals of high susceptibility to 399−406
　　chemically activated 432−434
　　detection of. See also Carcinogenicity, tests relation to
　　　biochemical phage induction assay 1−10
　　　calcium-independence of neoplastic cells
　　　　and 362−369
　　　differential killing test, for *E. coli* triple
　　　　mutant 290−294
　　　DNA repair synthesis assay 65−78
　　　DNA synthesis inhibition 59−64
　　　Drosophila assay 379−395
　　　fluctuation test 296−304
　　　by induction of resistant preneoplastic liver
　　　　cell 372−377
　　　by hamster embryo cell transformation 323−336
　　　hepatocyte primary culture/DNA repair
　　　　test 277−278
　　　by mammalian cell transformation *in vitro* 306−315
　　　nucleotide-permeable *E. coli* assay 140−157
　　　Salmonella/microsome assay 121
　　direct-acting 429−431
　　DNA-crosslinking 153, 485−486
　　enhancement of viral transformation by 350−360
　　environmental 121
　　enzyme induction/inhibition, effect on 439
　　human 109, 121
　　-induced mutagenesis, modifying agents for 449−455
　　-induced reactivation 21
　　metabolically activated 431−432
　　mutagenic 59−64, 111−112, 449−455, 490
　　nitrosation of 439 .
　　noncarcinogen, defined 112−113

nonmutagenic 114
　　organotropic 65−78
　　permissive influences on 439−444
　　tests for, unsolved problems with 474−481
　　transnitrosation of 439−440
　　trapping agents 439
　　ultimate 12, 65−78
　　xeroderma pigmentosum cell response to 78
Carcinogenesis
　　DNA repair, role in 20−21
　　promotion phases of. See Tumor(s) promoters
Carcinogenic potency, bacterial mutagenicity and 43
Carcinogenicity
　　of furyl-furamide 111
　　of nitrofurans 44
　　tests relation to
　　　alkaline elution of DNA assay 52
　　　DNA polymerase assay 130, 137
　　　Drosophila assay 391
　　　testicular DNA synthesis inhibition test 105
　　　in vitro tumor promotion assay 424−425
　　of Tris-BP 111
Catalase 435
Cell
　　culturing, automation in 256−260. See also under
　　　specific cell culture system
　　cycle, mutagen sensitivity and 165
　　membrane 489
Cellular capacity, for virus infection 27−28
Chemical(s)
　　as DNA-modifying agents 16−17
　　DNA synthesis inhibition test and 63
　　-induced
　　　DNA damage, *in vivo* 57
　　　transformation, of human cells 338−348
　　microsome-mediated activation of 137
　　mutagenesis, inhibition of 428−436
　　mutagens, plant genetic systems for 200−212
　　noncarcinogenic, defined 112−113
　　testicular DNA synthesis inhibition test and 98−101
　　viruses interaction with 350−360
Chinese hamster ovary cells 237−240
Chinese hamster V79 cells
　　alkaline elution of DNA assay and 49, 54−56
　　in vitro assay for tumor promoters and 423, 426
Chlorpromazine 31
Chromosome(s). See also Aneuploidy
　　aberration
　　　induction of, mutagenesis and 428−436
　　　de novo, in high-risk groups 493
　　　in meiotic material 208
　　　in plant root 201, 203−204
　　　somatic, in Tradescantia 204−205
　　abnormalities
　　　congenital 218, 230
　　　due to griseofulvin 114
　　breakage
　　　in Fanconi anemia family 224
　　　in hematologic cancer family 226
　　　in nevoid basal carcinoma family 227
　　　in retinoblastoma family 228
　　　spontaneous, and gentian violet-induced 219−233
　　　UV-induced, xeroderma pigmentosum and 218
　　harlequin. See Sister chromatid exchange test
　　"instability syndromes" 217−219
Philadelphia 188

Chromosomal mutations 187–189, 422
  chromosomal organization and 484–487
  X-chromosome 382
Chromosomal organization 484–487
Clastogen
  analysis, automation and 262
  gentian violet as 219–230
Cocarcinogenesis 439
Cocarcinogens
  detection of
    by mammalian cell transformation *in vitro* 306–315
    microbial mutagenicity assays for 438–445
  enzyme inducers/inhibitors and 441
  tryptophan metabolites as 441–442
Cockayne's syndrome 23, 26
Comutagens 116, 442
Congenital chromosomal abnormalities 218, 230
Cryopreservation, of Syrian hamster embryo
  cells 323–336
Cycasin 115
Cycloheximide 31
Cyclophosphamide 155, 245
Cysteine
  -N-methyl-N'-nitrosoguanidine interaction 430–431
  PM2 DNA, effect on 16–17
Cytogenetic tests
  for chemical mutagens 207–208
  for genetic instability, in humans 217–233
Cytogenetics
  automation in 255–262
  mammalian 492–495
Cytotoxicity studies, on human diploid cells 267–268

Daunomycin 155
Dibromochloropropane (DBCP) 408
Diethylstilbestrol
  E. coli differential killing test and 293
  Salmonella mutagenicity test and 109
Dikarion test, *Neurospora crassa* and 181–182
7,12-Dimethylbenz[a]anthracene (DMBA)
  enhancement of viral transformation by 354
  micronucleus assay for 245
Dimethylsulfoxide (DMSO) 86
DNA
  adduct formation with 484–485, 507
    in human cells 507
  alkaline elution of. See Alkaline elution of DNA assay
  binding, nucleotide-permeable *E. coli* model
    of 140–157
  crosslinking 153, 485–486
  damage
    due to chemically activated carcinogens 432–434
    due to direct-acting carcinogens 429–431
    due to metabolically activated carcinogens 431–432
    PM2 DNA assay for 15–16
    sucrose gradient assay for 90–93
    in vivo 57
  excision repair, antitumor agents with 140–157
  modifying agents. See Modifying agents
  nicking assay 13
  PM2. See PM2 DNA assay
  polymerase
    genotoxic activity determination with 127–139
    7-methylbenz[a]anthracene-5,6-oxide and 145–155
  repair
    carcinogenesis, role in 20–21

in diploid human cells 264–274
DNA polymerase and 127–139
tumor promoters, effect on 431–423
test, hepatocyte primary culture and 277–278,
  280–281
repair polymerization, inducing agents for 145–155
repair synthesis assays 65–78, 83–88, 428–436
  chemical mutagenesis inhibition and 428–436
  compounds evaluated with 68–72, 86
  vs., DNA replication 65–66
  in human cell lines 83–88, 428
  for humans use 507
  multi-well assay 83–88
  unsolved problems in 67, 73
  validity of 66–67
replication 65–66, 140–157
strand scissions in 486
synthesis inhibition test 59–64
  chemicals tested with 63
  principle of method 59–60
  results 61–64
  test organism and testing protocol for 60–61
  testicular. See Testicular DNA synthesis inhibition test
synthesis, replicative, inhibitors of 140–157
synthesis, unscheduled (UDS). See DNA repair synthesis
  assay
viruses. See Virus(es) and under specific virus name
Down syndrome 218
Drosophila
  assay 379–395
    agents tested 386–387
    Basc test and 382
    brooding technique in 384
    carcinogenicity relation to 391
    enzyme induction and 391–393
    genetic endpoint, choice of 379–380
    mutagen-sensitive strains and 393
    mutagenicity of precarcinogens in 386–389
    testing protocol 383–386
  mutagenic activity in vs. mammalian cytogenetics 495
Drugs 55
Dyes, aminoazo
  hepatocyte primary culture/DNA repair test and 280
  mutagenesis assay and 282

Embryo cells. See Hamster embryo cells (HEC)
Enhanced virus reactivation 28–30
Environmental carcinogens 121
Environmental mutagens 161–172
Environmentally induced aneuploidy 187–197
Enzyme inducers/inhibitors
  anticarcinogens, cocarcinogens and 441
  carcinogen, effect on 439
  Drosophila assay and 391–393
Enzymatic activity, of premutagens 491
Epithelial cell line, rat liver 279–285
Epoxides 146–147
Escherichia coli
  DNA polymerase of 127–130
  DNA polymerase deficient 137
  genetic markers of 141
  mutants
    nitrofuran-resistant (NFR) 39–41
    triple, differential killing test and 290–294
  nitroreductases of 39–40
  nucleotide-permeable 140–157

Ethyl methanesulfonate (EMS)
  DNA viruses reactivation and   31
  enhancement of viral transformation by   354
  PM2 DNA and   17
Ethylene dibromide   355
Ethylene oxide   125

Familial cancer syndromes   217, 226−228
Fanconi anemia   20, 25−26, 217−218
Fecalase   115
Fibroblasts. See Human cell lines: Xeroderma pigmentosum
  cell
Fish, melanophore system of   400−401
Fluctuation test   296−304
  agents tested with   300−303
  hepatocyte preparation for   296−298
  modifications of   301
  S9 fraction and   296−298
  statistical evaluation of   299−301
Fluorescent antibodies techniques   508−509
5-Fluorouracil   104
Furadantin   44
Furazolidone   44
Furyl-furamide   111

β-Galactosidase   2, 6
Gamma rays
  aneuploidy and   194, 196
  DNA viruses reactivation and   31
Gene(s)
  conversion, mitotic   164, 166
  mutations   187
  regulating   403
  tumor   403
Genetic analysis, of *Neurospora crassa* mutants   175,
    179−185
Genetic control, of *E. coli* nitroreductases   39
Genetic endpoint, in Drosophila assay   379−380
Genetic information, neoplastic transformation and   401
Genetic instability, in humans   217−233
Genetic markers, of *E. coli*   141
Genetic properties, of *Saccharomyces*
    cerevisiae   161−162
Genetic tests. See also DNA
  for humans   505−510
  in plants. See Plant genetic test systems
Genotoxic activity, determination of   127−139
Gentian violet, as clastogen   219−230
Glutathione   435
Glycosidases   115−117
Griseofulvin   114

Halo-alkanes   388−389
Halo-olefins   388−389
Hamster
  Chinese hamster ovary cells   237−240
  Chinese hamster V79 cells   49, 54−56, 423, 426
  embryo cell transformation assays   307−315,
      323−336, 351−360
    agents tested with   335−336
    bioassay   326−327
    with cryopreserved cells   323−336
    metabolic activation and   333
    morphologic criteria for   327−333
    for viruses   351−360
Haploid yeast reversion test   171−172. See also
    Saccharomyces cerevisiae

Hepatocyte
  activation, fluctuation test and   296−298
  primary culture/DNA repair test   277−278, 280−281.
    See also Liver culture systems
Herpesvirus, reactivation of
  chemical and physical agents affecting   31
  host-cell   21−27
Heterocyclic compounds, DNA, effects on   135
Heterokaryon test, Neurospora mutants and   180−181
Hordeum vulgare   207−208
Hormones, mutagenesis, role in   487−489
Host-cell reactivation (HCR)   21−27
  N-acetoxy-2-acetylaminofluorene and   23−24
  cell types used for   21−27
  nitrous acid and   23
  X-rays and   23
Human carcinogens   109, 121
Human cell lines
  chemically induced transformation of   338−348
    cell culture   341−342
    early transformation assay   345, 348
    malignant transformation assay   342−348
    markers of   338−341
    testing protocol   342−345
  chromosome aberration induction in   428
  diploid   83−88
    initiation of   265−267
    mutagenization of   264−273
    source of   265
  DNA repair synthesis in   83−88, 428
  DNA synthesis inhibition in   59−64
Human cells, neoplastic indices of   368
Human diseases, aneuploidy in   188−189, 197, 493
Human sperm assays   408−417
Human tumors   188−189, 197
Humans
  genetic instability in, test for   217−233
    blood culture and scoring procedure   220−221
    clinical information   220
    gentian violet use in   219−230
    statistical analysis   221
  genetic tests extended to   505−510
Huntington's chorea   23
Hydrazines   135
Hydrocarbons
  chlorinated   109
  polycyclic, aromatic
    alkaline elution of DNA assay for   54
    DNA-modifying activity of   134−136
    hepatocyte primary culture/DNA repair test for
        280
    mutagenesis assay for   282
    preneoplastic liver cells resistant to   376
    testicular DNA synthesis inhibition test for   98
N-Hydroxyurethan   129
Hypoxanthine   272
    -guanine phosphoribosyl transferase assay
        279−283

Ifosfamide   155
Infection, viral, cellular capacity for   27−28
Inorganics, DNA, effect on   136
Intercalation   484
Intercalating agents   132, 153, 156
Isoniazid   435

Karyotyping, automated   260−261

Lambda promoters   1–2
Lead chromate   355
Lesch-Nyhan syndrome   271
Leukemia
    aneuploidy in   188–189
    lymphatic, acute   226
Liver
    adult rat epithelial cell line (ARL)
        /hypoxanthine-guanine phosphoribosyl transferase
            assay   279–283
        transformation assays and   283–285
    cells
        neoplastic indices for   366
        preneoplastic, resistant, induction of   372–377
    culture systems, carcinogens detection and
        277–287
    homogenate, S9 in   115, 117
Lymphoproliferative neoplasm, multiple   225

Malignant transformation. See Transformation, malignant
Malondialdehyde   117
Mammalian cells. See also Human cell lines
    chemical mutagenesis inhibition in   428–436
    DNA synthesis inhibition test and   59–64
    nitroheterocyclics activation by   36–45
    nitroreductases of   38
    transformation of   306–315
Mammalian cytogenetics   492–495
    vs. mutagenic activity in Drosophila   495
Mammalian system, mutagenesis in   488
Melanoma, spontaneous and induced   401–406
Melanophore system, of Xiphophorus   400–401
Metabolic activation. See also S9 fraction, activation by
    in hamster embryo cell transformation assays   333
    of nitroheterocyclics   36–45
Metabolically activated carcinogens   431–432
Metabolism, mutagenesis, role in   487–489
Metal complexes   435
7-Methyl-benz[a]anthracene-5,6-oxide   144–145
N-Methyl-N′-nitrosoguanidine (MNNG)
    -cysteine interactions   430–431
    DNA repair synthesis assay for   74–75, 86
    mutagenic activity of   453
    PM2 DNA, effect on   16–17
    transformation of Syrian hamster cells by   314
8-Methoxypsoralen   31
Microbial assays. See also Bacteria; Escherichia coli
    for cocarcinogens and anticarcinogens   438–445
    DNA polymerase assay   127–139
    for modifying agents   449–455
    Salmonella mutagenicity test. See Salmonella
        mutagenicity test
Micronucleus test,
    for human use   508
    in vitro   250–254
        sister chromatid exchange test and   251–252
    in vivo   243–248
        agents tested with   245
        biology underlying   243–244
        scoring   248
        statistics   246
        technique   246–247
Microsomes. See also Salmonella/microsome assay
    -mediated activation of chemicals   137
    preparation of   115, 117
Miracil D   128, 130
Misonidasole   92–93

Mitomycin C   117
    E. coli differential killing test and   292
    micronucleus assay and   245
Mitotic aneuploidy   196
Mitotic gene conversion   164, 166
Mitotic recombination   162, 166
Modifying agents   134–139
    of carcinogen-induced mutagenesis   449–455
    chemicals as   16–17
    detection of, PM2 DNA and   12–19
Monkey kidney cell   27
Multi-well assay, for DNA repair synthesis   83–88
    cells harvesting and scintillation counting   84
    cells, prelabeling of   84–85
    compounds evaluated with   86
Murine sperm assays   414–417
Mutagenesis
    assay (ARL/HGPRT)   279–283
    carcinogen-induced   449–455
    chemical
        inhibition of   428–436
        metabolism, role in   487–489
    hormones and   487–489
    in mammalian system   488
    quantitative measures of   457–471
    studies, on diploid human cells   264–273
Mutagenic activity
    in Drosophila   495
    tests for, classification of   83
Mutagenic agents, Neurospora and   185
Mutagenic carcinogens   59–64, 111–112, 449–455, 490
Mutagenic efficiency, relative (RME)   464–467
Mutagenic potencies, of nitrofurans   44
Mutagenicity
    assays   109–117, 438–445
    bacterial, carcinogenic potency and   43
    of nitrofurazone   41
    of precarcinogens, in Drosophila assay   386–389
    screening   484
    selection   269–271
    testing, problems in   483–498
    tests relation to
        alkaline elution of DNA assay   52
        Salmonella mutagenicity test   109–117
        testicular DNA synthesis inhibition test   105
Mutagens. See also Comutagens
    detection of
        DNA polymerase assay   129–130
        DNA repair synthesis assay   68–72
        plant genetic test systems   200–212
        yeast as a test system for   164–165
    direct-acting, nitroheterocyclics as   36
    environmental   161–172
    premutagens. See Premutagens
    proximate   490, 492
    -sensitive Drosophila strains   393
    sensitivity to, cell cycle and   165
    ultimate   490
Mutants
    of E. coli
        nitrofuran-resistant (NFR)   39–41
        triple, differential killing test and   290–294
    specific locus, of Neurospora   175–186
Mutation(s)
    chromosomal. See Chromosomal mutations
    fixation   421
    gene   187

Mutation(s) [cont.]
  somatic, detection of   204–205, 211–212
  specific locus, in Neurospora   175–186
  stochastic description of   459–461
  in xeroderma pigmentosum   21
  yields   457, 461

Neoplastic indices   366–368. See also Tumor(s)
Neoplastic proliferation, calcium independence
    of   362–369
Neoplastic transformation. See Transformation, malignant
Neurospora crassa
  and environmentally induced aneuploidy   187–197
  specific locus mutations in   175–186
    agents tested   185
    general utility of   175–176
    genetic analysis   180–185
    heterokaryon tests and   180–181
    induction of   177–178
    test method   176–178
    trikaryon test and   182–184
Nevoid basal carcinoma syndromes   219, 227, 232
Nitrofuran(s)
  carcinogenic and mutagenic potencies of   44
  reduction of   37–39
  -resistant E. coli mutants   39
Nitrofurantoin   43–44
Nitrofurazone   41
Nitroheterocyclic compounds
  as antimicrobial agents   42–44
  as direct-acting mutagens   36
  metabolic activation of   36–45
    by bacteria and mammalian cells   36–45
    quantitative aspects of   42–44
    S9 fraction and   41–42
Nitroreductases
  of E. coli   37–40
  sensitivity to oxygen of   37–38, 44–45
Nitrosamines
  biochemical phage induction assay for   3
  DNA-modifying activity of   135
  Drosophila assay for   387
  hepatocyte primary culture/DNA repair test for
    280
  mutagenesis assay for   282
  testicular DNA synthesis inhibition test for   99
Nitrosation   439
N-Nitroso compounds   376
Nitrous acid
  DNA viruses reactivation by   23, 31
  PM2 DNA crosslinking by   16, 17
Norharmane   441–442
Nucleotide-permeable E. coli   140–157
  agents tested with   146–155
  results   142–154
  test organism   140–142
  testing protocol   142

Organotropic carcinogens   65–78
Osmium tetraoxide (OsO4)   151–152
Osteosarcoma   232
Oxidases   487
Oxides
  ethylene   125
  propylene   125, 355
Oxygen, nitroreductases sensitivity to   37–38,
    44–45

Permissive influences
  on carcinogen   439–444
  cocarcinogenesis, role in   439
  on responding tissue   444
  systemic   444
Phage. See Bacteriophage
Philadelphia chromosome   188
Pigment cells, of melanophore system   400–401
Plant(s)
  genetic tests systems for mutagens in   200–212
    in Allium species   201–203
    in Arabinopsis thaliana   209
    cytogenetic   207–208
    in Hordeum vulgare   207–208
    sister chromatid exchanges and   205
    in soybean   212
    in Tradescantia   204–205, 211–212
    in Vicia faba   203–204
  root, chromosome aberrations in   201, 203–204
  tumors of   212
Plate incorporation assay   128–132
PM2 DNA assay   12–19
  advantages and disadvantages of   17–18
  agents tested with   16–17
  classification of lesions in   19
  DNA-modifying agents and   12–19
  PM2 ^{3}H-labeled   14–15
Polymerase. See DNA polymerase
Precarcinogens
  DNA-repair synthesis assay for   65–78
  Drosophila assay for   386–389
Premutagens, activation of   165–166, 491
Preneoplastic, resistant, liver cell   372–377
Progeria   25, 26
Promoters, tumor. See Tumor promoters
Propanols, halogenated   131
Propyl gallate   454
Propylene oxide   125, 355
Proximate mutagens   490, 492
Pyrrolizidine alkaloids
  Drosophila assay for   388
  hepatocyte primary culture/DNA repair test for
    281

Quantitative measures
  of induced mutagenesis   457–471
  of nitroheterocyclics activation   42–44

Radiation
  enhanced virus reactivation, effect on   28
  mutagenic effects of, in Neurospora   185
  UV. See Ultraviolet light
  due to X-rays. See X-rays
Radiobiology, of yeast   162–164
Reactivation
  carcinogen-induced   21
  enhanced virus-   28–30
  host-cell   21–27
  UV-induced   21–27
  of viruses   20–32
Reductases. See Nitroreductases
Relative mutagenic efficiency (RME)   464–467
Retinoblastoma   219, 228, 232
Riboflavin   115, 117

S9 fraction
  activation by
    of diploid human cell lines   265

DNA repair test and   127
E. coli differential killing test and   291
fluctuation test and   296−298
of nitroheterocyclics   41−42
unsolved problems with   476
liver homogenate as a source for   115, 117
Saccharomyces cerevisiae
aneuploidy detection in   190
assay   161−172
haploid yeast reversion test   171−172
strains used in   166−170
excision-deficient   467−471
genetic properties of   161−162
nonreplicating test system   458
quantitative measures of induced mutagenesis
in   457−471
radiobiology of   162−164
Salmonella
/microsome assay   120−125
agents tested with   121−125
desiccator assay   121
mutagenicity assay   109−117
agents tested with   134−135
false positives and negatives   112−114
human carcinogens and   109
modifications of   114−116
mutagens found carcinogenic with   111−112
validation studies on   109−111
typhimurium
fluctuation test and   297
modifying agents and   449−455
nitroheterocyclics activation by   37−39, 44
Simian adenovirus   350−360. See also Adenovirus
Simian virus 40 (SV40)   31
Sister chromatid exchange (SCE)
in plants   205
test   236−241
Chinese hamster ovary cells use in   237−240
for humans   508
micronucleus assay and   251−252
protocol for   240−241
Somatic mutations, detection of   204−205, 211−212
Soybean assay system   212
Specific locus mutations   175−186
Sperm assays, human and murine   408−417
Spot test
for biochemical phage induction assay   4−5, 7
for DNA repair   127−128
Statistics
alkaline elution of DNA assay   51
cytogenetic test, in humans   221
fluctuation test   299−301
micronucleus assay   246
Stochastic description, of mutation   459−461
Sucrose gradient assay   90−93
Syrian hamster cells, transformation of   311−315,
323−336

Teratospermia   409
Testicular DNA synthesis inhibition test   94−106
agents tested with   98−101
carcinogenicity, mutagenicity and   105
false positives and negatives   101
pharmacokinetics, effect on   102−103
Tests, short-term
extended to humans   505−510
unsolved problems with   474−481
12-O-Tetradecanoylphorbol-13-acetate (TPA)   444−445

6-Thioguanine   266, 270
Tradescantia, somatic mutations in   204−205, 211−212
Transformation
assays   283−285, 306−315, 323−336
adult rat liver epithelial cells   283−285
hamster embryo cells   307−315, 323−336
bioassay   326−327
chemically induced, of human cells   338−348
malignant
assay, in human cells   342−348
calcium and   362−369
genetic information for   401
of pigment cells   400−401
of mammalian cells, *in vitro*   306−315
viral, enhancement of   350−360
Transnitrozation   439−440
Triazenes   388
Triaziquone   155
Triethylenethiophosphoramide   155
Trikaryon test, Neurospora and   182−184
Trimethypsoralen   31
Tris-BP   111−112, 131
Tryptophan, metabolites of   441−442
Tumor(s)
aneuploidy in   188−189, 197. See also Neoplastic
indices
gene   403
of plants   212
promoters   147, 420
DNA-modifying activity of   134−136
DNA repair and   421−423
mechanism of action of   444−445
in vitro assay for   420−426
second primary   232

Ultimate carcinogens   12, 65−78
Ultimate mutagens   490
Ultraviolet (UV) light
-enhanced reactivation (UVER)   28−30
-induced
chromosomal breakage   218
DNA repair   73
reactivation   21−27
mutagenic effects of
on diploid human cells   267
on yeast   467−471
PM2 DNA and   19
transformation of Syrian hamster cells by   313
Unscheduled DNA synthesis (UDS). See DNA repair
synthesis assay
Unsolved problems, in carcinogens testing   474−481
Ureas   136

V antigen, of adenovirus   22−23, 25, 30
Vicia faba   203−204
Vinyl acetate   355
Vinyl chloride   109
Viral DNA, UV-induced changes in   21−27
Viral enhancement assays   352−353
Viral transformation   350−360
Virus(es)
infection, cellular capacity for   27−28
reactivation   20−32
chemical and physical agents affecting   31
enhanced-   28−30
error-prone   30
host-cell   21−27

X-chromosome mutations   382
X-rays
   sensitivity to, chromosomal abnormalities and   218
   therapy, second primary tumor and   232
   viral reactivation and   23, 31
Xeroderma pigmentosum (XP)
   cell
      carcinogens, effect on   78

host-cell reactivation and   23
   mutagenesis studies and   268–271
   UV-induced changes of   218
DNA repair synthesis assay and   77
mutations leading to   21
Xiphophorus, melanophore system of   400–401

Yeast. See *Saccharomyces cerevisiae*